# Greenhouse Engineering

## Integrated Energy Management

İlhami Yıldız

CRC Press
Taylor & Francis Group
Boca Raton London New York

CRC Press is an imprint of the
Taylor & Francis Group, an **informa** business

First edition published 2021
by CRC Press
6000 Broken Sound Parkway NW, Suite 300, Boca Raton, FL 33487-2742

and by CRC Press
2 Park Square, Milton Park, Abingdon, Oxon, OX14 4RN

CRC Press is an imprint of Taylor & Francis Group, LLC



*Library of Congress Cataloging-in-Publication Data*
Names: Yildiz, İlhami, author.
Title: Greenhouse engineering : integrated energy management / İlhami Yildiz.
Description: First edition. | Boca Raton, FL : CRC Press, 2021. | Includes bibliographical references and index. | Summary: "Sustainable energy development concept requires and maintains multiple linkages among energy production, energy consumption, human well-being, and environmental quality. Book puts forward the concept of integrated energy management and modeling pertinent to greenhouses that will eventually help reduce the load on power grids, demand for fossil fuels and water, and also supply CO2 for the greenhouse production. It helps enhance the competitive position of the global greenhouse industry by introducing economically, environmentally and socially sustainable technologies and management strategies"-- Provided by publisher.
Identifiers: LCCN 2020049366 (print) | LCCN 2020049367 (ebook) | ISBN 9781498743136 (hardback) | ISBN 9781315117072 (ebook) Subjects: LCSH: Renewable energy sources. | Energy consumption. | Energy development. | Greenhouse gas mitigation. | Sustainable engineering.
Classification: LCC TJ808 .Y55 2021 (print) | LCC TJ808 (ebook) | DDC 621.31028/6--dc23
LC record available at https://lccn.loc.gov/2020049366
LC ebook record available at https://lccn.loc.gov/2020049367

ISBN: 978-1-498-74313-6 (hbk)
ISBN: 978-0-367-75413-6 (pbk)
ISBN: 978-1-315-11707-2 (ebk)

Typeset in Palatino
by SPi Global, India

Additional information is available for download as an eResource at: https://www.routledge.com/Greenhouse-Engineering-Integrated-Energy-Management/Yildiz/p/book/9781498743136.

*I dedicate this book,* the book of sustaining civilization under cover,
*to my dear wife Betül for her patience and the billion cups of medium roast coffee.*

İlhami Yıldız, Ph.D., MSME, MSAE, MSc.

# Contents

# *Preface*

We are using our renewable and nonrenewable environmental resources faster than they can be replenished. We are extracting minerals, oil, and ground water without sufficient concern for their limits or the need to recycle them. As a result, there is an existing shortage of some resources and there will be more shortage of other resources in the future. Therefore, we must learn how to sustain our resources so that they can continue to fuel human civilization and other living things on our planet within natural limits for many generations. Sustainability, therefore, means leaving our children and future generations a planet at least as rich as we live in today, and requires maintaining fully functional and healthy natural environment, which basically forms the foundation for our civilization. Due to the population increase, in reality, sustainability is a moving target; therefore, more resources will be required to maintain a healthy and functional status. Our choices, up to certain extent, determine both the local and global sustainable carrying capacity for the human population of our planet. For instance, more people will mean more energy use; and in addition, people in less developed parts of the world will be consuming more energy on per capita basis when they reach higher standards of living. The environmental impacts of fossil fuels are significant and may become even worse in the years to come due to the inevitable increase in their uses. Today, we optimistically believe that we have the required knowledge and technology to ensure a worry-free future, if we are to take the action now. Rather than relying on fossil fuels heavily, we have the option to build a sustainable energy future based on careful planning, innovative thinking, and a willingness to move from our dependency on fossil fuels.

Here is a good place to introduce the concept of integrated energy management. According to this concept, a series of options need to be employed, and the range will be determined locally. The sources of energy will have to consider both renewable and nonrenewable sources. With the integrated energy management concept, we would be able to provide diverse and reliable sources of energy including more local and decentralized production of energy, considering energy efficiency and conservation measures, protect our local or global environments, and make sure that our future generations live in a healthy and functional environment with rich enough resources. The biggest challenge for a sustainable energy future however is the human aspect – changing our total way of living. The overall goal of sustainable energy development requires and maintains multiple linkages among energy production, energy consumption, human well-being, and environmental quality. And for this, local production and locally designed systems are the key. As we are getting into the new renewable bio-economy era, bioenergy is expected to have a major share in local and global energy mixtures. However, non-food feedstock diversity and sustainability issues are important considerations.

One major factor hindering future expansion of greenhouse industry is the cost required for environmental control. Considerable effort is expended to conserve energy and look for alternative energy sources, especially environment-friendly renewable energy sources and technologies. Proper greenhouse and environmental management systems can significantly change the energy and moisture dynamics of greenhouse production systems. It is sincerely hoped that this book, *the book of sustaining civilization under cover*, will help enhance the competitive position of the global greenhouse industry by introducing economically, environmentally, and socially sustainable technologies and management strategies. The overall goal of this book is to help reduce the load on power grid, demand for fossil fuels and water, and also supply $CO_2$ for the greenhouse production. The use of heat pumps for heating and cooling greenhouses, and cogeneration or multigeneration systems, for instance, makes it possible to use local and renewable energy sources while reducing or totally eliminating $CO_2$ emissions. Heat pumps also make innovative semi-closed and closed greenhouse operations possible, which would totally eliminate the energy, water, and $CO_2$ losses due to ventilation, conserving all of these precious resources, and would make an insect-free greenhouse operation possible.

**İlhami Yıldız**
Montreal, QC, Canada
iyildiz@dal.ca

# Acknowledgements

The following individuals and organizations kindly supplied illustrations and provided many practical support and opportunities. Their contributions and the contributions of many others in many different forms are greatly acknowledged. In alphabetical order, they are at least:

ASABE, Standard EP406.2 (OCT95), Tables 6.1 and 6.2, used with permission
ASHRAE, Tables A.11 to A.16, used with permission
Elsevier, Table 10.16, used with permission
Mr. Arie van Heteren, formerly with BOM Greenhouses B.V., Monster, The Netherlands
Mr. Ben DeCloet, DeCloet Greenhouse Manufacturing Ltd., Simcoe, Ontario, Canada
Mr. Bram van der Kooy, P.L.J. BOM Greenhouses B.V., Naaldwijk, The Netherlands
Dr. Gagandeep Singh, CRC Press Taylor & Francis Group, and his team members
Dr. James E. Faust, Clemson University, Figures 7.1 to 7.11
Dr. Joanne Logan, The University of Tennessee, Figures 7.1 to 7.11
Dr. Murat Kaçıra, University of Arizona, Figure 6.7
Dr. Peter Ling, The Ohio State University, Figure 8.7
Ms. Thivya Vasudevan, SPi Global, and her team members

# Author Biography

**Dr. İlhami Yıldız** is a controlled environment systems engineer and has expertise in energy, environment, sustainability, and the relevant innovation areas covering environmental biotechnology, bioreactors, biofuels and industrial bioproducts production using waste streams, complex systems modeling, greenhouse engineering, heat pumps, combined heat and power generation, flue gas recovery and greenhouse gas mitigation, and in particular, hydronic heating, drying and hot water storage systems, as well as climate change mitigation and adaptation. He has provided extensive services at different capacities, and contributed to training of multiple HQPs internationally at various international institutions in Canada, Turkey, and the United States. He has also served as a member of editorial boards for several international journals, and review panels for a number of research granting agencies and institutions internationally, including the U.S. Secretary of Energy's review panel. He has published numerous books and book chapters, and peer-reviewed articles in scholarly journals and conference proceedings; organized several conferences and symposia; and delivered multiple keynote speeches and invited lectures internationally at a number of conferences, symposia, and institutions. He has also designed and built numerous environmentally friendly commercial energy systems transferring extensive research findings to commercial settings in Canada, Mexico, Russia, Turkey, and the United States. Dr. Yıldız is a recipient of a number of international honors and awards, and a firm human rights defender for all, everywhere, and all the time.

**Overview:** President - Istinye University, Istanbul, Turkey; Professor - Mechanical Engineering, Istinye University; Adjunct (Retired) Professor - Interdisciplinary PhD Program, Dalhousie University, Halifax, Canada; Professor - Engineering, Dalhousie University; Director - Bioenvironmental Engineering Centre, Dalhousie University; Professor, Director, and Department Chair - Engineering Programs, Dalhousie University - Truro Campus; Associate Professor - BioResource & Agricultural Engineering, Cal Poly - San Luis Obispo, California, USA; Assistant Professor - Earth & Environmental Sciences, University of Windsor, Ontario, Canada; PDF in Biological & Environmental Engineering, Cornell University, Ithaca, New York, USA; Ph.D. in Food, Agricultural and Biological Engineering, M.Sc. in Mechanical Engineering, and M.Sc. in Applied Meteorology & Climatology, The Ohio State University, Columbus, Ohio, USA; B.Sc. & M.Sc. in Agricultural Engineering, University of Ankara. He has been honored with the Grand Challenge Award by the U.S. Government (2008); Outstanding Faculty Award at Cal Poly (2008); Presidential Letter of Appreciation at Cal Poly (2008); Gold Medal Performance Award by the University of Windsor (2004); Postdoctoral Fellowship (1997) by the OECD Co-operative Research Programme; Graduate Fellowship (1985), Ministry of Education, Turkey; and Graduate Fellowship (1984) by the Scientific and Technological Research Council of Turkey (TUBITAK).

# 1

# Introduction

## 1.1 Trends in Global Energy

The story of humankind's success and his eventual downfall is one that rests largely on the shoulders of our creative exploitation and reimagining of energy and its uses. Throughout history, humankind has been able to utilize energy in ways other species have been unable to grasp. Ingenious use of energy has led to increased brain development, enabled us to travel great distances, allowed for the manufacturing of a variety of products at tremendous speed, and helped to power the machines, which influence everything in life from food to healthcare, to communication, and to science, research, and development. The rate at which energy has been exploited has however left humankind in a compromising position. Finite resources are rapidly being depleted; carbon emissions and competing water demands and pollution continue to cause large-scale environmental issues, increasing the attention to sustainability in all contexts. Renewable energy sources offer one answer to the problem and with increased implementation, they may one day power the planet. From the earliest uses of fire to the steam engine, to massive fossil fuel dependence, and to the beginning of the major shift toward renewable energy sources, humankind has continually innovated and adapted to his situation through desire or necessity. Advancements in energy production and usage improved social and economic aspects of life. Humankind has grown alongside energy and in a world where more and more demanded, continued growth and innovation will be required to sustain our societies and their ever-growing needs.

Energy is directly acquired from sunlight by photosynthetic organisms, including plants and microorganisms. The most primitive forms of energy, such as oil, natural gas, and coal are produced and accumulated underneath the earth due to natural processing of these ancient photosynthetic products in the past millions of years. These non-renewable fossil fuel resources, which have tremendous value for us now and for our future generations, are expected to be depleted due to the rising world population and industrial development (Omer, 2008; UN, 2014). With the emergence and development of agricultural and industrial activities, numerous products, including plastics, textiles, fertilizers, and other products from steel manufacturing and petrochemical industry, have been produced to meet our daily needs. Larger and more production activities mean more energy needs for processing and transportation. The world population is increasing rapidly, especially in the developing world, with an approximately 80 million more addition every year (U.S. Census Bureau, 2012). Lifestyles have been changing gradually, particularly in the developed world, and the developing nations are quickly following the trend. Energy makes daily work and life more convenient and efficient. The rising population and demand for convenient life styles require more goods and therefore more energy resources every passing day. Considering the imbalance between supply and demand due to depletion of non-renewable energy resources might eventually trigger an energy crisis; and indeed, open and undercover conflicts from an international perspective have already started taking place globally. Therefore, renewable energy exploitation and energy efficiency improvements are critical themes for sustainable energy supply, and play an essential role in providing a decent living standard for human beings (Omer, 2008). Between 1973 and 2010, the global energy consumption grew by 186% and supply for industrial use increased by 157% (IEA, 2012a; UN, 2014; REN21, 2016). Currently, the world's primary energy supplies are primarily contributed by fossil fuels, hydropower, nuclear energy, wind energy, solar

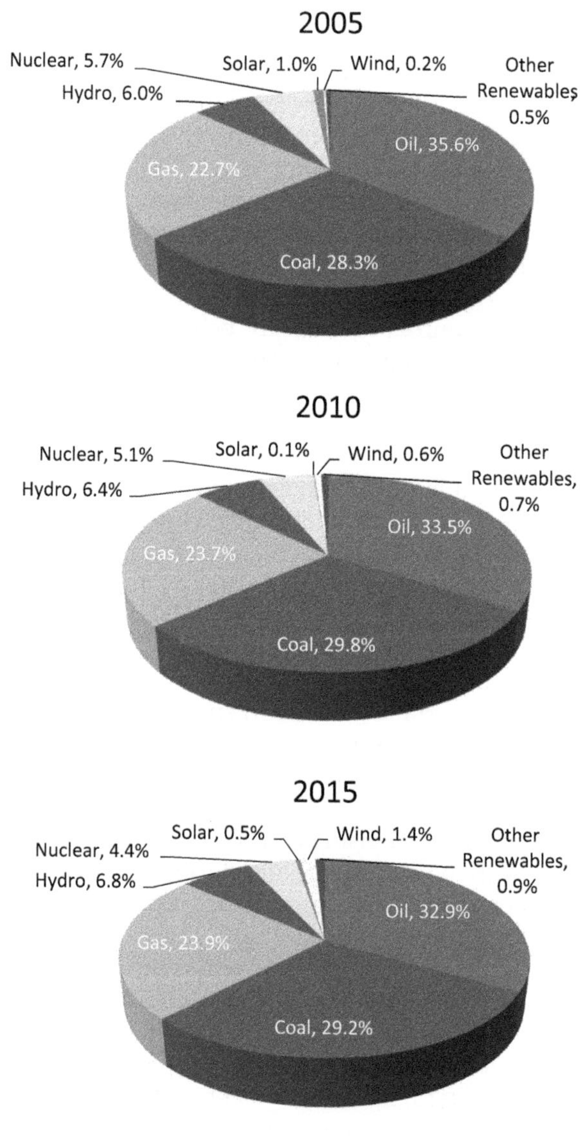


**FIGURE 1.1**
Global primary energy consumption. Data from WEC (2016).

**TABLE 1.1**
Projection of the World's Regional Energy Consumption 2012–2040 (EJ).

| Region | 2012 | 2025 | 2040 |
|---|---|---|---|
| OECD[a] | 251 | 274 | 298 |
| Americas | 121 | 135 | 146 |
| Europe | 85 | 92 | 101 |
| Asia | 41 | 47 | 51 |
| Non-OECD | 322 | 413 | 533 |
| Europe | 54 | 58 | 61 |
| Asia | 186 | 260 | 340 |
| Middle East | 34 | 47 | 65 |
| Africa | 34 | 32 | 46 |
| Americas | 33 | 39 | 50 |
| World Total | 591 | 710 | 860 |

Data from WEC (2016).
[a] Organization for Economic Cooperation and Development member countries. Europe: Austria, Belgium, Czech Republic, Denmark, Estonia, Finland, France, Germany, Greece, Hungary, Iceland, Ireland, Italy, Luxembourg, Netherlands, Norway, Poland, Portugal, Slovakia, Slovenia, Spain, Sweden, Switzerland, Turkey, United Kingdom. Other member countries: Australia, Canada, Chile, Israel, Japan, Mexico, New Zealand, South Korea, USA.

energy, and other renewables (see Figure 1.1). And the International Energy Outlook 2016 projects remarkable growth in worldwide energy consumption in different regions over the period of 2012–2040 (Table 1.1).

The World Energy Council (WEC) and International Energy Agency (IEA) reported that the world energy demand in 2020 is expected to rise by 50–80% from the 1990s and the number is projected to increase by more than one-third by the year 2035 compared to the present level. It is expected that approximately 60% of the growth in demand will mainly come from China, India, and the Middle East (IEA, 2013; WEC, 2013; UN, 2014). The global energy demand from all sources is expected to expand, including oil (13%), coal (17%), natural gas (48%), nuclear (66%), and renewables (77%) (UN, 2015). Global electricity demand as well is forecasted to increase by 70% by the year 2030 (IEA, 2013). The renewable energy contribution, including hydropower, geothermal, and solar, is expected to reach 30% of total electricity supply by 2035 (IEA, 2013).

Due to the environmental concerns and others, the contribution of fossil fuels is expected to drop with the rapid growth of clean and renewable energy sources. Fossil fuel reservoirs are not equally distributed around the world. Table 1.2 summarizes the fossil fuel reserves and productions in top five countries, and shows that they are mainly concentrated in the Middle East, North America, Russian Federation, and Asia. About 77% of total coal production was exploited from the top five reserves, while only 50% and 24% of global natural gas and oil productions, respectively.

The global capacity of primary renewable energy in top five countries in 2015 is summarized in Table 1.3 (REN21, 2015). Thermal electricity generation using coal, natural gas and nuclear will however keep contributing the power generation around the world, and sadly, coal will continue to dominate having the largest share in global electricity production (EIA, 2016).

**TABLE 1.2**

Oil, Natural Gas, and Coal Reserves in Top Five Countries and Their Productions in 2016.

| Oil | | | Gas | | | Coal | | |
|---|---|---|---|---|---|---|---|---|
| **Country** | **Reserves (Mt)** | **Production (Mt)** | **Country** | **Reserves (tcm)** | **Production (bcm)** | **Country** | **Reserves (Mt)** | **Production (Mt)** |
| Venezuela | 47,000 | 124.1 | Iran | 33.5 | 202.4 | USA | 281,582 | 521.1 |
| Saudi Arabia | 36,600 | 285.7 | Russian Fed. | 32.3 | 579.4 | China | 244,010 | 2,408.1 |
| Canada | 27,600 | 218.2 | Qatar | 24.3 | 181.2 | Russian Fed. | 160,364 | 275.4 |
| Iran | 21,800 | 216.4 | Turkmenistan | 17.5 | 66.8 | Australia | 144,818 | 427.6 |
| Iraq | 20,600 | 218.4 | USA | 8.7 | 749.2 | India | 94,769 | 401.3 |
| **Global** | **240,700** | **4,382.4** | **Global** | **186.6** | **3,551.6** | **Global** | **1,139,331** | **5,223.4** |

Data from BP (2017).

**TABLE 1.3**

Global Capacity of Renewable Energy in Top Five Countries (2015).

| Wind (GW) | | Hydropower (GW) | | Geothermal Power (GW) | |
|---|---|---|---|---|---|
| China | 145.4 | China | 296 | US | 3.6 |
| US | 75 | Brazil | 92 | Philippines | 1.9 |
| Germany | 45 | US | 80 | Indonesia | 1.4 |
| India | 25.1 | Canada | 79 | Mexico | 1.1 |
| Spain | 23 | Russian Fed. | 48 | New Zealand | 1.0 |
| **Global** | **433** | **Global** | **1,064** | **Global** | **13.2** |
| **Biofuels (GL)** | | **Solar PV (GW)** | | **Concentrating Solar Thermal Power (MW)** | |
| US | 62.1 | China | 43.5 | Spain | 2,300 |
| Brazil | 32.3 | Germany | 39.7 | US | 1,738 |
| Germany | 3.8 | Japan | 34.4 | India | 225 |
| Netherlands | 3.5 | US | 25.6 | Morocco | 180 |
| France | 3.4 | Italy | 18.9 | South Africa | 150 |
| **Global** | **130.7** | **Global** | **227.0** | **Global** | **4,755** |

Data from REN21 (2015).

Over the past two decades, hundreds of millions of people living in developing countries, particularly in China and India, have gained access to modern energy services due to their quickly growing urbanization and economic development (Olsson, 2012). However, there were still 1.3 billion people who did not have any access to electricity in the year 2010, and approximately 2.6 billion people still relied on traditional biomass energy, such as wood and other forms of biomass for cooking and heating purposes (UN, 2015; REN21, 2015). Moreover, it has been stated that about 9 billion people are projected to require affordable energy services by the year 2050 (IEA, 2012b). Fossil fuels have always had the major share in the global primary energy consumption. Even though the share of fossil fuel consumption has decreased in recent years and is predicted to decrease further, they will keep continuing to hold the major share in the primary energy mix in the foreseeable future as more unconventional fossil fuels are explored. Overall fossil fuel consumption increased approximately 51% in the period of 1995–2015, and it is predicted that the consumption will increase approximately 18% more in the period of 2015–2035. The renewable energy has recently experienced an exponential growth in the share of the energy markets; however, it still accounts for only a small portion of total primary energy supply, and their contribution is not expected to have enormous impacts in the near future. The WEC's Outlook Report in 2013 indicated that development of the renewables not only faces demand challenges, but also depends significantly upon financial and legal supports from governments in developing countries (WEC, 2013). In reality, the renewable energy development rate has been slower than what was predicted 20 years ago (WEC, 2013). A comparison of global primary energy consumption over the past 10 years is summarized in Figure 1.1. The findings clearly show that the majority of consumption was in the form of fossil fuels, even though a slight decline (0.61%) was observed from 2005 to 2015. Renewables only shared a small fraction that fluctuated from 12 to 14%, but it seems to have gone up steadily.

During the transition period from conventional to more sustainable future energy systems, it is primarily expected to improve energy systems' comprehensive performances; that is, energetic, exergetic, economic, and environmental performances. With the current developments in cutting-edge technologies and artificial intelligence applications, it is now possible to change the course of action in dealing with problematic energy issues, covering the entire energy spectrum: energy fundamentals and concepts, energy materials, energy production, energy conversion, and overall energy management. In the past, the energy consumption to gross domestic product ratio (energy intensity) was used as a development indicator – the countries consuming more energy were considered as more developed. Later, energy use efficiency became more meaningful; and now, almost everything has evolved into a more efficient and integrated smart management system.

## 1.2 Integrated Energy Management

The First Nations People believed in four sacred elements, namely the earth, air, water, and fire (meaning energy). It may not be so sacred for them, but the Sumerians, the Babylonians, and later, the ancient Greeks as well believed in the same four. The challenge we are facing today in our increasingly populated planet is how to live within such that its resources (the four sacred elements) can sustain human civilization and the natural environment in a healthy and functional state for many generations in the foreseeable future. How can we continue to improve human welfare within the limits of our planet's resources? The solution is sustainable development. In Our Common Future, also known as the Brundtland Report, the 1987 report of the World Commission on Environment and Development (1987), sustainable development was defined as development that "… meets the needs of the present without compromising the ability of future generations to meet their own needs." In other words, it is the use of renewable and nonrenewable resources in a manner that satisfies our current needs without compromising future availability of resources. Then, "What is the maximum number of people our planet can sustain?" That is, what is the sustainable human carrying capacity of our planet, sustaining the natural environment, its species, and its systems in a healthy and functional state? We are using our renewable and nonrenewable environmental resources faster than they can be replenished. We are extracting minerals, oil, and ground water without sufficient concern for their limits or the need to recycle them. As a result, there is an existing shortage of some resources and there will be more shortages of other resources in the future. Therefore, we must learn how to sustain our resources so that they can continue to fuel human civilization and other living things on our planet within natural limits for many generations. Sustainability, therefore, means leaving our children and future generations a planet at least as rich as we live in today, and requires maintaining fully functional and healthy natural environment, which basically forms the foundation for our civilization. Due to the population increase, in reality, sustainability is a moving target; therefore, more resources will be required to maintain a healthy and functional status. Our choices, up to certain extent, determine both the local and global sustainable carrying capacity for the human population of our planet. For instance, more people in the future will mean more energy use; and in addition, people in less developed parts of the world will be consuming more energy on per capita basis when they reach higher standards of living. Today, the environmental impacts of fossil fuels are significant and may become even worse in the years to come due to the inevitable increase in their uses. We optimistically believe that we have the required knowledge and technology to ensure a worry-free future, if we are to take the action now. Rather than relying on fossil fuels heavily, we have the option to build a sustainable energy future based on careful planning, innovative thinking, and a willingness to move from our dependency on fossil fuels. Here is a good place to introduce the concept of integrated energy management. According to this concept, a series of options need to be employed, and the range will be determined locally. The sources of energy will have to consider both renewable and nonrenewable sources. With the integrated energy management concept, we would be able to provide diverse and reliable sources of energy including more local and decentralized production of energy, considering energy efficiency and conservation measures, protect our local or global environments, and make sure that our future generations live in a healthy and functional environment with rich enough resources. The biggest challenge for a sustainable energy future however is the human aspect – changing our total way of living and thinking. The overall goal of sustainable energy development requires and maintains multiple linkages among energy production, energy consumption, human wellbeing, and environmental quality. And for this, local sustainability is the key.

Greenhouses are quite often considered as energy consuming structures; however, they also produce energy via biomass production and have even further potential to produce clean and renewable energy through innovative energy producing exterior and interior covering materials and a bio-agent that truly cares about the aforementioned four sacred elements, a non-food feedstock microalgae's energy and bioproducts potential within fully enriching cycles, and therefore, simply sustaining our civilization under cover.

## Bibliography


BP (2017) http://bp.com/statisticalreview Retrieved 21-10-2017.

EIA (2016) *United States Energy Information Administration, International Energy Outlook 2016.* Washington DC.

IEA (2012a) *International Energy Agency,* Key World Energy Statistics, OECD, Paris.

IEA (2012b) *International Energy Agency, World Energy Outlook 2012,* OECD, Paris.

IEA (2013) *International Energy Agency, World Energy Outlook 2013,* OECD, Paris.

Olsson, G (2012) *Water and Energy: Threats and Opportunities,* IWA Publishing, London, UK.

Omer, A (2008) Energy, environment and sustainable development, *Renewable and Sustainable Energy Reviews,* 12(9): 2265-2300.

REN21 (2015) *Renewable Energy Policy Network for the 21st Century, Renewables 2015, Global Status Report,* Paris.

REN21 (2016) *Renewable Energy Policy Network for the 21st Century, Renewables 2016 Global Status Report,* Paris.

U.S. Census Bureau (2012) *International Programs,* World Population.

UN (2014) *U.N. World Water Assessment Program, The United Nations World Water Development Report* 2014: Water and Energy, Paris.

UN (2015) U.N. World Water Assessment Program, *The United Nation World Water Development Report: Water for a Sustainable World,* UNESCO, Paris.

WEC (2013) *World Energy Council, 2013 Survey of World Energy Resources,* London, UK.

WEC (2016) *World Energy Council, 2016 Survey of World Energy Resources,* London, UK.

World Commission on Environment and Development (1987). *Our Common Future.* Oxford: Oxford University Press. ISBN 019282080X.

# 2

# Fundamentals

## 2.1 Introduction

This chapter establishes the essential groundwork for this book. Therefore, it will first start with covering the energy units, conversions, and dimensional analysis. The base and derived quantities, relationships between quantities, and quantity equations will be discussed. The multiples and submultiples in SI unit system will be presented, and an introduction will also be provided for dimensional analysis, which is quite a useful method for deriving algebraic relationships between different physical quantities. Furthermore, conversion factors for energy related units between SI and I-P will also be provided.

The chapter will then move into and briefly review thermodynamics, introducing the basic concepts: the laws of thermodynamics, and the energy and mass balance concepts before getting into different modes of heat transfer for different geometries. The chapter will then be relating the heat transfer analyses to different energy applications, for instance, heat transfers though walls or windows, which is essential for determining heating or cooling loads of buildings. Furthermore, an analysis of heat exchangers for numerous applications (also common in greenhouse operations) will be presented as well. And then, the chapter will introduce the study of psychrometrics, which deals with thermodynamic properties of moist air. Using these properties, we will analyze engineering situations and processes involving moist air, covering perfect gas relations and describing their use in thermal comfort problems, particularly in greenhouse environments. And then, another subsection will cover fuels, combustion of fuels, theoretical and excess air requirements for combustion, flue gas and carbon emissions, as well as the cost comparisons of different heating fuels.

After that, as the piping systems for water, natural gas, oil, etc., are all designed using the principles of fluid mechanics, the chapter will focus on the required essentials of fluid mechanics. A dedicated subsection will cover fundamental fluid mechanics, which is essential for handling general heating, ventilating, and air-conditioning processes.

And finally, as solar energy powers virtually everything in the Earth and atmosphere system and is a vitally important energy resource for greenhouse operations, this chapter will have a fully dedicated section on light, radiation, characteristics of solar radiation, and relevant data. The solar spectrum, direct and diffuse components of solar radiation, as well as the temporal and spatial distributions of solar energy will be studied in depth before the chapter is concluded.

Plenty of numerical examples are provided in each and every subsection of the chapter to demonstrate and grasp the relevant concepts easily.

## 2.2 Units, Conversions, and Dimensional Analysis

### 2.2.1 Introduction

When dealing with engineering and scientific relationships, in order to appreciate the magnitudes of physical quantities, it is essential to have a solid grasp of units, and recognize two types of equations, namely quantity equations and numerical equations. Both types are found in texts and reference books, and the concept of units and quantities is useful in understanding their respective features. In this chapter, we cover the main features of quantities and quantity equations, and provide the most important units and conversions relating to energy. Quantity equations are also called equations between quantities, or physical equations. And, numerical equations are alternatively called measure equations. We also introduce the technique of dimensional analysis, which is used to derive basic physical relationships without performing a full analysis of a system.

### 2.2.2 Quantities

In 1954, the 10th General Conference on Weights and Measures (CGPM) decided that an international system should be derived from six base units to provide for the measurement of temperature and optical radiation in addition to mechanical and electromagnetic quantities. Six base units recommended at this conference were: the meter, kilogram, second, ampere, degree Kelvin (later renamed Kelvin), and candela. In 1960, the 11th CGPM named the system the International System of Units, SI from the French name, Le Système International d'Unités (CGMP, 1961). Later, the seventh base unit, the mole, was added in 1971 by the 14th CGPM (CGMP, 1972). SI is the modern form of metric system, and today is the most widely used measurement system.

Therefore, the International System of Quantities (ISQ) is now a system based on seven base quantities, length, mass, time, thermodynamic temperature, electric current, luminous intensity, and amount of substance. Other quantities such as area, pressure, and electrical resistance are all derived from these base quantities. The ISQ defines quantity as any physical property that can be measured with the SI units (BIPM, 2012). A quantity may also be a physical constant, such as the gas constant, or the Planck's constant. Several hundred quantities are employed to describe and measure the physical world, and a few of these quantities are listed below (Wildi, 1995):

| | | | |
|---|---|---|---|
| length | viscosity | area | electromotive force |
| time | energy | luminance | entropy |
| mass | speed | angle | pressure |
| force | power | temperature | momentum |

#### 2.2.2.1 *Relationship Between Quantities*

The study of physics is to a great extent can be defined as the study of mathematical relationships among various physical properties. Physical quantities are defined, as above, when these properties allow a reasonable mathematical description. The relationship of all other quantities can be established in terms of a few base quantities selected properly, either by definition, by geometry, by physical law, or by a combination of the base quantities.

For instance, pressure is a quantity that is related, by definition, to a quantity force divided by a quantity area. Area, on the other hand, is a quantity related, by geometry, to the product of two quantities of length. Moreover, force is a quantity related (by Newton's Second Law) to the quantity mass times the quantity acceleration.

The relationships between quantities are expressed in the form of quantity equations. We can relate even an isolated quantity such as temperature to the quantities pressure, volume, and mass. We can further relate the quantities length and time by using the universal constant and the speed of light. Therefore, if we define our concepts correctly, we can relate any quantity to any other quantity. Thus, the equation area = length × width is a quantity equation, which states that the quantity (area of a rectangle) is equal to the quantity (length) times the quantity (width).

#### *2.2.2.2 Base Quantities*

In order to reduce a set of quantity equations, we have to first establish a number of so-called base quantities. Hence, base quantities are called the building blocks upon which we develop the entire structure and relationships of the physical world. As mentioned earlier, the international system of units, or SI, makes use of seven base quantities: mass (kg), length (m), time (s), temperature (K), electric current (A), luminous intensity (cd), and amount of substance (mol). The number of base quantities, as well as their choice, is quite an arbitrary choice; but, generally, we select quantities that are easy to understand and frequently used, and for which accurate and measurable standards can be established.

#### *2.2.2.3 Derived Quantities*

As mentioned in the relationship section earlier, using the selected base quantities as building blocks, derived quantities are expressed as those which can be deducted by definition, geometry, or physical law. Some derived quantity examples are area (equals the products of two lengths), velocity (equals length/time), and force (equals mass × acceleration), pressure, power, etc. We also have, what is called, supplementary units (as a class of derived units) namely the plane angle (radian = rad = m/m) and solid angle (steradian = sr = $m^2/m^2$).

#### *2.2.2.4 Multiples and Submultiples of Quantities*

Note that the magnitude of a quantity can have an extremely large range. In an effort to handle such a large range, the SI unit system generated 20 prefixes shown in Table 2.1.

#### *2.2.2.5 Types of Quantity Equations*

The energy of wind, the pressure at the bottom of an air or water column, the weight of an object, and the viscosity of a liquid are all physical quantities of nature. And, whether they are measured or not, these quantities are always there interacting with each other according to fundamental laws. Physicists often express these laws in terms of quantity equations because quantities conform to these laws. Quantity equations possess two important features; first, they show the relationship between quantities, and second, they can be used with any system of units.

**TABLE 2.1**

Multiples and Submultiples in SI Unit System.

| Prefix | Symbol | Multiplier | | Example |
|---|---|---|---|---|
| yotta | Y | $10^{24}$ | 1 000 000 000 000 000 000 000 000 | 5 Ym = 5 yottameters = $5 \times 10^{24}$ m |
| zetta | Z | $10^{21}$ | 1 000 000 000 000 000 000 000 | 2 Zm = 2 zettameters = $2 \times 10^{21}$ m |
| exa | E | $10^{18}$ | 1 000 000 000 000 000 000 | 7 Em = 7 exameters = $7 \times 10^{18}$ m |
| peta | P | $10^{15}$ | 1 000 000 000 000 000 | 6 PJ = 6 petajoules = $6 \times 10^{15}$ J |
| tera | T | $10^{12}$ | 1 000 000 000 000 | 5 TW = 5 terawatts = $5 \times 10^{12}$ W |
| giga | G | $10^{9}$ | 1 000 000 000 | 8 GJ = 8 gigajoules = $8 \times 10^{9}$ J |
| mega | M | $10^{6}$ | 1 000 000 | 2 MW = 2 megawatts = $2 \times 10^{6}$ W |
| kilo | k | $10^{3}$ | 1 000 | 3 km = 3 kilometers = $3 \times 10^{3}$ m |
| hecto | h | 100 | 100 | 6 hL = 6 hectoliters = 600 L |
| deka | da | 10 | 10 | 2 dam = 2 decameters = 20 m |
| deci | d | $10^{-1}$ | 0.1 | 3 dL = 3 deciliters = 0.3 L |
| centi | c | $10^{-2}$ | 0.01 | 5 cm = 5 centimeters = 0.05 m |
| milli | m | $10^{-3}$ | 0.001 | 9 mV = 9 millivolts = $9 \times 10^{-3}$ V |
| micro | μ | $10^{-6}$ | 0.000 001 | 5 μm = 5 micrometers = $5 \times 10^{-6}$ m |
| nano | n | $10^{-9}$ | 0.000 000 001 | 2 ns = 2 nanoseconds = $2 \times 10^{-9}$ s |
| pico | p | $10^{-12}$ | 0.000 000 000 001 | 3 pJ = 3 picojoules = $3 \times 10^{-12}$ J |
| femto | f | $10^{-15}$ | 0.000 000 000 000 001 | 6 fm = 6 femtometers = $6 \times 10^{-15}$m |
| atto | a | $10^{-18}$ | 0.000 000 000 000 000 001 | 5 aJ = 5 attojoules = $5 \times 10^{-18}$ J |
| zepto | z | $10^{-21}$ | 0.000 000 000 000 000 000 001 | 6 zJ = 6 zeptojoules = $6 \times 10^{-21}$ J |
| yocto | y | $10^{-24}$ | 0.000 000 000 000 000 000 000 001 | 8 yJ = 8 yoctojoules = $8 \times 10^{-24}$ J |

There are three basic types of quantity equations:

1. *Quantity equations developed from the laws of nature*; for instance, Newton's Second Law of motion

$$F = ma$$

where $F$ is the magnitude of the force, $m$ is the magnitude of the mass, and $a$ is the magnitude of the acceleration.

2. *Quantity equations developed from geometry*; for instance, area of a circle

$$A = \pi r^2$$

where $A$ is the magnitude of the area, $\pi$ is the coefficient based on the geometry of a circle, and $r$ is the magnitude of the radius.

3. *Quantity equations developed from a definition*; for instance, definition of pressure

$$p = F/A$$

where $p$ is the magnitude of the pressure, $F$ is the magnitude of the force, and $A$ is the magnitude of the area.

Many quantity equations can be developed as a combination of the basic quantity equations given above, and in all cases, we can use any units we want to describe the magnitudes of the relevant physical quantities.

### 2.2.3 Dimensional Analysis

Dimensional analysis is quite a useful method for deriving an algebraic relationship between different physical quantities, which relies on good physical intuition in choosing the different appropriate physical variables. The idea behind this analysis is that each variable is expressed in terms of its fundamental units of mass M, length L, and time T, etc., raised to some arbitrary index $a$, $b$, $c$, etc. These unknown indices are then determined by equating the indices of like units (Andrews & Jelley, 2007). One might also choose force, length, and mass as the base dimensions, with associated dimensions F, L, M; which corresponds to a different basis. It may sometimes be useful to choose one or another extended set of dimensional symbols. In electromagnetism, for instance, it may be advantageous to use dimensions of M, L, T, and Q, where Q is used to represent the dimension of electric charge. Another example is that, for instance, in thermodynamics, the base set of dimensions is often extended to include a dimension for temperature, θ.

Let's now perform a simple dimensional analysis to find an expression for the hydrostatic pressure in a fluid. The hydrostatic pressure is dependent on the density $\rho$, the gravitational acceleration $g$, and depth $h$. Now, let's assume a general algebraic equation in the form of

$$p = k\rho^a g^b h^c$$

where $k$ is a coefficient (dimensionless), and $a$, $b$, and $c$ are the indices (numbers) to be determined. Now, we can replace each symbol by its fundamental physical unit, and have

$$\mathrm{M^1L^{-1}T^{-2}} = \left(\mathrm{ML^{-3}}\right)^a \left(\mathrm{LT^{-2}}\right)^b \left(\mathrm{L}\right)^c$$

or

$$\mathrm{M^1L^{-1}T^{-2}} = \mathrm{M}^a\mathrm{L}^{-3a+b+c}\mathrm{T}^{-2b}$$

M, L, and T are all independent quantities; therefore, we can equate the indices on both sides, and have the following equations:

$$1 = a, -1 = -3a + b + c, -2 = -2b$$

Then, we can solve for and find that $a = b = c = 1$; consequently, the expression for hydrostatic pressure can be found as $p = k \rho g h$, where the coefficient $k$ cannot be determined from dimensional analysis because it is dimensionless.

More dimensional analysis examples are provided in the Examples section later.

### 2.2.4 Units and Conversions

This section, as modified after ASHRAE (1984) and (1989), references Standard for Metric Practice, ASTM Standard E 380-84 (ASTM, n.d.) as one of the basic standards for SI usage (National Bureau of Standards, n.d; CSA, n.d.; ASME, n.d.; IOS, n.d.). Table 2.2 provides conversion factors rounded to three or four significant figures for conversion between SI and I-P. And Table 2.3 provides conversion factors for different physical quantities related to energy further.

#### *Useful Units in Electricity*

*Coulomb*

In an electric circuit, the unit of electric charge in SI is coulomb, and has the symbol, C. An ampere, which has the symbol of A, is defined as the amount of charge transported through any cross section of a conductor in one second by a constant current of one ampere, and is equivalent to the amount of charge on about 6,241,510,000,000,000,000 electrons.

*Volt*

In an electric circuit, the unit of electric potential, potential difference, and electromotive force in SI is volt and has the symbol V. If and when we consider our house wiring as plumbing, volts can then be considered as a measure of the water pressure. One volt is the potential difference between two points on a conductor when the current flowing is one ampere and the power dissipated between the points is one watt.

The volt is a derived unit, and in terms of base units, it can be expressed as follows:

$$\text{Volt} = \text{Watt}/\text{Ampere} = \left(\text{m}^2.\text{kg}\right)/\left(\text{s}^3.\text{A}\right)$$

*Watt*

In an electric circuit, one watt (joules per second) is a current of one ampere at a pressure of one volt. In terms of base units,

$$\text{Watt} = \text{J}/\text{s} = \left(\text{m}^2.\text{kg}\right)/\text{s}^3$$

*Ohm*

In an electric circuit, the unit of electrical resistance (a derived unit) in SI is called ohm and has the symbol of Ω. One ohm is defined as the electrical resistance between two points on a conductor when a constant potential difference of one volt, applied to these points, produces in the conductor a current of one ampere. Ohm is a derived unit, and in terms of base units, it can be expressed as follows:

$$\text{Ohm}(\Omega) = \text{Volt}/\text{Ampere} = \left(\text{m}^2.\text{kg}\right)/\left(\text{s}^3.\text{A}^2\right).$$

**TABLE 2.2**

SI Energy Related Units and Conversions.

| Divide | By | To obtain | Divide | By | To obtain |
|---|---|---|---|---|---|
| ha | 0.405 | acre | J | 1.36 | ft·Ib$_f$ (work) |
| kPa | 100 | bar | J/kg | 2.99 | ft·Ib$_f$/Ib (Specific energy) |
| L | 159 | barrel (42 US gal, petroleum) | W | 0.0226 | ft·Ib$_f$/min (power) |
| m$^3$ | 0.159 | | L | 3.79 | gallon (US, 231 in$^3$) |
| kJ | 1.055 | Btu, IT | m$^3$ | 0.00379 | gallon |
| kJ/m$^3$; J/L | 37.3 | Btu/ft$^3$ | mL/s | 1.05 | gph |
| kJ/L | 0.279 | Btu/gal | L/s | 0.0631 | gpm |
| W/(m·K) | 1.731 | Btu·ft/h·ft$^3$·°F | mL/J | 0.0179 | gpm/ton refrigeration |
| | | Btu·in/(h· ft$^3$·°F) | g | 0.0648 | grain (1/7,000 Ib) |
| W/(m·K) | 0.144 | (thermal conductivity, k) | mg/L | 17.1 | gr/gal |
| W/(m·°C) | | | g/kg | 0.143 | gr/Ib |
| W | 0.293 | Btu/h | kW | 9.81 | horsepower (boiler) |
| kJ/m$^3$ | 11.4 | Btu/ft$^2$ | kW | 0.746 | horsepower (550 ft-Ib$_f$/s) |
| GJ/(y·m$^2$) | 0.0000114 | Btu/(y·ft$^2$) | mm | *25.4 | inch |
| W/m$^2$ | 3.15 | Btu/(h·ft$^2$) | kPa | 3.38 | in of mercury (60°F) |
| W/(m$^2$·K) | 5.68 | (overall heat transfer coef., *U*) | Pa | 249 | in of water (60°F) |
| | | (thermal conductance, *C*) | mm/m | 0.833 | in/100 ft, thermal expansion |
| kJ/kg | 2.33 | Btu/Ib | mN·m | 113 | in·Ib$_f$ (torque or moment) |
| kJ/(kg·K) | 4.19 | Btu/(Ib·°F) (specific heat, C) | mm$^2$ | 645 | in$^2$ |
| kJ/(kg·°C) | | | mL | 16.4 | in$^3$ (volume) |
| m$^3$ | 0.0352 | bushel | mL/s | 0.273 | in$^3$/min (SCIM) |
| J | 4.19 | calorie, gram | mm$^3$ | 16 400 | in$^3$ (section modulus) |
| kJ | 4.19 | calorie, kilogram; kilocalorie | mm$^4$ | 416 000 | in$^4$ (section moment) |
| mPa·s | *1.00 | centipoise, viscosity, μ | m/s | 0.278 | km/h |
| | | (absolute, dynamic) | MJ | *3.60 | kWh |
| mm$^2$/s | *1.00 | centistokes, kinematic viscosity, *ν*) | GJ/(y·m$^2$) | 0.0388 | kWh (y · ft$^2$) |
| Pa | *0.100 | dyne/cm$^2$ | J/L | 2.12 | kWh/100 cfm |
| W | 44.0 | EDR hot water (150 Btu/h) | N | 9.81 | kilopond (kg force) |
| W | 70.3 | EDR stream (240 Btu/h) | kN | 4.45 | kip (1,000 Ib$_f$) |
| COP | 0.293 | EER | MPa | 6.89 | kip/in$^2$ (ksi) |
| m | *0.3048 | ft | m$^3$ | *0.001 | liter |
| mm | *304.8 | ft | mPa | 133 | micron of mercury (60°F) |
| m/s | 0.00508 | ft/min, fpm | km | 1.61 | mile |
| m/s | *0.3048 | ft/s, fps | km | 1.85 | mile/nautical |
| kPa | 2.99 | ft of water | km/h | 1.61 | mph |

*(Continued)*

**TABLE 2.2 (Concluded)**

SI Energy Related Units and Conversions.

| Divide | By | To obtain | Divide | By | To obtain |
|---|---|---|---|---|---|
| kPa/m | 0.0981 | ft of water per 100 ft pipe | m/s | 0.44 | mph |
| $m^2$ | 0.0929 | $ft^2$ | kPa | *0.100 | millibar |
| $m^2$·K/W | | | kPa | 0.133 | mm of mercury (60°F) |
| $m^2$·°C/W | 0.176 | $ft^2$·h·°F/Btu | Pa | 9.80 | mm of water (60°F) |
| | | (thermal resistance, *R*) | kPa | 9.80 | meter of water |
| $mm^2$/s | 92 900 | $ft^2$/s, kinematic viscosity, $\nu$ | g | 28.3 | ounce (mass, avoirdupois) |
| L | 28.3 | $ft^3$ | N | 0.278 | ounce (force or thrust) |
| $m^3$ | 0.0283 | $ft^3$ | mL | 29.6 | ounce (liquid, US) |
| mL/S | 7.78 | $ft^3$/h, cfh | mN·m | 7.06 | ounce inch (torque or moment) |
| L/s | 0.472 | $ft^3$/min, cfm | g/L | 7.49 | ounce (avoirdupois) per gallon |
| L/s | 28.3 | $ft^3$/s, cfs | ng/(s·$m^2$·Pa) | 57.4 | perm (permeance) |
| N·m | 1.36 | ft·$Ib_f$ (torque or moment) | ng/(s·m·Pa) | 1.46 | perm inch (permeability) |
| mL | 473 | pint (liquid, US) | kg/$m^3$ | 16.0 | Ib/$ft^3$ (density, $\rho$) |
| | | pound | kg/$m^3$ | 120 | Ib/gallon |
| kg | 0.454 | Ib (mass) | mg/kg | *1.00 | ppm (by mass) |
| g | 454 | Ib (mass) | kPa | 6.89 | psi |
| N | 4.45 | $Ib_f$ (force or thrust) | EJ | 1.055 | quad |
| kg/m | 1.49 | Ib/ft (uniform load) | L | 0.946 | quart (liquid US) |
| mPa·s | 0.413 | $Ib_m$/(ft·h) viscosity | $m^2$ | 9.29 | square (100 sq ft) |
| | | (absolute, dynamic, $\mu$) | mL | 15 | tablespoon (approximately) |
| mPa·s | 1,490 | $Ib_f$/(ft·s) viscosity | mL | 5 | teaspoon (approximately) |
| | | (absolute, dynamic, $\mu$) | MJ | 105.5 | therm (US) |
| g/s | 0.126 | Ib/h | t (tonne); Mg | 1.016 | ton, long (2,240 Ib) |
| kg/s | 0.00756 | Ib/min | t (tonne); Mg | 0.907 | ton, short (2,000 Ib) |
| kW | 0.284 | Ib of steam per hour @ 212°F (100°C) | kW | 3.52 | ton, refrigeration (12,000 Btu/h) |
| | | | Pa | 133 | torr (1 mmHg@0°C) |
| Pa | 47.9 | $Ib_f$/$ft^2$ | W/$m^2$ | 10.8 | watt per square foot |
| mPa·s | 47,900 | $Ib_f$·s/$ft^2$ viscosity | m | *0.9144 | yd |
| | | (absolute, dynamic, $\mu$) | $m^2$ | 0.836 | $yd^2$ |
| kg/$m^2$ | 4.88 | Ib/$ft^2$ | $m^3$ | 0.765 | $yd^3$ |
| **To obtain** | **by** | **Multiply** | **To obtain** | **by** | **Multiply** |

* Conversion factor is exact.

**TABLE 2.3**

Conversion Factors.

| **Pressure** | | | | | | | |
|---|---|---|---|---|---|---|---|
| pascal | dyne/cm² | kg/cm² | bar | mm Hg | atm | in.Hg | psi |
| 1 | 10 | $1.0192 \times 10^{-5}$ | $10^{-5}$ | 0.00750 | $9.8692 \times 10^{-6}$ | $2.953 \times 10^{-4}$ | $1.45038 \times 10^{-4}$ |
| 0.100 | 1 | $1.01972 \times 10^{-6}$ | $10^{-6}$ | 0.000750 | $9.8692 \times 10^{-7}$ | $2.953 \times 10^{-5}$ | $1.45038 \times 10^{-5}$ |
| 98,066 | 980,665 | 1 | 0.98066 | 735.559 | 0.96784 | 28.959 | 14.223 |
| $10^5$ | $10^6$ | 1.01972 | 1 | 750.062 | 0.98692 | 29.530 | 14.5038 |
| 133.32 | 1,333.2 | 0.0013595 | 0.0013332 | 1 | 0.00131579 | 0.03937 | 0.0193368 |
| 101,325 | 1,013,250 | 1.03323 | 1.01325 | 760.0 | 1 | 29.921 | 14.6960 |
| 3,386.4 | 33,864 | 0.034532 | 0.33864 | 25.400 | 0.033421 | 1 | 0.491154 |
| 6,894.8 | 68,948 | 0.07030696 | 0.068948 | 51.715 | 0.068046 | 2.0360 | 1 |
| **Mass** | | kg | lb | | | | |
| | | 1 = | 2.20462 | | | | |
| | | 0.45359 = | 1 | | | | |
| **Volume** | | meter³ | liter | gal | ft³ | in³ | |
| | | 1 | 1,000 | 264.173 | 35.315 | 61,023.74 | |
| | | 0.001* | 1 | 0.264173 | 0.035315 | 61.02374 | |
| | | 0.0037854 | 3.7854 | 1 | 0.13368 | 231.0 | |
| | | 0.028317 | 28.317 | 7.48055 | 1 | 1,728* | |
| | | $1.63871 \times 10^{-5}$ | 0.0163871 | $4.329 \times 10^{-3}$ | $5.787 \times 10^{-4}$ | 1 | |

*(Continued)*

**TABLE 2.3 (Continued)**

Conversion Factors.

| | | | | | |
|---|---|---|---|---|---|
| **Energy** | Watt-sec | joule | calorie | fl·lb | Btu |
| | 1 | 1 | 0.2390 | 0.73756 | $9.4845 \times 10^{4}$ |
| | 4.184* | 4.184* | 1 | 3.08596 | $3.9683 \times 10^{3}$ |
| | 1.3558 | 1.3558 | 0.32405 | 1 | $1.2859 \times 10^{3}$ |
| | 1,054.35 | 1,054.35 | 251.9957 | 777.65 | 1 |
| **Density** | $kg/m^3$ (g/L) | $g/cm^3$ | lb/gal | $lb/ft^3$ | |
| | 1 | 0.001 | 0.008345 | 0.0624280 | |
| | 1,000 | 1 | 8.34538 | 62.4280 | |
| | 119.827 | 0.119827 | 1 | 7.48055 | |
| | 16.018463 | 0.016018 | 0.133680 | 1 | |
| **Specific Volume** | $m^3/kg$ (L/g) | $cm^3/g$ | gal/lb | $ft^3/lb$ | |
| | 1 | 1,000 | 119.827 | 16.018463 | |
| | 0.001 | 1 | 0.119827 | 0.016018 | |
| | 0.008345 | 8.34538 | 1 | 0.133680 | |
| | 0.0624280 | 62.4280 | 7.48055 | 1 | |
| **Specific Heat** | J/(g·K) | cal/(g·K) | Btu/lb·°F | | |
| **or Entropy** | 1 | 0.2390 | 0.2390 | | |
| | 4.184* | 1 | 1.0 | | |
| | 4.184* | 1 | 1 | | |
| **Entropy** | J/g | cal/g | Btu/lb | | |
| | 1 | 0.2390 | 0.43021 | | |
| | 4.184* | 1 | 1.8* | | |
| | 2.3244 | 0.5556 | 1 | | |

*(Continued)*

**TABLE 2.3 (Concluded)**

Conversion Factors.

| | | | | | | |
|---|---|---|---|---|---|---|
| **Thermal** | W/(m·K) | W/(cm·°C) | J/(s·cm·°C) | cal/(s·cm·°C) | Btu/h·ft·°F | |
| **Conductivity** | 418.4* | 1 | 1 | 0.2390 | 57.816 | |
| | 418.4* | 4.184* | 4.184* | 1 | 241.91 | |
| | 1.7296 | 0.017296 | 0.017296 | $4.1338 \times 10^{-3}$ | 1 | |
| **Viscosity** (1 poise = dyne-sec/cm² = 0.1 Newton-sec/m²) | | | | | | |
| | $Ib_m/(ft·s)$ | N·s/m² | kg/(m·s) | $Ib_f·hr/ft^2$ | $Ib_f·s/ft^2$ | Poise g/(cm·s) |
| | 1 | 1.4882 | 1.4882 | $8.6336 \times 10^{-6}$ | $3.1081 \times 10^{-2}$ | 14.8819 |
| | 0.0671955 | 1 | 1 | $5.8014 \times 10^{-6}$ | 0.020885 | 10 |
| | 115,827 | 172,369 | 172,369 | 1 | 3,600 | 1,723,689 |
| | 32.17405 | 47.88026 | 47.88026 | $2.7778 \times 10^{-4}$ | 1 | 478.8026 |
| | $6.71955 \times 10^{-2}$ | 0.1 | 0.1 | $5.8014 \times 10^{-7}$ | $2.0885 \times 10^{-3}$ | 1 |
| **Coefficient of** | W/(m²·K) | kcal/(h·m²·°C) | W/(cm²·°C) | cal/(h·cm²·°C) | Btu/h·ft²·°F | |
| **Heat Transfer** | 1 | 0.8598 | $1 \times 10^{-4}$ | $2.388 \times 10^{-5}$ | 0.1761 | |
| | 1.630 | 1 | $1.1630 \times 10^{-4}$ | $2.778 \times 10^{-5}$ | 0.2048 | |
| | 10,000 | 8,598 | 1 | 0.2388 | 1761.1 | |
| | 41,869 | 36,000 | 4.1869 | 1 | 7,373.5 | |
| | 5.6783 | 4.8823 | $5.6783 \times 10^{-4}$ | $1.3562 \times 10^{-4}$ | 1 | |

International Table Btu, cal, and kcal. Linear Temperature Difference – °F or °R, °C, or K.

### 2.2.5 Rules for Using SI Units

#### 2.2.5.1 Capitalization

The names of units start with a lowercase letter when writing the units out except for in a title or the beginning of a sentence. The only exception is "degree Celsius." Unless they come from an individual's name (in which case the first letter of the symbol is capitalized), lowercase is used in writing symbols for units. The only exception is L for liter. Symbols for numerical prefixes (multiplies and sub-multiplies) are also lowercase, except for those representing multipliers of $10^6$ or more, for instance, mega (M), giga (G), tera (T), peta (P), exa (E), zetta (Z), and yotta (Y). It means that all prefixes are written in lowercase when spelled out.

**Lowercase units:** m, kg, s, mol, etc. **Uppercase units:** A, K, Hz, Pa, C, etc.

Symbols rather than self-styled abbreviations should always be used to represent units.

**Correct usage:** A, s **Incorrect usage:** amp, sec

#### 2.2.5.2 Use of Plurals

Remember that symbols are never expressed as plural. That is, an "s" is never added to the symbol to denote plural. However, when the names of units are spelled out, they are made plural if the number to which they refer is greater than 1. Fractions, on the other hand, are always written as singular. Plurals are used as required when writing unit names. For example, henries is plural for henry. The following exceptions are noted:

**Singular:** lux, Hertz, Siemens **Plural:** lux, Hertz, Siemens

#### 2.2.5.3 Use of Hyphenation and Space

Also remember that a hyphen or a space is not used to separate a prefix from the name of the unit. A space, however, is left between a symbol and the number to which it refers, with the exception of the symbols for degree, minute, and second of angles, and for degree Celsius.

In three cases, the final vowel in the prefix is omitted: megohm, kilohm, and hectare.

**Example 2.1: Correct and Incorrect Usages of Units**

| Correct Usage | Incorrect Usage |
|---|---|
| 5 kg | 5 kgs |
| 5 kilograms | 5 kilogram |
| 5.57 kg | - |
| 5.57 kilograms | 5.57 kilogram |
| 0.57 kilogram | 0.57 kilograms |

**Example 2.2: Correct and Incorrect Usages of Units**

| Correct Usage | Incorrect Usage |
|---|---|
| 5 kg | 5kg |
| 40° 45′ 30″ | 40 ° 45 ′ 30 ″ |
| 30°C | 30 °C |
| 5 km | 5km |
| MJ | M J |
| 5 s | 5s |
| 5 milliseconds | 5 milli-seconds |

### 2.2.5.4 *Use of Numerals and Periods*

Remember that scientific and technical writing is different from any other writings such as newspaper, magazine, and other writings. In scientific and technical writing, numerals are used for all numbers expressing physical quantities; however, it is a common practice to write out the numbers from one to nine and use numerals for other numbers in newspapers. In ordinary books and magazines, for instance, whole numbers from one through ninety-nine, and any of these followed by "hundred," "thousand," "million," "billion," etc., are spelled out. Also, keep in mind that the associated number is written as numerals when the unit is represented by an abbreviation or symbol.

Periods are never used after SI symbols unless the symbol is at the end of a sentence.

### 2.2.5.5 *Use of Symbols for Mathematical Operations*

Units are represented by symbols, not by their spelled-out names when the units (SI) are used with symbols for mathematical operations.

Notes to remember:

1. When writing unit names as a product, always use a space (preferred) or a hyphen
   Correct usage: newton meter or newton-meter
2. When expressing a quotient using unit names, always use the word per and not a solidus (/). The solidus or slush mark is reserved for use with symbols.
   Correct usage: meter per second Incorrect usage: meter/second
3. When writing a unit name that requires a power, use a modifier, such as squared or cubed, after the unit name. For area or volume, the modifier can be placed before the unit name.
   Correct usage: millimeter squared or square millimeter
4. When denoting a quotient by unit symbols, any of the following are accepted form:
   Correct usage: m/s or $m.s^{-1}$

In more complicated cases, consider using negative powers or parentheses. For acceleration, use $m/s^2$ or $m\ s^{-2}$ but not m/s/s. For electrical potential, use $kg.m^2/(s^3.A)$ or $kg.m^2.s^{-3}.A^{-1}$ but not $kg.m^2/s^3/A$.

**Example 2.3: Correct and Incorrect Usages of Units**

| Correct Usage | Incorrect Usage |
|---|---|
| J/kg | |
| J $kg^{-1}$ | joules.$kg^{-1}$ |
| Joules per kilogram | joules/kilogram |
| N.m | newton.meter |
| Newton meter | |
| Newton-meter | |

### 2.2.6 Overall Examples

**Example 2.4: Area**

Show the unit of area in SI, and perform dimensional analysis.

**Solution**

Area equation is a quantity equation arising from geometry, for example, the area equation for a pipe is expressed as follows:

$$\text{Area}(A) = \pi r^2 = \pi (\text{m})^2$$
$$\text{Area}(A) = \text{m}^2$$

where $A$ is the magnitude of area in $m^2$, the magnitude of $\pi$ is 3.14 (dimensionless), and $r$ is the magnitude of radius in m.

Or in another example, the area for a rectangle is expressed as follows:

$$\text{Area}(A) = wl = \text{m} \times \text{m}$$
$$\text{Area}(A) = \text{m}^2$$

where $A$ is the magnitude of area in $m^2$, $w$ is the magnitude of width in m, and $l$ is the magnitude of length in m.

Let's now perform a simple dimensional analysis to find an expression for the area. The area is dependent on the dimensionless number $\pi$ and the radius. So,

$$A = k\, r^a$$

where $k$ is a dimensionless number, and $a$ is the number to be determined. Now, we can replace each symbol by its fundamental physical unit, and have

$$\text{L}^2 = \text{L}^a$$

L is an independent quantity; therefore, we can equate the indices on both sides, and have the following equation:

$$a = 2$$

Consequently, the expression for the area can be found as $A = k\, r^a$, where the coefficient $k$ cannot be determined from dimensional analysis because it is dimensionless; and, from geometry, we know that $k = \pi$.

**Example 2.5: Volume**

Show the unit of volume in SI and perform dimensional analysis.

**Solution**

Volume equation is a quantity equation arising from geometry, for example, the volume equation for a pipe is expressed as follows:

$$\begin{aligned} \text{Volume}(V) &= \pi\, r^2 L \\ &= \pi\left(\text{m}^2\right)(\text{m}) \\ \text{Volume}(V) &= \text{m}^3 \end{aligned}$$

where $V$ is the magnitude of volume in $\text{m}^3$, the magnitude of $\pi$ is 3.14 (dimensionless), and $r$ is the magnitude of radius in m.

Or in another example, the volume for a rectangular cross section is expressed as follows:

$$\text{Volume}(V) = w \times l \times h = \text{m} \times \text{m} \times \text{m}$$
$$\text{Volume}(V) = \text{m}^3$$

where $V$ is the magnitude of volume in $\text{m}^3$, $w$ is the magnitude of cross-sectional width in m, $l$ is the magnitude of cross-sectional length in m, and $h$ is the magnitude of height.

Let's now perform a dimensional analysis to find an expression for the volume having a tubular cross-sectional area. The volume is dependent on the dimensionless number $\pi$, the radius and length of the tube. So,

$$V = k\, r^a L^b$$

where $k$ is a dimensionless number, and $a$ and $b$ are the numbers to be determined. Now, we can replace each symbol by its fundamental physical unit, and have

$$\text{L}^3 = \text{L}^a \text{L}^b$$

L is an independent quantity; we can therefore equate the indices on both sides, and have the following equation:

$$a + b = 3$$

In the earlier example, it was determined that $a = 2$, so this leaves $b = 1$. Consequently, the expression for the volume can be found as $V = k\, r^a\, L^b$, where the coefficient $k$ cannot be determined from dimensional analysis because it is dimensionless; however, from geometry, we know that $k = \pi$; therefore, $V = k\, r^2\, L$.

**Example 2.6: Volume**

Determine the unit of volume ($m^3$) in SI for a given volume in I-P system.

**Solution**

Volume unit in I-P system is $ft^3$ and remember that 1 ft = 0.3048 m; then

$$\text{Volume}(V) = \text{ft}^3 (0.3048\ \text{m}/1\ \text{ft})^3$$
$$\text{Volume}(V) = 0.028317\ \text{m}^3$$

**Example 2.7: Mass**

Determine the unit of mass (kg) in SI for a given mass in I-P system.

**Solution**

Mass unit in I-P system is $lb_m$ and remember that 1 $lb_m$ = 0.45359 kg; then

$$\text{Mass}(m) = \text{lb}_\text{m} (0.45359\ \text{kg}/1\ \text{lb}_\text{m})$$
$$\text{Mass}(m) = 0.45359\ \text{kg}$$

**Example 2.8: Force**

Show that the unit of force is Newton (N) in SI, and perform dimensional analysis.

**Solution**

The unit of force in SI, defined as that force, which applied to a mass of 1 kilogram, gives it an acceleration of 1 meter per second per second. Newton's second law of motion, a quantity equation established from the laws of nature, is expressed as:

$$\begin{aligned} \text{Force}(F) &= \text{mass} \times \text{acceleration} = m\,a \\ &= (\text{kg})(\text{m/s}^2) = (\text{kg m/s}^2) \\ \text{Force}(F) &= \text{N} \end{aligned}$$

where $m$ is the magnitude of mass in kg, $a$ is the magnitude of acceleration in $m/s^2$, and $F$ is the magnitude of force in N.

The force is dependent on the mass and the acceleration. Now, let's assume a general algebraic equation for force in the form of

$$F = m^a a^b$$

where $a$ and $b$ are the indices (numbers) to be determined. Now, we can replace each symbol by its fundamental physical unit, and have

$$\text{M}^1\text{L}^1\text{T}^{-2} = (\text{M})^a \left(\text{LT}^{-2}\right)^b$$

or

$$M^1L^1T^{-2} = M^aL^bT^{-2b}$$

M, L, and T are all independent quantities; therefore, we can equate the indices on both sides, and have the following equations:

$$1 = a \text{ and } 1 = b$$

Then, we can solve for and find that $a = b = c = 1$; consequently, the expression for force can be found as $F = m.a$.

**Example 2.9: Force**

Show that the unit of force is Newton (N) in SI for a given force in I-P system.

**Solution**

Force unit in I-P system is $lb_f$ and remember that 1 $lb_m$ = 0.45359 kg, 1 ft = 0.3048 m, and gravitational acceleration $g$ is 32.174 $lb_m/s^2$; then

$$\begin{aligned}\text{Force}(F) &= 1\,\text{lb}_f = 1\,\text{lb}_m\left(32.174\,\text{ft/s}^2\right) = 32.174\,\text{lb}_m.\text{ft/s}^2 \\ &= \left(32.174\,\text{lbm.ft/s}^2\right)\left(0.45359\,\text{kg/lb}_m\right)\left(0.3048\,\text{m/ft}\right) \\ \text{Force}(F) &= 4.45\,\text{kg m/s}^2 = 4.45\,\text{N}\end{aligned}$$

where $g$ is the magnitude of gravitational acceleration, and $F$ is the magnitude of force.

Note that in I-P system, an acceleration of 9.80665 $m/s^2$ corresponds exactly to 32.174048 $ft/s^2$ as shown below:

$$g = \left[9.80665\,\text{m/s}^2\left(100\,\text{cm/1 m}\right)\right] / \left[\left(12\,\text{in/1 ft}\right)\left(2.54\,\text{cm/1 in}\right)\right] = 32.174048\,\text{ft/s}^2$$

**Example 2.10: Pressure**

Show the unit of pressure (Pascal) in SI, and perform dimensional analysis.

**Solution**

In solids, we deal with stresses; in liquids and gasses, however, we deal with pressure, which is defined as the normal component of force per unit area. Therefore, the unit for pressure is a derived unit, and has the symbol Pa (Pascal) in SI. One Pascal is the pressure resulting from a force of 1 Newton acting uniformly over an area of 1 square meter. So, the pressure equation is a quantity equation established form a definition, which is expressed as follows:

$$\begin{aligned}\text{Pressure}(p) &= \text{force / area} = F / A = m\,g / A \\ &= (\text{kg})(\text{m / s}^2) / \text{m}^2 = (\text{kg m / s}^2) / \text{m}^2 = \text{N / m}^2 \\ \text{Pressure}(p) &= \text{Pa}\end{aligned}$$

where $F$ is the magnitude of force (the mass times the acceleration) in Newton (N), and $A$ is the magnitude of area in $m^2$.

The pressure is dependent on the mass, the gravitational acceleration, and the area. Now, let's assume a general algebraic equation for pressure in the form of

$$p = m^a g^b A^c$$

where $a$, $b$, and $c$ are the indices (numbers) to be determined. Now, we can replace each symbol by its fundamental physical unit, and have

$$M^1L^{-1}T^{-2} = (M)^a\left(LT^{-2}\right)^b\left(L^{-2}\right)^c$$

or

$$M^1L^{-1}T^{-2} = M^aL^{b-2c}T^{-2b}$$

M, L, and T are all independent quantities; therefore, we can equate the indices on both sides, and have the following equations:

$$1 = a, -1 = b - 2c, -2 = -2b$$

Then, we can solve for and find that $a = b = c = 1$; consequently, the expression for pressure becomes $p = (m\,g)/A$.

**Example 2.11: Pressure**

Show that the unit of pressure is Pascal (Pa) in SI for a given force in I-P system.

**Solution**

Pressure unit in I-P system is psi ($lb_f/in^2$), and 1 $lb_f$ = 4.45 kg $m/s^2$ = 4.45 N; 1 in = 0.0254 m; then,

$$\begin{aligned}
\text{Pressure} &= \text{force/area} = F/A = p \\
&= \left[\left(1\,\text{lb}_f\right)\left(4.45\,\text{kg m/s}^2\right)/\left(1\,\text{lb}_f\right)\right]/\left\{\left[\text{in}^2\left(6.452\times10^{-4}\text{m}^2/\text{in}^2\right)\right]\right\} \\
&= 6{,}894.8\left(\text{kg}\right)\left(\text{m/s}^2\right)/\text{m}^2 \\
\text{Pressure}(p) &= 6{,}894.8\,\text{Pa} = 6.89\,\text{kPa}
\end{aligned}$$

where $F$ is the magnitude of force and $A$ is the magnitude of area.

**Example 2.12: Work**

Show that the unit of work is joule (J) in SI, and perform dimensional analysis.

**Solution**

The unit of work or energy in SI is joule, which has the symbol, J. The work done when the point of application of a force of 1 Newton is displaced 1 meter in the direction of the force. One watt-second is equal to 1 joule. The work equation is then a quantity equation established form a definition, which is expressed as follows:

$$\begin{aligned} \text{Work} &= \text{force} \times \text{distance} = (\text{mass} \times \text{acceleration}) \times \text{distance} = W = m\, a\, \Delta x \\ &= (\text{kg})(\text{m/s}^2)(\text{m}) \\ &= (\text{kg m/s}^2)\text{m} \\ \text{Work} &= \text{N.m} = \text{J} \end{aligned}$$

where $W$ is the magnitude of work in J, $m$ is the magnitude of mass in kg, and $a$ is the magnitude of acceleration in m/s$^2$, and $\Delta x$ is the distance travelled in m.

The work is dependent on the mass, the acceleration and the distance travelled, for instance. Let's now assume that a general algebraic equation for work is in the form of

$$W = \text{m}^a a^{\text{b}} \Delta x^{\text{c}}$$

where $a$, $b$, and $c$ are the indices (numbers) to be determined. Now, we can again replace each symbol by its fundamental physical unit, and have

$$\left(\text{M}^1\text{L}^1\text{T}^{-2}\right)\text{L}^1 = (\text{M})^{\text{a}}\left(\text{LT}^{-2}\right)^{\text{b}}(\text{L})^{\text{c}}$$

or

$$\text{M}^1\text{L}^2\text{T}^{-2} = (\text{M})^{\text{a}}\left(\text{L}^{\text{b}}\text{T}^{-2\text{b}}\right)\text{L}^{\text{c}}$$

M, L, and T are all independent quantities; therefore, we can equate the indices on both sides, and have the following equations:

$$1 = a,\ 1 = b \text{ and } 1 = c$$

Then, we can solve for and find that $a = b = \text{c} = 1$; consequently, the expression for force becomes $W = m\, a\, \Delta x$.

**Example 2.13: Energy**

Show that the unit of energy is joule (J) in SI, and perform dimensional analysis.

**Solution**

Energy is defined as the ability to perform work, and as expressed earlier, the unit of energy in SI is joule, which has the symbol, J. So, the energy equation is a quantity equation established form a definition, which is expressed as follows:

$$\begin{aligned}
\text{Energy} &= \text{work} = \text{force} \times \text{distance} = (\text{mass} \times \text{acceleration}) \times \text{distance} = E = m\,a\,\Delta x \\
&= (\text{kg})(\text{m}/\text{s}^2)(\text{m}) \\
&= (\text{kg.m}/\text{s}^2)\text{m} \\
\text{Energy} &= \text{N.m} = \text{J}
\end{aligned}$$

where $m$ is the magnitude of mass in kg, and $a$ is the magnitude of acceleration in m/s$^2$, and $\Delta x$ is the distance travelled in m.

In this case, the energy is dependent on the mass, the acceleration and the distance travelled. Let's now assume that a general algebraic equation for energy is in the form of

$$E = m^{\text{a}} a^{\text{b}} \Delta x^{\text{c}}$$

where $a$, $b$, and $c$ are the indices (numbers) to be determined. Now, we can again replace each symbol by its fundamental physical unit, and have

$$(\text{M}^1\text{L}^1\text{T}^{-2})\text{L}^1 = (\text{M})^{\text{a}} (\text{LT}^{-2})^{\text{b}} (\text{L})^{\text{c}}$$

or

$$\text{M}^1\text{L}^2\text{T}^{-2} = (\text{M})^{\text{a}} (\text{L}^{\text{b}}\text{T}^{-2\text{b}}) \text{L}^{\text{c}}$$

M, L, and T are all independent quantities; therefore, we can equate the indices on both sides, and have the following equations:

$$1 = a, 1 = b \text{ and } 1 = c$$

Then, we can solve for and find that $a = b = \text{c} = 1$; consequently, the expression for energy becomes $E = m\,a\,\Delta x$.

**Example 2.14: Power**

Show that the unit of power is watt (W) in SI, and perform dimensional analysis.

**Solution**

The unit of power in SI is watt and has the symbol, W. Power is defined as the rate at which energy is expended or work done. The watt in thermodynamics is defined as "the power which in one second gives rise to energy of one joule." In mechanical terms, however, a power of one watt can move a mass of one kilogram in one second, through a distance of one meter with such force that the kilogram mass' velocity at the end of the meter will be one meter per second greater than it was at the beginning. In an electric circuit, on the other hand, one watt is a current of one ampere at a pressure of one volt. So, the power equation is a quantity equation established form a definition, which is expressed as follows:

$$\text{Power} = \text{work/time, or energy generation/time, or energy consumption/time}$$

Since the unit of work or energy is J, then

$$\text{Power} = \text{J/s} = \text{W} = P = E/t = \text{energy/time} = m\, a\, \Delta x/t$$

where $t$ is the time in seconds. In this case, the power is dependent on the mass, the acceleration, the distance travelled, and the time taken to travel. Let's now assume that a general algebraic equation for power is in the form of

$$P = F^{a}.\Delta x^{b}.t^{-c}$$

where $a$, $b$, and $c$ are the indices (numbers) to be determined. Now, we can replace each symbol by its fundamental physical unit, and have

$$\left(\mathrm{M^1L^1T^{-2}}\right)\mathrm{L^1T^{-1}} = \left(\mathrm{M\,L\,T^{-2}}\right)^{a}\left(\mathrm{L}\right)^{b}\mathrm{T}^{-c}$$

or

$$\mathrm{M^1L^2T^{-3}} = \mathrm{M^aL^{a+b}T^{-2a-c}}$$

M, L, and T are all independent quantities; therefore, we can equate the indices on both sides, and have the following equations:

$$1 = a, 2 = a + b \text{ and } -3 = -2a - c$$

Then, we can solve for and find that $a = b = c = 1$; consequently, the expression for power can be found as $P = F\,\Delta x\, t^{-1} = m\, a\, \Delta x\, t^{-1}$.

**Example 2.15: Volt**

Express the unit of volt in terms of base units.

**Solution**

As mentioned earlier, in an electric circuit, the unit of electric potential, potential difference, and electromotive force in SI is volt and has the symbol V. One volt is the potential difference between two points on a conducting wire carrying a constant current of one ampere, and the power dissipated between the points is one watt.

The volt is a derived unit, and in terms of base units, it can be expressed as follows:

$$\begin{aligned} E &= P/I = \text{electromotive force} = \text{power/current} \\ \text{Volt} &= \text{Watt/Ampere} = \text{V} = \text{W/A} = \text{W/A} \\ &= (\text{J/s})/\text{A} \\ &= (\text{J/C}) \\ &= (\text{N.m})/\text{C} \\ &= \text{A.}\Omega \\ &= \text{A.}\left[\left(\text{kg.m}^2\right)/\left(\text{s}^3.\text{A}^2\right)\right] \\ \text{Volt} &= \text{kg.m}^2/\left(\text{A.s}^3\right) \end{aligned}$$

**Example 2.16: Radiant Flux Density**

Determine how many units of radiant flux density $I$ ($\text{W/m}^2$) in SI are in a given amount of $\text{Btu/h.ft}^2$.

**Solution**

Radiant flux density is defined as the amount of energy received on a unit surface in unit time.

$$\text{Radiant flux density}\left(\text{W/m}^2\right) = \left(\text{Btu/h.ft}^2\right)(1{,}054.35\ \text{J/Btu})(\text{h}/3{,}600\ \text{s})(\text{ft}/0.3048\ \text{m})^2$$

$$\text{Radiant flux density}\left(\text{W/m}^2\right) = 3.15\ \text{W/m}^2$$

**Example 2.17: Boiler Horsepower**

Determine how many kilowatts of power in SI are in 1 boiler horsepower (bhp) in I-P.

**Solution**

Remember that 1 boiler horsepower is the energy rate needed to evaporate 34.5 $\text{lb}_\text{m}$ of water at 212°F (100°C) in 1 h; therefore, it is equal to 33,475 Btu/h. A boiler horsepower is approximately13 times larger than mechanical horsepower (engine output).

Also remember that 1 Btu = 1,054.35 J, and 1 h = 3,600 s; therefore,

$$\text{Boiler power} = 1\ \text{bhp}\left[(33{,}475\ \text{Btu/h})\ /\ 1\ \text{bhp}\right](1{,}054.35\ \text{J}/1\ \text{Btu})(1\ \text{h}/3{,}600\text{s})$$

$$\text{Boiler power} = 9{,}804\ \text{W}$$

$$\text{Boiler power} = 9.8\ \text{kW}$$

**Example 2.18: Horsepower**

Determine how many kilowatts of power in SI are in 1 horsepower (hp) in I-P.

**Solution**

The horsepower (hp) is a unit in I-P system, sometimes used to express the rate at which mechanical energy is expended. It was originally defined as 550 foot-pounds per second (ft-$lb_f$/s).

Remember that 1 $lb_f$ = 4.45 kg m/$s^2$ = 4.45 N, and 1 ft = 0.3048 m; then

$$\text{Power} = (550\ \text{ft} - \text{lb}_f/\text{s})\left[(0.3048\ \text{m}/1\text{ft})(4.45\ \text{kg.m}/\text{s}^2)/1\,\text{lb}_f\right]$$
$$\text{Power} = 746\ \text{W} = 0.746\ \text{kW}$$

**Example 2.19: Refrigeration Ton**

Determine how many Btu/h and kWh of refrigeration is provided by 1 ton of refrigeration.

**Solution**

Remember that 1 Btu = 1,054.35 J, 1 h = 3,600 s, 1 $lb_m$ ice = 1 $lb_m$ of liquid water, latent heat of fusion = 144 Btu/$lb_m$, and 1 short ton = 2,000 $lb_m$.

Refrigeration is commonly rated in tons. 1 ton of refrigeration is the latent heat of fusion needed to melt 1 short ton (2,000 $lb_m$) of ice in 24 h. Therefore,

$$1\ \text{refrigeration ton} = (2{,}000\ \text{lb}_m\text{ice} \times 144\ \text{Btu}/\text{lb}_m)/24\ \text{h}$$
$$1\ \text{refrigeration ton} = 12{,}000\ \text{Btu}/\text{h}$$
$$1\ \text{refrigeration ton} = (12{,}000\ \text{Btu}/\text{h})(1{,}054.35\ \text{J}/\text{Btu})(1\ \text{h}/3600\ \text{s})(1\ \text{kJ}/1{,}000\ \text{J})$$
$$1\ \text{refrigeration ton} = 3.52\ \text{kW}$$

**Example 2.20: Energy**

Show how many MJ of energy 1 kWh is equal to.

**Solution**

Remember that 1 MJ of energy is equal to $10^6$ J of energy, and 1 kWh is equal to 1,000 J/s. Therefore,

$$1\ \text{kWh} = (1{,}000\ \text{J}/\text{s})(1\ \text{h})(60\ \text{min}/1\ \text{h})(60\ \text{s}/1\ \text{min}) = 3{,}600{,}000\ \text{J}$$
$$1\ \text{kWh} = (3{,}600{,}000\ \text{J})/(1{,}000{,}000\ \text{J}/1\ \text{MJ})$$
$$1\ \text{kWh} = 3.6\ \text{MJ}$$

**Example 2.21: Energy**

Show how many MJ and kWh of energy 1 therm (US) has.

**Solution**

The therm is a unit of heat energy equal to 100,000 Btu units. It is the energy equivalent of burning approximately 100 $ft^3$ (2.83 $m^3$) of natural gas. Natural gas meters measure volume rather than energy content; therefore, natural gas companies use a therm factor to convert the volume of gas used to its heat equivalent, and thence calculate the actual energy use. Please remember that natural gas with a higher than average concentration of butane, ethane, or propane has a higher therm factor, while impurities lower the therm factor.

$$1\,\text{therm} = (100{,}000\,\text{Btu})(1{,}054.35\,\text{J/Btu})(\text{MJ}/10^6\,\text{J}) = 105.5\,\text{MJ}$$
$$1\,\text{therm} = (105.5\,\text{MJ})(1\,\text{kWh}/3.6\text{MJ}) = 29.3\,\text{kWh}$$

**Example 2.22: Potential Energy**

Show that the unit of potential energy is Joule (J) in SI.

**Solution**

$$\text{Potential energy} = \text{force} \times \text{elevation} = (\text{mass} \times \text{gravitational acceleration}) \times \text{elevation}$$

$$\begin{aligned} E_{pe} &= m\,g\,h \\ &= (\text{kg})(\text{m/s}^2)(\text{m}) \\ &= (\text{kg m/s}^2)\,\text{m} \\ E_{pe} &= \text{N.m} = \text{J} \end{aligned}$$

where $m$ is the magnitude of mass in kg, and $g$ is the magnitude of gravitational acceleration in m/s$^2$, and $h$ is the elevation from a datum in m.

A dimensional analysis can easily be performed to find an expression for potential energy, as we did for work and energy examples earlier.

**Example 2.23: Kinetic Energy**

Show that the unit of kinetic energy is Joule (J) in SI.

**Solution**

$$\text{Kinetic energy} = 0.5 \times (\text{mass} \times \text{speed}^2)$$

$$\begin{aligned} E_{ke} &= 0.5\,m.u^2 \\ &= (\text{kg})(\text{m/s})^2 \\ &= (\text{kg.m/s}^2)\,\text{m} \\ E_{ke} &= \text{N.m} = \text{J} \end{aligned}$$

where $m$ is the magnitude of mass in kg, and $u$ is the magnitude of speed in m/s.

A dimensional analysis can easily be performed to find an expression for kinetic energy, as we did for work and energy earlier.

**Example 2.24: Pressure Energy**

Show that the unit of pressure energy is Joule (J) in SI.

**Solution**

$$\begin{aligned}
\text{Pressure energy} &= (\text{mass} \times \text{pressure difference})/\text{density} = m\,\Delta P/\rho \\
&= (\text{kg})(\text{Pa})/(\text{kg/m}^3) \\
&= (\text{kg})(\text{kg/m s}^2)/(\text{kg/m}^3) \\
&= (\text{kg.m/s}^2).(\text{m}) \\
\text{Pressure energy} &= \text{N.m} = \text{J}
\end{aligned}$$

where $m$ is the magnitude of mass in kg, and $\Delta P$ the magnitude of pressure difference between two points in Pa, and $\rho$ is the magnitude of density in kg/m$^3$.

Again, a dimensional analysis can easily be performed to find an expression for pressure energy, as we did for work and energy earlier.

**Example 2.25: Kinematic Viscosity**

Show the units of kinematic viscosity $\nu$ in SI.

**Solution**

The quantity equation for kinematic viscosity $\nu$ is given as follows:

$$\begin{aligned}
\nu &= \text{dynamic viscosity/density} = \mu/\rho = F\,t/\rho \\
&= (\text{Pa.s})/(\text{kg/m}^3) = (\text{kg/m.s}^2).\text{s}/(\text{kg/m}^3) \\
\nu &= \text{m}^2/\text{s}
\end{aligned}$$

where Pa (Pascal) is equal to kg/(m.s$^2$), $\mu$ is the magnitude of dynamic viscosity in Pa.s, and $\rho$ is the magnitude of density in kg/m$^3$.

Kinematic viscosity is dependent on the force $F$, the time $t$, and the density $\rho$. Now, let's assume a general algebraic equation in the form of

$$\nu = F^a t^b/\rho^c$$

where $a$, $b$, and $c$ are the indices (numbers) to be determined. Now, we can replace each symbol in kinematic viscosity by its fundamental physical unit, and overall, we would have

$$\text{L}^2\text{T}^{-1} = (\text{M L T}^{-2})^a (\text{T})^b (\text{M L}^{-3})^{-c}$$

where $a$, $b$, and $c$ are the indices (numbers) to be determined. Now, we can replace each symbol by its fundamental physical unit, and have

$$\text{M}^0\text{L}^2\text{T}^{-1} = \text{M}^{a-c}\,\text{L}^{a+3c}\,\text{T}^{-2a+b}$$

M, L, and T are independent quantities; therefore, we can equate the indices on both sides, and have the following equations:

$$0 = a - c, 2 = a + 3c \text{ and } -1 = -2a + b$$

Then, we can solve for and find that $a = c = ½$, and $b = 0$, which are the indices (numbers); consequently, the expression for kinematic viscosity can be found as $\nu = F\, t/\rho$.

**Example 2.26: Dimensionless Number**

Show that the Reynolds number Re is a dimensionless number.

**Solution**

The quantity equation for Reynolds number is given as follows:

$$\text{Re} = \rho u\, D/\mu = \left[\left(\text{kg/m}^3\right)\left(\text{m/s}\right)\left(\text{m}\right)\right] / \left[\left(\text{kg/m.s}^2\right).\text{s}\right] = \text{Dimensionless}$$

where $\rho$ is the magnitude of density of the fluid in kg/m$^3$, $u$ is the magnitude of average velocity of the fluid in m/s, $D$ is the diameter of the pipe in which the fluid flows in m, and $\mu$ is the magnitude of dynamic viscosity in Pa.s.

**Example 2.27: Pump Head**

Show that the pump head $h_p$ unit is m in SI.

**Solution**

$$\begin{aligned} h_p &= \left(\text{pumping energy}\right)/\left(\text{gravitational acceleration}\right) = E_p/g \\ &= \left[\left(\text{J/kg}\right)/\left(\text{m / s}^2\right)\right] = \left[\left(\text{N.m/kg}\right)/\left(\text{m/s}^2\right)\right] \\ &= \left[\left(\left(\text{kg.m/s}^2\right).\text{m/kg}\right)\left(\text{s}^2/\text{m}\right)\right] \\ h_p &= \text{m} \end{aligned}$$

where J is the unit of energy Joule in SI (= force × distance = N.m), N is the unit of force Newton (= kg.m/s$^2$), and $g$ is the magnitude of gravitational acceleration 9.81 m/s$^2$.

Pump head is dependent on the pumping energy $E_p$ and the gravitational acceleration $g$. Now, let's assume a general algebraic equation in the form of

$$h_p = E_p{}^{a} g^{-b}$$

where $a$, $b$, and $c$ are the indices (numbers) to be determined. Now, we can replace each symbol in kinematic viscosity by its fundamental physical unit, and overall, we would have

$$\text{M}^0\text{L}^1\text{T}^0 = \left(\text{M}^2\text{L}^1\text{T}^{-2}\text{M}^{-1}\right)^{a}\left(\text{L}^1\text{T}^{-2}\right)^{-b}$$

where $a$ and $b$ are the indices (numbers) to be determined. Now, we can replace each symbol by its fundamental physical unit, and have

$$M^0L^1T^0 = M^aL^{a-b}T^{-2a+2b}$$

M, L, and T are independent quantities; therefore, we can equate the indices on both sides, and have the following equations:

$$0 = a \text{ and } 1 = a - b$$

Then, we can solve for and find that $a = 0$ and $b = -1$, which are the indices (numbers); consequently, the expression for pump head can be found as $h_p = E_p/g$.

**Example 2.28: Pump Power**

Show that the pump power $\theta$ unit is W in SI.

**Solution**

$$\textit{Pump Power}\left(\theta\right) = \dot{m}\, E_p = \left(\frac{\text{kg}}{\text{s}}\right)\left(\frac{\text{J}}{\text{kg}}\right)$$

$$\textit{Pump Power}\left(\theta\right) = \frac{\text{J}}{\text{s}} = \text{W}$$

where $\dot{m}$ is the magnitude of mass flow rate in kg/s, and $E_p$ is the magnitude of pumping energy in J/kg.

If the pump has a conversion efficiency of $\eta$, then the pump input power $\theta_{bk}$ can be can be calculated as follows:

$$\text{Pump Input Power} = \text{pump break power}\left(\theta_{bk}\right) = \text{pump fluid power/efficiency} = \theta_{fl}/\eta$$

where $\theta_{fl}$ is the pump output (fluid) power, and the efficiency $\eta$ is in decimals (i.e. 0.6 is used for 60% conversion efficiency).

A dimensional analysis can easily be performed to find an expression for pump power, as we did for work and energy, and others earlier.

**Example 2.29: Pump Fluid Power**

Show that the pump fluid power $\theta_{fl}$ unit is W in SI.

**Solution**

$$\begin{aligned}\textit{Pump Fluid Power}\left(\theta_{fl}\right) &= \dot{m}\, g\, h_p = \left(\frac{\text{kg}}{\text{s}}\right)\left(\frac{\text{m}}{\text{s}^2}\right)\text{m} \\ &= \left(\frac{\text{kg m}}{\text{s}^2}\right)\left(\frac{\text{m}}{\text{s}}\right) = \text{N m/s} \\ \textit{Pump Fluid Power}\left(\theta_{fl}\right) &= \frac{\text{J}}{\text{s}} = \text{W}\end{aligned}$$

where $\dot{m}$ is the magnitude of mass flow rate in kg/s, $g$ is the magnitude of gravitational acceleration in m/s$^2$, and $h_p$ is the magnitude of pump head in m.

$$\theta_{fl} = F^a \Delta x^b t^{-c}$$

where $a$, $b$, and $c$ are the indices (numbers) to be determined. Now, we can replace each symbol by its fundamental physical unit, and have

$$\left(M^1L^1T^{-2}\right)L^1T^{-1} = \left(M\,L\,T^{-2}\right)^a (L)^b\, T^{-c}$$

or

$$M^1L^2T^{-3} = M^a\, L^{a+b}\, T^{-2a-c}$$

M, L, and T are all independent quantities; therefore, we can equate the indices on both sides, and have the following equations:

$$1 = a, 2 = a + b \text{ and } -3 = -2a - c$$

Then, we can solve for and find that $a = b = c = 1$; consequently, the expression for pump fluid power can be found as $\theta_{fl} = F\Delta x\, t^{-1} = \dot{m}\, g\, h_{pump}$.

**Example 2.30: Pump Break Power**

Show that the pump break power $\theta_{bk}$ unit is W in SI.

**Solution**

$$\text{Pump Break Power}\left(\theta_{bk}\right) = \Omega\,\omega = \left(\text{kg.m/s}^2\right)\left(\text{m/s}\right) = \text{N.m/s}$$
$$\text{Pump Break Power}\left(\theta_{bk}\right) = \text{J/s} = \text{W}$$

where Ω is the magnitude of torque in kg.m/s$^2$, and $\omega$ is the magnitude of angular velocity of the shaft in m/s.

A dimensional analysis can easily be performed to find an expression for pump break power, as we did in the previous example.

**Example 2.31: Temperature**

Show what the temperature is in SI for a thermodynamic temperature of 350 K.

**Solution**

By definition, the temperature in °C is the difference between the thermodynamic temperature and the thermodynamic temperature of 273.15 K. Note that, by definition, a temperature interval of 1°C is equal to a temperature interval of 1 K, and 0°C (also known as the ice point) corresponds to 273.15 K (Figure 2.1). Then, the temperature in SI (°C) can be expressed as follows:

Temperature (°C) = Thermodynamic temperature (K) − 273.15 = 350 K − 273.15

Temperature (°C) = 76.85°C

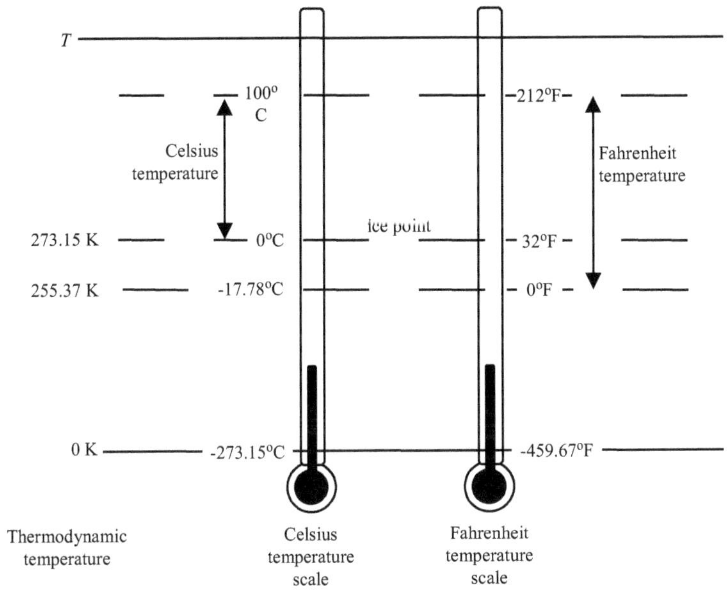


**FIGURE 2.1**
Relationships between thermodynamic temperature and temperature scales.

**Example 2.32: Temperature**

Determine the thermodynamic temperature equivalent of 550°C.

**Solution**

Thermodynamic temperature is expressed as follows:

$$\begin{aligned}\text{Thermodynamic temperature}(\text{K}) &= \text{Temperature}(^\circ\text{C}) + 273.15\,\text{K}\\ &= 550^\circ\text{C} + 273.15\,\text{K}\\ \text{Thermodynamic temperature}(\text{K}) &= 823.15\,\text{K}\end{aligned}$$

Note that K was substituted for °C because both units are identical as expressed before (Figure 2.1).

**Example 2.33: Temperature**

Determine the temperature in Fahrenheit for a given temperature of 21°C.

**Solution**

Note that ice point and boiling temperatures 0°C and 100°C temperatures on the Celsius scale corresponds to 32°F and 212°F, respectively. Therefore, the Celsius temperature range of (100°C – 0°C) = 100°C corresponds exactly to the Fahrenheit temperature range of (212°F – 32°F) = 180°F (Figure 2.1).

(212°F – 32°F) = (100°C – 0°C) = (180°F = 100°C); therefore, 1.0°C = 1.8°F, and we obtain K = 1.8°F.

Then, the relationship between the Celsius temperature (°C) and Fahrenheit temperature (°F) can be defined as follows:

$$\text{Temperature}\left({}^{\circ}\text{F}\right) = \left(1.8^{\circ}\text{F}/^{\circ}\text{C}\right)\left({}^{\circ}\text{C}\right) + 32^{\circ}\text{F}$$

where 32°F again is the freezing temperature for water in Fahrenheit temperature scale, which is equal to 0°C in the Celsius scale (Figure 2.1). Then, the temperature (°F) in this example can be determined as:

$$\text{Temperature}\left({}^{\circ}\text{F}\right) = \left(1.8^{\circ}\text{F}/^{\circ}\text{C}\right)\left(21^{\circ}\text{C}\right) + 32^{\circ}\text{F}$$

$$\text{Temperature}\left({}^{\circ}\text{F}\right) = 69.8^{\circ}\text{F} \approx 70^{\circ}\text{F}$$

The absolute temperature scale related to the Fahrenheit temperature scale is known as Rankine scale and is designated R. The relationship between the Rankine and Fahrenheit temperature scales is expressed as follows:

$$\text{Rankine} = {}^{\circ}\text{F} - 459.67$$

**Example 2.34: Temperature**

Determine the temperature in SI for a given temperature of 70°F.

**Solution**

The relationship between the Celsius temperature and the Fahrenheit temperature is defined as follows:

$$\text{Temperature}\left({}^{\circ}\text{C}\right) = \left({}^{\circ}\text{F} - 32^{\circ}\text{F}\right)(5/9)\left({}^{\circ}\text{C}/^{\circ}\text{F}\right)$$

where 32°F is the freezing temperature for water in Fahrenheit temperature scale, which is equal to 0°C in the Celsius scale. Then, the temperature in SI (°C) can be determined as:

$$\text{Temperature}\left({}^{\circ}\text{C}\right) = \left(70^{\circ}\text{F} - 32^{\circ}\text{F}\right)5/9\left({}^{\circ}\text{C}/^{\circ}\text{F}\right)$$

$$\text{Temperature}\left({}^{\circ}\text{C}\right) = 21.1^{\circ}\text{C} \approx 21^{\circ}\text{C}$$

**Example 2.35: Density**

Determine the unit (kg/m³) of density in SI if the unit is given in I-P system.

**Solution**

Remember that the density is defined as the mass *m* per unit volume *V*. So, the density equation is a quantity equation established form a definition, which is expressed in I-P system as follows:

$$\rho = m/V = \text{lb}_\text{m}/\text{ft}^3$$

Remember that 1 ft = 0.3048 m; and 1 $\text{lb}_\text{m}$ = 0.45359 kg

$$\rho = \left[\text{lb}_\text{m}\left(0.45359\,\text{kg}/1\,\text{lb}_\text{m}\right)\right]/\left[\text{ft}^3\left(0.3048\,\text{m}/1\,\text{ft}\right)^3\right]$$
$$\rho = 16.0184\,\text{kg}/\text{m}^3$$

Density is dependent on the mass *m* and the volume *V*. Now, let's assume a general algebraic equation in the form of

$$\rho = m^a/V^b$$

where *a* and *b* are the indices (numbers) to be determined. Now, we can replace each symbol in density by its fundamental physical unit, and overall, we would have

$$\text{M}\,\text{L}^{-3} = (\text{M})^a\left(\text{L}^{-3}\right)^b$$

L and T are independent quantities; therefore, we can equate the indices on both sides, and have the following equations:

$$1 = a \text{ and } -3 = -3b$$

Then, we can solve for and find that $a = b = 1$, which are the indices (numbers); consequently, the expression for kinematic viscosity can be found as $\rho = m^a/V^b$.

**Example 2.36: Specific Volume**

Determine the unit (m³/kg) of specific volume in SI if the unit is given in I-P system.

**Solution**

Remember that the specific volume is defined as the volume $\nu$ per unit mass *m*, is therefore a reciprocal of density. So, the specific volume equation is a quantity equation established form a definition, which is expressed in I-P system as follows:

$$v = 1/\rho = \text{ft}^3/\text{lb}_\text{m}$$

Remember that 1 ft = 0.3048 m; and 1 $\text{lb}_\text{m}$ = 0.45359 kg

$$v = \left[\text{ft}^3\left(0.3048\,\text{m}/1\,\text{ft}\right)^3\right]/\left[\text{lb}_\text{m}\left(0.45359\,\text{kg}/1\,\text{lb}_\text{m}\right)\right]$$
$$v = 0.062428\,\text{m}^3/\text{kg}$$

Now, we can replace each symbol in specific volume by its fundamental physical unit, and overall we would have

$$\left(M^{-1}\,L^{3}\right)=\left(M\right)^{a}\left(L\right)^{b}$$

where $a$ and $b$ are the indices (numbers), which are determined as $a$ = 3 and $b$ = −1, so specific volume has a dimension of $m^3/kg$ (in SI).

**Example 2.37: Specific Heat or Entropy**

Determine the unit (J/(kg.K)) of specific heat in SI if the unit is given in I-P system.

**Solution**

The constant pressure specific heat ($c_p$) and constant volume specific heat ($c_v$) are useful functions for particularly gases. Remember that the specific heat is defined as the heat energy required for raising the temperature of a unit mass of substance one degree. So, the specific heat equation is a quantity equation established form a definition, which is expressed in I-P system as follows:

$$c=\text{Btu}/\text{lb}_\text{m}{}^\circ\text{F}$$

Remember that 1 Btu = 1,054.35 J; 1 $lb_m$ = 0.45359 kg; K = 1.8°F

$$c=\left[\text{Btu}\left(1{,}054.35\ \text{J}/1\ \text{Btu}\right)\right]/\left[\text{lb}_\text{m}\left(0.45359\ \text{kg}/1\ \text{lb}_\text{m}\right)^\circ\text{F}\left(\text{K}/1.8^\circ\text{F}\right)\right]$$
$$c=4{,}184\ \text{J}/\left(\text{kg.K}\right)$$

So, in SI system, the specific heat is defined as the heat energy (J) required for raising the temperature of 1 kg substance 1K (or 1°C).

**Example 2.38: Enthalpy**

Determine the unit (J/kg) of enthalpy in SI if the unit is given in I-P system.

**Solution**

Remember that the enthalpy is defined as the heat energy available per unit mass of a substance. So, the enthalpy equation is a quantity equation established form a definition, which is expressed in I-P system as follows:

$$h=\text{Btu}\ /\ \text{lb}_\text{m}$$

Remember that 1 Btu = 1,054.35 J; and 1 $lb_m$ = 0.45359 kg

$$h=\left[\text{Btu}\left(1{,}054.35\ \text{J}/1\ \text{Btu}\right)\right]/\left[\text{lb}_\text{m}\left(0.45359\ \text{kg}/1\ \text{lb}_\text{m}\right)\right]$$
$$h=2{,}324.5\ \text{J}/\text{kg}$$

**Example 2.39: Entropy**

Determine the unit (J/g) of entropy in SI if the unit is given in I-P system.

**Solution**

Remember that the entropy is defined as a measure of the molecular disorder. The higher disorder of any system is, the greater is its entropy; conversely, a higher order status of a system gives a low entropy status. Boltzmann showed that there existed a simple relationship between the entropy of a given system of molecules and its probability (thermodynamic) of occurrence. So, the entropy equation is a quantity equation established form a definition, which has the symbol $s$, and is expressed in I-P system as follows:

$$s = \text{Btu}/\text{lb}_\text{m}$$

Remember that 1 Btu = 1,054.35 J; and 1 $\text{lb}_\text{m}$ = 0.45359 kg

$$s = \left[\text{Btu}\,(1{,}054.35\,\text{J}/1\,\text{Btu})\right] / \left[\text{lb}_\text{m}\left[(1{,}000\,\text{g}/1\,\text{kg})(0.45359\,\text{kg}/1\,\text{lb}_\text{m})\right]\right]$$
$$s = 2.3244\,\text{J}/\text{g}$$

**Example 2.40: Thermal Conductivity**

Determine the unit (W/(m·K)) of thermal conductivity in SI if the unit is given in I-P system.

**Solution**

Remember that the thermal conductivity is defined as the amount of heat that passes through a unit area when the temperature difference between the two sides is one degree per unit distance. So, the thermal conductivity equation is a quantity equation established form a definition, which is expressed in I-P system as follows:

$$k - \text{Btu}/\text{h.ft.}^\circ\text{F}$$

Remember that 1 Btu = 1,054.35 J; 1 ft = 0.3048 m; K = 1.8°F; 1 h = 3,600 s

$$k = \left[\text{Btu}\,(1{,}054.35\,\text{J}/1\,\text{Btu})\right] / \left[\text{h}\,(3{,}600\,\text{s}/1\,\text{h})\,\text{ft}\,(0.3048\,\text{m}/1\,\text{ft})^\circ\,\text{F}(\text{K}/1.8^\circ\text{F})\right]$$
$$k = (1.7296\,\text{J}/\text{s})/(\text{m.K})$$
$$k = 1.7296\,\text{W}/(\text{m.K})$$

**Example 2.41: Heat Transfer Coefficient**

Determine the unit (W/(m$^2$·K)) of heat transfer coefficient in SI if the unit is given in I-P system.

**Solution**

Remember that the heat transfer coefficient is defined as the amount of heat that is transferred from/to a unit area per unit time when the temperature difference between the surface and the ambient is one degree. So, the heat transfer coefficient equation is a quantity equation established form a definition, which is expressed in I-P system as follows:

$$h = \text{Btu/h.ft}^2.{}^\circ\text{F}$$

Remember that 1 Btu = 1,054.35 J; 1 $ft^2$ = 0.0929 $m^2$; K = 1.8°F; 1 h = 3,600 s

$$h = \left[\text{Btu}\left(1{,}054.35\,\text{J}/1\,\text{Btu}\right)\right] / \left[\text{h}\left(3{,}600\,\text{s}/1\,\text{h}\right)\text{ft}^2\left(0.0929\,\text{m}^2/1\,\text{ft}^2\right)^\circ\text{F}\left(\text{K}/1.8^\circ\text{F}\right)\right]$$

$$h = 5.68\,\text{W}/\left(\text{m}^2.\text{K}\right)$$

**Example 2.42: Thermal Transmittance**

Determine the unit (W/($m^2$·K)) of thermal transmittance in SI if the unit is given in I-P system.

**Solution**

Remember that thermal transmittance is the amount of heat that passes through an entire wall, ceiling, etc., section of a unit area per unit time when the temperature difference between the air on the warm side and the air on the cold side is one degree. So, the thermal transmittance equation is a quantity equation established form a definition, which is expressed in I-P system as follows:

$$U = \text{Btu/h.ft}^2.{}^\circ\text{F}$$

Remember that 1 Btu = 1,054.35 J; 1 $ft^2$ = 0.0929 $m^2$; K = 1.8°F; 1 h = 3,600 s

$$U = \left[\text{Btu}\left(1{,}054.35\,\text{J}/1\,\text{Btu}\right)\right] / \left[\text{h}\left(3{,}600\,\text{s}/1\,\text{h}\right)\text{ft}^2\left(0.0929\,\text{m}^2/1\,\text{ft}^2\right)^\circ\text{F}\left(\text{K}/1.8^\circ\text{F}\right)\right]$$

$$U = 5.68\,\text{W}/\left(\text{m}^2.\text{K}\right)$$

**Example 2.43: Thermal Resistance**

Determine the unit ($m^2$ K/W or $m^2$ °C/W) of thermal resistance in SI if the unit is given in I-P system.

**Solution**

Remember that thermal resistance is the resistance of one unit area to heat flow through a substance per unit time when the temperature difference between the two sides is one degree, and has the symbol of *R*. Thermal resistance is an additive quantity; that is, 2″ material has twice the R-value of 1″. And it does not include the boundary layer resistances. So, the heat transfer coefficient equation is a quantity equation established form a definition, which is expressed in I-P system as follows:

$$R = \text{h.ft}^2.{}^\circ\text{F/Btu}$$

Remember that 1 Btu = 1,054.35 J; 1 $ft^2$ = 0.0929 $m^2$; K = 1.8°C = 1.8°F; 1 h = 3,600 s

$$R = \left[\text{h}\left(3{,}600\,\text{s}/1\,\text{h}\right)\text{ft}^2\left(0.0929\,\text{m}^2/1\,\text{ft}^2\right)^\circ\text{F}\left(\text{K}/1.8^\circ\text{F}\right)\right] / \left[\text{Btu}\left(1{,}054.35\,\text{J}/1\,\text{Btu}\right)\right]$$

$$R = 0.176\,\text{m}^2.\text{K/W}$$

**Example 2.44: Wavelength**

Calculate the maximum wavelength $\lambda_{max}$ for the Sun's surface radiation.

**Solution**

Wien's Displacement Law states that the wavelength at which a blackbody emits its maximum amount of radiation is inversely proportional to its absolute temperature in Kelvin, $\lambda_{max} = a/T$, where $\lambda$ is in μm, $a$ is 2,897 μm K, and $T$ in K.

Assuming an average surface temperature of approximately 6,000 K, we would have its maximum emission at

$$\lambda_{max} = 2897\,\mu\text{m.K}/6{,}000\,\text{K}$$
$$\lambda_{max} = 0.48\,\mu\text{m} \approx 0.5\,\mu\text{m}$$

A wavelength of approximately 0.5 μm lies within the visible spectrum.

**Example 2.45: Frequency**

Determine the frequency of a radiational wavelength of 0.5 μm.

**Solution**

Wavelength ($\lambda$) and frequency ($v$) are related, and this relationship can formally be expressed through the speed of light $c$, as $c = \lambda\, v$, where $c$ is the speed of light, which has a value of $3 \times 10^8$ m/s, $\lambda$ is the wavelength (μm), and $v$ is the frequency (cycles/s, also known as Hz). Then, if one is known, the other can easily be determined using the speed of light, which is a constant.

$$\begin{aligned} c &= \lambda\, v \\ v &= c/\lambda \\ &= \left(3\times10^{8}\,\text{m/s}\right)/\left(0.5\,\mu\text{m}\right) \\ &= \left(3\times10^{14}\,\mu\text{m/s}\right)/\left(0.5\,\mu\text{m}\right) \\ &= 6\times10^{14}\,\text{cycles/s} \\ v &= 6\times10^{14}\,\text{Hz} \end{aligned}$$

**Example 2.46: Energy Content of a Single Wavelength Solar Radiation**

Determine the energy content $E$ of a solar radiation frequency of $6 \times 10^{14}$ Hz (or 0.5 μm wavelength).

**Solution**

Solar radiation consists of photons, which can be considered as packets of energy, which is related to frequency $v$ as $E = h\, v$, where $E$ is the energy content (J), and $h$ is the Planck constant ($6.626\times10^{-34}$ J.s).

$$\begin{aligned} E &= h\, v \\ &= \left(6.626\times10^{-34}\,\text{J.s}\right)\left(6\times10^{14}\,\text{cycles/s}\right) \\ E &= 3.98\times10^{-19}\,\text{J} \end{aligned}$$

## Nomenclature

| | |
|---|---|
| **a** | atto |
| *a* | index; acceleration; constant, 2,897 μm.K |
| **A** | unit of electric current, ampere |
| *A* | area |
| *b* | index |
| **Btu** | British thermal unit |
| **c** | centi |
| *c* | index; speed of light, $3 \times 10^8$ m/s; specific heat |
| **C** | unit of electric charge, coulomb; Celsius |
| **cd** | luminous intensity unit |
| **d** | deci |
| **da** | deka |
| **E** | exa |
| *E* | energy |
| **f** | femto |
| **F** | Fahrenheit |
| *F* | force |
| *g* | gravitational acceleration |
| **G** | giga |
| **h** | hector |
| *h* | height; pump head; enthalpy; heat transfer coefficient; Planck's constant, $6.626\times10^{-34}$ J.s |
| **hp** | unit of power in I-P, horsepower |
| *I* | radiant flux density |
| **J** | unit of energy and work, Joule |
| **k** | kilo |
| *k* | dimensionless coefficient; thermal conductivity |
| **kg** | unit of mass, kg |
| **kWh** | unit of energy, kilowatt-hour (= 3.6 MJ) |
| **K** | unit of thermodynamic temperature |
| *l* | length |
| *L* | length |
| **m** | milli; unit of length |
| *m* | mass |
| $\dot{m}$ | mass flow rate |
| **M** | mega |
| *M* | mass |
| **mol** | unit of amount of substance, mol |
| **n** | nano |
| **N** | Newton |
| **p** | pico |
| *p* | pressure |
| **P** | peta |
| *P* | power |
| **Pa** | Pascal |
| *r* | radius |

| | |
|---|---|
| $R$ | thermal resistance |
| **rad** | unit of plane angle, radian |
| $s$ | entropy |
| **sr** | unit of solid angle, steradian |
| **T** | tera |
| $T$ | time; temperature |
| **therm** | unit of heat energy, 105.5 MJ or 100,000 Btu |
| **ton** | refrigeration ton, 12,000 Btu/h or 3.52 kW |
| $U$ | thermal transmittance |
| **V** | Volt |
| $V$ | volume; electric potential; potential difference; electromotive force |
| $w$ | width |
| **W** | Watt |
| $W$ | work |
| $x$ | distance |
| **y** | yocto |
| **Y** | yotta |
| **z** | zepto |
| **Z** | zetta |

## Greek letters

| | |
|---|---|
| $\Delta$ | difference |
| $\eta$ | conversion efficiency |
| $\lambda$ | wavelength |
| µ | micro |
| $\mu$ | dynamic viscosity |
| $\nu$ | kinematic viscosity; specific volume |
| | frequency, cycles/s = hertz = Hz |
| $\theta$ | pump power |
| $\rho$ | density |
| $\Omega$ | unit of electric resistance, ohm |

## Subscripts

| | |
|---|---|
| *bk* | break power |
| *fl* | fluid power |
| *ke* | kinetic energy |
| **max** | maximum |
| *p* | pump; constant pressure |
| *pe* | potential energy |
| *v* | constant volume |

## Superscripts

| | |
|---|---|
| o | degree |
| ′ | minute (angle) |
| ″ | second (angle) |

## Bibliography

Andrews, J, and Jelley, N (2007) *Energy Science: Principles, Technologies, and Impacts*. Oxford University Press Inc., New York, pp. 13–14.

ASHRAE (1984) SI Guide for HVAC & R, 6th ed., American Society of Heating, Refrigeration and Air Conditioning Engineers, Inc. Atlanta, GA 30329.

ASHRAE (1989) ASHRAE Handbook of Fundamentals. American Society of Heating, Refrigeration and Air Conditioning Engineers, Inc. Atlanta, GA 30329.

ASME (n.d.) ASME Orientation and Guide for Use of Metric Units, American Society of Mechanical Engineers, 345 East 47th Street, New York, NY 10017.

ASTM (n.d.) Standards for Metric Practice, ASTM Standard E 380 (American Society for Testing and Materials), Philadelphia.

BIPM (2012), International Vocabulary of Metrology – Basic and Associated Terms (VIM). 3rd edition (2012).

CGMP (1961) Comptes Rendus de la 11e CGPM (1960), p.87.

CGMP (1972) Comptes Rendus de la 11e CGPM (1971), p.78.

CSA (n.d.) Metric Practice Guide, CSA Standard CAN 3-Z234-1 Canadian Standards Association, 178 Rexdale Blvd., Rexdale, Ontario M9W 1R3.

IOS (n.d.) SI Units and Recommendations for the Use of Their Multiples and of Certain Other Units, ISO Standard 1000 International Organization for Standardization, Geneva, Switzerland; available from American National Standards for Metric Institute, 1430 Broadway, New York, NY 10018

National Bureau of Standards (n.d.) The International System of Units (SI), National Bureau of Standards Special Publication 330. Superintendent of Documents, U.S. Government Printing Office, Washington, DC20402.

Wildi, T (1995) *Metric Units and Conversion Charts. A Metrication Handbook for Engineers, Technologists, and Scientists*. IEEE Press, New York, pp. 65-66.

Yildiz, I and Liu, Y (2018) Energy Units, Conversions, and Dimensional Analysis. In: Dincer, I. (ed.), *Comprehensive Energy Systems*. Vol. 1, Elsevier, Oxford, UK, pp. 1–23.

## 2.3 Thermal Energy

### 2.3.1 Introduction

In this section, first the laws of thermodynamics, which form the foundation of heat transfer analyses, are covered. Along the way, temperature is defined, and then, energy and mass balance concepts are introduced before covering different modes of heat transfer using different objects and geometries. Toward the end, a subsection is also provided on fossil fuels combustion and carbon sequestration. A number of numerical examples were included in the section to demonstrate and grasp the relevant concepts easily.

A system can have various forms of energy including chemical, electrical, kinetic, magnetic, nuclear, potential, or thermal energy. Potential and kinetic energy forms are the examples of what is called macroscopic forms, which are possessed by a system as a lump sum compared to a particular outside reference. The forms of energy related with the molecular structure and level of activity are however called as microscopic forms of energy, and the internal energy of a system is defined as the sum of different microscopic energy forms, which include sensible, latent, chemical, and nuclear energies. Sensible energy is associated with the kinetic energy of molecules (i.e. with a temperature change), whereas latent energy is associated with a change in phase in a system, while chemical and nuclear energy are related with the atomic bonds and the bonds within the atomic nucleus, respectively. Some of these concepts are covered further in depth in the following sections; however, we must first understand the laws of thermodynamics, and energy and mass balance concepts before moving any further.

### 2.3.2 Laws of Thermodynamics

Thermodynamics is defined as the science of energy that covers the entire characteristics of energy transformations. When two bodies at different temperatures are brought together, a transfer of energy occurs between these bodies until a thermal equilibrium is reached. The Zeroth Law of Thermodynamics states that when two bodies have temperature equality with a third body, then in turn, they have equal temperatures and are in thermal equilibrium with each other. This process of energy flow due the presence of a temperature difference is called heat transfer, and the rate of heat transfer is function of this existing temperature gradient. We use the conservation of energy principle in dealing with heat transfer analyses, which basically states that energy can neither be created nor destroyed; rather, it can only change its form. This principle is also called as the First Law of Thermodynamics, the study of which deals with quantifying the amount of energy transferred from one body or system to another; however, it does not specify the direction of energy transfer, which is actually defined by the Second Law of Thermodynamics. The second law states that energy does not only have a quantity, but also has a quality. And the direction of energy transfer is determined by the direction in which the quality of energy decreases. Furthermore, the second law dictates theoretical performance and efficiency limits of energy systems as well. When heat (energy in transition) is for instance added to a system, the system's thermal energy content rises and molecular activity in the system and therefore the temperature increase as a result. So the temperature, a property of a given system, mirrors the thermal energy content of the system, and is related to hotness or coldness of an object; however, it is difficult to give an exact definition of temperature. As explained above, the zeroth law explains the temperature relationships of two bodies with respect to a third body, while on the other hand, the Third Law of Thermodynamics, which defines thermodynamic temperature, states that a body or a system has its minimum possible energy content at the theoretically lowest temperature (i.e. absolute temperature). And at this point (absolute zero), the particle constituents of matter have minimal motion and can become no colder.

#### *2.3.2.1 First Law of Thermodynamics*

The first law, as introduced briefly above, states that energy can neither be created nor destroyed; rather, it can only change its form from one to another. Using the first law, we can analyze the process of energy transfer between a system or a body and its surroundings by simply employing the energy balance approach.

Overall, energy is transferred to or is extracted from a system in the form of heat, work, or mass flow. Employing a control volume and a system boundary, over which the energy gains or losses occur, we can easily analyze the process of energy transfer. As mentioned earlier, when heat is transferred to a system, it ends up with a rise in molecular energy and therefore in temperature. Hence, this process increases the internal energy of the system. If heat is transferred from a system to the surrounding, the internal energy of the system decreases due to the reduced molecular energy content. Energy transfer can also take place by performing work on a system or by a system performing work itself. If work is performed by the system itself, the energy of the system decreases, while the energy of the system increases when the work is done on the system. Furthermore, mass flow is another mechanism, through which energy transfer can take place over the boundary of a system. Similarly, when mass flows into a given control volume, the energy content of this volume increases as a result of the amount of energy introduced by the mass flowing into the system, whereas it decreases when the mass leaves the system.

###### *2.3.2.1.1 Energy Balance*

In any process, the difference between the total amount of energy entering and leaving a system is equal to the change in total energy of the system. This is called the energy balance and can be expressed as follows:

$$E_i - E_o = \Delta E_{sys} \tag{2.1}$$

where $E_i$ and $E_o$ are the amounts of energy entering and leaving the system, respectively, and $\Delta E_{sys}$ is the change in total energy of the system.

If we apply the energy balance concept per unit time basis, then we can write the energy balance equation as:

$$\dot{E}_i - \dot{E}_o = \frac{dE_{sys}}{dt} \tag{2.2}$$

where $\dot{E}_i$ and $\dot{E}_o$ are the amounts of energy entering and leaving the system per unit time, respectively.

We can apply the energy balance approach basically to any system for any process; however, we need to make sure that all forms of energy entering or leaving the system are properly identified and quantified. In heat transfer analyses, since we only consider the energy in the form of heat, the energy (heat) balance equation takes the following form:

$$Q_i - Q_o + Q_{gen} = \Delta E_{sys,thermal} \tag{2.3}$$

where $Q_i$ and $Q_o$ are the amounts of heat transfer into and out of the system, respectively, $Q_{gen}$ is the amount of heat generation, and $\Delta E_{sys,\,thermal}$ is the change in thermal energy content of the system.

###### *2.3.2.1.2 Energy Balance for Closed Systems*

A system is called a closed system, where there is no mass exchange with surroundings, that is, no inflow or outflow across its boundary. Since no mass enters or leaves close systems, we can only talk about work and heat transfer in such systems; and thence when applied to closed systems, Equation (2.1) can be written as:

$$(Q_i + W_i) - (Q_o + W_o) = \Delta E_{sys} \tag{2.4}$$

$$(Q_i - Q_o) + (W_i - W_o) = \Delta E_{sys} \tag{2.5}$$

In the systems that are stationary, which do not involve a change in kinetic or potential energies, total energy of the system comprises of its internal energy. Hence, the energy balance for such closed systems can be written as follows:

$$E_i - E_o = \Delta U \tag{2.6}$$

where $\Delta U$ is the change in internal energy of the system, which can be expressed in terms of the mass $m$, constant volume specific heat $c_v$, and the change in temperature $\Delta T$ as:

$$\Delta U = m c_v \, \Delta T \tag{2.7}$$

Thence, the heat balance equation for closed stationary systems can be written as:

$$Q_{net} = m c_v \, \Delta T \tag{2.8}$$

where $Q_{net}$ represents the net transfer of heat to or from the system.

###### 2.3.2.1.3 Energy and Mass Balance for Open Systems

An open system (or a control volume) is a system, where both mass and energy are allowed to cross the control boundary. In control volumes, energy can be transferred in the form of heat, work, and mass transfer; and compressors, heat exchangers, pumps, and turbines are just a few examples of such systems. For such systems, the principle of conservation of mass is particularly important, and is expressed as the difference between the total mass entering, $m_i$, and leaving, $m_o$, the system (control volume), $m_{cv}$, in a specific time interval:

$$m_i - m_o = \Delta m_{cv} \tag{2.9}$$

If it is considered on a per unit time basis, then the mass balance equation becomes:

$$\dot{m}_i - \dot{m}_o = \frac{dm_{cv}}{dt} \tag{2.10}$$

where $\dot{m}_i$ and $\dot{m}_o$ represent the rates of mass inflow and outflow, respectively.

In heat transfer applications, quite often, we encounter fluid flow within pipes and ducts. In such cases, we can determine the mass flow rate from the cross-sectional area of the flow passage, density, and velocity of the fluid. In one-dimensional approximations, the fluid properties are assumed to only change in the flow direction. The mass flow rate can then be expressed as:

$$\dot{m} = \rho u A \tag{2.11}$$

where $\rho$ is the density of the flowing fluid, $u$ is the average velocity of the fluid in the direction of flow, and $A$ is the cross-sectional area of the flow passage.

We can also express the mass flow rate $m$ in terms of the volumetric flow $Q$ rate as follows:

$$Q = u A \tag{2.12}$$

$$\dot{m} = \rho Q \tag{2.13}$$

where $Q$ is the volumetric flow rate of the fluid.

In steady-flow processes, the amount of mass within the control volume does not chance in time. Thence, the total mass inflow rates are equal to the total mass outflow rates:

$$\sum \dot{m}_i = \sum \dot{m}_o \tag{2.14}$$

And in many heat transfer applications, only one inlet and outlet conditions are encountered. Under such conditions, for steady-flow processes, the inlet and outlet mass flow rates are equal:

$$\dot{m}_i = \dot{m}_o \tag{2.15}$$

In open systems (control volumes), energy transfer takes place by mass transfer as well; therefore, it is necessary to determine the total energy of the fluid that enters or leaves the system. The total energy of a flow per unit mass entering or leaving a system can be expressed as follows:

$$E_{total} = Flow\,Energy + Internal\,Energy + Kinetic\,Energy + Potential\,Energy$$

$$E_{total} = \left(h + \frac{u^2}{2} + gz\right) \tag{2.16}$$

where the middle term on the right hand side of the equation represents the kinetic energy while the far right term represents the potential energy.

The energy balance equation for steady-flow processes can thus be written as:

$$\dot{E}_i - \dot{E}_o = \frac{dE_{sys}}{dt} \tag{2.17}$$

And for steady-flow processes, $\frac{dE_{sys}}{dt} = 0$; hence, $\dot{E}_i = \dot{E}_o$, then

$$\dot{Q}_i + \dot{W}_i + \sum \dot{m}_i\left(h + \frac{u^2}{2} + gz\right) = \dot{Q}_o + \dot{W}_o + \sum \dot{m}_o\left(h + \frac{u^2}{2} + gz\right) \tag{2.18}$$

where $\dot{Q}_i$ and $\dot{Q}_o$ are the rates of heat added to and removed from the system, respectively, and $\dot{W}_i$ and $\dot{W}_o$ are the rates of work done on and by the system, and the terms in the parenthesis were explained earlier.

In processes that do not include energy transfer by work, and also when the kinetic and potential energy changes are negligible, the energy balance equation for steady-flow conditions can be reduced to:

$$\dot{Q}_{net} = \dot{m}\left(h_o - h_i\right) = \dot{m}c_p\left(T_o - T_i\right) \tag{2.19}$$

where $\dot{Q}_{net}$ is the net rate of heat transfer, $c_p$ is the constant pressure specific heat, and $h$ and $T$ are the enthalpy and temperature of the flowing fluid, respectively.

#### 2.3.2.2 Second Law of Thermodynamics

When energy transfer happens between a system and its surroundings, the transfer occurs only in a certain direction. The first law of thermodynamics does not say anything about the direction of heat flow as it only deals with the transformation and conservation of energy. The second law of thermodynamics, on the other hand, employs entropy, which is a thermodynamic property that measures degree of randomization or disorder at the microscopic level, and defines the direction of an energy transfer or conversion process. Processes can be classified as reversible or irreversible. And the second law states that any process that would reduce the entropy of the isolated system is impossible. If the entropy is unchanged by the forward process, then it would also be unchanged by the reverse process. Since no entropy is produced, such a process can go in either direction without violating the second law and is called a reversible process. If the entropy is produced by the forward process, however, then the entropy would decrease for the reverse process; and hence the reverse process would be impossible. It is therefore said that the forward process is irreversible. Entropy is produced in irreversible processes, and all real processes are subjected to some measure of irreversibility. Recognition of the irreversibilities in real processes is quite important in engineering as they reflect an increase in the amount of disorganized energy (requires further work before it can be used effectively) at the expense of organized energy (can easily be put to practical use). In short, the second law of thermodynamics states that entropy can be produced but never destroyed. For the first law, we presented the energy balance approach; however, the second law deals with the entropy and exergy balance instead. And the entropy balance for a system can be expressed as follows:

$$S_i - S_o + S_{gen} = \Delta S_{sys} \tag{2.20}$$

where $S$ represents the entropy.

As discussed earlier, the first law deals with the quantity of energy, whereas the second law is concerned about the associated quality. At this point, we must introduce the concept of exergy, a property, which is known as a measure of the maximum useful work potential of a system in a particular reference environment. A system that is in equilibrium with its surroundings has zero exergy, and therefore, said to be at the dead state. In irreversible processes, exergy destruction takes place while entropy is being generated. Thence, the amount of destroyed exergy can be expressed as a function of entropy generation as follows:

$$Ex_{dest} = T_o S_{gen} \tag{2.21}$$

where $Ex_{dest}$ is the amount of exergy destroyed, $T_o$ is the temperature of the dead state (the system is in equilibrium with the surroundings, where no useful work can be obtained).

The exergy balance equation for a given system can then be developed as follows:

$$Ex_i - Ex_o - Ex_{dest} = \Delta Ex_{sys} \tag{2.22}$$

If we express it on a rate basis (per unit time), the above equation takes the following form:

$$\dot{E}x_i - \dot{E}x_o - \dot{E}x_{dest} = \frac{dEx_{sys}}{dt} \tag{2.23}$$

Just like we employ the energy balance analysis for control volumes, the exergy balance analysis can as well be used for control volumes. However, heat transfer analyses are based on the first law of thermodynamics and the energy balance concept. The following sections cover the types and mechanisms of heat transfer.

### 2.3.3 Heat Transfer and the Mechanisms of Heat Transfer

Heat is the form of energy in transition caused by a temperature difference. The process and study of energy flow due the presence of a temperature difference is called heat transfer, and the rate of heat transfer is function of this existing temperature gradient. As mentioned earlier, thermodynamic analysis deals with determining the amount of heat transferred between systems using the conservation of energy principle, whereas the study of heat transfer focuses on determining the rate of heat transfer.

All substances in principle can hold a certain amount of heat, and the property that signifies this amount of heat is known as heat (thermal) capacity. When heat is added to a system, the system's thermal energy content increases and molecular activity in the system and therefore the temperature rise as a result. When a solid substance is heated, for instance, its temperature keeps rising until it reaches the melting point of that substance, which is the highest temperature that can be achieved at the solid phase before the phase change from solid to liquid starts taking place. The heat that is absorbed by the solid until melting point is called sensible heat $Q_S$; however, the heat that is used to convert the solid to the liquid phase is called latent heat of fusion (melting) $Q_L$ (Figure 2.2).

Similarly, when a liquid is heated, the temperature of the liquid keeps rising until it reaches the boiling point, which is the highest temperature that can be reached at the liquid phase at the measured pressure. The heat that is absorbed by the liquid while increasing the temperature up until the boiling point is again sensible heat, while the heat that is used to convert the liquid to the vapor phase is called latent heat of vaporization.

The internal energy related to the kinetic energy of molecules as a result of the absorbed or released energy by a system, as discussed above, leads to an increase or decrease in

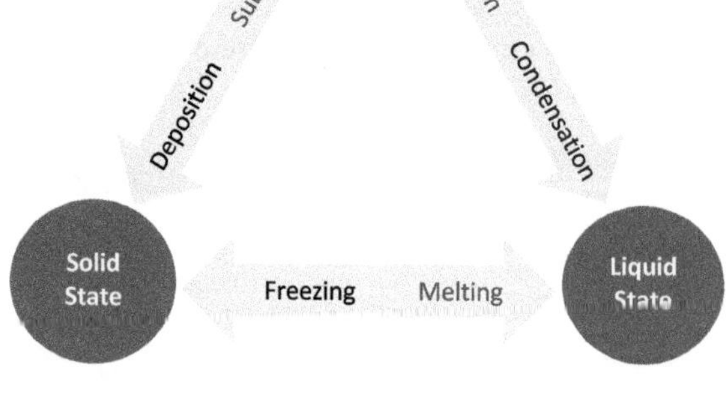


**FIGURE 2.2**
Phase changes and relevant processes.

temperature is known as sensible heat. Quantity of sensible heat associated with a temperature change, $Q_S$, can be expressed as follows:

$$Q_S = m c_p \left(T_2 - T_1\right) \tag{2.24}$$

where $Q_S$ is the total sensible heat of the system in J (Btu), $m$ is the mass of the system in kg (lb), $c_p$ is the specific heat in J/kg.°C (Btu/lb.°F), and the term $(T_2 - T_1)$ is the temperature change in °C (°F). Specific heat of common substances are provided in Table 2.4, which is the heat to raise 1 kg (1 lb) of substance, 1°C (1°F). For example, the specific heat of air is 1,004 J/kg.°C (0.24 Btu/lb.°F) while it is 4,186 J/kg.°C (1.00 Btu/lb.°F) for water.

Observable properties such as temperature and pressure characterize the state or phase of a system. Each and every specific property of a substance at a given state has a certain magnitude that is independent of the path of the substance arriving the given state. For instance, if enough heat is added to or removed from a substance at a given condition, the substance undergoes a state or phase change process. The substance temperature remains constant during this phase change process up until the process is completed. This is true for a solid to liquid phase change, liquid to vapor phase change, or vice versa. Figure 2.3 shows the typical diagram for the phase changes of water, where solid water (ice) is changed to liquid water after certain amount of heat addition, and the liquid water is changed to the water vapor after adding further heat. As shown in Figure 2.3, the temperature during these phase changes remains constant.

As discussed above, sensible energy is associated with the kinetic energy of molecules (i.e. with a temperature change), whereas latent energy is associated with a change in

**TABLE 2.4**

Specific heats of common substances.

| Material | Specific Heat (J/kg.°C) | Specific Heat (Btu/lb.°F) |
|---|---|---|
| Water | 4,186 | 1.00 |
| Aluminum | 900 | 0.22 |
| Iron/Steel | 448 | 0.12 |
| Copper | 387 | 0.093 |
| Concrete | 960 | 0.23 |
| Glass | 840 | 0.20 |
| White pine | 2,800 | 0.67 |
| Ice (0°C) | 2,090 | 0.50 |
| Air | 1,004 | 0.24 |
| Rock/Stone | 840 | 0.20 |
| Sand | 776 | 0.19 |
| Asphalt | 922 | 0.22 |
| Brick | 838 | 0.20 |

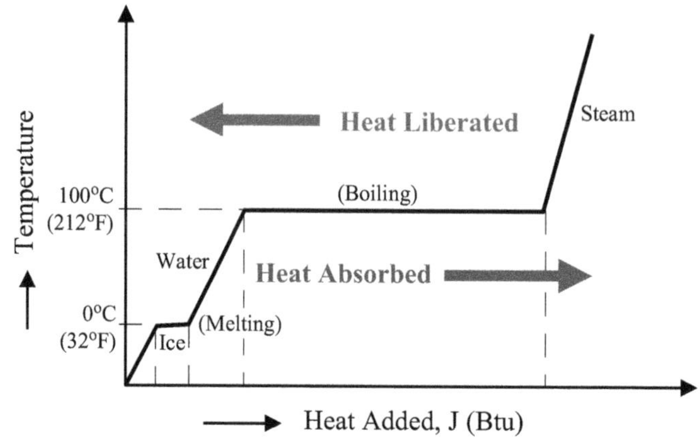


**FIGURE 2.3**
Phase changes of water.

phase in a system. The energy required to convert a unit mass of a particular substance from solid to liquid phase is known as latent heat of fusion; whereas, the energy required converting a unit mass of liquid of a particular substance from liquid to gaseous state is known as latent heat of vaporization. The temperature of a system remains constant during a phase change process until the process is complete. Quantity of latent heat associated with a phase change in a system, $Q_L$, can be expressed as follows:

$$Q_L = mh \tag{2.25}$$

where $Q_L$ is the total latent heat of the system in J (Btu), $m$ is the mass of the system in kg (lb), $h$ is the latent heat (enthalpy) of fusion or vaporization in J/kg (Btu/lb). Latent heat (enthalpy) of fusion and vaporization of some common substances are provided in Table 2.5.

When ice melts, for instance, latent heat of fusion is absorbed, and liquid water is formed while the temperature remains at 0°C (32°F). The latent heat of fusion of ice, $Q_L$, that is, the amount of energy that has to be supplied to change 1 kg of ice (at its melting point) to liquid water at the same temperature, is $3.39 \times 10^5$ J/kg. In imperial units, 1 lb of ice requires 144 Btu of energy to melt. The quantity of heat associated with a change in phase from liquid water to water vapor, latent heat of vaporization, $Q_L$, that is, the energy required to

**TABLE 2.5**
Latent heat (enthalpy) of fusion and vaporization of common substances.

| Material | Enthalpy of Fusion (J/kg) | Enthalpy of Fusion (Btu/lb) | Enthalpy of Vaporization (J/kg) | Enthalpy of Vaporization (Btu/lb) |
|---|---|---|---|---|
| Water | $3.39 \times 10^5$ | 144 | $2.26 \times 10^6$ | 1,044 |
| Ammonia | $3.33 \times 10^5$ | 143 | $1.36 \times 10^6$ | 583.2 |
| Alcohol-ethyl | $1.08 \times 10^5$ | 46 | $0.86 \times 10^6$ | 367.5 |
| Alcohol-methyl | $0.99 \times 10^5$ | 43 | $1.10 \times 10^6$ | 473.0 |

evaporate 1 kg liquid water without a change in temperature at 100°C is about $2.26 \times 10^6$ J. In imperial units, 1 lb of water requires 1,044 Btu of energy to evaporate.

**Example 2.47: Sensible Heat**

Determine the total amount of sensible heat required to increase the temperature of 1,000 kg of water from 4°C to 10°C.

**Solution**

From Table 2.5, specific heat $c_p$ of water is 4,186 J/kg.°C

Using Equation (2.24),

$$Q_S = m\, c_p\, (T_2 - T_1)$$

$$Q_S = 1{,}000 \text{ kg } (4{,}186 \text{ J/kg.°C})\ (40 - 4)\text{°C}$$

$$Q_S = 150{,}696{,}000 \text{ J } (\approx 142{,}840 \text{ Btu})$$

**Example 2.48: Latent Heat of Fusion**

Determine the total amount of latent heat of fusion of 1 ton (short) of ice melted in 24 h.

**Solution**

From Table 2.5, enthalpy of fusion $h$ of water is 144 Btu/lb

$$1 \text{ ton (short) of ice} = 2{,}000 \text{ lb}$$

Using Equation (2.25)

$$Q_L = m\, h$$

$$Q_L = 2{,}000 \text{ lb ice} \times 144 \text{ Btu/lb} = 288{,}000 \text{ Btu } (\approx 303{,}840 \text{ J})$$

It means 288,000 Btu of energy (latent heat of fusion) is needed for melting 2,000 lb (1 ton) of ice in 24 h. Then,

$$288{,}000 \text{ Btu/24 h} = 12{,}000 \text{ Btu/h } (12{,}660 \text{ J/h or } 3.516 \text{ J/s} = 3.516 \text{ W})$$

Additionally, in this example, we have also shown that 1 ton of refrigeration equals to 12,000 Btu/h (3.516 W).

**Example 2.49: Latent Heat of Vaporization**

Determine the total amount of latent heat of vaporization of 1,000 kg of water evaporated at 100°C.

**Solution**

From Table 2.5, enthalpy of fusion $h$ of water is $2.26 \times 10^6$ J/kg

Using Equation (2.25)

$$Q_L = (1{,}000\,\text{kg water})(2.26\times10^6\,\text{J/kg}) = 2.26\times10^9\,\text{J}\,(2.14\times10^6\,\text{Btu})$$

This section also covers the modes of heat transfer. Thermal energy is transferred from one region (or a body) to another by three modes namely conduction, convection, and radiation, Heat transfer is one of those transport phenomena, which includes mass transfer (discussed earlier), momentum transfer or fluid friction (will be discussed later in this chapter), and electrical conduction. Transport phenomena have similar rate equations, and the flux in each case is proportional to a potential difference. In heat transfer by conduction and convection, as we will see shortly, the potential difference is the temperature difference; and heat transfer takes place through matter due to this temperature gradient. In radiation heat transfer, we have a change in energy form internal energy to electromagnetic energy at the source for transmission, and then back to internal energy where it is received. The amount of thermal energy transferred by radiation increases quickly as the temperature increases.

In order to be able to handle heat transfer problems, some heat transfer equations have been developed from fundamentals; however, they are most of the time obtained from correlations of observed data. Typically, these correlations utilize certain dimensionless numbers developed from dimensional analysis or analogy, and Table 2.6 lists some dimensionless numbers important for heat transfer.

**TABLE 2.6**

Dimensionless Numbers Commonly Used in Heat Transfer[a].

| Name | Symbol | Value | Application |
|---|---|---|---|
| Nusselt Number | Nu | $\frac{hD}{k}$ or $\frac{hL}{k}$ | Natural or Forced Convection, Boiling or Condensing |
| Reynolds Number | Re | $\frac{\rho uD}{\mu}$ or $\frac{\rho uL}{\mu}$ | Forced Convection |
| Prandtl Number | Pr | $\frac{\mu c_p}{k}$ | Natural or Forced Convection, Boiling or Condensing |
| Stanton Number | St | $\frac{h}{\rho u c_p}$ or Nu/Re Pr | Forced Convection |
| Grashof Number | Gr | $\frac{L^3\rho^2\beta g(\Delta T)}{\mu^2}$ or $\frac{L^3\rho^2 g(\Delta T)}{T\mu^2}$ | Natural Convection (for ideal gases) |
| Rayleigh Number | Ra | Gr Pr | Natural Convection |
| Fourier Number | Fo | $\frac{\alpha\tau}{L^2}$ | Unsteady-State Conduction |
| Peclet Number | Pe | Re Pr | Forced Convection (small Pr) |
| Biot Number | Bi | $\frac{hL}{k}$ | Steady and Unsteady-State Conduction |
| Graetz Number | Gz | Re Pr *D/L* | Laminar Convection |

[a] A complete list of nomenclature appears at the end of this chapter.

### 2.3.3.1 Conduction Heat Transfer

Thermal conduction is the mode of heat transfer where energy is transferred from more energetic to less energetic neighboring particles or group of particles at the atomic level. Conduction occurs in solids, liquids, and gases. In liquids and electrically non-conducting solids, it is believed to occur by longitudinal lattice molecular oscillations as well as movement of free electrons; in gases, the molecules are under constant random motion, energy transfer occurs due to elastic collision and diffusion of molecules. In solid opaque bodies, thermal conduction is the mode of energy transfer because there is no material flow within the medium; and in flowing fluids, thermal conduction also dominates in the laminar flow region very close to a solid boundary where the flow is parallel to the surface and there is no eddy motion. And as a consequence of the second law, thermal energy transfer occurs in the direction of decreasing temperature (Figure 2.4), and conduction heat transfer can be classified into one-, two-, or three-dimensional. The classification obviously depends on the variation of temperature within the medium; and depending on the geometry, the Cartesian, cylindrical, or spherical coordinate system is utilized.

#### 2.3.3.1.1 Steady-State Heat Conduction

During the heat transfer process, if the surface temperature of the body in question does not change with time, the situation is then termed as steady-state condition, in which case the transfer process is analyzed without employing any differential equation simply by utilizing the thermal resistance concept.

*Steady State Heat Conduction in Walls*

If energy balance concept per unit time basis developed earlier in Equation (2.2) is applied to a plane wall, then we can write the energy balance equation as follows:

$$\dot{Q}_i - \dot{Q}_o = \frac{dE_w}{dt} \tag{2.26}$$

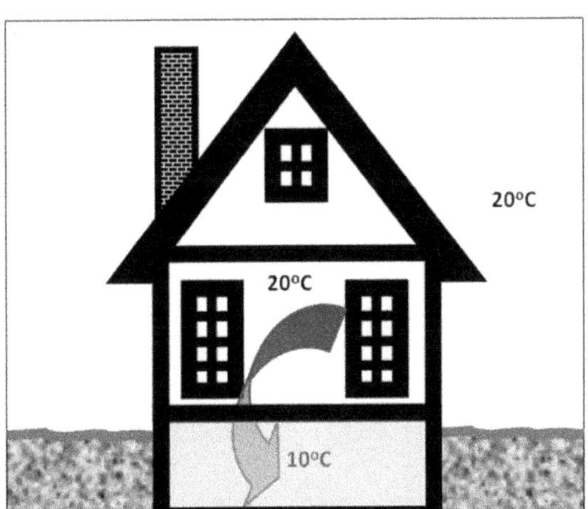


**FIGURE 2.4**
Heat flows when there is a temperature gradient.

Since energy content of the wall does not change with time in steady-state conditions, that is, $\frac{dE_w}{dt}$, then the energy balance equation simply reduces to:

$$\dot{Q}_i = \dot{Q}_o \tag{2.27}$$

For steady-state heat conduction in one dimension, the Fourier equation applies:

$$\dot{Q} = -k\,A\frac{dT}{dx} \tag{2.28}$$

where $\dot{Q}$ is the heat transfer rate (W or Btu/h), $k$ is the thermal conductivity (W/m.°C or Btu/h.ft.°F), $A$ is the cross-sectional area normal to the flow (m²), and $\frac{dT}{dx}$ is the temperature gradient (°C/m).

Since thermal conductivity values are given in different units, one must be very careful in using consistent units. Under steady-state conditions, since the conduction heat transfer rate through the plane wall is constant, and the thermal conductivity and the cross-sectional area do not vary, $\frac{dT}{dx}$ is a constant as well. Having a linear temperature distribution across the wall (Figure 2.5), the conduction heat transfer rate can be expressed as follows:

$$\dot{Q} = -k\,A\frac{(T_1 - T_2)}{L} \tag{2.29}$$

Fourier equation presented above states that the heat flow rate $\dot{Q}$ in the direction of $x$ is directly proportional to the temperature gradient $\frac{dT}{dx}$ and the cross-sectional area $A$ normal to the heat flow; and thermal conductivity $k$ is the proportionality constant in between (Tables 2.7 and 2.8). The minus sign on the right hand side of the equation states that the heat transfer is positive in the direction of decreasing temperature.

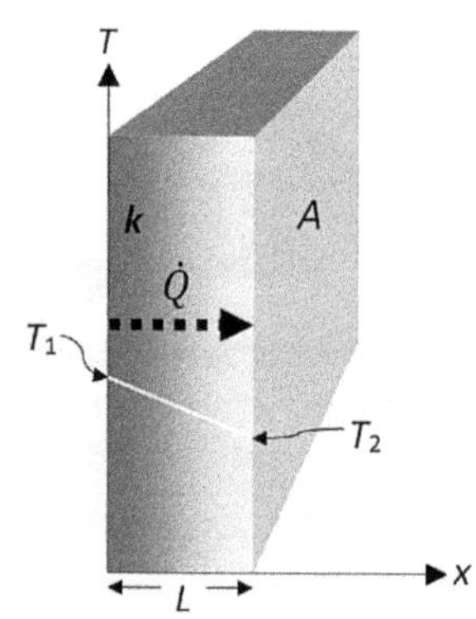


**FIGURE 2.5**
Heat conduction through a plain wall.

**TABLE 2.7**
Thermal Conductivities of Some Common Solid Materials.

| Material | Thermal Conductivity (W/m.°C) | Thermal Conductivity (Btu/h.ft.°F) |
|---|---|---|
| Cotton wool | 0.017 | 0.010 |
| Corkboard | 0.043 | 0.025 |
| Mineral wool | 0.045 | 0.026 |
| White pine | 0.112 | 0.065 |
| Balsa | 0.055 | 0.032 |
| Fir | 0.156 | 0.090 |
| Oak | 0.17 | 0.098 |
| Granit | 2.79 | 1.612 |
| Sandstone | 2.90 | 1.675 |
| Marble | 2.80 | 1.618 |
| Sand | 0.27 | 0.156 |
| Soil | 0.52 | 0.30 |
| Asphalt | 0.062 | 0.036 |
| Rubber – soft | 0.13 | 0.075 |
| Rubber – hard | 0.16 | 0.092 |
| Glass | 1.4 | 0.81 |
| Gypsum plaster | 0.519 | 0.30 |
| Common brick | 0.692 | 0.40 |
| Concrete (avg. house) | 1.385 | 0.80 |
| Mild steel | 45.0 | 26 |
| Wrought iron | 59.7 | 34.5 |
| Yellow brass | 90.0 | 52 |
| Aluminum | 204.2 | 118 |
| Copper | 381.0 | 220 |

*Sources:* Incropera and DeWitt (2001); Granet and Bluestein (2004)

**Example 2.50: Heat Conduction through a Plain Wall**

Consider a 20 cm thick concrete wall of 3 m × 6 m having a uniform thermal conductivity. Determine the total amount of steady-state conduction heat transfer rate when the inside surface temperature of the wall is 20°C and the outside surface temperature of the wall is −10°C. Assume steady-state conditions prevail and the heat transfer is one-dimensional.

**Solution**

$T_1 = 20^\circ$C and $T_2 = -10^\circ$C; $A = 3$ m × 6 m = 18 m$^2$; $L = 0.2$ m; $k = 1.385$ W/m.°C (Table 2.4)

For steady-state heat conduction in one-dimension, the Fourier equation, Equation (2.29) applies; then

$$\dot{Q} = -k\,A\,\frac{(T_1 - T_2)}{L}$$

*(Continued)*

**Example 2.50: (Continued)**

Thence, the rate of conduction heat transfer through the concrete wall is

$$\dot{Q} = -\left(1.385 \frac{\text{W}}{\text{m}^\circ\text{C}}\right)\left(18\,\text{m}^2\right)\frac{\left(20-(-10)\right)^\circ\text{C}}{0.2\,\text{m}} = -3{,}740\,\text{W}\ (\approx -12{,}763\ \text{Btu/h})$$

**TABLE 2.8**

Order of Magnitudes of Thermal Conductivities For Different Classes of Material.

| Material | Thermal Conductivity (W/m.°C) | Thermal Conductivity (Btu/h.ft.°F) |
|---|---|---|
| Gases | 0.009–0.035 | 0.005–0.02 |
| Insulating materials | 0.024–0.173 | 0.014–0.10 |
| Wood | 0.07–0.17 | 0.04–0.10 |
| Liquids (nonmetallic) | 0.09–0.69 | 0.05–0.40 |
| Brick, concrete, stone, plaster | 0.4–3.5 | 0.2–2.0 |
| Refractory materials | 0.87–17.3 | 0.50–10.0 |
| Metals and alloys | 17.3–415 | 10–240 |

*Sources:* Incropera and DeWitt (1996); Granet and Bluestein (2004)

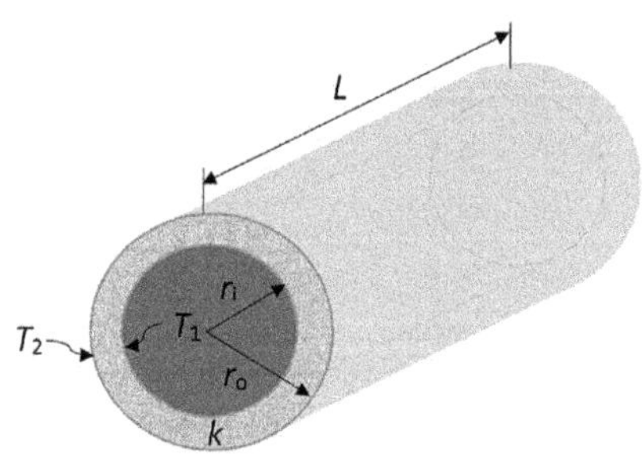


**FIGURE 2.6**
Heat conduction through a hollow circular cylinder.

### *Steady-State Heat Conduction in Cylinders*

In steady-state one-dimensional conduction heat transfer in a hollow cylinder (Figure 2.6) with a uniform wall thickness and inner radius $r_i$ and outer radius $r_o$ as well as constant and uniform inner and outer surface temperatures of $T_1$ and $T_2$, respectively, without any heat generation within the cylinder, the rate of heat conduction through the cylinder's wall and the cross-sectional area $A$ normal to the heat flow in the direction of temperature gradient, the Fourier's equation, takes the following form:

$$\dot{Q} = -k\,A\frac{dT}{dr} \tag{2.30}$$

where the surface area $A$ is equal to $2\pi rL$ at the distance $r$ from the center of the cylinder and for the length of the cylinder $L$.

Thermal conductivity being the proportionality constant in between, the equation above provides the following:

$$\dot{Q} = -2\pi L k \frac{(T_1 - T_2)}{\ln\left(\frac{r_o}{r_i}\right)} \tag{2.31}$$

**Example 2.51: Heat Conduction through a Cylinder**

Now consider a 51 mm steel tube with a 2.1 mm wall thickness and inside diameter of 46.7 mm with a uniform thermal conductivity of 17 W/(m.°C). Determine the total amount of steady-state conduction heat transfer rate when the inside surface temperature of the cylinder is 93°C and the outside surface temperature of the cylinder is 90°C. Assume the pipe length is 30 m, the steady-state conditions prevail and the heat transfer is one-dimensional.

**Solution**

$T_i = 93°C$ and $T_o = 90°C$; $r_i = 0.0467$ m and $r_{wall} = 0.0021$ m; $L = 30$ m; $k = 17$ W/m.°C

The rate of heat transfer through the pipe can then be determined using Equation (2.31) as follows:

$$\dot{Q} = -2\pi L k \frac{(T_1 - T_2)}{In\left(\frac{r_o}{r_i}\right)}$$

The outer radius of the pipe $r_o = r_i + r_{wall} = 0.0467 \text{ m} + 0.0021 \text{ m} = 0.0488 \text{ m}$

$$\dot{Q} = -2\pi(30\,\text{m})\left(17\frac{\text{W}}{\text{m°C}}\right)\frac{(93-90)^{\circ}\text{C}}{In\left(\frac{0.0488}{0.0467}\right)} = -218{,}442\,\text{W}\ (\approx -745{,}534\ \text{Btu/h})$$

*Steady-State Heat Conduction in Spheres*

In steady-state one-dimensional conduction heat transfer in a hollow sphere (Figure 2.7) with a uniform wall thickness and inner radius $r_i$ and outer radius $r_o$ as well as constant thermal conductivity, and inner and outer surface temperatures of $T_1$ and $T_2$, respectively, without any heat generation within the sphere, the rate of heat conduction through the sphere's wall and the cross-sectional area $A$ normal to the heat flow in the direction of temperature gradient, the Fourier's equation, takes the following form:

$$\dot{Q} = -kA\frac{dT}{dr} \tag{2.32}$$

where the surface area $A$ is equal to $4\pi r^2$.

After integrating Equation (2.32), the following equation is obtained:

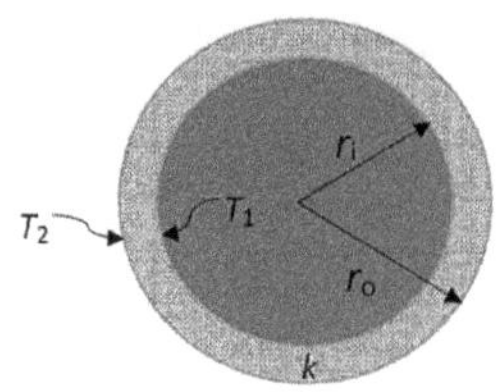


**FIGURE 2.7**
Heat conduction through a hollow sphere.

**Example 2.52: Heat Conduction through a Spherical Tank**

Let's now consider an aluminum spherical tank with a 20 mm wall thickness and inside diameter of 1.50 m with a uniform thermal conductivity. Determine the total amount of steady-state conduction heat transfer rate when the inside surface temperature of the sphere is 30°C and the outside surface temperature of the sphere is 10°C. Assume that the steady-state conditions prevail and the heat transfer is one-dimensional.

**Solution**

$T_i = 30°C$ and $T_o = 10°C$; $r_i = 1.50$ m and $r_{wall} = 0.02$ m; $k = 204.2$ W/m.°C (Table 2.7)

The outer radius of the spherical tank $r_o = r_i + r_{wall} = 1.50 + 0.02 \text{ m} = 1.52 \text{ m}$

Then, the rate of heat transfer through the sphere can be determined using Equation (2.33) as follows:

$$\dot{Q} = -\frac{4\pi k(r_i r_o)(T_1 - T_2)}{r_o - r_i}$$

$$\dot{Q} = -\frac{(4\pi)\left(204.2\frac{\text{W}}{\text{m}°\text{C}}\right)(1.50\text{ m})(1.52\text{ m})(30-10)°\text{C}}{(1.52-1.50)\text{ m}} = -5{,}847{,}635\text{ W}\ (\approx -19{,}957{,}797\text{ Btu/h})$$

$$\dot{Q} = -\frac{4\pi k(r_i r_o)(T_1 - T_2)}{r_o - r_i} \tag{2.33}$$

*2.3.3.1.2 Resistance Concept and Overall Heat Transfer*

In heat transfer applications, the concept of thermal resistance is useful to analyze steady-state problems. The steady-state heat conduction equation for a wall can be rearranged as

$$\dot{Q} = -k\,A_m\frac{(T_1 - T_2)}{L_m} = -\frac{\Delta T}{R_t} \tag{2.34}$$

where $A_m$ is the mean cross-sectional area normal to flow (m² or ft²), $L_m$ is the mean length of heat flow path (m or ft), $\Delta T$ is the overall temperature difference (°C or °F), and $R_t = \frac{L_m}{k\,A_m}$ is the thermal resistance (°C/W or °F.h/Btu), which is directly proportional to the mean length of the heat flow path $L_m$, and inversely proportional to the thermal conductivity $k$, and the mean cross-sectional area normal to the flow, $A_m$. In essence, it is the resistance to heat flow of a 1 m² (1 ft²) material when the temperature difference between the two sides is 1°C (1°F). Thermal resistance $R_t$ is an additive quantity, and does not include the boundary layer resistances; hence, a 10 cm material has twice the $R$-value of 5 cm same material.

Equation (2.34) is similar to Ohm's law for electrical circuits, and this analogy can be utilized for heat conduction in complex shapes that may sometimes not be easily solved by exact analytical means.

If the thermal resistance through the wall is expressed as $R_w$, then employing the resistance concept, Equation (2.29) takes the following form for the rate of heat conduction through the wall:

$$\dot{Q} = -\frac{(T_1 - T_2)}{R_w} \tag{2.35}$$

Thermal resistance equations for a few common shapes are also provided below. Resistance for radial flow through a circular cylinder shown in Figure 2.6 is:

$$R = \frac{\ln(r_o - r_i)}{2\pi k L} \tag{2.36}$$

And moreover, thermal resistance for a buried cylinder shown in Figure 2.8 is provided below:

$$R = \frac{\ln\left[\left(a + \sqrt{a^2 - r^2}\right)/r\right]}{2\pi k L} = \frac{Cosh^{-1}(a/r)}{2\pi k L} \quad (L >> 2r) \tag{2.37}$$

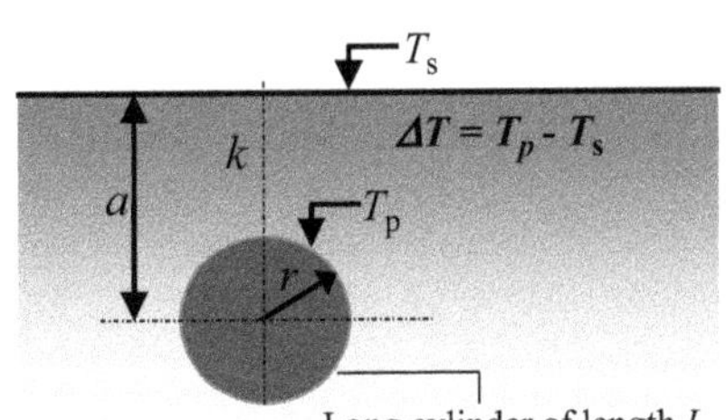


**FIGURE 2.8**
Heat conduction through a buried hollow cylinder.

**Example 2.53: Heat Conduction through a Buried Polyethylene Tube**

Now, consider a polyethylene tube with a diameter of 25 mm buried in a 15 cm deep dense concrete floor with a uniform thermal conductivity of 1.4 W/(m.K). Determine the total amount of conduction heat transfer rate when the floor surface temperature is 17.5°C and the surface temperature of the buried polyethylene tube is 32.5°C. Assume the tube length is 7.2 m, the steady-state conditions prevail and the heat transfer is one-dimensional.

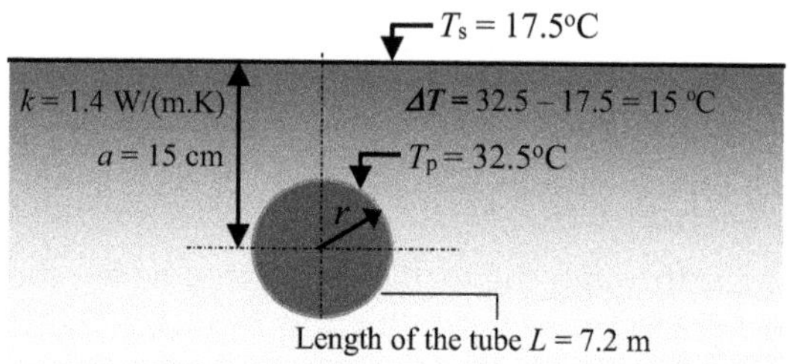


**Solution**

The tube radius $r = 0.025/2 = 0.0125$ m; the depth of the tube $a = 0.15$ m; $L = 7.2$ m >> $2r$ = 0.025 m.

The steady-state one-dimensional rate of heat transfer through the tube can then be determined using Equation (2.38) as follows:

$$R = \frac{\ln\left[\left(a + \sqrt{a^2 - r^2}\right)/r\right]}{2\pi k L} = \frac{\ln\left[\left(0.15\,\text{m} + \sqrt{\left(0.15\,\text{m}^2\right) - \left(0.0125\text{m}\right)^2}\right)/0.0125\,\text{m}\right]}{2\pi\left(1.4\dfrac{\text{W}}{\text{m.K}}\right)7.2\,\text{m}}$$

$$R = 0.0502\,\text{K/W}$$

Now using Equation (2.34), we can calculate the rate of heat transfer as follows:

$$\dot{Q} = -\frac{\Delta T}{R} = -\frac{15\,\text{K}}{0.0502\,\text{K/W}} = 299\,\text{W}\,(1{,}021\,\text{Btu/h})$$

Furthermore, resistance for radial flow through a sphere shown in Figure 2.6 can be expressed as:

$$R = \frac{\left(1/r_i - 1/r_o\right)}{2\pi k} \tag{2.38}$$

The thermal resistance concept is quite useful for heat transfer problems involving combine conduction, convection, and radiation. Actually, in most steady-state heat transfer problems, more than one mode of heat transfer is involved. And for such cases, different heat transfer coefficients can be combined into an overall heat transfer coefficient so that we can calculate the total heat transfer from the terminal temperatures simply by employing the thermal resistance concept.

Now consider from one fluid to another by a three-step steady-state heat transfer process: from a warmer fluid to a solid wall, through the wall, and then to a colder fluid. An overall coefficient of heat transfer $U$, based on the difference between the bulk temperatures of the two fluids, $T_i - T_o$, is defined as follows:

$$\dot{Q} = -\frac{(T_i - T_o)}{R_t} = -U\,A(T_i - T_o) \tag{2.39}$$

where $A$ is the surface area m² (ft²) normal to the heat flow, and $U$ is the overall heat transfer coefficient W/m².°C (Btu/h.ft².°F), which gives us the heat transfer per unit time through an entire wall, ceiling, etc., section of 1 m² (1 ft²) when the temperature difference between the warmer side fluid and the colder side fluid is 1°C (1°F). Here, the total resistance $R_t = \frac{1}{UA}$ and the overall heat transfer coefficient $U$ values include the inside and outside boundary layer resistances, $R_i$ and $R_o$, respectively, the values of which are provided in Table 2.9 for different flow directions and surface emissivities. The surface resistances combine the effects of conduction, convection, and radiation. The resistance for inside surfaces is usually for still air, and the outside surface resistances for winter and summer are based on 6.6 m/s (15 mph) and 3.3 m/s (7.5 mph) wind speeds, respectively.

**TABLE 2.9**

Surface Heat Transfer Resistances, $R_t$.[a]

| | | Surface Emissivity, ε | | | Surface Emissivity, ε | | |
|---|---|---|---|---|---|---|---|
| | | 0.90 | 0.20 | 0.05 | 0.90 | 0.20 | 0.05 |
| **Position of Surface** | **Heat Flow Direction** | **Heat Transfer Resistance (m².°C/W)** | | | **Heat Transfer Resistance (h.ft².°F/Btu)** | | |
| Horizontal (still air) | Upward | 0.107 | 0.194 | 0.232 | 0.61 | 1.10 | 1.32 |
| | Downward | 0.162 | 0.475 | 0.801 | 0.92 | 2.70 | 4.55 |
| Sloping 45° (still air) | Upward | 0.109 | 0.201 | 0.241 | 0.62 | 1.14 | 1.37 |
| | Downward | 0.134 | 0.294 | 0.391 | 0.76 | 1.67 | 2.22 |
| Vertical (still air) | Horizontal | 0.120 | 0.238 | 0.299 | 0.68 | 1.35 | 1.70 |
| 6.6 m/s (15 mph) wind (for winter). Any position | Any | | 0.030 | | | 0.17 | |
| 3.3 m/s (7.5 mph) wind (for summer). Any Position | Any | | 0.044 | | | 0.25 | |

[a] Adopted from the *2013 ASHRAE Handbook of Fundamentals*. Note that condensation can have a significant impact on surface emissivity. A surface emissivity of 0.90 can be assumed for ordinary building materials while a surface emissivity of 0.05 can be used for reflective materials such as aluminum foil.

**TABLE 2.10**

Heat Transfer Resistances of Some Common Materials.

| Material | Heat Transfer Resistance (m².°C/W) For 1 cm given thickness | Heat Transfer Resistance (h.ft².°F/Btu) For 1″ given thickness |
|---|---|---|
| Air (non-convecting) | 0.42 | 6 |
| Copper | $2.6 \times 10^{-5}$ | $3.7 \times 10^{-4}$ |
| Plywood | $8.6 \times 10^{-2}$ | 1.25 |
| Glass | $1.3 \times 10^{-2}$ | 0.18 |
| Brick | $7.7 \times 10^{-3}$ | 0.11 |
| Concrete | $1.0 \times 10^{-2}$ | 0.14 |
| Fiberglass batts | 0.25 | 3.6 |
| Polyurethane foam | 0.43 | 6.3 |

Heat transfer resistances of some common materials are provided here in Table 2.10; however, the readers of this book are referred to *2013 ASHRAE Handbook of Fundamentals* (ASHRAE, 2013) for a long list of typical thermal properties of common building and insulating materials. Remember that the units of resistances in both Table 2.4 and Table 2.5 are given in per unit area basis. The surface resistances combine the effects of conduction, convection, and radiation. Surface resistances increase with decreasing emissivity, and in buildings, the resistance for inside surfaces is usually taken for still air; and for moving outside air, forced convection is predominant; hence *R*-value is independent of the slope and heat flow direction.

***Heat Flow in Series:***

Now consider the overall heat transfer through a composite wall having several layers of different materials with different thicknesses as shown below in Figure 2.9:

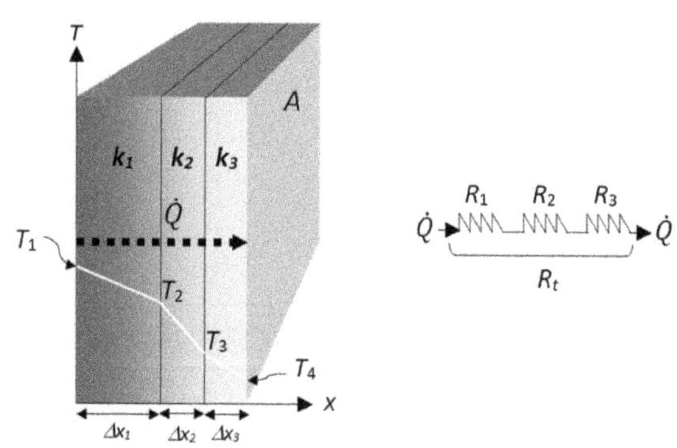


**FIGURE 2.9**
Thermal resistances in series.

The overall resistance of the thermal circuit becomes:

$$R_t = R_1 + R_2 + R_3 = \frac{\Delta x_1}{k_1\,A} + \frac{\Delta x_2}{k_2\,A} + \frac{\Delta x_3}{k_3\,A} \tag{2.40}$$

And, since the heat transfer $\dot{Q}_x$ through the circuit is constant, $\dot{Q}_x$ becomes:

$$\dot{Q}_x = -\frac{(T_1 - T_2)}{\dfrac{\Delta x_1}{k_1\,A}} = -\frac{(T_2 - T_3)}{\dfrac{\Delta x_2}{k_2\,A}} = -\frac{(T_3 - T_4)}{\dfrac{\Delta x_3}{k_3\,A}}$$

And then, using the overall thermal resistance and terminal temperatures, the overall heat transfer rate can be determined as follows:

$$\dot{Q}_x = -\frac{(T_1 - T_4)}{R_t} \tag{2.41}$$

A thermal resistance also exists for the heat flow from the surface of an object to the surrounding environment. The surface resistances combine the effects of conduction, convection, and radiation. We can determine surface resistance from Newton's law of cooling, which is provided below:

$$\dot{Q} = -h\,A(T_s - T_\infty) \tag{2.42}$$

$$R_i = \frac{1}{h_i\,A} = \frac{(T_s - T_\infty)}{\dot{Q}} \tag{2.43}$$

where $R_i$ is the surface thermal resistance, $h_i$ is the convective heat transfer coefficient, $A$ is the surface area normal to the direction of heat flow, $T_s$ is the surface temperature, and $T_\infty$ is the ambient temperature. The resistance for inside surfaces is usually for still air, and the outside surface resistances for winter and summer are based on 6.6 m/s (15 mph) and 3.3 m/s (7.5 mph) wind speeds, respectively (Table 2.6).

Now, let's consider the composite wall configuration given in Figure 2.9, one side facing to the indoor environment with a temperature of $T_i$ and the other side to the outdoor with a temperature of $T_o$ shown in Figure 2.10. Then, since the heat transfer $Q_x$ through the circuit is constant, $Q_x$ becomes:

$$\dot{Q}_x = -\frac{(T_i - T_1)}{\dfrac{1}{h_i\,A}} = -\frac{(T_1 - T_2)}{\dfrac{\Delta x_1}{k_1\,A}} = -\frac{(T_2 - T_3)}{\dfrac{\Delta x_2}{k_2\,A}} = -\frac{(T_3 - T_4)}{\dfrac{\Delta x_3}{k_3\,A}} = -\frac{(T_4 - T_o)}{\dfrac{1}{h_o\,A}} \tag{2.44}$$

where the overall resistance of the thermal circuit is:

$$R_t = R_i + R_1 + R_2 + R_3 + R_o = \frac{1}{h_i\,A} = \frac{\Delta x_1}{k_1\,A} = \frac{\Delta x_2}{k_2\,A} = \frac{\Delta x_3}{k_3\,A} = \frac{1}{h_o\,A} \tag{2.45}$$

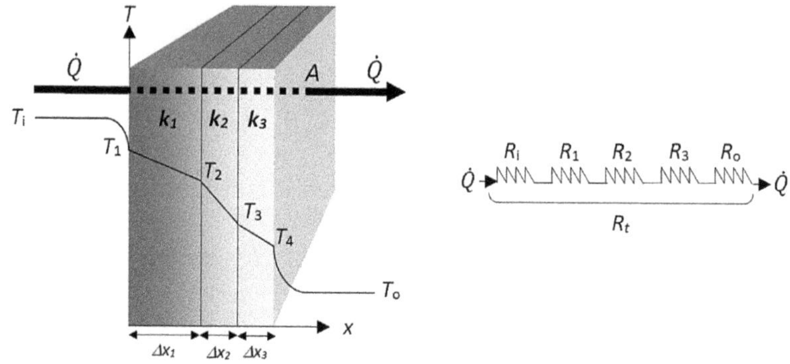


**FIGURE 2.10**
Thermal resistances in series including the surface resistances.

And when dealing with composite walls, it is easier to work with the overall heat transfer coefficient $U$, whose relationship with the total thermal resistance is given below:

$$U = \frac{1}{R_t A} = \frac{1}{h_i} + \frac{\Delta x_1}{k_1} + \frac{\Delta x_2}{k_2} + \frac{\Delta x_3}{k_3} + \frac{1}{h_o} \tag{2.46}$$

And then, using the overall thermal resistance $R_t$ or the overall heat transfer coefficient $U$ and the inside and outside temperatures, the heat transfer rate can be determined using Equation (2.39).

**Example 2.54: Heat Transfer through a Single-Pane Glass Glazing**

Consider a 2.2 m × 4.0 m and 5 mm thick single-pane glass sidewall. Determine the rate of overall heat transfer from a glasshouse having an indoor temperature of 20°C through the glass sidewall if the outside temperature is −10°C.

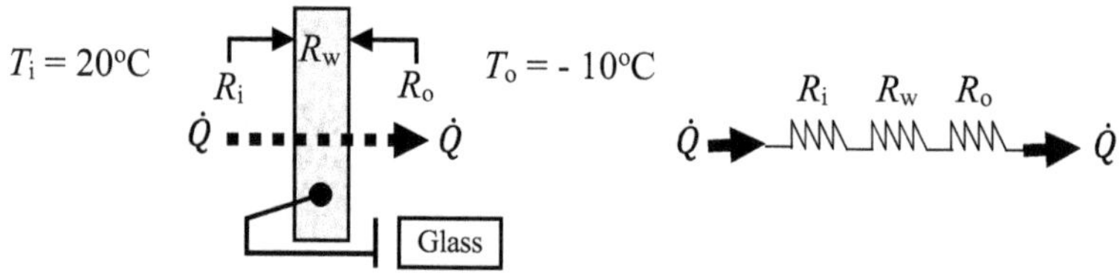


**Solution**

$T_i = 20°C$ and $T_o = -10°C$; $A = 2.2\ m \times 4.0\ m = 8.8\ m^2$; $R_i = 0.120\ m^2.°C/W$ and $R_o = 0.030\ m^2.°C/W$ (Table 2.6); $R_w = 1.3 \times 10^{-2}\ m^2.°C/W$ (Table 2.7).

Since the glass is 5 mm thick, the resistance $R_w = (1.3 \times 10^{-2}\ m^2.°C/W)/2 = 0.65 \times 10^{-2}\ m^2.°C/W$.

Then, the total resistance $R_T$ is:

$$R_t = R_i + R_w + R_o = 0.120\, m^2.°C/W + 0.65 \times 10^{-2} m^2.°C/W + 0.030\, m^2.°C/W$$
$$R_t = 0.1565\, m^2.°C/W$$

And the rate of heat transfer is:

$$\dot{Q} = -\frac{A}{R_t}(T_1 - T_2) = -\frac{8.8\ \text{m}^2}{0.1565\ \text{m}^2.°\text{C/W}}\left[20°\text{C} - (-10°\text{C})\right]$$

$$Q = -1{,}687\ \text{W}\left(\approx 5{,}758\frac{\text{Btu}}{\text{h}}\right)$$

Or using the overall heat transfer coefficient $U = \frac{1}{R_t A} = \frac{1}{0.1565\text{m}^2.°\text{C/W}} = 6.39\,\text{W/m}^2.°\text{C}$

$$\dot{Q} = -UA(T_1 - T_2) = -\left(6.39\frac{\text{W}}{\text{m}^2}°\text{C}\right)(2.2\ \text{m x } 4.0\ \text{m})\left[20°\text{C} - (-10°\text{C})\right]$$

$$\dot{Q} = -1{,}687\ \text{W}\left(\approx 5{,}758\frac{\text{Btu}}{\text{h}}\right)$$

*Heat Flow in Parallel*

Now consider the overall heat transfer through the parallel wall shown below in Figure 2.11, for instance, which has three sections side by side having the same thickness, but different materials and areas with different properties. The wall face temperatures are kept at $T_1$ and $T_2$ as shown below:

One-dimensional heat flow rate in this configuration basically has three different pathways, one through the first section, and another through the second section, and the other one through the third section; and the total heat flow rate is equal to the sum of the three. The individual wall areas normal to heat flow are $A_1$, $A_2$, and $A_3$. Here, we can simply calculate the total resistance by employing the parallel circuit approach. We can first add the individual resistances to obtain the total resistance $R_t$.

$$R_t = \left(\frac{1}{\frac{1}{\Delta x/k_1A_1} + \frac{1}{\Delta x/k_2A_2} + \frac{1}{\Delta x/k_3A_3}}\right) \tag{2.47}$$

And then, the rate of heat transfer through the composite wall can be determined using Equation (2.39).

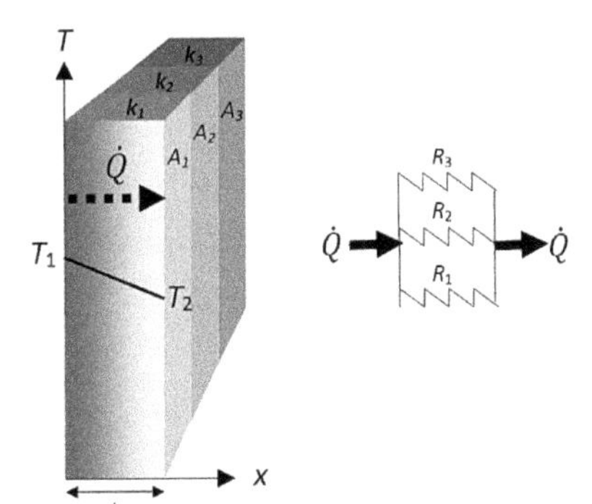


**FIGURE 2.11**
Thermal resistances in parallel circuits.

**Example 2.55: Heat Transfer through a Parallel Wall**

Consider the parallel wall provided in Figure 2.10, 2.2 m high, 6.0 m wide, and 10 cm thick. First 2 m of the wall is made of fir with a thermal conductivity of 0.156 W/m.°C, the next 2 m is made of sandstone with a thermal conductivity of 2.90 W/m.°C, and the last 2 m is made of common brick with a thermal conductivity of 0.692 W/m.°C (Table 2.4). Determine the rate of heat transfer from the wall if the left side is kept at a temperature of 20°C while the right side is kept at −20°C.

**Solution**

$$A_1 = A_2 = A_3 = 2.2\ \text{m} \times 2.0\ \text{m} = 4.4\ \text{m}$$

Let's first calculate the individual resistances for different wall sections. Thus,

$$R_{\text{fir}} = \frac{0.1\ \text{m}}{\left(0.156\ \text{W/m.}^\circ\text{C}\ 4.4\ \text{m}^2\right)} = 0.1457^\circ\text{C/W and} \frac{1}{R_{\text{fir}}} = 6.683$$

$$R_{\text{sandstone}} = \frac{0.1\ \text{m}}{\left(2.90\ \text{W/m.}^\circ\text{C}\ 4.4\ \text{m}^2\right)} = 0.00784^\circ\text{C/W and} \frac{1}{R_{\text{sandstone}}} = 127.551$$

$$R_{\text{brick}} = \frac{0.1\ \text{m}}{\left(0.692\ \text{W/m.}^\circ\text{C}\ 4.4\ \text{m}^2\right)} = 0.0328^\circ\text{C/W and} \frac{1}{R_{\text{brick}}} = 30.488$$

Then, the total resistance $R_t$ and $\dot{Q}$:

$$R_t = \left(\frac{1}{\frac{1}{R_{fir}} + \frac{1}{R_{sandstone}} + \frac{1}{R_{brick}}}\right) = \left(\frac{1}{6.663 + 127.551 + 30.488}\right)$$

$$= \left(\frac{1}{164.902}\right) = 0.006064^\circ\text{C/W}$$

$$\dot{Q} = -\frac{\Delta T}{R_t} = -\frac{\left[20-(-20)\right]^\circ\text{C}}{0.006064\ ^\circ\text{C/W}} = -6{,}596\ \text{W}\left(\approx -22{,}512 \frac{\text{Btu}}{\text{h}}\right)$$

As a check,

$$\dot{Q}_{fir} = -\frac{\Delta T}{R_{fir}} = -\frac{\left[20-(-20)\right]^\circ\text{C}}{0.1457\ ^\circ\text{C/W}} = -274.5\ \text{W}$$

$$\dot{Q}_{sandstone} = -\frac{\Delta T}{R_{sandstone}} = -\frac{\left[20-(-20)\right]^\circ\text{C}}{0.00784\ ^\circ\text{C/W}} = -5{,}102\ \text{W}$$

$$\dot{Q}_{brick} = -\frac{\Delta T}{R_{brick}} = -\frac{\left[20-(-20)\right]^\circ\text{C}}{0.0328\ ^\circ\text{C/W}} = -1{,}219.5\ \text{W}$$

$$\dot{Q}_{total} = -\left(\dot{Q}_{fir} + \dot{Q}_{sandstone} + \dot{Q}_{brick}\right) = -(274.5 + 5{,}102 + 1{,}219.5) = 6{,}596\ \text{W}\ (\approx 22{,}512\ \text{Btu/h})$$

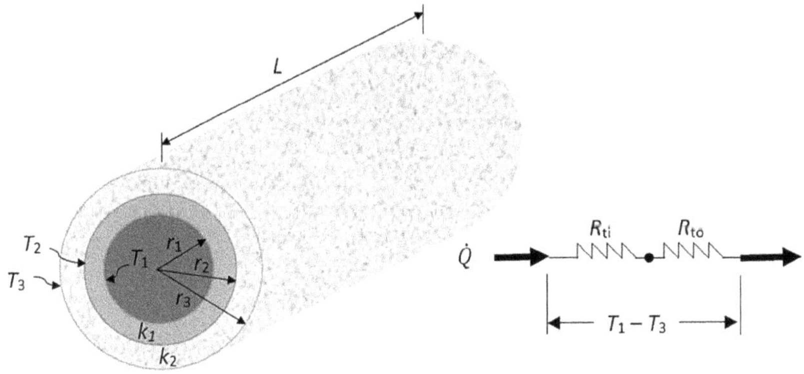


**FIGURE 2.12**
Heat transfer through a compound circular cylinder.

Furthermore, the result of Example 2.51 shows quite a large heat loss from a bare pipe. In order to decrease the loss from a bare pipe, we usually insulate the pipe with a material having a low thermal conductivity. When this is done, we have a situation of two resistances in a series thermal circuit, each having variable heat transfer areas. Applying Equation (2.31) to each of the cylinders in Figure 2.12 yields the following:

For the inner cylinder,

$$\dot{Q} = -2\pi k_1 L \frac{(T_1 - T_2)}{\ln\left(\frac{r_2}{r_1}\right)} \text{ and } R_{ti} = -\frac{\ln\left(\frac{r_2}{r_1}\right)}{2\pi k_1 L} \tag{2.48}$$

For the outer cylinder,

$$\dot{Q} = -2\pi k_2 L \frac{(T_2 - T_3)}{\ln\left(\frac{r_3}{r_2}\right)} \text{ and } R_{to} = -\frac{\ln\left(\frac{r_3}{r_2}\right)}{2\pi k_2 L} \tag{2.49}$$

We can now directly write that the overall resistance is the sum of individual resistances $R_{ti}$ and $R_{to}$. And using Equations (2.48) and (2.49), we can obtain the following:

$$\dot{Q} = -2\pi L \frac{(T_1 - T_3)}{\left(\frac{1}{k_1}\right)\ln\left(\frac{r_2}{r_1}\right) + \left(\frac{1}{k_2}\right)\ln\left(\frac{r_3}{r_2}\right)} \tag{2.50}$$

**Example 2.56: Heat Conduction through an Insulated Cylinder**

Now consider the same 51 mm steel tube with a 2.1 mm wall thickness and inside diameter of 46.7 mm with a uniform thermal conductivity of 17 W/(m.°C) insulated with 2.5 cm mineral wool (k = 0.045 W/(m.°C)). Determine the total amount of steady-state conduction heat transfer rate when the inside surface temperature of the cylinder is 93°C and the outside surface temperature of the insulation is 25°C. Assume the pipe length is 30 m, the steady-state conditions prevail and the heat transfer is one-dimensional.

**Solution**

$T_1 = 93^\circ\text{C}$ and $T_3 = 90^\circ\text{C}$; $r_i = 0.0467$ m; $r_{wall} = 0.0021$ m; $k_1 = 17$ W/m.°C; $k_2 = 0.045$ W/m.°C; $L = 30$ m;

The outer radius of the pipe $r_2 = r_1 + r_{wall} = 0.0467 + 0.0021 \text{ m} = 0.0509 \text{ m}$

The outer radius of the insulation $r_3 = r_2 + r_{ins} = 0.0509 + 0.025 \text{ m} = 0.0759 \text{ m}$

$$\left(\frac{1}{k_1}\right)\ln\left(\frac{r_2}{r_1}\right) = \left(\frac{1}{17\dfrac{\text{W}}{\text{m}^\circ\text{C}}}\right)\ln\left(\frac{0.0509 \text{ m}}{0.0467 \text{ m}}\right) = 0.005$$

$$\left(\frac{1}{k_2}\right)\ln\left(\frac{r_3}{r_2}\right) = \left(\frac{1}{0.045\dfrac{\text{W}}{\text{m}^\circ\text{C}}}\right)\ln\left(\frac{0.0759 \text{ m}}{0.0509 \text{ m}}\right) = 8.879$$

The rate of heat transfer through the compound pipe can then be determined using Equation (2.50) as follows:

$$\dot{Q} = -2\pi(30)\frac{(93-25)}{0.005+8.879} = -1,442 \text{ W} \left(\approx -4,922 \text{ Btu/h}\right)$$

#### *2.3.3.1.3 Mean Temperature Difference (Heat Exchangers)*

In industrial processes, if the fluids are involved in heat exchange without any mixing, the fluids are separated with heat exchanging surfaces known as heat exchangers. The heat flow in a heat exchanger typically involves conduction heat transfer within the separating wall between the two heat exchanging fluids, and convection heat transfer within each fluid in the heat exchanger. Heat exchangers have various shapes depending on the function, and are used as condensers, evaporators, economizers, air heaters, water heaters, reheaters, and so on. Typically, heat exchangers are designated by the geometric shape and the relative directions of the flow of the heat exchanging fluids. For example, Figure 2.13a shows a concentric-tube (or double-pipe) heat exchanger, the simplest type of heat exchangers, which contain two circular pipes with different diameters with fluids flowing in the same (parallel flow) or opposite (counter flow) directions at different temperatures in each pipe. Other common types of heat exchangers are shown in the reminder of Figure 2.13, which include shell and tube, plate-fin, and compact plate and frame type heat exchangers that are designed to achieve a greater heat transfer area per unit volume. Moreover, when a hot fluid flows through a porous storage mass, for instance, the porous storage mass stores the thermal energy temporarily until it is transferred back to a cold fluid. These types of heat exchangers are known as regenerative heat exchangers.

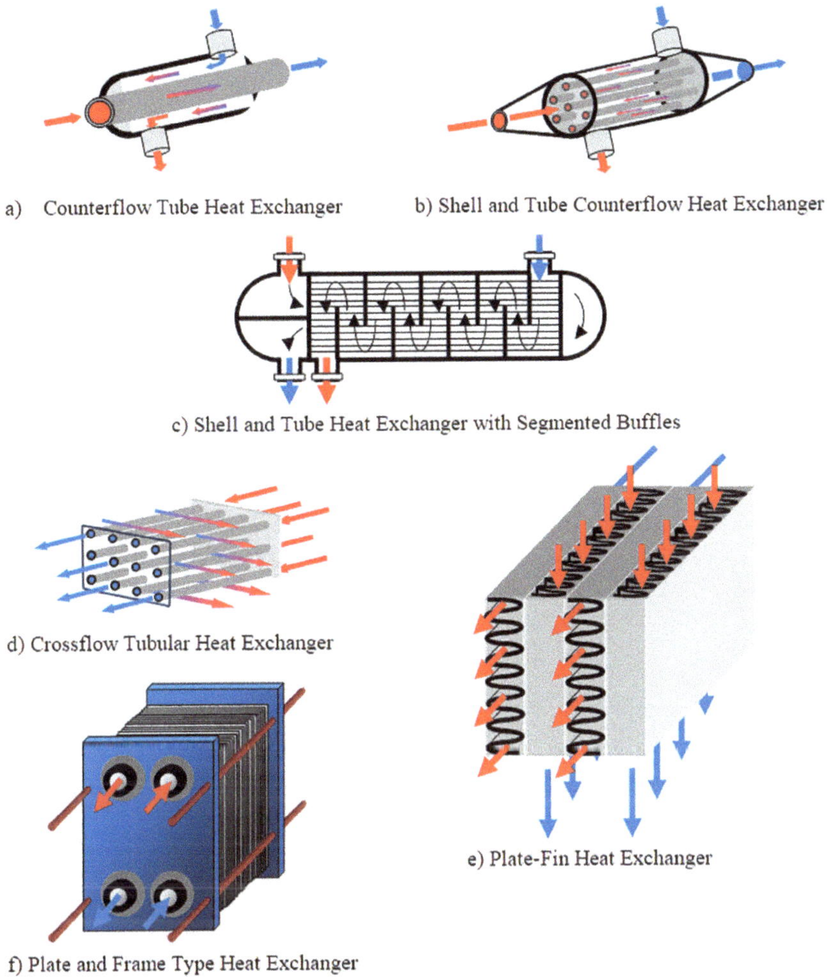


**FIGURE 2.13**
Schematic views of some heat exchanger types.

When heat is exchanged between two fluids flowing through a heat exchanger, the local temperature difference $\Delta T$, varies along the flow path. Heat transfer may be calculated using the following equation, which was developed for heat exchangers using Newton's law of cooling:

$$\dot{Q} = UA\Delta T_m \tag{2.51}$$

where $U$ is the overall heat transfer coefficient having the same units as the convection heat transfer coefficient $h$, W m$^{-2}$°C$^{-1}$ (Btu h$^{-1}$ ft$^{-2}$°F$^{-1}$); $A$ is the heat transfer surface area associated with the coefficient $U$ in m$^2$, and $\Delta T_m$ is an appropriate mean temperature difference in °C (°F). The value of $U$ in Equation (2.51) is not usually constant for all locations in the heat exchanger, and its local value is a function of the local fluid temperatures; however, it is a common practice to evaluate the individual heat transfer coefficients based on the arithmetic average of fluid temperatures. Similar to convection, we have a thermal resistance equal to $1/UA$, the use of which will be best illustrated in the following few example problems.

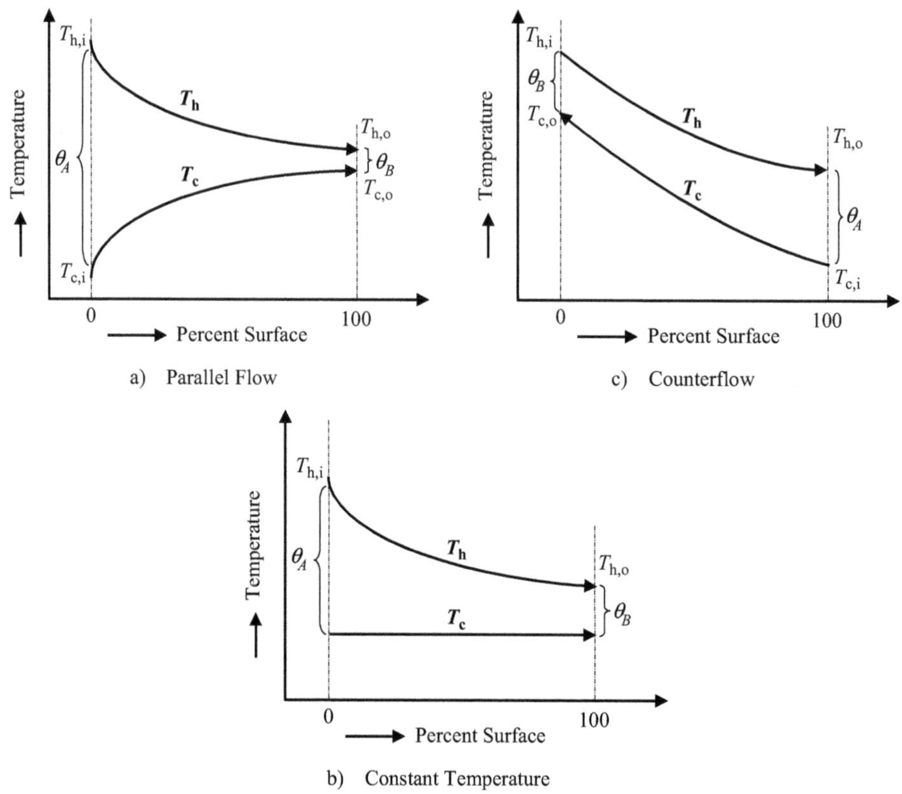


**FIGURE 2.14**
Fluid temperatures in different flow heat exchangers.

As it is seen in Figure 2.14, the temperature of one or both of the heat exchanging fluids varies continuously as the fluids proceed through the heat exchanger. That is why, the true mean temperature difference of fluids exchanging heat cannot be determined by simply subtracting temperatures. The true mean temperature difference can be obtained by assuming that the overall heat transfer coefficient $U$ and the specific heat of each fluid are constant over the entire heat exchanger, the fluid flows are in steady-state, and the exchange process is adiabatic (i.e. the heat losses are negligible). In Figure 2.14, the subscript $h$ is used for the hot fluid, whereas the subscript $c$ is used for the cold fluid. Furthermore, the subscript $i$ is used to denote the temperature at the entry of a fluid of a heat exchanger, and $o$ is used for the temperature of the fluid at the exit of a heat exchanger. The arrowheads on the temperature curves show the direction of the flow of each flow. The largest temperature difference between the fluids in the heat exchanger at either inlet or outlet is expressed as $\theta_A$, and the least temperature difference is designated as $\theta_B$. Now, we can develop $\Delta T_m$, also known as the logarithmic mean temperature difference, in Equation (2.51) for parallel flow, counter flow, and constant temperature heat exchangers as follows:

$$\Delta T_m = \frac{\theta_A - \theta_B}{\ln\left(\dfrac{\theta_A}{\theta_B}\right)} \tag{2.52}$$

Since we assumed that the heat exchange process is a steady-state and an adiabatic process, applying the energy balance approach, the rate of heat transfer $\dot{Q}$ from the hot fluid must be equal to that received by the cold fluid as shown below:

$$\dot{Q} = \dot{m}_h c_{p,h} \left(T_{h,o} - T_{h,i}\right) = \dot{m}_c c_{p,c} \left(T_{c,o} - T_{c,i}\right) \tag{2.53}$$

where $\dot{m}_h$ and $\dot{m}_c$ denote the mass flow rates of the hot and cold fluids, respectively; $c_{p,h}$ and $c_{p,c}$ represent the specific heats; and $T_{h,i}$, $T_{h,o}$, $T_{c,i}$, $T_{c,o}$ are the inlet and outlet temperatures of the hot and cold fluids. Generally, in the analysis of heat exchangers, the product of mass flow rate and specific heat, $C_h = (\dot{m}c_p)_h$ or $C_c = (\dot{m}c_p)_c$ is expressed as a single quantity known as the heat capacity rate.

The relationship developed above for the log mean temperature difference is specifically for parallel flow, counter-flow, and constant temperature heat exchangers. In the case of a cross-flow or shell and tube heat exchanger with multiple passes, correction factors have been developed to account for the deviation from parallel or counter-flow. The correction factor $F$ is applied to the log mean temperature difference to obtain the true temperature difference. The *log mean temperature difference* approach is quite useful when the inlet and outlet temperatures of the fluids and the overall heat transfer coefficient are known and the required heat transfer area needs to be determined. More specifically, in the case the inlet and outlet temperatures of the hot and cold fluids are known, the log mean temperature difference method can normally be employed because $\Delta T_m$ can easily be calculated. This type of problems is known as heat transfer design problems. Generally speaking, along with the inlet temperatures and mass flow rates of the hot and cold fluids, the desired temperature of the cold or hot fluid is known as well. After picking the type of heat exchanger, the heat transfer surface area that will provide the desired outlet temperature is determined, and we eventually decide the heat exchanger size accordingly. Using Equation (2.51), the heat transfer area $A$ can easily be determined. Naturally, for the same task, we can also use an alternate method, the *number of heat transfer units* (NTU) method, which is involved in the use of three non-dimensional parameters and discussed further in depth below. In using this method for solving the problem given above, we first determine the effectiveness $\varepsilon$ and the capacity ratio $Z$, and then find the $NTU$.

In many cases, however, the temperatures of the fluids leaving the exchanger are not known. For instance, if the inlet temperatures, heat exchanger type and size are known, but the heat transfer rate and the fluid outlet temperatures need to be determined (a performance type problem), even though the log mean temperature difference method can be employed to solve the problem, it may take some time as it requires trial and error approach. In such a case, the NTU method becomes quite handy, and the problem can be solved easily because it requires no iteration. If the heat exchanger type and dimensions as well as the mass flow rates are known, $NTU$ and $Z$ can be calculated first, and then using an appropriate diagram or an equation, $\varepsilon$ can be determined. As we will see shortly, since $\dot{Q}_{max}$ can be calculated, we can easily determine the actual heat transfer rate using Equation (2.54). And then, using Equation (2.53), we can determine the outlet temperatures of both fluids.

One of the non-dimensional parameters used in the NTU method is known as the exchanger heat transfer effectiveness $\varepsilon$, which is the ratio of the actual rate of heat transfer to the maximum possible rate of heat transfer, and can be expressed as:

$$\varepsilon = \frac{\dot{Q}}{\dot{Q}_{max}} \tag{2.54}$$

where $\dot{Q}$ is the actual rate of heat transfer, and $\dot{Q}_{max}$ represents the maximum possible rate of heat transfer. Employing the energy balance approach on either the hot or cold fluid, $\dot{Q}_{max}$ is determined from the maximum temperature difference between the hot and cold fluids and the minimum heat capacity rate of the fluids as follows:

$$\Delta T_{max} = \left(T_{h,i} - T_{c,i}\right) \tag{2.55}$$

$$C_h = \left(\dot{m}_h c_{p,h}\right) \text{ and } C_c = \left(\dot{m}_c\, c_{p,c}\right) \tag{2.56}$$

If we consider $C_c < C_h$, then cold fluid will experience higher temperature difference, and the temperature of the cold fluid will approach the inlet temperature of the hot fluid ($T_{c,\,o} = T_{h,\,i}$) as the heat exchanger length approaches infinity. Therefore, based on Equation (2.53),

$$\dot{Q}_{max} = C_c \Delta T_{max} \tag{2.57}$$

Similarly, if $C_h < C_c$, then hot fluid will experience higher temperature difference, and the temperature of the hot fluid will approach the inlet temperature of the cold fluid ($T_{h,\,o} = T_{c,\,i}$). Then, again based on Equation (2.53),

$$\dot{Q}_{max} = C_h \Delta T_{max} \tag{2.58}$$

Then, based on the discussions above, we can write the following general equation:

$$\dot{Q}_{max} = C_{min} \Delta T_{max} \tag{2.59}$$

where $C_{min}$ is either $C_c$ or $C_h$, whichever is smaller.

Then, with the help of Equations (2.53) and (2.59), we can develop

$$\varepsilon = \frac{C_h\left(T_{h,i} - T_{h,o}\right)}{C_{min}\left(T_{h,i} - T_{c,i}\right)} \tag{2.60}$$

or

$$\varepsilon = \frac{C_h\left(T_{c,o} - T_{c,i}\right)}{C_{min}\left(T_{h,i} - T_{c,i}\right)} \tag{2.61}$$

As mentioned earlier, the effectiveness $\varepsilon$ is a non-dimensional quantity, and has a value between zero and one ($0 \le \varepsilon \le 1$). If the effectiveness and the inlet and outlet temperatures of the hot and cold fluids are known, then the actual heat transfer rate can be determined from the following equation:

$$\dot{Q} = \varepsilon C_{min}\left(T_{h,i} - T_{c,i}\right) \tag{2.62}$$

Generally, for any heat exchanger, the heat transfer effectiveness can be expressed for a given exchanger as a function of the number of transfer units and the capacity rate ratio $Z = C_{min}/C_{max}$, the second non-dimensional quantity. Then, we can write the following relationship:

$$\varepsilon = f\left(NTU, \frac{C_{\min}}{C_{\max}}, \textit{flow arrangement}\right) \tag{2.63}$$

In the above relationship, the capacity rate ratio $Z$, depending on the heat capacity rates of the hot and cold fluids, can take the form of $C_c/C_h$ or $C_h/C_c$. The NTU, the third non-dimensional quantity, shown below is used extensively in heat exchanger analyses.

$$NTU = \frac{UA}{C_{\min}} \tag{2.64}$$

The effectiveness $\varepsilon$ is independent of the temperatures in the heat exchanger; however, it is dependent on the flow arrangement and geometry of the heat exchanger. Hence, the effectiveness relations are different for different types of heat exchangers.

For any heat exchanger in which the capacity rate ratio $Z$ is zero (for instance, if one fluid undergoes a phase change, e.g. in condenser or evaporator), the effectiveness becomes:

$$\varepsilon = 1 - \exp(-NTU) \tag{2.65}$$

And heat transfer rate can be determined from:

$$\dot{Q} = c_h\left(T_{h,o} - T_{h,i}\right) = c_c\left(T_{c,o} - T_{c,i}\right) \tag{2.66}$$

And the heat transfer rate in terms of entering fluid temperatures was developed and presented earlier in Equation (2.62). The proper mean temperature difference for the rate equation (Equation (2.51)) is then provided by:

$$\Delta T_m = \varepsilon \frac{\left(T_{h,i} - T_{c,i}\right)}{NTU} \tag{2.67}$$

Moreover, the effectiveness for parallel flow heat exchangers is:

$$\varepsilon = \frac{1 - e^{\left[-NTU(1+Z)\right]}}{1+Z}$$

If $Z = 1$ for a parallel flow heat exchanger, then the effectiveness equation becomes:

$$\varepsilon = \frac{1 - e^{(-2NTU)}}{2} \tag{2.68}$$

Furthermore, the effectiveness of a counter flow heat exchanger is expressed as:

$$\varepsilon = \frac{1 - e^{[-NTU(1-Z)]}}{1 - Z e^{[-NTU(1-Z)]}} \tag{2.69}$$

If Z = 1 for a counter-flow heat exchanger, then the effectiveness equation becomes:

$$\varepsilon = \frac{NTU}{1 + NTU} \tag{2.70}$$

Incropera and DeWitt (1996) and Kays and London (1964) show the relations of $\varepsilon$, $NTU$, and $Z$ for other flow arrangements. Afgan and Schlunder (1974) and these authors also present graphical representations for further convenience.

**Example 2.57: Irrigation Water Pre-heating with a Plate and Frame Type Counter-flow Heat Exchanger**

Irrigation water is to be pre-heated in a plate and frame type counter-flow heat exchanger from 4°C to 21°C at a mass flow rate of 12.5 kg/s. A hot stream of water from the boiler mixing group on manifold is used for this purpose. The temperature of the hot water at the inlet of heat exchanger is measured to be 95°C and it flows with a mass flow rate of 8.5 kg/s. The overall heat transfer coefficient of the plate and frame type heat exchanger is determined to be 5,750 W/m².K. Determine the total heat exchanger surface area required if the surface area per plate is 0.126 m². Assume that steady-state operation prevails; heat losses and gains to/from the surroundings as well as the fouling effect are negligible.

**Solution**

The log mean temperature difference approach is quite useful when the inlet and outlet temperatures of the fluids and the overall heat transfer coefficient are known and the required heat transfer area needs to be determined. Since the inlet and outlet temperatures of the hot and cold fluids are known in this problem, we will employ the log mean temperature difference method because $\Delta T_m$ can easily be calculated. As mentioned earlier, this type of problems is known as heat transfer design problems.

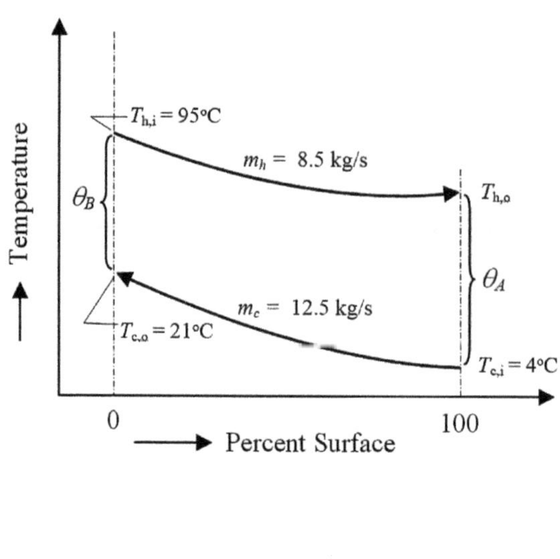

First, we can calculate the rate of heat transfer using Equation (2.53) as follows:

$$\dot{Q} = \dot{m}_c c_{p,c}\left(T_{c,o} - T_{c,i}\right) = \left(12.5\frac{\text{kg}}{\text{s}}\right)\left(4{,}180\frac{\text{J}}{\text{kg°C}}\right)(21-4)\text{°C} = 888{,}250\ \text{W} = 888.3\ \text{kW}$$

After the rate of heat transfer is calculated, we can use the energy balance approach to determine the outlet temperature of hot water side:

$$\dot{Q} = 888.3\ \text{kW} = \dot{m}_h c_{p,h}\left(T_{h,i} - T_{h,o}\right)$$

$$T_{h,o} = T_{h,i} - \frac{\dot{Q}}{\dot{m}_h\, c_{p,h}} = \left[95\text{°C}\right] - \frac{888.3\ \text{kW}}{\left(8.5\ \frac{\text{kg}}{\text{s}}\right)\left(4.18\frac{\text{kJ}}{\text{kg°C}}\right)} = 70\text{°C}$$

Now that the inlet and outlet temperatures are known, the log mean temperature difference can be determined using Equation (2.52) as follows:

$$\Delta T_m = \frac{\theta_A - \theta_B}{\ln\left(\frac{\theta_A}{\theta_B}\right)}$$

For a counter flow heat exchanger,

$$\theta_A = \left(T_{h,i} - T_{c,o}\right) = (95-21)\text{°C} = 74\text{°C}$$
$$\theta_B = \left(T_{h,o} - T_{c,i}\right) = (70-4)\text{°C} = 66\text{°C}$$

Thence,

$$\Delta T_m = \frac{\theta_A - \theta_B}{\ln\left(\frac{\theta_A}{\theta_B}\right)} = \frac{[74-66]\text{°C}}{\ln\left(\frac{74}{66}\right)} = 69.9\text{°C}$$

And now, the required heat transfer area can be calculated as:

$$A = \frac{\dot{Q}}{U\Delta T_m} = \frac{888{,}250\ \text{W}}{\left(5{,}750\ \frac{\text{W}}{\text{m}^2\text{°C}}\right)(69.9\text{°C})} = 2.21\ \text{m}^2$$

The required number of heat exchanger plates can therefore be calculated as:

$$No\ of\ plates = \frac{A}{Area\ per\ plate} = \frac{\left(2.21\ \text{m}^2\right)}{\left(0.126\ \text{m}^2\right)} = 17.5 \approx 18$$

**Example 2.58: Irrigation Water Pre-heating with a Double Pipe Counter-flow Heat Exchanger**

A double pipe counter-flow heat exchanger is used to heat irrigation water for a greenhouse operation. The inlet temperatures of the hot and cold streams are 35°C and 4°C, respectively. The cold stream of water (irrigation water) flows with a mass flow rate of 1.0 kg/s, and the hot stream flows with a mass flow rate of 1.5 kg/s. If the overall heat transfer coefficient is given as 800 W/m$^2$.°C, and the heat transfer area as 4 m$^2$, determine the outlet temperatures of both the hot and cold streams.

**Solution**

Here, we can use the NTU method to determine the outlet temperatures easily. Based on the givens in this problem (performance type), we will first calculate the *NTU* and the capacity ratio Z, and then using Equation (2.69) (or a diagram) for counter-flow heat exchanger, $\varepsilon$ can be determined. And as we will perform shortly, using Equation (2.59), we will first determine $\dot{Q}_{max}$ and after that we can easily determine the actual heat transfer rate using Equation (2.54). And then, using Equation (2.53), we will determine the outlet temperatures of both fluids.

The heat capacity rates for the cold and hot water can be calculated as follows:

$$C_c = \left(\dot{m}_c\, c_{p,c}\right) = \left(1.0\,\frac{\text{kg}}{\text{s}}\right)\left(4{,}180\,\frac{\text{J}}{\text{kg°C}}\right) = 4{,}180\,\frac{\text{W}}{\text{°C}}$$

$$C_h = \left(\dot{m}_h c_{p,h}\right) = \left(1.5\,\frac{\text{kg}}{\text{s}}\right)\left(4{,}180\,\frac{\text{J}}{\text{kg°C}}\right) = 6{,}270\,\frac{\text{W}}{\text{°C}}$$

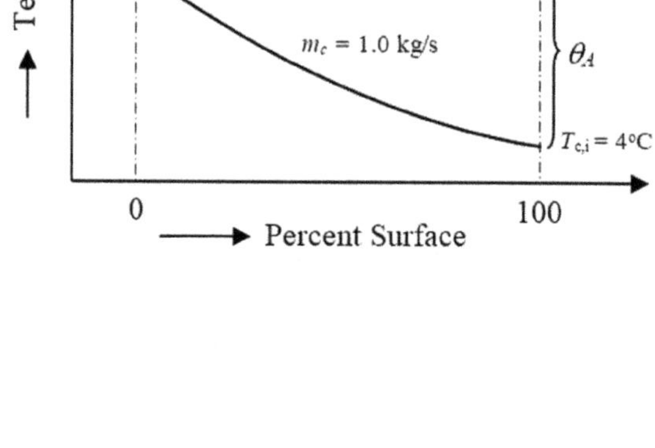

This shows that the minimum heat capacity rate $C_{min}$ is for the cold fluid. Now, we can first calculate the *NTU* using Equation (2.64),

$$NTU = \frac{UA}{C_{\min}} = \frac{\left(800 \frac{\mathrm{W}}{\mathrm{m}^{2\circ}\mathrm{C}}\right)\left(4\,\mathrm{m}^2\right)}{\left(4{,}180\ \frac{\mathrm{W}}{^\circ\mathrm{C}}\right)} = 0.77$$

and then the capacity ratio $Z = \frac{C_{\min}}{C_{\max}}$ as follows:

$$Z = \frac{C_{\min}}{C_{\max}} = \frac{\left(4{,}180 \frac{\mathrm{W}}{^\circ\mathrm{C}}\right)}{\left(6{,}270 \frac{\mathrm{W}}{^\circ\mathrm{C}}\right)} = 0.67$$

Knowing the *NTU* and *Z*, and using Equation (2.69) for counter-flow heat exchanger, the effectiveness $\varepsilon$ can be determined as:

$$\varepsilon = \frac{1 - e^{\left[-NTU(1-Z)\right]}}{1 - Z\,e^{\left[-NTU(1-Z)\right]}} = \frac{1 - \mathrm{e}^{\left[-0.77(1-0.67)\right]}}{1 - 0.67\,\mathrm{e}^{\left[-0.77(1-0.67)\right]}} = 0.47$$

Then, using Equation (2.59), the maximum rate of heat transfer can be determined as follows:

$$\dot{Q}_{\max} = C_{\min}\Delta T_{\max} = C_c\left(T_{h,i} - T_{c,i}\right) = 4{,}180\,\frac{\mathrm{W}}{^\circ\mathrm{C}}(35 - 4)^\circ\mathrm{C}$$

$$\dot{Q}_{\max} = 129{,}580\ \mathrm{W} = 129.6\ \mathrm{kW}$$

After that, we can easily determine the actual heat transfer rate using Equation (2.54) and the effectiveness $\varepsilon$:

$$\dot{Q} = \varepsilon\,\dot{Q}_{\max} = 0.47\left(129.6\ \mathrm{kW}\right) = 60.9\ \mathrm{kW}$$

Now, using Equation (2.53), we can determine the outlet temperatures of both fluids. First, the outlet temperature of the irrigation water is

$$T_{c,o} = \frac{\dot{Q}}{\dot{m}_c\, c_{p,c}} + T_{c,i} = \frac{60.9\ \mathrm{kW}}{4.18\,\frac{\mathrm{kW}}{^\circ\mathrm{C}}} + 4^\circ\mathrm{C} = 18.6^\circ\mathrm{C}$$

Then, similarly, the outlet temperature of the hot water stream is calculated as:

$$T_{h,o} = T_{h,i} - \frac{\dot{Q}}{\dot{m}_h\, c_{p,h}} = \left(35^\circ\mathrm{C}\right) - \frac{60.9\ \mathrm{kW}}{6.27\,\frac{\mathrm{kW}}{^\circ\mathrm{C}}} = 25.3^\circ\mathrm{C}$$

**Example 2.59: Finned Pipe Cross-flow Heat Exchanger for Recovering Waste Heat from Flue Gas**

Let's now consider a cross-flow finned pipe compact heat exchanger used to recover waste heat from boiler flue gas in a greenhouse operation. Heat exchanging body of the exchanger is made of aluminum, the pipe inside diameter is 13.8 mm, and all the other dimensions and overall configuration is shown in the figure provided below. The mass flow rate on the flue gas (hot) side is 1.25 kg/s normal to the heat exchanging pipes at 1 atm pressure and 550°C and the cross-sectional area of the heat exchanger is 0.20 m². The overall heat transfer coefficient based on the flue gas (hot) side is provided as 95 W/m².°C. If the mass flow rate of water is 5 kg/s and the inlet and outlet temperatures are 75°C and 95°C, respectively, determine the heat exchanger volume, the number of pipelines in the flue gas flow direction, and the outlet temperature of the flue gas. Assume that the flue gas has the thermal properties of air at the same temperature and pressure, and the ratio of flue gas (hot) side heat transfer surface area to heat exchanger volume, $\alpha$ is 269 m²/m³.

**Solution**

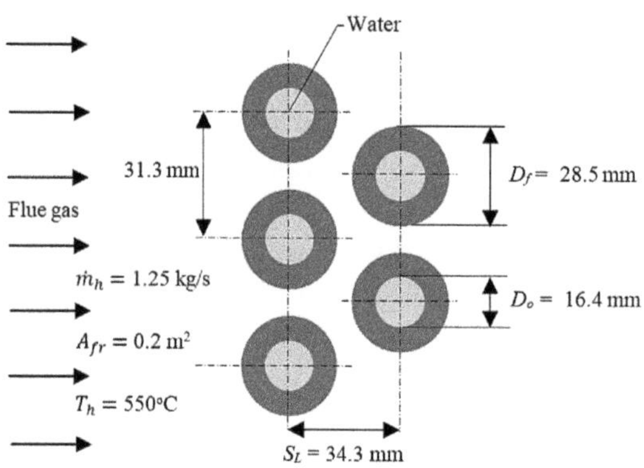


*Useful Information:*
Pipe outside diameter, $D_o$ = 16.4 mm
Fin diameter, $D_f$ = 28.5 mm
Number of fins per pipe: 275 per meter
Fin thickness: 0.254 mm
Free flue gas flow area/Heat exchanger cross-sectional area, $\sigma$ = 0.449
Heat transfer area/total volume, $\alpha$ = 269 m²/m³
Minimum free flue gas flow area is the area normal to the gas flow.

$$\text{Flue gas (air) at } \bar{T} = 430°\text{C: } c_p = 1{,}075\,\frac{\text{J}}{\text{kg}}$$

$$\text{Water at } \bar{T} = 85°\text{C: } c_p = 4{,}201\,\frac{\text{J}}{\text{kg°C}}$$

We will first calculate the heat capacity rates for the cold and hot fluids:

$$C_c = \left(\dot{m}_c\, c_{p,c}\right) = \left(5.0\,\frac{\text{kg}}{\text{s}}\right)\left(4{,}201\,\frac{\text{J}}{\text{kg°C}}\right) = 21{,}005\,\frac{\text{W}}{\text{°C}}$$

$$C_h = \left(\dot{m}_h c_{p,h}\right) = \left(1.25\,\frac{\text{kg}}{\text{s}}\right)\left(1{,}075\,\frac{J}{\text{kg°C}}\right) = 1{,}344\,\frac{\text{W}}{\text{°C}}$$

Then, we can calculate the required rate of heat transfer using Equation (2.53) as follows:

$$\dot{Q} = C_c\left(T_{c,o} - T_{c,i}\right) = \left(21{,}005\,\frac{\text{W}}{^\circ\text{C}}\right)[95 - 75]^\circ\text{C} = 420{,}100\ \text{W}$$

The minimum heat capacity rate $C_{min}$ is for the hot fluid (flue gas), so using Equation (2.59), the maximum rate of heat transfer can then be determined as follows:

$$\dot{Q}_{\max} = C_{\min}\Delta T_{\max} = C_h\left(T_{h,i} - T_{c,i}\right) = 1{,}344\,\frac{\text{W}}{^\circ\text{C}}(550 - 75)^\circ\text{C}$$

$$\dot{Q}_{\max} = 638{,}400\ \text{W}$$

Then, using Equation (2.54), the heat exchanger effectiveness can be calculated,

$$\varepsilon = \frac{\dot{Q}}{\dot{Q}_{\max}} = \frac{420{,}100\ \text{W}}{638{,}400\ \text{W}} = 0.658$$

And we can calculate the capacity ratio $Z = \dfrac{C_{\min}}{C_{\max}}$ as follows:

$$Z = \frac{C_{\min}}{C_{\max}} = \frac{\left(1{,}344\,\dfrac{\text{W}}{^\circ\text{C}}\right)}{\left(21{,}005\,\dfrac{\text{W}}{^\circ\text{C}}\right)} = 0.06$$

Using the effectiveness $\varepsilon$ equation provided below or the effectiveness $\varepsilon$, the capacity ratio $Z$ and the effectiveness graph for a non-mixing single pass cross-flow heat exchanger provided by Incropera and DeWitt (1996), the $NTU$ can be determined as 1.00.

$$\varepsilon = 1 - \exp\left[\left(\frac{1}{Z}\right)(NTU)^{0.22}\left\{\exp\left[-Z(NTU)^{0.78}\right] - 1\right\}\right]$$

$$NTU = \frac{U_h A_h}{C_{\min}} = 1.00$$

And then, using this $NTU$, the hot side heat transfer surface area $A_h$ is found,

$$A_h = \frac{1.00 \times 1{,}344\ \text{W}/^\circ\text{C}}{95\ \text{W}/\text{m}^2.^\circ\text{C}} = 14.15\ \text{m}^2$$

Since the hot (flue gas) side heat transfer surface area per heat exchanger volume $\alpha = \dfrac{A_h}{V}$ is already given as 269 $\text{m}^2/\text{m}^3$, we can easily determine the required heat exchanger volume $V$,

$$V = \frac{A_h}{\alpha} = \frac{14.15\ \text{m}^2}{269\ \text{m}^2/\text{m}^3} = 0.0526\ \text{m}^3$$

Since we know the heat exchanger volume $V$, we can now determine the heat exchanger dimensions in the flue gas flow direction using the following expression for $N_L$, which gives the number of pipe rows in the direction of flue gas flow:

$$N_{\mathrm{L}} \approx \frac{L - D_f}{S_L} + 1$$

where $L = \frac{V}{A_{\mathrm{fr}}} = \frac{0.0526\ \mathrm{m}^3}{0.20\ \mathrm{m}^2} = 0.263\ \mathrm{m}$, $D_f$ is the fin diameter, $S_L$ is the distance between the front and back row pipe centers.

$$N_{\mathrm{L}} \approx \frac{(263 - 28.5)\mathrm{mm}}{34.3\ \mathrm{mm}} + 1 = 7.84 \approx 8$$

And the outlet temperature of the flue gas is

$$T_{h,o} = T_{h,i} - \frac{\dot{Q}}{C_h} = (550°\mathrm{C}) - \frac{420{,}100\ \mathrm{W}}{1{,}344\ \frac{\mathrm{W}}{°\mathrm{C}}} = 237°\mathrm{C}$$

This calculated outside temperature shows that our previously assumed average flue gas (air) temperature of 430°C was not so bad an assumption.

#### 2.3.3.1.4 Transient Heat Conduction

So far, we have focused on steady-state conduction heat transfer processes, where the temperatures do not change over time. However, quite often, the heat transfer and temperature distribution under unsteady or transient conditions ought to be known. In transient conditions, the temperature of the body of interest at any point as well as its heat content varies temporally and spatially. Time-dependent transient heat transfer problems are encountered in many energy engineering applications. For instance, variation of external air temperature, and solar irradiation affecting the heating and cooling loads of a greenhouse or the temperature of different materials within would be an example for transient heat transfer conditions.

The fundamental equation for three-dimensional transient conduction in solids or fluids is:

$$\frac{\partial T}{\partial t} = \alpha\left(\frac{\partial^2 T}{\partial x^2} + \frac{\partial^2 T}{\partial y^2} + \frac{\partial^2 T}{\partial z^2}\right) \tag{2.71}$$

where $\alpha = \frac{k}{\rho c_p}$ is the thermal diffusivity, $k$ is the thermal conductivity, $\rho$ is the density, and $c_p$ is the specific heat. This equation tells us that in the case of high thermal conductivity, and low density and/or specific heat, we would have a large thermal diffusivity, that is, we would have fast heat diffusion.

In order to predict the rate of temperature change of a material being held at constant volume with uniform temperature, such as a well-stirred volume of fluid, temperature of which is changing due to a net rate of heat gain or loss, the proper equation can be written as:

$$m c_v \frac{\partial T}{\partial t} = -Q_{net} \tag{2.72}$$

where $m$ is the mass of the material, $c_v$ is the specific heat at constant volume. If the heating or cooling takes place at constant pressure, then $c_v$ is replaced by $c_p$. We should also keep in mind that, for liquids and solids, $c_v$ and $c_p$ are almost equal; therefore, we can use $c_p$ and the small error caused can be ignored. In Equation (2.72), $Q_{net}$ may include heat transfer due to conduction, convection, or radiation. When convective boundary conditions are needed in the solution for not so fast transient conditions, we often use the heat transfer coefficients ($h$ values) based on steady-state correlations.

### 2.3.3.2 Convection Heat Transfer

Convection heat transfer may involve energy transfer by conduction and eddy mixing and diffusion due to bulk fluid motion. The transferred thermal energy is usually sensible heat, and in some cases where a phase change between a liquid and vapor state takes place, then the convection heat transfer also include latent heat exchange. Let's now consider heat transfer to a flowing fluid inside a conduit. In the case we have a large enough Reynolds number, then there exist three flow regions: i) laminar sub-layer, the layer immediately adjacent to the wall where heat transfer mode is thermal conduction; ii) buffer layer, the transition region outside the laminar sub-layer, where both conduction and eddy mixing (convection) are significant; and iii) turbulent region, the region beyond the buffer layer extending to the center of the conduit, where the dominant mode of transfer is eddy mixing (convection).

The main body of fluid in most equipment is in turbulent mode, and a laminar layer is always present at the solid walls. In low velocity or viscous flows where low Reynolds numbers prevail, the entire flow might be laminar flow without having any transition or eddy region. Heat transfer by convection can be further categorized based on the nature of flow. When the flow is induced by buoyancy forces arising from density variations caused by temperature variations in the fluid, it is then termed as free or natural convection. On the other hand, if fluid currents are caused by external sources, for instance, a fan, pump, or atmospheric wind, then the mode of heat transfer is termed as forced convection.

The fundamental equation for convection heat transfer is known as Newton's law of cooling and is expressed as follows:

$$\dot{Q} = -h A_s \left(T_s - T_\infty\right) \tag{2.73}$$

where $h$ represents the convection heat transfer coefficient (W/m$^2$.°C or Btu/h.ft$^2$.°F), which includes all effects regarding the convection heat transfer and dependent upon the boundary layer conditions, that is, the nature of the fluid flow, thermal properties, surface geometry, and physical properties all impact the magnitude of convection heat transfer. Typical values of $h$ are provided in Table 2.11. In the above equation, $A_s$ is the surface area (m$^2$ or ft$^2$), $T_s$ is the surface temperature (°C or °F), and $T_\infty$ is the temperature of the fluid

**TABLE 2.11**

Typical Values of Convective Heat Transfer Coefficients.

| Material | Heat Transfer Coefficient (W/m².°C) | Heat Transfer Coefficient (Btu/h.ft².°F) |
|---|---|---|
| Gases (natural convection) | 0.6–28 | 0.1–5 |
| Flowing gases | 11–284 | 2–50 |
| Flowing liquids (nonmetallic) | 170–5,678 | 30–1,000 |
| Boiling liquids | 1,136–283,915 | 200–50,000 |
| Condensing vapors | 2,839–283,915 | 500–50,000 |

flowing over the surface (°C or °F). Newton's law of cooling equation does not take radiation heat transfer into account, which in many cases, is relatively small compared to conduction or convection heat transfer involving a surface and a fluid; however, in the case of high surface temperatures and natural convection, the rate of heat transfer due to radiation is comparable to that of the natural convection heat transfer.

When compared with Equation (2.34), we can write the convective resistance term (boundary layer resistance or film resistance) $R_c$ in Equation (2.73) as follows:

$$R_c = \frac{1}{h\,A_s} \tag{2.74}$$

The thermal resistance for convective heat transfer $R_c$ is treated in the same manner as the resistance concept in conduction heat transfer discussed earlier.

**Example 2.60: Heat Convection from a Circular Steel Tube**

Now assume that the 51 mm steel tube in Example 2.51 has a heat transfer coefficient on the outside due to natural convection to the surrounding air at 20°C of 5.1 W/m².°C. Determine the heat transfer from the given tube to the greenhouse air.

**Solution**

$T_i = 93°C$; $r_i = 0.0467$ m and $r_{wall} = 0.0021$ m; $L = 30$ m; $k = 17$ W/m.°C

The outer radius of the pipe $r_o = r_i + r_{wall} = 0.0467 + 0.0021\text{ m} = 0.0488\text{ m}$

The resistance of the pipe is:

$$R_{pipe} = \frac{\ln\left(\frac{r_o}{r_i}\right)}{2\pi k L} = \frac{\ln\left(\frac{0.0488}{0.0467}\right)}{2\pi(17\text{ W/m.°C})\times 30\text{ m}} = 1.3734\times 10^{-5}\text{ °C/W}$$

$$R_{convection} = \frac{1}{hA} = \frac{1}{(5.1\text{ W/m}^2.\text{°C})(\pi\times 0.0976\text{ m}\times 30\text{ m})} = 0.02133\text{ °C/W}$$

$$R_{total} = 6.025\times 10^{-4} + 0.02133 = 0.02134\text{ °C/W}$$

$$\dot{Q} = -\frac{(T_1 - T_2)}{R} - \frac{(93-20)\text{°C}}{0.02193\,\text{°C/W}} = -3{,}420\text{ W}\left(-11{,}673\frac{\text{Btu}}{\text{h}}\right)$$

Here, we have observed that the outside film resistance acted as an insulator, and the resistance of the pipe is negligible (see Example 2.51).

And furthermore, with the help of Nusselt number, also known as the dimensionless convection heat transfer coefficient, the heat transfer coefficient $h$ is non-dimensionalized and the relationship is defined in the following equation:

$$\text{Nu} = \frac{hD}{k} \text{ or } \frac{hL}{k} \tag{2.75}$$

where $D$ and $L$ are the characteristic length and diameter, respectively, and $k$ is the fluid thermal conductivity. Nusselt number in essence provides the ratio of the convective heat transfer through a fluid layer that has a temperature $T_1$ on one side and temperature $T_2$ on the other side of the fluid in the presence of fluid motion to the conductive heat transfer through the fluid when the fluid layer is stagnant:

$$\text{Nu} = \frac{\dot{Q}_{conv}}{\dot{Q}_{cond}} = \frac{h(T_1 - T_2)}{\dfrac{k(T_1 - T_2)}{L}} = \frac{hL}{k} \tag{2.76}$$

As it will be discussed further in the natural and forced convection sections below, using the techniques of dimensional analysis, the non-dimensionalized expression of Nusselt number reduces the total number of variables for natural and forced convection heat transfers. Once Nusselt number is determined using a proper correlation, then using the relationship expressed in Equation (2.76), we can easily determine the convective heat transfer coefficient.

Let's now recall that, earlier in this chapter, we considered a composite wall configuration, one side facing to the indoor environment with a temperature of $T_i$ and the other side to the outdoor with a temperature of $T_o$ shown in Figure 2.10, and determined the overall resistance of the thermal circuit considering the inside and outside boundary layer resistances as provided in Equation (2.45):

$$R_t = \frac{1}{h_i\,A} + \frac{\Delta x_1}{k_1\,A} + \frac{\Delta x_2}{k_2\,A} + \frac{\Delta x_3}{k_3\,A} + \frac{1}{h_o\,A}$$

And we also stated that, when dealing with composite walls, it is much easier to work with the overall heat transfer coefficient $U$, whose relationship with the total thermal resistance is provided earlier in Equation (2.46) as:

$$U = \frac{1}{R_t\,A} = \frac{1}{h_i} + \frac{\Delta x_1}{k_1} + \frac{\Delta x_2}{k_2} + \frac{\Delta x_3}{k_3} + \frac{1}{h_o}$$

And then, using the overall thermal resistance $R_t$ or the overall heat transfer coefficient $U$ and the inside and outside temperatures, the heat transfer rate can be easily determined using Equation (2.39).

#### 2.3.3.2.1 Natural Convection

As mentioned earlier, when the flow is induced by buoyancy forces arising from density variations caused by temperature variations in the fluid, natural convection heat transfer occurs. Using the techniques of dimensional analysis, the following non-dimensionalized expression that reduces the total number of variables have been developed for natural convection:

$$\frac{hD}{k} = \mathrm{Nu} = A(\mathrm{Gr})^a(\mathrm{Pr})^b \tag{2.77}$$

where Nu is the Nusselt number, Gr is the Grashof number, Pr is the Prandtl number, the details of which are provided in Table 2.6, and $A$, $a$, and $b$ are the constants whose values change depending on the system under consideration. We should remember that Nusselt, Grashof, and Prandtl numbers are dimensionless numbers, and consistent units must be employed throughout. The properties of fluid such as density, viscosity, specific heat, and thermal conductivity are usually evaluated at the film temperature of fluid. The characteristic length is the diameter $D$ of the tube, when a tube is involved; however, it is the length of the plane plate, if a plate type surface is involved.

The boundary layer of the fluid is either laminar or turbulent; hence, this situation will affect the values of constants in the above equation. Experimentally observed values suggest that the heat transfer coefficient increases (moving from laminar to turbulent) when the product of (Gr) (Pr) exceeds $10^8$. And heat transfer literature provides natural and forced heat transfer coefficients for different geometry and orientations. As mentioned earlier, natural convection is a consequence of density differences in the fluid. Therefore, since the warmer fluid molecules rise, the orientation of the hot surface impacts the rate of heat transfer. As Table 2.9 shows, a hot horizontal surface has lower thermal resistance compared to that of a vertical surface because the hot fluid molecules from the horizontal surface rise more easily and the flow becomes more readily turbulent. However, the thermal resistance increases if the heat transfer from the horizontal surface is downward, because the hot fluid molecules have the most difficult time rising from this orientation. Furthermore, fluid flow over hot surfaces experience frictional resistances that can slow down the fluid motion especially in forced convection, which is discussed in the following section.

#### 2.3.3.2.2 Forced Convection

Forced convection flow takes place between a solid surface and a moving fluid and can be either laminar or turbulent, inside or outside tubes, and involve phase changes such as boiling and condensation. For instance, heat exchangers, forced water and air heaters and coolers, forced water, and air condensers and evaporators all involve forced convection heat transfer. Before applying the Newton's law of cooling provided in Equation (2.73), as it is the case for natural convection as well, we first need to determine whether the boundary layer is laminar or turbulent, and then the heat transfer coefficient $h$ accordingly. For determining forced convection heat transfers, we can employ what's called as Nusselt–Reynolds correlations, which are provided in heat transfer literature for different fluids and flow regimes. The expressions for Nusselt and Reynolds numbers as are provided in Table 2.6.

As mentioned earlier, in the turbulent region, the dominant mode of transfer is eddy mixing, that is, turbulent mixing (convection). In the transition region, where we have the transition from laminar to turbulent, the value of Reynolds number is called the critical Reynolds number, which is a function of the surface geometry and fluid flow conditions. Because of the number of cases involved in forced convection, the heat transfer coefficient can be determined from the following generalized equation whose specific constants and shapes are provided in heat transfer literature:

$$\frac{hD}{k} = \text{Nu} = C(\text{Re})^a(\text{Pr})^b \tag{2.78}$$

where Nu is the Nusselt number, Re is the Reynolds number, Pr is the Prandtl number, the details of which are provided in Table 2.6, and $a$, $b$, and $C$ are the constants whose values vary depending on the system under consideration. It must be remembered that Nusselt, Reynolds, and Prandtl numbers are dimensionless numbers, and consistent units must be employed throughout. The properties of density, viscosity, specific heat, and thermal conductivity are usually evaluated at the film temperature of fluid. $D$ is the diameter of the tube, inside or outside, when a tube is involved; however, the characteristic length is the length of the plane plate, if a plate type surface is involved; for other shapes, the hydraulic diameter is used; and $u$ is the average velocity of the flowing fluid.

### *External Flow Forced Convection*

If a fluid flows over an external surface, for example, a flat surface, pipe, cylinder, and sphere, the flow is said to be an external flow. In such flows, in order to determine the heat transfer rates, we employ experimentally developed correlations between the Nusselt number, the Reynolds number, and the Prandtl number. The fluid properties used in determining these dimensionless parameters are typically evaluated at the film temperatures $T_f$, which is the average of the surface temperature and the bulk temperature of the fluid. Forced convection correlations for different geometries and different external flow regimes are covered in depth and provided in the heat transfer literature and therefore will not be repeated here.

**Example 2.70: External Flow Convective Heat Loss from a Flat Plate Collector**

Let's now consider a horizontal flat plate solar collector. As shown below, the top cover (3 m × 1 m) of the collector has a surface temperature of 50°C. If the wind speed is 6.5 m/s and the ambient air temperature is 4°C, assuming steady-state conditions, determine the rate of convective heat loss from the cover.

**Solution**

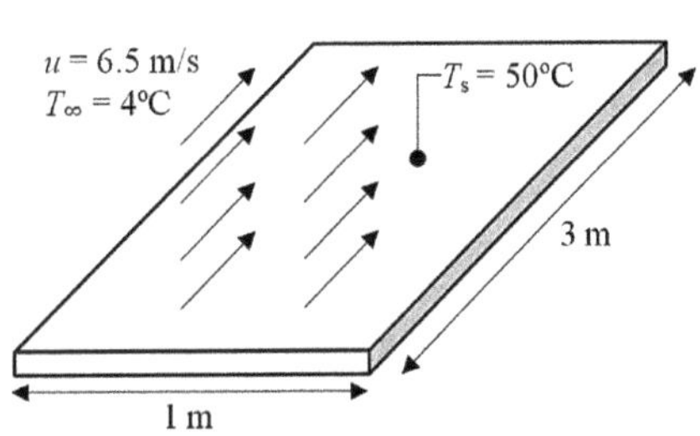

We first need to evaluate the properties of air at the film temperature $T_f$ needed for the dimensionless Re and Pr numbers. So, the film temperature $T_f$ is:

$$T_f = \frac{T_s + T_\infty}{2} = \frac{(50+4)^\circ\text{C}}{2} = 27^\circ\text{C}$$

Using $T_f$ = 27°C and Table 2.12, we can now determine the air density $\rho$ as 1.1765 kg/m$^3$, the kinematic viscosity $v$ as 15.76 × $10^{-6}$ m$^2$/s, the thermal conductivity as 0.0261 W/m.K, and Prandtl number Pr as 0.714.

The Reynolds number at the cover end can be determined as:

$$\text{Re} = \frac{uL}{v} = \frac{\left(6.5\frac{\text{m}}{\text{s}}\right)(3\,\text{m})}{\left(15.76\times10^{-6}\right)\text{m}^2/\text{s}} = 1,237,310$$

For the case Re ≤ $10^8$ and 0.6 < Pr < 60, the proper correlation from the literature is provided as:

$$\text{Nu} = \left(0.037\,\text{Re}^{0.8} - 871\right)\text{Pr}^{1/3}$$

**TABLE 2.12**

Thermophysical Properties of Air at Atmospheric Pressure.

| Temperature (°C) | $\rho$ (kg/m$^3$) | $c_p$ (J/kg.K) | $\mu \cdot 10^7$ (N.s/m$^2$) | $\nu \cdot 10^6$ (m$^2$/s) | $k \cdot 10^3$ (W/m.K) | $\alpha \cdot 10^6$ (m$^2$/s) | Pr |
|---|---|---|---|---|---|---|---|
| −40 | 1.5161 | 1,006 | 100.1 | 15.18 | 20.9 | 13.7 | 0.730 |
| −30 | 1.4535 | 1,005 | 108.1 | 15.70 | 21.7 | 14.9 | 0.727 |
| −20 | 1.3958 | 1,005 | 116.2 | 16.22 | 22.5 | 16.0 | 0.725 |
| −10 | 1.3426 | 1,005 | 124.6 | 16.73 | 23.3 | 17.3 | 0.722 |
| 0 | 1.2933 | 1,006 | 133.2 | 17.23 | 24.1 | 18.5 | 0.720 |
| 5 | 1.2699 | 1,006 | 137.6 | 17.47 | 24.5 | 19.2 | 0.719 |
| 10 | 1.2474 | 1,006 | 142.1 | 17.72 | 24.8 | 19.8 | 0.718 |
| 20 | 1.2047 | 1,006 | 151.1 | 18.21 | 25.6 | 21.1 | 0.716 |
| 30 | 1.1649 | 1,006 | 160.4 | 18.68 | 26.3 | 22.5 | 0.714 |
| 40 | 1.1275 | 1,007 | 169.8 | 19.15 | 27.1 | 23.9 | 0.712 |
| 50 | 1.0925 | 1,007 | 179.5 | 19.61 | 27.8 | 25.3 | 0.711 |
| 60 | 1.0597 | 1,008 | 189.3 | 20.06 | 28.5 | 26.7 | 0.709 |
| 70 | 1.0287 | 1,009 | 199.4 | 20.51 | 22.2 | 28.2 | 0.708 |
| 80 | 0.9995 | 1,010 | 209.6 | 20.95 | 29.9 | 29.7 | 0.707 |
| 90 | 0.9719 | 1.010 | 220.0 | 21.38 | 30.6 | 31.2 | 0.706 |
| 100 | 0.9458 | 1,011 | 230.6 | 21.81 | 31.3 | 32.7 | 0.705 |
| 110 | 0.9210 | 1,012 | 241.4 | 22.23 | 32.0 | 34.3 | 0.704 |
| 120 | 0.8975 | 1,013 | 252.3 | 22.65 | 32.6 | 35.9 | 0.703 |
| **Temperature (K)** | | | | | | | |
| 400 | 0.8711 | 1,014 | 230.1 | 26.41 | 33.8 | 38.3 | 0.690 |
| 450 | 0.7740 | 1,021 | 250.7 | 32.39 | 37.3 | 47.2 | 0.686 |
| 500 | 0.6964 | 1,030 | 270.1 | 38.79 | 40.7 | 56.7 | 0.684 |
| 550 | 0.6329 | 1,040 | 288.4 | 45.57 | 43.9 | 66.7 | 0.683 |
| 600 | 0.5804 | 1,051 | 305.8 | 52.69 | 46.9 | 76.9 | 0.685 |

Then,

$$\text{Nu} = \left(0.037 \times (1{,}237{,}310)^{0.8} - 871\right)0.714^{1/3} = 1{,}696$$

After determining the Nusselt number, we can then calculate the convective heat transfer coefficient $h$ using the following relationship:

$$h = \text{Nu}\frac{k}{L} = 1{,}696\,\frac{0.0261\ \text{W/m.K}}{3\ \text{m}} = 14.75\,\frac{\text{W}}{\text{m}^2.\text{K}}$$

Now that we have the transfer coefficient $h$, using the Newton's law of cooling (Equation (2.73)), the rate of convective heat loss from the cover can finally be determined as follows:

$$\dot{Q} = -h\,A\left(T_s - T_\infty\right) = -\left(14.75\,\frac{\text{W}}{\text{m}^2.\text{K}}\right)(3\ \text{m} \times 1\ \text{m})(50 - 4)^\circ\text{C} = -2{,}036\ \text{W}\left(\approx -6{,}948\ \text{Btu/h}\right)$$

**Example 2.71: External Flow Convective Heat Loss from a Hot Water Storage Tank**

The thermal energy generated during $CO_2$ dosing in a greenhouse operation is stored in a hot water storage tank for later use. Let's now consider that a cylindrical thermal energy storage tank with a diameter of 4 m and a height of 6 m is used to store the heat generated during the dosing. Assuming that the storage tank has a uniform and constant outside surface temperature of 50°C, and the steady-state conditions prevail, determine the rate of convective heat loss from the hot water storage tank if the average wind speed is 1.25 m/s and the ambient air temperature is −10°C. The efficiency of a thermal energy storage decreases as the thermal losses from the tank increase. If the overall efficiency of a hot water storage tank is defined as the ratio of the heat recovered to the heat input, determine the overall thermal efficiency of the hot water storage tank for a day for the conditions provided above if the total energy input was 33 GJ.

**Solution**

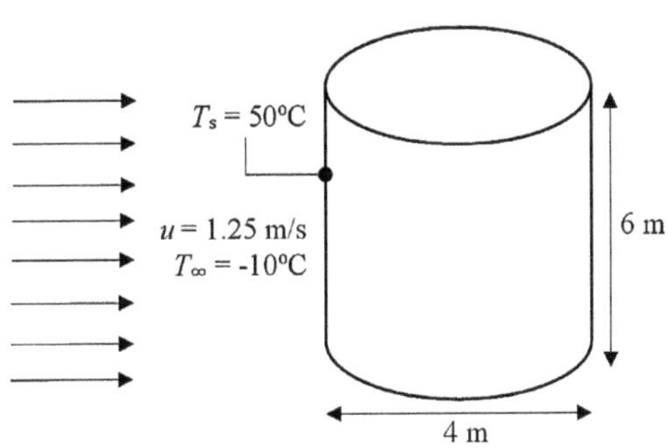


The outside surface temperature of the tank is given as 50°C. The velocity of cross-flow wind and temperature are known as well.

We need to determine the properties of air at the film temperature $T_f$. So, the film temperature is:

$$T_f = \frac{(T_s + T_\infty)}{2} = \frac{[50 - 10]^\circ\text{C}}{2} = 20^\circ\text{C}$$

At the temperature of 20°C, using Table 2.12, the air has a density $\rho$ of 1.2047 kg/m³, kinematic viscosity $v$ of $18.21 \times 10^{-6}$ m²/s, Prandtl number Pr of 0.716, and thermal conductivity $k$ of 0.0256 W/m.K.

We can now determine the Reynolds number using the tank diameter as the characteristic length:

$$\text{Re} = \frac{uD}{v} = \frac{\left(1.25\,\frac{\text{m}}{\text{s}}\right)(4\,\text{m})}{\left(18.21 \times 10^{-6}\,\frac{\text{m}^2}{\text{s}}\right)} = 274{,}574$$

For $40{,}000 < \text{Re} < 400{,}000$ and $\text{Pr} \geq 0.7$, the proper correlation from the literature is:

$$\text{Nu} = 0.027\,\text{Re}^{0.805}\,\text{Pr}^{1/3} = 0.027\left(274{,}574^{0.805}\right)\left(0.716^{1/3}\right) = 653.9$$

Using this Nusselt number, we can now determine the convective heat transfer coefficient $h$ as:

$$h = \text{Nu}\frac{k}{L} = (653.9)\frac{\left(0.0256\,\frac{\text{W}}{\text{mK}}\right)}{(4\,\text{m})} = 4.19\,\frac{\text{W}}{\text{m}^2.\text{K}}$$

Now, using the Newton's law of cooling, the rate of convective heat loss from the storage tank is determined,

$$\dot{Q} = -h\,A\left(T_s - T_\infty\right) = -4.19\frac{\text{W}}{\text{m}^2\text{K}}\left[\pi\,(4\,\text{m})(6\,\text{m})\right]\left[50 - (-10)\right]^\circ\text{C} = -18{,}923\text{W}\,(-64{,}582\,\text{Btu/h})$$

And, the overall thermal efficiency of the storage tank $\eta$ is expressed as:

$$\eta_{\text{overall}} = \frac{\textit{Heat recovered}}{\textit{Heat input}} = \frac{Q_{recovered}}{Q_{in}} = 1 - \frac{Q_{lost}}{Q_{input}}$$

where $Q_{lost}$ is the amount of heat lost during 24 h, and $Q_{input}$ is the total heat input to the storage tank during this time. The overall energy efficiency of the storage tank can then be calculated as:

$$\eta_{\text{overall}} = 1 - \frac{\left(18{,}923\,\text{J/s} \times 24\,\text{h} \times 3{,}600\,\text{s/h}\right)}{\left(33 \times 10^9\,\text{J}\right)} = 0.95$$

Thence, assuming that the given conditions remain constant, the overall efficiency of the tank is 95%.

### Internal Flow Forced Convection

The fluid flow in a conduit such as a pipe or duct is classified as internal flow. For determining internal flow forced convection heat transfers as well, we employ the experimental correlations, which are provided in heat transfer literature for different flow regimes. In determining the Reynolds and Nusselt numbers, we use the characteristic length; and when a tube is involved, the characteristic length is the internal diameter $D$ of the tube; however, for other shapes, it is the hydraulic diameter. After determining the Nusselt number, the heat transfer coefficient is calculated for a given situation.

**Example 2.72: Internal Flow Convective Heat Transfer from a Boiler Tube to Water**

Hot air is being used for generating hot water for hydronic heating in a hot water boiler. Approx. 550 K hot air enters a 6 m long 5 cm diameter tube with an average velocity of 8 m/s and boils water at 373 K. Assuming that steady-state conditions prevail, the outer surface temperature of the tube is uniform, constant, and equal to the boiling temperature with a negligible thermal resistance, determine the convective heat transfer coefficient of the air flowing in the tubes.

**Solution**

We know that the air inlet temperature is 550 K, but we do not know the exit temperature of the air; so, we will assume a mean air temperature of 500 K for determining the properties of air. At the end of our calculations, the air exit temperature will be calculated to see if the mean air temperature is in agreement with the assumed mean air temperature. If they are not in agreement, we will repeat the calculations using the new mean temperature.

At an assumed mean temperature $T_{avg}$ = 500 K, using Table 2.12, we can determine that the air density is 0.6964 kg/m$^3$, specific heat is 1,030 J/kg.K, kinematic viscosity is 3.88 × $10^{-5}$ m$^2$/s, thermal conductivity is 0.0407 W/m.K, and the Prandtl number is 0.684.

The average velocity of air through the tube is known to be 5 m/s, then the Reynolds number can be calculated:

$$\mathrm{Re} = \frac{uD}{v} = \frac{\left(8\,\frac{\mathrm{m}}{\mathrm{s}}\right)(0.05\,\mathrm{m})}{\left(3.88\times10^{-5}\,\frac{\mathrm{m}^2}{\mathrm{s}}\right)} = 10{,}309$$

So, the flow is turbulent. Assuming that the flow is fully developed as it flows through the tube, the literature provides the proper Nusselt number correlation as:

$$\mathrm{Nu} = 0.023\,\mathrm{Re}^{0.8}\,\mathrm{Pr}^{0.3} = 0.023(10{,}309)^{0.8}(0.684)^{0.3} = 33.3$$

Using this Nusselt number, we can now calculate the convective heat transfer coefficient $h$,

$$h = \mathrm{Nu}\frac{k}{D} = 33.3\frac{\left(0.0407\,\frac{\mathrm{W}}{\mathrm{m.K}}\right)}{(0.05\,\mathrm{m})} = 27.1\frac{\mathrm{W}}{\mathrm{m}^2.\mathrm{K}}$$

Now, using the heat transfer coefficient, we can determine the exit temperature of the air as follows:

$$T_o = T_s - (T_s - T_i)e^{\left(-\frac{hA}{\dot{m}c_p}\right)}$$

$$T_o = (373\,\text{K}) - (373\,\text{K} - 550\,\text{K})e^{\left[-\frac{\left(27.1\frac{\text{W}}{\text{m}^2.\text{K}}\right)\pi(0.05\,\text{m}\times 6\,\text{m})}{0.011\frac{\text{kg}}{\text{s}}\times 1{,}030\frac{\text{J}}{\text{kg.K}}}\right]} = 391.6\,\text{K}$$

We assumed that the surface temperature of the tube is uniform and constant; therefore, the log mean temperature difference can be calculated as:

$$\Delta T_m = \frac{(T_o - T_i)}{\ln\left(\frac{T_s - T_o}{T_s - T_i}\right)} = \frac{(391.6 - 550)\text{K}}{\ln\left(\frac{373 - 391.6}{373 - 550}\right)} = 70.3\,\text{K}$$

Thence, the rate of convective heat transfer can be determined as follows:

$$\dot{Q} = -h\,A\Delta T_m = -\left(27.1\frac{\text{W}}{\text{m}^2.\text{K}}\right)\pi\,(0.05\,\text{m}\times 6\,\text{m})(70.3\,\text{K}) = -1{,}795\,\text{W}\,(-6{,}126\,\text{Btu/h})$$

In the above calculations, we used the air properties determined at the initially assumed mean temperature of 500 K. The new mean temperature can now be calculated as $T_{avg} = \frac{391.6 + 550}{2} = 470.8\,\text{K}$, and we can see that it is not in agreement with the earlier assumed temperature of 500 K. Therefore, the calculations need to be repeated, this time evaluating the air properties at the new mean temperature of 470.8 K.

**Example 2.73: Internal Flow Convective Heat Loss from an Air Duct**

Waste heat is being used in a greenhouse heating, and the hot air at the source with a temperature of 50°C is directly sent to the greenhouse with an air blower at a flow rate of 5 m/s through a 40 m long rectangular duct having dimensions shown below. The temperature of the duct stays almost constant at 40°C. Assuming that the steady-state conditions prevail and the duct has a smooth inside surface, determine the rate of heat loss from the duct as the air flows through before reaching the greenhouse.

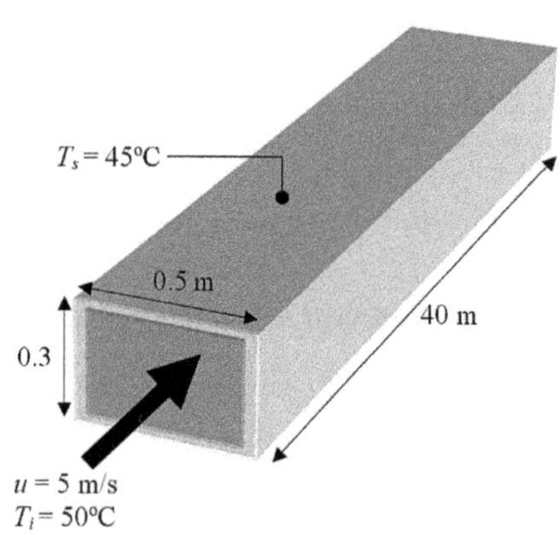

**Solution**

We know that the air inlet temperature is 50°C, but we do not know the outlet temperature; so, we will assume a mean air temperature of 40°C for determining the properties of air. At the end of our calculations, the air outlet temperature will be calculated to see if the mean air temperature is in agreement with the assumed mean air temperature. If they are not in agreement, we will repeat the calculations using the new mean temperature.

At the assumed temperature of 40°C, the air has a density of 1.1275 kg/m³, specific heat of 1,007 J/kg.K, thermal conductivity of 0.0271 W/m.K, kinematic viscosity of 1.915 × $10^{-5}$ m²/s, and Prandtl number of 0.712 (Table 2.12).

And since the duct is non-circular, we have to calculate the hydraulic diameter $D_e$ in order to be able to evaluate the Reynolds number. The hydraulic diameter $D_e$ is calculated either using the hydraulic radius $r_h$ or the cross-sectional area for flow $A_c$ and the total wetted perimeter $p$ as shown below:

$$D_e = 2\,r_h = \frac{4\,A_c}{p} = \frac{4(0.3\,\text{m} \times 0.5\,\text{m})}{2(0.3\,\text{m} + 0.5\,\text{m})} = 0.375\,\text{m}$$

Then, the Reynolds number can be calculated as:

$$\text{Re} = \frac{uD_e}{\upsilon} = \frac{\left(5\,\frac{\text{m}}{\text{s}}\right)(0.375\,\text{m})}{\left(1.915 \times 10^{-5}\,\frac{\text{m}^2}{\text{s}}\right)} = 94{,}937$$

So, the flow is turbulent, and the entry lengths are approximately equal to $10D_e = 3.75$ m. Since the duct is 40 m long, which is much longer than 3.75 m, the flow is fully developed as it flows through the duct. For the flow case $0.7 \le \text{Pr} \le 160$; turbulent; fully developed; and $\text{Re} \ge 10{,}000$; the literature provides the proper Nusselt number correlation as:

$$\text{Nu} = 0.023\,\text{Re}^{0.8}\,\text{Pr}^{0.3} = 0.023(94{,}937)^{0.8}(0.712)^{0.3} = 199.3$$

Using this Nusselt number and the hydraulic diameter, we can now calculate the convective heat transfer coefficient $h$,

$$h = \text{Nu}\frac{k}{D_e} 199.3 \frac{\left(0.0271\,\frac{\text{W}}{\text{m.K}}\right)}{(0.375\,\text{m})} = 14.4\,\frac{\text{W}}{\text{m}^2.\text{K}}$$

Now, using the heat transfer coefficient, we can determine the exit temperature of the air as follows:

$$T_o = T_s - (T_s - T_i)e^{\left(-\frac{hA}{\dot{m}c_p}\right)}$$

$$T_o = \left(45°C\right) - \left(45°C - 50°C\right) e^{\left[-\frac{\left(14.4 \frac{W}{m^2.K}\right)\left(2x\left(0.3\,m + 0.5\,m\right)x\left(40\,m\right)\right)}{0.846 \frac{kg}{s} x 1{,}007 \frac{J}{kg.K}}\right]} = 46.7°C$$

We assumed that the surface temperature of the duct is uniform and constant; therefore, the log mean temperature difference can be calculated as:

$$\Delta T_m = \frac{(T_o - T_i)}{\ln\left(\frac{T_s - T_o}{T_s - T_i}\right)} = \frac{(46.7 - 50)}{\ln\left(\frac{45 - 46.7}{45 - 50}\right)} = 3.1°C$$

Thence, the rate of convective heat loss from the duct can be determined as follows:

$$\dot{Q} = -h\, A \Delta T_m = -\left(16.3 \frac{W}{m^2.K}\right)\left(2 \times \left(0.3\,m + 0.5\,m\right) \times 40\,m\right)\left(3.1°C\right) = -3{,}191\,W\left(-10{,}891\,Btu/h\right)$$

In the above calculations, we used the air properties evaluated at the initially assumed mean temperature of 40°C. The new mean temperature can now be calculated as $T_{avg} = \frac{46.7 + 50}{2} = 48.4^{\circ}C$, and we can see that it is not in agreement with the earlier assumed temperature of 40°C. Therefore, the calculations need to be repeated, this time evaluating the air properties at the new mean temperature of 48.4°C.

#### *2.3.3.2.3 Boiling and Condensation Heat Transfer*

Many engineering problems involve boiling and condensation including problems in greenhouse environmental engineering. This section briefly and in general terms covers boiling and condensation phenomena and their types.

##### *Boiling Heat Transfer*

If vaporization takes place at a solid–liquid interface, then it is called boiling. And, if the temperature of a liquid at a certain pressure reaches and goes beyond the saturation temperature of that liquid, then it starts boiling. Since a fluid flow is involved in the boiling process, it is also considered to be a form of convective heat transfer, and boiling depends on the latent heat of vaporization of the liquid and the surface tension at the liquid–vapor interface as well. The heat transfer coefficient during boiling process is typically higher than those involving in the convective heat transfer processes with no phase changes. Using the Newton's law of cooling, the rate of heat transfer during a boiling process from a given surface to a fluid can be expressed as:

$$\dot{Q} = -h\, A_s\left(T_s - T_{sat}\right) \tag{2.79}$$

where $T_s$ and $T_{sat}$ are the surface and fluid saturation temperatures, respectively, $A_s$ is the heat transfer area, and $h$ is the boiling heat transfer coefficient.

If boiling takes place without any fluid flow, we then have a boiling known as pool boiling; whereas, if boiling occurs during a bulk fluid motion generally caused by external means, then we have what is called flow boiling. Heat transfer literature provides the equations for different boiling types.

*Condensation Heat Transfer*

Condensation is observed when the temperature of a given vapor decreases below its saturation temperature. If vapor comes in contact with a solid, free liquid, or gas surface that has a temperature lower than its saturation temperature, then it starts condensing. We have two types of condensation, and they are named as film condensation and dropwise condensation. In the case of film condensation, the vapor condenses as a film of liquid on the surface; whereas, in the case of dropwise condensation, the vapor condenses as liquid droplets on the surface. Typically, the rates of heat transfer in dropwise condensation are much higher than those of film condensation because the condensate droplets slide down the surface as they get bigger in size, allowing more vapor to be in contact with the condensing surface. In film condensation, however, the condensing surface is covered by the liquid film, which resists further exchange of heat between the vapor and the surface. The rate of condensation heat transfer is expressed in the following equation:

$$\dot{Q} = h_L\, A_s\left(T_{sat} - T_s\right) = \dot{m}\, h'_{fg} \tag{2.80}$$

where $h_L$ is the heat transfer coefficient, $A_s$ is the heat transfer area, $T_{sat}$ and $T_s$ are the saturated temperature of the vapor and the condensing surface temperature, respectively, $\dot{m}$ is the condensate mass flow rate, and $h'_{fg}$ is the modified latent heat of vaporization, which takes into consideration if the vapor is superheated and therefore requires cooling down to its saturation temperature before the condensation process begins, and if the liquid is cooled further below its saturation temperature. The expression for the modified latent heat of vaporization $h'_{fg}$ is provided below:

$$h'_{fg} = h_{fg} + 0.68\ c_{p,l}\left(T_{sat} - T_s\right) + c_{p,v}\left(T_v - T_{sat}\right) \tag{2.81}$$

where $h_{fg}$ is the latent heat of vaporization, $c_{p,\,l}$ is the specific heat of the liquid at the film temperature, and $c_{p,\,v}$ is the specific heat of the vapor at the average temperature of the vapor temperature $T_v$ and the saturation temperature $T_{sat}$. The heat transfer coefficient $h_L$ depends on a number of different variables including the properties of the fluids, nature of flow, and the surface geometry. Heat transfer literature provides the correlation equations for heat transfer coefficients for different geometries and condensation types.

#### 2.3.3.3 Radiation Heat Transfer

Radiation is a form of energy that is emitted by all objects having a temperature above absolute zero. Radiation is an electromagnetic phenomenon and differs from both conduction and convection as it does not require a medium to transfer heat, and can travel through a vacuum. A body that absorbs all incident radiation at all wavelengths is said to be a *blackbody*. The heat radiated by a body is composed of electromagnetic waves of many different wavelengths or frequencies. The radiation laws govern the relationships between a radiating surface and the heat radiated into the space in all directions. First of these laws is called the *Planck's law*, which basically states that the wavelength of emission from a perfectly black object, a blackbody that is a perfect emitter and absorber of radiation (*Kirchhoff's law*), depends on the absolute temperature of the emitting body.

$$E_\lambda^* = \frac{C_1}{\left\{\lambda^5\left[\exp\left(\frac{C_2}{\lambda T}\right)-1\right]\right\}} \tag{2.82}$$

where $E_\lambda^*$ is the amount of energy in W/m$^2$.µm (Btu/h.ft$^2$.µm) emitted at a single wavelength $\lambda$ (µm) (blackbody monochromatic radiation) at temperature $T$ in K (°R). The constants $C_1$ and $C_2$ have magnitudes of $3.74 \times 10^8$ W.µm$^4$/m$^2$ ($1.187 \times 10^8$ Btu.µm$^4$/h.ft$^2$) and $1.44 \times 10^4$ µm.K (25,898 µm.°R), respectively. Solar radiation, a very good approximation to that of a blackbody, which follows the Planck's curve, obtained using Equation (2.82) for an average outer surface temperature (6,000 K) of the Sun is presented in Figure 2.15. Actual (*nonblack*) bodies emit less than the theoretical blackbody emission, that is, $E_\lambda = \varepsilon_\lambda E_\lambda^*$ where $\varepsilon_\lambda$ is the emissivity, which is a measure of emission efficiency.

**Example 2.74: Blackbody Monochromatic Radiation**

Determine the blackbody monochromatic radiation at a single wavelength of 0.5 µm emitted by the Sun's surface, $E_\lambda^*$.

**Solution**

Using Equation (2.82) and assuming a surface temperature of 6,000 K, the blackbody monochromatic radiation at 0.5 µm is:

$$E_\lambda^* = \frac{3.74 \times 108\ \text{W.µm}^4/\text{m}^2}{\left\{(0.5\ \text{µm})^5\left[\exp\left(\frac{1.44 \times 104\ \text{µm.K}}{(0.5\ \text{µm})(6000\ \text{K})}\right)-1\right]\right\}}$$

$$E_\lambda^* = 99.3 \times 10^6\ \text{W/m}^2.\text{µm}$$

Or, it is simply 99.3 MJ per square meter solar surface area every second.

Absorption of radiation or photons is exact reverse of the photon emission process. A photon is absorbed as soon as it interacts with a molecule, an atom, or one of its electrons. And this interaction destroys the photon and brings the molecule or an atom to an excited state. The excited molecule or atom subsequently starts reemitting one or more photons, and continues the exchange of energy from radiant energy to atomic excitation and back again.

Two other particularly useful radiation laws, which can be derived from Equation (2.82) by differentiation and integration, respectively, are *Wien's law* for the wavelength of a maximum emission for a blackbody at a given temperature, and the *Stefan–Boltzmann law* for the total energy emitted by a blackbody.

*Wien's Displacement law* states that the wavelength at which a blackbody emits its maximum amount of radiation is inversely proportional to its absolute temperature in Kelvin (or in Rankine).

$$\lambda_{max} = \frac{a}{T} \tag{2.83}$$

where $\lambda$ is the wavelength in μm, $a$ is 2,897 μm.K (5,216 μm.°R), and $T$ in K (or in °R).

**Example 2.75: Wavelength of Maximum Radiation (Solar) Emission**

Determine the wavelength of maximum radiation emission $\lambda_{max}$ for the Sun's surface.

**Solution**

Using Equation (2.83) and assuming an average surface temperature of approximately 6,000 K, we would have its maximum emission at:

$$\lambda_{max} = \frac{u}{T}$$

$$\lambda_{max} = \frac{2897\ \mu m.K}{6{,}000\ K}$$

$$\lambda_{max} = 0.48\ \mu m$$

A wavelength of approximately 0.5 μm lies within the visible spectrum.

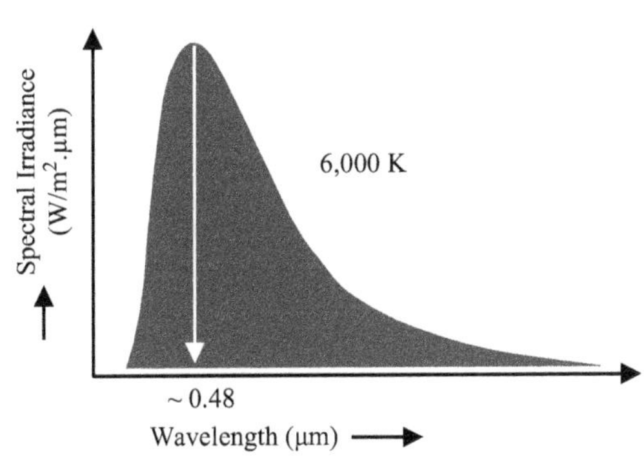


**FIGURE 2.15**
Schematic view of the standard blackbody curves for solar radiation assuming a surface temperature of 6,000 K. The values given by the Stefan–Boltzmann equation ($E^*$) is given by the area beneath the curve.

**Example 2.76: Wavelength of Maximum Radiation (Terrestrial) Emission**

Determine the wavelength of maximum radiation emission $\lambda_{max}$ for the Earth's surface.

**Solution**

Assuming an average surface temperature of 15°C (i.e. 288.15 K), we can determine the maximum emission,

$$\lambda_{max} = \frac{2897\ \mu m.K}{288.15\ K}$$

$$\lambda_{max} = 10.0\ \mu m$$

A wavelength of 10.0 μm lies within the infrared spectrum way beyond the visible spectrum. This shows that a hotter object emits a greater portion of its energy at shorter wavelengths than a cooler object.

For determining how much radiation a given object emits over all wavelengths combined, we need to sum up all the energy contributions of the photons at all wavelengths. Actually, this total energy is represented by the area between the object's blackbody curve and the *x*-axis as shown in Figure 2.15. That area, and therefore the radiant energy emission is given by the Stefan–Boltzmann law, given in the following equation:

$$E^{*} = -\sigma T^{4} \tag{2.84}$$

where $E^{*}$ is the total energy emitted in J (Btu) per unit time per unit area of black surface, $\sigma$ is known as Stefan–Boltzmann constant and has a magnitude of $5.679 \times 10^{-8}$ J/s.m$^2$.K$^4$ ($0.1714 \times 10^{-8}$ Btu/h.ft$^2$.R$^4$) and $T$ is the absolute temperature of the surface in K (R).

**Example 2.77: Sun's Total Radiation Emission**

Estimate the Sun's total radiation emission per unit time per unit surface area.

**Solution**

Using Equation (2.84) and assuming an average surface temperature of the Sun as 6,000 K, we can estimate that the Sun emits:

$$E^{*} = \left(5.679 \times 10^{-8}\,J/s.m^{2}.K^{4}\right)\left(6{,}000\,K\right)^{4}$$

$$E^{*} = 73.6\ MJ/s.m^{2}$$

Overall, in all directions with an approx. radius of $6.96 \times 10^{11}$ m, we can easily calculate the Sun's emission as approximately $4.5 \times 10^{20}$ MJ of energy every second.

When radiant energy falls on a surface, it can be absorbed, reflected, or transmitted through the material. Hence, from the first law of thermodynamics:

$$a + t + r = 1 \tag{2.85}$$

where $a$ is the fraction of incident radiation absorbed or *absorbance*, $t$ is the fraction of incident radiation transmitted or *transmittance*, and $r$ is the fraction of incident radiation reflected or *reflectance*.

Most solids behave as an opaque material in the infrared region, and for opaque materials, $t = 0$ and $a + r = 1$. For black surfaces, $a = 1$, $t = 0$, and $r = 0$.

The net interchange of heat by radiation between two bodies at different temperatures can be written as:

$$\dot{Q} = -F_E F_A A\,\sigma\left(T_1^4 - T_2^4\right) \tag{2.86}$$

where $F_E$ is the emissivity factor (Table 2.13), $F_A$ is the geometric (or view) factor (Table 2.13), $A$ is the area in m² (ft²), $\sigma$ is the Stefan–Boltzmann constant and has a magnitude of $5.679 \times 10^{-8}$ J/s.m².K⁴ ($0.1714 \times 10^{-8}$ Btu/h.ft².°R⁴), and $T_1$ and $T_2$ are the absolute temperatures of the surfaces in K (or °R).

**TABLE 2.13**

Radiation Between Solids: Factors For Use in Equation (2.86).

| Radiating surfaces | Area, $A$ | $F_A$ | $F_E$ |
|---|---|---|---|
| 1. Infinite parallel planes | $A_1$ or $A_2$ | 1 | $\dfrac{1}{\dfrac{1}{\varepsilon_1}+\dfrac{1}{\varepsilon_2}-1}$ |
| 2. Completely enclosed body, small compared with enclosing body (subscript 1 refers to enclosed body) | $A_1$ | 1 | $\varepsilon_1$ |
| 3. Completely enclosed body, large compared with enclosing body (subscript 1 refers to enclosed body) | $A_1$ | 1 | $\dfrac{1}{\dfrac{1}{\varepsilon_1}+\dfrac{1}{\varepsilon_2}-1}$ |

**TABLE 2.14**

Normal Total Emissivities of Various Surfaces.

| Surfaces | | | $\varepsilon$ |
|---|---|---|---|
| Metals | Aluminum | Polished plate | 0.04 |
| | | Rough plate, foil | 0.05 |
| | | Sheet | 0.12 |
| | Brass | Polished | 0.096 |
| | | Natural surface | 0.06 |
| | | Oxidized by heating | 0.59–0.61 |
| | Copper | Polished | 0.023 |
| | | Plate heated | 0.57 |
| | | Plate heated, with oxidized layer | 0.78 |
| | Iron & Steel | Cast iron, polished | 0.21 |
| | | Smooth sheet iron | 0.55–0.60 |
| | | Oxidized iron | 0.74 |
| | | Sheet steel | 0.66 |
| | | Galvanized steel | 0.25 |
| | | Oxidized steel | 0.79 |
| | | Steel plate, rough | 0.94–0.97 |
| | | Sheet steel, strong rough oxide layer | 0.80 |
| Building Materials | Brick | Red, rough | 0.93 |
| | | Silica, rough | 0.80–0.85 |
| | Glass, smooth | | 0.94 |
| | Oil paints, all colors | | 0.92–0.96 |
| | Wood | | 0.90 |
| | Water | | 0.95–0.96 |

**Example 2.78: Radiation Heat Loss from a Circular Steel Tube**

Now consider the 51 mm steel tube (in Examples 2.51 and 2.60), having an outside diameter of 0.0488 m, runs through a greenhouse whose walls are at 20°C. Determine the heat loss by radiation from 30 m of the tube if its outside surface temperature is 90°C. Assume that there is no crop canopy in the greenhouse yet.

**Solution**

From Table 2.13, case 2, $F_A = 1$, $A$ is the area of tube, and $F_e$ is the emissivity of the steel tube. From Table 2.14, for oxidized steel, $\varepsilon = 0.79$. Then, using Equation (2.87),

$$\dot{Q} = -F_e F_A A\,\sigma\left(T_1^4 - T_2^4\right)$$

$$\dot{Q} = -(0.79)(1)(\pi)(0.0488\text{ m x }30\text{ m})\left(5.679\text{x}10^{-8}\frac{J}{\text{s m}^2\text{K}^4}\right)\left[(90+273.15\text{ K})^4 - (20+273.15\text{ K})^4\right]$$

$$\dot{Q} = -2{,}064\text{ W}\left(-7{,}044\frac{\text{Btu}}{\text{h}}\right)$$

### 2.3.4 Air–Water Vapor Mixtures: Psychrometrics

The study of psychrometrics deals with thermodynamic properties of moist air, and using these properties, we analyze engineering and non-engineering situations and processes involving moist air. This section covers perfect gas relations and describes their use in thermal comfort problems.

*Atmospheric air* normally contains many gaseous components, some water vapor, and various contaminants such as smoke and pollen. The *dry-air* contains no water and contaminants; hence, its composition is relatively constant. Moisture, even though its amount in the air is relatively small, plays a significant role in thermal comfort for humans, plants, and animals as well as air-conditioning applications. Within the range of thermal comfort and air-conditioning applications, we can treat both dry and moist (atmospheric) air as perfect gases, with negligible errors. Therefore, the perfect-gas relation, $pV = nRT$, still holds, and the total pressure of atmospheric air (mixture) can be expressed in terms of the partial pressure as follows:

$$p = p_a + p_w \tag{2.87}$$

where $p_a$ is the partial pressure of dry air, and $p_w$ is the partial pressure of water vapor. Then, we can further write the following expressions:

$$p_a = \frac{n_a RT}{V} \quad (\text{dry air}) \tag{2.88}$$

$$p_w = \frac{n_w RT}{V} \quad (\text{water vapor}) \tag{2.89}$$

where $V$ is the total volume of the mixture, $n_a$ is the number of moles of dry air, $n_w$ is the number of moles of water vapor, $R$ is the universal gas constant (for SI: 8,314.41 J/kg$_{mol}$·K; and for I-P: 1,545.32 ft.lb$_f$/lb$_{mol}$·°R), and $T$ is the absolute temperature in Kelvin or Rankine.

At low pressures, the interaction between the air and gas molecules is quite modest, so the mixture obeys the perfect gas equation as well; and thence, we can write the following equation:

$$(p_a + p_w)V = (n_a + n_w)RT \tag{2.90}$$

where $(p_a + p_w)$ is the total mixture pressure (the sum of partial pressures), and $(n_a + n_w)$ is the total number of moles in the mixture.

The enthalpy of a mixture of perfect gases equals the sum of the individual partial enthalpies of the components, in this case, of the air and water vapor. Then,

$$H = H_a + H_g = m_a h_a + m_g h_g \tag{2.91}$$

Dividing the total enthalpy of the mixture by the mass of the dry air, we can obtain the enthalpy of the mixture per unit mass of dry air, which will be expressed as $h$ without any subscript:

$$\left(\frac{H}{m_a}\right) = h = h_a + \left(\frac{m_g}{m_a}\right) h_g = h_a + W\, h_g \tag{2.92}$$

It should be emphasized the new enthalpy of the mixture defined above, $h$, is the sum of the specific enthalpy of the dry air $h_a$ and the product of the specific enthalpy of the saturated water vapor $h_g$ and the ratio of the mass of water vapor $m_g$ to the mass of air $m_a$ (a.k.a. the humidity ratio or specific humidity $W$) in the mixture.

Within the range of interest, if we take 0°C as the reference temperature; then, having the constant-pressure specific heat for dry-air, $c_p$ = 1.006 kJ/kg.°C; we obtain:

$$h_a\left(\text{kJ/kg}\right) = c_p T \approx \left(1.006\,\text{kJ/kg.°C}\right)T \tag{2.93}$$

$$\Delta h_a\left(\text{kJ/kg}\right) = c_p \Delta T \approx \left(1.006\,\text{kJ/kg.°C}\right)\Delta T \tag{2.94}$$

where $T$ is the dry-bulb temperature in °C. The enthalpy of water vapor at 0°C is 2,501 kJ/kg. The average $c_p$ value of water vapor within the temperature range of interest can be taken as 1.86 kJ/kg.°C. Then, the enthalpy of water vapor can be determined from the following equation:

$$h_g\left(\text{kJ/kg}\right) \approx 2501 + 1.86T \tag{2.95}$$

Then, the enthalpy of the mixture (moist air) per unit mass of dry air is

$$h\left(\text{kJ/kg}_{\text{da}}\right) \approx 1.006T + W\left(2501 + 1.86T\right) \tag{2.96}$$

In I-P units:

$$h_a\left(\text{Btu/lb}\right) = c_p T \approx \left(0.240\,\text{Btu/lb.°F}\right)T \tag{2.97}$$

$$h_g\left(\text{Btu/lb}\right) \approx 1061 + 0.444T \tag{2.98}$$

where $T$ is the dry-bulb temperature in °F. The moist air enthalpy then becomes:

$$h\left(\text{Btu/lb}_{\text{da}}\right) \approx 0.240T + W\left(1061 + 0.444T\right) \tag{2.99}$$

#### *2.3.4.1 Some Definitions and Relationships*

*Humidity ratio, moisture content*, or *mixing ratio*, $W$, is the ratio of mass of water vapor to mass of air in a mixture. The advantage in dealing with humidity ratio is that if the amount of water vapor in the air remains unchanged, the humidity ratio remains constant. The humidity ratio is expressed in grams or kg of moisture (water vapor) per kilogram of dry air. Alternatively, in I-P units, it is given as $\text{lb}_{\text{water}}/\text{lb}_{\text{da}}$ or grains of water vapor per pound of dry air where 7,000 grains is equal to 1 lb of mass.

$$W\left(\frac{\text{kg}_{\text{water}}}{\text{kg}_{\text{da}}}\right) = \frac{m_w}{m_a} = 0.622\left(\frac{x_w}{x_a}\right) = 0.622\left(\frac{p_w}{p_a}\right) = 0.622\left(\frac{p_w}{p - p_w}\right) \tag{2.100}$$

where $m_w$ is the mass of water vapor, $m_a$ is the mass of dry air, 0.622 is the ratio of molecular masses (18.01528/28.9645 = 0.62198), $x_w$ is the mole fraction for water vapor, $x_a$ is the mole fraction for dry air, $x_w/x_a$ is the mole fraction ratio (by definition $x_a + x_w = 1$), $p_w$ is the partial pressure of water vapor, and $p$ is the total pressure ($p = p_a + p_w = 101{,}325$ Pa or 14.696 psia at sea level).

*Specific humidity*, $q$, is the ratio of mass of water vapor to the total mass of the moist air as shown below:

$$q\left(\frac{\text{kg}_{\text{water}}}{\text{kg}_{\text{air}}}\right) - \frac{m_w}{m_w + m_a} = \frac{W}{1+W} \tag{2.101}$$

The *degree of saturation*, $\mu$, is expressed as the air humidity ratio $W$ to the humidity ratio $W_s$ of saturated air at the same temperature and pressure as in the following equation:

$$\mu = \frac{W}{W_s} \tag{2.102}$$

where

$$W_s = 0.622\frac{p_{ws}}{p - p_{ws}} \tag{2.103}$$

where $p_{ws}$ is the *saturation pressure of water vapor* at the given temperature $T$. This is the amount of moisture air can hold in vapor form (a.k.a. the limit saturation) before condensation happens, required in determining a number of moist air properties (i.e. the saturation pressure of water vapor), is only function of temperature and can be found in Tables A.15 and A.16 as tabulated values or calculated using the following equations:

The saturation pressure $p_{ws}$ over ice for −100°C (−148°F) < $T$ < 0°C (32°F):

$$\ln(p_{ws}) = C_1/T + C_2 + C_3T + C_4T^2 + C_5T^3 + C_6T^4 + C_7\ln(T) \tag{2.104}$$

where the constants in SI units are:

$C_1 = -5.647\,535\,9$ E+03
$C_2 = 6.392\,524\,7$ E+00
$C_3 = -9.677\,843\,0$ E−03
$C_4 = 6.221\,570\,1$ E−07
$C_5 = 2.074\,782\,5$ E−09
$C_6 = -9.484\,024\,0$ E−13
$C_7 = 4.163\,501\,9$ E+00

For the I-P units, the constants are:

$C_1 = -1.021\,416\,462$ E+04
$C_2 = -4.893\,503\,01$
$C_3 = -5.376\,579\,44$ E−03
$C_4 = 1.920\,237\,69$ E−07

$C_5 = 3.557\ 583\ 16\ E{-}10$
$C_6 = -9.034\ 468\ 83\ E{-}14$
$C_7 = 4.163\ 501\ 9$

The saturation pressure $p_{ws}$ over liquid water for 0°C (32°F) < $T$ < 200°C (392°F):

$$\ln(p_{ws}) = C_8 T + C_9 + C_{10} T + C_{11} T^2 + C_{12} T^3 + C_{13} \ln(T) \tag{2.105}$$

where

$C_8 = -5.800\ 220\ 6\ E{+}03$
$C_9 = 1.391\ 499\ 3\ E{+}00$
$C_{10} = -4.864\ 023\ 9\ E{-}02$
$C_{11} = 4.176\ 476\ 8\ E{-}05$
$C_{12} = -1.445\ 209\ 3\ E{-}08$
$C_{13} = 6.454\ 967\ 3\ E{+}00$

For the I-P units, the constants are:

$C_8 = -1.044\ 039\ 708\ E{+}04$
$C_9 = -1.129\ 479\ 96\ E{+}01$
$C_{10} = -2.702\ 235\ 5\ E{-}02$
$C_{11} = 1.289\ 036\ 0\ E{-}05$
$C_{12} = -2.478\ 068\ E{-}09$
$C_{13} = 6.5459673$

In both Equations (2.104) and (2.105), ln is the natural logarithm, $p_{ws}$ is the saturation pressure in Pa for SI (and psia for I-P), and $T$ is the absolute temperature in Kelvin (K = °C + 273.15) for SI (and Rankine, °R = °F + 459.67 for I-P).

*Water vapor density* or *absolute humidity*, $d_v$, on the other hand, is the ratio of the mass of water vapor to the total volume, and can be expressed as follows:

$$d_v = \frac{m_w}{V} \tag{2.106}$$

And the *density* $\rho$ of the moist air mixture is the ratio of the total mass or air to the total volume of air as shown below:

$$\rho = \frac{m_a + m_w}{V} = \frac{1}{v}(1 + W) \tag{2.107}$$

where $v$ is the moist air volume in $m^3/kg_{da}$ (or $ft^3/lb_{da}$), $m_a$ is the total mass of dry air, $m_w$ is the total mass of water vapor, $V$ is the total volume of the mixed air, and $W$ is the humidity ratio.

The moist air *volume*, $v$, on the other hand, can be expressed as follows:

$$v = \frac{V}{m_a} = \frac{V}{28.9645(n_a)} \tag{2.108}$$

$$v = \frac{RT}{28.9645(p - p_w)} = \frac{R_a T}{(p - p_w)} = \frac{R_a T}{p}(1 + 1.6078W) \tag{2.109}$$

where 28.964 is the apparent molecular mass or weighted average molecular weight of all components for dry air, $n_a$ is the number of moles of dry air, $R$ is the universal gas constant, $R_a$ is the gas constant for dry air (For SI: 0.287055 kJ/kg·K; and for I-P: 53.352 ft.lb$_f$/lb$_m$.°R), $T$ is the absolute temperature, $p$ is the total pressure, $p_w$ is the partial pressure of water vapor, and $W$ is the humidity ratio.

The relative humidity, $\phi$, on the other hand, is the ratio of the mole fraction of water vapor, $x_w$, in moist air to the mole fraction $x_{ws}$ in the air, saturated at the same temperature and pressure, and expressed as:

$$\phi = \left.\frac{x_w}{x_{ws}}\right|_{T,p} = \left.\frac{p_w}{p_{ws}}\right|_{T,p} = \frac{\mu}{1 - (1 - \mu)\left(\frac{p_{ws}}{p}\right)} \tag{2.110}$$

Please note that both relative humidity $\phi$ and degree of saturation $\mu$ are zero for dry air and unity for saturated mixed air.

Also note that the standard temperature we are familiar with, a measure of internal energy, is called the *dry-bulb temperature*, $T$. We can also define *dew-point temperature*, $T_d$, which is the temperature of moist air at which water vapor condenses into liquid – Air is maximally saturated at the same pressure $p$, with the same humidity ratio $W$ as that of the given moist air, and it is defined as the solution $T_d$ ($p$, $W$) of the following equation:

$$W_s(p, T_d) = W \tag{2.111}$$

When the air temperature is lowered to the dew point at constant pressure, the relative humidity rises to 100% and condensation take place. $T_d$ is a direct measure of the humidity ratio $W$ because $W = W_s$ at $T = T_d$.

The dew point temperature $T_d$ expression for the dew point temperatures below 0°C (32°F):

$$T_d = 6.09 + 12.608\alpha + 0.4959\alpha^2 \tag{2.112}$$

where $T_d$ is the dew point temperature in °C, $\alpha = \ln p_w$, and $p_w$ is the water vapor partial pressure in kPa.

For the I-P units, the dew point temperature equation is provided as follows:

$$T_d = 90.12 + 26.142\alpha + 0.8927\alpha^2 \tag{2.113}$$

where $T_d$ is the dew point temperature in °F, $\alpha = \ln (p_w)$, and $p_w$ is the water vapor partial pressure in psia.

The dew point temperature $T_d$ expression for the dew point temperature range of 0°C (32°F) and 93°C (200°F):

$$T_d = \mathrm{a} + \mathrm{b}\alpha + \mathrm{c}\alpha^2 + \mathrm{d}\alpha^3 + \mathrm{e}(p_w)^{0.1984} \tag{2.114}$$

where

$T_d$ = dew point temperature in °C
a = 6.54
b = 14.526
c = 0.7389
d = 0.09486
e = 0.4569
$\alpha = \ln p_w$
$p_w$ = water vapor partial pressure in kPa.

And for the I-P units:

$$T_d = \mathrm{a} + \mathrm{b}\alpha + \mathrm{c}\alpha^2 + \mathrm{d}\alpha^3 + \mathrm{e}(p_w)^{0.1984} \tag{2.115}$$

where

$T_d$ = dew point temperature in °F
a = 100.45
b = 33.193
c = 2.319
d = 0.17074
e = 1.2063
$\alpha = \ln (p_w)$
$p_w$ = water vapor partial pressure in psia.

*Thermodynamic wet-bulb temperature*, $T_w$, is the temperature at which water, by evaporation into moist air at a given dry-bulb temperature $T$ and humidity ratio $W$, can bring air to saturation (100% relative humidity) adiabatically at the same temperature $T_w$ while the pressure $p$ is maintained constant.

In order to measure the wet-bulb temperature, we use a psychrometer, which consists of two thermometers; one of the thermometer's bulbs is covered by a thoroughly wetted wick. When the wet-bulb is left in an air stream, water evaporates from the wick and the bulb reaches an equilibrium temperature called the *wet-bulb temperature*. This process is not exactly an adiabatic saturation process, which defines the thermodynamic wet-bulb temperature discussed above, rather, is a simultaneous heat and mass transfer from the wet-bulb. The energy needed to evaporate liquid water (heat of vaporization) is taken from the air in the form of sensible heat and converted to latent heat, which lowers the temperature at constant enthalpy.

The equation provided below is used to calculate the wet-bulb temperature for SI, which requires iterative solving until we find the $T_w$ that satisfies the equation:

$$W = \frac{(2501 - 2.381T_w)W_{s@T_w} - 1.006(T - T_w)}{2501 + 1.805T - 4.186T_w} \tag{2.116}$$

where both $T_w$ and $T$ are in degree Celsius.

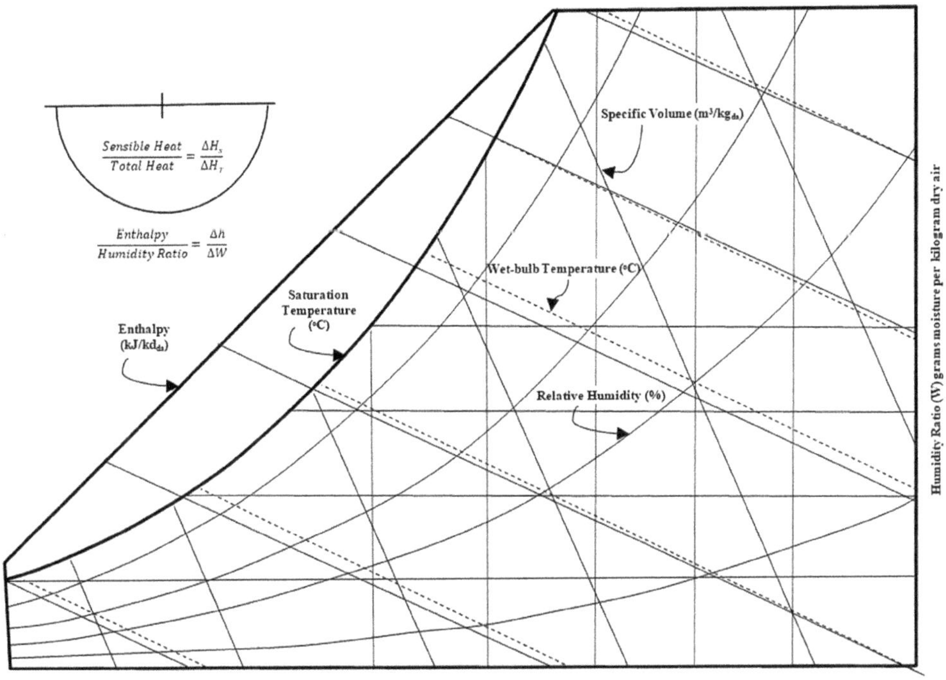


**FIGURE 2.16**
Components of a psychrometric chart – Normal Temperature Sea Level (SI).

For I-P units, the following equation is used, solving the equation iteratively until we find the $T_w$ that satisfies the equation:

$$W = \frac{(1093 - 0.556\,T_w)W_{s@T_w} - 0.240(T - T_w)}{1093 + 0.444T - T_w} \tag{2.117}$$

where $T_w$ and $T$ are in degree Farenheit.

In order to specify the state of moist air, we need three moist air properties, so that the others can be calculated either using the equations provided above for greater accuracy and automation or determined using the psychrometric charts (Figures 2.16 and 2.17). Total atmospheric pressure, $p$, is typically known; therefore, we only need to know two other properties. Thermodynamic properties of moist air at standard atmospheric pressure and thermodynamic properties of water at saturation were adopted from ASHRAE Handbook of Fundamentals (2013) and presented here in Tables A.15 and A.16.

**Example 2.79: Psychrometric Calculations**

Consider a moist air parcel having a dry-bulb temperature of 25°C with a 10°C dew point temperature. Calculate the following: a) the humidity ratio, $W$; b) degree of saturation, $\mu$; c) relative humidity, $\phi$; d) enthalpy, $h$; e) specific volume, $\nu$; f) density, $\rho$; and g) wet-bulb temperature, $T_w$.

**Solution**

a. Humidity ratio, $W$: Assuming $p$ = 101.325 kPa (sea level) and using Equation (2.100),

$$W\left(\frac{\text{kg}_{\text{water}}}{\text{kg}_{\text{da}}}\right)=0.622\left.\left(\frac{p_w}{p-p_w}\right)\right|_{@T=25^\circ C}$$

For a given $T_d$ = 10°C, the actual humidity ratio in the air $W$, is by definition the same as the saturation humidity ratio $W_s$, at an air temperature of 10°C, therefore,

$$W_{@T=25^\circ C}=W_{s@T=10^\circ C}=0.622\left.\left(\frac{p_{ws}}{p-p_{ws}}\right)\right|_{@T=10^\circ C}$$

From Table A.16, we find $p_{ws\,@\,10°C}$ = 1.2280 kPa; then,

$$W_{@T=25^\circ C}=W_{s@T=10^\circ C}=0.622\left.\left(\frac{1.2280\text{ kPa}}{101.325\text{ kPa}-1.2280\text{ kPa}}\right)\right|_{@T=10^\circ C}=0.00763\text{ kg}_{\text{water}}/\text{kg}_{\text{da}}$$

b. Degree of saturation, $\mu$, can be calculated using Equation (2.103); and for this, we need the saturation humidity ratio $W_s$ at 25°C as we just determined the humidity ratio $W$ at 25°C above:

$$W_{s@T=25^\circ C}=0.622\left.\left(\frac{p_{ws}}{p-p_{ws}}\right)\right|_{@T=25^\circ C}$$

From Table A.16, we find $p_{ws\,@\,25°C}$ = 3.1692 kPa; then,

$$W_{s@T=25^\circ C}=0.622\left.\left(\frac{3.1692\text{ kPa}}{101.325\text{ kPa}-3.1692\text{ kPa}}\right)\right|_{@T=25^\circ C}=0.02001\text{ kg}_{\text{water}}/\text{kg}_{\text{da}}$$

Then, using Equation (2.102), we can now calculate the degree of saturation as:

$$\mu=\frac{W}{W_s}=\frac{0.00763}{0.02001}=0.38$$

c. Relative humidity, $\phi$, can be calculated from Equation (2.110) as we now know both $p_w$ and $p_{ws}$:

$$p_{w@25^\circ C}=p_{ws@10^\circ C}=1.2280\text{ kPa}$$

$$\phi=\left.\frac{p_w}{p_{ws}}\right|_{T,p}=\frac{1.2280\text{ kPa}}{3.1692\text{ kPa}}=0.387=38.7\%$$

d. Enthalpy, $h$, now can be calculated using Equation (2.96) as follows:

$$h(\text{kJ}/\text{kg}_{\text{da}})\approx 1.006\,T+W(2501+1.86T)=1.006\,(25^\circ\text{C})+0.00763\big(2501+1.86\,(25^\circ\text{C})\big)$$

$$h\approx 44.6\text{ kJ}/\text{kg}_{\text{da}}$$

e & f. Specific volume, $v$, and density, $\rho$, can be calculated using Equations (2.109) and (2.107), respectively, with the gas constant for dry air $R_a$ = 0.287055 kJ/kg·K and temperature in Kelvin:

$$v=\frac{R_aT}{p}(1+1.6078W)=\frac{0.287055\,(25+273.15)}{101.325}\big(1+1.6078\,(0.00763)\big)=0.855\text{ m}^3/\text{kg}_{\text{da}}$$

$$\rho=\frac{1}{v}(1+W)=\frac{1}{0.855}(1+0.00763)=1.179\text{ kg/m}^3$$

g. And wet-bulb temperature, $T_w$, is the wet-bulb temperature $T_w$ that fits the following equation (Equation (2.116)), which requires iteration:

$$W = \frac{(2501 - 2.381\,T_w)W_{s@T_w} - 1.006(T - T_w)}{2501 + 1.805\,T - 4.186\,T_w} = 0.00763$$

where $T = 25^\circ$C, $T_w$ is unknown (°C), and $W_{s@T_w=?} = 0.622\left(\frac{p_{ws}}{p - p_{ws}}\right)\Bigg|_{@T_w=?}$

We just need to guess $T_w$ first, and determine $p_{ws}$ and calculate $W_s$ for that temperature, and then repeat the iteration until the calculated $W$ based on those values (and original temperature) in the equation above is equal to actual $W$ (0.00763 in our case). After satisfying this, the iteration process stops and we determine the wet-bulb temperature as follows:

$$T_w = 16.0^\circ\text{C}$$

**Example 2.80: Using the Psychrometric Chart**

Now consider a moist air parcel having a dry-bulb temperature of 24°C with a 50% relative humidity. Assuming $p = 101.325$ kPa (sea level) and this time using the psychrometric chart, determine the humidity ratio $W$, the saturation humidity ratio $W_s$, the wet bulb temperature $T_w$, the enthalpy $h$, the specific volume $\nu$, the density $\rho$, and the dew-point temperature $T_d$.

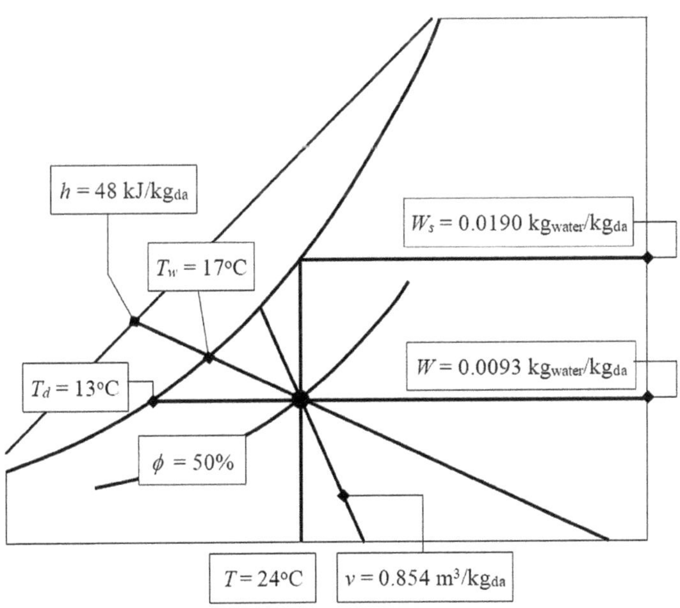


**Solution**

Using the psychrometric chart (ASHRAE, 2013) (see Figure 2.16 for the components), we can determine all the required quantities as follows:

#### 2.3.4.2 Psychrometric Processes

*Sensible Heating*

This is the heating process without changing the moisture content, that is, the humidity ratios at the initial (State 1) and final (State 2) conditions are the same as no moisture is added to or removed from the air as shown in the psychrometric sketch below (Figure 2.17).

Note that the relative humidity of air decreases during the sensible heating process. This is because the increased air temperature can hold more moisture and the moisture holding capacity of the air increases with increasing air temperature.

The conservation of mass reduces to:

$$\dot{m}_{a1} = \dot{m}_{a2} = \dot{m}_a \tag{2.118}$$

$$W_1 = W_2 \tag{2.119}$$

Neglecting any fan work and losses, the conservation of energy yields:

$$\dot{Q}_S = \dot{m}_a \left(h_2 - h_1\right) \tag{2.120}$$

where $\dot{Q}_S$ is the rate of sensible heating provided, $\dot{m}_a$ is the mass flow rate of dry air, $W_1$ and $W_2$ are the humidity ratios at the inlet and exit of the heating section, respectively, and $h_1$ and $h_2$ are enthalpies per unit mass of dry air at the inlet and exit of the heating section, respectively.

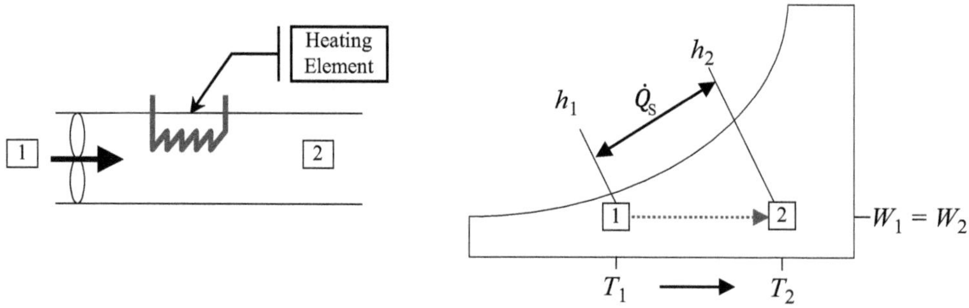


**FIGURE 2.17**
Sensible heating process.

**Example 2.81: Sensible Heating**

Now consider moist air saturated at 5°C entering a unit heater at a flow rate of 10 m³/s. Determine the rate of heating if the moist air must leave the unit heater at 30°C dry-bulb temperature.

**Solution**

Figure 2.17 schematically presents the solution. However, State 1 is located on the saturation curve, meaning that $T$, $T_d$ and $T_w$ are all equal to 5°C. From the psychrometric chart, $h_1 = 18.6\ \text{kJ/kg}_{da}$, $W_1 = 0.0054\ \text{kg}_{water}/\text{kg}_{da}$, and $v_1 = 0.795\ \text{m}^3/\text{kg}_{da}$. State 2 is located at the intersection of 30°C and $W_2 = W_1 = 0.0054\ \text{kg}_{water}/\text{kg}_{da}$. Hence, $h_2 = 45.0\ \text{kJ/kg}_{da}$.

Then, the mass flow rate is:

$$\dot{m}_a = \left(10\ \text{m}^3/\text{s}\right)/\left(0.795\ \text{m}^3/\text{kg}_{da}\right) = 12.6\ \text{kg}_{da}/\text{s}$$

And from Equation (2.120), the rate of heating $Q_s$:

$$\dot{Q}_S = \dot{m}_a\left(h_2 - h_1\right) = 12.6\left(45.0 - 18.6\right) = 332\ \text{kW}\left(= 1{,}133{,}106\frac{\text{Btu}}{\text{h}}\right)$$

## *Sensible Heating and Humidification*

Sometimes, in order to maintain the required relative humidity, we provide humidification to sensible heating process as shown in Figure 2.18. Humidification takes place by passing the air through a humidifying section. If steam is used in the humidifier, there will be an

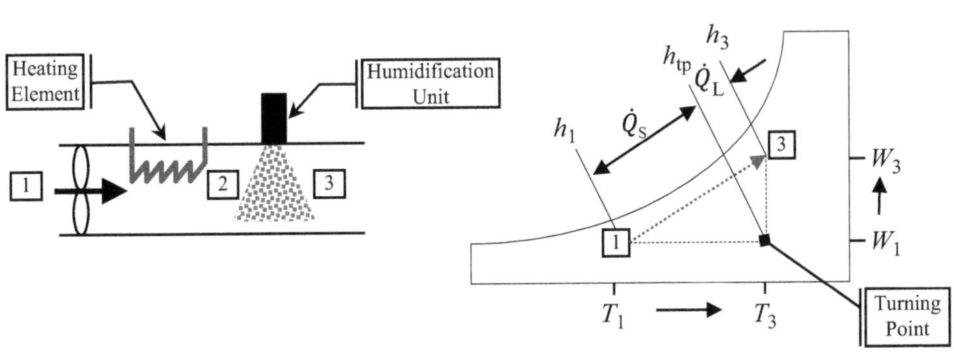


**FIGURE 2.18**
Sensible heating and humidification process.

additional heating; therefore, the temperature at Point 3 will be greater than the temperature at Point 2. However, if water is sprayed instead, part of the latent heat of vaporization will be absorbed from the air, which will result in some cooling of the air; therefore, the temperature at Point 3 will be lower than the temperature at Point 2.

In the sensible heating and humidification process, the conservation of mass reduces to:

$$\dot{m}_{a1} = \dot{m}_{a2} = \dot{m}_{a3} = \dot{m}_a \tag{2.121}$$

$$W_1 = W_2 \text{ and } W_1 < W_3 \tag{2.122}$$

$$\dot{m}_w = \dot{m}_a \Delta W = \dot{m}_a \left(W_3 < W_1\right) \tag{2.123}$$

where $\dot{m}_a$ is the mass flow rate of dry air, $\dot{m}_w$ is the rate of moisture addition to the moist air, and $W_1$ and $W_3$ are the humidity ratios at Points 1 and 3, respectively.

Neglecting any fan work and losses, the conservation of energy yields:

$$\dot{Q}_T = \dot{Q}_S + \dot{Q}_L \tag{2.124}$$

$$\dot{Q}_T = \dot{m}_a\left(h_{tp} - h_1\right) + \dot{m}_a\left(h_3 - h_{tp}\right) \tag{2.125}$$

where $\dot{Q}_T$ is the rate of heating provided, $\dot{Q}_S$ is the rate of sensible heating provided, $\dot{Q}_L$ is the rate of latent heating provided, $\dot{m}_a$ is the mass flow rate of dry air, $h_1$ and $h_3$ are the enthalpies per unit mass of dry air at the inlet and exit of the heating and humidification section, respectively, $h_{tp}$ is the enthalpy per unit mass of dry air at the turning point.

### *Sensible Cooling*

This is the cooling process without changing the moisture content of the air, that is, the humidity ratios at the initial (State 1) and final (State 2) conditions are the same as no moisture is added to or removed from the air as shown in the psychrometric sketch below (Figure 2.19).

Note that, as opposed to the sensible heating process, the relative humidity of air increases during the sensible cooling process. This is because the decreased air

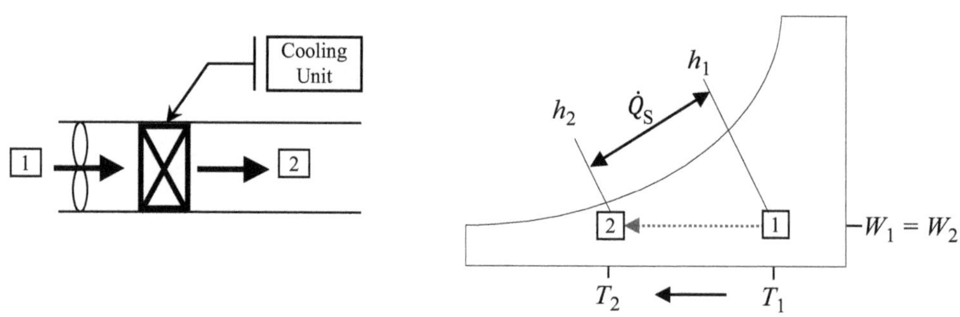


**FIGURE 2.19**
Sensible cooling process.

temperature can hold less moisture, and the moisture holding capacity of the air decreases with decreasing air temperature.

Just like in the sensible heating process, the conservation of mass reduces to:

$$\dot{m}_{a1} = \dot{m}_{a2} = \dot{m}_a \quad (2.126)$$

$$W_1 = W_2 \quad (2.127)$$

where $\dot{m}_a$ is the mass flow rate of dry air, and $W_1$ and $W_2$ are the humidity ratios at Points 1 and 2, respectively.

Neglecting any fan work and losses, the conservation of energy yields:

$$\dot{Q}_S = \dot{m}_a (h_2 - h_1) \quad (2.128)$$

where $\dot{Q}_S$ is the rate of sensible cooling provided, and $h_1$ and $h_2$ are the enthalpies per unit mass of dry air at the inlet and exit of the cooling section, respectively.

*Sensible Cooling and Dehumidification*

This is the cooling process with eventually the removal of some moisture from the air. In order to remove moisture from the air, we need to cool the air below its dew-point temperature. First, as the air passes through a cooling coil, the air temperature decreases and its relative humidity increases until the air temperature reaches its dew-point temperature. And then, any further cooling beyond the dew-point temperature condenses part of the moisture in the air while the air remains saturated and moves down the saturation curve during the entire condensation process as shown in Figure 2.20.

In the sensible cooling and dehumidification process, the conservation of mass reduces to:

$$\dot{m}_{a1} = \dot{m}_{a2} = \dot{m}_a \text{ and } W_1 \neq W_2 \quad (2.129)$$

$$\dot{m}_w = \dot{m}_a \Delta W = \dot{m}_a (W_2 - W_1) \quad (2.130)$$

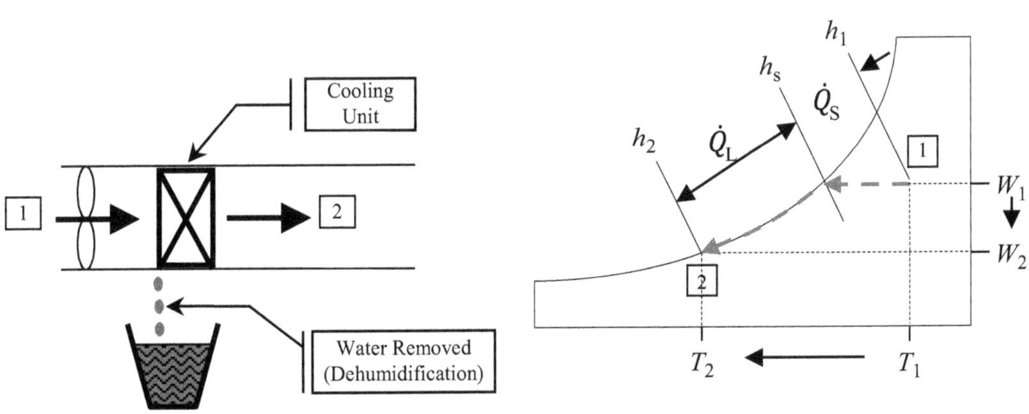


**FIGURE 2.20**
Sensible cooling and dehumidification process.

where $\dot{m}_a$ is the mass flow rate of dry air, $\dot{m}_w$ is the rate of moisture removal from the moist air, and $W_1$ and $W_2$ are the humidity ratios at State 1 and State 2, respectively.

Neglecting any fan work and losses, the conservation of energy yields:

$$\dot{Q}_T = \dot{Q}_S + \dot{Q}_L \quad (2.131)$$

$$\dot{Q}_T = \dot{m}_a (h_s - h_1) + \dot{m}_a (h_2 - h_s) = \dot{m}_a (h_2 - h_1) \quad (2.132)$$

where $\dot{Q}_T$ is the total cooling provided per unit time, $\dot{Q}_S$ is the rate of sensible energy removed, $\dot{Q}_L$ is the rate of latent energy removed, $h_1$ and $h_2$ are the enthalpies per unit mass of dry air at the inlet and exit of the heating and humidification section, respectively, and $h_s$ is the enthalpy per unit mass of dry air at the saturation point.

**Example 2.82: Sensible Cooling and Dehumidification**

Now consider moist air entering a chiller at 35°C dry-bulb temperature and 60% relative humidity with a flow rate of 0.425 $m^3/s$ and leaving the chiller and entering the conditioned space at a required dry-bulb temperature of 5°C and 100% relative humidity. Determine the total refrigeration capacity required and the total water removal rate from the moist air. Note: 1 ton of refrigeration equals a heat removal rate of 12,000 Btu/h.

**Solution**

Figure 2.20 schematically shows the solution. However, State 1 is located at the intersection of $T$ = 35°C and $\phi$ = 60%. Therefore, $h_1$ = 91.0 kJ/$kg_{da}$, $W_1$ = 0.0214 $kg_{water}/kg_{da}$, and $v_1$ = 0.903 $m^3/kg_{da}$. State 2 is located on the saturation curve at 5°C. Thus, $h_2$ = 19.0 kJ/$kg_{da}$, and $W_2$ = 0.0054 $kg_{water}/kg_{da}$.

The mass flow rate of dry air is:

$$\dot{m}_a = (0.425\,\text{m}^3/\text{s}) \Big/ \left(0.903 \frac{\text{m}^3}{\text{kg}_{\text{da}}}\right) = 0.471\,\text{kg}_{\text{da}}/\text{s}$$

Using Equation (2.132):

$$\dot{Q}_T = \dot{m}_a (h_2 - h_1) = 0.471(19.0 - 91.0) = -33.9\,\text{kW}\left(= -115{,}741 \frac{\text{Btu}}{\text{h}} = -9.65\text{ ton}\right)$$

Now, using Equation (2.130), we can determine the water removal rate as follows:

$$\dot{m}_w = \dot{m}_a \Delta W = \dot{m}_a (W_2 - W_1) = 0.425(0.0214 - 0.0054) = 0.0068\text{kg}_{\text{water}}/\text{s} \approx 24.5\text{kg}_{\text{water}}/\text{h}$$

### *Adiabatic Water Addition to Moist Air Stream*

Sometimes, steam or liquid water can be injected into a moist airstream to raise its humidity. Figure 2.21 shows a diagram of this common air-conditioning process.

As we did earlier, we can write conservation of mass for the steady-state process as:

$$\dot{m}_{a1} = \dot{m}_{a2} = \dot{m}_a \text{ and } W_1 \neq W_2 \quad (2.133)$$

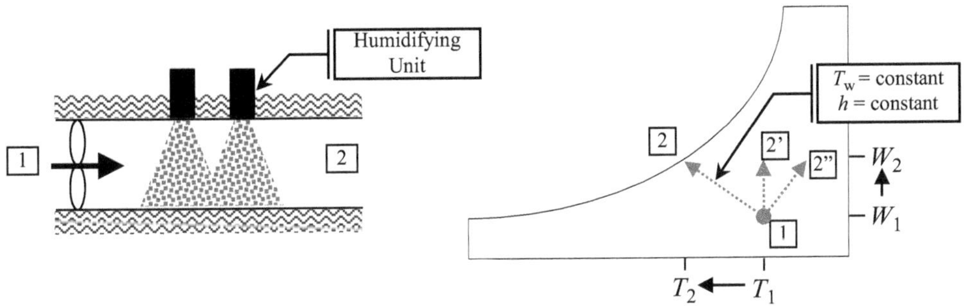


**FIGURE 2.21**
Adiabatic water addition to moist air stream.

$$\dot{m}_a W_1 + \dot{m}_w = \dot{m}_a W_2 \xrightarrow{yields} \dot{m}_w = \dot{m}_a \Delta W = \dot{m}_a \left(W_2 - W_1\right) \tag{2.134}$$

And assuming an adiabatic process, we can write the following energy balance equation:

$$\dot{m}_a h_1 + \dot{m}_w h_w = \dot{m}_a h_2 \tag{2.135}$$

Substituting Equation (2.134) in Equation (2.135) and dividing by $\dot{m}_a$, we obtain the following expression:

$$h_2 = h_1 + h_w \left(W_2 - W_1\right) \tag{2.136}$$

$$h_w = \frac{h_2 - h_1}{W_2 - W_1} = \frac{\Delta h}{\Delta W} \tag{2.137}$$

where $\dot{m}_a$ is the mass flow rate of dry air, $\dot{m}_w$ is the mass flow rate of water (any phase), $h_1$ and $h_2$ are the enthalpies of the moist air at State 1 and State 2, respectively, $h_w$ is the specific enthalpy of water (any state), and $W_1$ and $W_2$ are the humidity ratios at State 1 and State 2, respectively.

As shown in Figure 2.21, the direction of the final state (State 2) of the moist air is fixed by the specific enthalpy of the injected water, and drawn through the initial state (State 1) of the moist air.

If we spray cold water, the enthalpy of the water is relatively small compared to that of the moist air. As a result, the enthalpy of the moist air leaving the process (State 2) is equal to the enthalpy of the mixture entering (State 1). As water evaporates, the latent heat of vaporization is absorbed from the surrounding air and water body itself, causing a lower temperature. As a result, both air and water are cooled.

A simple energy balance gives the amount of cooling provided:

$$\dot{Q} = \dot{m}_a c_p \left(T_2 - T_1\right) = \dot{m}_a L_v \left(W_1 - W_2\right) \tag{2.138}$$

where $c_p$ is the specific heat of air at constant pressure (Table 2.12), $T_1$ and $W_1$ are the entering air dry-bulb temperature and humidity ratio at State 1, respectively, $L_v$ is the latent heat of vaporization or enthalpy of vaporization of water (Table A.16), and $T_2$ and $W_2$ are the leaving air dry-bulb temperature and humidity ratio at State 2, respectively.

This method of air-conditioning is quite an ancient and effective technique in dry climates. On the psychrometric chart, the process line 1 to 2 moves on a line of constant wet-bulb temperature $T_w$ and constant enthalpy $h$ on the psychrometric chart, which is termed evaporative cooling, essentially an adiabatic saturation process, from an initial state at low humidity ratio and high dry-bulb temperature to a final state at higher humidity and lower dry-bulb temperature. Since the sprayed water enthalpy is small compared to that of the moist air, lines of constant enthalpy are approximately parallel to lines of constant adiabatic wet-bulb temperature.

If hot water or steam with progressively higher enthalpy is sprayed into the moist air (the process line 1 to 2″), as Equation (2.136) indicates and Figure 2.21 shows, the enthalpy of the moist air at the end of the spraying process (State 2″) will increase. If, on the other hand, the enthalpy of the steam in the spray is the same as the enthalpy of the water vapor in the moist air entering the process (the process line 1 to 2′), the leaving moist air at State 2′ is at the same temperature as the entering mixture (Figure 2.21).

**Example 2.83: Evaporative Cooling**

Let's now consider a fan and pad type evaporative cooling system used for air-conditioning an institutional greenhouse. Assume that the outside air enters the evaporative cooling pad at 35°C dry-bulb temperature and 25% relative humidity with a flow rate of 4 m³/s. Assuming that the fan and pad evaporative cooler has a cooling efficiency of 80%, determine the supply $T$, $\phi$, total cooling capacity provided, and the rate of water used for this evaporative cooling. Note: 1 ton of refrigeration equals a heat removal rate of 12,000 Btu/h.

**Solution**

Figure 2.21 (the process line 1 to 2) schematically shows the solution. However, State 1 does not in this example lie on the saturation curve as the cooler is not 100% efficient. If it were 100% efficient, then the temperature at State 2 would be the $T_w$ of State 1, that is, State 1 is located at the intersection of $T$ = 35°C and $\phi$ = 25%, and $T_w$ would be approx. 20.2°C, which would at the same time be the temperature at State 2. With 80% efficiency, however, $T_2$ would be $T_1$ minus the 80% of $(T_1 - T_w)$, that is, $T_2$ = 35°C − 0.8 (35°C − 20.2°C) = 23.2°C, which provides the supply temperature to the greenhouse.

Using the intersection of $T_2$ = 23.2°C and the constant wet-bulb temperature line of $T_w$ = 20.2°C on the psychrometric chart, we can determine the humidity ratio at State 2 as $W_2 = 0.0137\ \text{kg}_{\text{water}}/\text{kg}_{\text{da}}$, the specific volume $v_2$ = 0.858 m³/kg, and the relative humidity at State 2 as $\phi$ = 77%.

Now, we can first calculate the mass flow rate of dry air using the specific volume $v_2$ at State 2 as the fan is positioned on the opposite wall (negative pressure or suction type ventilation). The mass flow rate of dry air is then:

$$\dot{m}_a = \left(4\,\text{m}^3/\text{s}\right)/\left(0.858\,\text{m}^3/\text{kg}_{\text{da}}\right) = 3.43\,\text{kg}_{\text{da}}/\text{s}$$

And now using Equation (2.138)

$$\dot{Q} = \dot{m}_a\, c_p \left(T_2 - T_1\right) = (3.43)(1.006)(23.2 - 35) = -40.7\,\text{kW}\left(-13{,}891\frac{\text{Btu}}{\text{h}} = -1.16\text{ ton}\right)$$

We can calculate the cooling capacity using the humidity ratios as well:

$$\dot{Q} = \dot{m}_a L_v (W_1 - W_2) = (3.43)(2,430)(0.0088 - 0.0137) = -40.8\,\text{kW}\,(-1.16\,\text{ton})$$

where we used a latent heat of vaporization $L_v$ of approx. 2,430 kJ/kg (at ~30°C) (Table A.16) and a humidity ratio at State 1, $W_1 = 0.0088\ \text{kg}_{\text{water}}/\text{kg}_{\text{da}}$ at the intersection of $T = 35°\text{C}$ and $\phi = 25\%$.

And now, using Equation (2.134), the theoretical rate of water use for the evaporative cooling process can be calculated as:

$$\dot{m}_w = \dot{m}_a (W_2 - W_1) = (3.43)(0.0137 - 0.0088) = 0.0168\text{kg}_{\text{water}}/\text{s} \approx 60.5\ \text{kg}_{\text{water}}/\text{h}$$

### *Adiabatic Mixing of Two Streams of Moist Air*

The adiabatic mixing of two moist airstreams, shown in Figure 2.22, is a very common practice in air-conditioning systems, which is governed by three equations namely the mass balance equation for the air, the mass balance equation for the water vapor, and the energy balance equation.

The mass balance equation for the entering and leaving air is:

$$\dot{m}_{a1} + \dot{m}_{a2} = \dot{m}_{a3} \tag{2.139}$$

The mass balance equation for the water vapor can be written as:

$$\dot{m}_{a1} W_1 + \dot{m}_{a2} W_2 = \dot{m}_{a3} W_3 \tag{2.140}$$

And similarly, assuming an adiabatic process, the energy balance equation using the mixture of enthalpies becomes:

$$\dot{m}_{a1} h_1 + \dot{m}_{a2} h_2 = \dot{m}_{a3} h_3 \tag{2.141}$$

After eliminating $\dot{m}_{a3}$ and substitution, we can develop the following relationships:

$$\dot{m}_{a1} h_1 + \dot{m}_{a2} h_2 = \dot{m}_{a3} h_3 \tag{2.142}$$

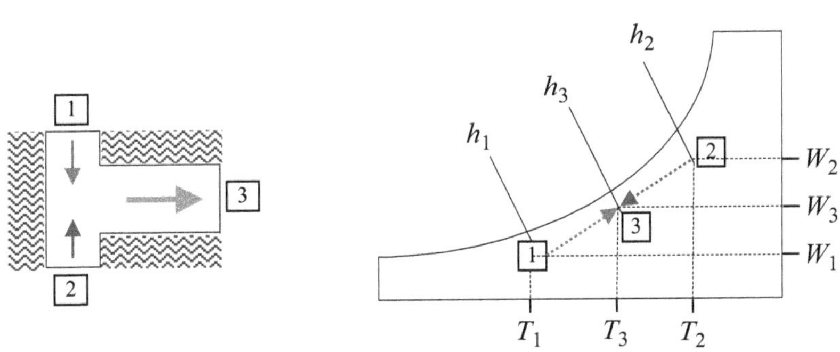


**FIGURE 2.22**
Adiabatic mixing of two streams of moist air.

Based on Equation (2.142), the state point of the resulting mixture (State 3) is always on the straight line connecting the state points of the streams 1 and 2 being mixed, and divides the line into two segments, with its location set by the same ratio as the masses of dry air in the streams 1 and 2.

### Absorption of Space Heat and Moisture Gains by Moist Air

In air-conditioning a space, we encounter moisture and energy gains almost all the time and it is required to remove given amounts of energy and water from the space at a specified condition. The energy gains consist of energy alone independent of the energy contributions due to the moisture gains, and we call this the sensible heat gain $\dot{Q}_s$. The net moisture gains $\sum \dot{m}_w$ can be through the system boundaries as well as from water sources within the space. Each and every unit of water mass injected into the space adds an amount of energy equal to its specific enthalpy. The system is basically governed by the mass and energy balance equations.

Assuming steady-state conditions, the mass balance equation can be written as:

$$\dot{m}_a W_1 + \sum \dot{m}_w = \dot{m}_a W_2 \text{ or } \sum \dot{m}_w = \dot{m}_a (W_2 - W_1) \tag{2.143}$$

which represents the total rate of net water addition to the space.

And similarly, the energy balance equation can be written as:

$$\dot{m}_a h_1 + \dot{Q}_s + \sum \dot{m}_w h_w = \dot{m}_a h_2 \text{ or } \dot{Q}_s + \sum \dot{m}_w h_w = \dot{m}_a (h_2 - h_1) \tag{2.144}$$

And the left side of this equation represents the total rate of energy addition to the space from all sources.

After eliminating $m_a$ and substitution, from Equations (2.140) and (2.141), we can develop the following relationships:

$$\frac{\Delta h}{\Delta W} = \frac{h_2 - h_1}{W_2 - W_1} = \frac{\dot{Q}_s + \sum(\dot{m}_w h_w)}{\sum(\dot{m}_w)} \tag{2.145}$$

Based on Equation (2.145), on the psychrometric chart and for a given state of the removed air, all possible states (conditions) for the supply air must lie on a straight line drawn through the state point of the removed air, which has a direction specified by the numerical value of $\left[\dot{Q}_s + \sum(\dot{m}_w h_w)\right] / \sum(\dot{m}_w)$. This line is the *condition line* for the given problem.

**Example 2.84: Absorption of Space Heat and Moisture Gains**

Let's now consider a research greenhouse from which moist air is removed at 30°C dry-bulb temperature and 22°C thermodynamic wet-bulb temperature. The sensible heat gain for the greenhouse is 20 kW. A rate of moisture gain of 10.0 kg/h occurs from the crop transpiration. This moisture is assumed as saturated water vapor at 28°C. Moist air is introduced to the greenhouse at a dry-bulb temperature of 20°C. Determine the supply air's thermodynamic wet-bulb temperature and the volumetric flow rate.

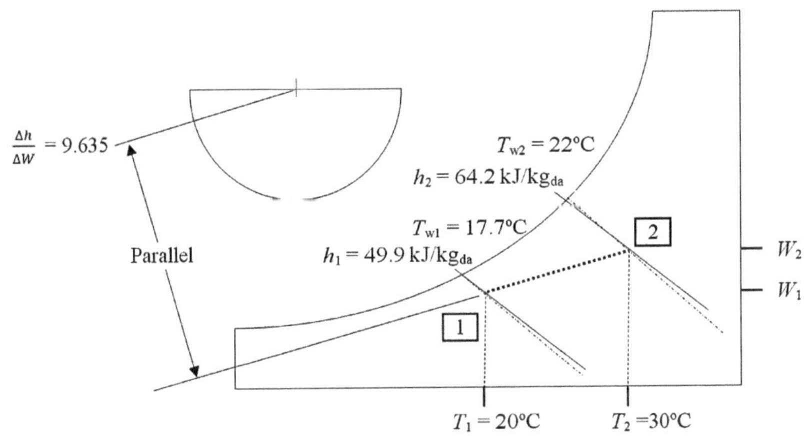


**Solution**

The psychrometric sketch above schematically shows the solution. State 2 is located on the psychrometric sketch at the intersection of 30°C dry-bulb temperature and 22°C thermodynamic wet-bulb temperature. Table A.16 provides the specific enthalpy of added water vapor at 28°C, $h_g$ as 2,434.55 kJ/kg$_{water}$.

From Equation (2.145):

$$\frac{\Delta h}{\Delta W} = \frac{20 + (10/3{,}600)(2434.55)}{\left(\frac{10}{3{,}600}\right)} = 9{,}635 \text{ kJ/kg}_{water}$$

With the $\Delta h/\Delta W$ protractor, we can now establish a reference line of direction $\Delta h/\Delta W$ = 9.635 kJ/g$_{water}$. Parallel to this reference line, a straight line on the psychrometric chart can be drawn through State 2. The intersection of this line with the 20°C dry-bulb temperature line fixes State 1. And thus, we can determine the wet-bulb temperature at State 1, $T_{w1}$ = 17.7°C.

Alternatively, we can also establish the condition line using the protractor's sensible-total heat ratio scale instead of the Δh/ΔW scale. On the protractor, the quantity $\Delta H_S/\Delta H_T$ is the ratio of the rate of sensible heat gain for the space to the rate of total energy gain for the space. Therefore,

$$\frac{\Delta H_S}{\Delta H_T} = \frac{\dot{Q}_s}{\dot{Q}_s + \Sigma(\dot{m}_w h_w)} = \frac{20}{20 + (10/3{,}600)(2434.55)} = 0.747$$

Observe that $\Delta H_S/\Delta H_T = 0.747$ on the protractor coincides closely with $\Delta h/\Delta W = 9.635$ kJ/g$_{water}$. The flow of dry air can now be calculated using either Equation (2.143) or (2.144). Using Equation (2.144),

$$\dot{m}_a = \frac{\left[\dot{Q}_s + \Sigma \dot{m}_w h_w\right]}{h_2 - h_1} = \left[20 + (10/3{,}600)(2434.55)\right]/(64.2 - 49.9)$$

$$\dot{m}_a = 1.872 \text{ kg}_{da}/\text{s}$$

At State 1, $v_1$ = 0.846 m$^3$/kg$_{da}$; therefore, the volumetric air flow rate = $\dot{m}_a v_1$ = (1.872) (0.846) = 1.584 m$^3$/s.

### 2.3.5 Fuels, Combustion Reactions and Calculations

Combustion is the chemical process where an oxidant reacts with a fuel rapidly to release the stored energy as thermal energy. The oxidant is typically oxygen in the air, and along with the thermal energy in the form of high temperature gases, usually small amounts of electromagnetic energy (as light), electric energy (as free electrons), and mechanical energy (as noise) are also released during the combustion process.

Hydrocarbon fuels are primarily composed of carbon and hydrogen, and most conventional fuels contain small amounts of sulfur as well. Flue gas is the product of complete or incomplete combustion including, if present, excess air. During *complete combustion* of hydrocarbon fuels, all the carbon (C) is oxidized to carbon dioxide ($CO_2$), all the hydrogen (H) to water ($H_2O$), and all the sulfur (S) to $SO_2$ or $SO_3$. Excess air, usually expressed as a percentage, is supplied beyond what is required theoretically for complete combustion to completely oxidize the fuel. *Stoichiometric* or *theoretical combustion* is the ideal combustion where fuel is burned completely. In stoichiometric combustion, fuel reacts with the exact amount of oxygen needed to completely oxidize all carbon, hydrogen, and sulfur in the fuel to $CO_2$, $H_2O$, and $SO_2$ or $SO_3$, respectively. Hence, the flue gas form stoichiometric combustion contains no excess or unreacted oxygen and no carbon monoxide.

The amount of carbon dioxide or sulfur oxides released during the stoichiometric combustion is the maximum achievable, and is called the *maximum theoretical percentage* or *ultimate amount*. In reality, stoichiometric combustion is rarely achieved due to imperfect mixing (of air and fuel) and finite reaction rates (fixed by fuel selection); therefore, most combustion equipment needs to operate with some excess air. In order to determine the *excess air* for a combustion system, we first determine the stoichiometric *air–fuel ratio*, which is the perfect ideal mixing ratio. When fuel is burned completely, all air is depleted without any left over. Heating equipment are rarely run on perfect mixing ratio. Boilers and high temperature furnaces typically incorporate a modest amount of excess air, approximately 10–20% more than what is required for burning the fuel completely. Natural gas-fired boilers may run as low as 5% excess air, whereas pulverized coal-fired boilers may run with 20% excess air.

If air provided to burner is not enough, *incomplete combustion* occurs, with all the non-oxidized components in the flue gas such as C, $H_2$, and CO. Incomplete combustion does not use fuel efficiently, a process which results in air pollution and can pose even further danger due to the carbon monoxide (CO) production and a potential for explosion. CO is an odorless, tasteless, and colorless toxic gas. Every combustion process has the potential to produce this toxic gas as a result of incomplete combustion. Carbon monoxide inhibits the blood's ability to carry oxygen to vital organs because it combines with the oxygen carrying hemoglobin in the blood. Health effects of carbon monoxide, its concentrations, and exposure hours are provided in Table 2.15. Typical sickness symptoms are reported as mild headache, fatigue, nausea, and dizziness; while an exposure of 1–3 min to 12–13,000 ppm CO concentration is deadly, a concentration of 1,600 ppm is fatal after 1 h. Combustion equipment is designed and operated in order to make sure we have the complete, not stoichiometric combustion so that we can eliminate or reduce the dangers of carbon monoxide production. Furthermore, air at higher altitudes has lesser density and, therefore, less oxygen per unit volume. Excess air must be adjusted accordingly; otherwise, we will have an incomplete combustion, which can eventually cease.

#### *2.3.5.1 Fuel Classification and Heating Values*

Typically, hydrocarbon fuels are classified based on their physical state, that is, gaseous, liquid, or solid. And, we will be analyzing the fuels in this section according to this classification.

**TABLE 2.15**
Health Effects of Carbon Monoxide, Its Concentrations, and Exposure Hours (Engineering ToolBox, 2003).

| | CO Concentration (ppm) | | |
|---|---|---|---|
| **Exposure (hour)** | **Perceptible** | **Sickness** | **Deadly** |
| ½ | 600 | 1,000 | 2,000 |
| 1 | 200 | 600 | 1,600 |
| 2 | 100 | 300 | 1,000 |
| 4 | 50 | 150 | 400 |
| 6 | 25 | 120 | 200 |
| 8 | 25 | 100 | 150 |

Fossil fuels have always had the major share in the global primary energy consumption. Overall fossil fuel consumption increased approximately 51% in the period of 1995–2015, and it is predicted that the consumption will increase approximately 18% more in the period of 2015–2035. It was observed that, in 1995, oil, natural gas, and coal had total primary energy consumption shares of 38%, 22%, and 26% (87% in total), respectively; that, in 2015, they had total primary energy consumption shares of 32%, 24%, and 29% (85% in total), respectively; whereas non-fossil energy consumption in 2015 included hydroelectric 6.8%, nuclear 4.4%, wind 1.4%, solar 0.45%, and other renewables amounting to 0.89% (Yildiz, 2018). And world energy consumption is growing. It is predicted that, by 2035, the shares will be 29%, 25%, and 24% (78% in total), respectively, having natural gas second place after oil, and that a 7% decrease in fossil fuels share in consumption is predicted in the global primary energy consumption. Even though the share of fossil fuel consumption has decreased and is predicted to decrease further, they will continue to hold the major share in the primary energy mix in the foreseeable future as more unconventional fossil fuels are explored. The highest change in fossil fuel consumption has happened and is predicted that it will continue to happen in natural gas, a cleaner fossil fuel compared to oil and coal; however, the extraction of and burning all fossil fuel forms have serious environmental consequences, which will require further and closer attention as more unconventional fossil fuel resources are explored and their overall consumption expand further.

As discussed earlier, combustion releases thermal energy or heat, and the amount of heat generated by complete combustion of a specific fuel is fixed and called as the *heating value, heat of combustion*, or *calorific value* of that fuel. The combustion process typically produces water vapor as well and we can recover the heat contained in this water vapor by condensing the water. When this water vapor is entirely condensed and the latent heat contained in the water vapor is recovered, the heating value of fuel is termed *higher heating value* (HHV), *higher calorific value* (HCV), *gross calorific value* (GCV), or *total heating value* (THV). Conversely, when the latent heat of vaporization is not recovered, the heating value of fuel is termed *lower heating value* (LHV), *lower calorific value* (LCV), *net calorific value* (NCV), or *net heating value* (NHV). Tables 2.16, 2.17, 2.18, 2.19, and 2.20 provide the heating values of different fuels. If the heating values of fuels are specified without designating higher or lower, they generally mean the HHVs, including those provided in this book as well.

#### *2.3.5.1.1 Gaseous Fuels*

In the past, a number of different gases have been used as energy sources; however, it is currently limited to natural gas and liquid petroleum gases (LPGs). Natural gas is an odorless and colorless hydrocarbon gas mixture mainly consisting of methane ($CH_4$), varying amounts of other higher alkenes such as ethane, propane, butane, pentane, hexane, and sometimes a small portion of carbon dioxide, oxygen, nitrogen, hydrogen sulfide, or helium. It is often referred to as "gas" when compared to other energy forms such as oil or coal; and is a non-renewable fossil fuel resource that is used as a source of energy for heating, cooking, electricity generation, etc. It is most commonly transported by pipeline; however, natural gas is increasingly being transported by ship in a liquefied form to meet greater global demand for the fuel. Natural gas (primarily methane) is also generated by the decomposition of municipal waste in landfills and manure from livestock production. Methane is a greenhouse gas that is approximately 22 times more potent than carbon dioxide. Capturing and burning the gas to produce usable heat and power prevents the methane from being released from the landfill or feedlot into the atmosphere directly. Before natural gas can be used as a fuel, most of the time, it is processed to remove impurities such as water to meet the specifications of marketable natural gas.

Heating values of natural gases differ considerably between markets from 33.6 to 44.8 MJ/m$^3$ (900–1,200 Btu/ft$^3$); however, typical range at sea level is 37.3–39.2 MJ/m$^3$ (1,000–1,050 Btu/ft$^3$). And for the purpose of safety, odorants are added to natural gas and LPG so that they can be noticed in the case of leaks. LPGs are usually obtained as a byproduct of oil refineries and typically consist of propane and butane, which are gaseous under normal atmospheric conditions; however, they are liquefied by applying moderate pressures at normal temperatures so that they can be transported conveniently. We basically have three commercially available LPGs namely *butane*, *propane*, and the *mixture* of the two.

At atmospheric pressure, commercial propane has a boiling point of approx. −40°C (−40°F), quite convenient for northern climates, mainly consists of propane, but usually contains 5–10% propylene as well; and it has a heating value of approximately 93.3 MJ/m$^3$ (2,500 Btu/ft$^3$) or 50.5 MJ/kg (21,669 Btu/lb). It is available in tank trucks, cylinders, and bottles. Commercial butane, on the other hand, has a relatively high boiling point of approx. 0°C (32°F) at atmospheric pressure, not convenient for northern climates unless it is maintained above 0°C, mainly consists of butane, but may contain up to 5% butylene as well; and it has a heating value of approximately 119.4 MJ/m$^3$ (3,200 Btu/ft$^3$) or 49.7 MJ/kg (21,321 Btu/lb). It is available in tank trucks and bottles, but not in cylinders. Commercial butane-propane mixtures, on the other hand, are available with varying ratios of butane and propane. Heating values of different gases and substances occurring in common fuels are provided in Table 2.16.

#### 2.3.5.1.2 Liquid Fuels

Liquid fuels are typically the mixtures of hydrocarbons derived from crude petroleum. After refining crude petroleum, we produce a variety of fuels and other useful products. Almost all lighter hydrocarbons are refined into gasoline, diesel fuels, jet fuels, kerosene, light heating oils, and LPGs. Heavy hydrocarbons, on the other hand, are refined into residual oils and some other products such as asphalt, petroleum coke, and lubricating oils. Biofuels, on the other hand, contain over 80% renewable materials such as biomass, which is derived from photosynthesis. Biofuels are essentially classified into first, second, third generations, and wastes. The first-generation biofuels are based on food biomass such as corn, sugar cane, plant, and vegetable oils. The second-generation biofuels are based on the non-food type biomass such as cellulose and hemicelluloses, whereas the third-generation biofuels primarily refer to algae and cyanobacteria. All of the above-mentioned biomass can be converted to alcohols, biodiesel, biogas, bio-oil, and bioproducts

**TABLE 2.16**

Heating Values of Gases and Substances Occurring in Common Fuels (ASHRAE, 1989; Engineering ToolBox, 2003).

| Gaseous Fuel | Density[a] | | Higher Heating Value (HHV) | | Lower Heating Value (LHV) | |
|---|---|---|---|---|---|---|
| | (kg/m³) | (lb/ft³) | (MJ/kg) | (Btu/lb) | (MJ/kg) | (Btu/lb) |
| Carbon (to CO) | | | 9.2 | 3,950 | 9.2 | 3,950 |
| Carbon (to $CO_2$) | | | 32.8 | 14,093 | 32.8 | 14,093 |
| Carbon monoxide | 1.184 | 0.074 | 10.1 | 4,347 | 10.1 | 4,347 |
| Acetylene | 1.097 | 0.070 | 49.9 | 21,453 | 48.4 | 20,769 |
| Ethylene | | | 50.4 | 21,636 | 47.2 | 20,275 |
| Propylene | 1.776 | 0.111 | 49.0 | 21,048 | 45.9 | 19,687 |
| Hydrogen | 0.090 | 0.0053 | 142.3 | 61,095 | 120.0 | 51,623 |
| Ethane | 1.280 | 0.080 | 52.0 | 22,323 | 47.6 | 20,418 |
| Butane | 2.528 | 0.158 | 49.7 | 21,321 | 45.9 | 19,678 |
| Methane | 0.672 | 0.042 | 55.6 | 23,875 | 50.0 | 21,495 |
| Propane | 1.920 | 0.120 | 50.5 | 21,669 | 46.5 | 19,937 |
| Natural gas (US market) | 0.777 | 0.049 | 52.3 | 22,446 | 47.2 | 20,262 |
| Sulfur (to $SO_2$) | | | 9.3 | 3,980 | 9.3 | 3,980 |
| Sulfur (to $SO_3$) | | | 13.8 | 5,940 | 13.8 | 5,940 |
| Hydrogen sulfide | 1.456 | 0.091 | 16.5 | 7,097 | 15.2 | 6,537 |

[a] @ 0°C and 1 bar (32°F and 29.92 in Hg).

using various processes: biochemical, thermo-chemical, and hydrothermal. Biofuels are promising alternatives to fossil fuels and have the capability of controlling the greenhouse gas emissions; therefore, they can have positive impacts on climate change. Bioethanol and biodiesel are currently the most widely used biofuels; however, the share of biofuels in the primary energy supply is still very small because the current costs of biofuels are not competitive with those of fossil fuels.

Fuel selection, particularly fuel oil grade selection for a specific application is usually a function of availability and economic factors, including the costs of fuel, preheating, handling, and equipment, as well as the environmental requirements. Heating values of a number of different liquid fuels are presented in Table 2.17.

### *2.3.5.1.3 Solid Fuels*

Solid fuels consist of coal, coke, crop residues, wood, and waste products of agricultural, forestry, and industrial operations. Out of these products, coal is the widely used fuel for heating and cooling applications. Following oil (32%), coal with a share of 29% is currently the world's second largest source of primary energy. Currently, about 40% of global electricity production is contributed by coal, but this number is expected to decline due to increased shares from renewable energy sources and natural gas. Global coal production in 2015 increased by 25.1% compared to that in 1995, and is projected to increase further in the next 20 years (8.4% based on the 2015 production). However, coal is primarily made up of carbon; therefore, it is a carbon-intensive energy source. Burning coal is the biggest single source of $CO_2$ emissions from human activity. Coal combustion produces almost double

**TABLE 2.17**

Heating Values of Liquid Fuels (ASHRAE, 1989; Engineering ToolBox, 2003; Demirel, 2018).

| Liquid Fuel | Density[a] (kg/l) | (lb/gal) | Higher Heating Value (HHV) (MJ/kg) | (Btu/lb) | Lower Heating Value (LHV) (MJ/kg) | (Btu/lb) |
|---|---|---|---|---|---|---|
| Acetone | 0.787 | 6.568 | 31.8 | 13,671 | 29.6 | 12,726 |
| Butanol | 0.810 | 6.759 | 37.3 | 16,036 | 34.4 | 14,789 |
| Ethanol | 0.789 | 6.584 | 29.7 | 12,769 | 26.7 | 11,479 |
| Methanol | 0.791 | 6.601 | 23.0 | 9,888 | 19.9 | 8,568 |
| Diesel | 0.846 | 7.060 | 45.6 | 19,604 | 42.6 | 18,315 |
| Biodiesel | 0.920 | 7.677 | 40.5 | 17,412 | 37.8 | 16,251 |
| Gasoline | 0.737 | 6.150 | 46.4 | 19,948 | 43.4 | 18,659 |
| Biooil (by pyrolisis) | 1.17 | 9.76 | 16–19 | 6,867–8,155 | | |
| Biooil (by liquefaction) | 1.17 | 9.76 | 34 | 14,592 | | |
| Fuel Oil Grade No. 1 | 0.816 | 6.813 | 46.9 | 20,109 | 44.5 | 19,507 |
| Fuel Oil Grade No. 2 | 0.854 | 7.128 | 46.4 | 19,893 | 44.8 | 19,220 |
| Fuel Oil Grade No. 4 | 0.910 | 7.592 | 45.5 | 19,507 | 44.4 | 19,050 |
| Fuel Oil Grade No. 5 (Light) | 0.936 | 7.813 | 44.7 | 19,199 | 43.8 | 18,789 |
| Fuel Oil Grade No.5 (Heavy) | 0.957 | 7.985 | 44.4 | 19,036 | 43.6 | 18,710 |
| Fuel Oil Grade No. 6 (Bunker C) | 0.989 | 8.250 | 44.0 | 18,897 | 42.7 | 18,339 |
| Kerosene | 0.821 | 6.851 | 46.2 | 19,862 | 43.0 | 18,487 |
| LNG | 0.428 | 3.572 | 55.2 | 23,732 | 48.6 | 20,894 |
| LPG | 0.537 | 4.481 | 49.3 | 21,195 | 45.5 | 19,561 |

[a] @ 15°C and 1 bar (60°F and 29.92 in Hg).

the greenhouse gas emissions as burning gas, for the same amount of energy. Although coal generated less than 30% of the world's energy supply, in 2013 for instance, it produced 46% of the global $CO_2$ emissions. It must be noted that coal mining also releases the potent greenhouse gas methane into the environment.

Even though there is no distinct demarcation between coal types, under proper conditions, as ancient plant material are exposed to high temperatures and pressures over geological time, metamorphic grade of the dead material increases successively into:

- *Peat*, which is considered to be a precursor of coal, which has industrial importance as a fuel in some regions.
- *Lignite*, also known as brown coal, is the lowest rank of coal, and almost solely used for electric power generation.
- *Sub-bituminous coal* whose properties range between those of lignite and bituminous coal, and is used primarily as fuel for steam-electric power generation.
- *Bituminous coal*, which is a dense and usually black (sometimes dark brown) sedimentary rock, which is mainly used as fuel in steam-electric power generation and for heat in manufacturing.
- *Steam coal*, which is a grade between bituminous coal and anthracite. It was once commonly used as a fuel for steam locomotives, and also used for domestic water heating.

- *Anthracite*, the highest rank of coal, which is a harder and glossy black coal mainly used for residential and commercial space heating. It is further divided into metamorphically altered bituminous coal and petrified oil.
- *Graphite*, one of the more difficult types of coal to ignite, which is not generally used as fuel.

Obviously, the solid fuel selection for a particular application is function of availability and economic factors, including the costs of fuel, handling, and equipment, as well as the environmental requirements. Heating values of various solid fuels, biomass, and waste material are presented in Tables 2.18, 2.19, and 2.20, respectively.

Wood, on the other hand, contains a lot of water when it is fresh. After it is cut and stacked for a year or two, the average moisture content of wood typically drops down to about 20%. During the combustion process, the temperature rises to flue gas temperature and water evaporates. Table 2.21 shows the average energy content of wood based on its moisture content when specific data is not available. Please note that by volume, wet wood has about 85% of the energy of oven-dry wood; whereas by weight wet, wood has less than half (42%) of the energy of oven-dry wood.

**TABLE 2.18**

Heating Values of Different Solid Fuels (ASHRAE, 1989; Engineering ToolBox, 2003).

| Solid Fuel | Higher Heating Value (HHV) (MJ/kg) | (Btu/lb) | Lower Heating Value (LHV) (MJ/kg) | (Btu/lb) |
|---|---|---|---|---|
| Petroleum coke | 31.3 | 13,460 | 29.5 | 12,685 |
| Coal (wet basis) | 24.0 | 10,304 | 22.7 | 9,773 |
| Coal (low bituminous) | 28.9 | 12,425 | | |
| Coal (bituminous, wet basis) | 27.3 | 11,723 | 26.1 | 11,230 |
| Coal (high bituminous) | 36.3 | 15,606 | | |
| Coal (anthracite) | 34.6 | 14,875 | | |
| Coke | 26.1 | 11,200 | | |
| Peat (dry) | 17.0 | 7,300 | | |
| Coal (lignite) | 14.0 | 6,000 | | |
| Coking coal (wet basis) | 29.9 | 12,840 | 28.6 | 12,300 |
| Charcoal (wood) | 34.7 | 14,918 | | |
| Farmed trees (dry basis) | 20.6 | 8,852 | 19.6 | 8,406 |
| Wood, oak | 18.9 | 8,126 | | |
| Wood, locust | 19.7 | 8,470 | | |
| Wood, Ponderosa pine | 20.0 | 8,599 | | |
| Wood, redwood | 20.7 | 8,899 | | |
| Herbaceous biomass (dry basis) | 18.1 | 7,791 | 17.2 | 7,399 |
| Corn stover (dry basis) | 17.4 | 7,487 | 16.4 | 7,038 |
| Grass (lawn cuttings) | 19.3 | 8,298 | | |
| Forest residue (dry basis) | 16.5 | 7,082 | 15.4 | 6,622 |
| Sugar cane bagasse | 16.4 | 7,031 | 15.1 | 6,474 |
| Newsprint | 18.6 | 7,997 | | |
| Cellulose | 17.3 | 7,438 | | |

**TABLE 2.19**

Heating Values of Various Biomass Material Used as Fuel (Engineering ToolBox, 2003).

| Biomass | Heating Value (MJ/kg) | Heating Value (Btu/lb) | Biomass | Heating Value (MJ/kg) | Heating Value (Btu/lb) |
|---|---|---|---|---|---|
| Alfalfa straw | 18.5 | 7,953 | Maize stalks | 18 | 7,738 |
| Charcoal | 30 | 12,897 | Paper, newspaper | 17 | 7,308 |
| Coconut husks | 10 | 4,299 | Peat | 14.5 | 6,234 |
| Coconut shells | 18 | 7,738 | Rice hulls | 15.5 | 6,663 |
| Coffee husks | 16 | 6,878 | Rice straw | 15 | 6,449 |
| Commercial wastes | 16 | 6,878 | Soybean stalks | 19 | 8,168 |
| Cotton hulls | 19.5 | 8,383 | Straw, harvested, baled | 15 | 6,449 |
| Cotton stalks | 17.5 | 7,523 | Sorghum bagasse | 19 | 8,168 |
| Domestic refuse | 9 | 3,869 | Sugar cane residues | 17 | 7,308 |
| Dung, dried | 16 | 6,878 | Wheat straw | 19 | 8,168 |
| Grass, fresh | 4 | 1,720 | Wood, green (60% moisture) | 6 | 2,579 |
| Groundnut shells | 20 | 8,598 | Wood, air-dried (20% moisture) | 15 | 6,449 |
| Maize cobs | 19 | 8,168 | Wood, oven-dried (0% moisture) | 18 | 7,738 |

**TABLE 2.20**

Heating Values of Different Waste Material (Engineering ToolBox, 2003).

| Waste | Heating Value (MJ/kg) | Heating Value (Btu/lb) | Waste | Heating Value (MJ/kg) | Heating Value (Btu/lb) |
|---|---|---|---|---|---|
| Fabric, nylon | 30.8 | 13,200 | Paper | 17.6 | 7,570 |
| Fabric, rubber-coated | 25.6 | 11,000 | Polyethylene, film | 44.5 | 19,100 |
| Fabric, vinyl-coated | 20.7 | 8,900 | Rags | 17.8 | 7,650 |
| Foam, scrap | 28.7 | 12,300 | Vinyl, scrap | 26.6 | 11,400 |
| Garbage | 19.8 | 8,480 | Wood | 20.1 | 8,610 |
| Glass tape, resin covered | 18.4 | 7,900 | | | |

**TABLE 2.21**

Moisture Content and Usable Energy Content of Wood (Engineering ToolBox, 2003).

| Wood | Energy by Volume (%) | Energy by Weight (%) |
|---|---|---|
| Oven-dry (0% moisture) | 100 | 100 |
| Air-dry (20% moisture) | 97 | 81 |
| Green (50% moisture) | 92 | 62 |
| Wet (100% moisture) | 85 | 42 |

#### 2.3.5.1.4 Comparing Costs of Heating Fuels

For convenience, average cost comparison formulas for heating fuels such as electricity, natural gas, liquid petroleum gas (LPG), fuel oil, and coal are provided below:

*Electricity*

$$C_E = P_E\left(\frac{100{,}000\,\text{Btu}}{3{,}412\,\text{Btu/kWh}}\right) \tag{2.146}$$

where $C_E$ is the cost of electricity per 100,000 Btu, $P_E$ is the price of electricity per kWh, and 1 kWh is equal to 3,412 Btu.

*Natural Gas*

$$C_{NG} = P_{NG}\left(\frac{100{,}000\,\text{Btu}}{1{,}036\,\text{Btu/ft}^3}\right) \tag{2.147}$$

where $C_{NG}$ is the cost of natural gas per 100,000 Btu, and $P_{NG}$ is the price of natural gas per cubic foot, and the average energy content of natural gas is assumed to be 1,036 Btu/ft$^3$.

*Liquid Propane Gas*

$$C_{LPG} = P_{LPG}\left(\frac{100{,}000\,\text{Btu}}{91{,}500\,\text{Btu/gal}}\right) \tag{2.148}$$

where $C_{LPG}$ is the cost of LPG per 100,000 Btu, and $P_{LPG}$ is the price of LPG per gallon, and the average energy content of LPG is assumed to be 91,500 Btu/gal.

*Fuel Oil*

$$C_{OIL} = P_{OIL}\left(\frac{100{,}000\,\text{Btu}}{139{,}000\,\text{Btu/gal}}\right) \tag{2.149}$$

where $C_{OIL}$ is the cost of fuel oil per 100,000 Btu, and $P_{OIL}$ is the price of fuel oil per gallon, and the average energy content of fuel oil is assumed to be 139,000 Btu/gal.

*Coal*

$$C_{COAL} = P_{COAL}\left(\frac{100{,}000\,\text{Btu}}{20{,}000{,}000\,\text{Btu/ton}}\right) \tag{2.150}$$

where $C_{COAL}$ is the cost of coal per 100,000 Btu, and $P_{COAL}$ is the price of coal per short ton, and the average energy content of Grade No. 2 coal used in medium capacity burner is assumed to be 20,000,000 Btu/ton.

*Wood*

$$C_{WOOD} = P_{WOOD}\left(\frac{100{,}000\,\text{Btu}}{20{,}000{,}000\,\text{Btu/cord}}\right) \tag{2.151}$$

where $C_{WOOD}$ is the cost of wood per 100,000 Btu, and $P_{WOOD}$ is the price of wood per cord, and the average energy content of dry wood varies considerably and is assumed to be 20,000,000 Btu/cord. The chord is the most commonly used measurement unit for purchasing firewood, which includes bark and air space, and can be calculated as follows:

In I-P units, 1 cord = (4 ft) (4 ft) (8 ft) = 128 $ft^3$; and in SI, 1 cord = (1.22 m) (1.22 m) (2.44 m) ≈ 3.63 $m^3$

It should be kept in mind that the air space in the cord can be as high as 40% due to the irregular shape of wood. In general, the net cord volume ranges somewhere between 80 and 100 $ft^3$, but for firewood it averages about 85 $ft^3$.

**Example 2.85: Calculating the Number of Cords**

Determine the number of cords in the pile.

**Solution**

A pile of firewood is 10 ft long, 5 ft wide, and 6 ft high.

$$\text{No. of Cords} = (10\text{ ft})\ (5\text{ ft})\ (6\text{ ft})/128\text{ ft}^3 = 2.34\text{ cord}$$

Dimensions for another pile of firewood are given as 3 m long, 1.5 m wide, and 1.8 m high.

$$\text{No. of Cords} = (3\text{ m})\ (1.5\text{ m})\ (1.8\text{ m})/3.63\text{ m}^3 \approx 2.23\text{ cord}$$

**Example 2.86: Comparing Costs of Fuels**

A greenhouse operation is trying to compare costs of fuel for a number of locally available fuels ignoring the handling costs, credits, and additional fees. Local fuel prices are listed as follows: $440 per cord of firewood; $0.0029 per $ft^3$ of natural gas; $3.59 per gallon of fuel oil; $0.125 per kWh of electricity; and $3.40 per gallon of propane. Determine the most cost effective fuel at this particular location.

**Solution**

Using the equations provided in the previous section, we can determine the fuel costs as follows:

$$C_E = 0.125\$/\text{kWh}\left(\frac{100{,}000\text{ Btu}}{3{,}412\text{ Btu/kWh}}\right) = \$3.66 \text{ per 100,000 Btu of electricity}$$

$$C_{NG} = 0.005\$/\text{ft}^3\left(\frac{100{,}000\text{ Btu}}{1{,}036\text{ Btu/ft}^3}\right) = \$0.48 \text{ per 100,000 Btu of natural gas}$$

$$C_{LPG} = 2.50\$/\text{gal}\left(\frac{100{,}000\text{ Btu}}{91{,}500\text{ Btu/gal}}\right) = \$2.73 \text{ per 100,000 Btu of LPG}$$

$$C_{OIL} = 2.45\$/\text{gal}\left(\frac{100{,}000\text{ Btu}}{139{,}000\text{ Btu/gal}}\right) = \$1.76 \text{ per 100,000 Btu of fuel oil}$$

$$C_{WOOD} = 325\$/\text{cord}\left(\frac{100{,}000\text{ Btu}}{20{,}000{,}000\text{ Btu/cord}}\right) = \$1.63 \text{ per 100,000 Btu of wood}$$

Based on the local fuel prices provided, the most cost effective fuel seems to be natural gas.

### 2.3.5.2 Combustion Reactions and Calculations

The exact amount of air required for complete combustion is termed theoretical or stoichiometric air. Table 2.22 provides the combustion reactions for complete combustion.

For instance, as Table 2.22 shows, 12 kg (or 12 lb) carbon is oxidized by 32 kg (or 32 lb) oxygen, 2 kg (or 2 lb) hydrogen is oxidized by 16 kg (or 16 lb) oxygen. One can also see the complete combustion reactions for other fuel constituents in Table 2.22.

Oxygen requirement for combustion is usually provided by air, a mixture of nitrogen, oxygen, water vapor, carbon dioxide, and some inert gases. For combustion calculations, we can use dry air content, which is composed of 20.95% oxygen and 79.05% inert gases (nitrogen, argon, etc.) by volume or 23.15% oxygen and 76.85% inert gases by weight. In combustion calculations, it is assumed that nitrogen does not change through the combustion process, even though small amounts of nitrogen oxides form.

The combustion process can in general terms be expressed as follows:

$$[C+H](fuel)+[O_2+N_2](air)\rightarrow[CO_2+H_2O+N_2](heat) \tag{2.152}$$

where C is carbon, H is hydrogen, O is oxygen, and N is nitrogen.

From the above relationship by weight, we can easily determine that the mass of dry air providing the required oxygen for complete combustion is 100/23.15 = 4.32 times the mass of oxygen. The mass of air provided in Table 2.22 were calculated based on this relationship. In determining the oxygen required for complete combustion, the oxygen contained in the fuel should be deducted from the required amount as this oxygen is already part of the fuel. Allowance should also be made for water vapor as it is always present in atmospheric air. We sometimes make the combustion calculations based on volume; in such

**TABLE 2.22**

Combustion Reactions and Stoichiometric Oxygen and Air Requirements of Common Fuel Constituents (ASHRAE, 1989).

| Fuel Constituent | Symbol | Combustion Reaction | $O_2$ kg/kg fuel[a] | Air kg/kg fuel[a] | $O_2$ m³/m³ fuel | Air m³/m³ fuel |
|---|---|---|---|---|---|---|
| Carbon | C | $C + 0.5O_2 \rightarrow CO$ | 1.33 | 5.75 | - | - |
| | | $C + O_2 \rightarrow CO_2$ | 2.66 | 11.51 | - | - |
| Carbon monoxide | CO | $CO + 0.5O_2 \rightarrow CO_2$ | 0.57 | 2.47 | 0.50 | 2.39 |
| Hydrogen | $H_2$ | $H_2 + 0.5O_2 \rightarrow H_2O$ | 7.94 | 34.28 | 0.50 | 2.39 |
| Methane | $CH_4$ | $CH_4 + 2O_2 \rightarrow CO_2 + 2H_2O$ | 3.99 | 17.24 | 2.00 | 9.57 |
| Ethane | $C_2H_6$ | $C_2H_6 + 3.5O_2 \rightarrow 2CO_2 + 3H_2O$ | 3.72 | 16.09 | 3.50 | 16.75 |
| Propane | $C_3H_8$ | $C_3H_8 + 5O_2 \rightarrow 3CO_2 + 4H_2O$ | 3.63 | 15.68 | 5.00 | 23.95 |
| Butane | $C_4H_{10}$ | $C_4H_{10} + 6.5O_2 \rightarrow 4CO_2 + 5H_2O$ | 3.58 | 15.47 | 6.50 | 31.14 |
| Ethylene | $C_2H_4$ | $C_2H_4 + 3O_2 \rightarrow 2CO_2 + 2H_2O$ | 3.42 | 14.78 | 3.00 | 14.38 |
| Propylane | $C_3H_6$ | $C_3H_6 + 4.5O_2 \rightarrow 3CO_2 + 3H_2O$ | 3.42 | 14.78 | 4.50 | 21.53 |
| Acetylene | $C_2H_2$ | $C_2H_2 + 2.5O_2 \rightarrow 2CO_2 + H_2O$ | 3.07 | 13.27 | 2.50 | 11.96 |
| Sulfur | S | $S + O_2 \rightarrow SO_2$ | 1.00 | 4.31 | - | - |
| | | $S + 1.5O_2 \rightarrow SO_3$ | 1.50 | 6.47 | - | - |
| Hydrogen sulfide | $H_2S$ | $H_2S + 1.5O_2 \rightarrow SO_2 + H_2O$ | 1.41 | 6.08 | 1.50 | 7.18 |

[a] Atomic masses: H = 1.008; C = 12.01; O = 16.00; S = 32.06.

cases, the volume of dry air providing the required oxygen for complete combustion is 100/20.95 = 4.78 times the mass of oxygen. The volume of required oxygen and air for specific fuel constituents are provided in the last two columns of Table 2.22, respectively.

Using the above relationships, we can develop the following equation for the theoretical mass of dry air required $W_a$ for stoichiometric combustion of a unit mass of any hydrocarbon fuel:

$$W_a = 0.0144\left(8\text{C} + 24\text{H} + 3\text{S} - 3\text{O}\right) \tag{2.153}$$

where C, H, S, and O are the mass percentages of carbon, hydrogen, sulfur, and oxygen in the fuel, respectively.

Sometimes, analyses of hydrocarbon fuels, as the case for gaseous fuels, are based on hydrocarbon components rather than the elemental components. In such cases, we can use the following relationship to determine the theoretical mass of dry air required $W_a$ for stoichiometric combustion of a unit mass of gaseous fuel:

$$\begin{aligned} W_a = {} & 2.47\,CO + 34.28\,H_2 + 17.24\,CH_4 + 16.09\,C_2H_6 + 15.68\,C_3H_8 \\ & + 15.47\,C_4H_{10} + 13.27\,C_2H_2 + 14.78\,C_2H_4 + 6.08\,H_2S - 4.32\,O_2 \end{aligned} \tag{2.154}$$

where CO, $H_2$, and others are the mass fractions of each constituent in the gaseous fuel.

However, if the fuel analysis is provided on volumetric basis, then it is better to calculate air requirements on volumetric bases. And then, we can convert this volumetric quantity to mass quantity if so desired. The theoretical volume of air required $V_a$ for stoichiometric combustion of a unit volume of gaseous fuel is determined as follows:

$$\begin{aligned} V_a = {} & 2.39\,CO + 2.39\,H_2 + 9.57\,CH_4 + 16.75\,C_2H_6 + 23.95\,C_3H_8 \\ & + 31.14\,C_4H_{10} + 11.96\,C_2H_2 + 14.38\,C_2H_4 + 7.18\,H_2S - 4.78\,O_2 \\ & + 30.47\,\text{illuminants} \end{aligned} \tag{2.155}$$

where CO, $H_2$, and others are the volumetric fractions of each constituent in the gaseous fuel. Illuminants, on the other hand, include a number of compounds not separated by common gas analysis. The main illuminants include ethylene ($C_2H_4$) and acetylene ($C_2H_2$) – both included in the above equation, and the dry air required for combustion, per unit volume of each gas, propylene, butylenes, pentene, benzene, toluene, and xylene. Because the illuminant percentages are small, the values can be lumped together and we can use an approximate value of 30 unit volume of dry air per unit volume of gas.

Usually, in engineering applications, only approximate amounts of air requirements are necessary; therefore, if the approximate amounts are enough, or if the fuel specific information is not available, then we can conveniently use the figures provided in Tables 2.23 and 2.24. Another commonly used approximation is that 0.9 ft$^3$ (0.0255 m$^3$) air is required for 100 Btu (105.5 kJ) of fuel.

In addition to the theoretical amount of air required for combustion, *excess air* must be provided to make sure the combustion is complete. Percent excess air can be calculated using the following equation:

$$\text{Excess Air}(\%) = \left(\frac{\textit{Supplied Air} - \textit{Theoretical Air}}{\textit{Theoretical Air}}\right) \times 100 \tag{2.156}$$

**TABLE 2.23**

Approximate Air Requirements for Stoichiometric Combustion of Fuels (ASHRAE, 1989).

| Type of Fuel | Air Required for Stoichiometric Combustion (lb/lb Fuel) | Air Required for Stoichiometric Combustion (ft³/Unit[a] Fuel) | Approximate Precision (%) |
|---|---|---|---|
| Solid | (Btu/lb) × 0.00073 | (Btu/lb) × 0.0097 | 3 |
| Liquid | (Btu/lb) × 0.00071 | (Btu/lb) × 0.0094 | 3 |
| Gas | (Btu/lb) × 0.00067 | (Btu/ft³) × 0.0089 | 5 |

[a] Units for solid and liquid fuels in Btu/lb; for gas in cubic feet.

**TABLE 2.24**

Theoretical Air Requirements and Flue Gas for Combustion of Various Fuels (Engineering ToolBox, 2003).

| Type of Fuel | Theoretical air for combustion[a] (m³/kg) | (ft³/lb) | Theoretical flue gas[a] (m³/kg) | (ft³/lb) |
|---|---|---|---|---|
| *Solid Fuels* | | | | |
| Anthracite | 9.4 | 150 | 9.5 | 152 |
| Semi anthracite | 8.4 | 135 | 8.6 | 137 |
| Bituminous coal | 6.9 | 110 | 7 | 112 |
| Charcoal | 8.4 | 134 | 8.4 | 135 |
| Coke | 8.4 | 134 | 8 | 135 |
| Lignite | 5.7 | 92 | 8.4 | 93 |
| Peat | 5.7 | 92 | 5.9 | 94 |
| Wood (dry) | 4.4 | 70 | 5 | 80 |
| *Liquid Fuels* | ($m^3_{air}/l_{fuel}$) | ($ft^3_{air}/gal_{fuel}$) | ($m^3_{air}/l_{fuel}$) | ($ft^3_{air}/gal_{fuel}$) |
| Gas oil | 9.8 | 1,570 | 10.4 | 1,670 |
| Heavy fuel oil | 10.8 | 1,730 | 11.6 | 1,860 |
| *Gaseous Fuels* | ($m^3_{air}/m^3_{fuel}$) | ($ft^3_{air}/ft^3_{fuel}$) | ($m^3_{gas}/m^3_{fuel}$)b | ($ft^3_{gas}/ft^3_{fuel}$)b |
| Natural gas | 9.5 | 9.5 | 10.5 | 10.5 |
| Propane | 23.8 | 23.8 | 25.8 | 25.8 |
| Butane | 31 | 31 | 33.4 | 33.4 |

[a] At standard pressure;
[b] Combustion of a gaseous fuel generates a flue gas with higher density than the fuel and combustion air.

The unit mass amount of dry air supplied per unit mass of fuel burned can be determined from the following equation:

$$\text{Dry Air Supplied} = \left(\frac{C(3.04\,N_2)}{CO_2 + CO}\right) \tag{2.157}$$

The value obtained from Equation (2.157) and the theoretical air value can be used in Equation (2.156) together to determine the amount of excess air.

The following equation, on the other hand, can be used for calculating excess air on a percentage basis for gaseous fuels from the unit volumes of stoichiometric combustion products and air, and volumetric analysis of the flue gas.

$$\text{Excess Air}(\%) = \left(\frac{U - CO_2}{CO_2}\right) \times 100 \times \left(\frac{P}{A}\right) \tag{2.158}$$

where $U$ is the ultimate carbon dioxide of flue gases resulting from stoichiometric combustion (%), $CO_2$ is the carbon dioxide content of flue gases (%), $P$ is the dry products resulting from stoichiometric combustion (unit volume per unit volume of gas burned), and $A$ is the air required for stoichiometric combustion (volume per unit volume of gas burned). The value of ($P/A$) is reported to be approximately 0.9 for most natural gases, hence, we can substitute 90 for the product of 100 ($P/A$) in the above equation.

Excess air calculations are most of the time made using the results of flue gas analyses, and the amounts of air required for stoichiometric combustion (theoretical air) are not always known, we may sometimes find the following relationship quite convenient for Equation (2.155):

$$\text{Excess Air}(\%) = \left(\frac{100\left[O_2 - \frac{CO}{2}\right]}{0.264 N_2 - \left[O_2 - \frac{CO}{2}\right]}\right) \tag{2.159}$$

where $O_2$, CO, and $N_2$ are the volumetric percentages obtained from the flue gas analysis.

The general practice for the excess air level is to provide 5–50% excess air; however, as discussed above, the exact amount depends on the fuel type. Typical amounts of excess air for various fuels are provided in Table 2.25.

*2.3.5.2.1 Boiler Room Air Supply*

As mentioned earlier, incomplete combustion generate carbon monoxide; and as a result of incomplete combustion, reignition may pose dangers for both personnel safety and property. Therefore, for complete combustion and safety, we provide boiler rooms with proper openings so that enough fresh air is ensured to be supplied. Temporary air intake openings such as doors and windows should definitely be avoided as they may be closed and hence cut off the required air supply. From experience, the ASME states that *1 square inch free area per 2,000 Btu/h* maximum fuel input of combined burners located in the boiler room works, and minimum 2 (two) *air supply openings must be provided and kept clear at all times.*

**TABLE 2.25**

Typical Values of Excess Air Provided for Various Fuels (Engineering ToolBox, 2003).

| Type of Fuel | Excess Air (%) | Type of Fuel | Excess Air (%) |
|---|---|---|---|
| Anthracite | 40 | Oil (No. 2 and No. 6) | 10–20 |
| Coke oven gas | 5–10 | Semi anthracite, hand firing | 70–100 |
| Natural Gas | 5–10 | Semi anthracite, with stoker | 40–70 |
| Coal, pulverized | 15–20 | Semi anthracite, with traveling grate | 30–60 |
| Coal, stoker | 20–30 | | |

### 2.3.5.2.2 Theoretical $CO_2$ Production

The theoretical $CO_2$ (a.k.a. the ultimate $CO_2$) concentration obtained in combustion of a hydrocarbon fuel with air is the amount of $CO_2$ released when the fuel is completely burned with the theoretically required air, that is, zero excess air. Using the following equation, theoretical $CO_2$ amounts released during combustion with excess air present can be calculated from the flue gas analyses as follows:

$$\text{Theoretical}\,CO_2(\%) = U = \frac{\%CO_2\,\textit{in flue gas sample}}{1-\left(\dfrac{\%O_2\,\textit{in flue gas sample}}{20.95}\right)} \tag{2.160}$$

Approximate theoretical $CO_2$ values and $CO_2$ values observed with different amounts of excess air for stoichiometric combustion of various common fuel types are provided in Table 2.26.

**TABLE 2.26**

Approximate Theoretical (Stoichiometric) $CO_2$ Values, and $CO_2$ Values for Various Fuels With Different Percentages of Excess Air (ASHRAE, 1989).

| Type of Fuel | Theoretical $CO_2$ (%) | Percent $CO_2$ at Given Excess Air Values | | |
|---|---|---|---|---|
| | | 20% | 40% | 60% |
| *Solid Fuels* | | | | |
| Anthracite | 20.2 | 16.8 | 14.4 | 12.6 |
| Bituminous coal | 18.2 | 15.1 | 12.9 | 11.3 |
| Coke | 21.0 | 17.5 | 15.0 | 13.0 |
| *Liquid Fuels* | | | | |
| No. 1 and 2 Fuel Oil | 15.0 | 12.3 | 10.5 | 9.1 |
| No. 6 Fuel Oil | 16.5 | 13.6 | 11.6 | 10.1 |
| *Gaseous Fuels* | | | | |
| Natural gas | 12.1 | 9.9 | 8.4 | 7.3 |
| Propane | 13.9 | 11.4 | 9.6 | 8.4 |
| Butane | 14.1 | 11.6 | 9.8 | 8.5 |
| Mixed Gas | 11.2 | 12.5 | 10.5 | 9.1 |

**Example 2.87: Calculating Theoretical $CO_2$ and Excess Air**

A greenhouse operation is burning natural gas for heating, and the flue gas analysis provides 12.0% $CO_2$, 3.0% $O_2$, and 85.0% $N_2$. Determine the theoretical percent $CO_2$ and the percent of excess air.

**Solution**

Using Equation (2.160), we can first determine the theoretical $CO_2$ as:

$$\text{Theoretical}\,CO_2(\%) = U = \frac{\%CO_2\,\textit{in flue gas sample}}{1-\left(\dfrac{\%O_2\,\textit{in flue gas sample}}{20.95}\right)} = \frac{12.0}{1-\left(\dfrac{3.0}{20.95}\right)} = 14.01\%$$

And then, using Equation (2.158), we can calculate the excess air:

$$\text{Excess Air}\,(\%) = \left(\frac{U - CO_2}{CO_2}\right) \times 100 \times \left(\frac{P}{A}\right) = \left(\frac{14.01 - 12.0}{12.0}\right) \times 90 = 15.08\%$$

Theoretical amount of $CO_2$ emission released during combustion can also be calculated from the following relationship. In order to calculate the $CO_2$ emission from a fuel, the carbon content of the fuel must be known and is multiplied with the ratio of molecular weight of $CO_2$ (44) to the molecular weight of C (12), which is 44/12 = 3.67. Then,

$$q_{CO_2} = \left(\frac{c_f}{h_f}\right)\left(\frac{M_{CO_2}}{M_C}\right) \tag{2.161}$$

where $q_{CO2}$ is the specific $CO_2$ emission ($kg_{CO2}$/kWh), $c_f$ is the specific carbon content of fuel ($kg_{carbon}/kg_{fuel}$), $h_f$ is the specific energy content of fuel ($kWh/kg_{fuel}$), $M_C$ is the molecular weight carbon ($kg_{carbon}/kmol_{carbon}$), and $M_{CO2}$ is the molecular weight of carbon dioxide ($kg_{CO2}/kmol_{CO2}$).

Specific carbon contents, energy contents, and carbon dioxide emissions from combustion of some common fuels are presented in Table 2.29.

**Example 2.88: Calculating and Comparing Theoretical $CO_2$ Emissions of Various Fuels**

Using Equation (2.103), calculate the theoretical $CO_2$ emissions from natural gas, bunker oil, bituminous coal, and wood. And compare your findings with the values provided in Table 2.27.

**Solution**

Using Equation (2.160), let's first determine the theoretical $CO_2$ emissions.

$$\text{Natural gas: } q_{CO_2} = \left(\frac{0.75}{15.4}\right)\left(\frac{44}{12}\right) = 0.18 kg_{CO_2}/kWh$$

$$\text{Bunker Oil: } q_{CO_2} = \left(\frac{0.85}{11.6}\right)\left(\frac{44}{12}\right) = 0.27 kg_{CO_2}/kWh$$

$$\text{Bituminous Coal: } q_{CO_2} = \left(\frac{0.65}{8.4}\right)\left(\frac{44}{12}\right) = 0.28 kg_{CO_2}/kWh$$

$$\text{Wood: } q_{CO_2} = \left(\frac{0.50}{4.6}\right)\left(\frac{44}{12}\right) = 0.40 kg_{CO_2}/kWh$$

From the calculations, we can see that wood has the highest $CO_2$ emission compared to the other fuels; however, wood is a biofuel, and considered as a carbon neutral fuel. Therefore, from the environmental point of view, it is still an environmentally friendly fuel. If, on the other hand, we compare the other three fossil fuels with respect to their $CO_2$ emissions, we can easily observe that natural gas is a relatively clean burning fuel. Considering the emissions of other flue gas constituents, natural gas is truly a clean burning fuel; and as a consequence, we use natural gas for $CO_2$ enrichment in commercial greenhouse operations, which will be covered further in depth later in this book.

**TABLE 2.27**

Emission of $CO_2$ from Combustion of Some Common Fuels.

| | Density | Specific Carbon Content | Specific Energy Content | | Specific $CO_2$ Emission (Amount of Fuel Basis) | | | Specific $CO_2$ emission (Amount of Energy Basis) | | |
|---|---|---|---|---|---|---|---|---|---|---|
| **Fuel** | kg/l | $kg_C/kg_{fuel}$ | $kWh/kg_{fuel}$ | $Btu/lb_{fuel}$ | $kg_{CO2}/kg_{fuel}$ | $kg_{CO2}/gal_{fuel}$ | $lb_{CO2}/gal_{fuel}$ | $kg_{CO2}/kWh$ | $kg_{CO2}/GJ$ | $lb_{CO2}/MBtu$ |
| Natural gas | | 0.75 | 15.4 | 23,900 | 2.75 | | | 0.18 | 50 | 115 |
| Propane | 0.510 | 0.82 | 13.8 | 21,300 | 2.99 | 5.78 | 12.7 | 0.22 | 60 | 140 |
| Butane | 0.564 | 0.83 | 13.6 | 21,100 | 3.03 | 6.47 | 14.3 | 0.22 | 62 | 144 |
| LPG | 0.537 | 0.82 | 13.7 | 21,200 | 3.01 | 6.12 | 13.5 | 0.22 | 61 | 142 |
| Gasoline | 0.737 | 0.90 | 12.9 | 19,900 | 3.30 | 9.20 | 20.3 | 0.26 | 71 | 165.3 |
| Kerosene (Jet) | 0.821 | 0.82 | 12.0 | 18,500 | 3.00 | 9.33 | 20.6 | 0.25 | 70 | 162.5 |
| Diesel | 0.846 | 0.86 | 12.7 | 19,605 | 3.15 | 10.1 | 22.3 | 0.25 | 69 | 160.8 |
| Heavy fuel oil (No.6/Bunker C) | 0.980 | 0.85 | 11.6 | 18,000 | 3.11 | 11.6 | 25.5 | 0.27 | 75 | 173.3 |
| Petroleum coke | | 0.89 | 9.4 | 14,500 | 3.26 | 14.7 | 32.4 | 0.35 | 97 | 225.1 |
| Anthracite | | 0.92 | 9.0 | 14,000 | 3.37 | | | 0.37 | 104 | 229.5 |
| Bituminous | | 0.65 | 8.4 | 13,000 | 2.38 | | | 0.28 | 79 | 231.7 |
| Subbituminous | | 0.40 | 6.8 | 10,500 | 1.47 | | | 0.22 | 60 | 233.9 |
| Lignite | | 0.30 | 3.9 | 6,000 | 1.10 | | | 0.28 | 79 | 236.1 |
| Coke | | 0.77 | 7.2 | 11,200 | 2.82 | | | 0.39 | 108 | 251.5 |
| Peat (dry) | | 0.52 | 4.7 | 7,300 | 1.91 | | | 0.40 | 112 | 260.7 |
| Bioethanol | 0.789 | 0.52 | 8.3 | 12,800 | 1.91 | 5.71 | 12.6 | 0.23 | 64 | 149.6 |
| Biomethanol | 0.791 | 0.37 | 5.5 | 8,500 | 1.37 | 4.11 | 9.10 | 0.25 | 70 | 162.2 |
| Biodiesel | 0.880 | 0.78 | 11.3 | 17,400 | 2.85 | 9.48 | 20.9 | 0.25 | 70 | 162.8 |
| Wood | | 0.50 | 4.5 | 7,000 | 1.83 | | | 0.40 | 113 | 263.1 |

### 2.3.5.2.3 Quantity of Flue Gas Produced

We need the mass of dry flue gas produced per mass of fuel for heat loss and efficiency calculations. And this mass consists of the fuel (minus the ash remaining in the furnace), theoretical air required for combustion, and excess air. This mass is calculated from the flue gas analysis for solid fuels as follows:

$$\text{Mass of Flue Gas Produced} = \frac{11CO_2 + 8O_2 + 7(CO + N_2)}{3(CO_2 + CO)} C \tag{2.162}$$

where $CO_2$, $O_2$, $CO$, and $N_2$ are the volumetric percentages from the flue gas analysis, and C is the mass of carbon burned per mass of fuel, corrected for carbon in the ash.

And for gaseous fuels, volume of dry flue gas per unit volume of gaseous fuel can be determined from the following equation:

$$\frac{\textit{Volume of Dry Flue Gas}}{\textit{Volume of Fuel Gas}} = \frac{(\textit{Unit volume of } CO_2 \textit{ produced per unit volume of gas burned} \times 100)}{\textit{Percent } CO_2 \textit{ by analysis}} \tag{2.163}$$

And the excess air quantity can be determined by subtracting the quantity of dry flue gases resulting from stoichiometric combustion from the total volume of flue gas. *Theoretical* flue gas values for combustion of various fuels are provided in Table 2.24.

**Example 2.89: Calculating the Amount of Required Dry Air, Flue Gas and its Constituents**

For the same flue gas analysis provided in Example 2.87, determine the m$^3$ of dry air required for combustion of m$^3$ of gaseous fuel, m$^3$ of each constituent in the flues gas, and the total volume of dry and wet flue gases. Fuel analysis also provided 85% $CH_4$, 10% $C_2H_6$, and 5% $N_2$ by volume.

**Solution**

Using Equation (2.155), we can first determine the volume of dry air required for combustion:

$$V_a = 9.57\,CH_4 + 16.75\,C_2H_6 = (9.57 \times 0.85) + (16.75 \times 0.10) = 9.81\,m^3_{\text{dry air}}/m^3_{\text{natural gas}}$$

Please note that this number is in agreement with the figure (9.5 $m^3_{\text{dry air}}/m^3_{\text{natural gas}}$) provided in Table 2.24.

And from Table 2.22, the constituents per cubic meter of flue gas are:

Carbon dioxide, $CO_2$:

| | |
|---|---|
| From Methane | : 0.85 ($CH_4$)(1.0) = 0.85 m$^3$ |
| From Ethane | : 0.10 ($C_2H_6$)(2.0) = 0.20 m$^3$ |
| Total $CO_2$ | : 1.05 m$^3$ |

Oxygen, $O_2$.

| | |
|---|---|
| $O_2$ in Excess Air | : 0.209 × 0.151 × 9.81 $m^3_{\text{dry air}}/m^3_{\text{natural gas}}$ = 0.31 m$^3$ |

Nitrogen, $N_2$:

| | |
|---|---|
| From Methane | : 0.85 $(CH_4)(9.57 - 2.0) = 6.43\ m^3$ |
| From Ethane | : 0.10 $(C_2H_6)(16.75 - 3.50) = 1.33\ m^3$ |
| $N_2$ in Fuel | : 0.05 $m^3$ |
| $N_2$ in Excess Air | : $0.791 \times 0.151 \times 9.81\ m^3{}_{dry\ air}/m^3{}_{natural\ gas} = 1.17\ m^3$ |
| Total $N_2$ | : 8.98 $m^3$ |

Water Vapor, $H_2O$:

| | |
|---|---|
| From Methane | : 0.85 $(CH_4)(2.0) = 1.70\ m^3$ |
| From Ethane | : 0.10 $(C_2H_6)(3.0) = 0.30\ m^3$ |
| Total $H_2O$ Vapor | : 2.00 $m^3$ |

Total volume of dry gases per cubic meter of natural gas: $1.05 + 0.31 + 8.98 = 10.34\ m^3{}_{dry\ flue\ gas}/m^3{}_{natural\ gas}$

Please note that this is in agreement with the figure ($10.5\ m^3{}_{dry\ flue\ gas}/m^3{}_{natural\ gas}$) provided in Table 2.24. The volume of dry flue gas per unit volume of gaseous fuel can also be computed using Equation (2.163).

Furthermore, total volume of wet gases per cubic meter of natural gas: $10.34 + 2.00 = 12.34\ m^3{}_{dry\ flue\ gas}/m^3{}_{natural\ gas}$.

## Nomenclature

| | |
|---|---|
| $a$ | absorbance; constant 2,897 μm.K (5,216 μm.°R) |
| $a$ to $e$ | constants in Equations (2.114) and (2.115) |
| $A$ | cross-sectional area |
| **Bi** | Biot number |
| $c$ | specific heat; specific carbon content |
| $C$ | heat capacity rate; cost; mass of carbon |
| $C_1$, $C_2$ | constants |
| $C_1$ to $C_{13}$ | constants in Equations (2.104) and (2.105) |
| $d_v$ | absolute humidity of moist air; mass of water per unit volume of mixture |
| $D$ | diameter |
| $E$ | amount of energy |
| $E^*$ | total radiation emission |
| $E_\lambda$ | real body (nonblack) radiation |
| $E_\lambda^*$ | blackbody monochromatic radiation |
| $Ex$ | amount of exergy |
| $F_A$ | geometric (or view) factor |
| $F_E$ | emissivity factor |
| **Fo** | Fourier Number |
| **GCV** | gross calorific value |
| **Gr** | Grashof Number |
| **Gz** | Graetz Number |
| $h$ | enthalpy; specific energy content |
| $h'_{fg}$ | modified latent heat of vaporization |
| $h_s@T_w$ | specific enthalpy of saturated moist air at thermodynamic wet-bulb temperature |

| | |
|---|---|
| $h_w@T_w$ | specific enthalpy of condensed water (liquid or solid) at thermodynamic wet-bulb temperature and pressure of 101.325 kPa |
| $H$ | rate of heat gain for space |
| **HCV** | higher calorific value |
| **HHV** | higher heating value |
| $k$ | thermal conductivity |
| $L$ | length |
| **LCV** | lower calorific value |
| **LHV** | lower heating value |
| **LPG** | liquid petroleum gas |
| $m$ | mass |
| $\dot{m}$ | mass flow rate |
| $M$ | mass; molecular weight |
| $n$ | number of moles in moist air sample |
| $N_L$ | number of pipe rows |
| **NCV** | net calorific value |
| **NHV** | net heating value |
| **NTU** | number of heat transfer units |
| **Nu** | Nusselt number |
| $p$ | pressure; wetted perimeter |
| $P$ | unit price |
| **Pe** | Peclet Number |
| **Pr** | Prandtl Number |
| $q$ | specific emission |
| $Q$ | amount of heat; volumetric flow rate |
| $\dot{Q}$ | rate of heat |
| $r$ | radius; reflectance |
| **R** | universal gas constant, 8,314.41 $J/(kg_{mole}{\cdot}K)$; gas constant for dry air or water vapor |
| **Re** | Reynolds number |
| $S$ | entropy |
| $S_L$ | distance between front and back row pipe centers |
| **St** | Stanton Number |
| $t$ | time, transmittance |
| $T$ | temperature; dry-bulb temperature of moist air |
| $T_d$ | dew-point temperature of moist air |
| $T_w$ | thermodynamic wet-bulb temperature of moist air |
| **THV** | total heating value |
| $u$ | average velocity |
| $U$ | internal energy; overall heat transfer coefficient; ultimate (theoretical) carbon dioxide |
| $v$ | specific volume |
| $V$ | heat exchanger volume, theoretical volume; total volume of moist air sample |
| $W$ | humidity ratio of moist air; mass of water per unit mass of dry air |
| $\dot{W}$ | rate of work |
| $W_a$ | theoretical mass of dry air |
| $W_s$ | humidity ratio of moist air at saturation at thermodynamic wet-bulb temperature |

| | |
|---|---|
| $x$ | mole fraction |
| $Z$ | capacity rate ratio |

**Greek letters**

| | |
|---|---|
| $\alpha$ | thermal diffusivity; transfer surface area per heat exchanger volume; ln ($p_w$), parameter used in Equations (2.112), (2.113), (2.114), and (2.115) |
| $\Delta$ | difference |
| $\varepsilon$ | effectiveness |
| $\phi$ | relative humidity, dimensionless |
| $\gamma$ | specific humidity of moist air, mass of water per unit mass of mixture |
| $\sigma$ | Stefan–Boltzmann constant $5.679 \times 10^{-8}$ J/s.m$^2$.K$^4$ ($0.1714 \times 10^{-8}$ Btu/h.ft$^2$.R$^4$) |
| $\lambda$ | wavelength |
| $\mu$ | dynamic viscosity; degree of saturation $W/W_s$ |
| $\nu$ | kinematic viscosity |
| $\eta$ | thermal efficiency |
| $\theta$ | temperature difference |
| $\rho$ | density |

**Subscripts**

| | |
|---|---|
| $a$ | air; dry air |
| $c$ | cold; convective |
| cv | control volume |
| C | carbon |
| $CO_2$ | carbon dioxide |
| $d$ | dew point |
| $E$ | electricity |
| $f$ | film; fuel; saturated liquid water |
| $g$ | water vapor |
| $h$ | hot; hydraulic |
| $i$ | in |
| $L$ | latent |
| LPG | liquid propane gas |
| $m$ | mean |
| max | maximum |
| NG | natural gas |
| $o$ | out |
| $p$ | constant pressure; pipe |
| $s$ | surface; saturated moist air |
| S | sensible |
| sat | saturation |
| $t$ | total |
| $T$ | total |
| $v$ | volume |
| $w$ | wall; window; water in any phase |
| $ws$ | saturated water vapor |
| 1, 2, 3 | numbers; state points |
| $\infty$ | ambient |

## Bibliography


Afgan, NH and Schlunder, EU (1974) *Heat Exchangers: Design and Theory Sourcebook*. McGraw-Hill Book Co., Inc., New York.

ASHRAE (1989) *Handbooke of Fundamentals*. American Society of Heating, Refrigeration and Air Conditioning Engineers, Inc. Atlanta, GA. pp. 15.1–15.14

ASHRAE (2013) *Handbook of Fundamentals*. American Society of Heating, Refrigeration and Air Conditioning Engineers, Inc. Atlanta, GA. pp. 6.1–6.17.

Brown, AI and Marco, SM (1958) *Introduction to Heat Transfer*, 3rd Ed., McFraw-Hill Book Co., Inc., New York.

Çengel, YA and Boles, MA (2002) *Thermodynamics: An Engineering Approach*. McGraw-Hill, Boston.

Çengel, YA and Ghajar, AJ (2015) *Heat and Mass Transfer: Fundamentals and Applications*. McGraw-Hill Education, New York.

Chaddock, JB (1965) Moist Air Properties from Tabulated Virial Coefficients. *Humidity and moisture measurement and control in science and industry* 3:273.

Demirel, Y (2018) Biofuels. In: Dincer, I. (ed.), *Comprehensive Energy Systems*, Vol. 1, Elsevier, Oxford, UK, pp. 875– 908.

Duffie, JA and Beckman, WA (2013) *Solar Engineering of Thermal Processes*. Wiley, New York.

Engineering ToolBox, (2003) Available at: https://www.engineeringtoolbox.com (Accessed 23.03.2020).

Goff, JA (1949) Standardization of Thermodynamic Properties of Moist Air. *Heating, Piping, and Air Conditioning* 21(11):118.

Goff, J.A., J.R. Anderson, and S. Gratch. 1943. Final Values of The Interaction Constant for Moist Air. *ASHVE Transactions* 49:269.

Goff, JA, Gratch, S (1945) Thermodynamic Properties of Moist Air. *ASHVE Transactions* 51:125.

Granet, I and Bluestein, M (2004) *Thermodynamics and Heat Power*. Pearson Prentice Hall, New Jersey.

Haines, RW (1961) How to Construct High Altitude Psychrometric Charts. *Heating, Piping, and Air Conditioning* 33(10):144.

Hardy, H.C., D. Telfair, and W.H. Pielemeier. 1942. The velocity of sound in air. *Journal of the Acoustical Society of America* 13:226.

Harrison, LP (1965) Fundamental Concepts and Definitions Relating To Humidity. *Humidity and moisture measurement and control in science and industry* 3:3. A. Wexler and W.A. Wildhack, eds. Reinhold Publishing, New York.

Hyland, RW and Wexler, A (1983a) Formulations for the thermodynamic properties of dry air from 173.15 K to 473.15 K, and of saturated moist air from 173.15 K to 372.15 K, at pressures to 5 MPa. *ASHRAE Transactions* 89(2A):520–535.

Hyland, RW and Wexler, A (1983b) Formulations for the thermodynamic properties of the saturated phases of H2O from 173.15 K to 473.15 K. *ASHRAE Transactions* 89(2A):500–519.

Incropera, FP and DeWitt, DP (1996) *Fundamentals of Heat Transfer*. John Wiley and Sons, Inc., New York.

Karig, HE (1946) Psychrometric Charts for High Altitude Calculations. *Refrigerating Engineering* 52(11):433.

Kays, WM and London, AL (1964) *Compact Heat Exchangers*. 2nd ed. McGraw-Hill Book Co., Inc., New York.

Keenan, JH and Kaye, J (1945) *Gas Tables*. John Wiley and Sons, New York.

Keenan, JH, Keyes, FG, Hill, PG and Moore, JG (1969) *Steam Tables*. John Wiley and Sons, New York.

Khademi, F and Yildiz, I (2018) Energy and Solid Wastes. In: Dincer, I. (ed.), *Comprehensive Energy Systems*, Vol. 1, Elsevier, Oxford, UK, pp. 980–1020.

Kuehn, TH, Ramsey, JW, and Threlkeld, JL (1998) *Thermal Environmental Engineering*, 3rd ed., p. 188. Prentice-Hall, Upper Saddle River, NJ.

Kusuda, T (1970) *Algorithms for psychrometric calculations. NBS Publication BSS21 (January) for sale by Superintendent of Documents*, U.S. Government Printing Office, Washington, D.C.

Mason, EA and Monchick, L (1965) Survey of the equation of state and transport properties of moist gases. *Humidity and moisture measurement and control in science and industry* 3:257. Reinhold Publishing, New York.
Midwest Plan Service (1987) *Structures and Environment Handbook*. Eleventh Edition. Ames, Iowa.
NASA (1976) U.S. Standard atmosphere, 1976. National Oceanic and Atmospheric Administration, National Aeronautics and Space Administration, and the United States Air Force. Available from National Geophysical Data Center, Boulder, CO.
NIST (1990) Guidelines for realizing the international temperature scale of 1990 (ITS-90). NIST *Technical Note* 1265. National Institute of Technology and Standards, Gaithersburg, MD.
Olivieri, J (1996) *Psychrometrics – Theory and Practice*. ASHRAE, Atlanta.
Palmatier, EP (1963) Construction of the normal temperature. ASHRAE Psychrometric Chart. *ASHRAE Journal* 5:55.
Peppers, VW (1988) Unpublished paper. Available from ASHRAE. Preston-Thomas, H. 1990. The international temperature scale of 1990 (ITS90). *Metrologia* 27(1):3-10.
Rohsenow, WM (1946) Psychrometric determination of absolute humidity at elevated pressures. *Refrigerating Engineering* 51(5):423.
Smithsonian Institution (1954) Smithsonian Physical Tables. 9th rev. ed. Available from the Smithsonian Institution, Washington, D.C.
Yildiz, I. (2018) Fossil Fuels. In: Dincer, I. (ed.), *Comprehensive Energy Systems*, Vol. 1, Elsevier, Oxford, UK, pp. 521– 567.

## 2.4 Essential Fluid Mechanics for Energy Conversion

### 2.4.1 Introduction

Fluid mechanics, the subcategory of mechanics, can overall be defined as the science that deals with the behavior of fluids at rest or flowing (or deforming) bodies. The piping systems for water, natural gas, oil, etc., are all primarily designed using the principles of fluid mechanics. Flowing fluids transfer mass or heat, and hence fluid flow and its mechanics are important for heating, ventilating, air-conditioning, and refrigeration systems engineers and designers as well. This section covers fluid mechanics fundamentals needed to handle general HVAC processes.

### 2.4.2 Types of Fluids and Fluid Properties

A substance can primarily exist in three phases: solid, liquid, and gas. It can also exist in plasma phase at very high temperatures. A substance in the liquid or gaseous phase however is referred to as a fluid. Liquids and gases behave differently with respect to the nature of molecular actions, particularly the degree of compressibility and liquid formation of an interface. When reaction of fluids to shearing is considered, fluids also separate themselves from solids. If a fluid is placed under a shear stress condition, it keeps deforming as long as the shear is present whereas a solid can resist an applied shear stress deforming just a finite amount. In fluid mechanics, a force per unit area, is called a stress; and the normal component of force acting on a surface (per unit area) is defined as normal stress (called pressure in a fluid at rest) while the tangential component per unit area is defined as shear stress (remember a fluid at rest would have zero shear stress). The normal stress can be either tensile or compressive, depending on whether it stretches or compresses the fluid; and it equals $F/A$, where $F$ is the normal force and $A$ is the area of the surface on which it

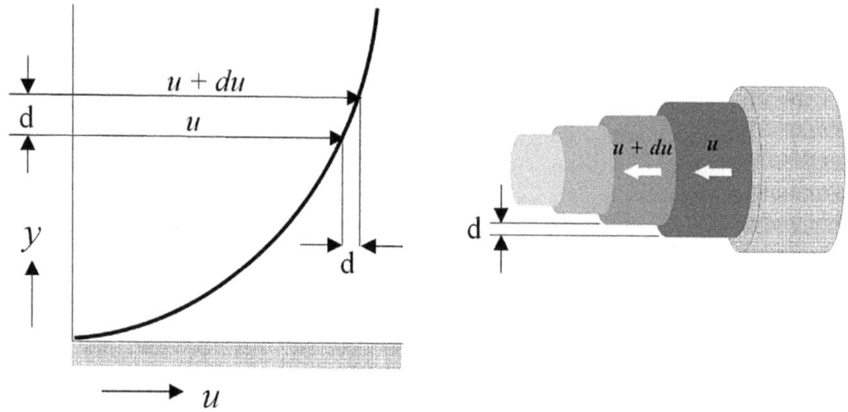


**FIGURE 2.23**
Velocity variation (non-linear profile) near a solid boundary.

acts. On the other hand, the shear stress $\tau$ equals *F/A*, where *F* is the tangential force and *A* is the area on which it acts. As the shear stress is directly proportional to the viscosity of the fluid, the importance of shear stress becomes quite obvious when the flow of viscous Newtonian fluids is considered. When fluid layers move relative to each other, a distance *dy* apart, move one over the other at different velocities, say *u* and *u+du* as shown in Figure 2.23, the viscosity together with relative velocity causes a shear stress acting between the fluid layers, and as a consequence, the slower layer offers resistance to and slows down the faster moving layer. The internal flow resistance is quantified by the fluid property called *viscosity*. All fluid flows have some degree of viscosity; therefore, there is no fluid flow called zero viscosity flow. The generated shear stress $\tau$ (that is, *F/A*) is proportional to the rate of change of velocity with respect to *y* (the Newton's law of viscosity), $\tau \propto (du/dy)$ or $\tau = \mu\ (du/dy)$, where $\mu$ is called the proportionality constant and known as the coefficient of *dynamic viscosity* or just *viscosity*, and $du/dy$ represents the rate of shear strain or rate of shear deformation or velocity gradient. Then, we have $\mu = \tau/(du/dy)$. Dynamic viscosity changes with the change of temperature and pressure. As temperature rises, viscosity of liquids decreases whereas viscosity of gases increases. In SI, the common unit of dynamic viscosity is [Pa s] – Pascal second, which may also be expressed as the [N s/m$^2$] or as the [kg/m s]. In fluid dynamics, viscosity $\mu$ quite often appears in ratio to fluid density $\rho$ as $\nu = \mu/\rho$, termed as *kinematic viscosity*, for which the unit in SI is [m$^2$/s]. And the fluid density $\rho$, or the mass per unit volume is another important fluid property, which is involved in most fluid models.

Fluid motion can be analyzed based on one of several models (Figure 2.24), the simplest being the ideal fluid model, which assumes that the fluid is incompressible and have no viscosity (resistance to shearing). In reality, no ideal fluid exists; rather, it is an imaginary fluid. While the ideal fluid model can be used and valid for quite a wide range of applications, it is frequently needed to include the effects of shearing (viscous action). A fluid, which possesses viscosity, is known as real fluid. All fluids in actual practice are indeed real fluids and possess some viscosity. If a real fluid obeys the Newton's law of viscosity, that is, the shear stress is directly proportional to the shear strain, the fluid is called as the Newtonian fluid. In HVAC applications, most fluids can be treated as Newtonian fluids, for which the deformation is directly proportional to the shearing stress. Examples would include water, air, and emulsions. Fluids that do not obey the Newton's law of viscosity are classified as non-Newtonian fluids. Moreover, if a fluid has a shear stress that is more than

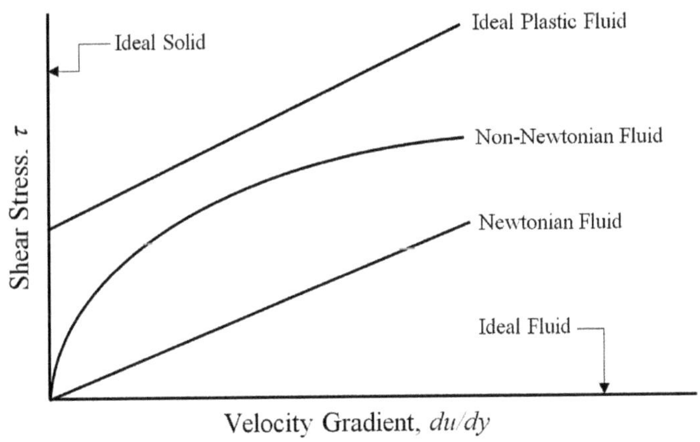


**FIGURE 2.24**
Velocity variation (non-linear profile) near a solid boundary.

the yield value, and shear stress is proportional to the velocity gradient (shear strain), then this fluid is called ideal plastic fluid.

### 2.4.3 Mass Continuity

Let's now consider a homogenous, constant property, and incompressible fluid flow (just like water flow almost), and apply conservation of matter to fluid flow in a conduit (Figure 2.25).

The flow equation is given by $Q = A\,u$ where $Q$ is the volumetric flow rate, $A$ is the cross-sectional area of the conduit, and $u$ is the average velocity of the fluid. If we have a conduit having varying cross sections as given in Figure 2.25, then the continuity relation states that except when branches occur, the flow rate $Q$ is the same at all sections along the conduit as no fluid can be gained or lost. Therefore,

$$Q = A_1u_1 = A_2u_2 = \text{constant} \tag{2.164}$$

This is called the continuity equation and it is true for any number of changes along the conduit.

Velocity of fluid in a conduit is not uniform across the cross-sectional area; thence, an average velocity that is calculated by the continuity equation for a steady flow is used.

$$u = \frac{Q}{A} = \frac{4Q}{\pi D^2} \tag{2.165}$$

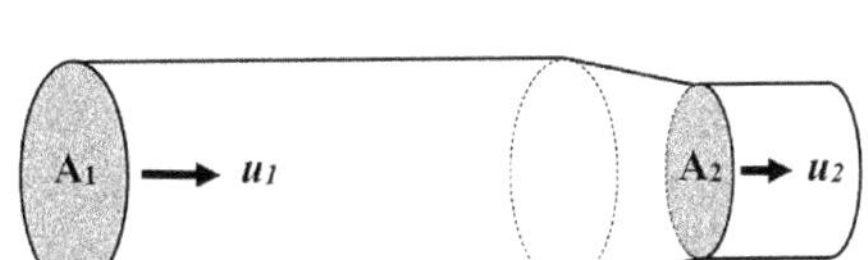


**FIGURE 2.25**
Fluid flow in a conduit having varying cross-sectional areas.

And the conduit (pipe) diameter can be determined easily when volumetric flow rate and velocity are known:

$$D = \sqrt{\left(\frac{4Q}{\pi u}\right)} \tag{2.166}$$

where $D$ is the internal pipe diameter, $Q$ is the volumetric flow rate, and $u$ is the velocity, and $A$ is the pipe cross-sectional area.

If mass flow rate is known instead, then the diameter can be calculated using the following equation:

$$D = \sqrt{\left(\frac{4\dot{m}}{\pi \rho u}\right)} \tag{2.167}$$

where $D$ is the internal pipe diameter, $\dot{m}$ is the mass flow rate, $\rho$ is the fluid density, and $u$ is the velocity, and $A$ is the pipe cross-sectional area.

### 2.4.4 Laminar and Turbulent Fluid Flows

In a conduit (pipe) flow, streamlines are in straight parallel lines if the fluid velocity is small; and the streamlines will keep being straight and parallel with the pipe wall up to a velocity – known as *critical velocity* – where the streamlines break into diffused patterns and start wavering, and after which the fluid flow involves random perturbations without any order or periodicity; the velocity of such flows varies temporally and spatially. When the velocity is lower than the critical velocity, the flow is called laminar flow, for which the flow velocity is highest on the pipe axis, and equal to zero on the pipe wall. The regime of flow is turbulent if the flow velocity is greater than the critical velocity. In turbulent flow, there is always a thin layer of fluid moving in laminar flow at the pipe wall, and this layer is known as the boundary layer.

The Reynolds number Re, a dimensionless quantity, giving the relative ratio of inertia to viscous forces, is used to determine flow regime. For flows through round pipes and tubes, the following form of Reynolds number is used:

$$\text{Re} = \frac{uD}{v} = \frac{\rho u D}{\mu} \tag{2.168}$$

where $u$ is the fluid velocity, $D$ is the internal pipe diameter, $\rho$ is the fluid density, $\nu$ is the kinematic viscosity, and $\mu$ is the dynamic viscosity. Laminar flow is generally expected if the Reynolds number is less than 2,000; for 2,000 < Re < 4,000, the flow is said to be in transition state and predictions cannot be reliable; and fully developed turbulent flow exists when Re > 4,000.

When Reynolds number is needed to be calculated for non-circular cross sections such as square, rectangular, or oval conduits, equivalent diameter approach ($D_{eq} = 4\ A/P_w$) is used, where $A$ is the flow area and $P_w$ is the wetted perimeter of the cross section. This approach is for instance what is used for air ducts, which are often rectangular in cross section.

**Example 2.90: Reynolds Number Calculation**

Water having a temperature of 30°C is flowing through a 30 m 2″ nominal steel pipe (Schedule 40) with a mass flow rate of 2 kg/s. Determine the Reynolds number for this fluid flow.

**Solution**

Internal diameter of 2″ nominal standard steel pipe (Schedule 40) (Table 2.28) = 0.0525 m

Water density $\rho$ at 30°C (Table A.17) = 995.7 kg/m$^3$

Dynamic viscosity $\mu$ at 30°C (Table A.17) = 798 × 10$^{-6}$ kg/m s = 798 × 10$^{-6}$ Pa s

From the continuity equation, $u(m/s) = \dfrac{Q(\text{m}^3/\text{s})}{A(\text{m}^2)}$ and $Q\left(\dfrac{\text{m}^3}{\text{s}}\right) = \dfrac{\dot{m}(\text{kg/s})}{\rho(\text{kg/m}^3)}$ So,

$$Q(\text{m}^3/\text{s}) = \frac{2(\text{kg/s})}{995.7(\text{kg/m}^3)} = 0.002\,\text{m}^3/\text{s}$$

Then,

$$u = \frac{4\left(0.002\dfrac{\text{m}^3}{\text{s}}\right)}{\pi(0.0525\,\text{m})^2} = 0.93\,\text{m/s}$$

And then,

$$\text{Re} = \frac{\rho u D}{\mu} = \frac{995.7\dfrac{\text{kg}}{\text{m}^3}0.93\dfrac{\text{m}}{\text{s}}0.0525\,\text{m}}{798\times10^{-6}\dfrac{\text{kg}}{\text{m s}}} = 60{,}813\,(\text{the flow is turbulent})$$

## 2.4.5 Bernoulli's Equation: Energy Conservation in an Ideal Fluid

Bernoulli's principle states that an increase in the speed of a fluid occurs simultaneously with a decrease in static pressure or a decrease in the fluid's potential energy. Although Daniel Bernoulli deduced that pressure decreases when the flow speed increases, it was Leonhard Euler who derived Bernoulli's equation in its usual form in 1752 (Darrigol and Frisch, 2008; Anderson, 2016). The principle can be applied to various types of fluid flow, and as a consequence, there are different forms of the equation for different types of flow. The simple form of Bernoulli's equation is used for incompressible flows (i.e. most liquid flows and gas flows moving at low Mach number).

Fluid energy is composed of kinetic, potential (due to elevation), and internal energies. Therefore, Bernoulli's equation can be derived from the principle of conservation of energy, which states that, in a steady flow, ignoring viscosity, compressibility, and thermal effects, the sum of all forms of energy in a fluid along a streamline is the same at all points on that streamline. This requires that the sum of kinetic, potential, and internal energy remains constant (Batchelor, 2000). Bernoulli's principle can also be derived directly from Newton's second law of motion. If a volume of fluid is flowing horizontally from a region of high

**TABLE 2.28**

Standard (Schedule 40) Steel Pipe Data (Frankland, 1984).

| Nominal Pipe Diameter (in) | Actual Outside in | Diameter m | Actual Inside in | Diameter m | Wall in | Thickness m | Inside $in^2$ | Area $m^2$ |
|---|---|---|---|---|---|---|---|---|
| ½ | 0.840 | 0.02134 | 0.622 | 0.01579 | 0.109 | 0.00278 | 0.304 | 0.000196 |
| ¾ | 1.050 | 0.02667 | 0.824 | 0.02093 | 0.113 | 0.00287 | 0.533 | 0.000344 |
| 1 | 1.315 | 0.03340 | 1.049 | 0.02644 | 0.133 | 0.00348 | 0.864 | 0.000549 |
| 1 ¼ | 1.660 | 0.04216 | 1.380 | 0.03505 | 0.140 | 0.00356 | 1.495 | 0.000964 |
| 1 ½ | 1.900 | 0.04826 | 1.610 | 0.04089 | 0.145 | 0.00369 | 2.035 | 0.001313 |
| 2 | 2.375 | 0.06033 | 2.067 | 0.05250 | 0.154 | 0.00392 | 3.354 | 0.002164 |
| 2 ½ | 2.875 | 0.07303 | 2.469 | 0.06271 | 0.203 | 0.00516 | 4.785 | 0.003087 |
| 3 | 3.500 | 0.08890 | 3.068 | 0.07793 | 0.216 | 0.00549 | 7.389 | 0.004767 |
| 3 ½ | 4.000 | 0.10160 | 3.548 | 0.09012 | 0.226 | 0.00574 | 9.882 | 0.006375 |
| 4 | 4.500 | 0.11430 | 4.026 | 0.10226 | 0.237 | 0.00602 | 12.724 | 0.008209 |
| 5 | 5.563 | 0.14130 | 5.047 | 0.12819 | 0.258 | 0.00656 | 19.996 | 0.012900 |
| 6 | 6.625 | 0.16828 | 6.065 | 0.15405 | 0.280 | 0.00712 | 28.876 | 0.018629 |
| 8 | 8.625 | 0.21908 | 7.981 | 0.20272 | 0.322 | 0.00818 | 50.002 | 0.032260 |
| 10 | 10.750 | 0.27305 | 10.020 | 0.25451 | 0.365 | 0.00927 | 78.814 | 0.050849 |
| 12 | 12.750 | 0.32385 | 12.000 | 0.30480 | 0.375 | 0.00953 | 113.040 | 0.072929 |
| 14 | 14.000 | 0.35560 | 13.250 | 0.33655 | 0.375 | 0.00953 | 137.817 | 0.088914 |
| 16 | 16.000 | 0.40640 | 15.250 | 0.38735 | 0.375 | 0.00952 | 182.562 | 0.117781 |
| 18 | 18.000 | 0.45720 | 17.250 | 0.43815 | 0.375 | 0.00953 | 233.587 | 0.150701 |
| 20 | 20.000 | 0.50800 | 19.250 | 0.48895 | 0.375 | 0.00953 | 290.892 | 0.187672 |
| 24 | 24.000 | 0.60960 | 23.250 | 0.59055 | 0.375 | 0.00953 | 424.342 | 0.273768 |
| 26 | 26.000 | 0.66040 | 25.250 | 0.64135 | 0.375 | 0.00953 | 500.487 | 0.322894 |
| 30 | 30.000 | 0.76200 | 29.250 | 0.74295 | 0.375 | 0.00953 | 671.617 | 0.433300 |
| 36 | 36.000 | 0.91440 | 35.250 | 0.89535 | 0.375 | 0.00953 | 975.412 | 0.629297 |
| 42 | 42.000 | 1.0668 | 41.250 | 1.04775 | 0.375 | 0.00953 | 1,335.727 | 0.861757 |

pressure to a region of low pressure, then there must be more pressure behind than that in front. This, as a result, gives a net force on the mentioned volume, moving it along the streamline.

Bernoulli performed his experiments on liquids, so as mentioned earlier his original equation is valid only for incompressible flow. A common form of Bernoulli's equation, valid at any arbitrary point along a streamline, can then be presented as:

$$\frac{u^2}{2}+gz+\frac{p}{\rho}=\text{constant} \tag{2.169}$$

where $u$ is the fluid flow speed at a point on a streamline, $g$ is the acceleration due to gravity, $z$ is the elevation (in the direction opposite to the gravitational acceleration) of the point above a reference plane, $p$ is the pressure at the chosen point, and $\rho$ is the density of the fluid at all points in the fluid. The constant on the right hand side of the equation depends only on the streamline chosen, whereas $v$, $z$, and $p$ depend on the particular point on that streamline. For this form of Bernoulli equation, the flow must be steady (i.e. the flow parameters velocity, density, etc., at any point cannot change with time), the flow must be incompressible (even though pressure varies), the density must remain constant along a streamline, and friction by viscous forces must be negligible.

Dividing Equation (2.169) by the gravitational acceleration $g$ produces the *head equation* derived from Bernoulli's principle, noting that each term can be described in the length dimension such as meters or feet:

$$\begin{aligned}&\frac{u^2}{2g}+z+\frac{p}{\rho g}=\text{constant};\\&\text{For two points}: \frac{u_1^{\,2}}{2g}+z_1+\frac{p_1}{\rho_1 g}=\frac{u_2^{\,2}}{2g}+z_2+\frac{p_2}{\rho_2 g}\end{aligned} \tag{2.170}$$

where the far left term $\frac{u^2}{2g}$ is called the *velocity head*, which represents the internal energy of the fluid due to its motion, the middle term $z$ is called the *elevation head*, which represents the potential energy of the fluid due to its elevation (with respect to a reference plain), and the far right term $\frac{p}{\rho g}$ is called the *pressure head*, which represents the internal energy of the fluid due to the exerted pressure on the container (pipe wall). So, as mentioned earlier, if friction losses are neglected and no energy is added to or taken from a piping system, the sum of these three heads give the total head.

By multiplying Equation (2.169) with the fluid density $\rho$, gives us an equation with three pressure terms:

$$\frac{\rho u^2}{2}+\rho gz+p=\text{constant} \tag{2.171}$$

It must be noted that, in this form of Bernoulli's equation, the total pressure is constant. If the middle term (the pressure due to elevation) is constant, and if the far right term (the static pressure) increases, then the far left term (the dynamic pressure) must have decreased. That is, an increase in the static pressure due to flow resistance results in a decrease in the flow speed if it is not due to elevation difference.

All three versions of Bernoulli equation given above are just simplified versions of energy balance on a system. And from Bernoulli equation, many other equations are derived for different energy losses and gains, which happen in the real world systems. In the real world, energy is added to or extracted from the fluid (e.g. via using pumps and turbines), and when the viscosity of the fluid is taken into account energy is lost due to friction due to the fluid rubbing against itself and the pipe and fitting walls, and by the flow turbulence; therefore, the total energy head presented in Equation (2.170) is no longer constant along the pipe. This lost energy is the energy converted to the sound and heat generated by fluids as they flow through the pipes. The amount of frictional loss is affected by the length, roughness, diameter, and changes in the shape or section of the pipe as well as the velocity and type of flow of the fluid. Fittings, valves, bends, etc., all increase frictional losses. We can then take Equation (2.170), the head equation for two points of one streamline in a fluid flow, modify it accordingly, and restore the equality taking all the above-mentioned terms into account as follows:

$$\frac{u_1^2}{2g} + z_1 + \frac{p_1}{\rho_1 g} + h_p = \frac{u_2^2}{2g} + z_2 + \frac{p_2}{\rho_2 g} + h_{f,major} + h_{f,minor} + h_t \tag{2.172}$$

where $u$ is the fluid flow speed at the chosen point on a streamline, $g$ is the acceleration due to gravity, $z$ is the elevation of the chosen point above reference plane, $p$ is the pressure at the chosen point, and $\rho$ is the density of the fluid at the chosen point of the fluid, $h_p$ is the pump head, $h_{f,\,major}$ is the head loss due to friction in the pipe, $h_{f,\,minor}$ is the frictional head losses due to fittings, valves, bends, etc., and $h_t$ is the turbine head. Head losses due to friction and fittings and turbine head are included as terms in Equation (2.172) in the right hand side because they all represent losses or extractions (in the presence of a turbine) of energy in the system, whereas the pump head is included in the left hand side because the pump adds mechanical energy to the system. When there is a flow in a pipe, it always creates an energy loss due to friction. Energy loss can be measured using two gauges just like static pressure in the direction of fluid flow. In practice, frictional head losses $h_{f,\,major}$ and $h_{f,\,minor}$ are calculated using tabulated values (typically provided by manufacturers) that require pipe type, diameter, length and flow rate, and type of fitting.

### 2.4.5.1 Pressure Loss due to Friction

General equation for pressure loss due to friction, also known as the Darcy–Weisbach equation or relation expressed in Pascals can also be used as follows:

$$\frac{\Delta P}{L} = f\frac{\rho}{2}\frac{u^2}{D} \text{ or } \Delta P = \rho f \frac{L}{D}\frac{u^2}{2} \tag{2.173}$$

Or head loss expressed in meters of fluid as follows:

$$\frac{\Delta h_{f,major}}{L} = f\frac{1}{2g}\frac{u^2}{D} \text{ or } h_{f,major} = f\frac{L}{D}\frac{u^2}{2g} \tag{2.174}$$

where $\Delta P$ is the pressure loss due to friction (Pa), $\Delta h_{f,\,major}$ or $h_{f,\,major}$ is the head loss due to friction (m), $f$ is the friction factor (unitless), $L$ is the pipe length (m), $D$ is the hydraulic diameter of the pipe (m), $g$ is the acceleration due to gravity (9.81 m/s$^2$), and $u$ is the mean flow velocity (m/s). Frictional head loss (Equation (2.174)) can be converted into frictional loss as a pressure (Equation (2.173)) by the relationship $\Delta P = h_{f,\,major}\,\rho\,g$.

Darcy–Weisbach equation can be used for both laminar and turbulent flows as well as for any liquid in a pipe. It can also be used for gases and vapors. The friction factor $f$ is some function of the Reynolds number and the roughness of the pipe and is almost constant for turbulent flow, varying only from about 0.01–0.05.

For fully developed laminar flow (also known as Hagen–Poiseuille flow) in a pipe, Re < 2,000, the friction factor is independent of relative roughness and depends only on Reynolds number; and in this case, we actually have an exact analytical solution as follows:

$$f = \frac{64}{\text{Re}} \tag{2.175}$$

where $f$ is the friction factor (dimensionless), and Re is the Reynolds number (dimensionless). Thence with laminar flow, the friction factor varies inversely with Reynolds number.

If Re > 4,000, then the flow is a turbulent flow, in this case the friction factor $f$ is function of the relative pipe roughness and the Reynolds number, and can be determined by using Moody's chart (Figure 2.26), one of the most famous and important charts or Colebrook equation or its approximations as shown below. In Moody's diagram, friction factor can be determined based on the values of Reynolds number and relative pipe roughness, which is equal to $\varepsilon/D$, where $\varepsilon$ is the surface roughness factor (also called absolute roughness, the height or measure of the roughness element) and $D$ is the internal diameter of the pipe, respectively. Surface roughness factors have been measured and tabulated for different materials used in piping systems (Table 2.29).

For very large Reynolds numbers, friction factor $f$ is independent of the Reynolds number. A very thin laminar sublayer exists for this completely turbulent flow along the pipe wall where the surface roughness completely dominates the character of the flow.

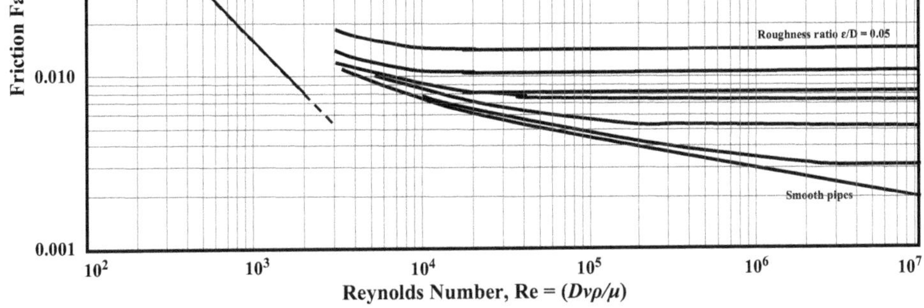


**FIGURE 2.26**
Friction factor as a function of Reynolds number and relative roughness of pipes.

**TABLE 2.29**
Surface Roughness Factor, $\varepsilon$, for Common Pipe Materials.

| Pipe Material | $\varepsilon$ (mm) | $\varepsilon$ (ft) |
|---|---|---|
| Glass, drawn brass, copper (new) | 0.0015 | 0.000005 |
| Seamless commercial steel (new) | 0.004 | 0.000013 |
| Commercial steel (enamel-coated) | 0.0048 | 0.000016 |
| Commercial steel (new) | 0.045 | 0.00015 |
| Wrought iron (new) | 0.045 | 0.00015 |
| Asphalted cast iron (new) | 0.12 | 0.0004 |
| Galvanized iron | 0.15 | 0.0005 |
| Cast iron (new) | 0.26 | 0.00085 |
| Wood stave (new) | 0.18–0.9 | 0.0006–0.003 |
| Concrete (steel forms, smooth) | 0.18 | 0.0006 |
| Concrete (good joints, average) | 0.36 | 0.0012 |
| Concrete (rough, visible, form marks) | 0.60 | 0.002 |
| Riveted steel (new) | 0.9–9.0 | 0.003–0.03 |
| Corrugated metal | 45 | 0.15 |

It should be remembered that there is a head loss in any pipe no matter how smooth the pipe surface is; and even for smooth pipes, the friction factor is not zero. This is because of the existing no-slip boundary condition that requires fluid to stick to the solid surface over which it flows. There is always some surface roughness that produces the no-slip behavior (therefore $f$ is not zero) on the molecular level. Such pipes are called *hydraulically smooth* pipes. For a hydraulically smooth pipe, the friction factor $f$ is approximated by Blasius (1911) equation:

$$f = (100\,\mathrm{Re})^{-1/4} \quad \text{for Re up to } 10^5 \tag{2.176}$$

$$f = 0.0032 + \frac{0.221}{\mathrm{Re}^{0.237}} \quad \text{for } 10^5 < \mathrm{Re} < 3\times10^6 \text{ at least} \tag{2.177}$$

**Example 2.91: Determining Friction Factor Using Moody Diagram**

Determine the relative roughness, $\frac{\varepsilon}{D}$, and friction factor $f$ using Moody diagram for Example 2.90.

**Solution**

Surface roughness factor $\varepsilon$ for new steel pipe (Table 2.29) = $4.5 \times 10^{-5}$ m.
Then,

$$\text{Relative Roughness} = \frac{\varepsilon}{D} = \frac{4.5\times10^{-5}\,\mathrm{m}}{0.0525\,\mathrm{m}} = 0.000857 \approx 0.0009$$

Now using the Moody Chart (Figure 2.26) with Re = 60,813 and the pipe relative roughness of 0.0009, we can find the friction factor $f$ as 0.023.

Colebrook combined all data for transition state and turbulent flow in smooth and rough pipes into a relation known as Colebrook equation, which is the most commonly used equation for calculating the friction coefficient for turbulent flow. The Colebrook equation is implicit in $f$, and determination of friction factor requires trial and error (iteration) approach as shown below:

$$\frac{1}{\sqrt{f}} = -2\log\left[\frac{\varepsilon}{3.7D} + \frac{2.51}{\text{Re}\sqrt{f}}\right] \tag{2.178}$$

We can also use an approximate explicit relation given by Swamee and Jain (1976) for $5{,}000 < \text{Re} < 10^8$ and $0.000001 < \varepsilon < 0.01$ as follows:

$$\frac{1}{\sqrt{f}} = -2\log\left[\frac{\varepsilon}{3.7D} + \frac{5.74}{\text{Re}^{0.9}}\right] \tag{2.179}$$

Or an approximate explicit relation for $f$ developed by Haaland (1983) can also be used for $4{,}000 < \text{Re} < 10^8$ and $0.000001 < \varepsilon < 0.05$ as follows:

$$\frac{1}{\sqrt{f}} = -1.8\log\left[\left(\frac{\varepsilon/D}{3.7}\right)^{1.11} + \frac{6.9}{\text{Re}}\right] \tag{2.180}$$

A transition region appears in Figure 2.26 for Reynolds numbers between 2,000 and 4,000, where the flow might be laminar, turbulent, or a mixture of both, and an interpolation approach is needed in this region for estimating the friction factor $f$.

**Example 2.92: Determining Friction Factor Using Colebrook Equation Approximations**

Find the friction factor $f$ using Equations (2.179) and (2.180) for Example 2.90.

**Solution**

**Step 1:** Find the Reynolds number

Reynolds number has already been determined as Re = 60,813.

**Step 2:** Classify flow

Since Re = 60,813 > 4,000, then the flow is turbulent flow.

**Step 3:** Determine friction factor, $f$

Equation (2.179) (Swamee equation) gives us:

$$\frac{1}{\sqrt{f}} = -2\log\left[\frac{\varepsilon}{3.7D} + \frac{5.74}{\text{Re}^{0.9}}\right]$$

$$\frac{1}{\sqrt{f}} = -2\log\left[\frac{4.5\times 10^{5}}{3.7\times 0.0525} + \frac{5.74}{60{,}813^{0.9}}\right] \Rightarrow f = 0.023$$

Similarly, using Equation (2.180) (Haaland equation) gives us:

$$\frac{1}{\sqrt{f}} = -1.8\log\left[\left(\frac{\varepsilon/D}{3.7}\right)^{1.11} + \frac{6.9}{\text{Re}}\right]$$

$$\frac{1}{\sqrt{f}} = -1.8\log\left[\left(\frac{4.5\times10^{-5}/0.0525}{3.7}\right)^{1.11} + \frac{6.9}{60{,}813}\right] \Rightarrow f = 0.023$$

So, both of the approximations calculated above and the friction factor determined earlier by using the Moody Chart are in agreement with each other.

#### 2.4.5.2 Valves and Fittings Head Loss, Resistance Coefficient K, and Equivalent Length L/D

Furthermore, most pipe systems consist of many more components than just straight pipes; therefore, the term $h_{f,\,minor}$ in Equation (2.172) needs to be calculated. As stated earlier, such losses are defined as minor losses, which might be due to the pipe entrance or exit, sudden or gradual expansions or contractions, bends, elbows, tees, and other fittings, open or partially closed valves. In some cases, the minor losses are greater than the major losses; for instance, a partially closed valve can have higher pressure drop or head loss than that of a long pipe. For most fluid dynamic designs, it is assumed that the fluid flow through valves and fittings is in turbulent range; therefore, almost all pressure drop or head loss data provided by manufacturers are reported for turbulent flow conditions, and are proportional to the velocity in valves and fittings.

The pressure or head loss data for valves and fittings are experimentally measured, tabulated, and provided by manufacturers. However, due to the time-consuming and costly nature of experimental testing, it is impossible to obtain test data for every size and type of valve and fitting. It is therefore desirable to provide a reliable means of extrapolating available test information to cover those items that have not been tested. Commonly used concepts for accomplishing this task are the "equivalent length *L/D*," "resistance coefficient *K*," and "flow coefficient $C_v$."

Since velocity in a pipe is obtained at the expense of static head, the decrease in static head due to velocity is $\frac{u^2}{2g}$, the far left term in Equation (2.170) is called the *velocity head.* The resistance coefficient *K* used for determining minor head losses or pressure drops in Equation (2.181) therefore is defined as the number of velocity heads lost due to a valve or fitting:

$$h_{f,minor} = \frac{\Delta P}{\rho g} = K\frac{u^2}{2g} \tag{2.181}$$

where *K* is a dimensionless resistance coefficient (also called the loss coefficient). Equation (2.181) is valid for computing head loss due to valves and fittings for all conditions of flow, including laminar flow, using resistance coefficient *K* as given in Table 2.30 or other relevant literature. The pressure or head loss due to valve and fitting resistance is always associated with the internal diameter in which the velocity in the term $\frac{u^2}{2g}$ occurs. The resistance

**TABLE 2.30**
Friction Losses for Standard Fittings.

| Fitting | Specifications | Resistance Coefficient, $K$ |
|---|---|---|
| *Elbows* | | |
| 45° Long radius | Flanged | 0.2 |
| 90° Long radius | Threaded | 1.0 |
| | Flanged | 0.2 |
| 45° Regular | Threaded | 0.4 |
| 90° Regular | Flanged | 0.5 |
| | Threaded | 1.5 |
| 90° Short radius | | 1.0 |
| 180° return bend | Flanged | 0.2 |
| | Threaded | 1.5 |
| *Tees* | | |
| Branch flow | Flanged | 1.0 |
| | Threaded | 1.5 |
| Line flow | Flanged | 0.2 |
| | Threaded | 1.0 |
| Cross flow | | 3.0 |
| True Y | | 1.5 |
| Union | Threaded | 0.8 |
| *Valves* | | |
| Angle valve | Fully open | 2 |
| Ball valve | Fully open | 0.05 |
| | ½ closed | 5.5 |
| | 2/3 closed | 210 |
| Diaphragm valve | Fully open | 2.3 |
| | ¼ closed | 2.6 |
| | ½ closed | 4.3 |
| Gate valve | Fully open | 0.15 |
| | ¼ closed | 0.26 |
| | ½ closed | 2.1 |
| | ¾ closed | 17 |
| Globe valve | Fully open | 10 |
| | ½ closed | 12.5 |
| Swing check | Backward flow | $\infty$ |
| | Forward flow | 2 |
| Entrance | Sharp | 0.5 |
| | Well-rounded | 0.05 |
| Contraction | Sharp $(D_2/D_1) = 0.5$ | 0.38 |
| Boiler | | 2.5 |

coefficient $K$ is considered as being independent of friction factor $f$ or Reynolds number, and may be treated as a constant for any given valve or fitting in a piping system under all flow conditions, including laminar flow. If the resistance coefficient is needed for a different internal pipe diameter from the diameter for which we can find the $K$ value from tables or diagrams, then we can use the relationship $K_a = K_b \left[\frac{D_a}{D_b}\right]^4$ where subscript "$a$" defines $K$ and $D$ with reference to the internal diameter of the pipe, and subscript "$b$" defines $K$ and $D$ with reference to the internal diameter of the pipe for which the values of $K$ were established and can be found in tables and diagrams.

The pressure or head loss due to friction in a straight pipe, on the other hand, is expressed by the Darcy–Weisbach equation (Equation (2.174)) as $h_{f,major} = \left(f\frac{L}{D}\right)\frac{u^2}{2g}$. Then it follows that $K = \left(f\frac{L}{D}\right)$, where the ratio $\frac{L}{D}$ is the equivalent length, in pipe diameters of straight pipe, that will cause the same pressure drop as the valve or fitting under the same flow conditions. Since the resistance coefficient $K$ is treated as a constant under all flow conditions, the value of equivalent length $\frac{L}{D}$ for any given valve or fitting must vary inversely with the change in friction factor for different flow conditions. When the equivalent length approach is used for determining the minor losses due to valves and fittings, the tabulated equivalent length $\left(\frac{L}{D}\right)$ values (Table 2.31) need to be employed instead of resistance coefficients.

Using the resistance coefficient $K$, the minor head losses, and therefore the total head loss due to friction can be calculated using the following equation:

$$h_{f,minor} = \sum_{i=1}^{n} K_i \left(\frac{u^2}{2g}\right); \text{then } h_{f,major} + h_{f,minor} = \left[f\frac{L}{D} + \sum_{i=1}^{n} K_i\right]\frac{u^2}{2g} \quad (2.182)$$

where $n$ is the total number of valves and fittings and the resistance coefficient $K$ needs to be determined for each valve and fitting from tables (Table 2.30).

The resistance coefficient $K$ due to sudden enlargements may be estimated using the following equation:

$$K = \left[1 - \frac{D_1^2}{D_2^2}\right]^2 \quad (2.183)$$

and the resistance coefficient due to sudden contractions as:

$$K = 0.5\left[1 - \frac{D_1^2}{D_2^2}\right] \quad (2.184)$$

where $D_1$ is the smaller internal diameter and $D_2$ is the larger internal diameter.

**TABLE 2.31**

Resistance of Valves and Fittings to Flow of Fluids in Terms of Equivalent Length of Straight Pipe.

| | | | Nominal Diameter | | | | | | | | | | |
|---|---|---|---|---|---|---|---|---|---|---|---|---|---|
| **Description of Fitting** | | **(in)<br>(mm)** | **½<br>15** | **¾<br>20** | **1<br>25** | **1 ¼<br>32** | **1 ½<br>40** | **2<br>50** | **2 ½<br>65** | **3<br>80** | **4<br>100** | **5<br>125** | **6<br>150** |
| *Globe Valve* | E.L. | ft | 13 | 16 | 26 | 35 | 40 | 55 | 65 | 80 | 32 | 134 | 164 |
| | | m | 4 | 5 | 8 | 11 | 12 | 17 | 20 | 24.5 | 24.6 | 40.2 | 49.2 |
| *Angle Valve* | E.L. | ft | 8 | 11 | 15 | 18 | 20 | 27 | 32 | 40 | 57 | 70 | 84 |
| | | m | 2.5 | 3.5 | 4.5 | 5.5 | 6 | 8.3 | 10 | 12 | 17 | 21 | 25.2 |
| *Butterfly Valve* | E.L. | ft | | | | | | 7.8 | 9.3 | 11.5 | 15.1 | 18.9 | 22.7 |
| | | m | | | | | | 2.3 | 2.8 | 3.5 | 4.5 | 5.7 | 6.8 |
| *Gate Valve* | E.L. | ft | 0.3 | 0.5 | 0.5 | 0.5 | 1 | 1 | 1.5 | 2 | 2.5 | 3 | 3 |
| | | m | 0.09 | 0.15 | 0.15 | 0.15 | 0.3 | 0.3 | 0.45 | 0.6 | 0.75 | 0.9 | 0.9 |
| *Swing Check Valve (Fully Open)* | E.L. | ft | 5.2 | 6.9 | 8.7 | 11.5 | 13.4 | 17.2 | 20.6 | 25.5 | 33.6 | 42.1 | 50.5 |
| | | m | 1.6 | 2.1 | 2.6 | 3.5 | 4.1 | 5.2 | 6.2 | 7.7 | 10.1 | 12.3 | 15.2 |
| *Elbow* | E.L. | ft | 1 | 2 | 2 | 3 | 4 | 5 | 6 | 8 | 11 | 13 | 17 |
| | | m | 0.3 | 0.6 | 0.6 | 0.9 | 1.2 | 1.5 | 1.8 | 2.5 | 3.5 | 4 | 5.2 |
| *Long Elbow* | E.L. | ft | 1 | 1.5 | 2 | 2.5 | 3 | 3 | 5 | 4 | 6 | 8 | 10 |
| | | m | 0.3 | 0.45 | 0.6 | 0.75 | 0.9 | 1 | 1.2 | 1.5 | 1.8 | 2.5 | 3 |
| *Run of Tee* →╨→ | E.L. | ft | 1 | 1.5 | 2.5 | 2.5 | 3 | 3 | 4 | 5 | 6 | 8 | 10 |
| | | m | 0.3 | 0.45 | 0.75 | 0.8 | 0.9 | 1 | 1.2 | 1.5 | 1.8 | 2.5 | 3 |
| *Run of Tee Reduced to ½* | E.L. | ft | 1 | 2 | 2 | 3 | 4 | 5 | 6 | 8 | 11 | 13 | 17 |
| | | m | 0.3 | 0.5 | 0.7 | 0.9 | 1.2 | 1.5 | 1.8 | 2.5 | 3.5 | 4 | 5.2 |

*(Continued)*

**TABLE 2.31 (Continued)**

Resistance of Valves and Ffittings to Flow of Fluids in Terms of Equivalent Length of Straight Pipe.

| Description of Fitting | | | | Nominal Diameter | | | | | | | | | | |
|---|---|---|---|---|---|---|---|---|---|---|---|---|---|---|
| | | | (in) | ½ | ¾ | 1 | 1 ¼ | 1 ½ | 2 | 2 ½ | 3 | 4 | 5 | 6 |
| | | | (mm) | 15 | 20 | 25 | 32 | 40 | 50 | 65 | 80 | 100 | 125 | 150 |
| *Branch of Tee* | ←╨→ | E.L. | ft | 3.5 | 5 | 6 | 8 | 10 | 13 | 15 | 18 | 24 | 30 | 35 |
| | | | m | 1.1 | 1.5 | 1.8 | 2.5 | 3 | 4 | 4.5 | 5.5 | 7.3 | 9 | 11 |
| *Close Return Bend* | | E.L. | ft | 2.6 | 3.4 | 4.4 | 5.8 | 6.7 | 6.6 | 10.3 | 12.8 | 16.8 | 21.0 | 25.3 |
| | | | m | 0.8 | 1.0 | 1.3 | 1.7 | 2.0 | 2.0 | 3.1 | 3.8 | 5.0 | 6.3 | 7.6 |
| *Sudden Enlargement* | d/D = ¼ | E.L. | ft | 1 | 2 | 2 | 3 | 4 | 5 | 6 | 8 | 11 | 13 | 17 |
| | | | m | 0.3 | 0.5 | 0.7 | 0.9 | 1.2 | 1.5 | 1.8 | 2.5 | 3.5 | 4 | 5.2 |
| | d/D = ½ | E.L. | ft | 1 | 1.5 | 2 | 2.5 | 3 | 3.5 | 4 | 5 | 7 | 9 | 11 |
| | | | m | 0.3 | 0.45 | 0.6 | 0.75 | 0.9 | 1.1 | 1.2 | 1.5 | 2.1 | 2.7 | 3.5 |
| | d/D = ¾ | E.L. | ft | 0.3 | 0.5 | 0.5 | 1 | 1 | 1 | 1.5 | 2 | 2.3 | 3 | 3 |
| | | | m | 0.09 | 0.15 | 0.15 | 0.25 | 0.3 | 0.35 | 0.45 | 0.6 | 0.75 | 0.9 | 1 |
| *Sudden Contraction* | d/D = ¼ | E.L. | ft | 0.8 | 1 | 1.2 | 1.5 | 2 | 2.5 | 3 | 4 | 5 | 6 | 8 |
| | | | m | 0.25 | 0.3 | 0.35 | 0.45 | 0.6 | 0.75 | 0.9 | 1.2 | 1.5 | 1.8 | 2.5 |
| | d/D = ½ | E.L. | ft | 0.5 | 0.8 | 1 | 1.2 | 1.5 | 2 | 2 | 3 | 4 | 5 | 6 |
| | | | m | 0.15 | 0.25 | 0.3 | 0.35 | 0.45 | 0.6 | 0.6 | 0.9 | 1.2 | 1.5 | 1.8 |
| | d/D = ¾ | E.L. | ft | 0.4 | 0.5 | 0.6 | 1 | 1 | 1.5 | 1.5 | 2 | 2.5 | 3 | 3.5 |
| | | | m | 0.12 | 0.15 | 0.18 | 0.3 | 0.3 | 0.4 | 0.5 | 0.6 | 0.8 | 0.9 | 1.1 |
| *Ordinary Entrance* | | E.L. | ft | 1 | 1 | 1.5 | 2 | 2.5 | 3 | 3.5 | 4.5 | 6 | 8 | 10 |
| | | | m | 0.3 | 0.3 | 0.45 | 0.6 | 0.75 | 0.9 | 11 | 1.4 | 1.8 | 2.5 | 3 |

**Example 2.93: Determining Major and Minor Head Losses**

Determine the major head loss $h_{f,\,major}$ and minor head loss $h_{f,\,minor}$ for the previous example assuming that the pipe is horizontal with two fully open gate valves and two regular flanged 90° elbows.

**Solution**

Using Equation (2.74) (Darcy–Weisbach equation), we can first calculate the major head loss $h_{f,\,major}$ as follows:

$$h_{f,major} = f\frac{L}{D}\frac{u^2}{2g} = 0.023\frac{30\text{ m}}{0.0525\text{ m}}\frac{\left(0.93\frac{\text{m}}{\text{s}}\right)^2}{2\times 9.81\text{m/s}^2}$$

$$h_{f,major} = 0.57\text{ m}$$

Then, using Equation (2.182), we can calculate the minor head loss $h_{f,\,minor}$ as follows:

$$h_{f,major} = \sum_{i=1}^{n} K_i\left(\frac{u^2}{2g}\right)$$

where we can substitute $K_1 = 0.15$ (gate valve), $K_2 = 0.15$ (gate valve), and $K_3 = 0.30$ (90° regular elbow-flanged).

$$h_{f,minor} = \sum_{i=1}^{n} K_i\frac{u^2}{2g} = (K_1 + K_2 + 2\times K_3)\left(\frac{u^2}{2g}\right)$$

$$h_{f,minor} = (0.15 + 0.15 + 2\times 0.30)\left(\frac{\left(0.93\frac{\text{m}}{\text{s}}\right)^2}{2\times 9.81\text{m/s}^2}\right)$$

$$h_{f,minor} = 0.04\text{ m}$$

Or similarly, using Equation (2.182), we can calculate both the major and minor head losses together as follows:

$$h_{f,major} + h_{f,minor} = \left[f\frac{L}{D} + \sum_{i=1}^{n} K_i\right]\frac{u^2}{2g}$$

$$h_{f,major} + h_{f,minor} = \left[0.023\frac{30\text{ m}}{0.0525\text{ m}} + (0.15 + 0.15 + 0.30\times 0.30)\right]\left(\frac{\left(0.93\frac{\text{m}}{\text{s}}\right)^2}{2\times 9.81\text{m/s}^2}\right)$$

$$h_{f,major} + h_{f,minor} = 0.61\text{ m}$$

#### 2.4.5.3 Flow Coefficient $C_v$ (or $K_v$), Pressure Drop, and Control Valve Flow Rate

The control valve industry has found convenient to express the valve capacity and the valve flow characteristics in terms of a coefficient, which is called the flow coefficient. The valve flow coefficient is basically a number, which represents a valve's ability to pass flow. In the United Kingdom and the United States, the coefficient $C_v$ is used, and is defined as the flow of water at the reference temperature of 60°F, in gallons per minute, at a pressure drop of one pound per square inch across the valve. The bigger the $C_v$ is, the more flow a valve can pass with a given pressure drop. For liquids, flow through a given valve (i.e. for a given $C_v$) is typically calculated from the following equation:

$$Q = C_v \sqrt{\left(\frac{\Delta p 62.4}{\rho}\right)} \tag{2.185}$$

where $Q$ is the valve flow rate in US gallons per minute (USGPM), $\Delta p$ is the difference between upstream and downstream pressure (psi), 62.4 is the conversion factor for fluids computed in relation to water, and $\rho$ is the density of fluid ($lb/ft^3$).

The equation above uses a conversion factor of 62.4, which is the density of water in $lb/ft^3$. The following equation is frequently utilized using a fluid's specific gravity $G$ where the relationship to water density is already considered.

$$Q = C_v \sqrt{\left(\frac{\Delta p}{G}\right)} \tag{2.186}$$

Likewise, $C_v$ for particular flow characteristics can be determined from the following equation:

$$C_v = Q \sqrt{\left(\frac{G}{\Delta p}\right)} \tag{2.187}$$

Similar equations are also available using metric units:

$$Q = K_v \sqrt{\left(\frac{\Delta p 1000}{\rho}\right)} \tag{2.188}$$

$$K_v = Q \sqrt{\left(\frac{\rho}{\Delta p 1000}\right)} \tag{2.189}$$

$$Q = K_v \sqrt{\left(\frac{\Delta p}{G}\right)} \tag{2.190}$$

$$K_v = Q \sqrt{\left(\frac{G}{\Delta p}\right)} \text{ or } K_v = 0.8646\, C_v \tag{2.191}$$

where $Q$ is the valve flow rate (m³/h), $K_v$ is the flow coefficient (primarily used in Europe), which describes the amount of water at the reference temperature of 15.5°C that flows through the valve at a pressure loss of 1 bar at a given valve position ($K_{vs}$ is used instead of $K_v$ when the flow that generates a pressure drop of 1 bar when the valve is fully open), $\Delta p$ is the difference between upstream and downstream pressure (bar), 1,000 is the conversion factor for fluids computed in relation to water, $\rho$ is the density of fluid (kg/m³), and $G$ is the specific gravity.

Liquid flow capacity of a valve in metric units can also be directly converted to $C_v$ using the following equation:

$$C_v = 0.0694\,Q\sqrt{\left(\frac{\rho}{\Delta p 999}\right)} \text{ or } C_v = 1.157\,K_v \tag{2.192}$$

where $Q$ is the flow rate (l/m), $\rho$ is the density (kg/m³), and $\Delta p$ is the pressure drop (bar).

The mixing valve can be sized by setting the maximum and minimum pressure drops over the valve. It is quite common to say that a mixing valve should have maximum and minimum pressure drops of 10 kPa = 0.1 bar (= 1.45 psi) and 3.5 kPa = 0.035 bar (0.5 psi), respectively. Then, using Equation (2.187) or Equation (2.191) and this pressure difference, a flow coefficient $C_v$ can be calculated and the correct mixing valve can be chosen easily.

**Example 2.91: Calculating Flow Coefficient and Choosing a Mixing Valve**

For a given volumetric flow rate of 300 USGPM of water and a maximum pressure drop of 1.45 psi, calculate the flow coefficient $C_v$ (and $K_v$).

**Solution**

Using Equation (2.187), we can calculate $C_v$ as follows:

$$C_v = Q\sqrt{\left(\frac{G}{\Delta p}\right)} = 300\text{ gpm}\sqrt{\left(\frac{0.999}{1.45\text{ psi}}\right)} = 249$$

Then, using this flow coefficient, we can choose the closest mixing valve from the manufacturer's list that has a minimum $C_v$ of 249. Then, using the chosen mixing valve's $C_v$ from the manufacturer's tables, we can go ahead and calculate the actual pressure drop $\Delta p$ to see if $\Delta p < 1.45$ psi.

And using Equation (2.191), we can do the same calculation in metric units for the water flow rate of 68.2 m³/h and a maximum pressure drop of 0.1 bar, and find $K_v$ as follows:

$$K_v = 68.2\text{ m}^3/\text{h}\sqrt{\left(\frac{0.999}{0.1\text{ bar}}\right)} = 216$$

Similarly, using this flow coefficient $K_v$, we can choose the closest mixing valve from the manufacturer's list that has a minimum $K_v$ of 216, and then, using the chosen mixing valve's $K_v$ from the manufacturer's tables, we can go ahead and calculate the actual pressure drop $\Delta p$ to see if $\Delta p < 0.1$ bar.

#### 2.4.5.4 Proper Sizing of Check Valves

Many difficulties are encountered with check valves, both lift and swing types. These difficulties are due to over-sizing, which ends up in noisy operation and premature wearing of moving parts. The minimum velocity required to lift the disc to the fully open and stable position has been determined by testing various types of check valves and provided by check valve manufacturers. Sizing a check valve on the basis of the specified minimum velocity for full disc lift often results in valves smaller in size than that of the pipe in which they are installed, but the pressure drop will be little, if any, higher than that of a full size valve with the disc not fully open. The losses due to sudden and gradual contraction and expansion, which will occur in such installations, can be easily calculated using the formulations provided earlier or tabulated data for fittings.

#### 2.4.5.5 Gas Pressure Regulator Capacity, Flow Coefficient $C_g$ (or $K_g$)

For gas pressure regulators operating in the range of inlet pressures up to 100 bar, the following equations for sub-critical and critical flows are used:

Sub-critical Flow:

$$Q = \frac{13.57}{\sqrt{S(T_u + 273)}} C_g \frac{p_u + p_b}{2} \sin\left[K_l \sqrt{\frac{p_u - p_d}{p_u + p_b}}\right] \tag{2.193}$$

Critical Flow:

$$Q = \frac{13.57}{\sqrt{S(T_u + 273)}} C_g \frac{p_u + p_b}{2} \tag{2.194}$$

where $Q$ is the gas flow rate (m³/h), $S$ is the gas relative density (for air $S = 1$), $T_u$ is the gas upstream temperature (°C), $C_g$ is the gas flow coefficient, $p_u$ is the gas upstream pressure (bar), $p_d$ is the gas downstream pressure (bar), $p_b$ is the ambient atmospheric pressure (bar), and $K_l$ is the valve body shape factor.

If $K_l < 130$ and $(p_u - p_d) > 0.1\,(p_u + p_b)$, then we can use the simplified calculation below:

Sub-critical flow rate where $(p_u - p_d) < 0.5\,(p_u + p_b)$:

$$Q = \frac{13.57}{\sqrt{S(T_u + 273)}} C_g \sqrt{(p_d + p_b)(p_u - p_d)} \tag{2.195}$$

Sub-critical flow rate where $(p_u - p_d) > 0.5\,(p_u + p_b)$:

$$Q = \frac{13.57}{\sqrt{S(T_u + 273)}} C_g \sqrt{\frac{p_u + p_b}{2}} \tag{2.196}$$

If desired, the equation above can also be expressed with the flow coefficient $K_g$ rather than $C_g$ as follows:

$$Q = K_g \sqrt{\frac{p_u + p_b}{2}} \tag{2.197}$$

Using the equations provided above, if the flow coefficient $C_g$ or $K_g$ provided by manufacturers is known, the regulator capacity $Q$ can be easily calculated for gas regulator's given upstream and downstream pressures. All the equations provided above are for regulators with fully open control element, as flow coefficients are measured for fully open regulators. Capacities for different positions of the control element in a regulator are calculated based on the curve of relevant regulator.

#### 2.4.5.6 Natural Gas Control Valve Flow Rate for Pressure Ratio at Normal Conditions

For natural gas control valves, it is a common practice to use the flow coefficient $KG$, which is defined for under the critical pressure ratio ($p_a/p_e > 0.54$) as follows:

$$KG = \frac{Q_n}{\sqrt{p_a(p_e - p_a)}} \tag{2.198}$$

For overcritical pressure ratio ($p_a/p_e < 0.54$):

$$KG = \frac{2Q_n}{p_e} \tag{2.199}$$

where $KG$ is the flow coefficient (m³/h), $Q_n$ is the flow rate (m³/h) at normal conditions ($p$ = 101,325 Pa and $T$ = 273.15 K), $p_a$ is the pressure after the control valve (bar), $p_e$ is the pressure in front of the control valve (bar).

For other gases, the following correction factor should be used:

$$Q_{n\,\mathrm{GAS}} = f\,Q_{n\,\mathrm{NATURAL\,GAS}} \tag{2.200}$$

$$f = \sqrt{\frac{0.78}{\rho_{\mathrm{GAS}}}} \tag{2.201}$$

where $Q_{n\,\mathrm{NATURAL\,GAS}}$ is the flow rate at normal conditions of natural gas ($m^3/h$), $Q_{n\,\mathrm{GAS}}$ is the flow rate of the other gas at normal conditions ($m^3/h$), and $f$ is the correction factor.

### 2.4.6 Lift, Circulation, and Pumping Energy Requirements

For a system containing a pump and a turbine, the Bernoulli equation was developed in Equation (2.172). If we take Equation (2.172) and multiply each side by $\rho\,g$, then we can obtain the pressure form of Bernoulli's equation as follows:

$$\frac{\rho u_1^2}{2} + \rho g z_1 + p_1 + p_p = \frac{\rho u_2^2}{2} + \rho g z_2 + p_2 + \rho g\left(h_{f,major} + h_{f,minor}\right) + p_t \tag{2.202}$$

For the application of Bernoulli's equation in certain situations, we can make the following simplifications and cancellations where applicable:

a. *Tanks, reservoirs, and nozzles exposed to the atmosphere*
   First of all, when the fluid at a reference point is exposed to the atmosphere, the pressure at that reference point is zero, and therefore the pressure head term can be

canceled from Bernoulli's equation. And second of all, the tank, from which the fluid is being drawn, can be assumed to be quite large compared to the size of the flow area inside the pipe. Since $u = Q/A$, the velocity at the surface of a reservoir or tank will be very small; and in velocity head, the velocity $u$ is squared, hence it produces much less value. In conclusion, the velocity head at the surface of a reservoir or tank can be considered as zero; therefore, it is canceled from the Bernoulli equation.

b. *When both reference points are in the same pipe*
Under the conditions of steady flow assumption, the velocity will be the same throughout the pipe due to an existing uniform flow area. Therefore, under steady flow conditions, when the two reference points for the Bernoulli equation are inside the same size pipe, then the velocity head terms on both sides of the equation are equal, and hence they are canceled.

c. *When elevations are equal at both reference points*
When both reference points for the Bernoulli equation are at the same elevation, the elevation head terms are equal as well; and therefore, they are canceled.

d. *Closed circuit conduit flow*
When both reference points for the Bernoulli equation are in the same closed circuit flow, the pressure head terms are equal at the both reference points; and therefore, they are canceled as well. Also, when both reference points for the Bernoulli equation are in the same closed circuit flow, the fluid lifting (elevation head) is compensated by the same amount of fluid falling; therefore, the elevation head terms are canceled.

The relationship between the pump head and the pump pressure $p_p$, on the other hand, is provided below:

$$h_p = \frac{p_p}{\rho g} \tag{2.203}$$

Basically, after calculating the pump pressure $p_p$, as explained below, we can determine the power delivered by the pump (pump fluid power) $\theta_{fl}$, the pump input power (pump break power) $\theta_{bk}$, the pump break horsepower (Bhp), and the pump's current requirement provided the voltage available.

The relationship between the pump fluid power (power delivered by the pump) $\theta_{fl}$ and the energy provided by the pump $p_p$ is provided below:

$$\theta_{fl} = p_p Q \tag{2.204}$$

And pumps have an efficiency ($\eta$), which represents the losses in the pump due to friction and electrical efficiency and is usually expressed in decimals (e.g. 0.6 is used for 60% conversion efficiency), and it is the ratio of the pump output power $\theta_{fl}$ to the pump input power (pump break power) $\theta_{bk}$. Therefore, we can write the following relationship:

$$\theta_{fl} = \eta \theta_{bk} \tag{2.205}$$

Then, combining Equations (2.203) and (2.204) gives the following:

$$p_p = \eta \theta_{fl} / Q \tag{2.206}$$

This expression can be substituted in the Bernoulli equation, Equation (2.202), and the required pump power or alternatively the system flow rate for a given pump power rate can be determined. And the power requirement that needs to be supplied to the pump, $\theta_{bk}$, then can be determined from the following equation:

$$\theta_{bk} = p_p Q / \eta \tag{2.207}$$

In other words, the power supplied to the pump $\theta_{bk}$ is equal to the electrical power sup plied, which is given by the following relationship:

$$\theta_{bk} = P_e = I.V \tag{2.208}$$

where $P_e$ is the electrical power (in *watt*), $I$ is the current (in *ampere*) and $V$ is the voltage (in *volt*). Typically, the voltage is known (e.g. 110 V, 220 V), so the current needs to be determined for cables, fuses, and controllers.

After determining the pump head for the system's required flow, we can develop the system curve (i.e. the flow capacity versus the system head) and plot it on the pump characteristic curves provided by pump manufacturers, and determine the operating point of the chosen point. Figure 2.27 depicts the typical efficiency curve, power curve, and rate of discharge curve for a centrifugal pump at constant operational speed, and the system curve plotted against and the pumps operating point accordingly.

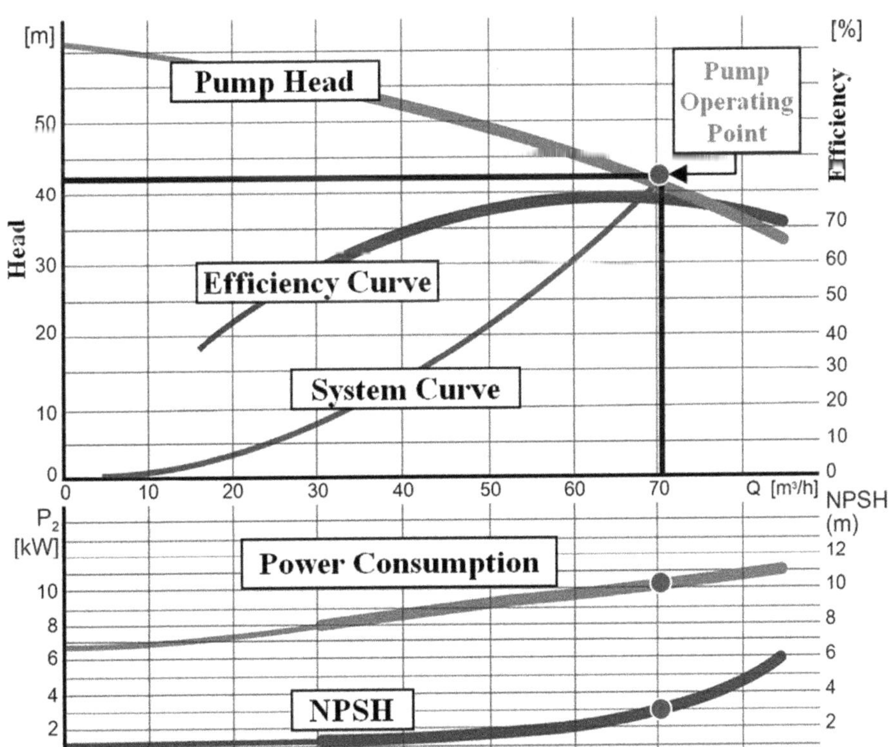


**FIGURE 2.27**
Characteristic curves of a centrifugal pump and a system curve.

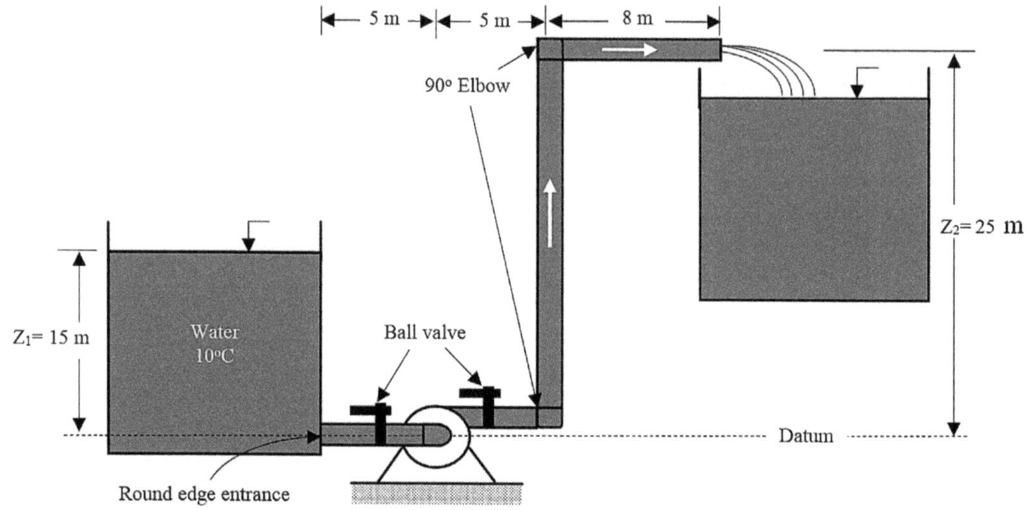


**FIGURE 2.28**
Water lifting between two reservoirs.

**Example 2.95: Water Lifting between Reservoirs**

Water at 10°C from a large reservoir is lifted with a pump to another reservoir at 1 kg/s through both a 1 ½ in and 2 in standard (Sch. 40) steel piping systems shown in Figure 2.28. Determine the total head loss and size the required pump.

**Solution**

Internal diameter of 1 ½″ standard (Schedule 40) steel pipe (Table 2.28) = 0.04089 m

Internal diameter of 2″ standard (Schedule 40) steel pipe (Table 2.28) = 0.05250 m

Surface roughness factor $\varepsilon$ for new steel pipe (Table 2.29) = $4.5 \times 10^{-5}$ m

Water density $\rho$ at 10°C (Table A.17) = 999.7 kg/m$^3$

Dynamic viscosity $\mu$ at 10°C (Table A.17) = $1{,}306 \times 10^{-6}$ kg/m.s = $1{,}306 \times 10^{-6}$ Pa.s

Using Equation (2.202) form of the Bernoulli equation, we can easily solve this problem as follows:

Remember that the reservoir at the left hand side and the discharge on the right hand side are exposed to the atmospheric pressure, so $p_1 = p_2$. And since the reservoir has a very large hydraulic diameter compared to the diameter of the discharge pipe, average water velocity at the top of the reservoir $u_1$ can be considered as zero ($u_1 = 0$). Since, there is no turbine in the system, the energy extracted by the turbine as pressure $p_t$ is zero as well.

$$\cancelto{0}{\frac{\rho u_1^2}{2}} + \rho g z_1 + \cancelto{0}{p_1} + p_p = \frac{\rho u_2^2}{2} + \rho g z_2 + \cancelto{0}{p_2} + \rho g(h_{f,\,major} + h_{f,\,minor}) + \cancelto{0}{p_t}$$

$$p_p = \rho\left[\frac{u_2^2}{2} + g\,(z_2 - z_1) + g(h_{f,\,major} + h_{f,\,minor})\right]$$

**Step 1:** Find the Reynolds number

From the continuity equation,

$$u\left(\mathrm{m/s}\right)=\frac{Q\left(\mathrm{m^3/s}\right)}{A\left(\mathrm{m^2}\right)}\text{ and }Q\left(\mathrm{m^3/s}\right)=\frac{\dot{m}\left(\mathrm{kg/s}\right)}{\rho\left(\mathrm{kg/m^3}\right)}$$

And then,

$$Q\left(\mathrm{m^3/s}\right)=\frac{1\left(\mathrm{kg/s}\right)}{999.7\left(\mathrm{kg/m^3}\right)}=0.0010\ \mathrm{m^3/s}$$

Then, for 1 ½″ pipe

$$u=\frac{4\left(0.001\frac{\mathrm{m^3}}{\mathrm{s}}\right)}{\pi\left(0.04089\ \mathrm{m}\right)^2}=0.76\ \mathrm{m/s}\text{ and for 2}''\text{pipe }u=0.46\ \mathrm{m/s}$$

And then, for 1 ½″ pipe

$$\mathrm{Re}=\frac{\rho uD}{\mu}=\frac{999.7\frac{\mathrm{kg}}{\mathrm{m^3}}0.76\frac{\mathrm{m}}{\mathrm{s}}0.04089\ \mathrm{m}}{1306\times10^{-6}\frac{\mathrm{kg}}{\mathrm{m\,s}}}=23{,}818\text{ and for 2}''\text{pipe Re}=18{,}579$$

**Step 2:** Classify flow

$$\mathrm{Re}=23{,}818>4000\text{ and Re}=18{,}551>4000,\text{ then both flows are turbulent.}$$

**Step 3:** Determine the friction factor, *f*

Using Equation (2.179), for 1 ½″ pipe

$$\frac{1}{\sqrt{f}}=-2\log\left[\frac{\varepsilon}{3.7D}+\frac{5.74}{\mathrm{Re}^{0.9}}\right]$$

$$\frac{1}{\sqrt{f}}=-2\log\left[\frac{4.5\times10^{-5}}{3.7\times0.04089}+\frac{5.74}{23{,}818^{0.9}}\right]$$

$$f=0.027\text{ and for 2}''\text{pipe }f=0.028$$

**Step 4:** Determine the pump pressure $p_p$

Using Equation (2.182), we first can calculate both the major and minor head losses together as follows:

$$h_{f,major}+h_{f,minor}=\left[f\frac{L}{D}+\sum_{i=1}^{n}K_i\right]\frac{u^2}{2g}$$

where $K_1$=0.05 (ball valve – fully open), $K_2$=1.50 (90° regular elbow – threaded), and the resistance coefficient due to sudden contractions is:

$$K_5 = 0.5\left[1 - \frac{D1^2}{D2^2}\right] = 0.5$$

For 1 ½″ pipe, then:

$$h_{f,major} + h_{f,minor} = \left[0.027\frac{(5+5+25+8)\text{m}}{0.04089\text{ m}} + (2\times0.05+2\times1.50+0.5)\right]\left(\frac{\left(0.76\frac{\text{m}}{\text{s}}\right)^2}{2\times9.81\text{ m/s}^2}\right)$$

$$h_{f,major} + h_{f,minor} = 0.838\text{ m} + 0.107\text{m} = 0.945\text{ m}$$

And for 2″ pipe:

$$h_{f,major} + h_{f,minor} = 0.287\text{m}$$

Then, using the velocity head, elevation head, and the head due to friction, the pump pressure can be calculated as follows:

$$p_p = \rho\left[\frac{u_2{}^2}{2} + g(z_2 - z_1) + g(h_{f,major} + h_{f,minor})\right]$$

For 1 ½″ pipe:

$$p_p = 999.7\text{ kg/m}^3\left[\frac{\left(0.762\frac{\text{m}}{\text{s}}\right)^2}{2} + 9.81\frac{\text{m}}{\text{s}^2}(25-15)\text{m} + 9.81\frac{\text{m}}{\text{s}^2}(0.945\text{ m})\right]$$

$$p_p = 107{,}627\text{ Pa}$$

And for 2″ pipe:

$$p_p = 100{,}993\text{ Pa}$$

**Step 5:** Determine the total pump head $h_p$
Dividing the pump pressure calculated in Step 4 by $\rho\, g$, we can easily determine the total pump head as follows:
For 1 ½″ pipe:

$$h_p = \frac{p_p}{\rho g} = \frac{107{,}627\text{ Pa}}{999.7\text{ kg/m}^3\ 9.81\text{ m/s}^2} = 11.0\text{ m}$$

For 2″ pipe:

$$h_p = 10.3\text{ m}$$

**Step 6:** Determine the required pump power $\theta_{fl}$

Using Equation (2.204) $p_p = \frac{\theta_{fl}}{Q}$ or $\theta_{fl} = p_p\, Q$, the required pump (fluid) power (the pump delivering) can be calculated as follows:

$$\theta_{fl} = p_p Q$$

For 1 ½″ pipe:

$$\theta_{fl} = (107{,}627\ \text{Pa})(0.0010\ \text{m}^3/\text{s}) = 107.7\ \text{W}$$

For 2″ pipe:

$$\theta_{fl} = 101.0\ \text{W}$$

**Step 7:** Determine the required pump input (break) power $\theta_{bk}$
Assuming a pump conversion efficiency of 60% and using Equation (2.205), we can calculate the pump input (break) power $\theta_{bk}$ as:

$$\theta_{fl} = \eta\theta_{bk}$$

For 1 ½″ pipe:

$$\theta_{bk} = \frac{\theta_{fl}}{\eta} = \frac{107.7\ \text{W}}{0.6} = 179.4\ \text{W}$$

For 2″ pipe:

$$\theta_{bk} = 168.4\ \text{W}$$

Since 1 hp = 746 W, the pump Bhp for 1 ½″ pipe is $179.4\ \text{W}\frac{1\ \text{hp}}{746\ \text{W}} = 0.24\ \text{hp}$ (and 0.23 hp for 2″ pipe). Assuming no electrical losses, the power supplied to the pump $\theta_{bk}$ is then equal to the electrical power supplied $P_e$, which has the following relationship:

$$\theta_{bk} = P_e = I.V$$

Then, the pump current requirement can be determined as:
For 1 ½″ pipe:

$$I = \frac{P_e}{V} = \frac{179.4\ \text{W}}{120\ V} = 1.50\ \text{A}$$

For 2″ pipe:

$$I = 1.40\ \text{A}$$

Using commercially available pump characteristic curves and the required pump capacity of 60 LPM with 11.0 m system head (15.9 USGPM with 36.0 ft system head) for the 1 1/2″ pipe system, and the required pump capacity of 60 LPM with 10.3 m system head (15.9 USGPM with 33.8 ft system head) for the 2″ pipe system, and we can easily choose pumps for this task.

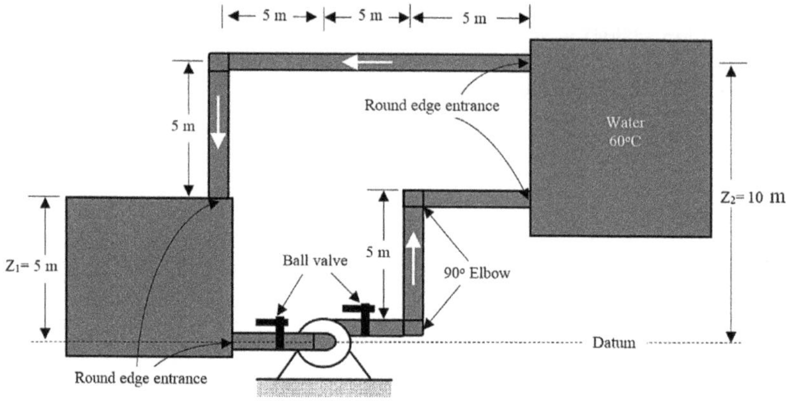


**FIGURE 2.29**
Closed loop water circulation.

**Example 2.96: Water Circulating in a Closed Loop Piping System**

Water at an average temperature of 60°C is circulated between two large reservoirs in a closed circuit piping system with a pump at 5 kg/s through a 2″ standard (Sch. 40) steel piping systems shown in Figure 2.29. Determine the total head loss and size the required pump.

**Solution**

Internal diameter of 2″ standard (Schedule 40) steel pipe (Table 2.28) = 0.05250 m

Surface roughness factor $\varepsilon$ for new steel pipe (Table 2.29 = $4.5 \times 10^{-5}$ m

Water density $\rho$ at 60°C (Table A.17) = 983.13 kg/m$^3$

Dynamic viscosity $\mu$ at 60°C (Table A.17) = $4.67 \times 10^{-4}$ kg/m.s = $4.67 \times 10^{-4}$ Pa.s

Using Equation (2.202) form of the Bernoulli equation, we can easily solve this problem as follows:

Remember that under steady flow conditions, the velocity head terms on both sides of the equation are equal, and hence they are canceled. When both reference points for the Bernoulli equation are in the same closed circuit flow, the pressure head terms are equal at the both reference points; and therefore, they are canceled as well. Also, when both reference points for the Bernoulli equation are in the same closed circuit flow, the fluid lifting is compensated by the same amount of fluid falling; therefore, the elevation head terms are canceled.

$$\cancelto{0}{\frac{\rho u_1^2}{2}} + \cancelto{0}{\rho g z_1} + \cancelto{0}{p_1} + p_p = \cancelto{0}{\frac{\rho u_2^2}{2}} + \cancelto{0}{\rho g z_2} + \cancelto{0}{p_2} + \rho g\left(h_{f,major} + h_{f,minor}\right) + \cancelto{0}{p_t}$$

$$p_p = \rho g\left(h_{f,major} + h_{f,minor}\right)$$

**Step 1:** Find the Reynolds number

From the continuity equation,

$$u(\text{m/s}) = \frac{Q(\text{m}^3/\text{s})}{A(\text{m}^2)} \text{ and } Q(\text{m}^3/\text{s}) = \frac{\dot{m}(\text{kg/s})}{\rho(\text{kg/m}^3)}$$

So,

$$Q\left(m^3/s\right) = \frac{5\left(kg/s\right)}{983.13\left(kg/m^3\right)} = 0.0051\,m^3/s$$

Then,

$$u = \frac{4\left(0.001\dfrac{m^3}{s}\right)}{\pi\left(0.05250\,m\right)^2} = 2.35\,m/s$$

And then,

$$Re = \frac{\rho u D}{\mu} = \frac{983.13\dfrac{kg}{m^3}1.07\dfrac{m}{s}0.05250\,m}{4.67\times10^{-4}\dfrac{kg}{m\,s}} = 259{,}791$$

**Step 2:** Classify flow
Re = 259,791 > 4,000; then, the flows is turbulent.

**Step 3:** Determine friction factor, $f$
Using Equation (2.179),

$$\frac{1}{\sqrt{f}} = -2\log\left[\frac{\varepsilon}{3.7D} + \frac{5.74}{Re^{0.9}}\right]$$

$$\frac{1}{\sqrt{f}} = -2\log\left[\frac{4.5\times10^{-5}}{3.7\times0.05250} + \frac{5.74}{259{,}791^{0.9}}\right]$$

$$f = 0.020$$

**Step 4:** Determine the pump pressure $p_p$
Using Equation (2.182), we can calculate both the major and minor head losses together as follows:

$$h_{f,major} + h_{f,minor} = \left[f\frac{L}{D} + \sum_{i=1}^{n} K_i\right]\frac{u^2}{2g}$$

where $K_1$=0.05 (ball valve – fully open), $K_3$=0.3 (90° regular elbow – flanged), and the resistance coefficients due to sudden contraction and sudden expansion, respectively, are:

$$K_5 = 0.5\left[1 - \frac{D_1^2}{D_2^2}\right] = 0.5K = \left[1 - \frac{D_1^2}{D_2^2}\right]^2 = 1.0$$

$$h_{f,major} + h_{f,minor} = \left[0.027\frac{(5+5+25+8)m}{0.04089\,m} + \left(2\times0.05 + 3\times0.3 + 2\times\left(0.5+1.0\right)\right)\right]\left(\frac{\left(0.76\dfrac{m}{s}\right)^2}{2\times9.81\,m/s^2}\right)$$

$$h_{f,major} + h_{f,minor} = 4.298m + 1.126m = 5.425\,m$$

Then, using the velocity head, elevation head, and the head due to friction, the pump pressure can be calculated as follows:

$$p_p = \rho g \left( h_{f,major} + h_{f,minor} \right)$$

$$p_p = 983.13 \text{ kg/m}^3 \, 9.81 \frac{\text{m}}{\text{s}^2} (5.425 \text{ m})$$

$$p_p = 52,318 \text{ Pa}$$

**Step 5:** Determine the total pump head $h_p$
Dividing the pump pressure calculated in Step 4 by $\rho g$, we can easily determine the total pump head as follows:

$$h_p = \frac{p_p}{\rho g} = \frac{52,318 \text{ Pa}}{983.13 \text{ kg/m}^3 \; 9.81 \text{ m/s}^2} = 5.425 \text{ m}$$

**Step 6:** Determine the required pump power $\theta_{fl}$
Using Equation (2.204) $p_p = \frac{\theta_{fl}}{Q}$ or $\theta_{fl} = p_p Q$, the required pump (fluid) power (the pump delivering) can be calculated as follows:

$$\theta_{fl} = p_p Q$$

$$\theta_{fl} = (52,318 \text{ Pa})(0.0051 \text{ m}^3/\text{s}) = 266.1 \text{ W}$$

**Step 7:** Determine the required pump input (break) power $\theta_{bk}$
Assuming a pump conversion efficiency of 60% and using Equation (2.205), we can calculate the pump input (break) power $\theta_{bk}$ as:

$$\theta_{fl} = \eta \, \theta_{bk}$$

$$\theta_{bk} = \frac{\theta_{fl}}{\eta} = \frac{224.6 \text{ W}}{0.6} = 443.5 \text{ W}$$

Since 1 hp = 746 W, the pump Bhp is $374.4 \text{ W} \frac{1 \text{ hp}}{746 \text{ W}} = 0.59 \text{ hp}$. Assuming no electrical losses, the power supplied to the pump $\theta_{bk}$ is then equal to the electrical power supplied $P_e$, which has the following relationship:

$$\theta_{bk} = P_e = I.V$$

Then, the pump current requirement can be determined as:

$$I = \frac{P_e}{V} = \frac{443.5 \text{ W}}{120 \text{ V}} = 3.70 \text{ A}$$

Using commercially available pump characteristic curves and the required pump capacity of 305.1 LPM with 5.4 m system head (80.6 USGPM with 17.8 ft system head), we can easily choose a pump for this task.

## Nomenclature

| | |
|---|---|
| $A$ | cross-sectional area |
| $C$ | flow coefficient |
| $D$ | diameter |
| $f$ | friction factor; correction factor |
| $g$ | acceleration due to gravity |
| $G$ | specific gravity |
| $h$ | head; head loss |
| $I$ | current |
| $K$ | resistance coefficient |
| $KG$ | flow coefficient for natural gas control valves |
| $K_l$ | valve body shape factor |
| $K_v$ | flow coefficient |
| $L$ | length |
| $m$ | mass flow rate |
| $p$ | pressure |
| $P$ | power; perimeter |
| $Q$ | volumetric flow rate |
| Re | Reynolds number |
| $S$ | gas relative density |
| $T$ | temperature |
| $u$ | average velocity |
| $V$ | voltage |
| $z$ | elevation of the point above a reference plane |

### Greek letters

| | |
|---|---|
| $\varepsilon$ | surface roughness factor |
| $\mu$ | dynamic viscosity |
| $\nu$ | kinematic viscosity |
| $\eta$ | pump efficiency |
| $\theta_{bk}$ | pump input power (pump break power) |
| $\theta_{fl}$ | pump output power (fluid power) |
| $\rho$ | density |
| $\tau$ | shear stress |

### Subscripts

| | |
|---|---|
| $a$ | after the control valve |
| $A$ | cross-sectional area |
| $b$ | ambient |
| $e$ | electrical; in front of the control valve |
| $f$ | frictional |
| g | gas |
| $eq$ | equivalent |

| | |
|---|---|
| *n* | normal conditions |
| *p* | pump |
| *t* | turbine |
| *u* | upstream |
| *v* | valve |
| *w* | wetted |
| **1, 2, 3** | numbers |

## Bibliography

Anderson, JD (2016) Some reflections on the history of fluid dynamics. In: Johnson, R.W. (ed.), *Handbook of Fluid Dynamics* (2nd ed.), CRC Press, ISBN 9781439849576

Batchelor, GK (2000) *An Introduction to Fluid Dynamics*. Cambridge University Press, Cambridge, UK, ISBN 978-0-521-66396-0.

Beck, S and Collins, R (2008) Moody Diagram. Available on: Wikipedia Commons https://commons.wikimedia.org/wiki/File:Moody_diagram.jpg (Retrieved on 21.03.2020)

Blasius, H (1911) Das Ähnlichkeitsgesetz bei Reibungsvorgängen. *Phys. Zeischr.* 12:1175–1177

Cengel, YA and Cimbala JM (2017) *Fluid Mechanics Fundamentals and Applications*. 4th Edition. McGraw-Hill Education, New York

Darrigol, O and Frisch, U (2008) From Newton's mechanics to Euler's equations. *Physica D: Nonlinear Phenomena*, **237** (14–17): 1855–1869, doi:10.1016/j.physd.2007.08.003

Frankland, TM (1984) *The Pipe Fitter's and Pipe Welder's Handbook*. Revised Edition. Glencoe Publishing Co., Mission Hills, CA. ISBN 0-02.802500-8

Haaland, SE (1983) Simple and Explicit Formulas for the Friction Factor in Turbulent Pipe Flow. *Journal of Fluids Engineering*, 105, 89–90. http://dx.doi.org/10.1115/1.3240948

Swamee, DK and Jain, AK (1976) Explicit Equations for Pipe Flow Problems. *Journal of the Hydraulics Division*, 102, pp. 657–664.

## 2.5 Light, Radiation, and Characteristics of Solar Radiation

### 2.5.1 Introduction

Solar energy powers virtually everything in the Earth and atmosphere system. This energy warms the air and the Earth's surfaces, and drives the winds, currents, evaporation, clouds, rain, etc. Essentially, the process starts when the Sun's energy in the form of electromagnetic radiation (radiation, in short) enters the atmosphere. The seasonal distribution of this energy obviously depends on the orbital characteristics of the Earth revolving around the Sun. Solar radiation reaching the Earth is significantly impacted by the inclination of the Earth's axis. This axis (the line joining the two poles through the center of the globe) is tilted 23.5° from the perpendicular. The axis maintains the same orientation with respect to the galaxy; therefore, the amount of incoming solar radiation at the top of the atmosphere, hence at the surface of the Earth, varies considerably creating seasons (by impacting the duration of daylight and elevation of the Sun in the sky with respect to time).

The solar spectrum covers the range of radiation from very short wavelengths to very long wavelengths. The spectrum belonging to a blackbody at 6,000 K is pretty close to the incident radiation from the Sun at the top of the Earth's atmosphere. When the Sun's radiation passes through the atmosphere, it is reflected, scattered, and absorbed by dust particles, gas molecules, ozone, and water vapor. The magnitude of the solar radiation's attenuation at a given time and location is determined by atmospheric composition and length of atmospheric pathway the solar radiation travels. Compared to scattering, atmospheric absorption of solar radiation is relatively small. Ozone's absorption of ultraviolet radiation is a significant component, which is vital for sustaining life on Earth. In addition to the ultraviolet absorption, at other wavelengths, absorption by nitrous oxide, carbon dioxide, oxygen, ozone, and water vapor takes place as well.

The solar radiation reaching a surface on the Earth has both *direct* and *diffuse* components. Energetically, they act in the same way. Since diffuse radiation comes from the entire sky, it is difficult to predict its intensity as it varies as moisture and pollutant contents of the atmosphere change throughout any given day at any location.

This radiant energy entering the Earth and atmosphere system is eventually transformed into a variety of other energy forms. As the radiation laws are already covered in the Section 2.3, we will begin this section with the introduction of light and solar physics, and then, move to the characteristics of solar radiation and cover it in depth with relevant numerical examples.

### 2.5.2 Light and Solar Physics

Newton was among the first to realize that a sunbeam is the composite of beams of different colors that comprise what is termed as the visible spectrum. We have spent the last three centuries for understanding the nature of light, and now we have a fairly complete model of the nature of light, or more generally speaking, of radiation.

Light is considered to behave as the composite waves of different wavelengths. *Wavelength* is the distance between two given points (let's say the peak) on one wave and the same point (the next peak) on the next wave in the same wavelength. The wavelengths of, let's say, the visible light, is extremely short, therefore, we use relatively small unit of wavelength: the micrometer. A micrometer is one-millionth of a meter, and is abbreviated as μm where μ stands for one-millionth, and m indicates "meter." In the visible spectrum, the red light has the longest wavelengths while the violet has the shortest. As will be discussed later in depth, the atmosphere scatters light differently based on its wavelength. In the visible spectrum, for instance, shorter wavelength blue and violet light are more readily scattered compared to longer wavelength lights. Actually, this is the reason for us to see the sky in blue (as the human eye is not so sensitive to violet wavelengths).

Performed experiments also indicated that the light energy behaves like a stream of particles. For instance, it is difficult to understand how sunlight travels through an empty space, as waves require a medium to travel in or propagate their energy. Ocean and sound waves for instance have water and air media to travel using the water and air molecules to propagate their energy, respectively. Without the mentioned media, it would not be possible for them to exist as there would be no molecules to vibrate. Light waves, however, require no medium to propagate. Light is considered as a stream of particles of radiant energy, also known as *photons*. And they have no mass, occupy no space, and travel at the speed of light; in this sense, they are quite different from particles of matter. As will be presented further in this section, the amount of energy carried by an individual photon is extremely small and inversely related to wavelength. The shorter the light's wavelength is, the higher the energy content of its photons. In the visible spectrum, for instance, violet light photons have more energy content than those of red light photons.

As presented above, each of the wave and particle model describes certain behavior of light that the other model cannot. So, light is only explained by this wave-particle duality model. Another important aspect is that, the photons are different from particles of matter. Countless photons are created and destroyed at any time; however, these creations and destructions do not violate the law of conservation of energy, as they simply change forms through emission and absorption. Energized particles of matter (a molecule or electron), before discharging their excess energy fully or partially and returning to their lower energy levels, remain at energized or excited stage for a very short time. This discharging or shedding of excess energy is done by releasing a photon. Obviously, more energy interactions result in higher energy photons, therefore emissions at shorter wavelengths. The energy source leading to photon production can be in any form, such as nuclear, combustion, and electricity. In summary, a photon can be described as a bundle of energy emitted by an excited molecule, atom, or electron as it returns from a high energy level to a lower energy level. Molecules, atoms, and electrons (that is, the basic particles of matter) create and destroy photons, which form the basic particles of radiant energy.

Solar energy comes from the thermonuclear reactions in the core of the Sun, where most of the solar mass is concentrated (Figure 2.30). In the core, the temperature at a pressure of about $10 \times 10^6$ bar is estimated to be approximately $15 \times 10^6$ K. This energy moves outward by radiative diffusion in the interior section; and from there; it is transferred by convection to the photosphere. Then from this outer surface zone, the energy, mostly shortwave radiation, is emitted into space in all directions by radiation. The outer surface or photosphere has an average temperature of about 6,000 K (more accurately 5,780 K); however, it fluctuates due to the sunspots, which are relatively cooler regions having temperatures closer to

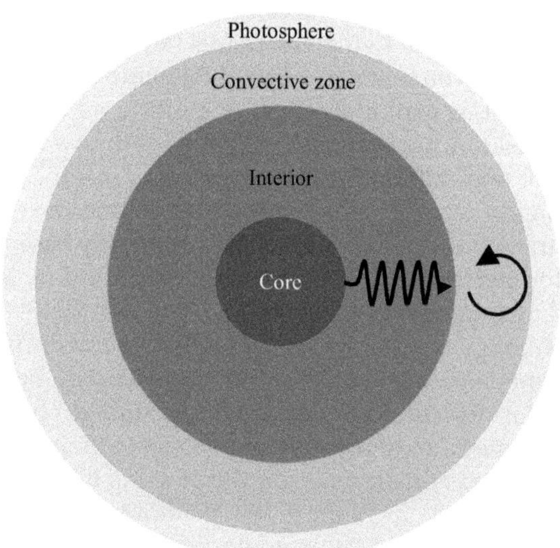


**FIGURE 2.30**
Main radial zones of the Sun (Modified after Taylor (2005)).

4,000 K. When it strikes a material, the radiation can be absorbed, transmitted, or reflected. The full discussion on the laws of radiation can be found in Section 2.3.

The Sun has a radius (696,000 km) approximately 100 times larger than that of the Earth. And its mass is estimated as $2 \times 10^{30}$ kg which is 300 million times bigger than that of the Earth. Just like any other planet, it is believed that the Sun is 4.5 billion years old, and most likely to last for 5.5 billion more years.

As noted earlier, solar energy comes from the thermonuclear reactions in the core of the Sun, where most of the solar mass is concentrated. Spectroscopic measurements of sunlight reaching the Earth from the photosphere (Figure 2.30) dictates that the solar mass is mainly composed of two elements, hydrogen (H) – which makes up about 70% of solar mass, and helium (He) – about 27%; and the remaining 3% of solar mass is made up of all the other elements (McAlester, 1983). The reaction is:

$$4\left({}^{1}\mathrm{H}\right) \rightarrow 4\mathrm{He} + \text{energy} + 2\,\text{neutrinos}$$

The conversion of H into He through solar fusion serves as the source of solar radiation (electromagnetic radiation) received on the Earth. Every second, approximately 630 ton of H are estimated to be converted to 625 ton of He and 5-tonne mass equivalent energy ($E = m\,c^2$), which gives an approximate energy release of $4 \times 10^{20}$ MJ every second.

### 2.5.3 Characteristics of Solar Radiation

This section starts with the solar radiation reaching the outer space of the atmosphere, and analyzes the spectral distribution of this radiation. And then, variations in the amount of this radiation are studied covering the impacts of different variables, such as time, solar geometry, and atmospheric effects.

#### *2.5.3.1 The Solar Constant*

It is pretty useful to know a few solar energy terms. The first one is the amount of solar radiation reaching onto a horizontal surface, which is called the *insolation* – short for "incoming solar radiation." The second term is the rate of energy flow, which is also known as power. As presented earlier, the unit of power used in SI units is Watt. One Watt is an energy flux of one Joule per second; and therefore, the unit of time is built into the definition of Watt. The third term is the amount of insolation reaching the top of the Earth's atmosphere, that is, the amount of solar energy passing thorough unit area every second at the mean distance of the Earth (perpendicular to the Earth–Sun) line is called the *solar constant*. The term, solar constant, provides a convenient way of referring to the flux of energy radiated by the Sun. Measurements over approximately the past 100 years show that the Sun's energy output is nearly constant. The measured flux is virtually the same. The quantity of the solar constant, the amount of solar radiation incident per second on one square meter surface oriented perpendicular to the beam and positioned at the top of the atmosphere at Earth's mean distance from the Sun varies only slightly with time, and expressed in Watt, has a mean measured value of 1368 W/m$^2$. And this unit, Watt, is related to the unit in which people buy electric energy for their households or greenhouses, which are typically given in the unit of kilowatt-hour (kWh).

**Example 2.97: Calculating the Solar Constant**

Using the measured solar surface temperature of 5,780 K, calculate the solar constant.

**Solution**

$$E^* = \left(5.679 \times 10^{-8}\,\mathrm{W/m^2K^4}\right)\left(5{,}780\,\mathrm{K}\right)^4$$
$$E^* = 63{,}384{,}522\,\mathrm{W/m^2}$$

And using a surface area of the Sun ($= 4\,\pi\,R_S^2 = 6.09 \times 10^{12}\,\mathrm{km^2}$), we can calculate $E^*$ as follows:

$$E^* = \left(63{,}384{,}522\,\mathrm{W/m^2}\right)\left(6.09 \times 10^{18}\,\mathrm{m^2}\right)$$
$$E^* = 3.86 \times 10^{26}\,\mathrm{W}$$

And dividing this by the surface area of a sphere at the mean distance of Earth to the Sun ($4\,\pi\,R_{SE}^2 = 2.83 \times 10^{17}\,\mathrm{km^2} = 2.83 \times 10^{23}\,\mathrm{m^2}$), we can calculate the solar constant as:

$$\begin{aligned} \textit{Solar Constant} &= \frac{E^*}{4\,\pi\,R_{SE}^2} \\ &= \frac{3.86 \times 10^{26}\,\mathrm{W}}{2.83 \times 10^{23}\,\mathrm{m^2}} \\ \textit{Solar Constant} &= 1{,}364\,\mathrm{W/m^2} \end{aligned}$$

This calculated solar constant is pretty close to the mean measured value (by satellites) of 1,368 $\mathrm{W/m^2}$ at an average distance (Earth orbit radius), $R_{SE}$, of $1.495 \times 10^8$ km. Note that we employ the inverse-square law here (that is, radiation emitted from a spherical source decreases with the square of the distance from the center of the sphere, $E_2 = E_1 (R_1/R_2)^2$, where $R$ is the radius from the center of the sphere).

If we multiply, let's say, the measured solar constant 1,368 $\mathrm{W/m^2}$ by the projected area of the Earth ($\pi\,R_E^2$, where $R_E$ is the mean radius of the Earth, 6,371 km), then using the area of a disk rather than the surface area of a sphere, we can find the intercepted solar radiation by the Earth as a whole at the top of the atmosphere every second approximately $1.74 \times 10^{11}$ MJ. In reality, incoming radiation is the solar constant minus reflected sunlight, times the area of the intercepting surface; that is, $[(1\text{-}A)(\textit{Solar Constant})(\pi\,R_E^2)]$, where $A$ is the global albedo (0.3), which will be discussed further in depth later. Overall, the top of the atmosphere receives about $4.5 \times 10^{-10}$ of the total energy output of the Sun.

**Example 2.98: Calculating the Intercepted Solar Radiation by the Earth**

Determine the intercepted solar radiation by the Earth as a whole at the top of the atmosphere every second.

**Solution**

As discussed above,

$$\textit{Intercepted Solar Radiation} = (1 - A)(\textit{Solar Constant})\left(\pi R_E^2\right)$$

where albedo $A$ is 0.3, the measured solar constant is 1,368 W/m$^2$, and the mean radius of the Earth is 6.371 × 10$^6$ m.

$$Intercepted\ Solar\ Radiation = (1-0.3)(1368\ \mathrm{W/m^2})\left(\pi\left[6.371\times10^6\,\mathrm{m}\right]^2\right)$$

$$Intercepted\ Solar\ Radiation = 1.22\times10^{11}\,\mathrm{MJ}$$

Ignoring the losses from the solar beam as it goes through the atmosphere; we can find that a one square meter area of the Earth's surface oriented toward the solar beam at about 45° angle receives 1.368 kilowatt-hours of solar energy every hour. Considering that a kilowatt-hour is a 1,000-Watt energy flow lasting for an hour, we can now imagine the amount of solar energy reaching a greenhouse roof or wall surface exposed to the Sun during a day.

#### 2.5.3.2 The Solar Spectrum

Figure 2.31 shows the complete set of possible wavelengths, also known as the electromagnetic spectrum. The solar spectrum covers the range of radiation from very short wavelengths (high frequency) to very long wavelengths (low frequency), as shown in Figure 2.32. The smooth spectrum (dashed line) in Figure 2.32 belongs to a blackbody at 6,000 K. And this spectral shape is pretty close to the incident radiation from the Sun at the top of the atmosphere. The observed deviations are mostly in the high-energy (ultraviolet) region ($\lambda < 0.4$ μm), where the molecular bands responsible for ozone production are found.

The wavelength of peak radiation emission at approx. 0.5 μm actually suggests the Sun to have a blue-green color; however, we see the Sun as yellow-white due to the radiation's interaction with the Earth's atmosphere and relative sensitivity of the human eye (has a maximum sensitivity at about 0.55 μm). And as shown in Figure 2.32, the effect of the Earth's atmosphere is to reduce the total insolation reaching the top of the atmosphere to the magnitudes reaching sea level. The solar radiation emission mainly

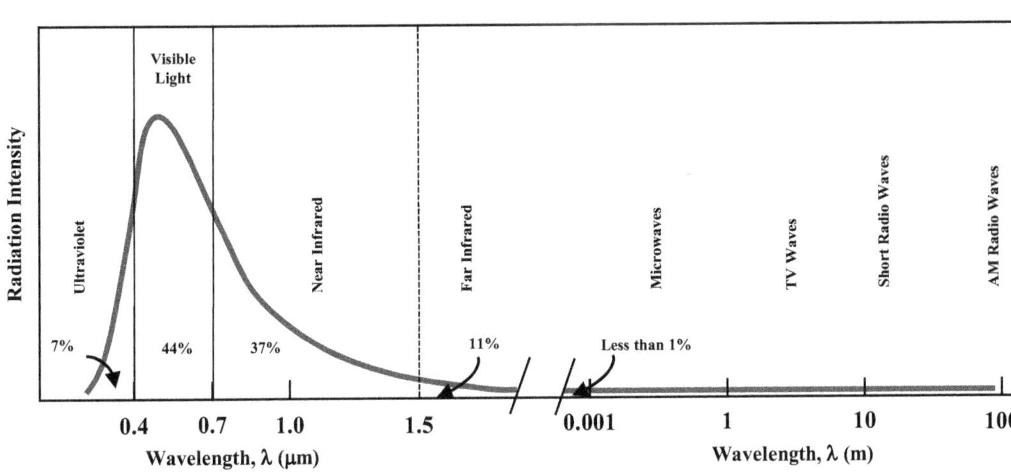


**FIGURE 2.31**
Electromagnetic spectrum: The complete set of possible wavelengths.

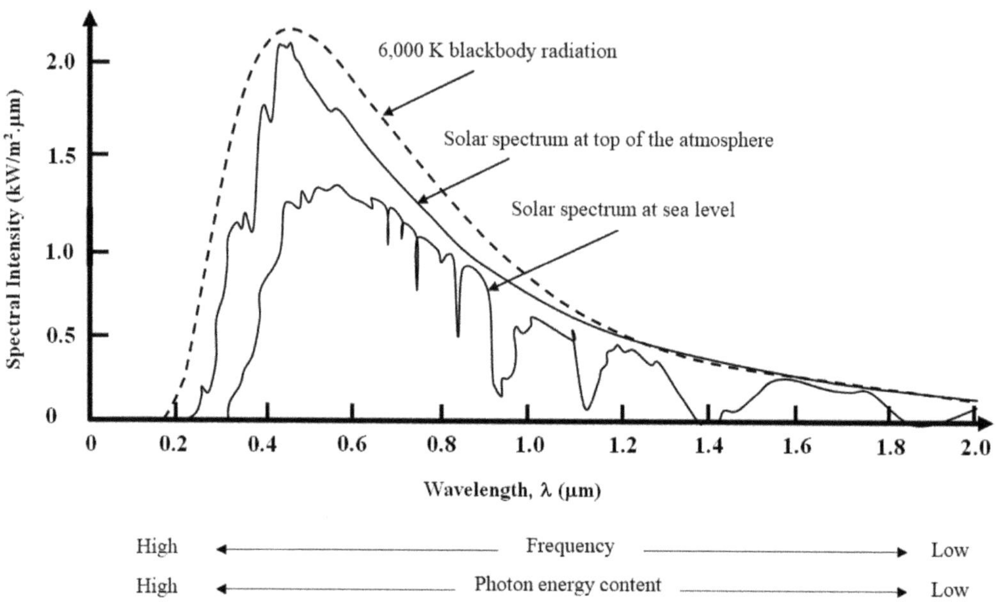


**FIGURE 2.32**
Solar spectrum. The smooth dashed line is the spectrum of the Sun would have if it were a blackbody at the "best fit" temperature of 6,000 K.

falls in the wavelength region of 0.15–4.0 µm, with about 9% in the form of ultraviolet radiation ($\lambda < 0.40$ µm), 38% in the visible (0.40–0.70 µm), and 53% in the near infrared (0.7–4.0 µm) regions. In Figure 2.32, we see that at very short wavelengths, the Sun emits relatively small amount of radiation; and as we move to the right toward the visible spectrum, the emission increases sharply reaching a maximum at about the middle of the visible range. However, the visible spectrum is still just a segment in a much broader range of solar radiation. Earlier we called this electromagnetic radiation because they both have electric and magnetic properties. The properties of light presented earlier in this section apply to radiation throughout the electromagnetic spectrum or solar spectrum. Therefore, moving to the left in Figures 2.31 and 2.32 also means moving toward radiation of shorter wavelengths and hence more energetic photons. Moving to the right, on the other hand, means that we move toward longer wavelengths, and therefore less energetic photons. As will be discussed further later, the photons of ultraviolet radiation, for instance, are highly energetic and the stratospheric ozone plays an important role in shielding us at the Earth's surface from these photons. The Sun also emits shorter wavelength X-rays and gamma rays, and longer wavelength infrared (4.0–100 µm), microwave (0.1–10 µm) and radio (> 1cm) photons; however, these do not make any significant contribution to the total solar energy reaching to the Earth and atmosphere system. As mentioned earlier, the smooth spectrum (dashed line) in Figure 2.32 belongs to a blackbody at 6,000 K. Blackbody is the term employed to describe a perfectly efficient radiator (not referring to its actual color). Therefore, the blackbody radiation given in Figure 2.32 represents the maximum radiation the Sun at a given average temperature of 6,000 K can emit. And we should also remember that most solids, liquids

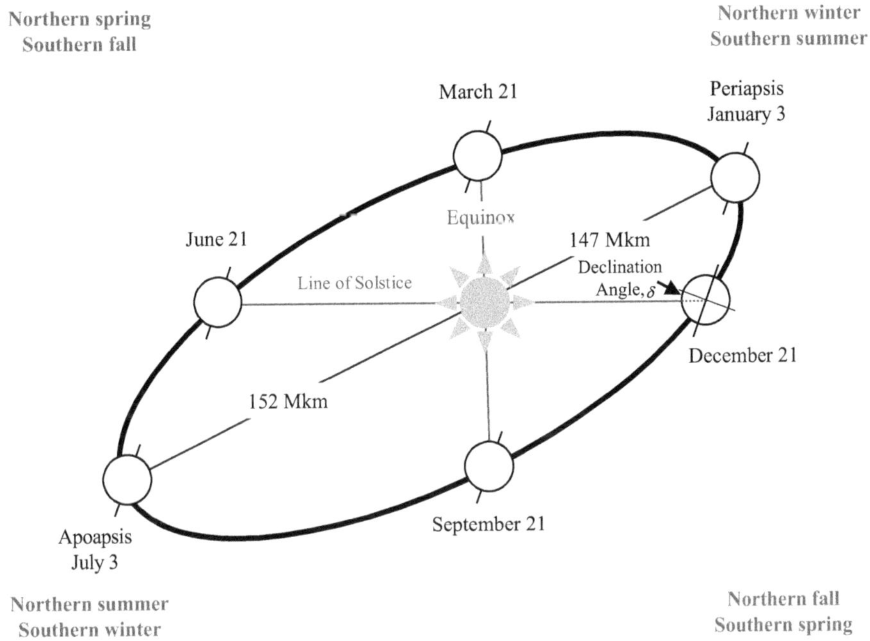


**FIGURE 2.33**
Earth's motion around the Sun (time of orbit is one revolution of 365.242 days).

and dense gases radiate very close to the blackbody radiation and have, what we call, continuous spectra of radiation.

Air molecules, as in the atmosphere, however, are quite separated and have much less interaction with each other compared to those molecules of denser substances such as liquids and solids. And each air molecule has distinct mass and structure, vibrates, rotates, and goes through different processes more or less independent from the surrounding molecules. Substances (such as $CH_4$, $N_2O$, $O_2$ and $O_3$, $CO_2$, and $H_2O$) which interact with only certain discrete energy level photons and wavelengths are defined as selective absorbers and emitters. These substances are not considered as blackbodies because at certain wavelengths they emit (or absorb) either very little or no radiation – line spectra of radiation. It is because of this reason, the Stefan–Boltzmann equation (which gives the total amount of radiation emitted or absorbed) is not applicable for a selective emitter or absorber like the atmosphere, therefore the solar spectrum at sea level has a different shape from the other two on Figure 2.32.

As Figures 2.31 and 2.32 show, wavelength ($\lambda$) and frequency ($v$) are related, and this relationship can formally be expressed through the speed of light, $c$, as follows:

$$c = \lambda v \tag{2.209}$$

where $c$ is the speed of light, which has a value of $3 \times 10^8$ m/s, $\lambda$ is the wavelength (μm), and $v$ is the frequency (cycles/s, also known as Hz). If one is known, then the other can easily be determined using the speed of light, which is a constant.

**Example 2.99: Calculating the Frequency of Radiation**

Determine the frequency for a radiational wavelength of 0.5 μm.

**Solution**

Using Equation (2.209),

$$c = \lambda v$$
$$v = c/\lambda$$
$$v = 3\times10^{8}\text{m/s}/(0.5\mu\text{m}) = 3\times10^{14}\mu\text{m/s}/(0.5\mu\text{m}) = 6\times10^{14}\text{cycles/s} = 6\times10^{14}\text{Hz}$$

And similarly, a radiational wavelength of 10 μm has a frequency, $v$, of $3 \times 10^{13}$ cycles/s (or $3 \times 10^{13}$ Hz).

Solar radiation consists of photons, which can be considered as packets of energy, which is related to frequency $v$ as:

$$E = h\,v \tag{2.210}$$

where $E$ is the energy content (J), and $h$ is the Planck constant ($6.626\times10^{-34}$ J.s). This shows that shorter wavelengths have higher frequencies and higher energy contents compared to those of longer wavelengths (lower frequencies), as shown in Figures 2.31 and 2.32.

**Example 2.100: Calculating the Energy Content of Light at a Certain Frequency**

Determine the energy content, $E$, of light at a frequency of $6 \times 10^{14}$ Hz (or 0.5 μm wavelength) .

**Solution**

Using Equation (2.210),

$$E = h\,v$$
$$E = \left(6.626\times10^{-34}\text{J.s}\right)\left(6\times10^{14}\text{cycles/s}\right)$$
$$E = 3.98\times10^{-19}\text{J}$$

And similarly, a radiation frequency of $3 \times 10^{13}$ Hz (or 10 μm wavelength) has an energy content, $E$, of approximately $1.99\times10^{-20}$ J. This clearly shows the above-mentioned difference between the energy contents of two different wavelengths.

### *2.5.3.3 Solar Time*

Orbital velocity of the Earth varies throughout the year, as well so does *Apparent Solar Time* (AST), which is determined by a sundial, changes in small amount from the *Mean Time*, which is kept by a clock running at a constant rate. This change is called the *Equation of Time* (ET). The position of the Sun in the sky is dictated by *Local Solar Time* (LST), which is determined by adding the Equation of Time to the *Local Civil Time*. Local Civil Time, on the other hand, is calculated from *Local Standard Time* (either by adding to or subtracting from) the longitude correction of 4 min/° difference between the local longitude (LON) and the *longitude of the standard time meridian* (LSTM) for the location in hand (e.g. 75° for the Eastern Standard Time in Canada and the United States).

The relationship between AST and Local Standard Time is provided in Equation (2.211):

$$\text{AST} - \text{LTS} + \text{ET} + 4(\text{LSTM} + \text{LON}) \tag{2.211}$$

where ET is the Equation of Time in minutes of time (Table 2.32), LSTM is the Local Standard Time Meridian in ° of arc (in Canada and the United States, the LSTs are 60° for Atlantic Standard Time (ST); 75° for Eastern ST; 90° for Central ST; 105° for Mountain ST; 120° for Pacific ST; 135° for Yukon ST; 150° for Alaska-Hawaii ST), LON is the LON in ° of arc, and 4 is the minutes of time required for the Earth's 1.0° rotation.

In the next section, we will see that the equatorial plane of the Earth is tilted at an angle of 23.5° to the orbital plane (Figure 2.33); and therefore, the solar declination $\delta$ (the angle between the Earth–Sun line and the equatorial plane varies throughout the year (Figure 2.33 and Table 2.32). And this variation, with their non-uniform periods of daylight and darkness, is the reason behind having different the seasons.

### 2.5.3.4 Solar Geometry

We know that the planets in the solar system have elliptical orbits around the Sun. The main features of the orbital geometry of the Earth are shown in Figure 2.33. The ellipticity in the solar system is relatively small; therefore, The Earth's motion around the Sun as well follows a nearly circular elliptical orbit with a period of 365-1/4 days. Kepler found that, for circular orbits, the time period ($Y$) of the orbit is related to the distance ($R$) of the planet

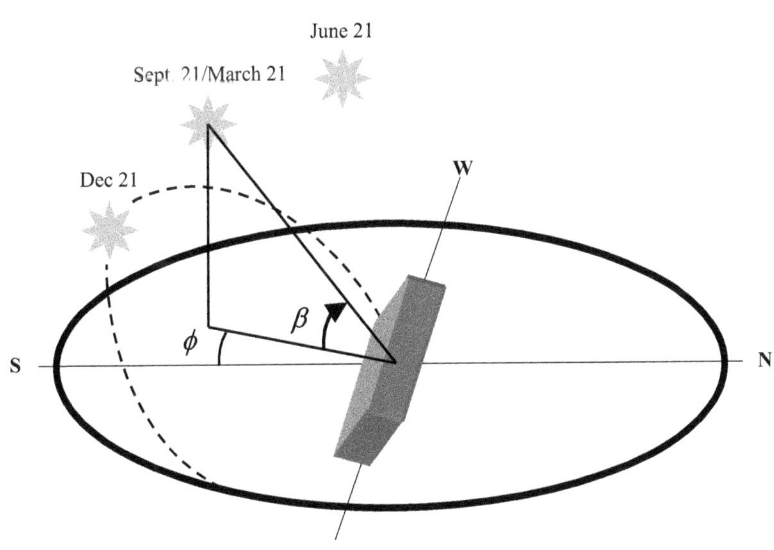


**FIGURE 2.34**
Annual and hourly changes in the Sun's position for 40°N. ($\beta$ is the solar altitude – angle above the horizon, and $\phi$ is the solar azimuth – angle from the true south) (Modified after Hinrichs and Kleinbach, 2006).

**TABLE 2.32**

Equation of Time, Declination Angle, Apparent Insolation (A) at Zero Air Mass, and Related Data for the 21st Day of Each Month (Adopted from ASHRAE, 1989).

| Month | Equation of Time (min) | Solar Declination (degrees) | A ($W/m^2$) | B (Dimensionless Ratios) | C (Dimensionless Ratios) |
|---|---|---|---|---|---|
| January | −11.2 | −20.0 | 1,228.5 | 0.142 | 0.058 |
| February | −13.9 | −10.8 | 1,212.8 | 0.144 | 0.060 |
| March | −7.5 | 0.0 | 1,184.4 | 0.156 | 0.071 |
| April | 1.1 | 11.6 | 1,134.0 | 0.180 | 0.097 |
| May | 3.3 | 20.0 | 1,102.5 | 0.196 | 0.121 |
| June | −1.4 | 23.45 | 1,086.8 | 0.205 | 0.134 |
| July | −6.2 | 20.6 | 1,083.6 | 0.207 | 0.136 |
| August | −2.4 | 12.3 | 1,105.7 | 0.201 | 0.122 |
| September | 7.5 | 0.0 | 1,149.8 | 0.177 | 0.092 |
| October | 15.4 | −10.5 | 1,190.7 | 0.160 | 0.073 |
| November | 13.8 | −19.8 | 1,219.1 | 0.149 | 0.063 |
| December | 1.6 | −23.45 | 1,231.7 | 0.142 | 0.057 |

from the Sun; that is, $Y = a\,R^{3/2}$. Parameter $a = 0.1996\ d/(\text{Gm})^{3/2}$, where $d$ is Earth days and Gm is Gigameters ($10^9$ m).

**Example 2.101: Calculating the Earth's Orbital Period**

Calculate the Earth's orbital period, $Y$.

**Solution**

Using an average Sun to Earth distance, $R = 149.6$ Gm.

$$Y = aR^{3/2}$$

$$Y = 0.1996\ d/(\text{Gm})^{3/2}(149.6\ \text{Gm})^{3/2}$$

$$Y = 365.2\ \text{days}$$

Using a leap year in every four years, the shortcoming of 365 days we use per year is corrected accordingly.

Figure 2.33 shows that the Earth is closest to the Sun when it is winter (January) in the Northern Hemisphere, and farthest away in July.

As mentioned earlier, solar radiation reaching the Earth is significantly impacted by the inclination of the Earth's axis. This axis (the line joining the two poles through the center of the Earth) is tilted 23.5° from the perpendicular. And, the axis maintains the same orientation with respect to the galaxy, therefore, the amount of incoming solar radiation at the top of the atmosphere, hence at the surface of the Earth, varies considerably creating seasons (by impacting the duration of daylight and elevation of the Sun in the sky with respect to time). The incoming solar radiation reaching a unit horizontal area at any specific location on the Earth's surface, therefore, even though the Sun's output energy is nearly constant, varies between 0 and 1,050 W/m$^2$, depending on the latitude, the season, the time of the day, the degree of cloudiness and air pollution. And this section addresses the variations due to the latitude, the season, and the time.

Figure 2.34 shows the Sun's apparent paths across the sky on different dates. It is called the apparent path because it is not the actual motion of the Sun; rather, it is the Earth's rotation about its own axis one per day. As the Earth rotates, different regions start receiving the sunlight and end again. On September 21 (and on March 21), anywhere on the Earth, the Sun seems to rise from the eastern side of the sky, move toward the west across the sky, and set 12 h later in the west. However, the actual apparent path varies considerably based on the latitude. Solar beam is actually very rarely perpendicular to the Earth's surface at any given location. Figure 2.34 shows that the smaller the angle between the incoming solar beam and the Earth's surface is, the larger the area the solar energy spreads; as a result, less solar energy is received per unit area. In late September, for instance, at the top of the atmosphere over equatorial regions, more solar energy is received per unit area compared to that over other regions; however, a location directly on the equator will receive the solar beam directly downward for a very short time around noon.

Daily path of the Sun is not constant throughout the year; rather it shifts slightly and steadily every day. And this has a huge impact on insolation. The latitude and the season are a result of the solar geometry. For instance, the North Pole has a tilt toward the Sun when it is summer in the Northern Hemisphere, and is away from the Sun when it is winter. Consequently, the Northern Hemisphere has more daylight hours in the summer, reaching a maximum value on *summer solstice*, June 21 (the position on the left side of Figure 2.33) when the North Pole's tilt toward the Sun is greatest, and hence, the amount of solar radiation reaching a horizontal surface is at the maximum in the summer. In the Northern Hemisphere, the Sun does not directly shine down on the equator; but rather, on latitude 23.5° N. On summer solstice, the daily path is highest in the sky, and the length of daylight is the longest of the year. In the Southern Hemisphere, on the other hand, the situation is the reverse, and the hemisphere is tilted farthest from the Sun on or about June 21 and receives the least amount of sunlight and onset of winter. In the Northern Hemisphere's winter, however, due to the tilt angle (and the North Pole) facing away, that is, farthest from the Sun, the solar radiation needs to go through a longer path resulting in more absorption and scattering in the atmosphere; therefore, less radiation reaches to the Earth's surface when the need for heat is the greatest. At this time of the year, the Sun is directly overhead at latitude 23.5°S, which is called the *winter solstice*; and at this time, the Southern Hemisphere starts enjoying the beginning of summer with long hours of the Sun high in the sky. Figure 2.34 schematically shows the annual and hourly changes in the Sun's position and relevant solar angles for latitude, approximately let's say, 40° North latitude. At the extreme, the noontime Sun is directly overhead at either the *Tropic of Cancer* (summer solstice) or the *Tropic of Capricorn* (winter solstice). At winter solstice, the areas north of the Arctic Circle have 24 h darkness, while the areas south of the Antarctic Circle have 24 h daylight. When the Sun is directly overhead at the equator (*spring* and *autumnal equinoxes* – on March 21 and September 21, respectively), everywhere on the Earth, there are 12 h of daylight; that is, all latitudes have equal amounts of day and night (Figure 2.33).

Figure 2.34 shows that as summer moves into fall, and then into winter, the sunrise and sunset times of the Sun's motion across the sky gradually move southward. As a result, the day lengths get shorter and the solar path gets lower in the sky. In December, the Sun rises and sets quite a bit south of east and south of west, respectively. Figure 2.34 depicts the Sun's annual and hourly position only for 40°N, note that the Sun's summertime daily path never goes below the horizon; therefore, these regions experience continuous sunlight (24 h of daylight each and every day) during summer months, which gives them the highest amount of insolation at the top of the atmosphere, at least, at that time of the year. We must also note that there is a lack of significant variation in insolation values at the top of the atmosphere throughout the year along the equator, compared to seasonal changes at higher latitudes. We also need to recognize that January insolation values decrease drastically as we go from the equator toward the North Pole, but remain constant south of the equator. In June, however, this pattern is reversed, with northern latitudes exposing little change.

#### 2.5.3.5 Solar Radiation

As it will be discussed in depth later in this section, some solar radiation scattered by air molecules and dust reaches the Earth in the form of *diffuse radiation*, $I_d$. The intensity of diffuse radiation is a difficult task to complete as it comes from all directions in the sky, and varies as moisture and dust content of the atmosphere change throughout any given day. For instance, on an overcast day, the diffuse component is all the solar radiation reaching the Earth.

The *total solar radiation* reaching a terrestrial surface, $I_t$, is composed of the *direct solar radiation* $I_D$, the *diffuse sky radiation* $I_d$, and the *solar radiation reflected from the surrounding surfaces* $I_r$ (Robinson and Sellers, 1999). The direct solar radiation $I_D$, is the product of the *direct normal solar radiation* $I_{DN}$ and the cosine of the angle of incidence $\theta$ between the incoming solar rays and a line normal (perpendicular) to the surface (ASHRAE, 1989):

$$I_t = I_{DN}\cos\theta + I_d + I_r \tag{2.212}$$

#### 2.5.3.6 Solar Angle

The position of the Sun in the sky is expressed in terms of the *solar altitude* and the *solar azimuth*. The height of the Sun, the elevation of the Sun, is usually given in terms of the solar altitude $\beta$ (Figure 2.34). This is the angular distance between the Sun's rays and the horizon, and is given by Equation (2.213). And the solar azimuth $\phi$ is the angle measured from the true south (Figure 2.34), given by Equation (2.214) (ASHRAE, 1989).

$$\sin\beta = \cos L \cos\delta \cos H + \sin L \sin\delta \tag{2.213}$$

$$\cos\phi = (\sin\beta \sin L - \sin\delta)/(\cos\beta \cos L) \tag{2.214}$$

where these angles depend on the *local latitude* $L$; the *solar declination* $\delta$ (the angle between the Earth–Sun line and the equatorial plane (Figure 2.33 and Table 2.32), which is a function of the time of year, and therefore varies from +23.5° on June 21 to −23.5° on December 21; and the AST, expressed as the *hour angle* $H$, where $H = 0.25 \times$ number of minutes from local solar noon), in degrees. Hour angle is zero at local noon and increases in magnitude by $\pi/12$, (15°), for every hour before or after noon.

Further, assuming that the Earth's orbit is circular, the solar declination angle for any day of the year can be approximated using $\delta = \Phi \cos [C\,(d - d_r)/d_y]$, where $\Phi$ is the tilt angle of 23.5°, C is full 360°, $d$ is the Julian Day, $d_r$ is the Julian day for summer solstice on June 21, and it is 172 for non-leap years, $d_y$ is the number of days per year, that is, 365 days (use 366 on a leap year).

**Example 2.102: Calculating the Solar Declination Angle**

Given $\Phi$ = 23.5°; C = 360°; $d_r$ = 172 days; $d_y$ = 365 days. Calculate the solar declination angle on March 21.

**Solution**

Assume: Not a leap year.

Using $\delta = \Phi \cos[C(d - d_r)/d_y]$,

$$\delta = 23.5^\circ \cos\left[360^\circ\left(80\,\text{days} - 172\,\text{days}\right)/365\,\text{days}\right]$$
$$\delta = -0.3^\circ$$

On spring equinox (March 21), the declination angle should be zero; so our calculation is a good approximation. It is still winter before March 21, and the declination angle should be negative; and the values in spring and summer are positive.

**Example 2.103: Calculating the Solar Azimuth and Altitude**

Determine the solar azimuth and altitude at 0830 Atlantic Standard Time (ST) on March 21 at 45° North latitude and 63° West longitude.

**Solution**

Using Equation (2.211), Local Time is 0830 + 4 (60° − 63°) = 0818. Table 2.32 gives the Equation of Time as −7 min, so AST = 0818 − 7 = 0811 or 229 min from local solar noon; therefore, the hour angle, $H$ = 0.25 × 229 min = 57.3°. As calculated in Example 2.102 and also provided in Table 2.32, the solar declination on March 21 is 0°. Therefore, using Equation (2.213), the solar altitude, $\beta$, can be calculated as follows:

$$\sin\beta = \cos 45^\circ \cos 0^\circ \cos 57.3^\circ + \sin 45^\circ \sin 0^\circ$$
$$\sin\beta = 0.382; \text{therefore}, \beta = 22.5^\circ$$

And, the solar azimuth, $\phi$, can be calculated using Equation (2.214) as:

$$\cos\phi = \left(\sin 22.5^\circ \sin 45^\circ - \sin 0^\circ\right)/\left(\cos 22.5^\circ \cos 45^\circ\right)$$
$$\cos\phi = 0.414;\ \text{therefore},\ \phi = 65.5^\circ$$

Figure 2.35 shows the solar angles and incident angles for horizontal and vertical surfaces. Line OV is perpendicular to the horizontal plane in which the solar azimuth, angle HOS and the surface azimuth, angle POS ($\Psi$) are located. Angle HOP is the surface solar azimuth ($\gamma$) and is given by

$$\gamma = \phi - \Psi \tag{2.215}$$

The solar azimuth angle $\phi$ is negative for morning hours and positive for afternoon hours. The absolute value of the surface solar azimuth $\gamma$ is used in Equation (2.216). The surface is considered in the shade if $\gamma$ is greater than 90° or less than 270°.

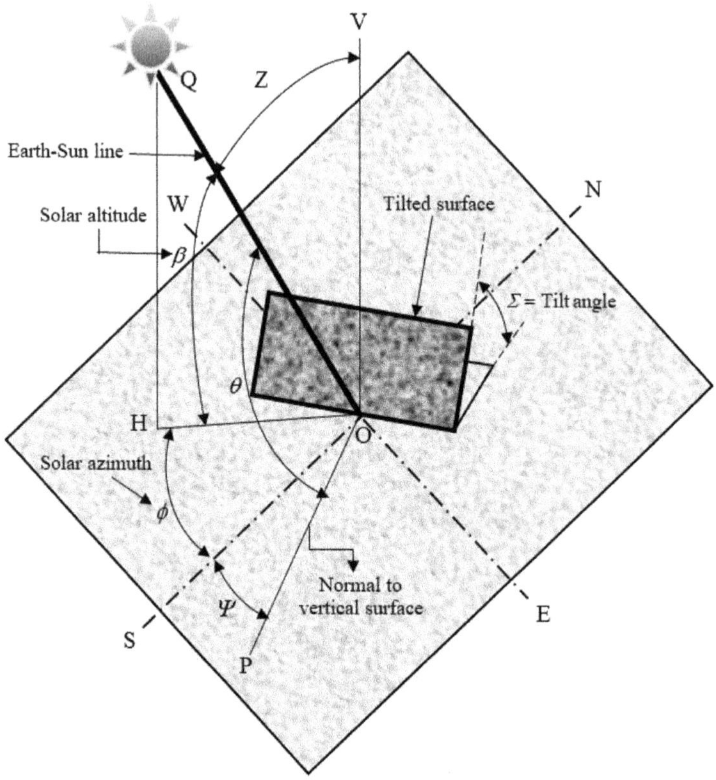


**FIGURE 2.35**
Solar angles ($\beta$ is the solar altitude – angle above the horizon, and $\phi$ is the solar azimuth – angle from the true south) (Modified after ASHRAE, 1989).

For any surface, the angle between the incoming solar rays and a line normal to the surface is called as the angle of incidence $\theta$. For the horizontal surface given in Figure 2.35, the angle of incidence $\theta_H$ is QOV; the incident angle for the vertical surface $\theta_V$ is QOP. The incident angle $\theta$ for any surface is related to the solar altitude $\beta$, the surface solar azimuth $\gamma$, and the tilt angle of the surface from the horizontal $\Sigma$ as shown below:

$$\cos\theta = \cos\beta \cos\gamma \sin\Sigma + \sin\beta \cos\Sigma \tag{2.216}$$

where $\Sigma$ is the tilt angle of the surface from the horizontal plane. When the surface is horizontal $\Sigma$ is equal to zero degrees, and $\cos\theta_H = \sin\beta$. When the surface is vertical, however, $\Sigma$ is equal to 90°, and $\cos\theta_V = \cos\beta \cos\gamma$.

### *2.5.3.7 Atmospheric Effects*

In the previous section, we mainly focused on the distribution of insolation at the top of the atmosphere at different times and locations. In this section, we will trace a beam of solar energy and explore its interactions with the atmosphere. It will be assumed that the beam has 100 units of energy as it enters the atmosphere. The fate of this 100 units of energy can be tracked by referring to Figure 2.36 as we read. From our daily observations, it would be

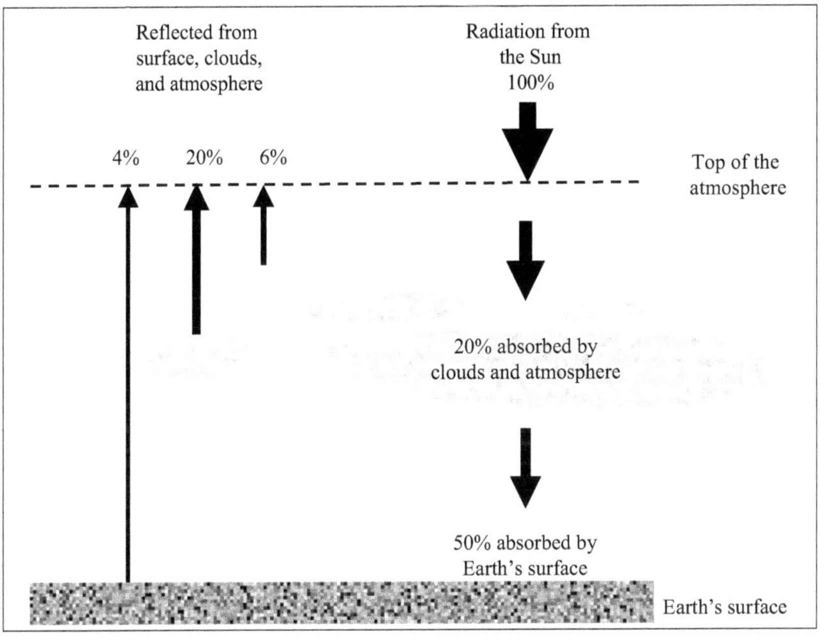


**FIGURE 2.36**
Schematic distribution of what happens to solar radiation after it falls on the Earth (Modified after Taylor, 2005).

**Example 2.104: Calculating the Incident Angle**

For the conditions of Example 2.103 given above, calculate the incident angle at a vertical surface facing southeast.

**Solution**

As we have determined that the surface azimuth angle is to the east (AST < 1,200), and the surface azimuth is also to the east (Table 2.33), they both must be negative, that is, $\phi = -65.5^\circ$ and $\Psi = -45^\circ$. Therefore, the surface solar azimuth, $\gamma$, can be calculated using Equation (2.215) as:

$$\gamma = -65.5^\circ - \left(-45^\circ\right) = -20.5^\circ$$

The negative surface solar azimuth angle that we calculated above indicates that the Sun at the given time is east of the normal to the surface.

As mentioned above, when the surface is vertical, $\Sigma$ is equal to 90°, and the incident angle, $\theta_v$, can then be determined using the relationship mentioned earlier, $\cos\theta_v = \cos\beta \cos\gamma$. Therefore,

$$\cos\theta_V = \cos 22.5^\circ \cos 20.5^\circ = 0.865$$

Then, the incident angle, $\theta_v$, the angle between the incoming solar rays and a line normal to the surface:

$$\theta_V = 30.1^\circ$$

**TABLE 2.33**

Surface Orientations and Azimuth Angles, Measured from the South (Adopted from ASHRAE, 1989).

| Orientation | N | NE | E | SE | S | SW | W | NW |
|---|---|---|---|---|---|---|---|---|
| Surface Azimuth Angle, $\Psi$ | 180° | −135° | −90° | −45° | 0° | 45° | 90° | 135° |

clear that, if the air is free of clouds and pollution, the solar beam would reach the ground with little interference from the atmosphere. The solar disk, on the other hand, becomes completely obscured under cloudy or dirty sky conditions. Clear air is quite transparent to sunlight. Normally, when the Sun's radiation passes through the Earth's atmosphere, it is reflected, scattered, and absorbed by dust particles, gas molecules, ozone, and water vapor. The magnitude of the solar radiation's attenuation at a given time and location is determined by atmospheric composition and length of atmospheric pathway the solar radiation travels. The length of atmospheric path is given in terms of the air mass $m$, which is the ratio of the atmospheric air mass in the actual Earth–Sun path to the mass which would exist if the Sun were overhead at sea level ($m = 1.0$). Obviously, above the atmosphere, $m$ is equal to zero. Almost for all purposes, the air mass $m$ at any given time and location equals the cosecant of the solar altitude $\beta$ (Figure 2.35), multiplied by the ratio of the existing barometric pressure to standard pressure.

Interactions of solar radiation within the atmosphere takes place almost simultaneously; however, these interactions are separate processes with different consequences; therefore, they must be treated separately.

Whenever a photon hits a particle or an object without being absorbed, scattering of radiation takes place. As soon as radiation gets into the atmosphere, it starts interacting with small particles, and this process is called as *scattering*. When however radiation hits a larger object, there is a complete change in the photon's travel direction, and this special condition is then called as *reflection*. This change in travel direction can be in any or multiple directions.

Air molecules have sizes of somewhere between 0.0001 and 0.001 μm, which indeed are much smaller than the wavelength of the visible band (0.4–0.7 μm). Small particles compared to the wavelength of solar radiation, such as atmospheric gas molecules, generate something called as *Rayleigh scattering*. How much scattering generated is directly proportional to the fourth power of the wavelength; therefore, the scattering of blue light ($\lambda \approx 0.4$ μm) within a cloudless atmosphere without pollution is about ten times greater than that for red light ($\lambda \approx 0.7$ μm). As a result of this Rayleigh scattering process, we see the daytime sky in blue. In the evening time, however, we observe a much redder Sun and reddish sky. This is because of the very long pathway the radiation travels through the atmosphere, which causes most of the visible wavelengths to be scattered many more times leaving only the red wavelength, some of which is also scattered to form a reddish sky.

When larger particles, such as water droplets and pollution particles exist in the atmosphere, that is, the scattering particles and the wavelength of the radiation have similar sizes; then a different type of scattering process takes place, which is called and simplified as *Mie scattering*. In this case, all wavelengths are scattered almost uniformly, and the scattering is function of both the particle size and the wavelength of the radiation. When the atmosphere is polluted or overcast, no wavelength is preferentially scattered; thence, it creates a light blue/grayish sky.

Air molecules and other small particles deflect some of the incoming solar beam in all directions. And under clear sky conditions, 6% of the original 100 units are scattered back to space.

Compared to scattering, atmospheric absorption of solar radiation is relatively small. Ozone's absorption of ultraviolet radiation is a significant component, which is vital for sustaining life on Earth. In addition to the ultraviolet absorption, at other wavelengths, absorption by nitrous oxide, carbon dioxide, oxygen, ozone, and water vapor takes place as well. Overall, as shown in Figure 2.36, approximately 20% of the energy reaching the top of the atmosphere is absorbed by the clouds (2%, i.e. 2 units of our sample 100 units) and the atmosphere (18%, i.e. 18 units of our sample 100 units). As mentioned above, some of the atmospheric absorption takes place due to the molecules of ozone and monatomic oxygen in the upper atmosphere. Actually, it is this absorbed energy for the cause of the high temperatures observed in the stratosphere and thermosphere. As a result of the scattering, and the absorption by the clouds and the atmosphere, only 74 units of the sample 100 units of energy are able to reach the Earth's surface as direct beam or diffuse solar radiation (which will be discussed further in the following section). Figure 2.36 shows that the clouds alone significantly reduce the amount of solar energy reaching the Earth's surface. As mentioned above, clouds absorb 2%, and reflect 20% back to space. All these atmospheric processes (absorption, reflection, and scattering) leave only 54% of the original energy, which eventually reach the Earth's surface. It is obvious that the areas of maximum radiation receipt are the desert regions of the Earth, while minimum radiation is received in the polar regions. It should also be remembered that Figure 2.36 represents the long-term, global averages; therefore, the local values would differ drastically. In high latitudes, where the solar angle is low, for instance, the solar beam has to pass through a much longer atmospheric path than that in low latitudes; therefore, the solar beam is more likely to be scattered or absorbed further.

As briefly mentioned above, the solar radiation reaching a surface on the Earth has both direct (the radiation that casts a shadow) and diffuse (radiation scattered from clouds, particles, and air molecules, coming from the entire sky) components. Energetically though, they act in the same way. Since diffuse radiation comes from the entire sky, it is difficult to predict its intensity as it varies as moisture and pollutant contents of the atmosphere change throughout any given day at any location. Since cloud cover and atmospheric pollution vary considerably and is hard to predict, the best thing we can do is to average the solar radiation received at a location over a number of recent years, and assume that the same amount will on average be received in the future. At the Earth's surface on clear days, solar radiation is approximately 85% direct and 15% diffuse. On a completely overcast day, all solar radiation reaching the Earth's surface is diffuse radiation.

#### 2.5.3.7.1 Direct Normal Solar Radiation

As mentioned earlier, main portion of the solar radiation reaching the Earth's surface on clear days is direct normal radiation or solar intensity $I_{DN}$, which is determined by the following equation:

$$I_{DN} = \frac{A}{\left[e^{\left(\frac{B}{\sin\beta}\right)}\right]} \tag{2.217}$$

where $A$ is the apparent incoming solar radiation (insolation) at air mass = 0 (Table 2.32), and $B$ is the atmospheric extinction coefficient, a dimensionless ratio (Table 2.32). Both values vary during the year due to the seasonal changes in atmospheric pollution and water vapor contents and the Earth–Sun distance.

Using Equation (2.217) and Table 2.32, values for direct normal radiation or solar intensity $I_{DN}$, at the Earth's surface on a clear day for any given latitude, can be calculated and tabulated for the daylight hours for the 21st day of each month.

**Example 2.105: Calculating the Direct Component of Insolation on a Horizontal Surface**

For a clear day, determine the direct component of insolation, $I_{DH}$, on a horizontal surface for the conditions given in our original example.

**Solution**

We have already determined that the solar altitude, $\beta = 22.5^\circ$, and that $\sin\beta = 0.382$. Using Equation (2.217), and the $A$, $B$, and $C$ values provided in Table 2.32, we can determine the direct normal irradiation or solar intensity, $I_{DN}$ as:

$$I_{DN} = A/\left[\exp\left(B/\sin\beta\right)\right] = \left(1184.4\ \text{W/m}^2\right)/\left[\exp\left(0.156/0.382\right)\right]$$
$$I_{DN} = 787.3\ \text{W/m}^2$$

And then, the direct component of insolation, $I_{DH}$, on a horizontal surface becomes:

$$I_{DH} = I_{DN}\sin\beta = \left(787.3\,\text{W/m}^2\right)\sin 22.5^\circ$$
$$I_{DH} = 301.3\ \text{W/m}^2$$

**Example 2.106: Calculating the Direct Component of Insolation on an Inclined Surface**

For a clear day, determine the direct component of insolation, $I_D$, on an inclined surface which has a tilt angle of 45° and facing southeast for the aforementioned conditions used in the examples above. The solar azimuth and altitude are at 0830 Atlantic Standard Time (ST) on March 21 at 45° North latitude and 63° West longitude.

**Solution**

Earlier, the Local Time was determined as 0818. And Table 2.32 provided the Equation of Time as −7 min, so AST was determined as 0811 or 229 min from local solar noon; therefore, the hour angle was then calculated as 57.3°. As also determined previously and provided in Table 2.32, the solar declination on March 21 is 0°. Then, using the solar altitude, $\beta$, and solar azimuth were calculated as $\beta = 22.5^\circ$ and $\phi = 65.5^\circ$, respectively. We also determined that the surface azimuth is to the east (Table 2.33), it must be negative, $\Psi = -45^\circ$ Therefore, the surface solar azimuth, $\gamma$, was calculated using Equation (2.215) as −20.5°. The negative surface solar azimuth angle obviously indicates that the Sun at the given time is east of the normal to the surface. Finally, the incident angle,$\theta_v$, the angle between the incoming solar rays and a line normal to the surface was calculated as $\theta_v = 30.1°$.

In Example 2.105, the direct normal irradiation or solar intensity, $I_{DN}$, was calculated as 787.3 W/m². As it was expressed earlier, the total solar radiation reaching a terrestrial surface, $I_t$, is composed of the direct solar radiation $I_D$, the diffuse sky radiation $I_d$, and the solar radiation reflected from the surrounding surfaces $I_r$.

The direct solar radiation $I_D$, is the product of the direct normal solar radiation $I_{DN}$, the cosine of the angle of incidence $\theta$ between the incoming solar rays and a line normal (perpendicular) to the surface, and also the surface solar azimuth $\gamma$ (as the Sun is east of the inclined surface):

$$I_D = I_{DN}\cos\theta\cos\gamma = \left(787.3\ \mathrm{W/m^2}\right)\cos 30.1°\cos 20.5°$$

$$I_D = 638\ \mathrm{W/m^2}$$

### 2.5.3.7.2 Diffuse Solar Radiation

The diffuse solar radiation reaching a surface on Earth may come from the sky and reflected solar radiation from adjacent surfaces. On a clear sky, a simplified relation for the diffuse solar radiation reaching on any surface from the sky is given by Equation (2.218):

$$I_{ds} = C\,I_{DN}F_{ss} \tag{2.218}$$

where $C$ is the diffuse radiation factor (dimensionless), $F_{ss}$ is 0.5 and 1.0 for vertical and horizontal surfaces, respectively, and $I_{DN}$ is the sky radiation falling on a horizontal surface, and $F_{ss}$ is the angle factor between the surface and the sky (dimensionless) determined by Equation (2.219) for other surfaces.

$$F_{ss} = \left(1.0 + \cos\Sigma\right)/2 \tag{2.219}$$

where $\Sigma$ is the tilt angle measured upward from the horizontal plane (Figure 2.35).

Solar radiation reflected by ground has the components of diffuse sky and direct solar radiation falling on a horizontal surface. The amount of total solar radiation reaching the ground is determined by Equation (2.220):

$$I_{tH} = I_{DN}\left(C + \sin\beta\right) \tag{2.220}$$

where $I_{DN}$ times $\sin\beta$ gives the direct radiation falling on a horizontal surface. Then, the ground-reflected diffuse solar radiation on any surface can be estimated by Equation (2.221):

$$I_{dg} = I_{tH}\varsigma_g F_{sg} \tag{2.221}$$

where $\varsigma_g$ is the reflectance of the foreground (dimensionless), and $F_{sg}$ is the angle factor between the surface and the ground (dimensionless). Obviously, the sum of the angle

**TABLE 2.34**

Solar reflectance values of various surfaces (Adopted from Threlkeld, 1962).

| | Incident Angle | | | | | |
|---|---|---|---|---|---|---|
| **Foreground Surface** | **20°** | **30°** | **40°** | **50°** | **60°** | **70°** |
| New Concrete | 0.31 | 0.31 | 0.32 | 0.32 | 0.33 | 0.34 |
| Old Concrete | 0.22 | 0.22 | 0.22 | 0.23 | 0.23 | 0.25 |
| Bright Green Grass | 0.21 | 0.22 | 0.23 | 0.25 | 0.28 | 0.31 |
| Crushed Rock | 0.20 | 0.20 | 0.20 | 0.20 | 0.20 | 0.20 |
| Bitumen and Gravel Roof | 0.14 | 0.14 | 0.14 | 0.14 | 0.14 | 0.14 |
| Bituminous Parking Lot | 0.09 | 0.09 | 0.10 | 0.10 | 0.11 | 0.12 |

### Example 2.107: Calculating the Overall Total Solar Radiation Reaching a Solar Collector's Surface

Determine the diffuse solar radiation incident on a solar collector, and then determine the overall total solar radiation reaching the collector's surface with a 45° slope that faces southeast, at 0830 Atlantic time on March 21, 45° North latitude.

#### Solution

Equation (2.213) for the local latitude of $L = 45^\circ$, the declination angle of $\delta = 0^\circ$ (Table 2.32), and the hour angle of $H = 57.3^\circ$ gave the solar altitude of $\beta$ as 22.5°. And then, we calculated $I_{DN}$ as 787.3 W m$^{-2}$ for 45° North latitude at 0830 Atlantic time (i.e. at AST of 0811). And Table 2.32 gives $C = 0.071$ for March 21. Using Equation (2.222) we can calculate the angle factor, $F_{sg}$, from the collector to the ground as:

$$F_{sg} = \left(1.0 - \cos 45^\circ\right)/2 = 0.146$$

Remember that, if the surface is exposed to only the ground and the sky, then $F_{ss} = (1.0 - F_{sg})$. Therefore, the angle factor from the collector to the sky, $F_{ss}$, is calculated as:

$$F_{ss} = \left(1.0 - 0.146\right) = 0.854$$

As mentioned earlier, solar radiation reflected by ground has the components of diffuse sky and direct solar radiation falling on a horizontal surface. Then, the diffuse solar radiation reaching on the collector from the sky can be determined using Equation (2.218) as:

$$I_{ds} = 0.071\left(787.3\ \text{W}/\text{m}^2\right)0.854$$
$$I_{ds} = 47.7\ \text{W}/\text{m}^2$$

The amount of total solar radiation reaching on the ground is calculated using Equation (2.220):

$$I_{tH} = \left(787.3\ \text{W}/\text{m}^2\right)\left(0.071 + \sin 22.5^\circ\right)$$
$$I_{tH} = 357.2\text{W}/\text{m}^2$$

If the ground is crushed rock having a solar reflectance of 20% (Table 2.34), the ground-reflected diffuse radiation, $I_{dg}$, falling on the collector can then be easily determined using Equation (2.221) as:

$$I_{dg} = \left(357.2\,\mathrm{W/m^2}\right)(0.20)(0.146)$$
$$I_{dg} = 10.4\,\mathrm{W/m^2}$$

The total diffuse solar radiation falling on the collector is, then, determined as the sum of diffuse sky radiation and the ground-reflected diffuse radiation, that is, 47.7 + 10.4 = 58.1 W/m².

And then, the overall total solar radiation reaching a collector surface with a 45° slope that faces southeast, at 0830 Atlantic time on March 21, 45° North latitude can be determined by Equation (2.212) as:

$$I_t = I_D + I_d + I_r$$
$$I_t = 638 + 47.7 + 10.4$$
$$I_t = 696.1\,\mathrm{W/m^2}$$

factors equals 1.0, and the angle factor for surface to ground is determined by Equation (2.222):

$$F_{sg} = \left(1.0 - \cos\Sigma\right)/2 \tag{2.222}$$

And if the surface is exposed to only the ground and the sky, then $F_{ss} = (1.0 - F_{sg})$. The reflectance values for different ground surfaces and foreground surfaces are provided by Threlkeld (1962) and reported by ASHRAE (1989).

### 2.5.4 Insolation Levels at the Earth's Surface

As explored in the previous section, on a long-term, global average, only 54% of the solar energy passing through the top of the atmosphere reaches the Earth's surface (Figure 2.36). Just a portion of this energy is absorbed at the Earth's surface, while remaining is reflected. The reflected portion compared to the incoming radiation is defined as the albedo.

$$Albedo, A = \left(Reflected\ Solar\ Radiation\right)/\left(Incoming\ Solar\ Radiation\right) \tag{2.223}$$

Albedo is normally expressed as a percentage, and satellite measurements indicate that the Earth surface has an average albedo value of 8%; that is, on average, 8% of the insolation reaching the Earth's surface is reflected back. Albedos for typical surfaces are presented in Table 2.35. When both sides of Equation (2.223) are multiplied by the incoming solar radiation, the expression becomes:

$$Reflected\ Solar\ Radiation = A \times \left(Incoming\ Solar\ Radiation\right) \tag{2.224}$$

If the albedo concept is applied to the whole Earth and atmosphere system, then we can obtain the term so called the *planetary albedo*, which has a long-term average value of 30%, that is, 30 units (20 from clouds, 6 from atmosphere, and 4 from the Earth's surface) of the sample 100 units of solar energy is reflected back to space (Figure 2.36). The remaining 70%

**TABLE 2.35**

Albedos of Typical Surfaces (Adopted from Stull, 2000).

| Surface | *A* | Surface | *A* | Surface | *A* | Surface | *A* |
|---|---|---|---|---|---|---|---|
| Fresh Snow | 75–95 | Sandy Soil | 20–25 | Asphalt Road | 5–15 | Tobacco | 19 |
| Old Snow | 35–70 | Pet Soil | 5–15 | Dirt Road | 18–35 | Potatoes | 19 |
| Gray Ice | 60 | Lime | 45 | Concrete | 15–37 | Alfalfa | 23–32 |
| Deep Water | 5–20 | Gypsum | 55 | Buildings | 9 | Cotton | 20–22 |
| Dark Wet Soil | 6–8 | Lava | 10 | Mean Urban | 15 | Sorghum | 20 |
| Light Dry Soil | 16–18 | Granite | 12.18 | Fallow Field | 5–12 | Coniferous Forest | 5–15 |
| Red Soil | 17 | Stones | 20–30 | Wheat | 10–23 | Deciduous Forest | 10–25 |
| Wet Clay | 16 | Tundra | 15–20 | Rice Paddy | 12 | Green Grass | 26 |
| Dry Clay | 23 | Sand Dune | 20–45 | Sugar Cane | 15 | Green Meadow | 10–20 |
| Wet Loam | 16 | Thick Cloud | 70–95 | Winter Rye | 18–23 | Savanna | 15 |
| Dry Loam | 23 | Thin Cloud | 20–65 | Corn | 18 | Steppe | 20 |

**Example 2.108: Calculating the Reflected Solar Radiation from a Surface**

Using Equation (2.224) and an average surface albedo of 8% (that is, 0.08), determine the reflected solar radiation.

**Solution**

$$\begin{aligned} \textit{Reflected Solar Radiation} &= A \times (\textit{Incoming Solar Radiation}) \\ &= 0.08 \times 54 \text{ units} \\ \textit{Reflected Solar Radiation} &= 4.32 \text{ units} (\approx 4 \text{ units}) \text{ reflected (see Figure.2.36)} \end{aligned}$$

of the solar energy reaching the Earth and atmosphere system is absorbed and transformed into heat.

Earlier, we defined insolation as the amount of solar radiation reaching onto a horizontal surface – short for "incoming solar radiation." Basically, that means how much sunlight is shining down on us. By knowing the insolation levels of a particular region, we can actually determine the size of a solar collector required for a particular application, and how much energy it can produce. Obviously, an area with poor insolation levels will need a larger collector area than those with high levels.

For application purposes, insolation level is generally expressed in kWh/m$^2$.day, and is the amount of solar energy that strikes a square meter of the Earth's surface in a single day. Btu or MJ may also be used, in which case the conversion is: 1 kWh/m$^2$.day = 317.1 Btu/ft$^2$.day = 3.6 MJ/m$^2$.day. The raw energy conversions are 1 kWh = 3,412 Btu = 3.6 MJ.

As discussed earlier in depth, insolation levels change throughout the year, having the lowest levels in winter and the highest in summer. And, close to the equator, the difference throughout the year is minimal whereas at high latitudes winter can be just a fraction of summer levels. A very high summer value, as we would see in a hot desert area is 7 kWh/m$^2$.day. For comparison purposes, the readers of this book are referred to the average annual insolation value of 2.27 kWh/m$^2$.day for Oslo, Norway (considered as very low), and of 5.26 kWh/m$^2$.day for Miami, Florida (considered as very high).

Tables A.3–A.8 list the average insolation values for main cities in each region of the world, which can be used in solar energy project designs (Taylor, 2005). The National Aeronautics and Space Administration (NASA) Surface meteorology and Solar Energy (SSE) data set consisting of resource parameters were developed for assessing and designing renewable energy systems. The monthly average amount of the total solar radiation incident on a horizontal surface at the surface of the Earth for a given month, averaged for that month over a 10-year period. Each monthly averaged value was evaluated as the numerical average of 3-hourly values for the given month. Renewable energy technologies range in complexity from the introduction of solar ovens and simple photovoltaic panels into rural communities to the construction of commercial buildings with integrated photovoltaics and large thermal and wind generating power plants. The availability of accurate, global solar radiation and meteorology data is extremely important for successful greenhouse and other renewable energy projects. NASA's Earth Science Enterprise (ESE) makes the SSE data set available free-of-charge over the Internet (NASA, 2016).

Historically, climatological profiles of insolation and meteorology parameters calculated from ground measurements have been used for determining the viability of greenhouse and other renewable energy projects. Although ground measurement data have been used successfully, ground measurement stations are situated mainly in populated regions. In remote areas, where many greenhouse and other renewable energy projects are being implemented, measurement stations are quite limited. Also, at some stations, available data can be sporadic and incomplete; and, data inconsistencies may occur as well. In contrast to ground measurements, the SSE data set is a continuous, consistent, and long-term global insolation data. Although the SSE data within a particular grid cell are not necessarily representative of a particular microclimate, or point, within the cell, the data are considered to be the average over the entire area of the cell. The SSE data set is however not intended to replace ground measurement data. NASA emphasizes that it is prepared to fill the gap where ground measurements do not exist, and to augment areas where ground measurements are available. In utilizing the SSE data set, the renewable energy resource potential can be determined for any location on the globe and is considered to be accurate for preliminary feasibility studies for greenhouse and other renewable energy projects. Detailed insolation information (including direct solar normal and diffuse components, and more) for any particular location or time can be directly obtained from NASA. Table A.9, on the other hand, lists the estimates of annual solar resource availabilities and ranks around the world, and may prove to be useful for potential greenhouse and other renewable energy projects.

**Example 2.109: Determining a Flat Plate Collector Area Required for Water Heating**

Assuming that the flat plate collector (FPC) efficiency is 50% and ignoring the ground reflected radiation falling on the collector, determine the collector size to be tilted from the horizontal surface facing south (for both Los Angeles (34° North latitude), California, USA, and Halifax (45° North latitude), Nova Scotia, Canada) required to heat a total of 150 L of hot water per day (needed for 4 people in a family operated greenhouse) from 5°C to 45°C in March.

**Solution**

We know that the amount of heat required for the given temperature difference can be calculated using Equation (2.24), that is, $Q_S = m\, c_p\, (T_2 - T_1)$, where $m$ is the mass of water, $c_p$ is the specific heat of water (Table 2.4), and $Q_S$ is the heat required.

$$Q_S = 150\,\text{kg}\left(4186\,\text{J/kg.}^\circ\text{C}\right)\left(45-5\right)^\circ\text{C}$$
$$Q_S = 25{,}116{,}000\,\text{J/d}\left(\approx 25.1\,\text{MJ/d}\right)$$

The heat available from the FPC is $Q_S$ = (*Insolation*) × (*FPC Area*) × (*Efficiency*). Table A.8 provides the average intensity of the insolation falling on the ground for March in Los Angeles as 5.09 kWh/m².day. Then, the heat available from the FPC is:

$$25.1\,\text{MJ/day} = \left(5.09\,\text{kWh/m}^2\text{.day}\right)\left(3.6\,\text{MJ/kWh}\right)\times \textit{FPC Area}\left(\text{m}^2\right)\times\left(0.5\right)$$
$$\textit{FPC Area} = 2.7\,\text{m}^2$$

As a useful rule of thumb, the collector should be tilted 44°–45° from the horizontal surface facing south for Los Angeles (34° N local latitude plus 10°), California, USA.

Table A.5 provides the average intensity of the insolation falling on the ground for March in Halifax as 3.46 kWh/m².day. Then, the heat available from the FPC is

$$25.1\,\text{MJ/day} = \left(3.46\,\text{kWh/m}^2\text{.day}\right)\left(3.6\,\text{MJ/kWh}\right)\times \textit{FPC Area}\left(\text{m}^2\right)\times 0.5$$
$$\textit{FPC Area} = 4.0\,\text{m}^2$$

Again, the collector should be tilted 55° from the horizontal surface facing south for Halifax (45° N local latitude plus 10°), Nova Scotia, Canada. As more insolation is available in Los Angeles compared to that in Halifax, much less (~32%) collector area is required to heat the same amount of water in Los Angeles, California.

This example gives us rough approximations for the required collector areas. For a detailed analysis, we would use the average monthly direct and diffuse components of solar radiation per square meter per day, or daily or hourly direct and diffuse components of solar radiation per square meter when available, and perform detailed calculations for the average direct, diffuse, and ground-reflected solar radiation reaching the collector surface for a given collector orientation and tilt angle. Then, we would decide what the design collector area is for this application.

#### Example 2.110: Determining a Flat Plate Collector Area Required for a Small Greenhouse Heating

Assuming that the FPC efficiency is 50%, determine the collector size required to heat a small greenhouse for one day when the heating load is 5.86 kW. Use the mean daily insolation falling on the ground as 2.25 kWh/m².day (or 8.1 MJ/m².day) and the local latitude as 40°N.

#### Solution

Remember that $Q_S$ = (*Insolation*) × (*FPC Area*) × (*Efficiency*). The thermal energy which is needed for a day will be 5.86 kW × (24 h/d) (3,600 s/h) = 506,304 kJ/d = 506.3 MJ/d.

The solar energy collected in one day will be (8.1 MJ/m².day) × *FPC Area* × *Efficiency*. Therefore,

$$\textit{FPC Area} = \left(506.3\,\text{MJ/day}\right)/\left[\left(8.1\,\text{MJ/m}^2\text{.day}\right)\times 0.5\right]$$
$$\textit{FPC Area} = 125\,\text{m}^2$$

NOTE: A useful rule of thumb is that the collector for space heating should be south facing and inclined at an angle (collector tilt angle from the horizontal) equal to the local latitude plus 10°. Therefore, we would have the collector in the example facing south at 40° plus 10° = 50°. This example gives us a rough approximation for the required collector area. For a detailed analysis, we would use the average monthly direct and diffuse components of solar radiation per square meter per day, or daily or hourly direct and diffuse components of solar radiation per square meter when available, and perform detailed calculations for the average direct, diffuse, and ground-reflected solar radiation reaching the collector surface for a given collector orientation and tilt angle. Then, considering a thermal storage device to be used for night time heating, we would decide what the design collector area is for this application.

### Example 2.111: Determining the Size of Hot Water Thermal Storage

Now, using the FPC (125 m² surface area with an efficiency of 50%) and the greenhouse in the previous example (located at 40°N latitude and having a heating load is 5.86 kW), calculate how many tons of water would be necessary to store enough thermal energy for let's say 3 days of space heating. Assume the water in the storage tank begins providing heat at 65°C and has a lower temperature limit of 30°C (i.e. $\Delta T$ = 35°C).

#### Solution

The thermal storage system is required to provide heat for space heating; so the 3-day capacity of the thermal storage system would be:

$$Q = (506.3\ \text{MJ/day}) \times (3\ \text{day}) = 1{,}518.9\ \text{MJ}$$

For a given temperature difference, using Equation (2.24), $Q_S = m\, c_p\, \Delta T$, where $Q_S$ is the heat taken from the thermal storage system, and $\Delta T$ is the temperature change after lowering its temperature, $c_p$ is the specific heat of water (4,186 J/kg.°C; Table 2.4), and $m$ is the mass of water required for thermal storage. Then,

$$Q_S = m\, c_p\, \Delta T$$

$$1{,}518.9\ \text{MJ} = m(\text{kg})(4186\ \text{J/kg.°C})(35\text{°C})$$

$$m = 10{,}367\ \text{kg} = 10.4\ \text{ton}$$

### Example 2.112: Determining Solar Collector Area for Power Generation

Assuming a collector conversion efficiency of 8% and ignoring the ground reflected radiation falling on the collector, determine the collector size to be tilted from the horizontal surface facing south (for both Los Angeles (34° North latitude), California, USA, and Halifax (45° North latitude), Nova Scotia, Canada) needed to provide power for a greenhouse community, which has an average annual power requirement of 5,000,000 kWh.

#### Solution

The required daily power output for the community can be calculated as:

$$(5{,}000{,}000\ \text{kWh/year})(3.6\ \text{MJ/kWh})/(1\ \text{year}/365\ \text{day})$$

$$\textit{Required Daily Power Output} = 49{,}315\ \text{MJ/day}$$

The power available from the collector is (*Insolation*) × (*Collector Area*) × (*Efficiency*). Table A.8 provides the annual average of daily intensity of insolation falling on the ground in Los Angeles as 5.40 kWh/m².day (=19.44 MJ/m².day). Then, the collector area required for the daily power output of 49,315 MJ/day in Los Angeles, California, is:

$$Collector\ Area\left(m^2\right) = \left(49{,}315\ MJ/day\right) / \left[\left(19.44\ MJ/m^2.day\right)\left(0.08\right)\right]$$
$$Collector\ Area = 31{,}710\ m^2 = 3.1\ ha$$

As a useful rule of thumb, the collectors should be tilted 44°–45° from the horizontal surface facing south for Los Angeles (34° N local latitude plus 10°), California, USA.

Table A.5 provides the annual average of daily intensity of insolation falling on the ground in Halifax as 3.40 kWh/m².day (= 12.24 MJ/m².day). Then, the collector area required for the daily power output of 49,315 MJ/day in Halifax, Nova Scotia, is:

$$Collector\ Area\left(m^2\right) = \left(49{,}315\ MJ/day\right) / \left[\left(12.24\ MJ/m^2.day\right)\left(0.08\right)\right]$$
$$Collector\ Area = 50{,}363\ m^2 = 5.0\ ha$$

Again, the collector should be tilted 55° from the horizontal surface facing south for Halifax (45° N local latitude plus 10°), Nova Scotia, Canada. As more insolation is available in Los Angeles compared to that in Halifax, much less (~38%) collector area is required to generate the same amount of power required. For a detailed analysis, we would use the average monthly direct and diffuse components of solar radiation per square meter per day, or daily or hourly direct and diffuse components of solar radiation per square meter when available.

## Nomenclature

| | |
|---|---|
| a | constant (2,897 μm.K); or (0.1996 d/(Gm)$^{3/2}$) |
| *A* | albedo; apparent incoming solar radiation at zero air mass |
| **AST** | apparent solar time (minutes) |
| **B** | dimensionless ratio |
| *c* | speed of light (3 × $10^8$ m/s); constant; dimensionless ratio |
| *C* | diffuse radiation factor (dimensionless), degree Celsius |
| *d* | day; Julian day (number); number of days per year (365) |
| **E** | amount of energy; east |
| **ET** | equation of time (minutes) |
| *F* | angle factor (dimensionless) |
| *FPC* | flat plate collector |
| **Gm** | Gigameter |
| *h* | Planck constant (6.626×$10^{-34}$ J.s) |
| *H* | hour angle (° of arc) |
| **Hz** | hertz (cycle/s) |
| *I* | amount of solar radiation per unit area per unit time |
| **J** | joule |
| **K** | absolute temperature Kelvin |
| **L** | local latitude |
| **LON** | local Longitude |

**LSTM** longitude of the standard time meridian
**LST** local solar time
*m* mass; meter
**M** mega
**N** north
**NE** northeast
**NW** northwest
*Q* amount of heat energy
*QOP* incident angle for the vertical surface
*QOV* angle of incidence
**R** radius
**S** south
**SE** southeast
**ST** standard time
**SW** southwest
*T* temperature
**W** Watt; west
**Y** time period (days)

## Greek letters

$\beta$ solar altitude
$\delta$ solar declination
$\Delta$ difference
$\varepsilon$ emissivity
$\phi$ solar azimuth angle
$\Phi$ tilt angle of the Earth
$\gamma$ surface solar azimuth
$\lambda$ wavelength
$\mu$ micro
$v$ frequency
$\theta$ angle of incidence
$\sigma$ Stefan–Boltzmann constant ($5.679 \times 10^{-8}$ W/m$^2$.K$^4$)
$\Sigma$ tilt angle of the surface
$\varsigma$ reflectance
$\Psi$ surface azimuth angle

## Subscripts

*d* diffuse
*D* direct
*DH* direct component of insolation on a horizontal surface
*DN* direct normal
*dg* ground reflected diffuse
*ds* diffuse solar
*E* Earth
*g* ground
*H* horizontal
*max* maximum

| | |
|---|---|
| *r* | ground-reflected; summer solstice |
| **S** | Sun |
| *sg* | between the surface and the ground |
| *ss* | between the surface and the sky |
| *SE* | Earth to the Sun |
| *t* | direct solar radiation |
| *V* | vertical |
| $\lambda$ | at a certain wavelength (monochromatic) |
| **1, 2** | number |

**Superscripts**

| | |
|---|---|
| * | Blackbody |

## Bibliography

Ahrens, C. Donald (2004) *Essentials of Meteorology: An Invitation to the Atmosphere*. 4th Edition. Thomson Learning, Inc., Pacific Grove, CA.

ASHRAE (1989) *Handbook of Fundamentals*. American Society of Heating, Refrigeration and Air Conditioning Engineers, Inc. Atlanta, GA 30329. pp. 27.1–27.38.

Hinrichs, RA and Kleinbach, M (2006) *Energy: Its Use and the Environment*. 4th edition. Thomson Brooks/Cole. Belmont, California. pp. 160–204.

McAlester, AL (1983) *The Earth: An Introduction to the Geological and Geophysical Sciences*. Prentice-Hall, Englewood Cliffs.

NASA (2016) Surface meteorology and Solar Energy Data Set. https://asdc.larc.nasa.gov/data/Retrieved November 11, 2016.

Robinson, PJ and Sellers, AH (1999) *Contemporary Climatology*. 2nd edition. Pearson Education Ltd., Essex, England. pp. 17–38.

Sen, Z (2008) Solar Energy Fundamentals and Modeling Techniques. Springer-Verlag London Limited. pp. 47–98.

Stull, RB (2000) *Meteorology for Scientists and Engineers*. 2nd edition. Thomson Brooks/Cole. Pacific Grove, California. pp. 23–38.

Taylor, FW (2005) Elementary Climate Physics. Oxford University Press Inc. New York. pp. 29–39.

Threlkeld, JL (1962) *Thermal Environmental Engineering*. Prentice Hall, New York. p. 321.

Yildiz, I (2018) Solar Energy. In: Dincer, I. (ed.), *Comprehensive Energy Systems*. Vol.1, Oxford: Elsevier. pp. 638–664.

# 3

# *Greenhouse Structures*

## 3.1 Introduction

Greenhouses are built for a variety of purposes, but in the first place, to provide a suitable environment for the intensive production of a variety of crops. This environment is provided for not only optimal plant growth but also efficient work or visits to be performed safely. Therefore, greenhouse structures and all parts should in the first place be designed and built choosing a proper site and safely supporting all loads, including dead load, without exceeding the allowable stresses for the materials from which the greenhouse is constructed. The topic of load analyses and requirements for designing greenhouse structures and their components, however, is not the focus of this book or chapter; therefore, they will not be dealt with in depth here. Rather, the readers of this book are referred to National Greenhouse Manufacturers Association (NGMA) Structural Design Manual (2004) for design loads and greenhouse structural designs. The manual provides guidelines for engineers through the building code provisions for production and commercial greenhouse structures. It includes the provisions for roof live load, snow load, wind loads, and seismic and collateral loads as contained in the building code or reference standards, and also provides structural design examples. The loads specified by the NGMA are to be used in conjunction with the allowable stresses recommended in current design specification for aluminum, steel, wood, glass, concrete, or any other conventional structural material used in the construction of greenhouses.

Overall, this chapter will briefly cover site selection and layout as well as design loads for greenhouses, and then provide in order typical foundations and floors, framing materials and structural forms, covering materials, curtain systems, and conclude with insect screens.

## 3.2 Site Selection and Layout

Choosing a good greenhouse site is the first and foremost important consideration before a greenhouse complex is designed and built as it affects a number of vital requirements of a greenhouse operation. Two of these requirements are dependable supply of energy and quality water.

Dependable supply of energy means the energy in the form of electricity and fuel for heating if needed. Electricity can be obtained from the grid or from a cogeneration system built within operation complex if the cost is justified. In the case it relies on the grid, we need to make sure that the bringing cost of power will be reasonable, that is, access to the grid will be easy, so that it will not put excessive burden on the operation. Another consideration for providing electricity is related to the environment. It means that we also need to check the utility company's energy mix to see whether it is using clean, renewable and local energy sources for power generation or not; or at least, we need to know what percent of the mix comes from such sources. If greenhouse heating will be required in the region, then we also need to know what kinds of energy sources are available for heating at a reasonable cost. In the case the heating fuel will be outsourced, we also need to make sure that the operation has an easy access to, for instance, a natural gas line or an all-season public road so that the operation has an uninterruptable supply of heating fuel. Having an easy access to a public road is also important for other supplies reaching the operation and plants or produce being delivered to market with little or no problems.

Dependable supply of quality water is a vitally important requirement. There will not be any plant production unless there is enough quality water; therefore, availability of municipal water, underground and surface water potential, relevant regulations and its quality must be known before an intensive crop production is attempted. River and stream waters are fine as well; however, they should not be the sole source of water due to the extensive damage they could possess in the case of a contamination. Today, there are many greenhouse communities around the world, which have been doing very well with respect to the production; however, they are having extremely difficult times with respect to their operational expansion due to municipal, ground, or surface water shortages. Further, the quantity of available water may not be an issue or problematic, but the quality. Therefore, the quality of available water must be tested as well. Table 3.1 provides water quality classification, which divides water quality for greenhouse use into three classes: Class 1 water is good for all purposes; Class 2 water can be used in soil culture and substrate where leaching can be performed; however, it should not be used for nutrient film technique (NFT) plants or when nutrient solution recirculation is considered; and Class 3 should not be used for crops that are sensitive to salt.

When no municipal water is available, then well water and surface water can be used, again depending on the availability and quality. Additionally, building a pond or a cistern for collection and storage of rainwater might be considered; however, this as a sole source will not supply enough water for year-round crop production. It should be kept in mind that the irrigation system must be able to provide in average 8 L/m$^2$.day water for a fully grown crop. Some fully grown crops such as cucumbers are reported to require water up to 5 L/plant.day during sunny days (See Figure 3.1).

**TABLE 3.1**

Classification of Water Quality (OMAFRA, 2001).

| Class | Electrical Conductivity (mS/cm) | Sodium (ppm) | Chloride (ppm) | Sulfate (ppm) |
|---|---|---|---|---|
| 1 | 0.5 | 30 | 50 | 100 |
| 2 | 0.5–1.0 | 30–60 | 50–100 | 100–200 |
| 3 | 1.0–1.5 | 60–90 | 100–150 | 200–300 |

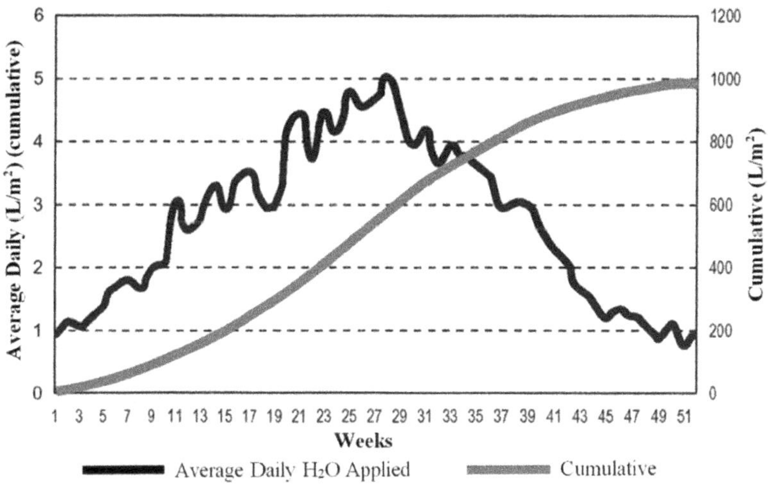


**FIGURE 3.1**
Average daily water application to greenhouse vegetables in Harrow, ON, Canada (OMAFRA, 2001).

Surface and subsurface drainage as well as ground slope are all important. A well-drained potential greenhouse site will make the construction much easier in the first place and help prevent the environmental stress on the greenhouse structure in the long run. A landscape that has a south-facing slope in northern hemisphere (a north-facing slope in the southern hemisphere) would provide an excellent greenhouse orientation for proper solar radiation in winter months and protect greenhouses from strong northerly winds in the northern hemisphere (southerly winds in the southern hemisphere) providing a natural windbreak as well as an adequate drainage. High winds drain the greenhouse interior heat and increase the operational costs. If a windbreak is provided naturally by choosing a sloping landscape in the planning stage, that is perfect; if not however, choosing a site that allows for planting trees as a windbreak, which will not cause too much shadowing, is still important.

In addition to the above-mentioned requirements, internet and telephone services are other essential services, which are required for a successful greenhouse operation, therefore, a proper site selection.

We know that greenhouse structures are built in many different variations to meet grower and other site-specific conditions and functional needs. NGMA (2004) defines a greenhouse as a *production greenhouse* if it is occupied for growing plants on a production or research basis, not normally occupied (without public access). Incidental use such as tending or moving products and other incidental uses is considered production activities, and therefore, not considered as occupied. Any other greenhouse structure used for the display and sale of horticultural crops and supplies is called as a *commercial greenhouse*. These two groups of greenhouses function in different ways and hence their functional and orientation needs are different, which are sometimes are dictated by the existing buildings and other structures. Hence, the best location, layout, and orientation (the direction of the greenhouse facing) for a greenhouse strongly depends on how the greenhouse is planned to be used and the position of the property with respect to a nearby public road and as mentioned, other nearby buildings and structures. Gutters oriented in a north–south orientation give a constantly changing shadow while the Sun moves from east to west throughout the day. Gutters oriented in an east–west orientation however cast a

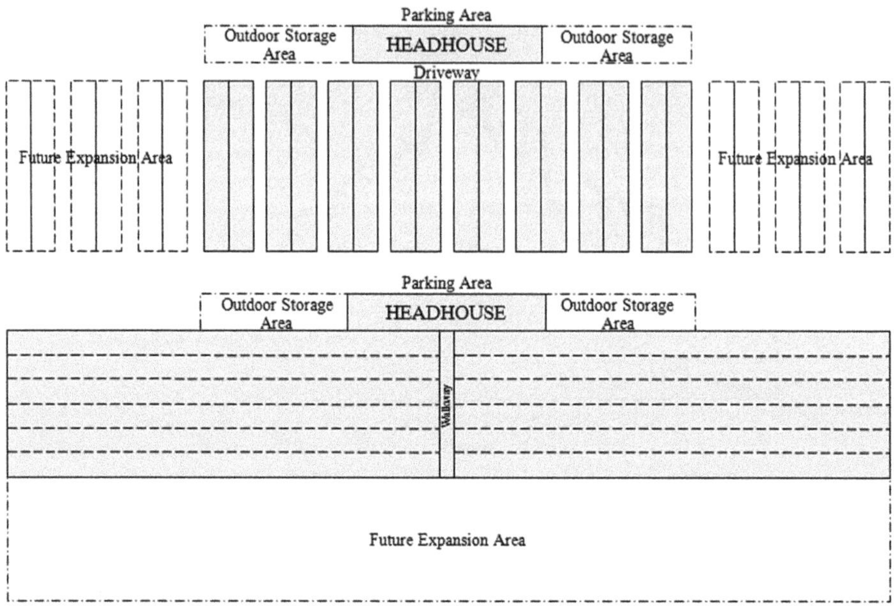


**FIGURE 3.2**
Layout examples for individual (or twin) greenhouses with separate headhouse (***above***) and gutter connected greenhouse with attached headhouse (***below***).

relatively permanent shadow as the Sun moves throughout the day, which creates a relatively poorer light uniformity. In any case, the layout of the facility should consider future expansion locating the headhouse (the work and indoor storage area where boiler rooms are typically located) to the north of the crop production area to prevent shading (Figure 3.2); separating the commercial area from the production area, and the customer and supplier traffic and parking; providing convenient access to people with wheelchairs and those assisted by walkers as well; and permitting windbreaks as discussed above and privacy to a residence, if applicable.

## 3.3 Design Loads

Greenhouse structures and all structural parts are designed and built to support four general types of loads, the weight of the structure (dead loads), the loads associated with building use (live loads), and snow and wind loads without exceeding the allowable stresses for the materials. Sometimes, a greenhouse is built as a lean-to or attached even-span greenhouse, which is actually added to an existing building. In such a case, we need to make sure that the existing structure is properly strengthened to withstand existing loads and the additional loads imposed by the lean-to or attached greenhouse. It should be kept in mind that the most unfavorable effect of the load combination in the structure, foundation, or member under consideration will govern the situation.

### 3.3.1 Dead Loads

These loads are the weight of all permanent construction material including but not limited to floor, walls, roofs, framing and glazing materials, and fixed service equipment. The actual weights of relevant building materials are used in estimating dead loads. These weights are typically tabulated and provided in handbooks for building designs.

### 3.3.2 Live Loads

These are the temporary loads generated by the use and occupancy of the greenhouse, and do not include the dead load or wind and snow loads. Live loads, both static and dynamic loads, can be exterior or interior loads resulting from the use and occupancy of the greenhouse. The static loads result from the weight, pressure from equipment, plants, stored products and the materials used for maintenance activities. The dynamic loads, on the other hand, result from the dynamic effect of material handling equipment such as forklifts. Exterior live loads may include the temporary loads of workers performing work on greenhouse roofs, and other temporary equipment such as scaffolds. Interior live loads, on the other hand, are temporary loads imposed on greenhouse structure by hanging objects, etc., for less than 30 days. If a live load is imposed on greenhouse structure for a continuous period of 30 days or more, then such loads are handled as permanent loads; and hence they are included as part of the dead load. Live dynamic loads due to materials handling and irrigation equipment are also treated as live loads.

### 3.3.3 Snow Loads

Snow load is a vertical load, which is based on expected ground snow accumulation at the greenhouse site, roof slope, covering material, and type of greenhouse structure, that is, whether the greenhouse is a single span or a gutter connected greenhouse, and whether it is a heated or an unheated greenhouse. Snow ground loads are based on the 50-year mean recurrence interval (with an annual probability of 0.02), and the readers of this book are referred to consult local sources for pertinent data. The nearby buildings should also be taken into account as they influence snow sliding from the roof of these buildings and snow drifting. Greenhouse gutter slopes should be designed carefully so that snowmelt and rain can smoothly run-off without accumulating. Enough space (at least 3 m or 10 ft) between greenhouses should also be left for snow accumulation. This space also prevents the risk of sidewall crushing due to snow sliding off greenhouse roof.

### 3.3.4 Wind Loads

The design wind loads for greenhouse structures as a whole or for individual components and glazing are based on wind speed data with a 50-year mean recurrence interval (with an annual probability of 0.02) and determined by an analytical procedure, which covers the velocity pressure, a gust response factor, and appropriate pressure or force coefficients. The readers of this book are referred to consult local sources for pertinent wind speed data. Wind design pressure, that is, the equivalent static pressure, is used for determining wind loads on greenhouses. The pressure due to wind load is assumed to act in a direction normal to the surface under consideration, either as a positive pressure directed toward the surface or as a negative pressure (suction) directed away from the surface. In calculating the design wind loads for components and covering materials of greenhouses, the pressure difference between opposite faces of the surface are taken into consideration. An exposure category is also determined for the site where the greenhouse is to be constructed, and reflects variations in surrounding ground surface roughness due to natural topography and vegetation as well as existing construction. Greenhouses have very small deadweights, that is, there is not much to resist the uplift force created by the wind; therefore, we need to make sure that enough anchorage to the ground is provided against the uplift.

It must be noted that no cookie cutter approach will work for determining the greenhouse design loads as each and every design project is a unique engineering project and each structural element must be designed with great care. The readers of this book are referred to MWPS (1987), NGMA (1996), and ASABE Standards (2018) for further information on loads.

## 3.4 Foundations and Floors

As for any building, a proper foundation is essential for the usefulness and life of a greenhouse. A greenhouse foundation resists the forces acting onto itself such as the weight of the greenhouse as a building, snow, wind, and its contents. A good foundation is able to distribute these forces so that movement of the greenhouse is small and uniform. The greenhouse foundations may be flagpole type placed directly in the ground, spread footings, or continuous concrete footings type foundations. Figure 3.3 shows some typical greenhouse foundations.

Flagpole footings are designed for vertical and horizontal loads. The footing must be able to handle the vertical loads and distribute the lateral bearing to the soil produced by a lateral load applied to the column. Sometimes, the flagpole type footings are backfilled with concrete or other filling material. The footings are typically the unrestrained type; however, a restrained flagpole footing design can also be employed when there is a slab on grade surrounding the flagpole footing. Spread footings as well are designed to distribute vertical and horizontal loads to the soil. The size of the footing is determined by the vertical load acting on the footing and the bearing pressure of the soil. Continuous footings as well are frequently used for building greenhouses, and often constructed as a base for the glazing. The footings are rest on level and undisturbed soil, and set below frost depths or to a minimum depth of 60 cm (24 in) below ground surface as shown in Figure 3.2. Concrete masonry, reinforced concrete, galvanized steel, or treated wood can be used for piers; and the wall between the piers can be built using poured concrete, precast concrete, masonry, or any decay resistant material. A continuous concrete or masonry foundation wall is built on poured concrete footings. When 6 m (24 ft) span greenhouses are in consideration, then a 15-cm (6 in) wall is considered sufficient, whereas a 20 cm (8 in) wall should be used for wider span greenhouses (Aldrich & Bartok, 1994). Pipe or tube foundations driven into the ground without footing should not be used for permanent greenhouse structures, as they do not provide much resistance to uplift in high wind locations. When they are used for temporary (less than a year) greenhouses such as pipe frame (hoop) greenhouses, then a minimum of 75 cm ground insert from grade should be used.

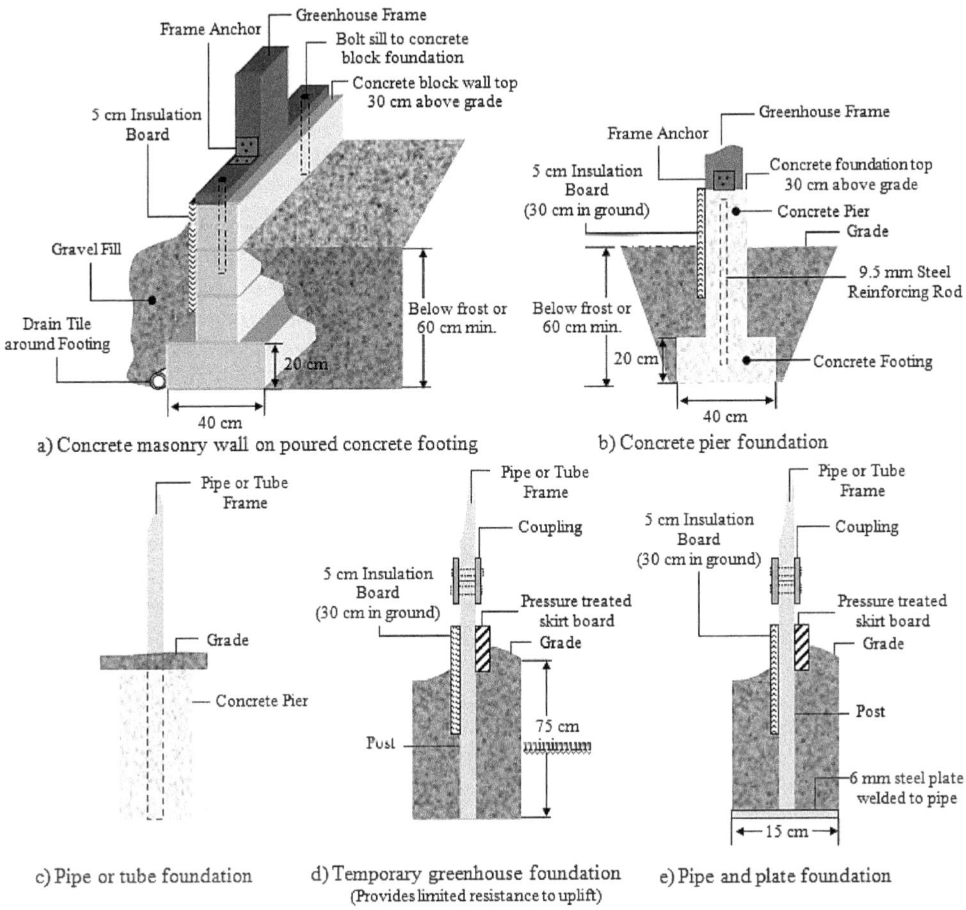


**FIGURE 3.3**
Typical greenhouse foundations.

The best greenhouse flooring materials are gravel, pea stone, and trap rock, which are designed specifically for greenhouses. A thickness of 15–20 cm (6–8 in) allows for good drainage, while helping to block out weeds. They can easily be sanitized as well with a mild bleach solution in case of disease. When, however, a smoother and harder flooring surface is desired, then a 5–7.5 cm (2–3 in)-thick porous concrete, which is made from uniform-sized aggregate and a cement–water paste can be used. Porous concrete flooring should only be used in areas where personnel and light vehicles such as garden tractors operate; however, they are not recommended for soil mixing areas where small particles have the potential to clog the pores of concrete and therefore prevent downward movement of water. The flooring in heavy traffic areas and aisles should be asphalt or concrete whose thickness depends on the traffic load; however, a flooring thickness of 7.5–10 cm (3–4 in) is recommended in such areas (ASABE Standards, 2018). Floor surfaces need to have proper slope for assuring surface drainage, and for the safety purposes, the final surface should have a rough broom finish for comfortable walking and riding materials handling machinery.

## 3.5 Frame Materials and Structural Forms

### 3.5.1 Frame Materials

Aluminum, reinforced concrete, steel, and wood are the materials typically used for building greenhouse frames either individually or in combination. Reinforced concrete is used for foundations and low walls only. Aluminum and galvanized steel are used for building almost all structural members and usually require no maintenance; however, they should be kept from direct contact with the ground so that they can be protected from corrosion. Painting the indoor surfaces of all frame materials naturally improve the light reflection in greenhouses. Wood, on the other hand, needs to be preserved against decay, especially when in contact with soil; however, the chemical preservatives used should pose no danger to plants or humans.

### 3.5.2 Structural Forms

The most common structural form of greenhouses is a straight sidewall and a gable roof. Post and beam, and post and truss arches are employed for forming the greenhouse gable structure. Any of these forms can easily be used to build a single span, a twin span, or a large gutter-connected greenhouse complex. Figure 3.4 shows some typical greenhouse shapes.

## 3.6 Covering Materials and Insect Screens

### 3.6.1 Covering Materials

Covering materials (also called glazing) have huge impact on plant production in a greenhouse system. Different glazing materials have different radiative properties; therefore, they influence the amount and spectral distribution of solar radiation eventually reaching the plant canopy, which directly affects plant growth. Furthermore, the type of covering

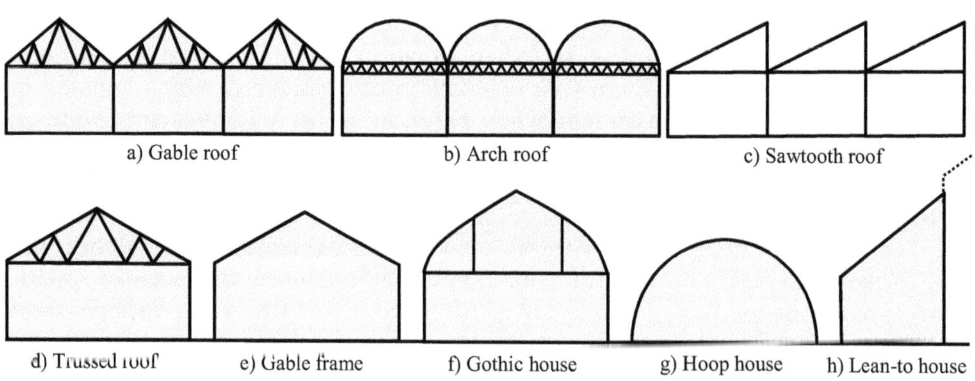


**FIGURE 3.4**
Common greenhouse shapes.

material used directly influences condensation on the surface of covering material itself, creating unfavorable environments for plants, and indirectly affects moisture and carbon dioxide concentrations in a greenhouse.

Giacomelli (1999) states that if the goal is to produce high quality plant products within efficient and cost-effective controlled environment systems, then the greenhouse design must consider (i) the type of crop and its associated cultural demands, and (ii) the production system, which will support the growth of the crop. And all other greenhouse systems including the greenhouse covering should properly support these two essential needs. This is quite a fair statement, and there are many commercially available transparent covering materials that include rigid materials (glass derivatives), thin films (normally double layer, air inflated polyethylene (PE), ethylene vinyl acetate (EVA), polyvinyl chloride (PVC), polyvinyl fluoride (PVF) and polyester films), and flexible plastics (such as fiberglass reinforced plastic (FPR), acrylic (PMMA), polycarbonate (PC), polypropylene (PP), PVC, and polyethylene terephthalate glycol (PETG) panels and the multiwall type materials such as polycarbonate, acrylics, polypropylene, PETG and laminar composites). Statistics Canada (2019), for instance, reported that the greenhouse industry in Canada in 2018 was a \$3 billion industry having a total greenhouse area of 2,336.5 ha, and 57% of which were plastic greenhouses while the remaining 43% were glass-covered greenhouses. This distribution was 62% and 38%, respectively, in the province of Ontario, where 65% of the Canada's greenhouses are located. Labor and heating costs are the major costs of greenhouse operations particularly in mid- and high-latitude countries including Canada. Annual labor costs typically range from 30 to 35% of the total cost of production, while heating costs typically ranging from 20 to 35% of the total production cost depending on the fuel prices. The covering material not only directly influences plant growth but also the cost of heating, cooling, and ventilation in a greenhouse due to their varying radiative properties and insulation effects.

As covered earlier in Chapter 2, when radiant energy falls on a surface, including a greenhouse covering, it is absorbed, reflected, or transmitted through the material. Thence, from the first law of thermodynamics, this relationship was expressed in Equation (2-85) as $a + t + r = 1$. As mentioned above, there exists a range of different greenhouse covering materials having different radiative properties, and Table 3.2 lists the transmissivities of different greenhouse glazing materials. However, there is no universal covering material, which would essentially provide the ideal environment for all plants in all climatic conditions. What is critical for plant growth and development is that the visible portion of the spectrum, that is, the photosynthetically active radiation (PAR) range, which covers the wavelengths from 0.4 to 0.7 μm. And the right covering material for a greenhouse is not only determined by the plant light requirements, but also the climate, longevity, handling requirements, and budget, and sometimes by the personal preferences. Plant requirements dictate the amount of PAR that the covering material would provide; however, the climate and budget determine the number of glazing involved for any type of glazing material used.

As Table 3.2 shows, the shortwave (solar) and longwave (thermal) transmissivities of covering materials are different. For instance, PE film is quite transparent to thermal radiation, which is particularly important on clear sky nights rather than overcast conditions, because greenhouse interiors will under such conditions have direct radiation exchange with the sky transmitting more thermal energy. In double PE greenhouses, the infrared (IR) film that blocks thermal radiation is typically installed as the inner layer to block the inside IR radiation and reduce heat losses throughout the night. In heated greenhouses, the savings have been reported to total from 10 to 20% depending on the sky conditions (Bartok,

**TABLE 3.2**

Transmissivities and Estimated Life of Some Common Greenhouse Glazing Materials. (Aldrich & Bartok, 1994).

| Material | | Shortwave Transmittance (%) | Longwave Transmittance (%) | Estimated Lifetime (Years) |
|---|---|---|---|---|
| Glass | Single (double strength) | 88 | 3 | 25+ |
| | Insulated Units | 75–80 | <3 | 25+ |
| | Solatex | 91–94 | <3 | 25+ |
| Acrylic | Single Wall Plexiglass | 93 | <5 | 20+ |
| | Lucite | 93 | <5 | 20+ |
| | Acrylite | 93 | <5 | 20+ |
| | Double Wall Exolite | 83 | <5 | 20+ |
| | Acrylic SDP | 83 | <5 | 20+ |
| Polycarbonate | Single Wall Dynaglass | 91–94 | <3 | 10–15 |
| | Single Wall Lexan Corrugated | 91–94 | <3 | 10–15 |
| | Macrolux Corrugated | 91–94 | <3 | 10–15 |
| | Double Wall Macrolux | 83 | 23 | 10–15 |
| | PolyGal | 83 | 23 | 10–15 |
| | Lexan Dripgard | 83 | 23 | 10–15 |
| Fiberglass Reinforced Polyester | Lascolite | 90 | <3 | 10–15 |
| | Excelite | 90 | <3 | 10–15 |
| | Filon | 90 | <3 | 10–15 |
| | Double Wall Roof Panels | 60–80 | 7–12 | 7–12 |
| Polyethylene Film | Tufflite | <85 | 50 | 3+ |
| | Standard UV | <85 | 50 | 3+ |
| | Tufflite Dripless | <85 | 50 | 3+ |
| | Fog Block | <85 | 50 | 3+ |
| | Sun Saver | <85 | 50 | 3+ |
| | Tufflite Infrared | <85 | 50 | 3+ |
| | Dura-Therm | <85 | <20 | 3+ |
| PVC Corrugated | Bio 2 | 84 | <25 | 10+ |
| Polyester Film | Llumar | 85–88 | 30 | 7–10 |
| | Mylar | 85–88 | 30 | 7–10 |
| | Melinex | 85–88 | 30 | 7–10 |
| PVF Film | Tedlar | 92 | 21 | 10+ |

2010). The IR film costs only a few more cents per square meter compared to the ordinary PE film; however, it provides huge savings with only a few months of payback periods.

As discussed earlier in Chapter 2, the overall rate of heat transfer (U-value) coefficient (Table 3.3) of a greenhouse glazing is important for the heat loss; and the lower the tabulated U-value is, the higher the resistance to heat loss is. The commercially used "R-value" concept (=1/U) expresses the resistance to heat transfer. It means the higher the R-value of a glazing material is, the higher the resistance to heat transfer is. Air infiltration is another important component of heat loss, which is not part of the U-value calculation procedures, but it must also be dealt with. Greenhouse covering systems vary quite a bit in air infiltration performances, which is included in heat loss calculations separately. Air exchange values for greenhouses with different glazings are provided in Table 3.4.

Single layer covering usually provides higher total solar (shortwave) radiation, therefore higher PAR, compared to multiple layer covering (used in many high latitude greenhouses); however, the latter provides more insulation against heat loss, which is quite a significant consideration especially in colder climates. If the greenhouse is operated during the heating season, inflated double PE films and acrylic or PC panels are quite

**TABLE 3.3**

Overall Heat Transfer Coefficients of Some Common Greenhouse Glazing and Building Materials (ASABE, 2018).

| Greenhouse Glazing | U (W/m$^2$.°C) | U (Btu/h.ft$^2$.°F) |
|---|---|---|
| Glass-single layer | 6.2 | 1.1 |
| Glass-single layer (low emissivity) | 5.4 | 0.95 |
| Glass-double layer (sealed) | 3.7 | 0.65 |
| Glass-triple layer (1/4″ space)[a] | 2.8 | 0.50 |
| PE film-single layer | 6.2 | 1.1 |
| PE film-double layer (separated) | 4.0 | 0.7 |
| PE film-double layer (IR inhibited) | 2.8 | 0.5 |
| Fiberglass-single layer (corrugated) | 5.7 | 1.0 |
| Polycarbonate-single layer (corrugated) | 6.2–6.8 | 1.1–1.2 |
| Rigid polycarbonate-double layer | 3.2–3.6 | 0.56–0.63 |
| Rigid acrylic-double layer | 3.2 | 0.56 |
| Rigid acrylic-double layer (filled w/poly pellets) | 0.57 | 0.10 |
| PE film-double layer (over glass) | 2.8 | 0.50 |
| PE film-double layer + internal thermal blanket | 2.5 | 0.44 |
| Glass-single layer + internal thermal blanket | 4.0 | 0.70 |
| **Wall Materials** | | |
| Concrete block, 20 cm (8″) | 2.9 | 0.51 |
| Poured concrete, 15 cm (6″) | 4.3 | 0.75 |
| Concrete block, 20 cm (8″) + 5 cm (2″) foamed urethane board[b] | 0.4 | 0.07 |
| Concrete block, 20 cm (8″) + 5 cm (2″) foamed polystyrene board[b] | 0.6 | 0.10 |
| Poured concrete, 15 cm (6″) + 5 cm (2″) foamed urethane board[b] | 0.4 | 0.07 |
| **Perimeter** | **(W/m.°C)** | **(Btu/h.ft.°F)** |
| Perimeter (uninsulated) | 1.4 | 0.80 |
| Perimeter (insulated)[c] | 0.7 | 0.40 |

[a] Roberts et al. (1989)
[b] Aldrich and Bartok (1994)
[c] Insulation equivalent to 5 cm (2″) of foamed polystyrene extending 60 cm (24″) into the ground.

**TABLE 3.4**

Air Exchange Values for Greenhouses (Aldrich & Bartok, 1994).

| Greenhouse Glazing | Air Exchange (1/hour)[a] |
|---|---|
| New Construction – Glass or Fiberglass | 0.75–1.50 |
| New Construction – Double PE Film | 0.5–1.0 |
| Old Construction – Glass (Good Condition) | 1.0–2.0 |
| Old Construction – Glass (Poor Condition) | 2.0–4.0 |

[a] Air exchange is the internal air volume exchange per hour. Low wind speeds or protection against reduced air exchange.

popular in high latitude regions. For normal operation conditions, a single layer PE film is quite enough; however, the double layer reduces heat loss at night by about 40% (Bartok, 2010).

Normally, covering materials influence the spectral distribution of solar radiation reaching the plant canopy, and high light transmission is absolutely essential to plant growth. Light passing through a greenhouse glazing is split into direct and diffused radiation. Direct light creates shadows while diffusion eliminates shadows and ensures a more uniform light distribution in the greenhouse so that it can even reach the lower parts of the plants, prevent burnings and offer a moderate cooling effect. Therefore, manufacturers have added the light diffusion property to their glazing materials and this property enhances the amount of diffused solar radiation reaching the plants and lower parts of the plant canopy while keeping the PAR levels the same, which is particularly important for tall vegetable crops. Increased diffused solar radiation in the lower parts is important for reducing fungal development and insect propagation as well.

Where plenty of solar radiation is available, photoselective films reflecting or absorbing specific wavelengths of the solar spectrum that is not required for plant growth may decrease greenhouse temperatures quite a bit, hence lowering the need for greenhouse cooling and irrigation, and improving the thermal comfort in work environment. Ultraviolet (UV)-blocking films, which absorb UV-radiation up to 390 nm, for instance, reduce some insects and control some fungal diseases as they block out a significant part of the visible range of insects. However, if a greenhouse operation is using the bumblebees to pollinate plants, then using a film that transmits at least some UV light through should be considered as the bees need the UV light for navigational purposes.

Moreover, the PE films have a typical tendency to attract dust particles, which can accumulate on the PE film over a long period of exposure and lead to reduction in light transmission, which has a negative impact on plant growth and yield. Manufacturers now use special anti-dust agents in the film, which prevent dust particles from sticking on the film.

Furthermore, in greenhouses, condensation is quite a common problem, whose droplets reduces light transmission and can commonly lead to disease problems particularly when they drip onto plants. Anti-dripping (AD) films, also referred to as anti-condensation (AC) films, prevent early morning and late afternoon fog formation in the greenhouse. The anti-condensation agent reduces the surface tension and allows condensation to flow rather than forming droplets. The additives are slowly washed out by water in time; therefore, the effect lasts for up to 2 years or so.

Mechanical strength (e.g., hail and impact resistance, resistance to windborne debris, tear resistance of plastic films), longevity (the usable life of covering materials to function effectively internally and externally in a commercial greenhouse operation), flammability, and chemical and material compatibility are also important considerations in choosing an adequate glazing material. Sometimes, the decrease in light transmission rather than the loss of physical strength determines the useful life of the glazing material.

Generally speaking, while there is no magic glazing material, any of the discussed materials would perform very well in greenhouses. The specific choice for any given situation however depends on the crop light requirement, cost, climate, and sometimes, personal preferences.

### 3.6.2 Curtain Systems

See Chapter 11 for curtain systems and relevant discussions.

### 3.6.3 Insect Screens

Integrated pest management (IPM) of greenhouse crops does not depend upon pesticide application and includes screening to physical control flying insect pests moving into the greenhouse from outside. Common greenhouse pests such as aphids, thrips, and whiteflies as well as less-common pests like tarnished plant bug are all included in this control. Benefits of insect screens have been demonstrated, and when they are properly chosen, installed, and used, reduced insect and pest levels as well as pesticide use and exposure, and improved pest control effectiveness are reported. Disadvantages of using insect screens include difficulties in sizing and fastening to structure, reduced ventilation, and screen maintenance. For instance, smaller hole-size screens exclude more pests than larger hole-size screens; however, they are much more resistant to airflow, causing higher static pressure drops, insufficient air exchange, higher greenhouse temperatures, higher energy consumption by the fans, and excessive wear on the fan motors. The insufficient air exchange, that is, the reduced ventilation can however be addressed by increasing the vent surface area and building a screened framework around the vent opening so that the vents provide sufficient air exchange to allow proper cooling of the greenhouse.

There exist however many insect screen types in the market today, and each may have its own merits depending on the application. Basically, the challenge is to match the proper type of insect screen to the need, and the mesh size of the screen is determined by the size of the target pests. Large mesh sizes of the household screens are not adequate for excluding most major greenhouse pests; and for common greenhouse pests, smaller mesh sizes are needed. Typically, growers have multiple pests of concerns; therefore, it would be wise to choose a mesh size, which would exclude the smallest of the concerned pests, usually thrips or whiteflies. Table 3.5 shows the screen mesh sizes needed to exclude major greenhouse pest species.

The installation and type of screening structure to be installed depends upon a number of considerations including the ventilation, duration of use, cost, durability, and maintenance. Figure 3.5 shows some common insect screen installations. Generally speaking, passively ventilated greenhouses are more difficult for screened operations and the ventilation in such greenhouses are not as effective as in forced ventilation systems. And some applications may, on the other hand, only need screens during certain

**TABLE 3.5**

Screen Mesh Sizes Needed to Exclude Major Greenhouse Pest Species (Bethke, 1994).

| Pest | Minimum Mesh Size[a] (microns) |
|---|---|
| Serpentine leafminer (*Liriomyza trifolii*) | 608 |
| Citrus leafminer (*Phyllocnistis citrella*) | 435 |
| Green peach aphid (*Myzus persicae*) | 434 |
| Melon aphid (*Aphis gossypii*) | 355 |
| Greenhouse whitefly (*Trialeurodes vaporariorum*) | 288 |
| Silverleaf whitefly (*Bemisia argentifolii*) | 239 |
| Western flower thrips (*Frankliniella occidentalis*) | 215 |

[a] 1 millimeter = 1,000 microns (μ)

a) Ridge and sidewall type

b) Box type

c) Lean-to type

**FIGURE 3.5**
Common insect screen installations. (a) Ridge and sidewall type. (b) Box type. (c) Lean-to type.

times or seasons of the year depending on when the concerned pests appear. Sometimes, heavy-duty, long-term use considerations are required, and in such cases, stronger screens should be used.

There are mainly four types of screens available on the market based on the materials used: stainless steel and brass, polyethylene, acrylic, and nylon. The first of these is the longest-lasting, however, the most costly option. Polyethylene screens come in different shapes such as monofilament (each thread is a solid strand), multifilament (strands made of many fibers make a strong screen), and low-cost film (with punched microholes restricting airflow). The acrylic yarns are multifilament screens as well, while nylon is a low-cost, less-durable, shorter-term, and light-duty option. Most commercial screens are reported to have 3–5 years life span.

Small pore size of screens means that they are quite susceptible to blockage with dust and other deposits, which reduce air flow through the screens and may cause excessively high temperatures in greenhouse if they are not maintained properly. Hence, it is extremely important to keep the screens clean.

## Nomenclature

| | |
|---|---|
| *a* | absorptivity |
| **EVA** | ethylene vinyl acetate |
| **FPR** | fiberglass reinforced plastic |
| **IR** | infrared |
| **PAR** | photosynthetically active radiation |
| **PC** | polycarbonate |
| **PE** | polyethylene |
| **PMMA** | acrylic |
| **PP** | polypropylene |
| **PVC** | polyvinyl chloride |
| **PVF** | polyvinyl fluoride |
| *r* | reflectivity |
| *t* | transmissivity |
| **U** | overall heat transfer coefficient |
| **UV** | ultraviolet |

## Bibliography

Aldrich RA and Bartok, JW (1994) *Greenhouse Engineering. NRAES-33. Northeast Regional Agricultural Engineering Service*. Cornell University, Ithaca, NY, USA.

ASABE Standards (2018) Standards, Engineering Practices, and Data. *V21(1)*. St. Joseph, MI, USA.

ASAE (1989) ASAE Engineering Practice. *Heating, ventilating and cooling greenhouses*. Hahn, R.H. and E.E. Rosentreter (eds.), 36th edition, pp. 452–455, ASAE, St. Joseph, MI.

Baker JR, Bell, ML, Shearin EA (1997) Insect Screening. Information Note 104: Ornamentals and Turf, Integrated Pest Management. North Carolina Cooperative Extension Service. NCSU, p. 4

Bartok JW (2010) Choose the right plastic film for your needs, *Greenhouse Management*, May 2010, Valley View, Ohio, USA.

Bell ML and Baker JR (1997) Choose a greenhouse screen based on its pest exclusion efficiency. *North Carolina Flower Growers' Bulletin*. 42(2):7–13.

Bethke JA (1994) Considering Installing Screening? This is what you need to know. *Greenhouse Manager*, April 1994, 34–36.

BP (2017) http://bp.com/statisticalreview. Accessed October 21, 2017.

EIA (2016) *United States Energy Information Administration, International Energy Outlook 2016*. Washington DC.

Friend, HH and Decoteau, DR (1990) Spectral transmission properties of selected row cover materials and implications in early plant development. *Proceedings of the 22nd National Agricultural Plastics Congress, Montreal*, Quebec, Canada, pp. 1–6.

Giacomelli GA (1999) *Greenhouse Glazings: Alternatives Under the Sun. Department of Bioresource Engineering, Cook College, Rutgers University*. Available at https://ceac.arizona.edu/sites/default/files/G%20Systems.pdf. Accessed May 10, 2020.

Godbey, LC, Bond, TE and Zornig, HF (1979) Transmission of solar and long-wavelength energy by materials used as covers for solar collectors and greenhouses. *Transactions of the ASAE* 22(5):1137-1144.

Goldammer T (2019) *Greenhouse Management: A Guide to Operations and Technology*. Apex Publishers, Centreville, Virginia, USA

Gray HE (1992) Greenhouse glazing uncovered. *Greenhouse Grower* 10(8):5457.

IEA (2012) *International Energy Agency*, Key World Energy Statistics, OECD, Paris.

MWPS (1987) *Midwest Plan Service, Structures and Environment Handbook*. 11th Edition. Ames, Iowa, USA.

NGMA (1996) *National Greenhouse Manufacturers Association Standards for Design Loads in Greenhouse Structures*. Structural Loads, Harrisburg, PA, USA.

NGMA (2004) *National Greenhouse Manufacturers Association Structural Design Manual*. NGMA, Harrisburg, PA, USA.

OMAFRA (2001) Growing Greenhouse Vegetables. *Ontario Ministry of Agriculture, Food and Rural Affairs*. Publication 371. ISSN: 1492-6601. Toronto, Canada. p. 77.

OMAFRA (2003) Production Recommendations for Greenhouse Floriculture. *Ontario Ministry of Agriculture, Food and Rural Affairs*. Publication 370. ISSN: 1492-6512. Toronto, Canada. p. 39.

Roberts WJ, Bartok JW, Fabian EE, Simpkins J (1989) *Energy Conservation for Commercial Greenhouses. NRAES-3. Northeast Regional Agricultural Engineering Service*. Cornell University, Ithaca, NY, USA.

Sase S and Christianson LL (1990) Screening Greenhouses – Some Engineering Considerations. Proceedings of the 1990 Northeast Agricultural/Biological *Engineering Conference*; pp. 1–13.

Statistics Canada (2019) *CANSIM Tables: 32-10-0019-01; 32-10-0022-01; 32-10-0023-01; 32-10-0026-01; 32-10-0245-01*. Ottawa, Canada.

# 4

# Greenhouse Environment

## 4.1 Introduction

The greenhouse industry already has a major share in the economies of some countries, and it is growing rapidly in other places as well. Countries having major greenhouse productions can be listed as China with a total greenhouse area of over 2.7 million ha, South Korea with more than 55,000 ha, Spain with more than 52,000 ha, Japan close to 50,000 ha, and Turkey with close to 40,000 ha. One major factor hindering future expansion of this industry however is the cost required to provide proper environmental control for plant growth. Environment is defined as the total of all external conditions surrounding plants, which comprise physical, chemical, and biological factors; that is, no plant can exist independently from an environment. Factors such as light and radiation, temperature, relative humidity, carbon dioxide ($CO_2$), water, nutrition, insect, and disease stresses are the key environmental components affecting optimum plant growth and yield. And greenhouses provide a suitable environment for the intensive production of a variety of plants. They are designed to provide optimal control of solar radiation, temperature, relative humidity, $CO_2$ levels, and air movement in the aerial environment. And in hydroponic greenhouses, nutrient levels and root temperatures can also be controlled.

From the engineering point of view, the environment provides the initial and boundary conditions in plant biosystems; therefore, an engineer is concerned with i) how much food, feed, or fiber will plants produce in a specific environment; ii) if the environment is not ideal, what needs to be changed and how much; iii) if the cost of changing the environment be compensated for by a profitable increase in output; and iv) what the pertinent social (if any) and environmental impacts are. In order to answer these questions, an engineer needs to have a basic understanding of the physiological bases for plant's responses to the environment and the physiological aspects of the environment and their effects on the functioning of a whole plant. Hence, modification of the abovementioned environmental parameters provides ample opportunities for engineering applications in greenhouses. For instance, low light conditions at higher latitudes in winter are the limiting factor and not usually adjusted except where artificial lighting is used as supplemental lighting at seedling or plant raising stages. Therefore, greenhouses are built with proper glazing materials in the first place to maximize natural light penetration so that plant growth can also be maintained during especially low light seasons. Large energy costs are frequently incurred to maintain the required thermal and radiant environments in greenhouses during both winter and summer seasons. During the winter season, supplemental heat is needed at night, and it may be needed during the day as well. Due to high solar radiation loads, shade curtains, whitewashing the greenhouse exterior, or some other cooling must be provided to maintain an appropriate environment during warm weather.

In order to provide economically optimal microenvironments for plant growth, producers can use or control the number of glazing layers, insulation curtains or screens to reduce longwave radiation losses at night, reduced ventilation rates, evaporative coolers, and different shading mechanisms to control incoming solar radiation. Actually, almost all of these mechanisms have successfully been used by commercial growers, and especially evaporative cooling and supplemental lighting have extended the greenhouse production to almost year-round operation.

However, while lower ventilation rates reduce the heating requirements, they may cause high humidity levels in greenhouses and low $CO_2$ levels if $CO_2$ enrichment is not provided. Consequently, low ventilation rates may adversely affect plant responses. If conventional ventilation rates are used to maintain reasonable relative humidity levels within greenhouses, they result in higher heating energy requirements (high operating costs).

Similarly, installation of curtains or screens reduces longwave radiation losses at night and lowers operating costs; however, it would introduce some additional cost, which must eventually be justified. Shading devices reduce warm weather cooling requirements; though justifiable, most of them also increase both initial and pertinent operating costs.

In addition, heat pumps can as well be used effectively for winter heating and warm weather cooling in greenhouses, including closed and semi-closed loop greenhouse operations. These systems however involve complex tradeoffs between initial and operating costs for cooling and heating, plant responses to various environmental factors, and the strategies used to regulate temperature, humidity and $CO_2$ levels in the greenhouse. Although the use of heat pumps to heat and cool greenhouses permit the use of low ventilation rates, a semi-closed or a closed loop operation that would totally eliminate the ventilation heat loss, it also results in higher initial cost. Special attention must also be given to the pertinent operational (dehumidification) strategies in maintaining acceptable relative humidity levels within greenhouses.

This chapter will briefly cover basic plant physiology first, and then move to the impact of major environmental parameters and their interactions on greenhouse plant production. The parameters covered include light and radiation, temperature, relative humidity, $CO_2$, and pollutants and air movement.

## 4.2 Basic Plant Physiology

Plant physiology is defined as the science of normal function that can be investigated at the cellular level, the functioning of discrete organs, and the functioning of whole plant. For engineers, however, in order to be able to put scientific knowledge into practical uses, it is more important to understand the basic physiology of plants from the viewpoint of its role in the whole plant system as influenced by the parameters of environment.

Green plants are photosynthetic autotrophs that manufacture (synthesize) their own food from the inorganic raw materials (carbon dioxide from the air and water from the root zone) by photochemical and biochemical processes with the help of chlorophyll (green pigment that absorbs light) and using the energy absorbed from light, which is called *photosynthesis*. The classical formulation of photosynthesis is provided below, which shows that the plant stores energy in the form of carbohydrates:

$$6CO_2 + 12H_2O \xrightarrow[(2{,}880\,kJ/mol\,glucose)]{Light} C_6H_{12}O_6\,(sugars) + 6H_2O + 6O_2$$

The stored sugars are used in the process of growth and production of biomass. Therefore, maintaining optimum light, temperature, $CO_2$, water, and nutrient regimes ensures that a desired yield or form of plant is obtained. A vegetative plant, for instance, produces more leaves than flowers and fruits; therefore, their environmental requirements are different from a generative plant, which typically has smaller leaves, plenty of flowers and fruits compared to the size of the plant. However, for optimum plant development, the environmental parameters need to be adjusted so that the plant can have the vegetative phase at certain times of the year and then move toward the generative phase.

Plants obtain their carbon dioxide and oxygen by the diffusion of gas through their stomata (tiny openings on leaves); and the gas exchange ($O_2$ uptake and $CO_2$ release) with the environment with a release of energy, related to the breakdown of reserves (sugars), is called *respiration*. And the classical formulation of respiration is provided below:

$$C_6H_{12}O_6\,(sugars) + 6O_2 + 6H_2O \xrightarrow{\left(-2{,}880\frac{kJ}{mol}\,glucose\right)} 6CO_2 + 12H_2O$$

The released energy is used for supporting different plant processes such as root, leaf, and bud formation and development. Respiration common to all plants is called *dark respiration*, which does not depend on light, and hence, occurs during 24 hours, day and night. Some plants, however, have an additional respiration mechanism known as *photorespiration*, which occurs only during the time when light is available. The green plants can therefore be classified into three major groups on the basis of their photosynthetic mechanisms: i) $C_4$ plants whose stomata are open during the day and dark respiration occurs during the day and night (e.g. corn, sugar cane, sorghum); ii) $C_3$ plants whose stomata are open during the day and dark respiration occurs during the day and night, and additionally, photorespiration takes place during daytime (e.g. virtually all other species); iii) CAM plants whose stomata are open at night and dark respiration occurs during the day and night (e.g. pineapple). When $O_2$ concentration increases, the rate of net photosynthesis in $C_3$ plants is inhibited due to the increased photorespiration (*Warburg Effect*); therefore, modulating the $CO_2$ concentration in the aerial environment ($CO_2$ enrichment) helps tremendously for developing a highly productive plant. This topic will be explored further in the chapter later.

Under low light and temperature conditions (early winter months), for instance, assuming that proper watering and nutrient regimes are provided, a greenhouse plant typically has a vegetative growth having large leaves and flowers, and relatively small fruits. Under such conditions, if the temperature is elevated under a controlled fashion, then a better plant growth is achieved under decreased relative humidity conditions. We can modulate the environmental parameters further by manipulating the day and night temperatures. If a higher temperature is maintained during the day, for instance, the plant manufactures more sugar and grows faster. The temperature is lowered at night for translocating the sugars manufactured (during the day) in the fruits, leaves, stems, and the growing points so that a thicker and heavier plant can be obtained. Not only adjusting the temperatures and determining when to make these adjustments, but also modulating the other environmental parameters are the tools engineers always use in designing control systems so that growers can obtain a well-balanced and highly productive plant in their greenhouses.

## 4.3 Light and Radiation

Light energy is required directly to sustain the growth of all green plants. As we have seen earlier in the classical formulation of photosynthesis, light must always be present for green plants to perform photosynthesis, but temperature and $CO_2$ concentration may modify its influence.

### 4.3.1 Photosynthetically Active Radiation Range

Light intensity, light quality, and light duration all influence plant development, and plants respond primarily to the levels and duration of light within *photosynthetically active radiation* (PAR) range (400–700 nm), the visible portion of the spectrum. Figure 4.1 shows the "McCree Curve", that is, the relative photosynthetic response of typical plants. As we can see, blue and red are the most photosynthetically efficient wavelengths, but other colors such as green and yellow are quite useful as well. With the help of developments in LED technology, every passing day, we are learning more about how light quality influences plant growth, development, flowering, antioxidant, and vitamin contents.

#### *4.3.1.1 Red Light*

Red light (600–700 nm) is quite effective at regulating plant growth and development. It is the most efficient light for photosynthesis as chlorophylls have absorbance peaks in the red band of the spectrum. Plants primarily use two photoreceptors termed as cryptochromes and phytochromes that mediate flowering of plants with a photoperiodic flowering

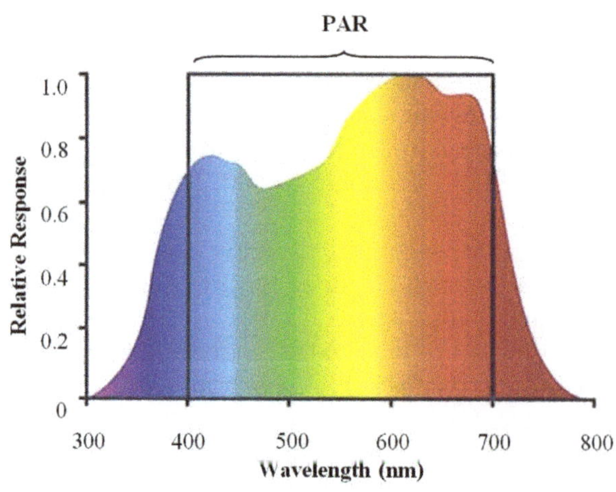


**FIGURE 4.1**
Relative photosynthetic response of typical plants.

response (photoperiodic lighting is used at a low intensity to inhibit flowering of short-day plants and promote flowering of long-day plants), which are highly sensitive to the blue and UV-A light regions and the red and far-red light (700–800 nm) regions, respectively, to determine the length of the day for *photoperiodism* (a response based on the portion of light to darkness in a 24-hour cycle). Phytochromes are used as temperature sensors as well as for photomorphogenesis in seed germination. Plants are overly sensitive to low intensities of red light, which can inhibit flowering of short-day plants and promote flowering of some long-day plants when provided at night. In short, it can be stated that red light is the most dominant color for both photosynthetic and photoperiodic lighting. Longer wavelengths including red light can penetrate the growth media such as soil as well; and therefore, they are used by seedlings to determine in what direction to grow.

However, plants grown under red light only (e.g. LEDs) have been reported not having adequate growth characteristics (e.g. thin and large leaves, taller plants) due to extension growth. When a small amount of blue light is added, however, the extension growth is inhibited. In greenhouses, on the other hand, plants receive the full solar spectrum and therefore the color provided by supplemental lighting does not have much impact on extension growth. The addition of far-red radiation as well has been reported to be effective for stimulating flowering of many long-day plants.

#### 4.3.1.2 Blue Light

Blue light (400–500 nm) as well is one of the most essential light ranges for photosynthesis and high-quality vegetative growth and flowering. Blue light also promotes the contents of antioxidants and certain vitamins. And usually, a low blue light intensity is sufficient for fully functional photosynthesis. Extension growth is generally suppressed by blue light; and plants grown with blue light have a narrow internodal spacing and stronger stems along with smaller, thicker, and darker green leaves. Moreover, it has been reported that blue light can promote flowering of long-day plants and inhibit flowering of short-day plants at higher intensities; however, it does not regulate flowering of most daylength-sensitive crops at low light intensities. Besides, blue light regulates the stomatal openings (tiny pores on leaves), which control transpiration (water loss) and carbon dioxide uptake for growth. Blue light contributes in leaf coloration via anthocyanin production as well, and plants as a result produce appealing vegetation. Furthermore, blue range influences plant morphology through plant cryptochromes, which are sensitive to blue light and tell the plant to grow toward the light.

#### 4.3.1.3 Green Light

Figure 4.1 shows that plants, even though not as efficiently as blue or red light, also use green light (500–600 nm) very well. Chlorophylls do not absorb green light that well and plants reflect more green light compared to other wavelengths (that's why we see green plants in green). Green light, on the other hand, helps mitigate the performance of lower leaves because it penetrates further into the canopy, and as well can penetrate deeper in the leaf mesophyll and energize chloroplasts and therefore further enhances photosynthesis. Green light has high energy transfer efficiency (quantum yield), which means that more energy reaches the reaction centers and can be used for photosynthesis under high light intensity when the chlorophylls and pigments on the upper leaf surface are already saturated by blue and red lights, much of which is absorbed on the upper leaf surface by flavonoids and carotenoids before reaching the reaction centers and therefore leaving chlorophylls and pigments deeper in the leaf not saturated. Green light is particularly useful under high light intensity conditions.

As discussed earlier, phytochromes and cryptochromes mediate flowering of plants with a photoperiodic flowering response. These pigments absorb green light as well, and recent research findings have shown that in many plants, green light is as effective at regulating flowering of long-day plants as the same intensity of red plus far-red radiation.

### 4.3.2 Non-PAR Range

Both solar radiation (see Figure 2.31) and supplemental lighting (see Chapter 7) provide PAR and non-PAR in greenhouses. Non-PAR from 750 nm to 3,000 nm, for instance, contributes to the air and soil temperatures in greenhouses while increasing the plant temperatures. Some non-PAR may contribute to crop responses as well.

#### *4.3.2.1 Ultraviolet Radiation*

A non-PAR range, ultraviolet (UV) radiation is defined as the part of the electromagnetic spectrum from 1 nm to about 400 nm (Figure 2.31), a biologically active radiation that regulates plant growth. As discussed in Chapter 2 earlier, the Earth's atmosphere (molecular oxygen and ozone) filters out UV radiation with a wavelength smaller than 280 nm, which is called the UV-C range. UV-A is defined as the UV radiation with a wavelength between 315 and 400 nm. Regular optical (glazing) materials such as glass and acrylic do not transmit UV radiation with a wavelength smaller than 360 nm effectively. And chlorophyll *a* has an absorption peak around 415 nm. Also, plants have a photoreceptor called UVR8 (responsible for triggering the stress response in plants), which is quite sensitive to UV-B light covering a range from 280 nm to 315 nm centered at about 285 nm. The stress response in plants can be manipulated for growing stronger plants if done carefully. It has also been shown that low doses of UV-B can, for instance, increase tetrahydrocannabinol (THC) levels, the main active ingredient, in cannabis as well as other secondary metabolites such as terpenes.

Overall, UV light increases leaf coloration, thickness, and waxiness and promotes shorter stems and smaller leaves inhibiting extension growth. Recent research has also shown that plants grown under some UV light usually have less insect feeding on leaves (higher resistance to pests) and are less vulnerable to fungal pathogens (suppressed under UV light) compared to plants grown under little or no UV. Therefore, it seems that there is a potential to use particularly UV-B and UV-C wavelengths to kill pathogens on leaves. UV-C and sometimes UV-B are also used for disinfecting water, which is another important potential use of UV light in greenhouse operations.

Furthermore, as discussed earlier in the previous chapter, UV is also used by different insects including bumble bees to help navigate to flowers. If a greenhouse environment is totally free of UV light, pollinators like bumble bees and pests may have navigation problems in locating flowers.

#### *4.3.2.2 Far-Red Light*

Another non-PAR range is far-red light, which covers a range approximately from 700 to 800 nm. As discussed earlier, plant photoreceptors are quite sensitive to red and far-red light, which are used by plants to evaluate their growth environment. For instance, in full sunlight, the proportion of red light to far-red inside a greenhouse is high, and as it is well known, plants respond to this very well. When the solar *daily light integral* (DLI) inside a greenhouse is considerably low, then supplemental lighting is effective at increasing plant growth, and including some far-red light might be useful for accelerating flowering of especially long-day plants. Far-red light does not have any influence on flowering of day-neutral or short-day plants. When the DLI is low, however, providing some red and far-red light during long nights can accelerate flowering of some species such as tomatoes. Under overcast or shade conditions, however, the proportion of far-red to red light is high, which initiates a phenomenon known as the shade avoidance response in plants. Under such conditions, in order to reach into the full sun, plants use all their energy for growing tall, therefore, neglecting vegetative growth. This obviously leads to an undesirable tall and weaker plant growth. The proportion of red to far-red light is manipulated by the photoperiodic plants for determining the time of year, so that they would know when to start flowering. Also, as the Sun sets, the proportion of far-red increases; and using this information, photoperiodic plants determine the length of the day. Once the plant, either short-day or long-day, determines that the length of day is adequate, it then performs flowering. The effectiveness for photoperiod response in plants typically peak at wavelengths of 660 nm (red) and 730 nm.

### 4.3.3 Photoperiodism

At this point, it would be wise to talk about photoperiodism little further. Daylength (photoperiod) influences the performance of some plants exerting a basic effect on flowering, fruit growth, and stem elongation. There exists four basic daylength groups of plants: i) short-day plants, which require a period of darkness longer than a critical length, usually in fall (e.g. chrysanthemum, poinsettia, aster); ii) long-day plants, which require a period of darkness shorter than a critical length, usually in summer (e.g. lettuce, most grains); iii) Daylength-intermediate plants, which flower only within a narrow range of daylength, usually between 10 and 14 hours (e.g. sugarcane, coleus); and iv) day-neutral plants, which are insensitive to the length of the day (e.g. tomato, carnation, sunflower, corn).

In summary, depending on this classification, an artificial single light source (e.g. sodium lamps) as the sole source of lighting may not be adequate in greenhouse environments; therefore, the environment may require the added blue, red, or infrared emissions, which will be discussed further in Chapter 7.

### 4.3.4 Light Intensity and Uniformity

Light intensity influences plant development, and as Figure 4.2 shows, the rate of photosynthesis increases as the light intensity increases; however, beyond a maximum intensity, light does not increase the photosynthetic rate. If the light available to a plant is too low, its energy production from photosynthesis may be equaled by its energy output in respiration. This is called the *light compensation point*. At this point, the light becomes a limiting factor.

In mid-winter months, there is not much light energy available for plant growth compared to the summer months, for instance; therefore, it takes twice as long for plants to perform the same amount of growth in mid-winter as in mid-spring or early summer. It is therefore particularly important to effectively manage the scheduling of greenhouse crop production. In order to have a cost-effective production operation, the scheduling must be well aligned with the prevailing light conditions.

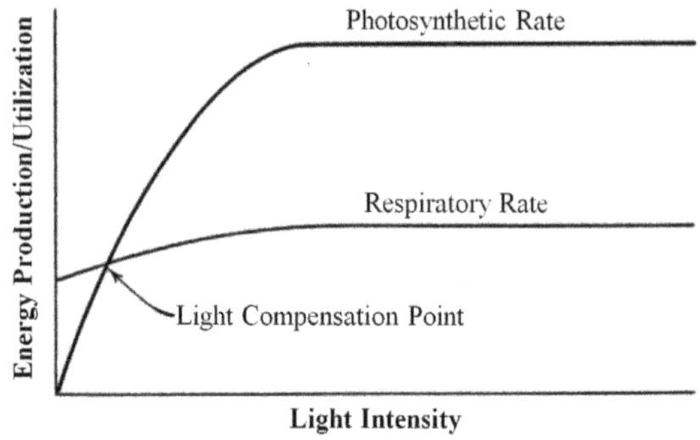


**FIGURE 4.2**
Schematic representation of photosynthetic and respiratory rates versus light.

Plants on the other hand can adapt themselves to different light levels depending on the age of plant or leaves and the acclimatization length. Under low light intensity environments, plants develop larger leaf surface area for intercepting more light so that the adverse effects of low light intensity can be compensated without impacting the photosynthetic rate per unit surface area of these leaves. Plants also arrange the light capturing chloroplasts accordingly depending on the light levels. Under low light conditions, they align chloroplasts perpendicular to the incoming radiation so that they can capture more light. Under high light conditions, however, the chloroplasts are aligned parallel to the incoming radiation.

As discussed earlier in Chapter 3, the amount of light that is received by plants in a greenhouse is directly affected by greenhouse covering material. Glass and double PE are the most used covering materials, and their transmissivities as well were provided earlier. Modern roof designs and elevated gutter heights enable improved light transmission to the crop canopy. Only 60–75% of incoming light reaching the cover's outside surface is transmitted into the greenhouse. There exists a theoretical maximum of transmissivity for each type of glazing because of the refractive index at each glazing–air interface (~4%) and absorptivity of each glazing material. Therefore, choosing the right glazing material in the first place is extremely important for providing an adequate light intensity for the plant grown in a greenhouse.

Especially, in the production of flowering ornamentals and some other specialty crops, the common goal of all growers is to consistently grow marketable high-quality and uniform plants. There might be different reasons for plant nonuniformity, and one of these reasons might be nonuniform light distribution in a greenhouse. Therefore, we must make sure that greenhouse lighting is evenly distributed as well to ensure uniform crop response.

### 4.3.5 Light and Carbon Dioxide Interaction

While several environmental stressors such as watering and nutrient availability also influence photosynthesis, particularly the $CO_2$ concentration and temperature affect photosynthesis dramatically (Figures 4.3 and 4.4).

Figure 4.3 indicates that, at low light intensities, the rate of photosynthesis depends on the quantity of incident light, whereas at high light intensities, that is, beyond the *light*

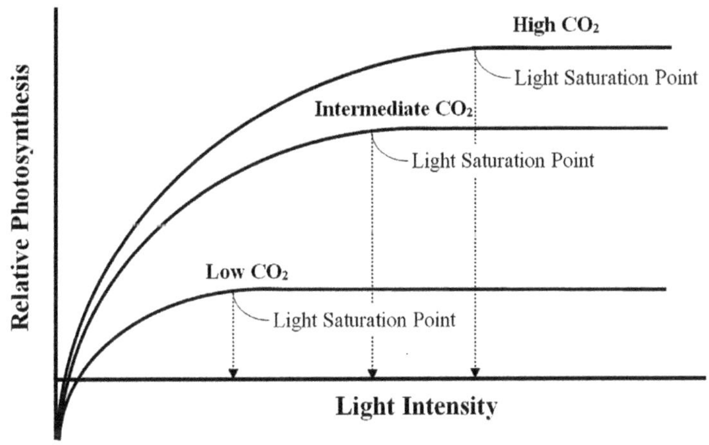


FIGURE 4.3
Schematic illustration of light and carbon dioxide interaction on photosynthesis. Plants use more light and perform more photosynthesis when the $CO_2$ concentration increases. Beyond the light saturation point at each $CO_2$ concentration, increased light does not influence photosynthesis.

*saturation point* (the intensity where further increases in light do not increase photosynthesis), the amount of $CO_2$ is rate limiting. Figure 4.3 also illustrates that low $CO_2$ concentration decreases the light saturation point reducing the rate of photosynthesis while high $CO_2$concentration levels increase the light saturation point, resulting in more effective utilization of light by the plants.

### 4.3.6 Light Pollution

See Section 11.3.4.

## 4.4 Temperature

Temperature affects a number of plant processes, and all of which influence the overall plant growth. For instance, photosynthesis and respiration, development of roots, leaves, and flowers, size and shape of leaves, and obviously the relative humidity in a greenhouse are all influenced by temperature, which is an extremely important environmental parameter for engineers to manipulate in their designs.

As a result of the net effect of photosynthesis and respiration processes, plants develop and accumulate dry matter in different forms, and the photosynthesis and respiration processes are affected by temperature; however, the optimum temperature for maximum development depends on the intensity of light.

For the development of roots, leaves, and flowers, and therefore for the whole plant development, 24-hour average temperature is important; and it is reported that there is a 10% increase in leaf and flower formation for each degree Celsius increase between 10°C and 30°C. The higher the light intensity is, the higher the optimum temperature becomes; that's why the 24-hour average temperature is raised under high light conditions. And

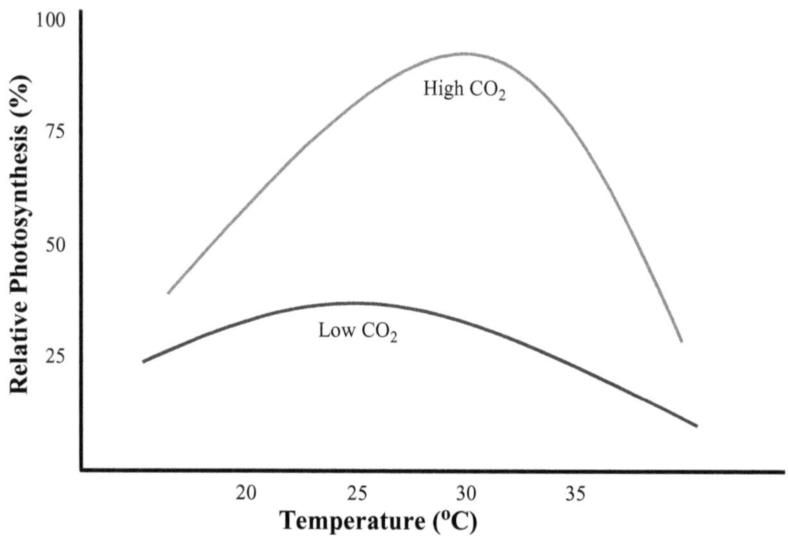


**FIGURE 4.4**
Schematic illustration of temperature and carbon dioxide interaction on photosynthesis. Photosynthesis of plants increases as temperature increases until a maximum and then decreases; and same thing is true for $CO_2$ increase as well. Also, the temperature for maximum photosynthesis increases when the $CO_2$ concentration increases.

under low light conditions, however, this 24-hour average temperature is lowered, and a lower temperature might increase foliage quality; but, plants develop more slowly and it takes much longer for them to develop fully. Figure 4.4 shows that photosynthesis increases as temperature increases until a maximum, and then decreases; however, as clearly seen on the figure, the temperature influence is also function of $CO_2$ concentration. The temperature for maximum development increases when the $CO_2$ concentration increases.

Besides, too high or low 24-hour temperature is no good for plant development. High temperatures promote thinner leaves and stems, and small flowers, while low temperatures promote thicker leaves and stems as well as large flowers, and damage flowers or delay harvest.

Additionally, the day–night temperature differentials influence development of thermoperiodic plant species, which have different optimum day and night temperatures for each stage of their growth. When there is no day–night temperature differential, plants produce more vegetation with shorter stems, and thicker and larger leaves; and when there is a large temperature differential, plants then turn into a generative production with longer stems, and thinner and smaller leaves. Many plants grow better when the night-time temperature is lower than the day-time temperature. In practice, generally a more even day–night temperature control is provided in early development of a plant, and then a temperature differential is used after the plants flower and move into the fruit setting stages.

Moreover, temperature and relative humidity of the ambient air are intimately linked. When temperature rises, then relative humidity drops; and when temperature drops, then relative humidity rises. In short, relative humidity is part of the plant's thermal environment and directly affects plant transpiration rate and the overall water status of the plant. Generally, plants produce longer stems and larger leaves in high relative humidity conditions.

**TABLE 4.1**

Recommended Air Temperatures for Three Major Greenhouse Crops (OMAFRA, 2001).

| Plant | | Temperature (°C) | | | | |
|---|---|---|---|---|---|---|
| | | Germination | Transplanting | Raising | Harvesting Stage | Full Harvest |
| Tomato | Day | 25 | 24 | 19–21 | 19 | 20–22 |
| | Night | 25 | 24 | 19–21 | 19 | 17 19 |
| Cucumber | Day | 28 | 25 | 24 | 23 | 23 |
| | Night | 28 | 25 | 21 | 21 | 17–21 |
| Pepper | Day | 25 | 21 | 21 | 21 | 21 |
| | Night | 25 | 24 | 23 | 24 | 18 |

Overall, plants have significantly varying minimum, optimum, and maximum temperatures, which heavily depend on light intensity, the stage of plant growth, $CO_2$ concentration, and season. The optimum temperatures for major greenhouse plants such as lettuce, tomatoes, peppers, and cucumbers are generally reported to be 17–18°C, 19°C, 19°C, and 21°C, respectively; however, further specifics of temperature requirements for these and other crops are provided in Table 4.1 and Table A.10.

As discussed earlier, many crops grow better when the night temperature is lower than the day temperature. They have different optimum day and night temperatures for each stage of their growth, and recommended night temperatures for a list of greenhouse crops can be found in Table A.10 as provided by ASHRAE (1997) and OMAFRA (2001).

## 4.5 Relative Humidity

Relative humidity is defined earlier in Chapter 2 as the ratio of the mole fraction of water vapor in moist air (the actual amount of water vapor present) to the mole fraction in the air saturated (the maximum amount of water vapor that can be held) at the same temperature and pressure. Relative humidity influences transpiration rate, latent heat transfer, and certain diseases, especially bacterial and fungal. Normal plant growth takes place at relative humidity levels between 20% and 80% if plants have well-developed root systems; however, the relative humidity in most greenhouses typically range between 40% and 100% (not desirable and cause many problems). The actual water vapor contents per unit volume of air for various relative humidity levels at a given temperature were provided in Figure 4.5.

Also, the water holding capacity per unit volume of air (saturation humidity) is correlated with air temperature, and as Figure 4.6 shows, water holding capacity of air increases tremendously with the increasing temperature, almost doubling for every 10°C increase in air temperature.

It should be kept in mind that relative humidity in a greenhouse changes with the change in air temperature if the absolute humidity of air is not adjusted via ventilation or so; and as shown in Figure 4.7, this change is typically about 5% increase or decrease in relative humidity for each degree Celsius decrease or increase, respectively. This is

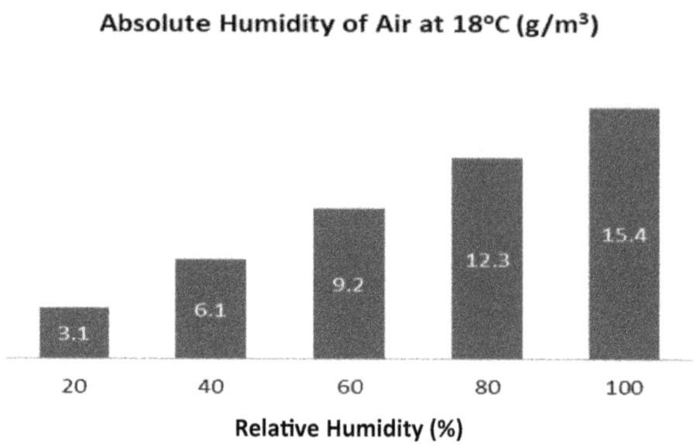


**FIGURE 4.5**
Actual water content per unit volume of air at 18°C for various relative humidity levels.

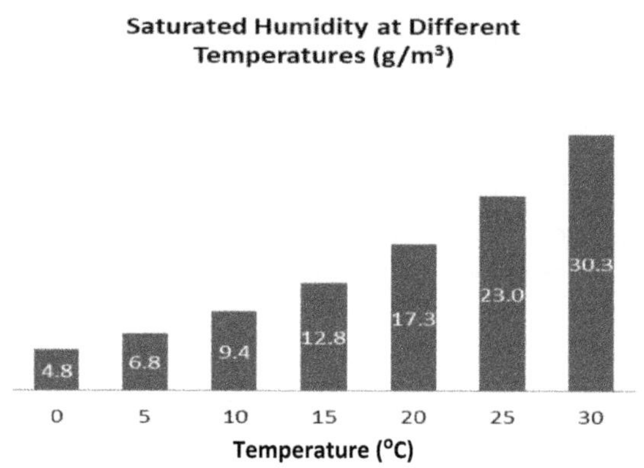


**FIGURE 4.6**
Effect of temperature on the water holding capacity of air.

particularly important at the times when greenhouse set point temperatures are changed during day and night.

In greenhouse operations, we typically use relative humidity to express the moisture content of greenhouse air even though it gives an idea about the moisture status only within a narrow temperature operation. It is not sufficient to know relative humidity as it fluctuates quite a bit with temperature changes (see Figure 4.7). Figure 4.8 further clarifies why relative humidity does not provide sufficient information about the actual moisture content of greenhouse air. Actual water contents and total water holding capacities of air with 70% relative humidity at various temperatures were presented in this figure; and the meaning of 70% relative humidity at each temperature with respect to the actual water content changes as the water holding capacity increases with rising temperature. For instance, while the air at 30°C temperature and 70% relative humidity has much higher actual moisture content (21.2 g/m$^3$) compared to the lower temperatures presented in Figure 4.8, the air still has quite a bit room (water vapor deficit of 9.1 g/m$^3$) before it gets saturated with water vapor. This means that the air at a lower temperature has a lower

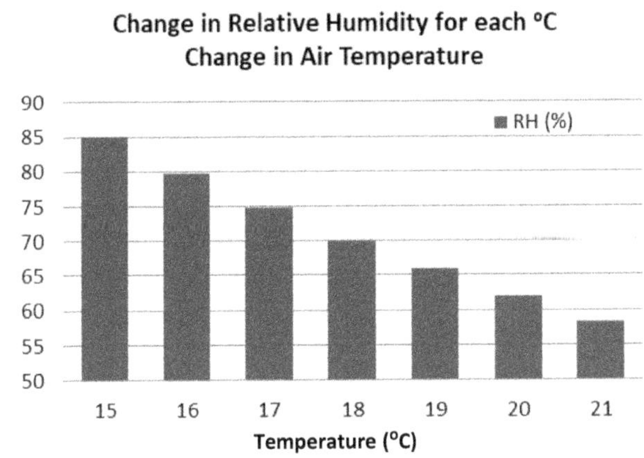


**FIGURE 4.7**
Change in relative humidity for each degree Celsius change in air temperature. ($T$ = 18°C and $T_d$ = 12.5°C (i.e. 70% RH) was used as the starting point).

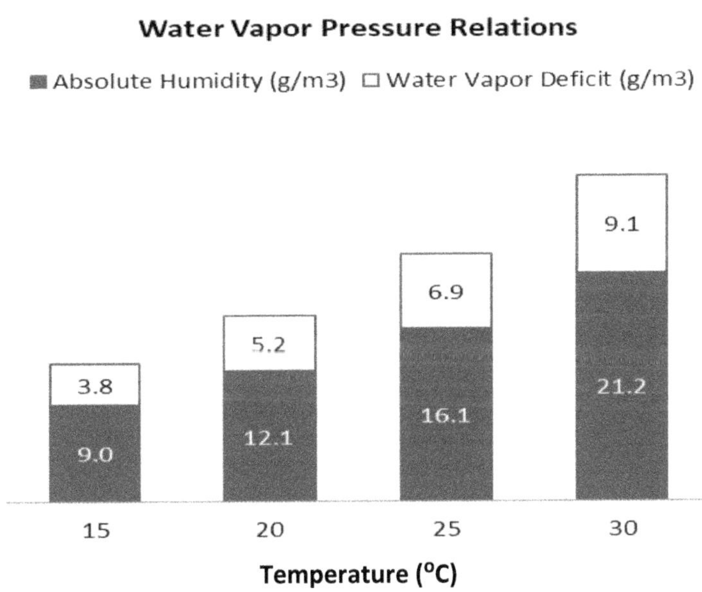


**FIGURE 4.8**
Actual water contents and water holding capacities of air with 70% relative humidity at various temperatures.

water holding capacity; and therefore, it becomes saturated faster. A high water vapor deficit, on the other hand, means that there exists a higher forcing function for transpiration; and therefore, transpiration will increase. Tiny pores on plant leaves termed as stomatal openings biologically control transpiration by opening and closing. When there is abundant water, or light, or low internal $CO_2$ concentration, stomatal openings will open; and

**TABLE 4.2**

Vapor Pressure Deficit as Influenced by Temperature and Relative Humidity (mbar) (OMAFRA, 2001).

| | Relative Humidity (%) | | | | | | | | | |
|---|---|---|---|---|---|---|---|---|---|---|
| **Temperature (°C)** | **50** | **55** | **60** | **65** | **70** | **75** | **80** | **85** | **90** | **95** |
| 15 | 8.5 | 7.7 | 6.8 | 6.0 | 5.1 | 4.3 | 3.4 | 2.6 | 1.7 | 0.9 |
| 16 | 9.1 | 8.2 | 7.3 | 6.4 | 5.5 | 4.6 | 3.6 | 2.7 | 1.8 | 0.9 |
| 17 | 9.7 | 8.7 | 7.8 | 6.8 | 5.8 | 7.5 | 3.9 | 2.9 | 1.9 | 1.0 |
| 18 | 10.3 | 9.3 | 8.3 | 7.2 | 6.2 | 5.2 | 4.1 | 3.1 | 2.1 | 1.0 |
| 19 | 11.0 | 9.9 | 8.8 | 7.7 | 6.6 | 5.5 | 4.4 | 3.3 | 2.2 | 1.1 |
| 20 | 11.7 | 10.5 | 9.4 | 8.2 | 7.0 | 5.9 | 4.7 | 3.5 | 2.3 | 1.2 |
| 21 | 12.4 | 11.9 | 9.9 | 8.7 | 7.5 | 6.2 | 5.0 | 3.7 | 2.5 | 1.2 |
| 22 | 13.2 | 11.9 | 10.6 | 9.3 | 7.9 | 6.6 | 5.3 | 4.0 | 2.6 | 1.3 |
| 23 | 14.1 | 12.6 | 11.2 | 9.8 | 8.4 | 7.0 | 5.6 | 4.2 | 2.8 | 1.4 |
| 24 | 14.9 | 13.4 | 11.9 | 10.4 | 9.0 | 7.5 | 6.0 | 4.5 | 3.0 | 1.5 |
| 25 | 15.8 | 14.3 | 12.7 | 11.1 | 9.5 | 7.9 | 6.3 | 4.8 | 3.2 | 1.6 |
| 26 | 16.8 | 15.1 | 13.4 | 11.8 | 10.1 | 8.4 | 6.7 | 5.0 | 3.4 | 1.7 |
| 27 | | | | 12.5 | 10.7 | 8.9 | 7.1 | 5.4 | 3.6 | 1.8 |
| 28 | | | | 13.2 | 11.3 | 9.5 | 7.6 | 5.7 | 3.8 | 1.9 |
| 29 | | | | | 12.0 | 10.0 | 8.0 | 6.0 | 4.0 | 2.0 |
| 30 | | | | | 12.7 | 10.6 | 8.5 | 6.4 | 4.2 | 2.1 |

*Note:* The green area is the optimum range for most greenhouse crops; 1 mbar = 0.1 kPa = 0.0145 psi; 0.1 kPa = 0.7 g/m$^3$ = 3% RH.

in reverse, when there is a lack of water or light, or when high internal $CO_2$ concentration conditions prevail, then stomatal openings will close. Stomatal cavities in plant leaves are normally fully saturated with water vapor while the greenhouse air is usually not. And the water vapor on each side exerts a pressure in the air whether it is greenhouse air or the air in stomatal cavity. So, the difference in between these two water vapor pressures, also known as vapor pressure deficit (VPD), regulates the plant water loss through transpiration. Table 4.2 presents the VPD values for various air temperature and relative humidity combinations, and the VPD values between 4 and 8 mbar are considered optimal for optimal nutrient uptake and transpiration supported by proper carbohydrate production and transfer, therefore for a healthy and well producing plant. Too low or too high VPDs have detrimental effects on plant growth and productivity. Engineers manipulate VPD information in their designs for controlling the greenhouse environment for a safe, healthy, and productive greenhouse operation.

Due to the incoming solar radiation, leaf temperatures are typically higher than the air temperatures during the day. However, at night, in the absence of solar radiation under especially clear sky conditions, leaf temperatures may drop several degrees due to radiational cooling depending on the greenhouse glazing and whether night curtain is used or not. Under such conditions, if the inside relative humidity is close to or higher than approx. 95–96%, then condensation would take place on leaf and/or other surfaces. Lower leaves, which stay moist longer, are typically more susceptible to condensation, and therefore, to bacterial and fungal diseases. Transpiration from the leaves raises the relative humidity in interfoliage air; thence, the still air surrounding the plant canopy usually tends to be moister. However, manipulating especially the rail pipe, underbench, or grow pipe heating systems coupled with ventilation for induced convection within the canopy along with several other measures, we can remove the risk of condensation on plant and other surfaces whenever it occurs.

Not only high relative humidity levels, but also low or extremely low relative humidity levels are no good for plants either. Rapid drops in relative humidity due to excessive ventilation may cause foliage burning and/or reduced fruit quality and yield in some greenhouse crops. Relative humidity also impacts pollen transfer and germination for fertilization and proper fruit size. Tomatoes, for instance, require the relative humidity level be maintained between 70% and 75% for proper pollination.

## 4.6 Carbon Dioxide and Air Pollutants

### 4.6.1 Carbon Dioxide

Today, the atmospheric $CO_2$ concentration is about 0.04% (400 ppm) and burning fossil fuels keeps adding significant amounts of $CO_2$ to the atmosphere every passing day. As it was discussed earlier, $CO_2$ taken up by plants through their stomatal openings is an essential ingredient for photosynthesis, which responds to the changes in $CO_2$ concentration as it is the main constituent of plant dry matter, approximately 45%, and essential for the plant growth and yield. The amount of $CO_2$ available to terrestrial plants is usually not limiting, although at high altitudes, it may well become a limiting factor. But, the amount of $CO_2$ available in greenhouses can drop well below the ambient levels quickly and become limiting especially during sunny periods depending on the light, temperature, ventilation, and supplemental $CO_2$ enrichment levels if there is any. Plants use more light and perform more photosynthesis when the $CO_2$ concentration in the ambient air increases (Figure 4.3). At low light intensities, the rate of photosynthesis depends on the quantity of incident light, whereas at high light intensities the amount of $CO_2$ becomes rate limiting assuming that the other management practices are adequately performed. Elevated $CO_2$ concentrations coupled with elevated daytime temperatures and high light conditions shortens the time between germination and harvest as much as 50 percent.

As Figure 4.9 schematically shows, the driving force for $CO_2$ is no more than 0.04 kPa (0.16 in. $H_2O$) (the case of zero concentration of $CO_2$ inside plant tissues). At 25°C, the outward driving force for loss of water to dry air from plant tissues is 3.2 kPa (12.9 in. $H_2O$) (80 times as high as the inward driving pressure for $CO_2$).

Due to the low diffusion gradient of $CO_2$ into the plant, normally it is difficult for plants to obtain the necessary $CO_2$ from the ambient air without incurring relatively high water loss by transpiration through the stomatal openings. Because of this unfavorable situation, plants have very high water requirements. In most plants, the stomatal openings are located primarily in the lower epidermis – the side of the leaf usually turned away from the Sun's rays, where the drying tendency is less severe. When plants run short of moisture, they block most water loss by simply closing the stomatal openings at the expense of growth.

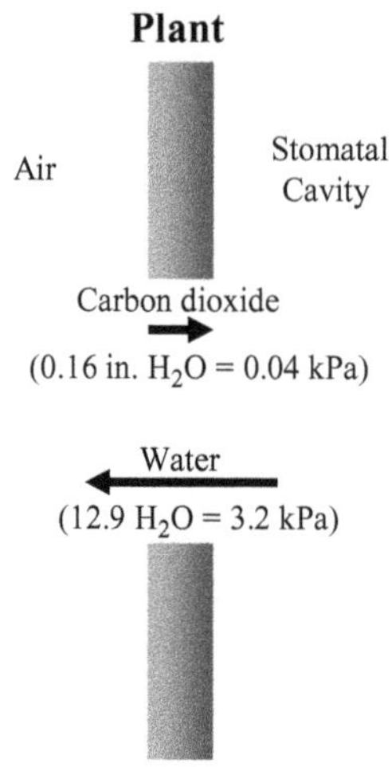


**FIGURE 4.9**
Partial pressure differences across stomatal openings.

It is possible however to increase the rate of plant growth by increasing $CO_2$ concentrations several times above normal values. Plant species differ in their photosynthetic responses to $CO_2$, with $C_3$ plants exhibiting increases in net photosynthetic rate up to about 1,000 ppm (*$CO_2$ saturation point*) and $C_4$ plants exhibiting the maximum rate at about 360 ppm (*$CO_2$ saturation point*) (Figure 4.10. As discussed earlier, $C_3$ plants have an additional light-dependent respiratory mechanism known as photorespiration; and the normal $O_2$ concentration of 21% inhibits $CO_2$ uptake by the plant and increases the loss of carbon due to photorespiration. Since the partial pressure of $O_2$ (21.3 kPa = 85.6 in. $H_2O$) in the air is normally much greater than that of $CO_2$ (0.04 kPa = 0.16 in. $H_2O$), the rate of net photosynthesis in $C_3$ plants is inhibited due to the increased photorespiration (*Warburg Effect*) as a result of the competition between ambient $CO_2$ and $O_2$ to be fixed by the enzyme ribulose diphosphate carboxylase. The oxygen inhibition of photosynthesis decreases with $CO_2$

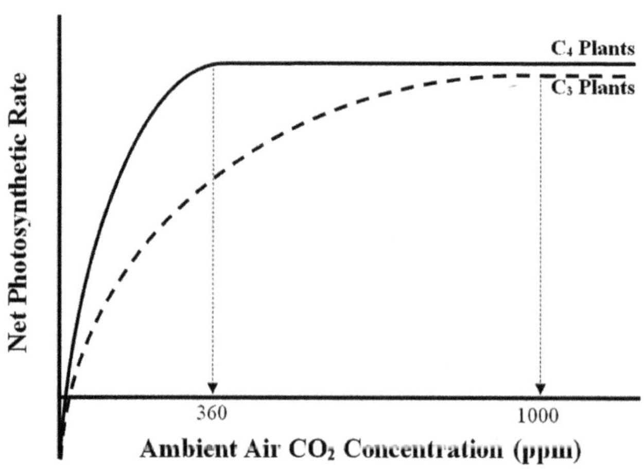


**FIGURE 4.10**
Schematic representation of photosynthetic response of different plant species to $CO_2$ concentration.

enrichment, and is almost eliminated at a $CO_2$ concentration of about 900 ppm or so; and consequently, the net photosynthesis in plants increases (Figure 4.10). Elevated $CO_2$ concentrations also increase the optimal temperature for growth (Figure 4.4). Pot plants, cut flowers, vegetables, and forest plants expose quite positive responses to $CO_2$ enrichment with increased plant quality and number of flowers.

Research findings have shown that plant growth is improved by 15–20% in greenhouses when $CO_2$ concentrations are increased to 800–1,300 ppm during the day. Continuous $CO_2$ enrichment during the light period has been found to be superior to intermittent $CO_2$ application. The literature suggest that the optimal $CO_2$ concentration for plant growth is somewhere between 600 ppm and 1,000 for most species; and in general, the optimum $CO_2$ coupled with temperature and PAR intensity is about 400 ppm on cloudy days or vents are opened more than 10%, and 1,000 ppm on sunny days.

Furthermore, elevating the $CO_2$ concentration leads to higher internal $CO_2$ concentrations in plant tissues closing the stomatal openings, hence reducing the transpiration of plants by 20–40%. Water consumption is therefore significantly reduced by $CO_2$ enrichment while photosynthesis is increased simultaneously. We know that increasing the air relative humidity, let's say, from 60% to 95% decreases plant transpiration by about 50% or so; and research has shown that $CO_2$ enrichment at 95% relative humidity decreases transpiration and water uptake by some plants further by about 30% increasing the water-use efficiency (amount of water used per unit dry weight produced) by about 30%.The decreased transpiration might however cause higher leaf temperature due to the reduced evaporative cooling performed by plants themselves, especially under high light conditions.

### 4.6.2 Air Pollutants

See Section 11.3.3.

## 4.7 Air Movement and Altitude

Air movement affects evaporation, plant transpiration, and the $CO_2$ availability to plants. An air speed of 0.5–0.7 m/s (100–140 fpm) is reported to be optimum in greenhouses. Air speeds exceeding 1 m/s (200 fpm) inhibit plant growth while 4.5 m/s (900 fpm) can cause physical damage. The direction of air movement does not seem to affect results.

Altitude, on the other hand, affects plant growth indirectly due to the decreased average temperatures, wavelength distribution of solar radiation, humidity, average wind speed, etc. At high altitudes, greenhouse heating costs for providing an adequate plant thermal environment rise due to decreased outside temperatures and increased wind speeds; partial pressures of atmospheric $CO_2$ and $O_2$ concentrations drop as well, and unless proper measures are taken, they may be limiting.

Furthermore, combustion equipment is designed and operated in order to make sure we have complete combustion. If air provided to burner is not enough, incomplete combustion occurs, with all the nonoxidized components in the flue gas such as C, $H_2$, and CO. Incomplete combustion does not use fuel efficiently, a process which results in air pollution and can pose even further danger due to CO production and a potential for explosion.

CO is an odorless, tasteless, and colorless toxic gas. As discussed earlier, every combustion process has the potential to produce this toxic gas as a result of incomplete combustion. Air at higher altitudes has lesser density and therefore less oxygen per unit volume. Excess air must be adjusted accordingly; otherwise, we can have incomplete combustion, or even combustion can eventually cease.

Other plants, animals, and insects (if there is any), bacteria, fungi, and viruses form the biotic environment in a greenhouse, must be managed below an economic injury level in an integrated fashion using cultural, physical, biological, and chemical control strategies.

## Nomenclature

**LED** light emitting diode
**nm** nanometer
**PAR** photosynthetically active radiation
**ppb** parts per billion
**ppm** parts per million
**RH** relative humidity
$T$ temperature
**THC** tetrahydrocannabinol
**UV** ultraviolet
**VPD** vapor pressure deficit

### Subscripts

$d$ dew point

## Bibliography

Abeles FB (1973) *Ethylene in Plant Biology*. Academic Press, New York, London.

Aldrich RA and Bartok JW (1994) *Greenhouse Engineering. NRAES-33. Northeast Regional Agricultural Engineering Service*. Cornell University, Ithaca, NY, USA.

Anderson L (1978) Gas-fired heaters cause crop damage. *Grower*, 89: 574–575.

Anderson LS and Mansfield TA (1979) The effects of nitric oxide pollution on the growth of tomato. *Environ. Pollut.*, 20:113–121.

Ashenden TW, Mansfield TA and Harrison RM (1977) Generation of air pollutants from kerosene combustion in commercial and domestic glasshouses. *Environ. Pollut.*, 14: 93–100.

ASHRAE (1997) Handbook of Fundamentals. American Society of Heating, Refrigeration and Air Conditioning Engineers, Inc. *Atlanta, GA* 30329. pp. 10.25–10.29.

Calvert A and Slack G (1975) Effect of carbon dioxide enrichment on growth, development and yield of glasshouse tomatoes. I. Responses to controlled concentrations. *J. Hortic. Sci.*, 50: 61–71.

Calvert A and Slack G (1976) Effect of carbon dioxide enrichment on growth, development and yield of glasshouse tomatoes. II. The duration of daily periods of enrichment. *J. Hortic. Sci.*, 51: 401–409.

Capron TM and Mansfield TA (1975) Generation of nitrogen oxide pollutants during $CO_2$ enrichment of a glasshouse atmosphere. *J. Hortic. Sci.*, 50: 233–238.

Capron TM and Mansfield TA (1976) Inhibition of net photosynthesis in tomato in air polluted with NO and $NO_2$. *J. Exp. Bot.*, 27: 1181–1186.

Capron TM and Mansfield TA (1977) Inhibition of growth in tomato by air polluted with nitrogen oxides. *J. Exp. Bot.*, 28: 112–116.

Dutta Gupta S (2017) *Light Emitting Diodes for Agriculture*. Springer Singapore.

Ehleringer JR (1979) Photosynthesis and photorespiration. Biochemistry, physiology, and ecological implications. *HortScience*, 14: 217–222.

Enoch HZ and Hurd RG (1977) Effect of light intensity, carbon concentration, and leaf temperature on gas exchange of spray carnation plants. *J. Exp. Bot.*, 28: 84–95.

Enoch HZ and Hurd RG (1979) The effect of elevated $CO_2$ concentrations in the atmosphere on plant transpiration and water use efficiency. A study with potted carnation plants. *Int. J. Biometeorol.*, 23: 343–351.

Enoch HZ, Rylski I and Samish Y (1970) $CO_2$ enrichment to cucumber, lettuce and sweet pepper plants grown in low plastic tunnels in a subtropical climate. *Israel J. Agric. Res.*, 20: 63–69.

Enoch HZ, Rylski I and Spigelman M (1976) $CO_2$ enrichment of strawberry and cucumber plants grown in unheated greenhouses in Israel. *Scientia Hortic.*, 5: 33–41.

Forrester ML, Krotkow G and Nelson CD (1966) Effect of oxygen on photosynthesis, photorespiration and respiration in detached leaves. I. Soybean. *Plant Physiol.*, 41: 422–427.

Freeman R (1983) Better crops through carbon dioxide injection. *Grower Talks*, 47: 42–48.

Goldammer T (2019) *Greenhouse Management: A Guide to Operations and Technology*. Apex Publishers, Centreville, Virginia, USA

Hanan JJ (1973) Ethylene pollution from combustion in greenhouses. *HortScience*, 8: 23–24.

Hand DW (1971) $CO_2$ from hydrocarbon fuels. *A.D.A.S.Q. Rev.*, 1: 18–23.

Hand DW (1979) Injury to crops in glasshouse polluted with nitrogen oxids. *A.D.A.S.Q. Rev.*, 33: 134–143.

Hand DW (1982) $CO_2$ enrichment, the benefits and problems. *Sci. Hortic.*, 33: 14–43.

Hand DW and Cockshull KE (1975) Roses. I: The effects of $CO_2$ enrichment on winter bloom production. *J. Hortic. Sci.*, 50: 183–192.

Hand DW and Postlethwaite JD (1971) The response to $CO_2$ enrichment of capillary-watered single-truss tomato at different plant densities and seasons. *J. Hortic. Sci.*, 46: 461–470.

Hand DW, Slack G and Sweeney DG (1981) Lettuce: crop responses to controlled levels of $CO_2$ enrichment. *Rep. Glasshouse Crops Res. Inst.*, 1979: 58–59.

Haukeness MO, Maginnes EA, Green GH and Brooks EE (1978) Using the heat and carbon dioxide from turbine exhaust gases for the production of greenhouse tomatoes. *Hort. Science*, 13: 292–293.

Heij G and Uffelen JAM van (1984) Effects of $CO_2$ concentration on growth of glasshouse cucumber. *Acta Hortic.*, 162: 29–36.

Hesketh J (1967) Enhancement of photosynthetic $CO_2$ assimilation in the absence of oxygen, as dependent upon species and temperature. *Planta (Berl.)*, 76: 371–374.

Hill AC and Bennett JH (1970) Inhibition of apparent photosynthesis by nitrogen oxides. *Atmos. Environ.*, 4: 341–348.

Holley WD and Goldsberry KL (1963) $CO_2$ and temperature recommendations. *Colo. Flower Grower Assoc. Bull.*, 164: 3.

Holley WD, Goldsberry KL and Juengling C (1964) Effects of $CO_2$ concentration and temperature on carnation. *Colo. Flower Grower Assoc. Bull.*, 174: 1–5.

Janes BE (1970) Effect of carbon dioxide, osmotic potential of nutrient solution, and light intensity on transpiration and resistance to flow of water in pepper plants. *Plant Physiol.*, 45: 95–103.

Joliffe PA and Tregunna EB (1968) Effect of temperature, $CO_2$ concentration, and light intensity on oxygen inhibition of photosynthesis in wheat leaves. *Plant Physiol.*, 43: 902–906.

Jones P, Allen LH, Jones JW and Valle R (1985) Photosynthesis and transpiration responses of soybean canopies to short- and long-term $CO_2$ treatments. *Agron. J.*, 77:119–126.

Kimball BA and Mitchell ST (1979) Tomato yields from $CO_2$ enrichment in unventilated and conventionally ventilated greenhouses. *J. Am. Soc. Hortic. Sci.*, 4: 515–520.

Kozai T, Fujiwara K, Runkle ES (2016) *LED Lighting for Urban Agriculture*. Springer Singapore.

Laing WA, Ogren WL and Hageman RH (1974) Regulation of soybean net photosynthetic $CO_2$ fixation by the interaction of $CO_2$, $O_2$, and ribulose 1,5-diphosphate carboxylase. *Plant Physiol.*, 54: 678–685.

Lopez R and Runkle ES (2017) *Light Management in Controlled Environments*. Meister Media Worldwide, Willoughby, OH, USA.

Matsumaru T, Yoneyama T, Totsuka T and Shiratori K (1979) Absorption of atmospheric $NO_2$ by plants and soils. (I) Quantitative estimation of absorbed $NO_2$ in plants by 15N method. *Soil Sci. Plant Nutr.*, 25: 255–265.

McCree, KL (1972) The action spectrum, absorptance and quantum yield of photosynthesis in crop plants. *Agric. Meteorol.*, 9: 191–216.

Monson RK, Stidham MA, Williams, GJ III, Edwards GE and Uribe EG (1982) Temperature dependence of photosynthesis in Agropyronsmithii Rydb. *Plant Physiol.*, 69: 921–928.

Morison JIL (1985) Sensitivity of stomata and water use efficiency to high $CO_2$. *Plant Cell Environ.*, 8: 467–474

Morison JIL and Gifford RM (1984) Ethylene contamination of $CO_2$ cylinders. *Plant Physiol.*, 75: 275–277.

Mortensen LM (1983a) Air pollution and plant injuries in greenhouses. Dep. Floriculture and Greenhouse Crops, Agricultural University of Norway, *Rep.* 281: pp. 1–22 (in Norwegian).

Mortensen LM (1983b) Growth responses of some greenhouse plants to environment. VIII. Effect of $CO_2$ on photosynthesis and growth of Norway spruce. *Meld. Nor. Landbrukshoegsk.*, 62: 1–13.

Mortensen LM (1983c) Growth responses of some greenhouse plants to environment. IX. Effect of $CO_2$ on photosynthesis of Chrysanthemum morifolium Ramat. at different light, temperature and $O_2$ levels. Meld. Nor. *Landbrukshoegsk.*, 62: I–12.

Mortensen LM (1985a) Effects of $CO_2$ enrichment and supplementary light on growth and flowering of poinsettia, Euphorbia pulcherrima Willd. *Meld. Nor. Landbrukshoegsk.*, 64: 1–8.

Mortensen LM (1985b) Nitrogen oxides produced during $CO_2$ enrichment. I. Effects on different greenhouse plants. *New Phytol.*, 101: 103–108.

Mortensen LM (1985c) Nitrogen oxides produced during $CO_2$ enrichment. II. Effects on different tomato and lettuce cultivars. *New Phytol.*, 101: 411–415.

Mortensen LM (1986a) $CO_2$ during ventilation to cucumber and roses. *Gartneryrket*, 76:140 (in Norwegian).

Mortensen LM (1986b) Effect of intermittent as compared to continuous $CO_2$ enrichment on growth and flowering of Chrysanthemum × morifolium Ramat. and Saintpauliaionantha H. Wendl. *Scientia Hortic.*, 29: 283–289.

Mortensen LM (1986c) Nitrogen oxides produced during $CO_2$ enrichment. III. Effects on tomato at different photon flux densities. *New Phytol.*, 104: 653–660.

Mortensen LM (1987) Review: $CO_2$ enrichment in greenhouses. Crop responses. *Scientia Horticulturae*, 33: 1–25, Elsevier Science Publishers B.V., Amsterdam, The Netherlands.

Mortensen LM and Moe R (1983a) Growth responses of some greenhouse plants to environment. IV. Effects of carbon dioxide on photosynthesis and growth of Chrysanthemum morifolium Ramat. *Meld. Nor. Landbrukshoegsk.*, 61: 1–11.

Mortensen LM and Moe R (1983b) Growth responses of some greenhouse plants to environment. V. Effect of $CO_2$, $O_2$ and light on net photosynthetic rate in Chrysanthemum morifolium Ramat. *Scientia Hortic.*, 19: 133–140.

Mortensen LM and Moe R (1983c) Growth responses of some greenhouse plants to environment. VI. Effect of $CO_2$ and artificial light on growth of Chrysanthemum morifolium Ramat. *Scientia Hortic.*, 19: 141–147.

Mortensen LM and Moe R (1983d) Growth responses of some greenhouse plants to environment. VII. The effect of $CO_2$ on photosynthesis and growth of roses. *Meld. Nor. Landbrukshoegsk.*, 62: 1–11.

Mortensen LM and Ulsaker R (1985) Effect of $CO_2$ concentration and light levels on growth, flowering and photosynthesis of Begonia × hiemalis Fotsch. *Scientia Hortic.*, 27: 133–141.
OMAFRA (2001) Growing Greenhouse Vegetables. *Ontario Ministry of Agriculture, Food and Rural Affairs*. Publication 371. ISSN: 1492-6601. Toronto, Canada. p.77.
OMAFRA (2003) Production Recommendations for Greenhouse Floriculture. *Ontario Ministry of Agriculture, Food and Rural Affairs*. Publication 370. ISSN: 1492–6512. Toronto, Canada. p. 39.
Rogers HH, Sionit N, Cure JD, Smith JM and Bingham GE (1984) Influence of elevated carbon dioxide on water relations of soybeans. *Plant Physiol.*, 74: 233–238.
Runkle ES (2011) The R to FR Ratio. *Greenhouse Product News Magazine*, May 2011. Sparta, MI, USA. p. 46.
Runkle ES (2012) Maximizing photosynthesis to improve growth. *Greenhouse Product News Magazine*, December 2012. Sparta, MI, USA. p. 66.
Runkle ES (2015a) Light wavebands and their effects on plants. *Greenhouse Product News Magazine*, March 2015. Sparta, MI, USA. p. 42.
Runkle ES (2015b) Interactions of light, $CO_2$ and temperature on photosynthesis. *Greenhouse Product News Magazine*, June 2015. Sparta, MI, USA. p. 54.
Runkle ES (2017a) Effects of blue light on plants. *Greenhouse Product News Magazine*, February 2017. Sparta, MI, USA. p. 38.
Runkle ES (2017b) Growing plants with green light. *Greenhouse Product News Magazine*, June 2017. Sparta, MI, USA. p. 58.
Runkle ES (2019) Is Green Light Useful to Plants? *Greenhouse Product News Magazine*, June 2019. Sparta, MI, USA. p. 50.
Salisbury FB and Ross C (1969) *Plant Physiology*. Wadsworth, Belmont, CA, p. 194
Saxe H (1981) Growth reduction by $NO_x$ during $CO_2$ enrichment in the greenhouses. *Ugeskr. Jordbruk*, 28:406 409 (in Danish).
Saxe H and Christensen OV (1985) Effects of carbon dioxide with and without nitric oxide pollution on growth, morphogenesis and production time of pot plants. *Environ. Pollut.*, 38: 159–169.
Schapendonk AHCM and Gaastra P (1984) A simulation study on $CO_2$ concentration in protected cultivation. *Scientia Hortic.*, 23: 217–229
Slack G and Hand DW (1985) The effect of winter and summer $CO_2$ enrichment on the growth and fruit yield of glasshouse cucumber. *J. Hortic. Sci.*, 4: 507–516.
Tolley LC and Strain BR (1984) Effects of $CO_2$ enrichment and water stress on growth of Liquidambar stipaciflua and Pinus taeda seedlings. *Can. J. Bot.*, 62: 2135–2139.
Yoneyama T, Sasakawa H, Ishizuka S and Totsuka T (1979) (II) Nitrite accumulation, nitrite reductase activity and diurnal change of $NO_2$ absorption in leaves. *Soil Sci. Plant Nutr.*, 25: 267–275.
Zeevaart AJ (1976) Some effects of fumigating plants for short periods with $NO_2$. *Environ. Pollut.*, 11: 97–108.

# 5

# Greenhouse Heating

## 5.1 Introduction

Labor and heating costs are the major costs of greenhouse operations particularly in mid- and high-latitude countries. Large energy costs are incurred to maintain the required thermal and radiant environments in greenhouses particularly during the winter season. While heating is mostly required at night, it may also be needed during the day. Annual labor costs typically range from 30% to 35% of the total cost of production, while heating costs typically range from 20% to 35% of the total production cost depending on the location and the local fuel prices. Many heating and air circulation systems are commercially available today, and these systems include old steam heating, hot water heating, radiant heating, spaced unit heaters, central heating, underbench heating, floor heating, perforated tube air distribution, and horizontal air circulation.

This chapter goes over estimating heating loads in greenhouses first, and then moves to the passive and unheated greenhouse systems. The chapter then introduces the reverse return principle and covers in depth the spaced unit heaters and central hot water (hydronic) heating systems using step-by-step approach. Actual boiler room (including the boiler loops, manifolds, transport groups, expansion systems, vent stack and fresh air inlets) and heat distribution system designs, that is, the main supply and return lines, pipe rail heating, underbench heating, in-floor heating, sidewall and gable end heating, and snowmelt and overhead heating systems designs are included. Flue gas condensers and low grade heating systems, water pre-heating, combined heat and power (CHP) system integration, heat buffer system design, and carbon dioxide dosing systems integration are covered in depth as well. Real-world design examples are provided in each and every step in order to fully demonstrate and grasp the abovementioned design concepts.

## 5.2 Estimating Heating Loads

Applying the energy balance equation (Equation (2.1) discussed earlier in Chapter 2) to a greenhouse and ignoring the internal storage, we can simply write the following energy balance equations:

$$\sum E_{gains} = \sum E_{loss} \tag{5.1}$$

$$\dot{Q}_t + \dot{Q}_{internal} = \dot{Q}_{cond} + \dot{Q}_{inf} + \dot{Q}_{evap} \tag{5.2}$$

where $\dot{Q}_t$ is the rate of supplemental heat provided by the furnace, $\dot{Q}_{cond}$ is the rate of heat loss from the greenhouse envelop, $\dot{Q}_{inf}$ is the rate of heat loss due to air exchange between the inside and outside air, $\dot{Q}_{evap}$ is the rate of heat loss by evaporating water, and $\dot{Q}_{internal}$ is the rate of total internal heat gains as follows:

$$\dot{Q}_{internal} = \dot{Q}_{equipment} + \dot{Q}_{lighting} + \dot{Q}_{solar} \tag{5.3}$$

In a steady-state analysis, heat gains must be the average for the period in question, not for the peak values. In particular, solar radiation must be based on averages, not peak values. Only when the outdoor temperature drops below the balance point temperature (see Chapter 9 for further discussions), supplemental heat is required to maintain the indoor design temperature. Above the balance point temperature, internal heat gains (solar, equipment and lights, when available) are able to maintain the indoor temperature at or above the indoor design temperature; therefore, no supplemental heating is required. Supplemental heat by furnace is however determined for night time heating when there is no sun. Furthermore, lighting or other equipment gains should never be included in the furnace capacity sizing because full heating may be required when lighting or other equipment may not be operational. The rate of heat loss by evaporating water is ignored as well. Therefore, the supplementary heating equation takes the following form:

$$\dot{Q}_t = \dot{Q}_{cond} + \dot{Q}_{inf} \tag{5.4}$$

$$\dot{Q}_{cond} = \Sigma UA\left(T_i - T_o\right)$$

where $\Sigma UA$ is the total conduction heat loss term (Btu/h.°F in I-P; W/°C in SI). See Table 3.3 for U values of different glazing and building materials, $A$ is the area, $T_i$ is the night-time temperature of the greenhouse (see Table A.10 for recommended night temperatures for a list of greenhouse crops), and $T_o$ is the winter design dry-bulb temperature (99%) (see Tables A.11–A14 for climatic conditions for selected locations).

$$\dot{Q}_{inf} = \text{ACH}\, V \rho c_p \left(T_i - T_o\right) \quad \text{(I-P)} \tag{5.5}$$

where ACH is the number of air changes per hour (see Table 3.4), $V$ is the volume of the greenhouse (ft$^3$), $\rho$ is the density (lb/ft$^3$) and $c_p$ is the specific heat of greenhouse inside air (Btu/lb.°F).

$$\dot{Q}_{inf} = \text{ACH}\left(\frac{1}{3{,}600}\right) V \rho c_p \left(T_i - T_o\right) \quad \text{(SI)} \tag{5.6}$$

where ACH is the number of air changes per hour, $V$ is the volume of the greenhouse (m$^3$), 1/3,600 is for unit conversion (1 h = 3,600 s), $\rho$ is the density (kg/m$^3$) and $c_p$ is the specific heat of greenhouse inside air (J/kg.°C).

In order to determine the overall heat loss from a greenhouse or the required supplementary heating for a greenhouse at a particular location, Equation (5.4) should be multiplied by an applicable wind factor provided in Table 5.1.

Greenhouse heating loads can be reduced using opaque and transparent insulation, retractable night curtains, reduced ventilation rates, etc., as covered in depth in Chapters 9

**TABLE 5.1**

Recommended Wind Factors as a Function of Wind Velocity.

| Wind Velocity | | |
|---|---|---|
| (mph) | (m/s) | Factor |
| 15 | 7 | 1.00 |
| 20 | 9 | 1.04 |
| 25 | 11 | 1.08 |
| 30 | 13 | 1.12 |
| 35 | 15 | 1.16 |

and 11. However, when determining the furnace capacity, these energy saving measures are not taken into account and are totally ignored.

## 5.3 Passive and Unheated Systems

Where possible, greenhouse operators or potential operators should definitely consider unheated or passively heated greenhouses. Many times, especially in low latitudes, the amount of incoming solar radiation to the greenhouse exceeds the total energy required to keep the indoor temperature at or above design temperatures. Since it is readily available and free, passive solar heating is the most cost effective and environmentally friendly method of heating greenhouses. Using proper greenhouse glazing materials (see Chapter 3) and curtain systems (see Chapter 11) are integral part of proper passive greenhouse designs. Furthermore, passive solar greenhouses may include solar heat storage systems (rocks, concrete, water, etc.) as well, and they can keep greenhouses warm during cloudy days or relatively cold nights.

## 5.4 Greenhouse Heating Systems

Forced air spaced unit heaters are quite simple and convenient heating systems for small greenhouse operations. They can be vented gas-fired, unvented gas-fired, oil-fired, electric, steam, or hot water unit heaters. For large operations, however, we typically use central heating systems. Central heating systems include a standard bare steel, black iron, and copper or aluminum pipe or finned pipe (with minimal footage and water volume) containing hot water or steam, hot air distributed in sheet metal or plastic tubing, and electrical resistance strips. Steam heating is no longer a favorable heating system due to the required excessive maintenance, initial cost, lack of installers, etc. Today, the most popular and common greenhouse heating systems are gas-fired unit heaters with one way gas piping and hot water (hydronic) heating systems; therefore, this chapter will focus mainly on gas-fired unit heater and central hot water heating systems. For a uniform temperature distribution in greenhouses, horizontal air circulation fans (HAFs) are integral part of spaced unit heaters. Even though HAFs are desirable with hot water systems, they readily provide quite a uniform temperature distribution and require custom engineering designs and quite complex installations for each and every successful application. They are designed to operate at different flow rates and temperatures. Bare piping, finned piping, or even spaced unit heaters can be used in central hot water heating systems. The piping or tubing in central heating systems may be placed as pipe rails between the crop rows, underbench, in-ground, along the outside walls, overhead, or directly above plant bed surfaces for root zone heating. Pipe rail heating alone can usually provide approximately 75-80% or more of the total heat loss of a greenhouse depending on the location and number of heating lines provided per bay. The remaining heat requirement is typically provided by sidewall, gable end, and overhead heating. Induced air circulation usually provides uniform temperatures within the plant canopy. Usually though, HAFs, the specifics of which will be discussed later, are desirable to improve the naturally induced circulation patterns. Overhead perforated plastic tubes with unit heaters are also used for greenhouse heat distribution. In these systems, the discharge holes are arranged in such a way that air is not blown directly onto plants. Several other discharge units are available with or without additional circulation fans to obtain better temperature distribution within the canopy. The unit employed may differ depending on the architectural characteristics of the plants grown in the greenhouse.

Another common practice for heat distribution is floor heating if plant containers can be set directly on the floor. Even though it varies from location to location and from one greenhouse type to another, this system can, for instance, provide up to 25% of the heat requirement for a double glazed greenhouse structure. A number of studies also reported that soil, floor, and bench-top heating systems for greenhouses conserve energy compared to more conventional natural or forced convection heated air systems (Boersma and Rykbost, 1975; Gosselin and Trudel, 1982; Seager, 1982; Zeroni and Gale, 1982).

Another heating method available today is radiant infrared heating; when properly designed and installed, they transfer the heating energy quite well even without HAFs, and provide adequate leaf temperatures while permitting the use of lower air temperature set points. Several studies compared infrared radiant heating systems with hot water systems in terms of the energy consumption. Knies et al. (1984) reported a net energy consumption reduction of 12% and 6% for two different configurations of infrared systems compared to hot water systems.

Casault et al. (1989), on the other hand, compared three gas-fired greenhouse heating systems (infrared emitters, condensing boilers, and unit heaters) in cold climates and reported that each system produced relatively uniform air temperature distributions both horizontally and vertically. However, the infrared emitters permitted equivalent leaf temperatures with lower air temperature set points (at least 3°C), while condensing boilers providing the best energetic performance because of their high steady-state efficiency and better sizing with respect to the heating load.

Alternative energy sources such as solar and wind energy are also receiving greater attention for heating, cooling, and power generation applications. In addition, the environmental concerns and the future of fossil fuels have advanced the use of heat pumps for heating and cooling applications. In order to improve the heat pump COP and replace the fossil fuel energy source, the idea of combining the heat pump and solar energy in beneficial ways has also been developed. The principal advantage of using solar radiation as a heat pump energy source is that, it provides heat at a higher temperature level than other sources, resulting in an increase in COP. Typical solar heat pumps can be operated in two basic configurations – direct and indirect. In a direct or conventional system, the refrigerant passes through the evaporator tubes in a solar collector (usually a flat plate type), where it is evaporated by incident solar energy. The heat pump extracts energy and delivers it to the load at the required temperature. In this system, the collector and evaporator functions are combined into one unit. Substantial energy savings can be realized using solar heat pump systems; however, the economics may not be favorable when initial and operating costs are considered together.

### 5.4.1 Reverse Return Principle

In piping systems having multiple branches and loops, the flow takes the path having the least resistance. In direct return systems with no controls shown in Figure 5.1 where the return header is connected to the load closest to the pump, there exists an inherent difference in flows. And this may be due to many factors, including pipe length and size, pipe roughness, material, fittings, bends, and many more.

Some may not have heard of this simple approach, but designers and operators have an option called the reverse return system. A reverse return system is a closed loop system where the return header is connected to the most hydraulically remote load, as shown in Figure 5.2. The basic principle is the last in, first out (the acronym LIFO is used to

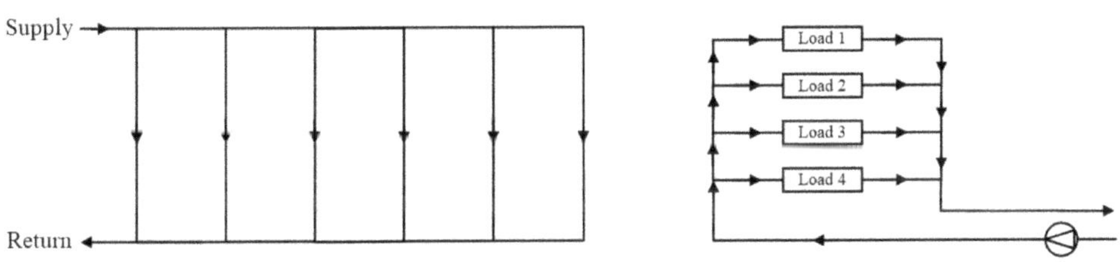


**FIGURE 5.1**
Different configurations of direct return closed loop system.

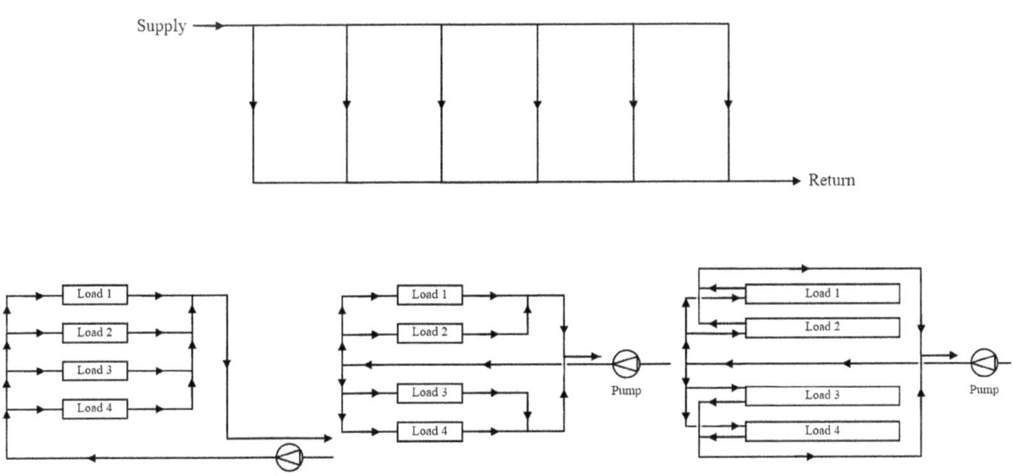


**FIGURE 5.2**
Different configurations of reverse return closed loop system.

remember easily). Compared to the direct return system, the reverse return system distributes the flows and pressures more evenly across the system, making it inherently balanced. The reverse return principle has been employed extensively in the heating, ventilation, and air conditioning (HVAC) industry to help balance flows. HVAC has many identical devices that require equal amounts of flow, such as boilers and chillers along with their respective heat emitters such as pipe rail, underbench, in-floor, overhead, and sidewall and gable end heat distribution systems.

### 5.4.2 Unit Heater Systems and Design

As mentioned earlier, unit heaters are the most commonly used greenhouse heating systems (see Figure 5.3). They can be vented gas-fired, unvented gas-fired, oil-fired, electric, steam, or hot water unit heaters. Greenhouse operations usually start small and expand in time; and as they expand, the operators keep adding more unit heaters to meet the heating requirement of their greenhouses. They are simple and cost effective, and normally rated on their input capacity not the output capacity (i.e. the capacity they deliver). When sizing an adequate unit heater, the output capacity must be considered in order to meet the heating demand; and the unit heaters have typically 80% efficiency.

**FIGURE 5.3**
Installed unit heater with a fresh-air louver on the back.

The two main types are the vented and unvented unit heaters. Unvented gas-fired unit heaters use the greenhouse air for combustion and exhausts the flue gas back into the greenhouse environment; therefore, all the energy after the natural gas or propane is combusted stays inside and is used to heat the greenhouse air. Some operators preferred this type of gas-fired unit heaters assuming that they could also provide $CO_2$ enrichment in their greenhouses; however, this argument has no justification because most of the time heating is required when $CO_2$ is not. Another byproduct of combustion is water vapor, which leads to condensation; therefore, it is not desirable in greenhouse environments either. Also, when the combustion is incomplete, a toxic gas CO is released, which must be monitored closely. Vented gas-fired unit heaters, on the other hand, exhausts the flue gas along with some heat to the outside air. Figures 5.4 and 5.5 show different (straight and 90° elbow) vented gas-fired unit heater connections.

In order to have proper combustion, each unit heater must be provided with sufficient fresh air supply so that its maximum efficiency can be achieved (see Figure 5.3 for the fresh air inlet louver on the back of the unit heater). As discussed earlier in Chapter 2, the ASME states that ***1 square inch free area per 2,000 Btu/h*** maximum fuel input of combined burners *must be provided and kept clear at all times*. Proper inlet louver sizing exercises will take place later in this chapter.

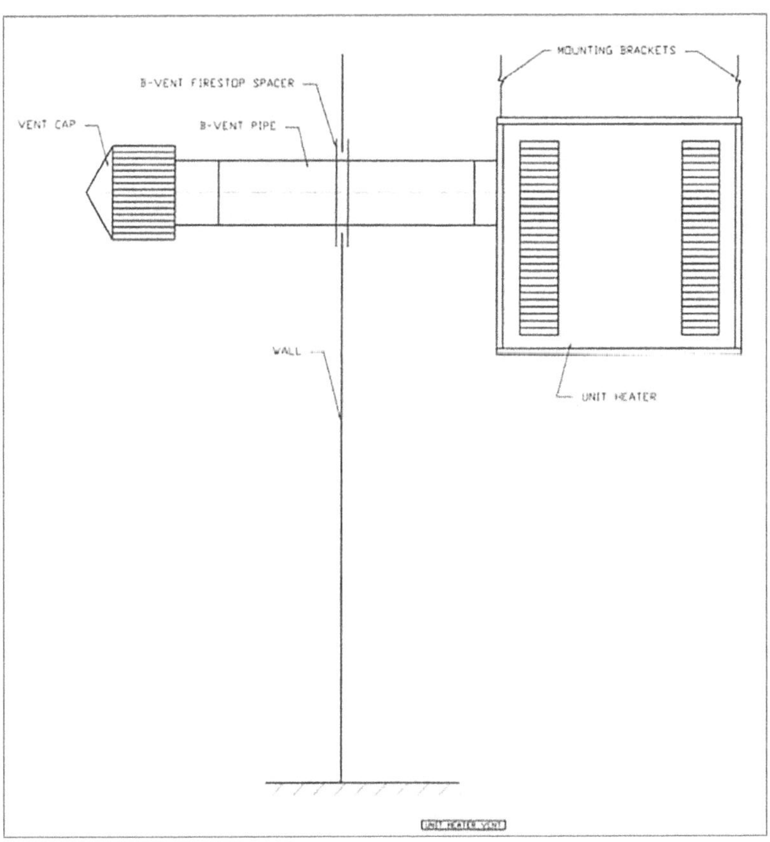


**FIGURE 5.4**
Unit heater with straight B-Vent connection.

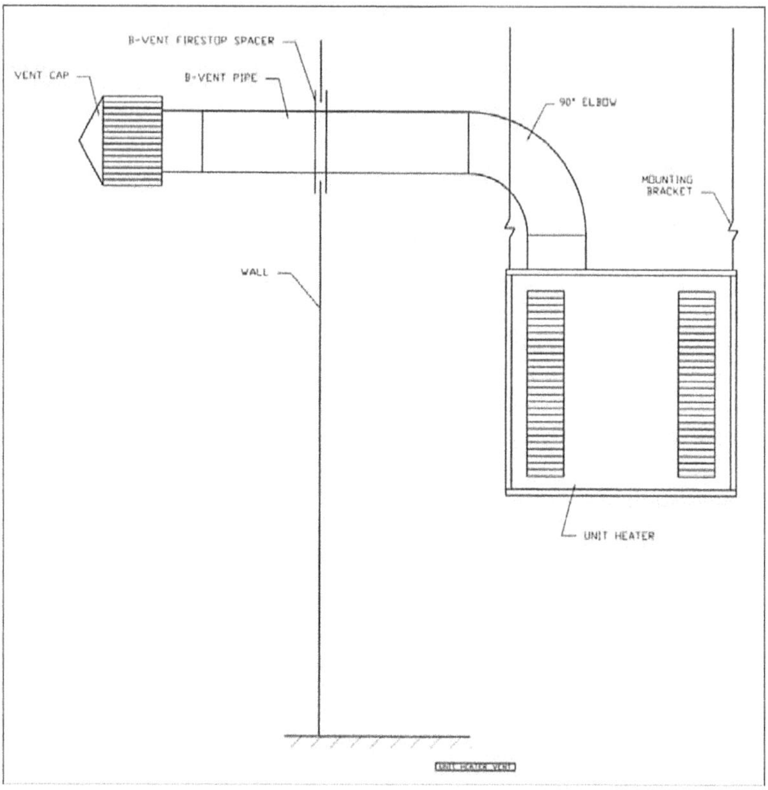


**FIGURE 5.5**
Unit heater with 90° elbow B-Vent connection.

Figure 5.6 shows the connection details of another type of unit heater, a hot water unit heater. These unit heaters are hooked up to a hot water or steam boiler system and require further engineering expertise for piping and controls.

### 5.4.3 Central Hot Water Heating Systems and Design

Central heating systems have a central boiler, which is either a hot water or a steam boiler, and a heat distribution system made up of the standard black pipe (natural convection), finned-pipe (natural convection) or hot water or steam unit heaters (forced convection). Boilers come in many different shapes and sizes, and are classified by operating temperature and pressure, fuel type (fuel oil, natural gas, coal, wood, etc.), and whether they are condensing flue gas or not. They are also described by their heat exchanger designs: cast-iron (oldest designs), fire-tube, steel water tube, and copper water tube boilers. In a fire-tube boiler, the hot flue gas moves through the tubes surrounded by hot water, whereas in a water tube boiler, whether it is steel or copper, water moves through tubes, which are surrounded by the hot flue gas this time. Condensing boilers can either be fire tube or water tube boilers, which condense the water vapor in the flue gas and extract the waste heat before the flue gas is exhausted into the atmosphere.

Hot water systems are as well classified by temperature and pressure, piping arrangement, and mode of circulation. Low temperature systems are generally open to the atmosphere through vent pipes and expansion tanks, and are called as low pressure systems.

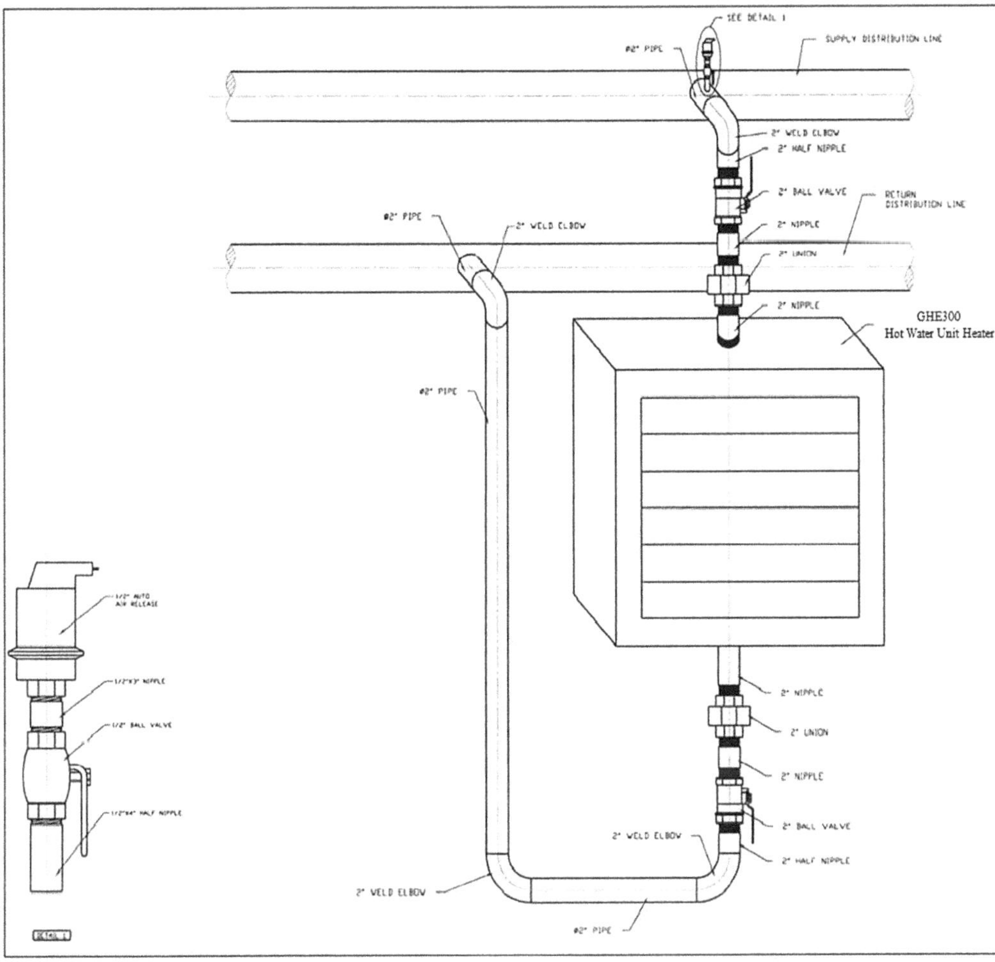


**FIGURE 5.6**
Hot water unit heater connection details.

They typically operate with flow water temperatures of less than 90°C (195°F). Modern systems today use lightly pressurized diaphragm type or nitrogen expansion vessels and have no open vent pipes as in the previous low temperature systems. These systems can operate at temperatures up to 120°C (250°F). These systems are known as medium temperature or pressure systems. Those systems that operate at temperatures above 120°C (250°F) are called high temperature or pressure systems. As mentioned above, natural gas, fuel oil, coal, or wood is burned in boilers to heat water and generate transportable heat. Boilers are rated in boiler horsepower based on the output capacity (1 bhp = 33,479 Btu/h). The hot water generated by a boiler is pumped and distributed in the greenhouse for heating purposes.

### *5.4.3.1 Boiler Room*

#### *5.4.3.1.1 The Heat Source*

The heat source in a hydronic heating system is typically a hot water boiler situated in the boiler room. The heating load of a greenhouse, as calculated using the estimating

**Example 5.1: Sizing Unit Heaters for a Gutter Connected Single Glass Greenhouse Complex**

Now let's consider a 10-bay gutter connected single glass gable roof greenhouse complex in Columbus, Ohio, USA, with an uninsulated floor, 21 ft (6.4 m) span, 130 ft (39.6 m) long, 16 ft (4.9 m) wall height, and 5.25 ft (1.6 m) gable height. Assuming that the headhouse is pre-heated, determine the heating load and design the heating system for the greenhouse using natural gas-fired unit heaters. See the layout of the greenhouse complex below.

**Solution**

| GREENHOUSE DIMENSIONS | | |
|---|---|---|
| **COMPONENT** | **(ft)** | **(m)** |
| Wall Height | 16 | 4.9 |
| House Width | 21 | 6.4 |
| House Length | 130 | 39.6 |
| Rafter Length | 11.5 | 3.5 |
| Gable Height | 5.25 | 1.6 |
| Number of Bays | 10 | 10 |

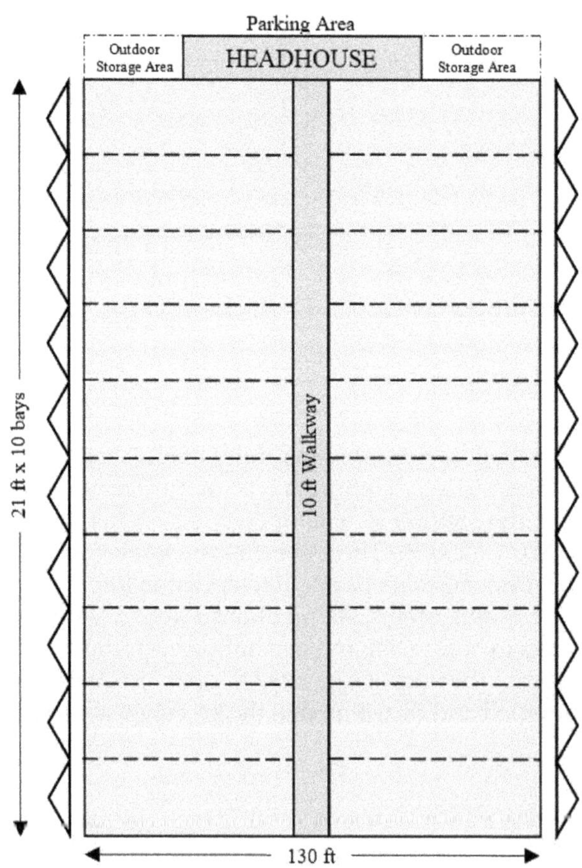


Layout of the gutter connected glass greenhouse for Example 5.1.

| OTHER PROVIDED or CALCULATED INPUTS | | | | |
|---|---|---|---|---|
| PARAMETER | (I-P) | | (SI) | |
| $T_i$ | 65 | °F | 18 | °C |
| $T_o$ | 6 | °F (99%); Table A.11 | −14 | °C (99%); Table A.12 |
| $\Delta T$ | 59 | °F | 32 | °C |
| Greenhouse Area | 27,300 | ft² | 2,536 | m² |
| | 0.63 | acre | 0.25 | ha |
| Greenhouse Volume | 508,463 | ft³ | 14,389 | m³ |
| Air Change per Hour | 1.00 (Table 3.4) | | | |
| Wind Factor | 1.00 (Table 5.1) | | | |
| Location | Columbus, Ohio, USA | | | |

| | AREA | | *U* VALUES | | *UA* VALUES & TRANSMISSION LOSSES | | | |
|---|---|---|---|---|---|---|---|---|
| COMPONENT | *A* (ft²) | *A* (m²) | *U* (Btu/h.°F.ft²) | *U* (W/°C.m²) | *UA* (Btu/h.°F) | *UA* Δ*T* (Btu/h) | *UA* (W/°C) | *UA* Δ*T* (W) |
| Single Material Wall | 10,880 | 1,011 | 1.1 | 6.2 | 11,968 | 706,112 | 6,267 | 200,533 |
| Gable End | 1,103 | 102 | 1.1 | 6.2 | 1,213 | 71,552 | 635 | 20,321 |
| Gable Roof | 29,900 | 2,778 | 1.1 | 6.2 | 32,890 | 1,940,510 | 17,222 | 551,098 |
| Perimeter Length | 680 | 207 | 0.8 | 1.4 | 544 | 32,096 | 290 | 9,285 |
| TOTAL | | | | | **46,615** | **2,750,270** | **24,414** | **781,237** |

* See Table 3.3 for determining the *U* values.

| HEAT LOSSES | | |
|---|---|---|
| | Btu/h | W |
| $Q_{cond} = UA \times \Delta T$ | 2,750,270 | 781,237 |
| $Q_{inf} = \rho \times C_p \times \Delta T \times Vol \times ACH$ | 599,986 | 155,023 |
| $Q_t = Q_{cond} + Q_{inf}$ | **3,350,256** | **936,260** |
| | **100 BHP** | |

The required heating load of 3,350,256 Btu/h (936 kW) can be supplied by using 10 (ten) 80% efficient gas-fired GHE200 (156,000 Btu/h output each) and 10 (ten) 80% efficient gas-fired GHE230 (184,000 Btu/h output each) providing a total of 3,400,000 Btu/h. The unit heaters come with 5 in. (12.7 cm) round flue gas connection on the back and can be installed with straight B-Vent connection as shown in Figure 5.4 and diagonally in each bay as shown below. GHE200 and GHE230 have ½ in. and ¾ in. gas piping sizes, respectively, and natural gas plumbers can easily connect the individual unit heaters to a single natural gas supply pipe along each gable end.

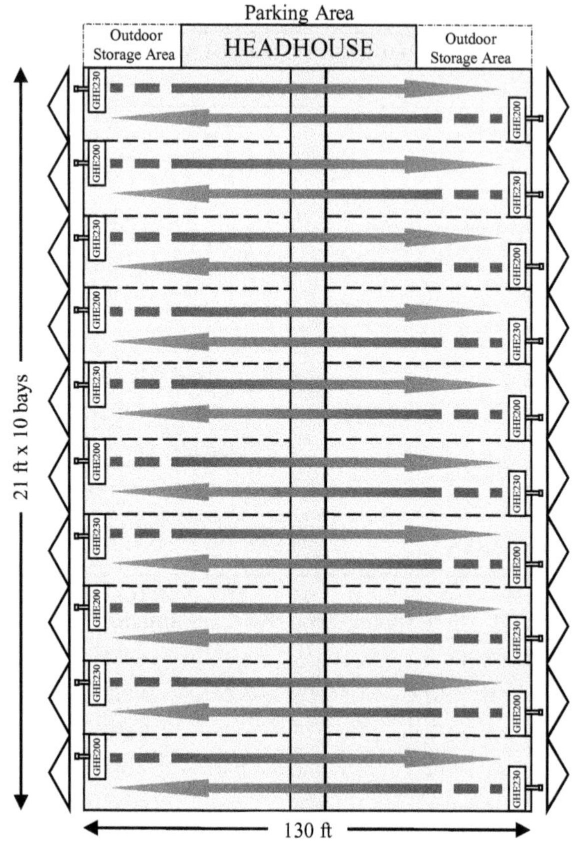


Layout of the unit heater heating system for Example 5.1.

In this project, we have used 10 (ten) 80% efficient GHE200 (156,000 Btu/h output each) and 10 (ten) 80% efficient GHE230 (184,000 Btu/h output each) providing a total of 3,400,000 Btu/h). This makes (10 × 156,000 Btu/h + 10 × 184,000 Btu/h)/0.8 = 4,250,000 Btu/h (input total). As mentioned earlier, the ASME states that ***1 square inch free area per 2,000 Btu/h*** maximum fuel input of combined burners *must be provided and kept clear at all times.* Then,

*Total Free Area Required* = (4,250,000 Btu/h input total) × (1 $in^2$/2,000 Btu/h) = 2,125 $in^2$

*Free Area per Unit Heater* = 2,125/20 heaters ≈ 106 $in^2$

Inlet louvers typically have 60% free area; therefore, 106 $in^2$/0.6 =177 $in^2$ louver area per unit heater needs to be provided. Then, for the purpose of providing fresh air here, 20 GHE13 motorized inlet shutters having opening dimensions of 13 ½ in. × 13 ½ in. each (which provides a louver area of 182 $in^2$) would meet the requirement. Instead, we could have sized the inlet louvers separately for GHE200 and GHE230 heaters. Please note that while the specifications are real, the manufacturer model is a generic model.

For this heating system, we also need to install adequate number and proper sizes of horizontal air flow (HAF) fans, which will be dealt with in Example 5.14 in Section 5.4.5 later in this chapter.

**Example 5.2: Sizing Unit Heaters for a Gutter Connected Double Poly Greenhouse Complex**

This time, let's consider a 10-bay gutter connected, IR inhibited double poly and curved roof greenhouse complex again in Columbus, Ohio, USA, with an insulated floor, 21 ft (6.4 m) span, 130 ft (39.6 m) long, 16 ft (4.9 m) wall height, and 5.25 ft (1.6 m) gable height. Assuming that the headhouse is pre-heated, determine the heating load and design the heating system for the greenhouse using natural gas-fired unit heaters. See the layout of the greenhouse complex below.

**Solution**

| GREENHOUSE DIMENSIONS | | |
|---|---|---|
| **COMPONENT** | **(ft)** | **(m)** |
| Wall Height | 16 | 4.9 |
| House Width | 21 | 6.4 |
| House Length | 130 | 39.6 |
| Length of Arch | 24.4 | 7.4 |
| Gable Height | 5.25 | 1.6 |
| Number of Bays | 10 | 10 |

| OTHER PROVIDED or CALCULATED INPUTS | | | | |
|---|---|---|---|---|
| **PARAMETER** | (I-P) | | (SI) | |
| $T_i$ | 65 | °F | 18 | °C |
| $T_o$ | 6 | °F (99%); Table A.11 | -14 | °C (99%); Table A.12 |
| $\Delta T$ | 59 | °F | 32 | °C |
| Greenhouse Area | 27,300 | ft$^2$ | 2,536 | m$^2$ |
| | 0.63 | acre | 0.25 | ha |
| Greenhouse Volume | 532,350 | ft$^3$ | 15,066 | m$^3$ |
| Air Change per Hour | 0.75 (Table 3.4) | | | |
| Wind Factor | 1.00 (Table 5.1) | | | |
| Location | Columbus, Ohio, USA | | | |

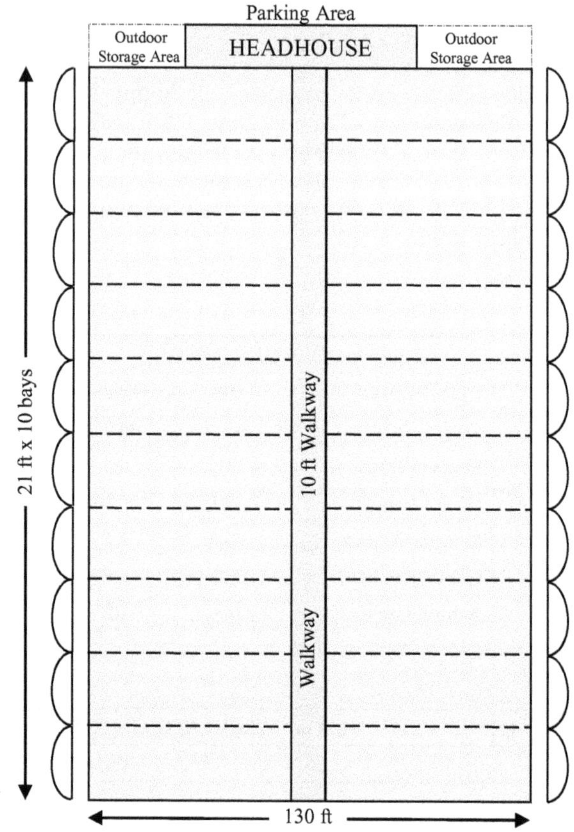


Layout of the gutter connected double poly greenhouse for Example 5.2.

| | AREA | | U VALUES | | UA VALUES & TRANSMISSION LOSSES | | | |
|---|---|---|---|---|---|---|---|---|
| COMPONENT | *A* (ft²) | *A* (m²) | *U* (Btu/h.°F.ft²) | *U* (W/°C.m²) | *UA* (Btu/h.°F) | *UA* Δ*T* (Btu/h) | *UA* (W/°C) | *UA* Δ*T* (W) |
| Single Material Wall | 10,880 | 1,011 | 0.5 | 2.8 | 5,440 | 320,960 | 2,830 | 90,563 |
| Curved End | 1,433 | 133 | 0.5 | 2.8 | 717 | 42,281 | 373 | 11,930 |
| Curved Roof | 31,720 | 2,947 | 0.5 | 2.8 | 15,860 | 935,740 | 8,251 | 264,032 |
| Perimeter | 680 | 207 | 0.4 | 0.7 | 272 | 16,048 | 145 | 4,643 |
| **TOTAL** | | | | | **22,289** | **1,315,029** | **11,599** | **371,168** |

* See Table 3.3 for determining the *U* values.

| HEAT LOSSES | | |
|---|---|---|
| | **Btu/h** | **W** |
| $Q_{cond} = UA \times \Delta T$ | 1,315,029 | 371,168 |
| $Q_{inf} = \rho \times C_p \times \Delta T \times Vol \times ACH$ | 471,130 | 121,729 |
| $Q_t = Q_{cond} + Q_{inf}$ | **1,786,159** | **492,898** |
| | | **53 BHP** |

The required heating load of 1,756,159 Btu/h (493 kW) can be supplied by using 4 (four) 80% efficient gas-fired GHE100 unit heater (80,500 Btu/h output each) and 10 (ten) 80% efficient gas-fired GHE115 unit heater (92,000 Btu/h output each) providing a total of 1,794,000 Btu/h). The unit heaters come with 4 in. (10.2 cm) round flue gas connection on the back and can be installed with straight B-Vent connection as shown in Figure 5.4 and diagonally in each bay as shown below. Both GHE100 and GHE115 have ½ in. gas piping sizes, and natural gas plumbers can easily connect the individual unit heaters to a single natural gas supply pipe along each gable end of the greenhouse complex.

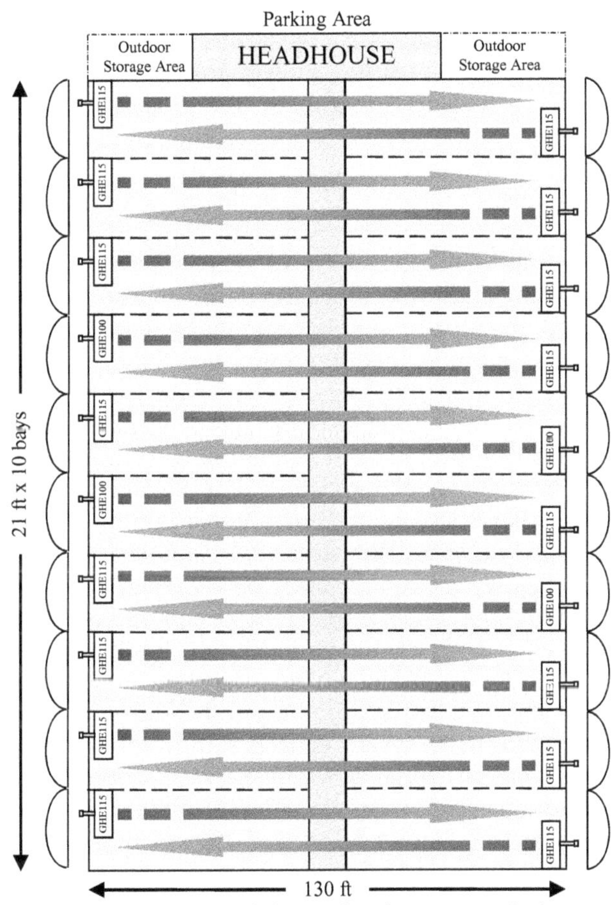


Layout of the unit heater heating system design.

We have used 4 (four) 80% efficient GHE100 (80,500 Btu/h output each) and 10 (ten) 80% efficient GHE115 (92,000 Btu/h output each) unit heaters. This makes (4 × 80,500 Btu/h + 10 × 92,000 Btu/h)/0.8 = 1,552,500 Btu/h (input total). And again, ***1 square inch free area per 2,000 Btu/h*** maximum fuel input of combined burners *must be provided and kept clear at all times*. Then,

$$\textit{Total Free Area Required} = (1{,}552{,}500 \text{ Btu/h input total}) \times (1 \text{ in}^2/2{,}000 \text{ Btu/h}) \approx 776 \text{ in}^2$$

$$\textit{Free Area per Unit Heater} = 776/(14 \text{ heaters}) \approx 56 \text{ in}^2$$

Inlet louvers typically have 60% free area; therefore, 56 $in^2$/0.6 = 92.4 $in^2$ louver area needs to be provided. Then, for the purpose of providing fresh air here, 14 GHE10 motorized inlet shutters having opening dimensions of 10 in. × 10 in. each (which provides a louver area of 100 $in^2$) would meet the requirement in this project. Instead, we could have sized the inlet louvers separately for GHE100 and GHE115 heaters. Please note that while the specifications are real, the manufacturer model is a generic model.

For this heating system, we also need to install adequate number and proper sizes of HAF fans, which will be dealt with in Example 5.15 in Horizontal Air Circulation section later in this chapter.

methodology introduced earlier in this chapter, determines the size of the heat source, a boiler for instance. As mentioned earlier, boilers are rated based on the output boiler horsepower; therefore, we first need to convert the greenhouse heating load to boiler horsepower. In greenhouse applications, we divide the total boiler horsepower requirement (e.g. 200 bhp) by 2 (two) to determine the capacity of each boiler (100 bhp) to be installed if we decide to install at least two boilers. In greenhouse applications, it is important to provide a back-up capacity of approximately 50 per cent. This means, if one of the boilers fails, then we would still have a 50% capacity to keep the greenhouse inside air temperature above freezing and protect the crop while the system is being fixed. If future expansion is considered, then the boiler room should be designed accordingly to accommodate the future boilers, pumps, mixing groups, etc. Figure 5.7 shows the connection details of a two-boiler central heating system where future expansion is not considered; however, connection details of alternative boiler room systems with two or more boilers and potential expansions are provided in Appendix B. While the sizes will vary from one boiler to another, typical connection details of a boiler relief valve and a 12 in. dia. top-connected B-Vent boiler chimney are provided in Figures 5.8 and 5.9, respectively. More connection details with different sizes and configurations are also provided in Appendix B.

The boiler lines, the manifold and the transport groups and mains are insulated with at least 30 mm thick insulation and jacketed with 0.8 mm aluminum sheet. The temperature and pressure sensors must be located where the volumetric flow is fully developed, which means that the sensors are installed at a minimum distance of 5 times the diameter of a pump or a valve. The air within the system collects at the highest point, typically where the flow direction changes from horizontal to vertical downward or from vertical upward to horizontal. The air should be bled by auto air release systems or other forms of air release systems at these locations (see Figures 5.7, 5.12, 5.13, and B.24). Should a water filtering assembly needs to be installed, connection details of two different water filtering systems are also provided in Appendix B. Thermal expansion of transport pipes, especially the long lines, should also be accommodated adequately so that transport line and/or structural damages can be prevented. As an example, the detail of a transport main flex joint connection against thermal expansion of pipes is provided in Appendix B as well (see Figure B.31). Overall, the boiler room must comply with the applicable local codes and regulations.

##### *5.4.3.1.2 The Manifold*

In a hot water heating system, the mixing groups, which will be discussed shortly, sit on two large diameter pipes (supply and return) called as the manifold (see Figure 5.7, the right hand side; and Figure 5.10 below for the details). As shown in Figure 5.11, the supply

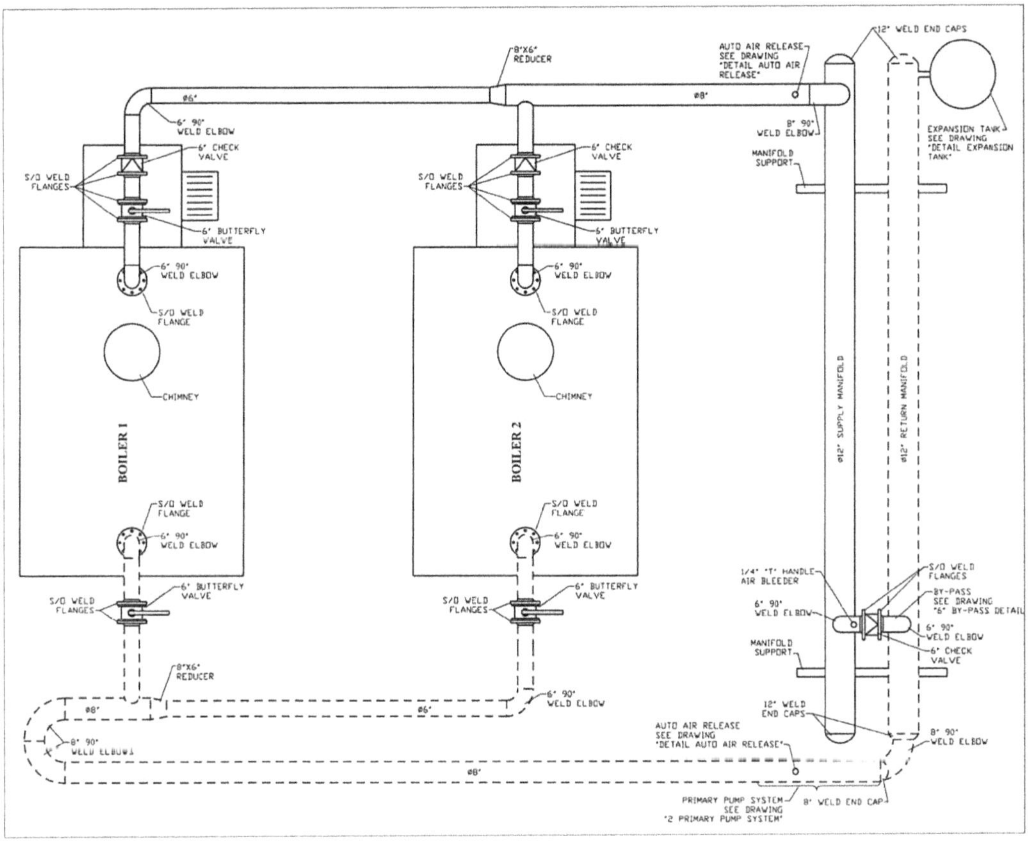


**FIGURE 5.7**
Connection details of a two-boiler central heating system.

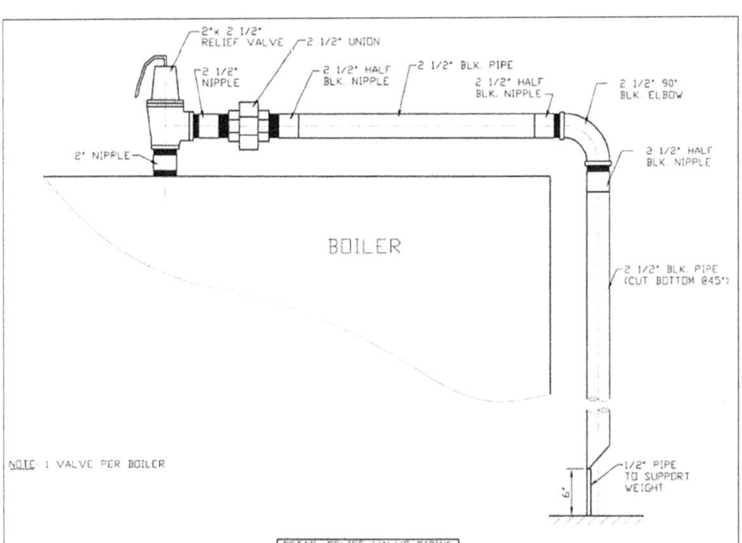


**FIGURE 5.8**
Connection details of a boiler relief valve.

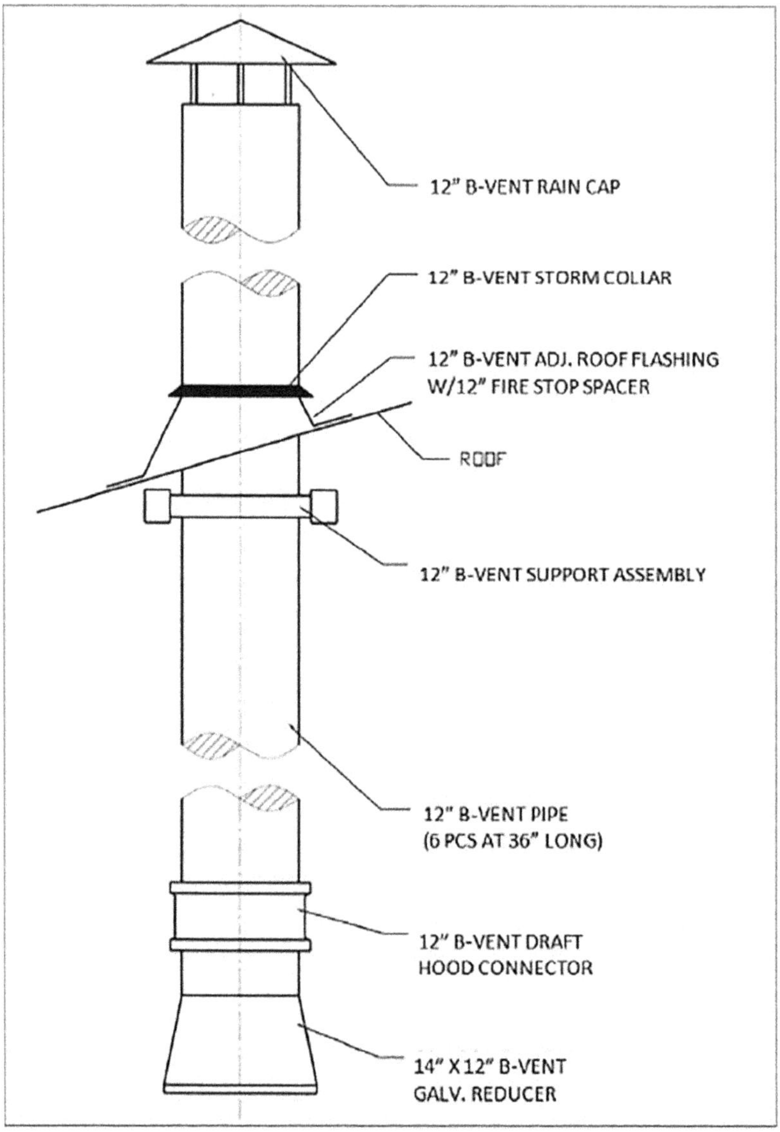


**FIGURE 5.9**
Connection details of a 12 in. dia. top-connected B-Vent boiler chimney.

and return manifolds are connected via a check valve preventing the back flow. The manifold simply provides extra water volume for pumps so that they would not "starve" for water and have access to enough hot water to transport heat. The manifold size is determined using the following relationship:

$$D = \sqrt{2\left(M_1^2 + M_2^2 + M_3^2 + . . . . + M_n^2\right)} \tag{5.7}$$

where $D$ is the manifold diameter in mm, $M_1$, $M_2$, ..., and $M_n$ are the size of mixing groups in mm.

**Example 5.3: Determining the Manifold Diameter**

Considering the manifold will feed 2 (two) 125 mm mixing groups now and 2 (two) 100 mm mixing groups in the future, determine the supply and return manifold sizes.

**Solution**

Using Equation (5.7), we can easily determine the manifold size as follows:

$$D = \sqrt{2\left(125^2 + 125^2 + 100^2 + 100^2\right)} = 320\,\text{mm} \approx 12''\left(\text{See Table 2.28}\right)$$

Since nominal size pipes come with 21 ft (6.4 m) lengths, we can use a full length of 12″ steel pipe to make both supply and return manifolds (10′–6″ or 3.2 m long each). Tubes come between 6 m (20 ft) and 9 m (30 ft) lengths.

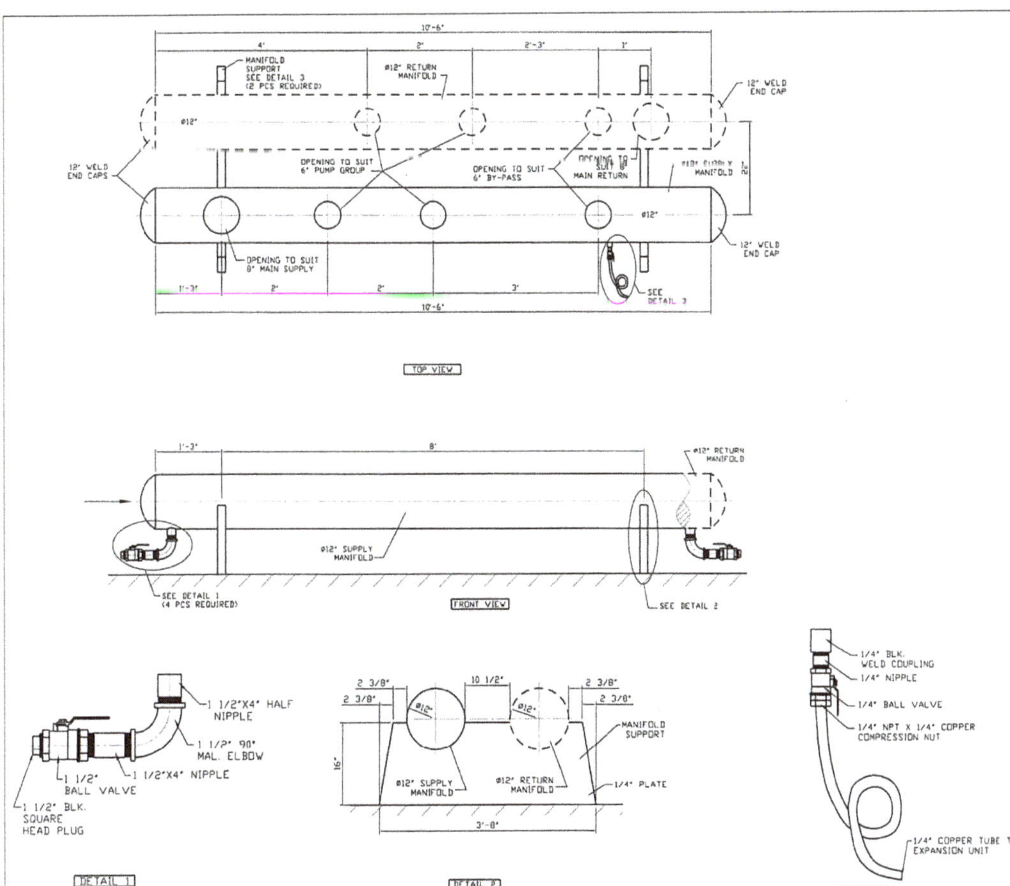


**FIGURE 5.10**
Details of a 12 in. manifold with two 6 in. pump connection openings.

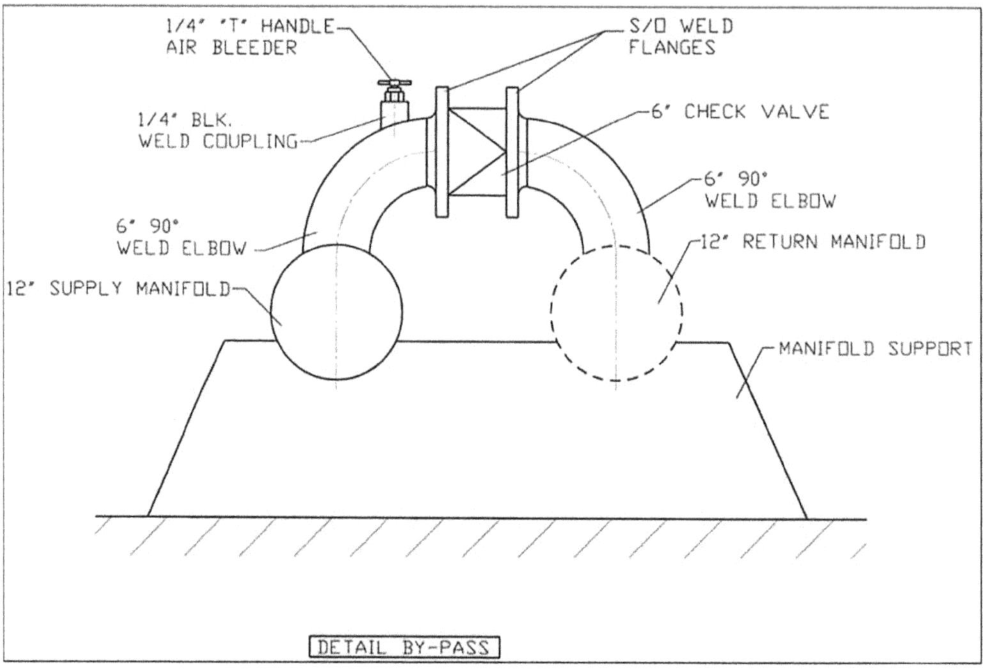


**FIGURE 5.11**
Connection details of a 6 in. transport main by-pass system.

#### 5.4.3.1.3 Boiler Main Lines

The boiler main lines are the pipe lines that connect the boilers to the manifold and are designed according to the reverse return principle discussed earlier in this chapter. Remember that the basic principle is the LIFO (see Figure 5.7). The boiler main lines can be designed using the methods introduced in Section 2.4. Connection details of boiler main lines for more boilers can be found in Appendix B.

#### 5.4.3.1.4 The Primary Loop

Most boilers require continuous flow over, and this is provided by the primary loop system. Circulation pumps are installed in the main return line and they keep circulating the water through the boilers and the supply and return manifolds, which are typically connected with a check valve to prevent backflow in the case of a pump failure. The loop is equipped with temperature and pressure gauges along with an auto air release system installed at the loop's highest point in order to release the air within the loop. If one of the circulation pumps fails, then the other pump will provide enough water circulation so that the greenhouse can still be provided with some heat in order to prevent the crops from freezing. Figure 5.12 shows the connection details of a two-pump primary group.

The capacity of the pump is calculated using Equation (2.24) and keeping in mind that the hot water boiler typically delivers 95°C (203°F) hot water with a return water temperature of 80°C (176°F) having therefore a $\Delta T$ of 15°C (27°F). After determining the pump capacity in l/h (or in gpm), the boiler main line pipe sizes and fittings and therefore, as stated above, the resistances are determined and the main line is designed. It should be kept in mind that the main line resistance not only includes the pipe, valve, and fitting

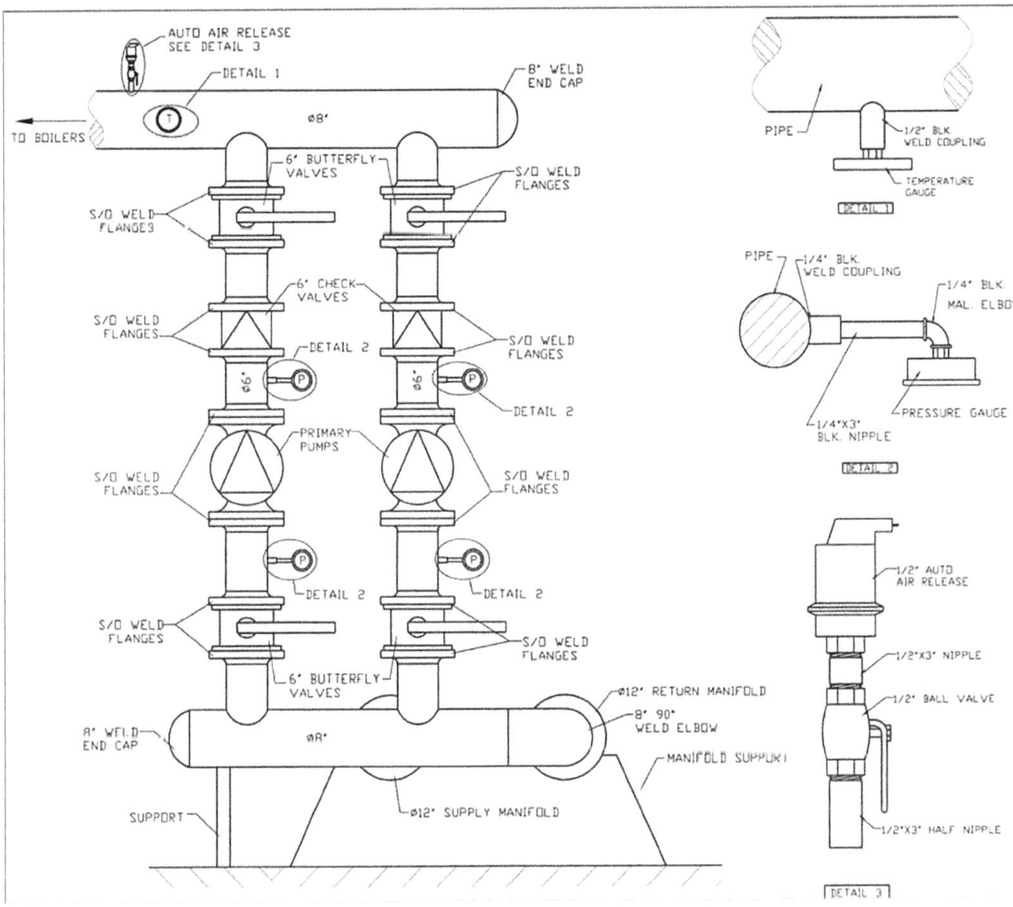


**FIGURE 5.12**
Connection details of a two-pump primary group.

resistances, but also the boiler's internal resistance. Boiler manufacturers typically provide the required resistance (head loss) information of the boiler; however, if the information is not provided, the experience has shown that a boiler head loss of 750 mm $H_2O$ (2.5 ft $H_2O$) can be assumed and used safely, and should be added to the line resistance in the primary loop. The selected circulation pump should obviously be able to deliver the required volumetric capacity (l/h or gpm) at the designed head (mm $H_2O$ or ft $H_2O$). An actual primary loop design work in step by step fashion will be shown later in this chapter.

#### *5.4.3.1.5 The Transport or Mixing Group*

As shown in Figure 5.13, insulated and normally jacketed pipes and manifolds deliver the hot water to the greenhouse zone being heated. The mixing group connected to the manifold adjusts the water temperature based on the need by mixing the supply water (typically 90°C or 194°F) with the return water (typically 75°C or 167°F) having a $\Delta T$ of 15°C (27°F). If the greenhouse zone calls for more heat, then the supply water is mixed with less return water; and if the zone requires less heat, then the supply water is mixed with more return water coming at a lower temperature. The transport pumps circulate the hot water

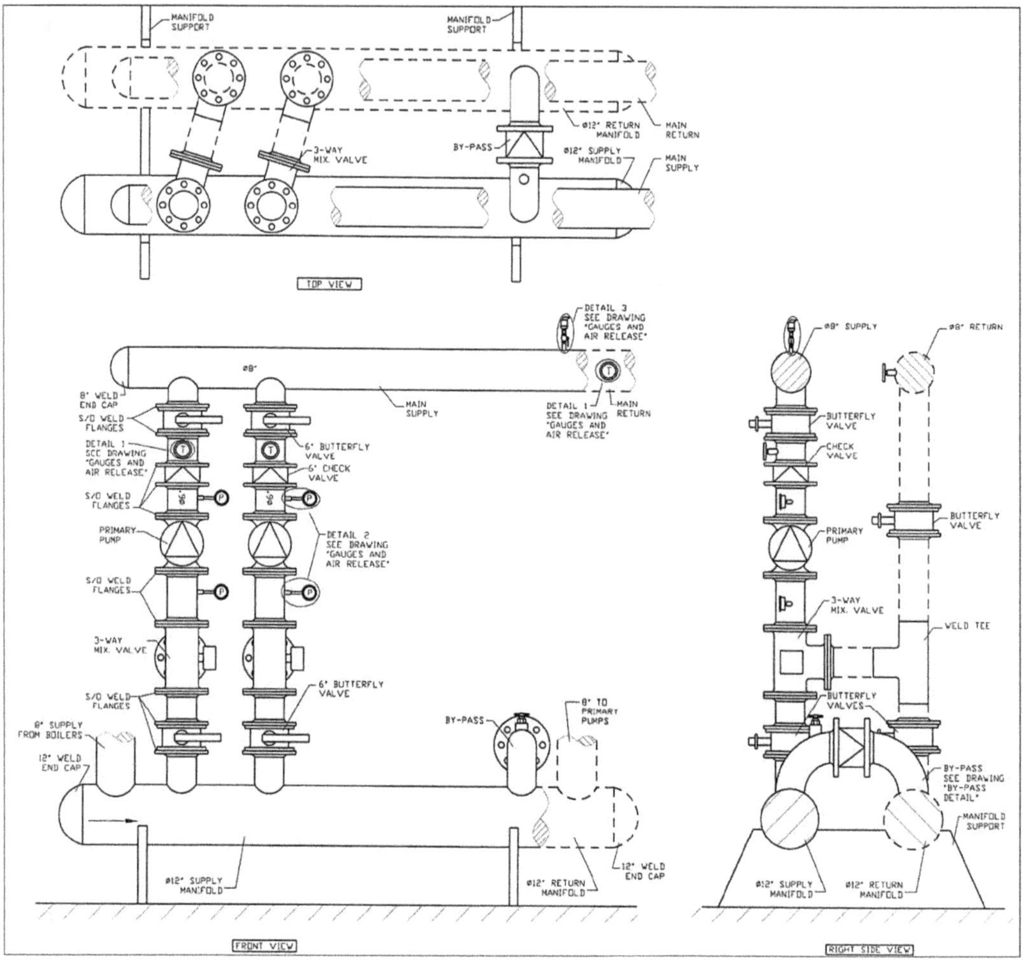


**FIGURE 5.13**
Connection details of a two-pump transport group.

through the main transport lines in the greenhouse and brings back to the mixing group located on the manifold. Figure 5.13 shows all the connection details of a two-pump transport group, each connected to the return line with a dedicated mixing valve (a 3-way mixing valve shown in the figure). As discussed earlier in Chapter 2, the mixing valve is sized using Equation (2.187) or Equation (2.189) for I-P or SI, respectively; and the relationship between the flow coefficients $C_v$ and $K_v$ is given by Equation (2.191). In order to determine the flow coefficient for the mixing valve, the flow rate (in proper units) and a pressure drop of 1.45 psi (or ~1,020 mm $H_2O$) are used. And the mixing valve is sized based on the calculated flow coefficient ($C_v$ or $K_v$). Then, using the flow coefficient of the actual mixing valve, the actual pressure loss across the mixing valve is calculated and used in determining the transport pump's head. An actual design work will be shown later in this chapter.

The capacity of the transport pump is calculated using Equation (2.24) and typically a supply water temperature of 90°C (194°F) with a return water temperature of 75°C (167°F). After determining the pump capacity either in l/h or in gpm, the main transport line pipe sizes and fittings and therefore, the resistances are determined and the transport line is designed.

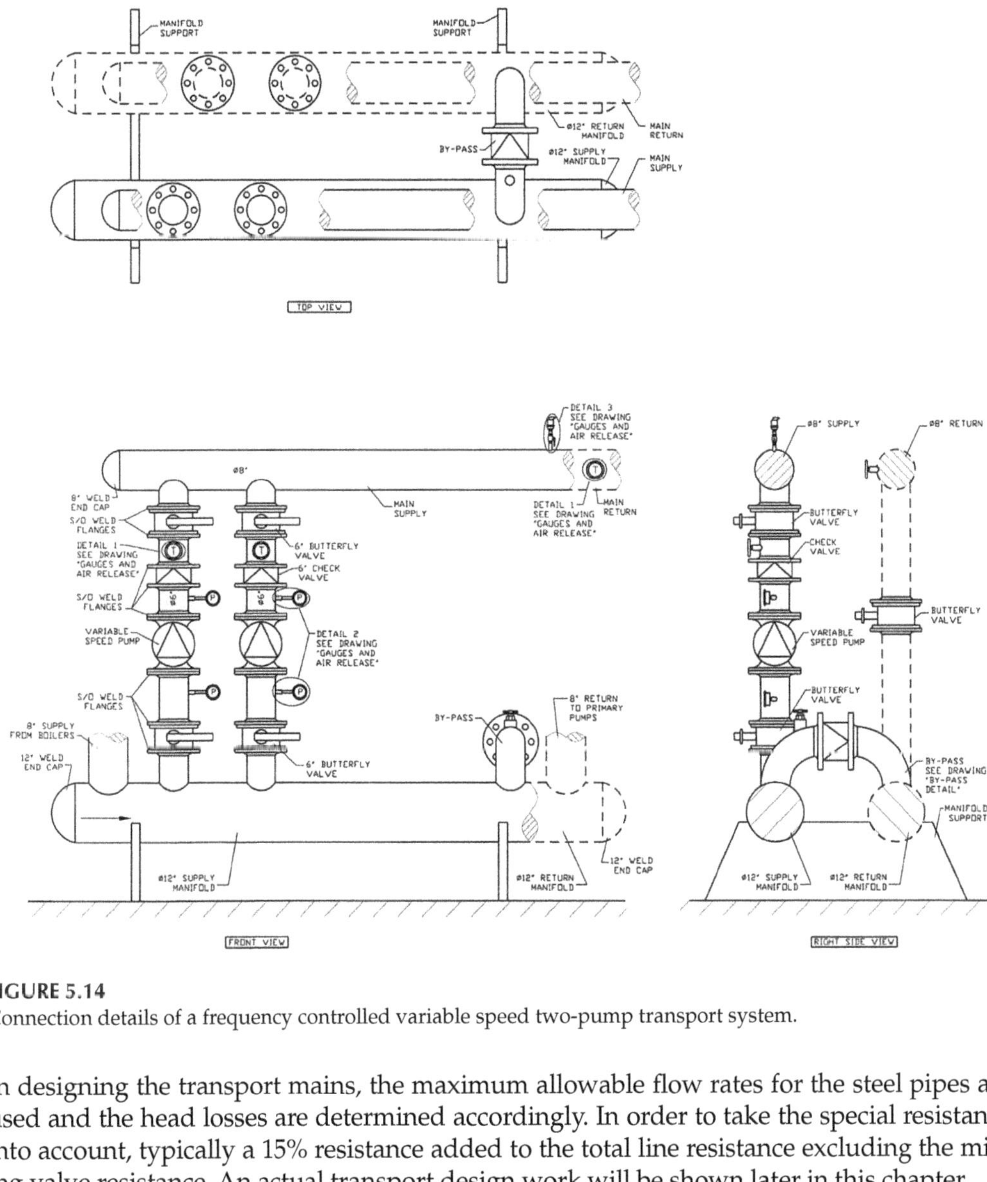


**FIGURE 5.14**
Connection details of a frequency controlled variable speed two-pump transport system.

In designing the transport mains, the maximum allowable flow rates for the steel pipes are used and the head losses are determined accordingly. In order to take the special resistance into account, typically a 15% resistance added to the total line resistance excluding the mixing valve resistance. An actual transport design work will be shown later in this chapter.

Figure 5.14, on the other hand, shows the connection details of a frequency controlled variable speed two-pump transport system. The design procedure is exactly the same as in the previous case except for the mixing valve. No mixing valve is needed or used in a frequency controlled transport line circuit; therefore, the resistance or the pump head is now smaller. The benefits of using a frequency controlled variable speed pumps are discussed further with numerical examples later in Chapter 11.

### *5.4.3.1.6 Expansion System*

The expansion and contraction of the water in a hot water heating system is controlled by the expansion vessel. When the system is heated up, the expansion vessel stores the increased volume of the system's water content and prevents the piping system from

**TABLE 5.2**

Density of Water as a Function of Temperature.

| Temperature °C (°F) | 4 (39.2) | 10 (50) | 20 (68) | 30 (86) | 40 (104) | 50 (122) | 60 (140) | 70 (158) | 80 (176) | 90 (194) |
|---|---|---|---|---|---|---|---|---|---|---|
| Density kg/m$^3$ (lb/ft$^3$) | 1,000 (62.43) | 999.8 (62.42) | 998.3 (62.32) | 995.8 (62.17) | 992.3 (61.95) | 988.1 (61.69) | 983.2 (61.38) | 977.7 (61.04) | 971.4 (60.64) | 965.1 (60.25) |

bursting. When it cools and shrinks down, the stored water is sent back to the system. The size and location of the expansion vessel are important. The mass of the water in the installation does not change; however, the volume of the water increases as temperature increases, that is, the density of the water decreases. Table 5.2 lists the density of water at different temperatures.

Water has its maximum density at 4°C or 39.2°F ($\rho$ = 1,000 kg/m$^3$). When water is heated from 4°C to 90°C, the volume of water increases by 1,000/965 = 1.036 times, having an expansion of 3.6%. In practice, an expansion of typically 4–5% is used.

Expansion systems are typically categorized under six different types: i) open expansion vessel; ii) expansion vessel with nitrogen cushion of varying pressure; iii) expansion vessel with constant pressure nitrogen cushion; iv) open pump-expansion vessel based on the overflow principle; v) closed pump-expansion vessel with pressure-controlled pump; and vi) membrane expansion vessel.

An open expansion vessel openly communicates with the environment. A secondary function of the open expansion vessel is to protect the heating system against excessive overpressure. The vessel is connected to the hydronic heating system by a non-closable safety pipe. Excess water is discharged outside through an overflow. A glass gauge indicates the water level in the vessel, and the position of the vessel determines the pressure in the installation. Obviously, the pressure variations are caused by the varying water level in the vessel, and the higher the vessel is placed, the higher the static pressure in the system. When the boiler room height is considered, installing this kind of an expansion vessel may be problematic. Therefore, the open expansion vessels are installed horizontally, creating a very large water–air separation surface. It means that the water is in contact with air; hence, the water absorbs plenty of air, and the absorbed oxygen leads to corrosion in the hydraulic system. Unless it is released properly, the air accumulates at the highest point of a plumbing section and partially blocks the circulation.

In the expansion vessel with nitrogen cushion of varying pressure, the nitrogen volume above the water surface is compressed as the water expands; and when the water shrinks, the nitrogen cushion expands. The pressure of the nitrogen cushion and therefore the heating system pressure vary between 0.5 and 2 bar. The lost nitrogen is supplemented from a permanently connected nitrogen bottle.

In the expansion vessel with constant pressure nitrogen cushion, when the system's water is warmed up, the water expands, the water level in the vessel rises, and the nitrogen pressure increases. The compressor starts running and presses the excess nitrogen back into the expansion vessel. The maximum nitrogen pressure in the storage tank is 5 bar. When the water in the hot water heating system cools down, the water shrinks and the pressure in the expansion vessel drops. An on/off control valve installed between the expansion valve and the nitrogen supply vessel is opened; and the nitrogen pressure in the expansion vessel increases until the pressure goes back to the set point pressure. This expansion system includes a nitrogen bottle as well in order to fill the expansion system and compensate for any losses.

The open pump expansion vessel based on the overflow principle stores and replenishes the water of a hot water heating system. The vessel is connected to the heating system via a pressure booster pump with a pressure regulator. Above the vessel, there exists air at about atmospheric pressure. If the pressure is too high, the relief valve discharges the air; and a vacuum valve prevents the observance of underpressure in the vessel. A continuously running booster pump creates the overpressure in the system, and the pressure increase is set and controlled by a pressure regulator. This regulator is incorporated in a shunt or overflow pipe and maintains the necessary continuous flow through the overflow pump. In this system, provisions must be made to ensure that the booster pump remains in continuous operation.

Instead of a continuously operating pressure boosting pump and a safety valve, in a closed pump expansion vessel with pressure-controlled pump, the pump and solenoid valve are controlled by a pressure switch. When the pressure in the heating system increases, the solenoid valve opens and the water flows to the expansion vessel. When the system pressure drops, this time the water is pumped from the expansion vessel to the hot water heating system.

In a membrane or diaphragm type expansion tank, a membrane separates the water and the compressed air that accommodates the expanding water. When the system's water expands, the excess water moves into the membrane and the air in the tank is compressed; and when it shrinks during the cooling process, the air volume in the closed section expands.

*Determination of the Content of a Membrane Type Expansion Reservoir:*

If the content of the expansion reservoir is expressed as $I$, $\Delta V$ is the maximum volume increase of water due to expansion, and the minimum and maximum pressures in the system are $p_{min}$ and $p_{max}$ (absolute pressures), respectively; then from Boyle's law, we have the following relationship for the air volume:

$$I \times p_{\min} = (1 - \Delta V) \times p_{\max} \tag{5.8}$$

Then,

$$I = \Delta V \times \left( \frac{p_{\max}}{p_{\max} - p_{\min}} \right) \tag{5.9}$$

**Example 5.4: Determining the Size of a Membrane Type Expansion Vessel**

Determine the size of a membrane type expansion vessel if the water volume increase or expansion is 2,000 liters.

**Solution**

Using Equation (5.9), we can easily determine the tank size as follows:

The initial and final pressures are 0.5 bar (1.5 bar absolute) and 3 bar (= 4 bar absolute), respectively. Then the content of the expansion tank becomes:

$$I = (2{,}000 \text{ liter}) \times \left( \frac{4}{4 - 1.5} \right) = 3{,}200 \text{ liter}$$

**Location of the expansion vessel:**
The expansion vessels are located on the return side (see Figures 5.7 and 5.15) of the hydronic heating system so that the whole system can operate at or above the system pressure. Hydronic systems typically operate a system pressure of 12–15 psi (0.83–1.03 bar), which is controlled by the use of expansion vessels.

### *5.4.3.1.7 Boiler Room Air Supply*

As discussed earlier in Chapter 2, for complete combustion and safety, boiler rooms must be provided with proper openings so that enough fresh air is supplied. Temporary air intake openings such as doors and windows should definitely be avoided as they may be

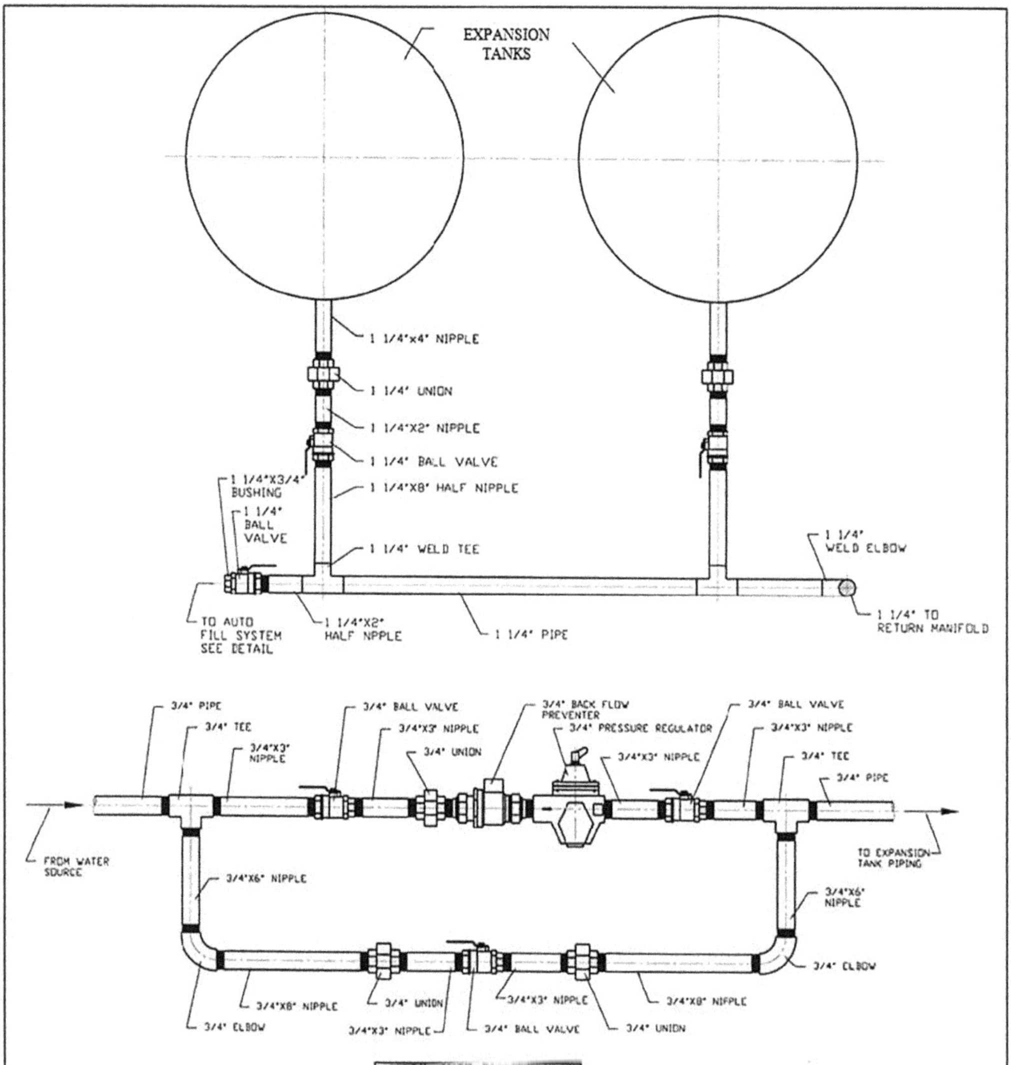


**FIGURE 5.15**
Connection details of two-tank membrane type expansion vessels and auto fill system.

closed and hence cut off the required air supply. From experience, the ASME states that ***1 square inch free area per 2,000 Btu/h*** maximum fuel input of combined burners located in the boiler room works, and minimum 2 (*two) air supply openings must be provided and kept clear at all times.*

It should also be kept in mind that inlet louvers typically have 60% free area; therefore, the actual (gross) louver area per unit heater will be larger. Properly designed motorized inlet shutters will do the job perfectly fine, and an actual design example will be shown later in this chapter.

### *5.4.3.2 Heat Distribution Systems*

As mentioned earlier, the insulated transport mains and manifolds deliver the hot water to the greenhouse zone that is being heated. The heat distribution systems are normally thin-walled 51 mm OD pipe rail heating and 33 mm OD grow pipe heating systems for the row vegetable crops or cut flowers; 51 mm OD pipe underbench heating systems for the seed germination, propagation, or general growth of crops grown on the bench tops; 25 mm OD (1″) polyethylene tube in-floor or on floor heating systems for the plants in containers directly placed on the floor; 51 mm OD pipe overhead and snowmelt systems, if needed; and again 51 mm OD pipe sidewall and gable end heating systems. In each case, the temperature of the circulating water is controlled via a zone mixing valve to match the heat requirement of a particular greenhouse zone. The heating zones are typically determined by the ventilation zones; it means, the number of ventilation zones determine the number of heating zones. Hot water heating systems often provide an even heat distribution, which results in a better temperature uniformity. When designed and installed properly, observing both the energy and mass balances, maximum spatial (horizontal) temperature difference observed in hot water greenhouse heating systems is approximately 1°C. In order to keep the spatial temperature differences at minimal levels, the greenhouse heating zones are designed accordingly. For instance, in small applications, let's say, if the length or width of a greenhouse section to be heated is less than 100 m, a single heating zone is used, where the mixing groups and transport lines are directly connected to the manifold in the boiler room; and in large applications, if the length or width is greater than 100 m, typically the section is split into more heating zones, where the main transport line is connected to the pre-mix group in the boiler room and the zone mixing groups are typically connected to the main transport lines within the greenhouse space. Other considerations are also taken into account for a uniform heat distribution within the greenhouse. For instance, aluminum painted pipes typically emit 30% less heat compared to black pipes; vertical pipes emit 50% less heat compared to the same size horizontal pipes; and plastic hoses half horizontal plus half vertical emit 3/4$^{th}$ of a horizontal plastic hose. Considering all these fine differences and taking them into account, we can design a well-functioning gable end and sidewall heating systems, for instance.

#### *5.4.3.2.1 Pipe Rail and Grow Pipe Heating System Designs*

*Pipe Rail Heating Systems*

As mentioned earlier, rail pipe heating systems are used quite effectively for row vegetable crops or row cut flowers. In order to distribute heat in such systems, thin-walled 51 mm OD pipe loops are used (see Figure 5.16). Depending on the bay width, 4, 5, or 6 loops of 51 mm OD pipes are installed. The rail loops start 0.7 m (2.3 ft) from the gable ends and run to the walkway, and are connected to the distribution headers along the gable end via 1.5–2

**FIGURE 5.16**
Pipe rail heating system.

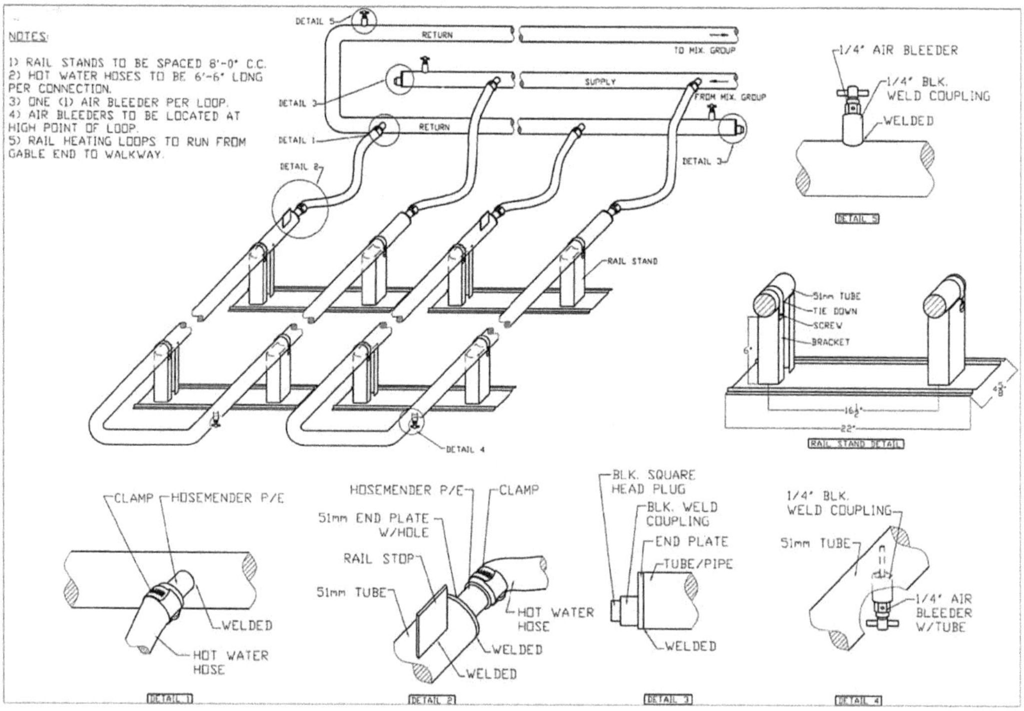


**FIGURE 5.17**
Connection details of pipe rail heating system.

m (5–6 ½ ft) 19 mm or 25 mm ID black hot water heating hoses (see Figures 5.17 and 5.18). If the water flow in the loop is less than 800 l/h (3.5 gpm), a 19 mm ID black hot water hose is used, whereas a 25 mm ID black hot water hose is used if the flow is greater than 800 l/h (3.5 gpm). Each pipe rail loop has a return bend, the older versions of which have 42 cm c.t.c. (50 cm total) widths, and relatively newer ones have 55 cm c.t.c. (60 cm total) widths. And each loop is supplied with an air bleeder with tube, and each distribution header is supplied with a regular air bleeder at the highest points (see Figures 5.17 and 5.19).

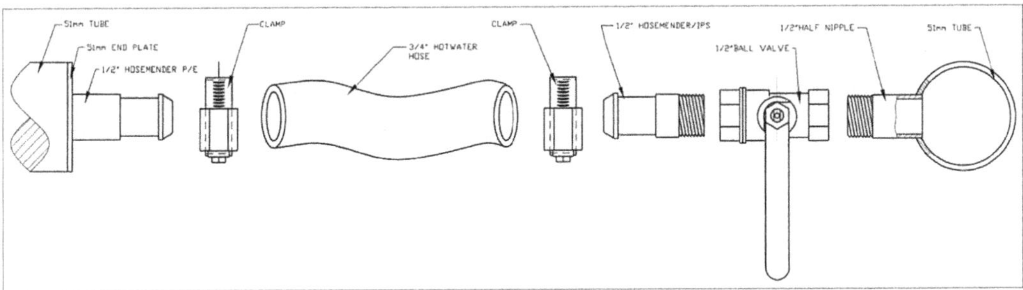


**FIGURE 5.18**
Details of pipe rail heat distribution system hose and valve connection to distribution headers.

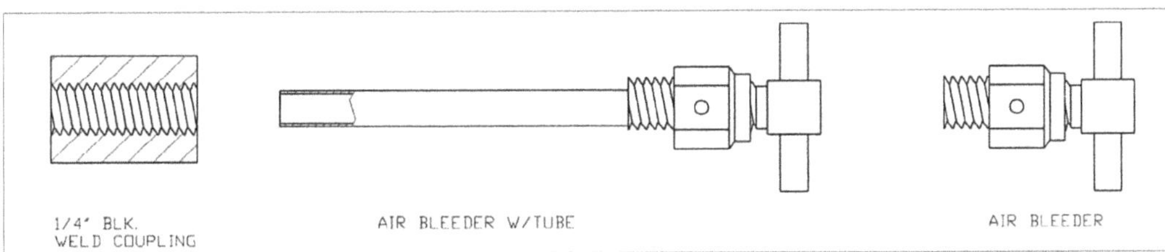


**FIGURE 5.19**
Details of air bleeders (see Detail 4 in Figure 5.17).

The distribution headers are designed according to the reverse return principle with a maximum flow resistance limit of 10 mm $H_2O$ (0.033 ft $H_2O$) and connected to a rail pipe heating zone mixing group as discussed earlier (see Figures 5.20 and B.32 for an alternative connection). The average pipe size of the distribution header, whether it is supply or return, is 0.8 times the maximum pipe size in the distribution line. See Example 5.6 for a rail heating system design example.

With respect to the heat loss from the greenhouse, the heat loss from the space 0.7 m (2.3 ft) from the gable end wall in the bay is treated as part of the heat loss from the gable end, and the heat demand is supplied by the distribution headers, the loops' plastic hose connections, and the additional 51 mm OD pipes, if needed (see Figure 5.21). Maximum pipe rail heating system temperature used is typically 90°C (194°F) whereas the return temperature is 75°C (167°F). In order to get these system temperatures, the flow temperatures to the pipe rail heating zone mixing group are 91°C (~196°F) and 74°C (~165°F) for supply and return, respectively.

The maximum flow resistance limit used in the heat distribution lines as well as in the loops is 10 mm $H_2O$ (0.033 ft $H_2O$) and a minimum of 250 mm $H_2O$ (0.83 ft $H_2O$) loop resistance from the supply header to return header is typically used over the 51 mm OD loop, including the connections. A special resistance is built in within this magnitude for the hose, the hosemenders, the nipple, and the valve.

### *Grow Pipe Heating Systems*

Grow or vegetation pipe heating systems are used with pipe rail heating systems mainly for getting rid of the excess moisture within the canopy of row crops so that the risks of condensation on plant surfaces and therefore the pertinent diseases are reduced or eliminated. When grow pipe heating is used, not only the excess moisture is removed, but also

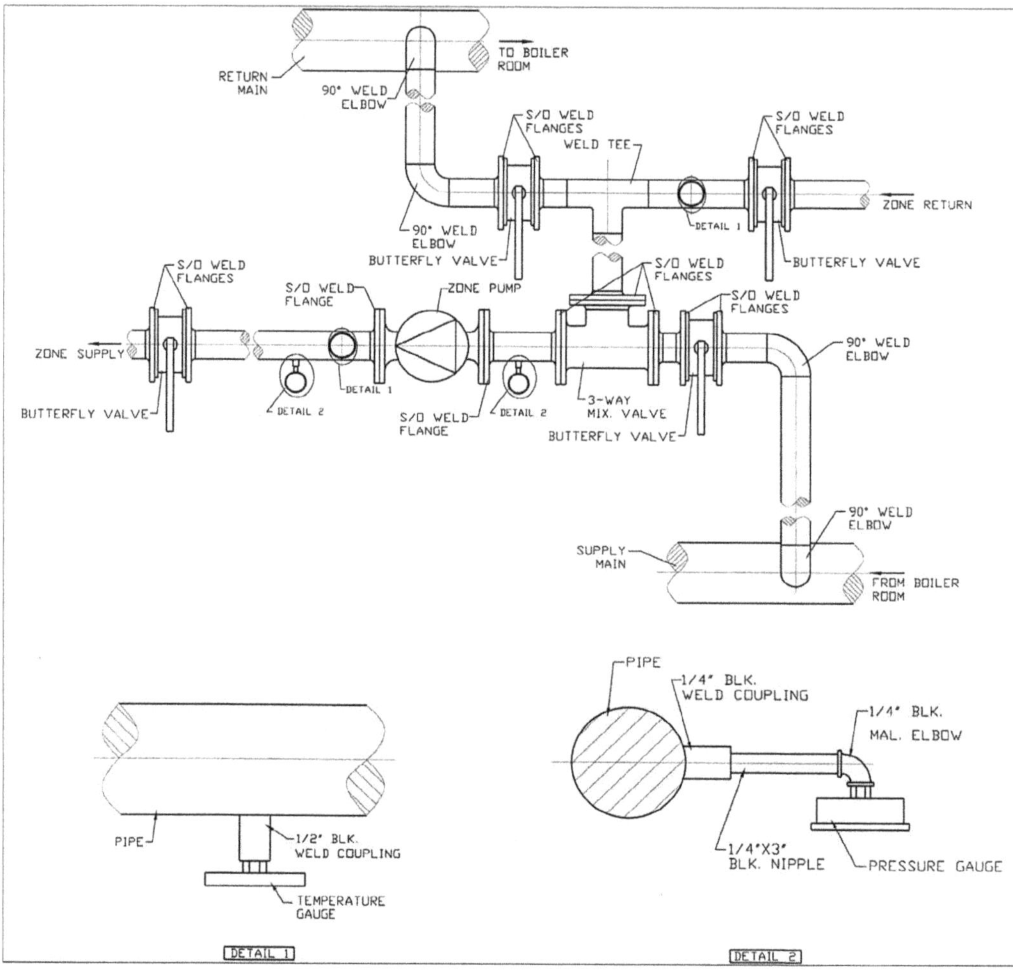


**FIGURE 5.20**
Details of a pipe rail heating zone mixing group connection.

a modest amount of heat is contributed within the immediate environment of plants at a relatively lower pipe temperature compared to those of the rail heating system. The heat released by the connecting arching pipes and the plastic hoses of grow pipe heating can also be used to cover the deficiency of the rail pipe heating along the walkways; therefore, the need for on-ground 51 mm OD heating along the walkway on both sides is reduced. Depending on the climate, sometimes one or two lengths of 51 mm OD pipe on each side of the walkway are required in order to have a balanced heating along the walkways as well.

As shown in Figures 5.22 and 5.23, grow pipe heating systems have separate distribution headers along the gable end walls and therefore separate mixing groups as well. A total of 2 ½ loops (5 × 33 mm OD, 1″ or ¾″ pipes) per bay, that is, one pipe length per pipe rail loop is used. The 33 mm OD, 1″ or ¾″ pipes are installed about 2.1 m (7 ft) above the floor and 0.3 m (1 ft) below the plant level.

Just like pipe rail loops, the grow pipe loops start 0.7 m (2.3 ft) from the gable ends and run to the walkway, and are connected to the distribution headers along the gable end via

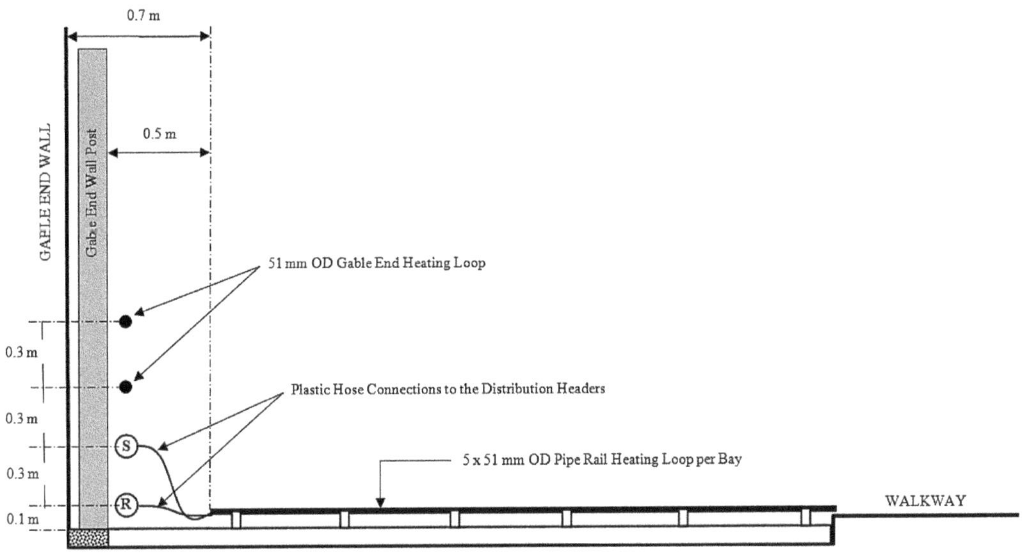


**FIGURE 5.21**
Distribution headers and the pipe rail heating loop (side view).

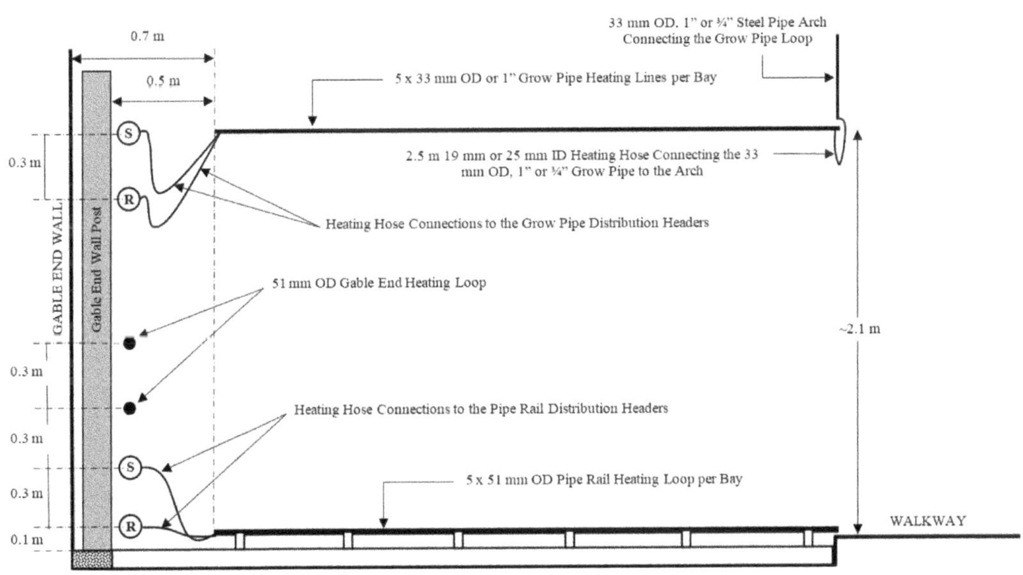


**FIGURE 5.22**
Distribution headers and the grow pipe heating loop along with the pipe rail (side view).

1.5–2 m (5–6 ½ ft) 19 mm or 25 mm ID black hot water heating hoses (see Figure 5.22). As in the pipe rail heating system, for connecting to the distribution headers, a 19 mm ID black hot water heating hose is used if the water flow in the grow pipe loop is less than 800 l/h (3.5 gpm), whereas a 25 mm ID black hot water heating hose is used if the flow is greater than 800 l/h (3.5 gpm). A same size pipe is used for the arch connecting the 33 mm OD, 1″ or ¾″ pipe lengths via 2.5 m long 19 mm or 25 mm OD black hot water heating hoses on each side in between (see Figure 5.23). And each loop and each distribution

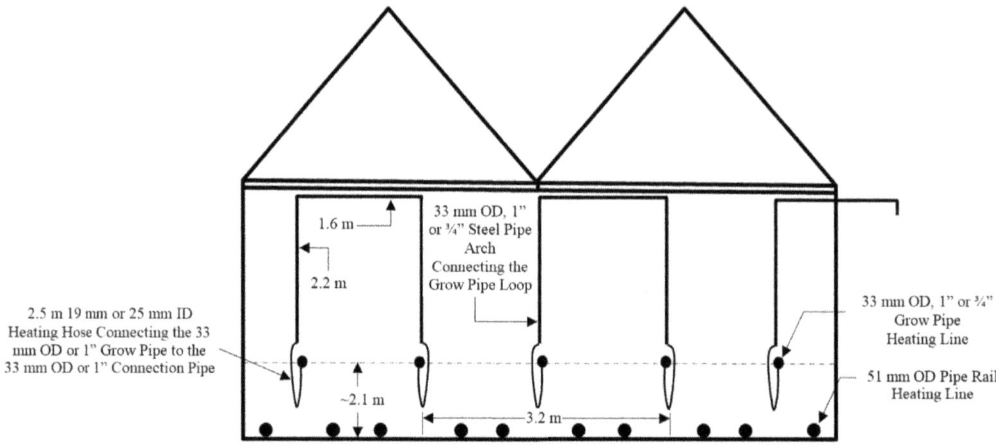


**FIGURE 5.23**
2 ½ loops (5 pipes) per bay grow pipe heating connections along with 5 × 51 mm OD pipe rail loops per bay (view from the walkway).

header is supplied with a regular air bleeder at the highest points. The distribution headers are designed according to the reverse return principle with a maximum flow resistance limit of 10 mm $H_2O$ (0.033 ft $H_2O$) and connected to a grow pipe heating zone mixing group (see Figure 5.20). The average pipe size of the distribution header, whether it is supply or return, is 0.8 times the maximum pipe size in the distribution line. Maximum grow pipe heating system temperature used is typically 70°C (158°F) whereas the return temperature is 55°C (131°F). In order to get these system temperatures, the flow temperatures to the grow pipe heating zone mixing group are 71°C (~160°F) and 54°C (~129°F) for supply and return, respectively. See Example 5.7 for a grow pipe heating system design example.

Since the horizontal pipe has 100% of the horizontal pipe heat emission, this rate should be used for the 1.6 m horizontal section; and the vertical pipe has the 50% of the horizontal pipe emission, 50% of the horizontal pipe emission should be used for the 2 (two) 2.2 m vertical pipes. Since the total pipe length over the arch is 6 m, the heat emission factor for the arch is calculated as follows:

$$\text{Factor for the Arch} = \left( \frac{1.6\,\text{m} \times 100/100 + 2 \times 2.2\,\text{m} \times 50/100}{6\,\text{m}} \right) = 0.63$$

*5.4.3.2.2 Underbench Heating System Design*

As mentioned earlier in this section, thin-walled 51 mm OD pipe underbench heating systems are used for the seed germination, propagation, or general growth of crops grown on the bench tops (see Figure 5.24). Four or five loops of 51 mm OD pipes per bay are installed under benches depending on the bay width and the need. Also, either 2 or 3 lengths of 51 mm OD pipes are installed under benches. As in pipe rail heating, the underbench loops start 0.7 m (2.3 ft) from the gable ends and run to the walkway, and are connected to the distribution headers along the gable end via 1.5–2 m (5–6 ½ ft) 19 mm or 25 mm ID black hot water heating hoses (see Figures 5.25 and 5.26). If the water flow in the loop is less than

**FIGURE 5.24**
Three-tube underbench heating system.

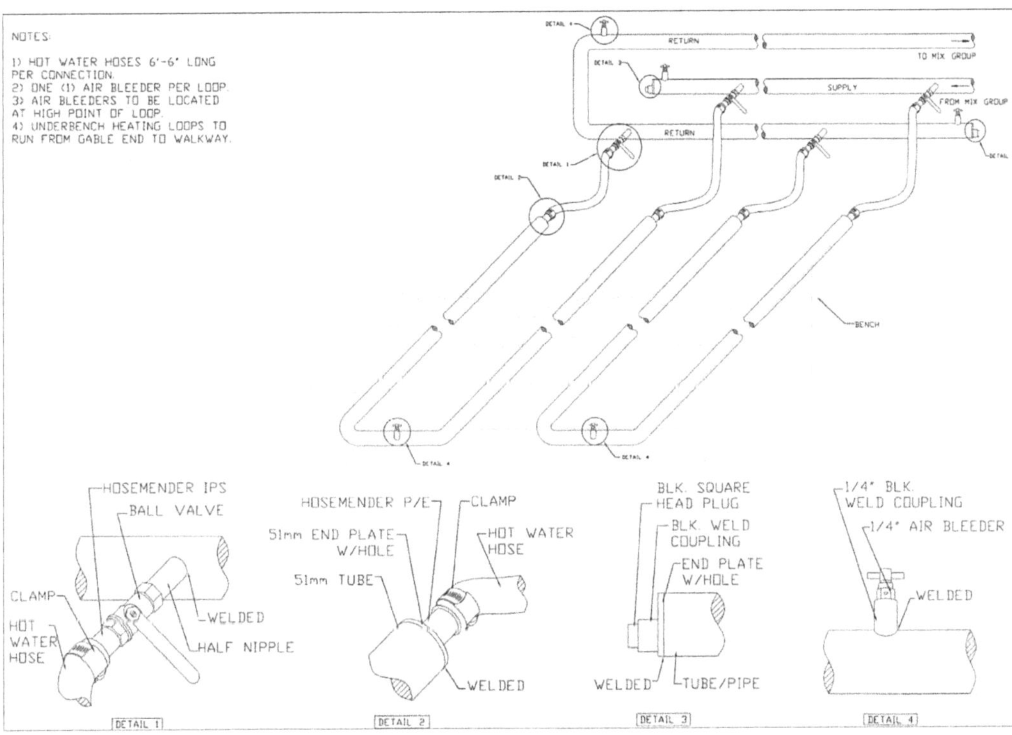


**FIGURE 5.25**
Connection details of two-tube underbench heating system.

800 l/h (3.5 gpm), a 19 mm ID plastic hose is used, whereas a 25 mm ID plastic hose is used if the flow is greater than 800 l/h (3.5 gpm). Each underbench loop has a return bend, ball vales for supply and return lines; and each loop and distribution header is supplied with a regular air bleeder at the highest point (see Figures 5.17 and 5.19). Underbench heating pipes (51 mm OD) emit 20% less heat compared to the same size uncovered horizontal pipes; therefore, just like pipe rail and grow pipe heating system, it is vital for the underbench heating systems to have an engineered design for adequate performance. The distribution headers are designed according to the reverse return principle with a maximum flow resistance limit of 10 mm $H_2O$ (0.033 ft $H_2O$) throughout and connected to a dedicated underbench heating zone mixing group (see Figure 5.20). The average pipe size of the distribution header, whether it is supply or return, is again 0.8 times the maximum pipe size in the distribution line.

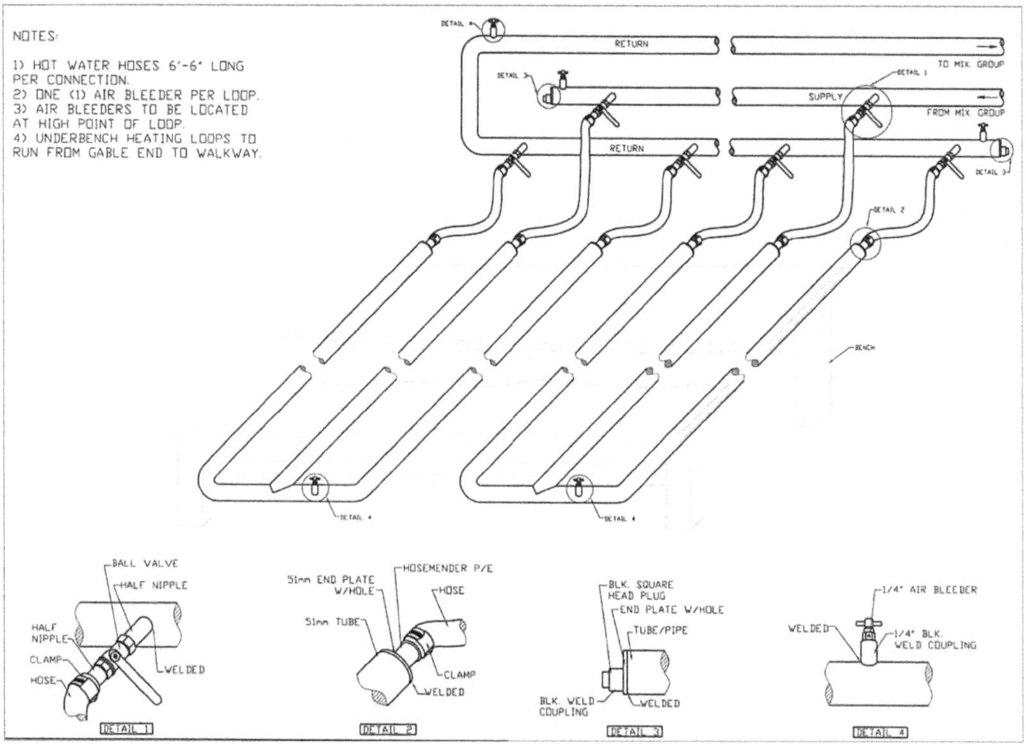


**FIGURE 5.26**
Connection details of three-tube underbench heating system.

#### *5.4.3.2.3 Floor and Overhead Heating System Designs*

##### *Floor (Low Grade) Heating Systems*

In-floor or on-floor heating systems are used very effectively for plants in bags and containers grown on the floor or even for vegetable crops grown directly in the soil (see Figure 5.27). In order to distribute heat in such systems, 25 mm OD (1″) thermoplastic tubes are used (see Figure 5.28). For on-ground heating, 25 mm OD smooth polyethylene tubes with a wall thickness of 1.9 mm is used while 25 mm OD smooth polyethylene tubes with a wall thickness of 2.7 mm is used for in-floor heating whether it is in sand, soil or concrete. Depending on the purpose, either 20 cm c.t.c. (8″ c.t.c.) (for propagation and growth) or 30 cm c.t.c. (12″ c.t.c.) (for packing sheds, pathways, etc.) tube spacings are used. Maximum loop lengths are maintained somewhere in between 125 m (415 m) and 150 m (500 ft). The floor heating loops start almost immediately, 0.2 m (~0.65 ft) from the gable ends and run to the walkway, and both supply and return lines are directly connected via ball valves to the distribution headers along the gable end (see Figures 5.27 and 5.28). Each polyethylene tube is fastened and secured to a steel mesh with tie wraps every 30 cm (12″) for integrity and longevity (see Figure 5.28). When concrete floors are used, in order to get uniform floor temperatures, a minimum concrete slab thickness of 10 cm (4″) is used and the floor should be insulated around the perimeter and as deep as the front line with typically 5 cm (2″) extruded styrene board insulation. A zone supply temperature of 35°C (95°F) and a return temperature of 30°C (86°C) are used providing a $\Delta T$ of 5°C (9°F). The distribution headers

**FIGURE 5.27**
In-floor heating system.

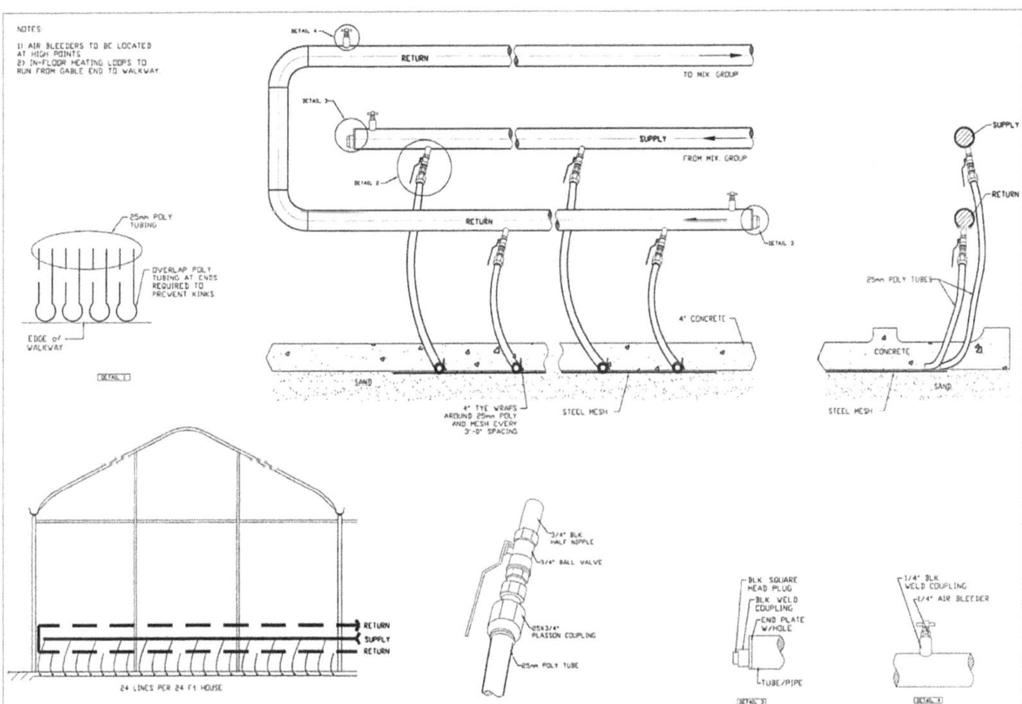


**FIGURE 5.28**
Connection details of in-floor heating system.

are designed according to the reverse return principle with a maximum flow resistance limit of 10 mm $H_2O$ (0.033 ft $H_2O$) and connected to a dedicated floor heating zone mixing group with a by-pass (see Figure 5.29). The average pipe size of the distribution header, whether it is supply or return, is 0.8 times the maximum pipe size in the distribution line. The by-pass in the mixing group is essential, because in the absence of the by-pass if the

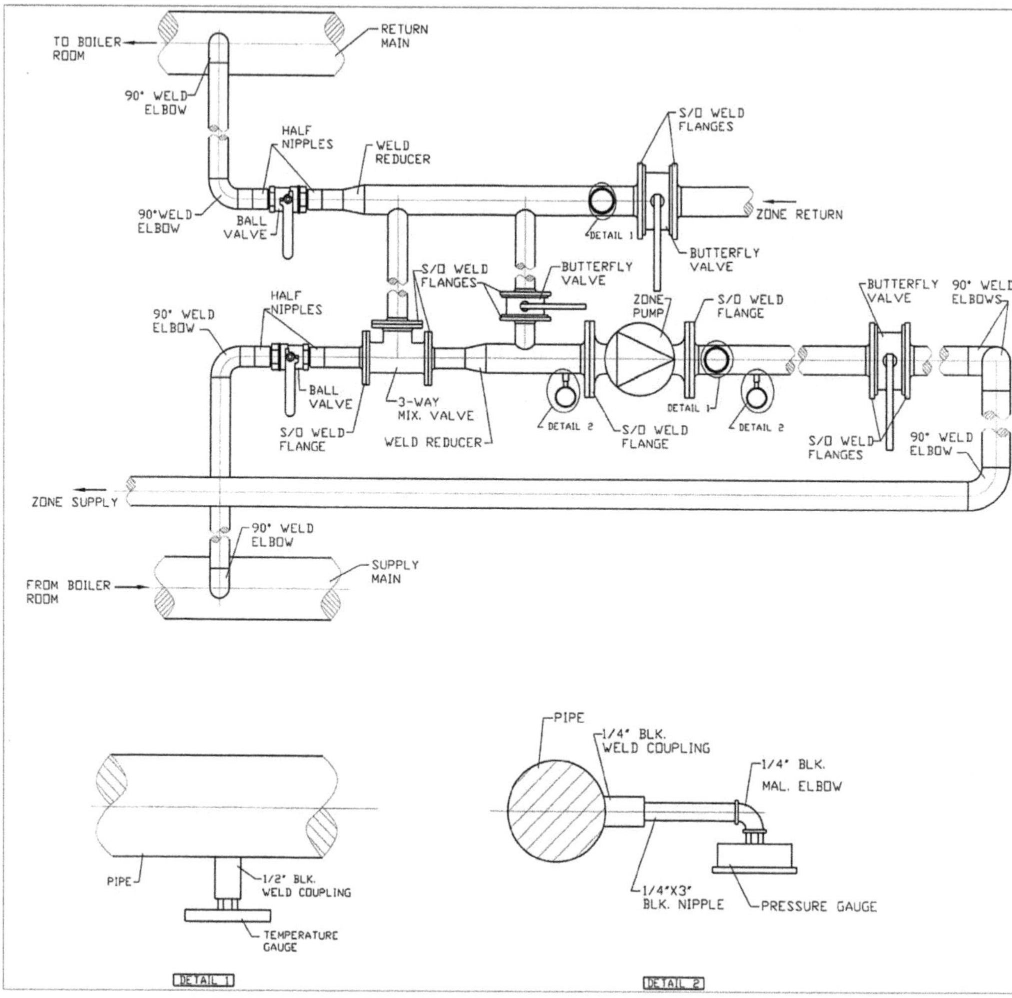


**FIGURE 5.29**
Connection details of floor heating zone mixing group.

mixing valve is opened by hand or if the mixing valve actuator fails, then the water temperature in the polyethylene tubes may go beyond the critical temperature of 60°C (140°F) and melt the tubes (see Example 5.5). And as an industry standard, typically 25 kcal/h.m$^2$ = 29 W/m$^2$ (100 Btu/h.m$^2$) floor and 50 kcal/h.m$^2$ = 58 W/m$^2$ (200 Btu/h.m$^2$) floor heating are provided for plant covered floors and uncovered floors (sheds, walkways, etc.), respectively. During the polyethylene tube installation, before they are covered with concrete, the tubing should be filled with water to check for and fix leakages, to make it more impact resistant, and to prevent them from floating. When designed properly, hot-water floor heating systems provide the most uniform heat distribution horizontally and within the crop canopy as well. Except for the cool-season crops such as lettuce and spinach, in addition to the floor heating, some type of supplemental heating will be required and it is typically provided by overhead, gable, end wall and sidewall heating systems. See Example 5.8 for an in-floor heating system design example.

**Example 5.5: Determining the Size of the By-pass Line in a Floor (Low Grade) Heating Mixing Group**

Determine the maximum zone supply temperature and the size of the by-pass line if the floor heating mixing group operates directly on the boiler water temperature of 95°C, and if the floor supply and return temperatures are 35°C and 30°C, respectively.

**Solution**

In this given case, the floor zone supply temperature of 35°C has to be made by mixing the floor zone return water at 30°C with the water coming from the boiler at 95°C. Then, the mixing valve without the by-pass works only on:

$$\frac{\Delta T_{\text{Floor Zone}}}{\Delta T_{\text{Mixing Group}}} = \left(\frac{35°\text{C} - 30°\text{C}}{95°\text{C} - 30°\text{C}}\right) 100\% = 7.7\% \text{ of } 95°\text{C water coming from the boiler room.}$$

This is fine; however, if the mixing valve is opened by hand or if the mixing valve actuator fails, then the water temperature in the polyethylene tubes may go beyond the critical temperature of 60°C (140°F) and melt the tubes. Therefore, a by-pass line in the mixing group as shown below is essential (see the figure below).

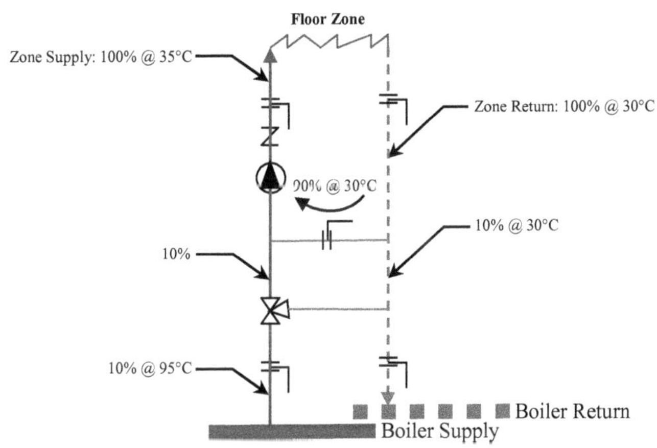


Then, the maximum water temperature in the case (with a by-pass line) would be:

$$100X = 10(95°\text{C}) + 90(30°\text{C})$$

$$X = \frac{10(95°\text{C}) + 90(30°\text{C})}{100} = 36.5°\text{C}$$

Now, the tubes are safe and protected. However, we still need to determine the flow and the velocity through the by-pass line; and after that, using this information and pipe flow and velocity tables, we can pick a proper pipe size for the by-pass flow so that we can ensure that 90% the flow comes from the zone return at 30°C. Alternatively, one can use the following equation rather than the pipe tables for determining the pipe size:

$$d = \sqrt{\left(\frac{5.0(Q)}{\sqrt{\Delta P}}\right)}$$

where $d$ is the pipe size (mm), $Q$ is the flow rate (l/h), and $\Delta P$ is the pressure drop (mm $H_2O$).

*Overhead and Snowmelt Heating Systems*

In overhead heating systems, just like in the pipe rail and underbench heating systems, thin-walled 51 mm OD pipes are used in order to meet the deficiency of pipe rail, underbench, or floor heating systems (see Figure 5.30). The number of 51 mm OD pipe lengths per bay is a function of the climate and the type of the greenhouse installed. The 51 mm OD pipes are installed in trusses and run from gable end to gable end, from supply into return from alternating sides (see Figures 5.31 and 5.32) or sometimes they are installed in loops and run from gable end to gable end, from the supply header on gable end to the return header on the same gable end using a return bend. Overhead heating lengths start 0.15 m (6″) from the gable ends and run to the to the other gable end, and are connected to the distribution headers along the gable end via an average of 3.5 m (~12 ft) 19 mm or 25 mm ID clamped black hot water heating hoses (see Figure 5.31) or 27 mm (¾″) welded steel pipes (see Figure 5.32). If the water flow in the loop is less than 800 l/h (3.5 gpm), a 19 mm ID black hot water heating hose is used, whereas a 25 mm ID black hot water heating hose is used if the flow is greater than 800 l/h (3.5 gpm). Each 51 mm OD pipe length is supplied with a regular air bleeder at the highest point (see Figures 5.31 and 5.32). The distribution headers are designed according to the reverse return principle with a maximum flow resistance limit of 10 mm $H_2O$ (0.033 ft $H_2O$) throughout and connected to a dedicated overhead heating zone mixing group (see Figure 5.20). The average pipe size of the distribution header, whether it is supply or return, is again 0.8 times the maximum pipe size in the distribution line. See Example 5.8 for an overhead heating system design example.

In addition to the overhead heating, snowmelt systems are installed as well at locations where considerable snow loads are expected. Typically, two lengths of thin-walled 51 mm OD pipes per bay are used for snow melting, and they are installed immediately under gutters and rest on trusses and run from gable end to gable end, from supply into return or sometimes they are installed in loops and run from gable end to gable end, from the supply header on gable end to the return header on the same gable end using a return bend. The line connections to the headers are done via plastic hoses or 27 mm (¾″) welded steel

**FIGURE 5.30**
Overhead heating system.

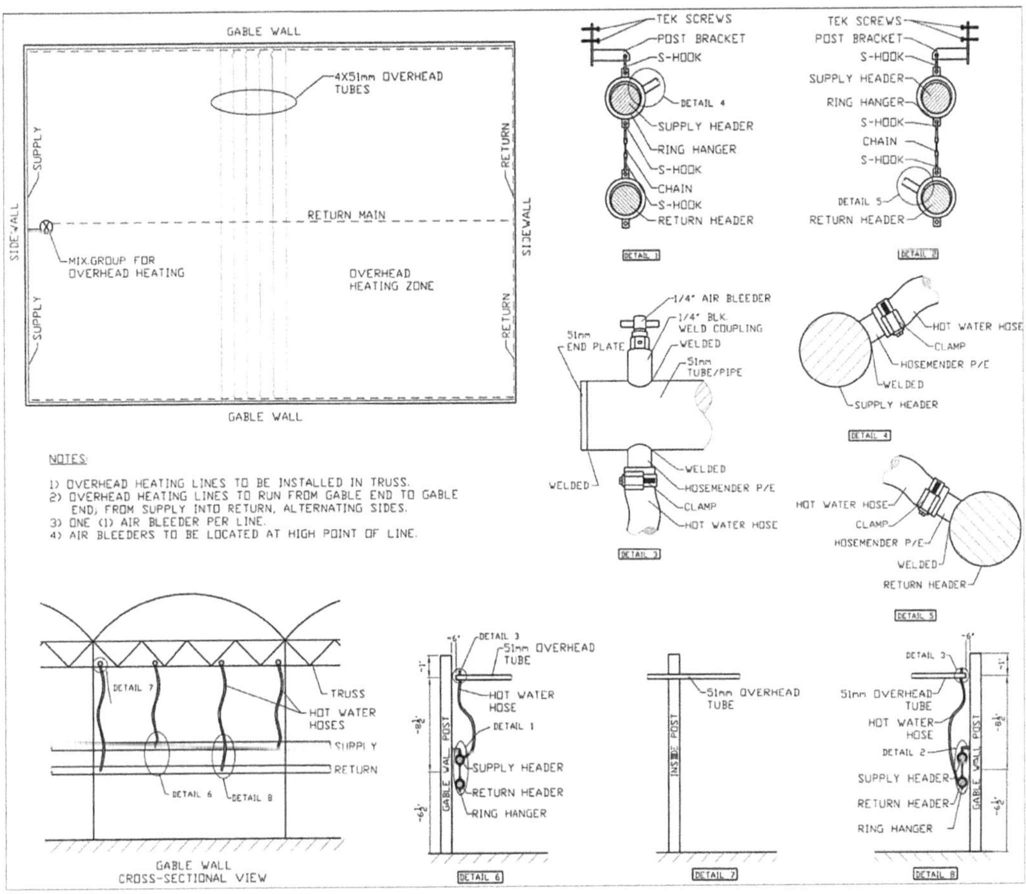


**FIGURE 5.31**
Connection details of plastic hose connected overhead heating system.

pipes. Each 51 mm OD pipe length is supplied with a regular air bleeder at the highest point. The distribution headers are designed according to the reverse return principle with a maximum flow resistance limit of 10 mm $H_2O$ (0.033 ft $H_2O$) throughout and connected to a dedicated snowmelt zone mixing group, which is a separate mixing group from the overhead heating, and provides further flexibility. The average pipe size of the distribution header, whether it is supply or return, is again 0.8 times the maximum pipe size in the distribution line.

#### *5.4.3.2.4 Sidewall and Gable End Heating System Designs*

In sidewall and gable end heating system, just like in the pipe rail and underbench heating systems, thin-walled 51 mm OD pipes are used in the case there is still a deficiency after heating provided by the distribution headers (see Figure 5.33).

The number of 51 mm OD pipe lengths per sidewall or gable end is a function of the climate and the type of the greenhouse installed (see Figures 5.21, 5.22, and 5.27). The 51 mm OD pipes run from the supply header on gable end to the return header on the same gable end. Sidewall or gable end heating lengths start almost immediately 0.15 m (6″) from the gable end or the sidewall, respectively and run along the gable end or sidewall, and are

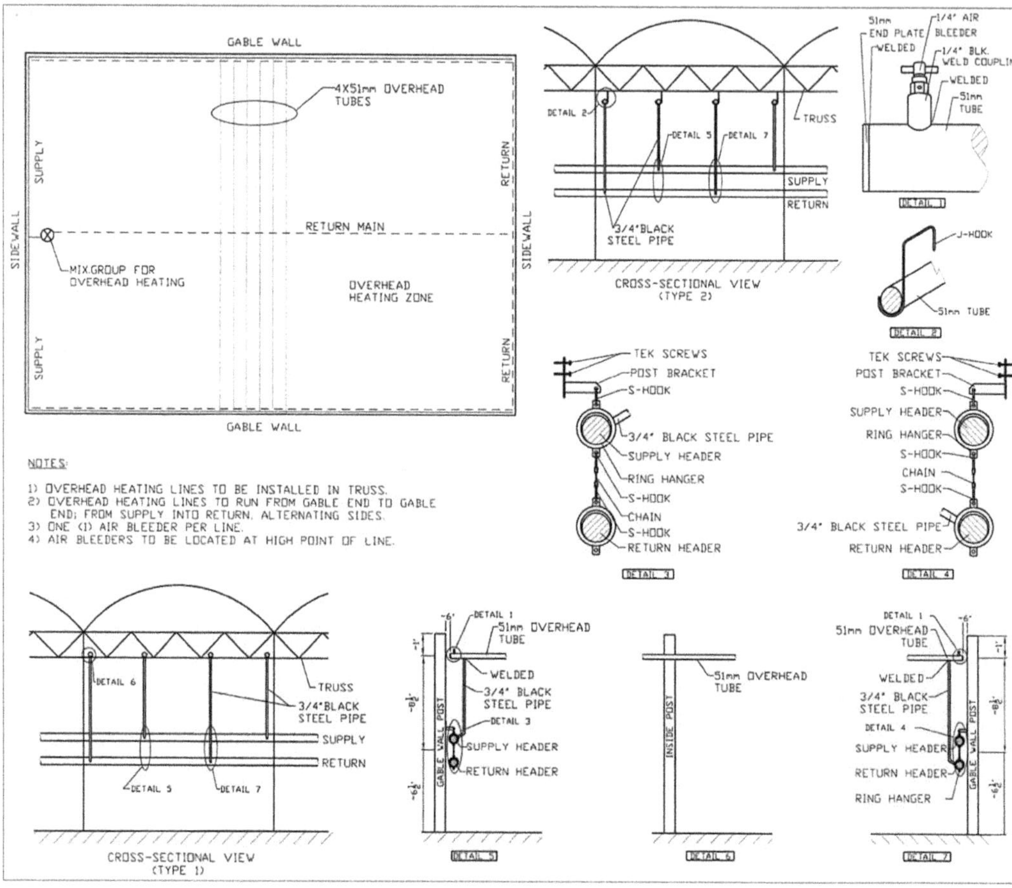


**FIGURE 5.32**
Connection details of steel pipe connected overhead heating system.

**FIGURE 5.33**
Steel pipe connected gable end wall and sidewall heating system.

connected to the distribution headers typically via 19 mm or 25 mm ID clamped black hot water heating hoses or welded steel pipes. They are typically connected to the main hot water heating zone mixing group, that is, the pipe rail heating, the underbench, or the overhead heating.

**Example 5.6: Designing a Central Hot Water System for a Gutter Connected Double Poly Greenhouse Complex**

Let's now consider a 10-bay gutter connected, separated double poly and curved roof greenhouse complex to be built in Columbus, Ohio, USA, with an insulated floor, 21 ft (6.4 m) span, 240 ft (72 m) long, 16 ft (4.8 m) wall height, and 5.3 ft (1.6 m) gable height. Minimum night-time temperature requirement of the plant is given as 20°C (68°F). Assuming that the headhouse is pre-heated, determine the heating load and design a hot water pipe rail heating system for the greenhouse using natural gas-fired hot water boilers. See the layout of the greenhouse complex below.

| GREENHOUSE DIMENSIONS | | |
|---|---|---|
| **COMPONENT** | **(ft)** | **(m)** |
| Wall Height | 16.0 | 4.8 |
| House Width | 21 | 6.4 |
| House Length | 240 | 72 |
| Width of Center Walkway | 12.0 | 3.6 |
| Average Roof Height | 19.3 | 5.8 |
| Length of Arch | 24.4 | 7.4 |
| Gable Height ($ft^2$ or $m^2$) | 5.3 | 1.6 |
| Roof Area per Unit Bay Length ($ft^2/ft$ bay or $m^2/m$ bay) | 82.2 | 7.4 |
| Number of Bays | 10 | 10 |

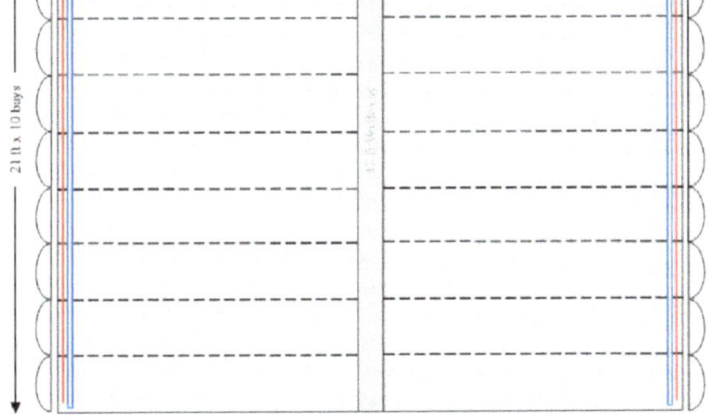

Gutter connected double poly greenhouse layout for Example 5.6.

**Solution**

Remember that the maximum pipe rail heating system temperature is typically 90°C (194°F) whereas the return temperature is 75°C (167°F) with an average water temperature of 82.5°C (180.5°F). We will design a single zone of pipe rail heating system here.

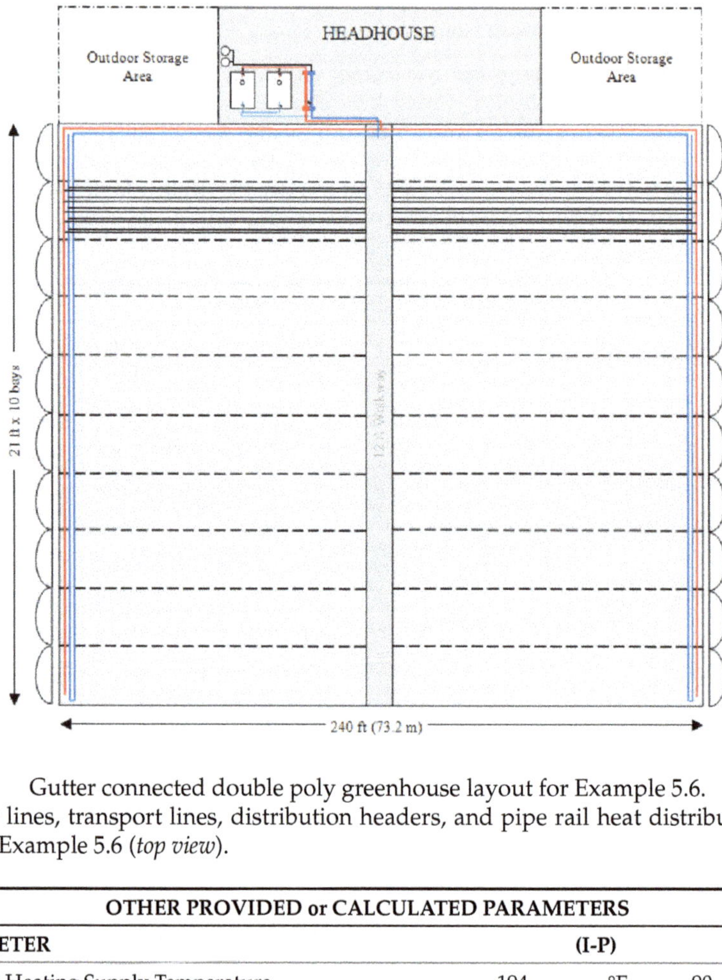


Gutter connected double poly greenhouse layout for Example 5.6.
Boiler lines, transport lines, distribution headers, and pipe rail heat distribution system for Example 5.6 (*top view*).

| OTHER PROVIDED or CALCULATED PARAMETERS | | | | |
|---|---|---|---|---|
| **PARAMETER** | **(I-P)** | | **(SI)** | |
| Pipe Rail Heating Supply Temperature | 194 | °F | 90 | °C |
| Pipe Rail Heating Return Temperature | 167 | °F | 75 | °C |
| $\Delta T_{water}$ | 27 | °F | 15 | °C |
| Average Water Temperature, $T_{avg}$ | 180.5 | °F | 82.5 | °C |
| Greenhouse Inside Design Temperature, $T_i$ | 68 | °F | 20 | °C |
| Outside Design Temperature, $T_o$ (Table A.13 or A.14); (99% Level) | 6 | °F | −14 | °C |
| $\Delta T = (T_i - T_o)$ | 62 | °F | 34 | °C |

*(Continued)*

| OTHER PROVIDED or CALCULATED PARAMETERS | | | | |
|---|---|---|---|---|
| **PARAMETER** | | **(I-P)** | | **(SI)** |
| Over Temperature ($T_{avg} - T_i$) | | 112.5 °F | | 62.5 °C |
| Greenhouse Roof $U$-value (Provided or Table 3.3) | 0.7 Btu/h.ft².°F | 3.5 kcal/h.m².°C | | 4.0 W/m².°C |
| Greenhouse Wall $U$-value (Provided or Table 3.3) | 0.7 Btu/h.ft².°F | 3.5 kcal/h.m².°C | | 4.0 W/m².°C |
| Bay Width | | 6.4 m | | |
| Roof Area per Unit Bay Length (m²/m bay) | | 7.4 m²/m bay | | |
| Wall Height | | 4.8 m | | |
| Average Roof Height | | 4.8 m + 1 m = 5.8 m | | |
| Air Change per Hour | | 0.75 (Table 3.4) | | |
| Wind Factor | | 1.00 (Table 5.1) | | |
| Location | | Columbus, Ohio, USA | | |
| **Heat Loss Through the Roof** | | | | |
| Transmission Heat Losses per ° Temp. Diff. per Bay Length, $Q_c$ | | (3.5 kcal/h.m².°C) × (7.4 m²/m bay) × (1°C) = 25.9 kcal/h.m$_{roof.}$°C | | |
| Infiltration Heat Losses per ° Temp. Diff. per Bay Length, $Q_{inf}$ | | 6.4 m × 5.8 m × 1 m × 1.293 × 0.24 × 0.75 ach × 1°C = 8.64 kcal/h.m$_{roof}$.°C | | |
| Total Heat Loss per ° Temp. Diff. per Bay, $Q_t = Q_c + Q_{inf}$ | | 34.5 kcal/h.m.roof.°C | | |
| **Heat Supply of the Pipes** | | | | |
| Heat Supply by the 51 mm Pipes @ 62.5°C Over Temp. (Table A.18) | | 122 kcal/h.m | | |
| **Calculation of the Heat Supply by Pipe Rail Heating** | | | | |
| 10 lengths × 51 mm Pipes per Bay: 10 × 122 kcal/h.m | | 1,220 kcal/h.m | | |
| Pipe Rail Heating Temperature Lift = Heat Supply by 51 mm OD/($Q_t/\Delta T$) | | (1,220 kcal/h.m$_{roof}$)/(34.5 kcal/h.m$_{roof}$.°C) = 35.4°C | | |
| **Water Flow Rate per 51 mm Pipe Rai Heating Pipe** (one side only) = Number of Pipe Runs × Length × Pipe Heat Supply/$\Delta T_{water}$ | | (2 length × 34 m × 122 kcal/h.m)/15°C = 553 l/h | | |
| Water Flow Rate for Pipe Rail Heating (one side only) = 50 loops × 553 l/h = 27,650 l/h | | 27,650 l/h (122 gpm) | | |
| Maximum Header Size (Table A.21 for Flow Rate per Section & max resist. 10 mm $H_2O$) | | 108 mm | | |
| Avg. Distribution Header Pipe Size = 0.8 × Max Size = 0.8 × 108 mm | | 89 mm | | |

Now using the total water flow figures, the maximum limit of 10 mm $H_2O$ resistance, and Table A.21, we can design the pipe rail heating distribution headers as follows:

| GABLE WALL SECTION (m) | | DISTRIBUTION HEADER SIZE | | | |
|---|---|---|---|---|---|
| | | **SUPPLY Flow (l/h)** | **SUPPLY OD (mm)** | **RETURN Flow (l/h)** | **RETURN OD (mm)** |
| 0.00 | 3.20 | 27,650 | **108** | 383 | **27** |
| 3.20 | 6.40 | 25,268 | **102** | 1,766 | **42** |

*(Continued)*

| | | | | | |
|---|---|---|---|---|---|
| 6.40 | 9.60 | 23,885 | **102** | 3,148 | **51** |
| 9.60 | 12.80 | 22,503 | **102** | 4,531 | **57** |
| 12.80 | 16.00 | 21,120 | **95** | 5,913 | **60** |
| 16.00 | 19.20 | 19,738 | **95** | 7,296 | **70** |
| 19.20 | 22.40 | 18,355 | **89** | 8,678 | **70** |
| 22.40 | 25.60 | 16,973 | **89** | 10,061 | **76** |
| 25.60 | 28.80 | 15,590 | **89** | 11,443 | **76** |
| 28.80 | 32.00 | 14,208 | **83** | 12,826 | **83** |
| 32.00 | 35.20 | 12,825 | **83** | 14,208 | **83** |
| 35.20 | 38.40 | 11,443 | **76** | 15,591 | **89** |
| 38.40 | 41.60 | 10,060 | **76** | 16,973 | **89** |
| 41.60 | 44.80 | 8,678 | **70** | 18,356 | **89** |
| 44.80 | 48.00 | 7,295 | **70** | 19,738 | **95** |
| 48.00 | 51.20 | 5,913 | **60** | 21,121 | **95** |
| 51.20 | 54.40 | 4,530 | **57** | 22,503 | **102** |
| 54.40 | 57.60 | 3,148 | **51** | 23,886 | **102** |
| 57.60 | 60.80 | 1,765 | **42** | 25,268 | **102** |
| 60.80 | 64.00 | 383 | **27** | 27,650 | **108** |

And remember that the average size of the distribution header piping is 0.8 times the maximum pipe size. And the pipe rail heating piping starts 0.5 m from one gable end and runs to the center walkway (see Figure 5.21). Keeping this in mind, we can now determine the heat loss and the supply along the gable end, sidewall, and center walkway as follows:

| GABLE WALL, SIDEWALL, and CENTER WALKWAY HEAT DEMAND and SUPPLY | | |
|---|---|---|
| **Heat Loss Through the Gable End Wall** | | |
| Transmission Heat Loss for Avg. Gable Wall Height | $Q_c$ = (3.5 kcal/h.m².°C) × 5.8 m × 34°C | 690 kcal/h.m |
| Transmission Heat Loss from 0.7 m Bay Length | = 0.7 m × 1,220 kcal/h.$m_{roof}$/6.4 m | 133 kcal/h.m |
| Total Heat Loss from the Gable End Wall | | **823 kcal/h.m** |
| **Heat Supply of the Distribution Headers along the Gable End Wall** | | |
| Average Distribution Header Pipe Size: 0.8 × The Max Pipe Size | 0.8 × 108 mm ≈ 89 mm | |
| Heat Supply by the 89 mm Pipes @ 62.5°C Over Temp. (Table A.18) | 196 kcal/h.m | |
| Heat Supply by 2 Distribution Headers = 2 × 196 kcal/h.m | 392 kcal/h.m | |
| Heat Supply by 1.5 m 19 mm ID Heating Hose = (3/4 × 46.5 kcal/h.m × 10 hoses × 1.5 m)/6.4 m | 65 kcal/h.m | |
| Heat Supply by 108 mm Reverse Return Piping = 1 pipe × 198 kcal/h.m × (75.20)°C | 198 kcal/h.m | |

*(Continued)*

| | | |
|---|---|---|
| Total Heat Supply Along the Gable End Wall | | **655 kcal/h.m** |
| Heat Balance Along the Gable End Wall = Supply − Demand = (655 − 823) kcal/h.m | | **−168 kcal/h.m** |
| **NOTE:** The heat deficiency of -168 kcal/h.m needs to be removed by introducing additional 51 mm OD tubing along the gable end wall; in this case, we can introduce valve connected 2 × 51 mm OD tubing. See below: | | |
| Heat Supply by 2 × 51 mm Pipes @ 62.5°C Over Temp. (Table A.18) = 2 × 122 kcal/h.m | | 244 kcal/h.m |
| New Heat Balance Along the Gable End Wall = (392 + 65 + 198 + 244) kcal/h.m | | **899 kcal/h.m** (perfectly fine) |
| **NOTE:** If the 108 mm reverse return pipe is painted (aluminized), then 0.7 × 198 kcal/h.m = 138.6 kcal/h.m. Total heat supply will then be 839 kcal/h.m. The new heat balance would be 16 kcal/h.m, which is perfectly fine. | | |
| **Heat Loss Through the Sidewalls** | | |
| Transmission Heat Loss for the Sidewall (each) | $Q_c$ = (3.5 kcal/h.m$^2$.°C) × 4.8 m × 34°C | **571 kcal/h.m** |
| **Heat Supply of the Pipe Rail Heating along the Sidewall** | | |
| Maximum Distribution Header Pipe Size | | 108 mm |
| Heat Supply of the 108 mm Pipe @ 62.5°C Over Temp., ($T_{Supply} - T_i$) (Table A.18) | | 232.5 kcal/h.m |
| **SIDEWALL-1:** | | |
| Heat Supply of 2 × 108 mm Pipes (Supply & Return) = (2 × 232.5 kcal/h.m) × 0.70 (painted) | | 326 kcal/h.m |
| Heat Supply by 2 × 51 mm return pipe @ 62.5°C Over Temp. (Table A.18) = 2 × 122 kcal/h.m | | 244 kcal/h.m |
| Total Heat Supply Along Sidewall-1 | | **570 kcal/h.m** |
| Heat Balance Along Sidewall-1 = Supply − Demand = (570 − 571) kcal/h.m | | **1 kcal/h.m** |
| **NOTE:** The heat balance along Sidewall-1 is fine. | | |
| **SIDEWALL-2:** | | |
| Heat Supply by 5 × 51 mm return pipe @ 62.5°C Over Temp. (Table A.18) = 5 × 122 kcal/h.m | | 610 kcal/h.m |
| Heat Balance Along Sidewall-2 = Supply − Demand = (610 − 571) kcal/h.m | | **39 kcal/h.m** |
| **NOTE:** The heat balance along the Sidewall-2 is fine as well. | | |
| **Heat Loss Along the Center Walkway** | | |
| Width of the Center Walkway = 3.6 m | | |
| Transmission Heat Losses | 3.6 m × (1,220 kcal/h.m)/6.4 m bay width | **686 kcal/h.m$_{path}$** |

*(Continued)*

**Heat Supply Above the Center Walkway**

| | |
|---|---|
| Heat Supply by the 51 mm Return Bends @ 62.5°C Over Temp. (Table A.18); 5 lengths of 51 mm; 5 × 0.5 m × 122 kcal/h.m path/6.4 m = 48 kcal/h.m path | 48 kcal/h.m |

*Note:* Remember that for center walkway heating, we can use side path heating by introducing 2 lengths of 51 mm pipe along both sides of the walkway. Then, 4 × 122 kcal/h.m = 488 kcal/h.m path heat supply would be provided. Plus, in-floor heating can be provided: 3.6 $m^2$ × 50 kcal/h.$m^2$.m = 180 kcal/h.m. So, these two heat supplies would provide total heat supply of 668 kcal/h.m path. Then 668 kcal/h.m + 48 kcal/h.m = 716 kcal.h.m. The heat balance would then be perfectly fine. Also remember that the grow pipe heating is typically installed with the pipe rail heating, and it contributes the heat supply along the center walkway, the gable end walls and the sidewalls; therefore, no additional 51 mm loop heating might be required. Hence, for this project, we will consider it as covered either introducing several lengths of 51 mm pipe along both sides of the walkway or in combination of grow pipe heating contribution or in-floor heating.

Now that the heating (energy balance) side is covered, the mass balance side, that is, the flow requirement and the resistances for the pipe rail heating can be worked on as follows:

**PIPE RAIL HEATING FLOW REQUIREMENT and RESISTANCES**

**NOTE:** Remember that in order to meet the heat demand, 2 lengths of 51 mm pipe along the gable end wall, 2 lengths of 51 mm pipe along Sidewall-1, and 5 lengths of 51 mm pipe along Sidewall-2 have been introduced; hence, in order to have a mass balance, the flow requirement needs to be adjusted accordingly; and it is done as follows:

| | | | |
|---|---|---|---|
| Pipe Rail Loop Flow Total | | | 27,650 l/h |
| 51 mm loops along the end gable wall: 2 pipes × 64 m × 122 kcal/h.m/15°C | | | 1,041 l/h |
| 51 mm loops along Sidewall-1: 2 pipes × 34 m × 122 kcal/h.m/15°C | | | 553 l/h |
| 51 mm loops along Sidewall-2: 5 pipes × 34 m × 122 kcal/h.m/15°C | | | 1,383 l/h |
| **Adjusted Total Flow Rate (one side only):** | | | **30,627 l/h** |
| **Total Flow Rate (both sides):** | | | **61,254 l/h** |
| Flow Rate: **61,254 l/h** | Main Pipe Size: **140 mm** (Table A.21) | Resistance, *R*: **9.5 mm $H_2O$/m** (Table A.21) | Velocity: **1.21 m/s** (Table A.21) |
| **Calculation of the Resistance in the Overhead Heating Circuit** | | | |
| Greenhouse Width, *W*: 64 m | Rail Pipe Heating Depth, *L*: 34 m | | |
| Resistance of the Loop = 0.26 mm $H_2O$/m (Table A.21 for 553 l/h × (*L* = 34) × 2 length = 0.26 mm $H_2O$/m × 34 m × 2 lengths | | | **18 mm $H_2O$** |
| Resistance over the 19 mm Hose Connection = (2 lengths × 1.5 m × 22 mm $H_2O$) + 144 mm $H_2O$/m)<br>Note that 22 mm $H_2O$/m is the resistance for 553 l/h in 19 mm hose; 144 mm $H_2O$ is the special resistance for that (Table A.25) | | | **210 mm $H_2O$** |
| Distribution Header Resistance = (2 × *W*) × *R* × 1.10 = (2 × 64 m) × 10 mm $H_2O$/m × **1.10** (10% extra) | | | **1,408 mm $H_2O$** |

*(Continued)*

| | | |
|---|---|---|
| Resistance over the Transport Mains outside the Greenhouse:<br>2 × 20 m (assumed) × 9.5 mm $H_2O$/m × **1.50** (50% extra for fittings) | | **570 mm $H_2O$** |
| $K_v$ is calculated using Equation (2.189) for $\Delta P$ =0.1 bar (1.45 psi) allowed;<br>and 125 mm 3-way Mixing Valve with $K_v$ = 250 is chosen.<br>Then, the Resistance over the 3-way Mixing Valve,<br>$\Delta P = [(61.2\ m^3/h)/250]^2 \times 10{,}000$ mm $H_2O$ = 599 mm $H_2O$ < 700 mm $H_2O$ | | **599 mm $H_2O$** |
| Other Resistances to be Included (boiler loop, etc., if there is any)? | | 0 |
| **Total Resistance** | | **2,805 mm $H_2O$** |
| **Pump Size** | **~270 gpm @ 9.3 ft Head** | **61,254 l/h @ 2.8 m Head** |

Using the adjusted total water flow rate and the gable end and sidewall requirements, we can now redesign the pipe rail heating distribution headers as follows:

| GABLE WALL SECTION (m) | | DISTRIBUTION HEADER SIZE | | | |
|---|---|---|---|---|---|
| | | SUPPLY Flow (l/h) | SUPPLY OD (mm) | RETURN Flow (l/h) | RETURN OD (mm) |
| 0.00 | 3.20 | 30,627 | **108** | 2,977 | **48** |
| 3.20 | 6.40 | 27,651 | **108** | 4,359 | **57** |
| 6.40 | 9.60 | 26,268 | **102** | 5,742 | **60** |
| 9.60 | 12.80 | 24,886 | **102** | 7,124 | **64** |
| 12.80 | 16.00 | 23,503 | **102** | 8,507 | **70** |
| 16.00 | 19.20 | 22,121 | **102** | 9,889 | **76** |
| 19.20 | 22.40 | 20,738 | **95** | 11,272 | **76** |
| 22.40 | 25.60 | 19,356 | **95** | 12,654 | **83** |
| 25.60 | 28.80 | 17,973 | **89** | 14,037 | **83** |
| 28.80 | 32.00 | 16,591 | **89** | 15,419 | **89** |
| 32.00 | 35.20 | 15,208 | **89** | 16,802 | **89** |
| 35.20 | 38.40 | 13,826 | **83** | 18,184 | **89** |
| 38.40 | 41.60 | 12,443 | **83** | 19,567 | **95** |
| 41.60 | 44.80 | 11,061 | **76** | 20,949 | **95** |
| 44.80 | 48.00 | 9,678 | **76** | 22,332 | **102** |
| 48.00 | 51.20 | 8,296 | **70** | 23,714 | **102** |
| 51.20 | 54.40 | 6,913 | **64** | 25,097 | **102** |
| 54.40 | 57.60 | 5,531 | **60** | 26,479 | **108** |
| 57.60 | 60.80 | 4,148 | **57** | 27,862 | **108** |
| 60.80 | 64.00 | 2,766 | **48** | 30,627 | **108** |

Note that the maximum and the average size of the distribution header are the same as before. Next, let's check whether the total flow rate provided actually meets the total greenhouse requirement or not.

| FLOW RATE CHECK | |
|---|---|
| **Total Greenhouse Heat Loss (Demand)** | |
| Heat Loss from the Roof = 10 bays × 72 m × 34.5 kcal/h.m.°C × [20 °C – (–14 °C)] | 844,560 kcal/h |
| Heat Loss from the Gable End Walls = 2 gable end walls × 64 m × 690 kcal/h.m (for the pure gable end wall portion only) | 88,320 kcal/h |
| Heat Loss from the Sidewalls = 2 sidewalls × 72 m × 571 kcal/h.m | 82,224 kcal/h |
| **Total Greenhouse Heat Loss** | **1,015,104 kcal/h** |
| **Total Check Flow Rate = Total Greenhouse Heat Loss/ ($\Delta T_{water}$ + 5°C) = (1,015,104 kcal/h)/(20°C)** | **50,755 l/h** |
| Since **50,755 l/h** (the minimum) < **55,300 l/h** (what is provided), we are perfectly fine. | |

*Notes:*

1. $\Delta T$ over the mixing group is limited to $\Delta T_{loop}$ + 5°C; therefore, 20°C was used for calculating the check flow rate.
2. Check flow rate is the minimum flow rate (pump capacity), which must be less than the total flow rate provided earlier. If not, using the check flow rate calculated (which becomes the new flow rate) as the new flow rate and we do the calculations again and continue doing it until it meets the requirement. Check flow rate must always be less than the flow rate provided (used).

The above calculations show that the provided total flow rate meets the total greenhouse flow requirement.

| OVERALL GREENHOUSE HEAT LOSS & THE BOILER SIZE | | | |
|---|---|---|---|
| | **Btu/h** | **kcal/h** | **W** |
| Total Greenhouse Heat Loss | 4,025,564 | 1,015,104 | 1,180,566 |
| Boiler Size | | **120 BHP** | |
| Required Boilers | | **2 × 60 BHP (Rated Output)** | |

The above table shows that total required boiler capacity is 120 bhp; however, in the case the boiler fails, approximately a 50% back-up capacity should be considered in greenhouse applications. Therefore, instead of using one 120 bhp boiler, two 60 bhp boilers should be used. This means, if one of the boilers fails, then the overall capacity will drop down to 50%; however, the remaining boiler will be able to maintain the greenhouse temperature above freezing until the failed boiler gets back to work. Normally, the greenhouse boiler room is designed considering the future expansion; however, for this design application, we will design the boiler room using two natural gas-fired GHE 60 bhp (rated output: sea level to 900 m (3,000 ft)) hot water boilers. Please note that while the manufacturer model is a generic model, the specifications of the boiler are real. The gas-fired GHE 60 bhp hot water boiler comes with a 10″ diameter vent stack, 4″ water outlet and return connection sizes, and has approx. 1,572 liter (420 US gal) flooded water capacity. We can now go ahead and design the boiler room as follows:

The boiler main lines are the pipe lines that connect the boilers to the manifold and are designed according to the reverse return principle discussed earlier in this chapter (see Figure 5.7). The following sketch shows the boiler loop designed accordingly based on the flow requirements of the boilers.

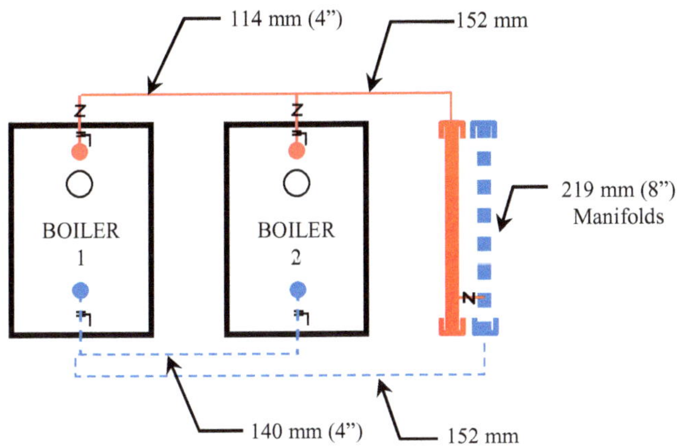


BOILER LOOP

As discussed earlier, most boilers require continuous flow over, and this is provided by the primary loop system. Circulation pumps are installed in the main return line and they keep circulating the water through the boilers and the supply and return manifolds, which are typically connected with a check valve to prevent backflow in the case of a pump failure. The loop is equipped with temperature and pressure gauges along with an auto air release system installed at the loop's highest point in order to release the air within the loop. If one of the circulation pumps fails, then the other pump will provide 50% water circulation so that the greenhouse can still be provided with some heat in order to prevent the crops from freezing. Figure 5.12 and the sketch below shows the connection details of a two-pump primary group.

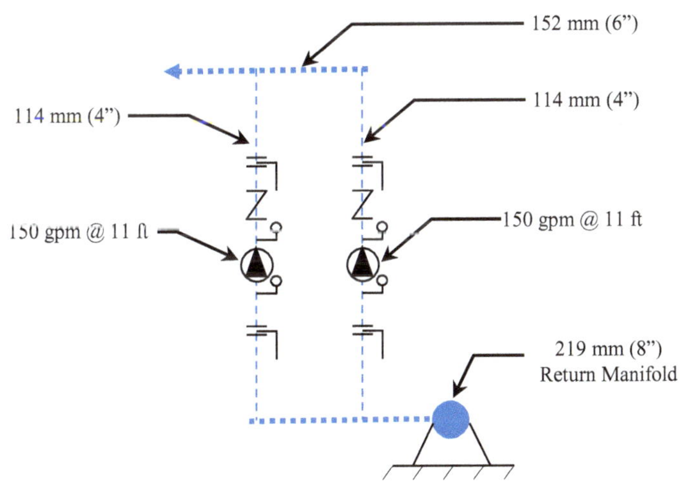


PRIMARY GROUP

The capacity of the pump is calculated using Equation (2.24) and the hot water boiler delivers 95°C (203°F) hot water with a return water temperature of 80°C (176°F) having therefore a $\Delta T$ of 15°C (27°F). Then, the capacity for each pump can be calculated as 34,069 l/h (~150 gpm). After determining the pump capacity, the boiler main line pipe sizes and fittings and therefore, as stated above, the resistances are determined and the main line is designed. It should be kept in mind that the main line resistance not only includes the pipe, valve and fitting resistances, but also the boiler's internal resistance. A boiler head loss of 750 mm $H_2O$ (2.5 ft $H_2O$) is provided by the boiler manufacturer and added to the line resistance here in the primary loop. Therefore, a head for each pump was determined as approx. 3.3 m $H_2O$ (11 ft $H_2O$). Then, the primary loop circulation pumps were determined to be two pumps with 34,069 l/h @ 3.3 m $H_2O$ head each (~150 gpm @ 11 ft $H_2O$).

The manifold size is determined by using Equation (5.7) and the mixing group sizes as follows:

$$D = \sqrt{2\left(M_1^2 + M_2^2 + M_3^2 + \cdots + M_n^2\right)} = \sqrt{2\left(100^2 + 100^2\right)} = 200 \text{ mm} \approx 8''$$

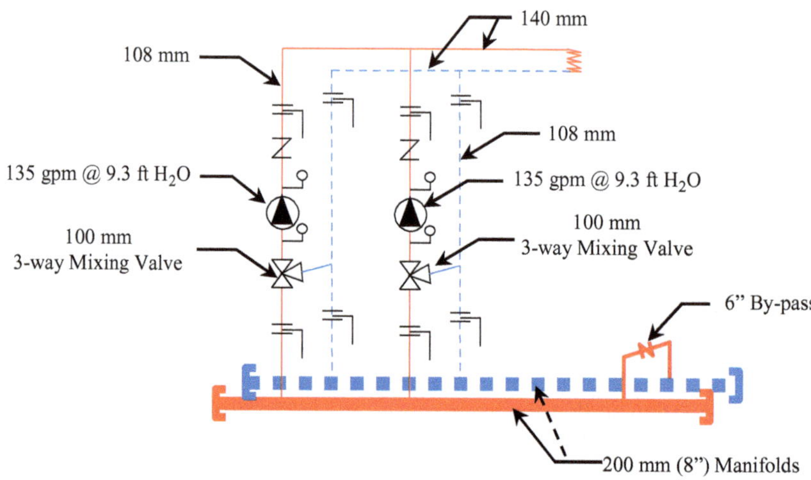


MIXING GROUP

Note that rather than using a single pump for the pipe rail heating zone here, two pumps (50% of the full capacity) were used because in the case one pump fails, then the other pump would still provide 50% water circulation so that the zone can still be provided with 50% heat in order to prevent the crops from freezing. New mixing valve sizes were calculated and the same head was kept for the half capacity pumps (see the above sketch for the mixing group). Please also note that, in North America, we can use readily available normal pipe sizes (i.e. Sch. 40) in boiler room piping design (boiler loop, primary group, and transport group) as well as the transport piping.

Now, let's determine the size of a membrane type expansion vessel for our system in hand. Using Equation (5.9), we can easily determine the tank size as follows:

Total water content of the system was calculated as 18,240 liter (4,754 gal), and 4.5% expansion volume was determined as 821 liter (214 gal) accordingly. Then,

$$I = \Delta V \text{x} \left( \frac{p_{\max}}{p_{\max} - p_{\min}} \right)$$

The initial and final pressures are 0.5 bar (1.5 bar absolute) and 3 bar (= 4 bar absolute), respectively. Then the content of the expansion tank becomes:

$$I = (821 \text{ litre}) \text{x} \left( \frac{4}{4 - 1.5} \right) = 1{,}314 \text{ litre}$$

For this volume, we can use 1 × GHEXT800 (800 liter) + 1 × GHEXT600 (600 liter) providing a combined total capacity of 1,400 liter. Please note that while the manufacturer model is a generic model, the specifications of the expansion tanks are real. See Figure 5.15 for two expansion tank system connection details.

As discussed earlier in Chapter 2, for complete combustion and safety, boiler rooms must be provided with proper openings so that enough fresh air is supplied. From experience, the ASME states that ***1 square inch free area per 2,000 Btu/h*** maximum fuel input of combined burners located in the boiler room works, and minimum 2 (*two*) *air supply openings must be provided and kept clear at all times.* It should also be kept in mind that inlet louvers typically have 60% free area; therefore, the actual (gross) louver area per unit heater will be larger.

Note that 2 natural gas-fired GHE 60 bhp (rated output: sea level to 900 m (3,000 ft)) hot water boilers were chosen for this particular project. These boilers have rated input capacities of 2,511,000 Btu/h; then, the total required free air inlet area can be calculated as follows:

$$(2\text{ Boilers}\times 2{,}511{,}000\text{ Btu/h})/(2{,}000\text{ Btu/h}) = 2{,}511\text{ in}^2$$

And, since a minimum of two louvers need to be used, the required free air inlet area per louver would be 2,511 in$^2$/2 = 1,255.5 in$^2$. Also, keep in mind that metal louvers have 60% free area. For instance, each GHE50 motorized shutter has opening dimensions of 50″ × 50″ and provides a free inlet area of 1,500 in$^2$; hence, using two GHE50 motorized inlet shutters will perform the required job perfectly fine providing a minimum of two air supply openings and a total free air inlet area of 3,000 in$^2$.

**Example 5.7: Designing a Hot Water Grow Pipe Heating System**

Now, let's design a hot water grow pipe heating system for the following Venlo greenhouse to be built in Vancouver area, British Columbia, Canada.

| GREENHOUSE DIMENSIONS | | |
|---|---|---|
| **COMPONENT** | **(ft)** | **(m)** |
| Wall Height | 16.4 | 5 |
| House Width | 26.2 | 8 |
| House Length | 672.6 | 205 |
| Width of Center Walkway | 12.0 | 3.6 |
| Average Roof Height | 18.7 | 5.7 |
| Gable Area (ft$^2$ or m$^2$) | 53.8 | 5 |
| Roof Area per Unit Bay Length (ft$^2$/ft bay or m$^2$/m bay) | 92.6 | 8.6 |
| Number of Bays | 30 | 30 |

**Solution**

Remember that the maximum grow pipe heating system temperature is typically 70°C (158°F) whereas the return temperature is 55°C (131°F).

| OTHER PROVIDED or CALCULATED PARAMETERS | | | |
|---|---|---|---|
| **PARAMETER** | **(I-P)** | | **(SI)** |
| Greenhouse Inside Design Temperature, $T_i$ | 68 | °F | 20 °C |
| Outside Design Temperature, $T_o$ (Table A.13 or A.14); (99% Level) | 24 | °F | −4.4 °C |
| $\Delta T$ | 44 | °F | 24.4 °C |
| Grow Pipe Supply Water Temperature, $T_{supply}$ | 158 | °F | 70 °C |
| Grow Pipe Return Water Temperature, $T_{return}$ | 131 | °F | 55 °C |
| $\Delta T_{water}$ | 27 | °F | 15 °C |
| Average Water Temperature, $T_{avg}$ | 144.5 | °F | 62.5 °C |
| Over Temperature ($T_{avg} - T_i$) | 108.5 | °F | 42.5 °C |
| Greenhouse Roof $U$-value (Provided or Table 3.3) | 1.0 Btu/h.ft$^2$.°F | 5 kcal/h.m$^2$.°C | 5.8 W/m$^2$.°C |
| Greenhouse Wall $U$-value (Provided or Table 3.3) | 1.0 Btu/h.ft$^2$.°F | 5 kcal/h.m$^2$.°C | 5.8 W/m$^2$.°C |
| Number of 32 mm Pipes per Bay | | 5 | |
| Air Change per Hour | | 0.5 (Table 3.4) | |
| Wind Factor | | 1.00 (Table 5.1) | |
| Location | | Vancouver, British Columbia, Canada | |
| **Heat Loss Through the Roof** | | | |
| Transmission Heat Losses per ° Temp. Diff. per Bay Length, $Q_c$ | 92.6 Btu/ h.ft roof.°F | 43 kcal/ h.m roof.°C | 49.9 W/ m roof.°C |
| Infiltration Heat Losses per ° Temp. Diff. per Bay Length, $Q_{inf}$ | 15.4 Btu/ h.ft roof.°F | 7.08 kcal/ h.m.roof.°C | 8.21 W/ m roof.°C |
| Total Heat Loss per o Temp. Diff. per Bay, $Q_t$ | 108.0 Btu/ h.ft roof.°F | 50.1 kcal/ h.m.roof.°C | 58.11 W/m roof.°C |
| **Heat Supply of the Grow Pipe Loop Pipes** | | | |
| Heat Supply by the 33 mm Pipes @ 42.5°C Over Temp. (Table A.18) | 64 Bth/h.ft | 53 kcal/h.m | 61.5 W/m |
| Heat Supply per Pipe Length per Bay (# Pipes × Pipe Heat Supply) | 320 Bth/ h.ft.bay | 265 kcal/ h.m.bay | 307.5 W/m. bay |
| Temperature Lift Provided by the Grow Pipes, Pipe Heat/Supply/$Q_t$ | 2.9°F | 5.3°C | 5.3°C |
| **Water Flow Rate per Grow Pipe Loop** = Number of Pipe Runs × Length × Pipe Heat Supply/$\Delta T_{water}$ | 3.11 gpm | 707 l/h | 707 l/h |
| Total Water Flow Rate for Grow Pipe (2.5 loops/bay & 30 bays) | 233.5 gpm | 53,025 l/h | 53,025 l/h |

Now using the total water flow figures, the maximum limit of 10 mm $H_2O$ resistance mentioned earlier, and Table A.21, we can design the distribution headers as follows:

| | | DISTRIBUTION HEADER SIZE | | | |
|---|---|---|---|---|---|
| **GABLE WALL SECTION (m)** | | **SUPPLY Flow (l/h)** | **SUPPLY OD (mm)** | **RETURN Flow (l/h)** | **RETURN OD (mm)** |
| 0.00 | 6.40 | 53,025 | **133** | 707 | **33** |
| 6.40 | 12.80 | 51,611 | **133** | 2,121 | **42** |

*(Continued)*

| | | | | | |
|---|---|---|---|---|---|
| 12.80 | 19.20 | 50,197 | **133** | 3,535 | **51** |
| 19.20 | 25.60 | 48,783 | **133** | 4,949 | **57** |
| 25.60 | 32.00 | 47,369 | **127** | 6,363 | **63** |
| 32.00 | 38.40 | 45,955 | **127** | 7,777 | **70** |
| 38.40 | 44.80 | 44,541 | **127** | 9,191 | **70** |
| 44.80 | 51.20 | 43,127 | **127** | 10,605 | **76** |
| 51.20 | 57.60 | 41,713 | **121** | 12,019 | **76** |
| 57.60 | 64.00 | 40,299 | **121** | 13,433 | **83** |
| 64.00 | 70.40 | 38,885 | **121** | 14,847 | **83** |
| 70.40 | 76.80 | 37,471 | **121** | 16,261 | **89** |
| 76.80 | 83.20 | 36,057 | **114** | 17,675 | **89** |
| 83.20 | 89.60 | 34,643 | **114** | 19,089 | **95** |
| 89.60 | 96.00 | 33,229 | **114** | 20,503 | **95** |
| 96.00 | 102.40 | 31,815 | **114** | 21,917 | **102** |
| 102.40 | 108.80 | 30,401 | **108** | 23,331 | **102** |
| 108.80 | 115.20 | 28,987 | **108** | 24,745 | **102** |
| 115.20 | 121.60 | 27,573 | **108** | 26,159 | **102** |
| 121.60 | 128.00 | 26,159 | **102** | 27,573 | **108** |
| 128.00 | 134.40 | 24,745 | **102** | 28,987 | **108** |
| 134.40 | 140.80 | 23,331 | **102** | 30,401 | **108** |
| 140.80 | 147.20 | 21,917 | **102** | 31,815 | **114** |
| 147.20 | 153.60 | 20,503 | **95** | 33,229 | **114** |
| 153.60 | 160.00 | 19,089 | **95** | 34,643 | **114** |
| 160.00 | 166.40 | 17,675 | **89** | 36,057 | **114** |
| 166.40 | 172.80 | 16,261 | **89** | 37,471 | **121** |
| 172.80 | 179.20 | 14,847 | **83** | 38,885 | **121** |
| 179.20 | 185.60 | 13,433 | **83** | 40,299 | **121** |
| 185.60 | 192.00 | 12,019 | **76** | 41,713 | **121** |
| 192.00 | 198.40 | 10,605 | **76** | 43,127 | **127** |
| 198.40 | 204.80 | 9,191 | **70** | 44,541 | **127** |
| 204.80 | 211.20 | 7,777 | **70** | 45,955 | **127** |
| 211.20 | 217.60 | 6,363 | **63** | 47,369 | **127** |
| 217.60 | 224.00 | 4,949 | **57** | 48,783 | **133** |
| 224.00 | 230.40 | 3,535 | **51** | 50,197 | **133** |
| 230.40 | 236.80 | 2,121 | **42** | 51,611 | **133** |
| 236.80 | 240.00 | 707 | **33** | 53,025 | **133** |

Remember that the average size of distribution header piping is 0.8 times the maximum pipe size. Keeping this in mind, we can now determine the heat loss and the supply along the gable end as follows:

*(Continued)*

| GABLE END WALL HEAT DEMAND and SUPPLY | | |
|---|---|---|
| **Heat Loss Through the Gable End Wall** | | |
| Transmission Heat Loss for Avg. Gable Wall Height | $Q_c$ = (5 kcal/h.m².°C) × 5.7 m × 5.3°C | 151 kcal/h.m |
| Transmission Heat Losses for the 0.7 m roof | 0.7 m × (265 kcal/h.m)/ 8 m bay width | 23 kcal/h.m |
| Total Heat Demand of the Gable Wall per Unit Wall Length, $Q_t$ | | **174 kcal/h.m** |
| **Heat Supply of the Grow Pipe along the Gable End Wall** | | |
| Average Distribution Header Pipe Size: 0.8 × The Max Pipe Size | 0.8 × 133 mm = 106.4 ≈ 108 mm | |
| Heat Supply by the 108 mm Pipes @ 42.5°C Over Temp. (Table A.18) | 144 kcal/h.m | |
| Heat Supply by 2 Distribution Headers = 2 × 144 kcal/h.m | 288 kcal/h.m | |
| Heat Supply by the 19 mm Heating Hoses @ 42.5°C Over Temp. (Table A.19) | 30 kcal/h.m | |
| Heat Supply by 19 mm Heating Hoses = 2 m Hose × (3/4) × (30 kcal/h.m)/ (8 m Bay Width) | 28 kcal/h.m | |
| Total Grow Pipe Heating Supplied along the Gable End Wall = (288 + 28) kcal/h.m | **316 kcal/h.m** | |
| Heat Balance along the Gable End Wall = Supply − Demand = (316 − 174) kcal/h.m | **142 kcal/h.m** | |

*Note:* The total gable end wall heating provided by the grow pipe heating (316 kcal/h.m) along the gable end wall is to be added to the rail pipe heating supplied along the gable end wall. If there is still shortage of heating, then extra loops of 51 mm heating is provided until the total heat provided matches the heat loss through gable end wall. However, it seems that the grow pipe heating provides more heat than what is demanded along the gable end wall; therefore, no extra loops of 51 mm tubing is required.

Keeping in mind that, in this design exercise, the supply piping runs along the sidewall (both sides of the center walkway) and the return piping runs along the opposite sidewall (again, both sides of the center walkway) Let's now determine the heat loss and the supply along the sidewalls as follows:

| SIDEWALL HEAT DEMAND and SUPPLY | | |
|---|---|---|
| **Heat Loss Through the Sidewall** | | |
| Transmission Heat Loss for the Sidewall (each) | $Q_c$ = (5 kcal/h.m².°C) × 5.3 m × 5.3°C | 140 kcal/h.m |
| **Heat Supply of the Grow Pipe along the Sidewall** | | |
| Maximum Distribution Header Pipe Size | 133 mm | |
| Heat Supply of the 133 mm Main Supply Pipes @ 50°C Over Temp., ($T_{Supply} - T_i$) (Table A.18) | 210 kcal/h.m | |
| Heat Supply of the 133 mm Main Return Pipes @ 35°C Over Temp., ($T_{Return} - T_i$) (Table A.18) | 136 kcal/h.m | |

*Note:* The total sidewall heating provided by the grow pipe heating supply main (210 kcal/h.m) along the first sidewall is more than what is required; therefore, an aluminum painted piping can be used along this sidewall in order to reduce heat supply by the main supply piping. The return main of the grow pipe heating system along the opposite sidewall almost meets the demand.

*(Continued)*

Remember that the grow pipe heating system provides heat through arch piping and hot water heating hoses used in connecting the loops along the center walkway. Let's now determine the heat loss and the supply along the center walkway as follows:

| CENTER WALKWAY HEAT DEMAND and SUPPLY | | |
|---|---|---|
| **Heat Loss Along the Center Walkway** | | |
| Width of the Center Walkway = Full Width/2 (in this design case as one side of the grow pipe is responsible for one half of the greenhouse) – 5/2 = 2.5 m | | |
| Transmission Heat Losses from the 2.5 m roof | 2.5 m × (265 kcal/h.m)/8 m bay width | **83 kcal/h.m path** |
| **Heat Supply Above the Center Walkway** | | |
| Heat Supply by the 133 mm Pipe @ 35°C Over Temp., ($T_{Return} - T_i$) (Table A.18) | 136 kcal/h.m | |
| Heat Supply by 33 mm Arch Pipe @ 42.5°C Over Temp. (Table A.19)<br>Arch Factor: 0.63; 2.5 loops of 33 mm pipes; Total Length of 33 mm: 6 m/loop; Bay Width: 8 m | 53 kcal/h.m | |
| Total Heat Supply by 33 mm Arch Pipe = 53 kcal/h.m × 6 m × 0.63 × 2.5 loops/8 m | 63 kcal/h.m | |
| Heat Supply by 19 mm Arch Pipe Heating Hoses @ 42.5°C Over Temp. (Table A.19)<br>Hose Length: 2.5 m/pipe; 5 hoses/bay; Hose Factor: ¾; Bay Width: 8 m | 30 kcal/h.m | |
| Total Heat Supply by 19 mm Hose = 30 kcal/h.m × 2.5 m × ¾ × (5 hoses/bay)/8 m | 35 kcal/h.m | |
| Total Grow Pipe Heating Supplied along the Center Walkway = (136 + 63 + 35) kcal/h.m | **234 kcal/h.m path** | |
| Heat Balance Along the Center Walk = Supply – Demand = (234 – 83) kcal/h.m | **151 kcal/h.m path** | |

*Note:* The total center walkway heating provided by the grow pipe heating (234 kcal/h.m) along the walkway is to be added to the rail pipe heating supplied along the center walkway. If there is still shortage of heating, then extra loops of 51 mm heating is provided until the total heat provided matches the heat loss along the center walkway. However, it seems that the grow pipe heating provides more heat than what is demanded along the center walkway; therefore, no extra lengths of 51 mm tubing along the walkway might be required. If 2 lengths of 51 mm tubing are provided along the sides of the walkway; then we may need to reduce it to one length on each side, or totally eliminate it depending on the heat demand.

Now that the heating (energy balance) side is covered, the mass balance side, that is, the flow requirement and the resistances can be worked on as follows:

| GROW PIPE HEATING FLOW REQUIREMENT and RESISTANCES | | | |
|---|---|---|---|
| Flow Rate: **53,025 l/h** | Main Pipe Size: **133 mm** (Table A.21) | Resistance, *R*: **9.5 mm $H_2O$/m** (Table A.21) | Velocity: **1.17 m/s** (Table A.21) |
| **Calculation of the Resistance in the Circuit (Including the Loop)** | | | |
| Greenhouse Width, *W*: 240 m | Grow Pipe Heating Depth, *L*: 100 m | | |
| Resistance of the Loop = 5.4 mm $H_2O$/m (Table A.21 for 31.8 mm used here) × ($L$ = 100) × 2 = 5.4 mm $H_2O$/m × 100 × 2 | | | **1,080 mm $H_2O$** |
| Resistance of the Arch = (6 m 31.8 mm pipe used here) × (5.4 mm $H_2O$/m) = 32 mm $H_2O$<br>Resistance of the 19 mm Hose = 33.8 mm $H_2O$/m (Table A.21) for ~700 l/h; Special Resistance = 227.8 mm $H_2O$ for ~700 l/h | | | **429 mm $H_2O$** |

*(Continued)*

| | | |
|---|---|---|
| Total Resistance of the 19 mm Hose<br>= (2 × 2.5 m × 33.8 mm $H_2O$) + 227.8 mm $H_2O$/m) = 397 mm $H_2O$<br>Total Resistance of the Arch Pipe and the Hose = 32 mm $H_2O$ + 397 mm $H_2O$ | | |
| Resistance over the 19 mm Hose on the Distribution Header Side = (2 × 2 m × 33.8 mm $H_2O$) + 227.8 mm $H_2O$/m) | | **363 mm $H_2O$** |
| Resistance over the Mains = (2 × $W$ + 2 × $L$) × $R$ × 1.10<br>= (2 × 240 m + 2 × 100 m) × 9.5 mm $H_2O$/m × **1.10** (10% extra) | | **7,106 mm $H_2O$** |
| Resistance over the Transport Mains outside the Greenhouse<br>= 2 pipes × 20 m × 9.5 mm $H_2O$/m × **1.50** (50% extra) | | **570 mm $H_2O$** |
| $K_v$ = 168 is calculated using Equation (2.189) for<br>$\Delta P$ = 0.1 bar (1.45 psi) allowed; the 125 mm 3-way Mixing Valve has a $K_v$ = 250<br>Then, the Resistance over the 3-way Mixing Valve,<br>$\Delta P = [(53\ m^3/h)/250]^2 \times 10{,}000$ mm $H_2O$ = 450 mm $H_2O$ < 700 mm $H_2O$ | | **450 mm $H_2O$** |
| Other Resistances to be Included (boiler loop, etc., if there is any)? | | 0 |
| **Total Resistance** | | **9,998 mm $H_2O$** |
| **Pump Size** | **~234 gpm @33 ft Head** | **53,025 l/h @10 m Head** |

In this design example, we have chosen to connect the grow pipe heating mixing group directly to the boiler room manifold and eliminated the transport pump; and the required grow pipe mixing group pump size is approximately 53,025 l/h with 10 m head (or 234 gpm with 33 ft head). We will need to match this requirement with the pump manufacturer's supplied pump curves and choose a high efficiency pump (see Figure 2.27). This grow pipe heating system will take care of one side of the greenhouse. The layout of the grow pipe heating and the pump for the other side is identical, and the mixing group directly connects to the manifold in the boiler room. However, if the system has additional resistances, it should be added to the above pump head.

#### Example 5.8: Designing an In-floor and Overhead Heating System

In this example, let's design an in-floor and overhead heating system for the following double poly greenhouse facility to be built in Leamington, Ontario, Canada.

| GREENHOUSE DIMENSIONS | | |
|---|---|---|
| **COMPONENT** | **(ft)** | **(m)** |
| Wall Height | 16 | 4.8 |
| House Width | 24 | 7.2 |
| House Length | 252 | 75.6 |
| Width of Center Walkway | 12 | 3.6 |
| Average Roof Height | 19.3 | 5.8 |
| Gable Area ($ft^2$ or $m^2$) | 88.3 | 8.2 |
| Number of Bays | 17 | 17 |

#### Solution

Remember that the maximum in-floor heating system supply temperature is 35°C (95°F), whereas the return temperature is 30°C (86°C) providing a $\Delta T$ of 5°C (9°F). We will design a single zone in-floor heating system here.

| OTHER PROVIDED or CALCULATED PARAMETERS | | | |
|---|---|---|---|
| **PARAMETER** | **(I-P)** | | **(SI)** |
| Greenhouse Inside Design Temperature, $T_i$ | 70 | °F | 21 °C |
| Outside Design Temperature, $T_o$ (Table A.13 or A.14); (99% Level) | 5 | °F | −15 °C |
| $\Delta T$ | 95 | °F | 35 °C |
| In-Floor Heating Supply Water Temperature, $T_{supply}$ | 95 | °F | 35 °C |
| In-Floor Heating Return Water Temperature, $T_{return}$ | 86 | °F | 30 °C |
| $\Delta T_{water}$ | 9 | °F | 5 °C |
| Average Water Temperature, $T_{avg}$ | 90.5 | °F | 32.5 °C |
| Over Temperature ($T_{avg} - T_i$) | 108.5 | °F | 11.5 °C |
| Greenhouse Roof *U*-value (Provided or Table 3.3) | 0.7 Btu/h.ft².°F | 3.5 kcal/h.m².°C | 4.0 W/m².°C |
| Greenhouse Wall *U*-value (Provided or Table 3.3) | 0.7 Btu/h.ft².°F | 3.5 kcal/h.m².°C | 4.0 W/m².°C |
| Number of 25 mm OD PE Hose Loops per Bay | | 18 | |
| Air Change per Hour | | 1.0 (Table 3.4) | |
| Wind Factor | | 1.00 (Table 5.1) | |
| Location | | Leamington, Ontario, Canada | |
| **Heat Supply of the In-Floor Heating PE Tubes** | | | |
| Heat Supply by the 25 mm Hoses @ 11.5°C Over Temperature | 100 Bth/h.m² | 25 kcal/h.m² | 29 W/m² |
| Heat Supply by the 25 mm Hoses per Bay Length (per 7.2 m²/m bay) | 720 Btu/h.m bay length | 180 kcal/h.m bay length | 208.8 W/m bay length |
| Heat Supply in One Side of the Greenhouse (17 × 7.2 m² = 4,406.4 m²) | 440,640 Bth/h | 110,160 kcal/h | 127,786 W |
| Σ Greenhouse Heat Loss per Bay (see Overhead Heating Section) | 5,949 Btu/h.m bay | 1,500 kcal/h.m bay | 1,740 W/m bay |
| Σ Greenhouse Heat Loss per ° Temperature Diff., $Q_t$; $\Delta T$ = (21+15)) = 36°C | 91.8 Btu/h. m².°F | 41.7 kcal/h.m².°C | 48.3 W/ m².°C |
| Temperature Lift Provided by the In-Floor Heating, Hose Heat Supply per Bay/ ($Q_t$/36°C) | 7.8°F | 4.3°C | 4.3°C |
| **Water Flow Rate for In-Floor Heating per Side** = Heat Supply in One Side of the Greenhouse/$\Delta T_{water}$ = (110,160 kcal/h)/5°C | 97 gpm | 22,032 l/h | 22,032 l/h |
| Σ Water Flow Rate for In-Floor Heating | 194 gpm | 44,064 l/h | 44,064 l/h |
| In-Floor Heating Loops per Side = 17 bays × 18 loops/bay) | | 306 | |
| Σ In-Floor Heating Loops | | 612 | |
| Water Flow Rate per Loop | 0.317 gpm | 72 l/h | 72 l/h |

*(Continued)*

| | | | |
|---|---|---|---|
| Σ Greenhouse Width = 17 bays × Bay Width | 408 ft | 122.4 m | 122.4 m |
| Σ Number of In-Floor Heating Sections per Bay = (Σ Greenhouse Width)/3.20 m | 38.25 | | |
| Flow Rate per Section = Water Flow Rate per Side/Σ Number of Sections | 2.54 gpm | 576 l/h | 576 l/h |
| Maximum Header Size (Table A.21 for Flow Rate per Section & max resist. 10 mm $H_2O$) | 102 mm | | |
| Avg. Distribution Header Pipe Size = 0.8 × Max Size = 0.8 × 102 mm | 83 mm | | |

Now using the total water flow figures, the maximum limit of 10 mm $H_2O$ resistance, and Table A.21, we can design the distribution headers for in-floor heating system as follows:

| GABLE WALL SECTION (m) | | DISTRIBUTION HEADER SIZE | | | |
|---|---|---|---|---|---|
| | | **SUPPLY Flow (l/h)** | **SUPPLY OD (mm)** | **RETURN Flow (l/h)** | **RETURN OD (mm)** |
| 0.00 | 3.20 | 22,032 | **102** | 576 | **38** |
| 3.20 | 6.40 | 21,888 | **102** | 1,152 | **38** |
| 6.40 | 9.60 | 21,312 | **95** | 1,728 | **45** |
| 9.60 | 12.80 | 20,736 | **95** | 2,304 | **45** |
| 12.80 | 16.00 | 20,160 | **95** | 2,880 | **51** |
| 16.00 | 19.20 | 19,584 | **95** | 3,456 | **51** |
| 19.20 | 22.40 | 19,008 | **95** | 4,032 | **57** |
| 22.40 | 25.60 | 18,432 | **95** | 4,608 | **57** |
| 25.60 | 28.80 | 17,856 | **89** | 5,184 | **57** |
| 28.80 | 32.00 | 17,280 | **89** | 5,760 | **63** |
| 32.00 | 35.20 | 16,704 | **89** | 6,336 | **63** |
| 35.20 | 38.40 | 16,128 | **89** | 6,912 | **63** |
| 38.40 | 41.60 | 15,552 | **89** | 7,488 | **70** |
| 41.60 | 44.80 | 14,976 | **89** | 8,064 | **70** |
| 44.80 | 48.00 | 14,400 | **83** | 8,640 | **70** |
| 48.00 | 51.20 | 13,824 | **83** | 9,216 | **70** |
| 51.20 | 54.40 | 13,248 | **83** | 9,792 | **76** |
| 54.40 | 57.60 | 12,672 | **83** | 10,368 | **76** |
| 57.60 | 60.80 | 12,096 | **83** | 10,944 | **76** |
| 60.80 | 64.00 | 11,520 | **76** | 11,520 | **76** |
| 64.00 | 67.20 | 10,944 | **76** | 12,096 | **83** |
| 67.20 | 70.40 | 10,368 | **76** | 12,672 | **83** |
| 70.40 | 73.60 | 9,792 | **76** | 13,248 | **83** |
| 73.60 | 76.80 | 9,216 | **70** | 13,824 | **83** |
| 76.80 | 80.00 | 8,640 | **70** | 14,400 | **83** |

*(Continued)*

| | | | | | |
|---|---|---|---|---|---|
| 80.00 | 83.20 | 8,064 | **70** | 14,976 | **89** |
| 83.20 | 86.40 | 7,488 | **70** | 15,552 | **89** |
| 86.40 | 89.60 | 6,912 | **63** | 16,128 | **89** |
| 89.60 | 92.80 | 6,336 | **63** | 16,704 | **89** |
| 92.80 | 96.00 | 5,760 | **63** | 17,280 | **89** |
| 96.00 | 99.20 | 5,184 | **57** | 17,856 | **89** |
| 99.20 | 102.40 | 4,608 | **57** | 18,432 | **95** |
| 102.40 | 105.60 | 4,032 | **57** | 19,008 | **95** |
| 105.60 | 108.80 | 3,456 | **51** | 19,584 | **95** |
| 108.80 | 112.00 | 2,880 | **51** | 20,160 | **95** |
| 112.00 | 115.20 | 2,304 | **45** | 20,736 | **95** |
| 115.20 | 118.40 | 1,728 | **45** | 21,312 | **95** |
| 118.40 | 121.60 | 1,152 | **38** | 21,888 | **102** |
| 121.60 | 122.40 | 576 | **38** | 22,032 | **102** |

Remember that the average size of distribution header piping is 0.8 times the maximum pipe size. Keeping this in mind, we can now determine the heat loss and the supply along the gable end, sidewall, and center walkway as follows:

| GABLE WALL, SIDEWALL, and CENTER WALKWAY HEAT DEMAND and SUPPLY | | |
|---|---|---|
| **Heat Loss Through the Gable End Wall** | | |
| Transmission Heat Loss for Avg. Gable Wall Height | $Q_c$ = (3.5 kcal/h.m².°C) × 5.8 m × 4.3°C | **87 kcal/h.m** |
| **Heat Supply of the In-Floor Heating Headers along the Gable End Wall** | | |
| Average Distribution Header Pipe Size: 0.8 × The Max Pipe Size | 0.8 × 102 mm = 81.6 ≈ 83 mm | |
| Heat Supply by the 83 mm Pipes @ 11.5°C Over Temp. (Table A.18) | 24.5 kcal/h.m | |
| Heat Supply by 2 Distribution Headers = 2 × 24.5 kcal/h.m | **49 kcal/h.m** | |
| Heat Balance along the Gable End Wall = Supply − Demand = (49 − 87) kcal/h.m | **-38 kcal/h.m** | |
| **NOTE:** The total gable end wall heating provided by the in-floor heating distribution headers (49 kcal/h.m) along the gable end wall is less than the heat loss through gable end wall. See the overhead heating section to see how this deficiency is removed and a heat balance is maintained. | | |
| **Heat Loss Through the Sidewall** | | |
| Transmission Heat Loss for the Sidewall (each) | $Q_c$ = (3.5 kcal/h.m².°C) × 4.8 m × 4.3°C | **72 kcal/h.m** |
| **Heat Supply of the Grow Pipe along the Sidewall** | | |
| Maximum Distribution Header Pipe Size | 102 mm | |
| Heat Supply of the 102 mm Main Supply Pipes @ 14°C Over Temp., ($T_{Supply} - T_i$) (Table A.18) | 37 kcal/h.m | |

*(Continued)*

| GABLE WALL, SIDEWALL, and CENTER WALKWAY HEAT DEMAND and SUPPLY | | |
|---|---|---|
| Heat Supply of the 102 mm Main Return Pipes @ 9°C Over Temp., ($T_{Return} - T_i$) (Table A.18) | 22 kcal/h.m | |

**NOTE:** The total sidewall heating provided by the in-floor heating supply main (37 kcal/h.m) along the first and second sidewalls are less than what are required; however, see the amount of heating supplied by the overhead heating main supply and return pipes to see if there is a heat balance over each sidewall. If not, additional loops of 51 mm OD piping will be introduced along the sidewalls.

**Heat Loss Along the Center Walkway**

| | | |
|---|---|---|
| Width of the Center Walkway = 3.6 m | | |
| Transmission Heat Losses | 3.6 m × (180 kcal/h.m)/7.2 m bay width | **90 kcal/h.m path** |

**Heat Supply Above the Center Walkway**

| | |
|---|---|
| Heat Supply by the 127 mm Return Pipe @ 9°C Over Temp., ($T_{Return} - T_i$) (Table A.18) | **27 kcal/h.m** |
| Heat Balance Along the Center Walk = Supply − Demand = (27 − 90) kcal/h.m | **−63 kcal/h.m path** |

**NOTE:** The total center walkway heating provided by the in-floor heating return pipe (27 kcal/h.m) along the walkway is not sufficient to meet the demand; however, the shortage is typically covered by the other pipes along the center walkway. See the overhead heating section to see what is provided and if it meets the requirement.

Now that the heating (energy balance) side is covered, the mass balance side, that is, the flow requirement and the resistances can be worked on as follows:

| IN-FLOOR HEATING FLOW REQUIREMENT and RESISTANCES | | | |
|---|---|---|---|
| Flow Rate: **44,064 l/h** | Main Pipe Size: **127 mm** (Table A.21) | Resistance, *R*: **6.5 mm $H_2O$/m** (Table A.21) | Velocity: **0.96 m/s** (Table A.21) |
| Water Flow through the Mixing Valve = 10% × 44,064 l/h | | | 4,406 l/h |
| Water Flow through the By-pass = 90% × 44,064 l/h | | | 39,658 l/h |

**By-pass Line**

| | |
|---|---|
| Flow over the By-pass = 39,658 l/h; We have 2 T-branch flows and 1 ball valve (see Table 2.30 for resistance coefficients; 1.5 for T-branch flow; ~1.0 for the valve). Then, Σ Resistance Coefficient over the By-pass = (2 × 1.5 + 1) = | 4 |
| Mixing Valve Resistance (40 mm 3-way mixing valve with $K_v = 25$) = [ (4.4 m³/h)/25]² × 10,000 mm $H_2O$ = 310 mm $H_2O$ < 700 | 310 mm $H_2O$ |
| And using Table A.27 for the total resistance coefficient of 4 and theoretical resistance of the by-pass 1,310 mm $H_2O$ (310 +1,000 mm $H_2O$), we can find the theoretical velocity (as 2.6 m/s). | 2.6 m/s |

By-pass Size = 83 mm By-pass Valve Size = 65 mm = 2 ½″

Resistance over 0.5 m by-pass (using Table A.21 for 83 mm pipe and 39,658 l/s = 70 mm $H_2O$) = 35 mm $H_2O$; $V$ = 2.35 m/s.
And using Table A.27 for the total resistance coefficient of 4 & $V$ = 2.35 m/s, Z = 1,090 mm $H_2O$. Actual Total Resistance = (35 + 1,090) mm $H_2O$ = 1,125 mm $H_2O$, which is less than the theoretical resistance of 1,310 mm $H_2O$. However, that's why we have a throttling valve over the by-pass; it means, we can introduce more resistance by throttling the valve.

**Calculation of the Resistance in the Circuit (Including the In-Floor Loop)**

| | |
|---|---|
| Width of the Section, *W*: 122.4 m | In-Floor Heating Depth, *L*: 75.6/2 = 37.8 m ≈ 38 m |

*(Continued)*

| IN-FLOOR HEATING FLOW REQUIREMENT and RESISTANCES | |
|---|---|
| Actually, Table A.24 provides a resistance of 2 mm $H_2O$ for minimum 149 l/h; then, if the flow rate is halved, then the resistance for 72 l/h becomes the square of the half times the resistance, that is, $(1/2)^2 \times 2$ mm $H_2O$ = 0.5 mm $H_2O$. Then, Resistance of the Loop = 0.5 mm $H_2O$/m (for 72 l/h) $\times$ ($L$ = 36 m) $\times$ 2 = 0.5 mm $H_2O$/m $\times$ 36 m $\times$ 2 = 36 mm $H_2O$ | **36 mm $H_2O$** |
| Resistance of the Loop Connection: ½″ Valve; Special Resistance = 25 mm $H_2O$ | **25 mm $H_2O$** |
| Resistance over the Mains = $(2 \times W + 2/2 \times L) \times 10 \times$ **1.10** = (2 × 122.4 m + 1 × 75.6 m) × 10 mm $H_2O$/m × **1.10** (10% extra) | **3,524 mm $H_2O$** |
| $K_v$ is calculated using Equation (2.189) for $\Delta P$ =0.1 bar (1.45 psi) allowed; the chosen 40 mm 3-way Mixing Valve has a $K_v$ = 25. Then, the Resistance over the 3-way Mixing Valve, $\Delta P = [(4.4\ m^3/h)/25]^2 \times 10{,}000$ mm $H_2O$ = 310 mm $H_2O$ < 700 mm $H_2O$ | **310 mm $H_2O$** |
| Other Resistances to be Included (transport line, boiler loop, etc., if there is any)? None here. | 0 |
| **Total Resistance** | **3,895 mm $H_2O$** |
| **Pump Size** **~194 gpm @13 ft Head** | **44,064 l/h @4 m Head** |

In this design example, we have chosen to connect the in-floor heating mixing group directly to the boiler room manifold and eliminated the transport pump; and the required in-floor heating mixing group pump size is approximately 44,064 l/h with approx. 4 m head (or 194 gpm with 13 ft head). We will need to match this requirement with the pump manufacturer's supplied pump curves and choose a high efficiency pump (see Figure 2.27). This in-floor heating system will take care of one side of the greenhouse. The layout of the in-floor heating and the pump for the other side is identical, and the mixing group directly connects to the manifold in the boiler room. However, if the system has additional resistances, it should be added to the above pump head.

Now, for the same greenhouse in this example, let's design an overhead heating system for the remaining heat requirement using three overhead heating zones as shown below.

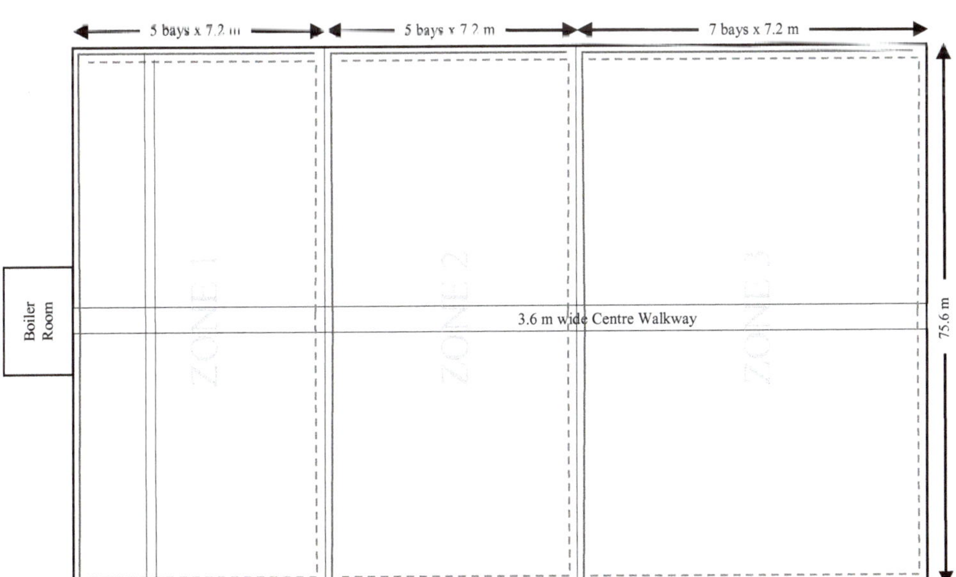

| OTHER PROVIDED or CALCULATED PARAMETERS | | | | |
|---|---|---|---|---|
| **PARAMETER** | **(I-P)** | | **(SI)** | |
| Overhead Heating Supply Temperature | 203 | °F | 95 °C | |
| Overhead Heating Return Temperature | 176 | °F | 80 °C | |
| $\Delta T_{water}$ | 27 | °F | 15 °C | |
| Average Water Temperature, $T_{avg}$ | 189.5 | °F | 87.5 °C | |
| Greenhouse Inside Design Temperature, $T_i$ | 70 | °F | 21 °C | |
| Outside Design Temperature, $T_o$ (Table A.13 or A.14); (99% Level) | 5 | °F | -15 °C | |
| $\Delta T = (T_i - T_o)$ | 64.8 | °F | 36 °C | |
| Over Temperature $(T_{avg} - T_i)$ | 119.5 | °F | 66.5 °C | |
| Greenhouse Roof $U$-value (Provided or Table 3.3) | 0.7 Btu/h.ft$^2$.°F | 3.5 kcal/h.m$^2$.°C | 4.0 W/m$^2$.°C | |
| Greenhouse Wall $U$-value (Provided or Table 3.3) | 0.7 Btu/h.ft$^2$.°F | 3.5 kcal/h.m$^2$.°C | 4.0 W/m$^2$.°C | |
| Bay Width | 7.2 m | | | |
| Roof Area per Unit Bay Length (m$^2$/m bay) | 8.2 m$^2$/m bay | | | |
| Wall Height | 4.8 m | | | |
| Average Roof Height | 4.8 m + 1 m = 5.8 m | | | |
| Air Change per Hour | 1.0 (Table 3.4) | | | |
| Wind Factor | 1.00 (Table 5.1) | | | |
| Location | Leamington, Ontario, Canada | | | |
| **Heat Loss Through the Roof** | | | | |
| Transmission Heat Losses per Bay Length, $Q_c$ | (3.5 kcal/h.m$^2$.°C ) × (8.2 m$^2$/m bay) × (36°C) = 1,033 kcal/h.m roof | | | |
| Infiltration Heat Losses per Bay Length, $Q_{inf}$ | 7.2 m × 5.8 m × 1 m × 1.293 × 0.24 × 1.0 ac × 36°C = 467 kcal/h.m.roof | | | |
| Total Heat Loss per Bay, $Q_t = Q_c + Q_{inf}$ | 1,500 kcal/h.m.roof | | | |
| **Heat Supply of the Pipes** | | | | |
| Heat Supply by the 51 mm Pipes @ 66.5°C Over Temp. (Table A.18) | 132 kcal/h.m | | | |
| **Calculation of the Number of 51 mm Pipes Required for Overhead Heating** | | | | |
| $= Q_t - Q_{Floor\ Heat\ Supplied}$ = (1,500 – 180) kcal/h.m roof/132 kcal/h.m | 10 pipes/bay | | | |
| Overhead Heating Temperature Lift = Heat Supply by 51 mm OD/($Q_t$/ΔT) | (1,320 kcal/h.m.roof )/(1,500 kcal /h.m.roof/36°C) = 31.5°C | | | |
| **Water Flow Rate per 51 mm Overhead Heating Pipe** = Number of Pipe Runs × Length × Pipe Heat Supply/$\Delta T_{water}$ | (1 length × 75 m × 132 kcal/h.m)/15°C = 660 l/h | | | |
| However, due to the return line running between Zone 1 & 2, we will use 4 lengths of 51 mm less (along the zone separation), that is, a total of 46 lengths, approx. 4 × 0.72 m ≈ 2.9 m from Zone 2. Then, the **Adjusted Flow Rate** per 51 mm overhead pipe is; | 717 l/h | | | |
| Total Water Flow Rate for Overhead Heating Zone 1 | 33,000 l/h (145 gpm) | | | |

*(Continued)*

| OTHER PROVIDED or CALCULATED PARAMETERS | | |
|---|---|---|
| PARAMETER | (I-P) | (SI) |
| Water Flow Rate for One Side of Center Walkway of Zone 1 | 16,500 l/h (72.5 gpm) | |
| Maximum Header Size (Table A.21 for Flow Rate per Section & max resist. 10 mm $H_2O$) | 89 mm | |
| Avg. Distribution Header Pipe Size = 0.8 × Max Size – 0.8 × 89 mm | 70 mm | |

Now, using the total water flow figures, the maximum limit of 10 mm $H_2O$ resistance, and Table A.21, we can design the overhead heating distribution headers as follows:

| | | DISTRIBUTION HEADER SIZE | | | |
|---|---|---|---|---|---|
| GABLE WALL SECTION (m) | | SUPPLY Flow (l/h) | SUPPLY OD (mm) | RETURN Flow (l/h) | RETURN OD (mm) |
| 0.00 | 3.60 | 16,500 | **89** | 359 | **27** |
| 3.60 | 7.20 | 14,707 | **83** | 2,152 | **45** |
| 7.20 | 10.80 | 12,913 | **83** | 3,946 | **57** |
| 10.80 | 14.40 | 11,120 | **76** | 5,739 | **60** |
| 14.40 | 18.00 | 9,326 | **70** | 7,533 | **70** |
| 18.00 | 21.60 | 7,533 | **70** | 9,326 | **70** |
| 21.60 | 25.20 | 5,739 | **60** | 11,120 | **76** |
| 25.20 | 28.80 | 3,946 | **57** | 12,913 | **83** |
| 28.80 | 32.40 | 2,152 | **45** | 14,707 | **83** |
| 32.40 | 33.10 | 359 | **27** | 16,500 | **89** |

Again, the average size of the distribution header piping is 0.8 times the maximum pipe size. And the overhead heating piping starts 0.2 m from one gable end and runs to the opposite gable end (0.2 m from that gable end as well). Keeping this in mind, we can now determine the heat loss and the supply along the gable end, sidewall, and center walkway as follows:

| GABLE WALL, SIDEWALL, and CENTER WALKWAY HEAT DEMAND and SUPPLY | | |
|---|---|---|
| **Heat Loss Through the Gable End Wall** | | |
| Transmission Heat Loss for Avg. Gable Wall Height | $Q_c$ = (3.5 kcal/h.m².°C) × 5.8m × (36 – 40.5)°C | 640 kcal/h.m |
| Transmission Heat Loss from 0.2 m Bay Length | = 0.2 m × (1,500 – 180) kcal/h.m roof/7.2 m | 37 kcal/h.m |
| Total Heat Loss from the Gable End Wall | | **677 kcal/h.m** |
| **Heat Supply of the Distribution Headers along the Gable End Wall** | | |
| Average Distribution Header Pipe Size: 0.8 × The Max Pipe Size | 0.8 × 89 mm ≈ 70 mm | |
| Heat Supply by the 70 mm Pipes @ 66.5°C Over Temp. (Table A.18) | 173 kcal/h.m | |
| Heat Supply by 2 Distribution Headers = 2 × 173 kcal/h.m | 346 kcal/h.m | |

*(Continued)*

| GABLE WALL, SIDEWALL, and CENTER WALKWAY HEAT DEMAND and SUPPLY | | |
|---|---|---|
| **Heat Loss Through the Gable End Wall** | | |
| Heat Supply by 27 mm Connection Pipe = (½× 77 kcal/h.m × 10 pipes × 3.5 m)/7.2 m | 187 kcal/h.m | |
| Total Heat Supply Along the Gable End Wall | **533 kcal/h.m** | |
| Heat Balance Along the Gable End Wall = Supply − Demand = (533 − 677) kcal/h.m | **−144 kcal/h.m** | |

**NOTE:** The total gable end wall heat balance for the in-floor heating is −38 kcal/h.m); therefore, −33 − 144 = −177 kcal/h.m deficiency needs to be removed by introducing additional 51 mm OD tubing along the gable end wall; in this case by a valve connected 2 × 51 mm OD tubing. See below:

| | | |
|---|---|---|
| Heat Supply by 2 × 51 mm Pipes @ 66.5°C Over Temp. (Table A.18) = 2 × 132 kcal/h.m | 264 kcal/h.m | |
| New Heat Balance Along the Gable End Wall = (533 + 264 − 677 − 33) kcal/h.m | **87 kcal/h.m** (perfectly fine) | |
| **Heat Loss Through the Sidewall** | | |
| Transmission Heat Loss for the Sidewall (each) | $Q_c$ = (3.5 kcal/h.m$^2$.°C) × 4.8 m × 31.5°C | **529 kcal/h.m** |
| **Heat Supply of the Grow Pipe along the Sidewall** | | |
| Maximum Distribution Header Pipe Size | 89 mm | |
| Heat Supply of the 89 mm Main Supply Pipes @ 74°C Over Temp., ($T_{Supply} - T_i$) (Table A.18) | 244 kcal/h.m | |
| Heat Supply by 2 × 51 mm Pipes @ 74°C Over Temp. (Table A.18) = 2 × 152 kcal/h.m | 304 kcal/h.m | |
| Total Heat Supply Along the Sidewall | **548 kcal/h.m** | |
| Heat Balance Along the Sidewall = Supply − Demand = (548 − 529) kcal/h.m | **19 kcal/h.m** | |

**NOTE:** The total sidewall heating provided by the in-floor heating along the first sidewall was 35 kcal/h.m less than what is required; however, the overhead heating now provides 19 kcal/h.m extra heat. And the remaining 16 kcal/h.m is quite negligible and will be handled by the overall heating.

| | | |
|---|---|---|
| **Heat Loss Along the Center Walkway** | | |
| Width of the Center Walkway = 3.6 m | | |
| Transmission Heat Losses | 3.6 m × (1,320 kcal/h.m)/ 7.2 m bay width | **660 kcal/h.m path** |
| **Heat Supply Above the Center Walkway** | | |
| Heat Supply by the 114 mm Return Pipe @ 51°C Over Temp., ($T_{Return} - T_i$) (Table A.18); 1 pipe | 227 kcal/h.m | |
| Heat Supply by the Overhead Pipes Crossing the Walkway @ 66.5°C Over Temp. (132 kcal/h.m; Table A.18) = (132 kcal/h.m) × (10 pipes/bay) × (3.6 m width of the walkway)/7.2 m bay width | 660 kcal/h.m | |
| Total Heat Supply Above the Center Walkway | **887 kcal/h.m path** | |

**NOTE:** The total center walkway heating provided is too much (887 − 660 − 63 = 164 kcal/h.m) even considering the in-floor heating deficiency of 63 kcal/h.m. We can now use aluminum painting to reduce the heat contributions. Remember that we also have the insulated transport lines for the other zones running along the center walkway, so we are fine.

Now that the heating (energy balance) side is covered, the mass balance side, that is, the flow requirement and the resistances for the overhead heating can be worked on as follows:

| OVERHEAD HEATING FLOW REQUIREMENT and RESISTANCES | | | |
|---|---|---|---|
| Flow Rate: **33,000 l/h** | Main Pipe Size: **114 mm** (Table A.21) | Resistance, $R$: **8.3 mm $H_2O$/m** (Table A.21) | Velocity: **0.99 m/s** (Table A.21) |
| **Calculation of the Resistance in the Overhead Heating Circuit** | | | |
| Greenhouse Width, $W$: 36 m | Grow Pipe Heating Depth, $L$: 37.8 m ≈ 38 m | | |
| Resistance of the Loop = 0.44 mm $H_2O$/m (Table A.21 for 717 l/h × ($L$ = 75) × 1 length = 0.44 mm $H_2O$/m × 75 m × 1 length | | | **33 mm $H_2O$** |
| Resistance over the 27 mm Connecting Pipe = (2 × 3.5 m × 17.6 mm $H_2O$) + 60.1 mm $H_2O$/m) Note that 17. 6 mm $H_2O$/m is the resistance for 717 l/h in 27 mm pipe; 60.1 mm $H_2O$ is the special resistance for that. | | | **183 mm $H_2O$** |
| Resistance over the Mains = (2 × $W$ + 2 × ½ × $L$) × $R$ × 1.10 = (2 × 36 m + 2 × ½ × 75 m) × 10 mm $H_2O$/m × **1.10** (10% extra) | | | **1,617 mm $H_2O$** |
| Resistance over the Transport Mains outside the Greenhouse: Since it is right ne to the boiler room, no transport line! | | | **0** |
| $K_v$ is calculated using Equation (2.189) for $\Delta P$ =0.1 bar (1.45 psi) allowed; and 100 mm 3-way Mixing Valve with $K_v$ = 160 is chosen. Then, the Resistance over the 3-way Mixing Valve, $\Delta P = [(33\ m^3/h)/160]^2 \times 10{,}000$ mm $H_2O$ = 420 mm $H_2O$ < 700 mm $H_2O$ | | | **425 mm $H_2O$** |
| Other Resistances to be Included (boiler loop, etc., if there is any)? | | | 0 |
| **Total Resistance** | | | **2,258 mm $H_2O$** |
| **Pump Size** | | **~145 gpm @8 ft Head** | **53,025 l/h @2.5 m Head** |

Similarly, Zones 2 and 3 are designed and connected to the boiler room. For the same reasons, in overhead heating, 2 × 4 =8 lengths less (5 × 10 – 8 = 42 lengths in total) of 51 mm OD tubing and 2 + 2 × 2 = 6 lengths less (7 × 10 – 6 = 64 lengths in total) of 51 mm OD tubing are used in the Zones 2 and 3, respectively.

### *5.4.3.3 Carbon Dioxide Dosing Systems*

#### *5.4.3.3.1 Plant Responses to Carbon Dioxide Enrichment*

As discussed earlier in Chapter 4 in depth, research findings have shown that plant growth is improved by 15–20% in greenhouses when $CO_2$ concentrations are increased to 700–1,200 ppm during the day. Continuous $CO_2$ enrichment during the light period has been found to be superior to intermittent $CO_2$ application. The literature suggest that the optimal $CO_2$ concentration for plant growth is somewhere between 600 and 1,000 ppm for most species; and in general, the optimum $CO_2$ coupled with temperature and PAR intensity is about 400 ppm on cloudy days or vents are opened more than 10%, and 1,000 ppm on sunny days.

Furthermore, elevating the $CO_2$ concentration leads to higher internal $CO_2$ concentrations in plant tissues closing the stomatal openings hence reduces the transpiration of plants by 20–40%. Water consumption is therefore significantly reduced by $CO_2$ enrichment while photosynthesis is increased simultaneously. It is well known that increasing the air relative humidity, let's say, from 60% to 95% decreases plant transpiration by about 50% or so; and research has shown that $CO_2$ enrichment at 95% relative humidity decreases transpiration and water uptake by some plants further by about 30% increasing the water-use efficiency (amount of water used per unit dry weight produced) by about 30%.

Regardless, because of the overall benefits of $CO_2$ and as being one of the basic nutrients for plants, $CO_2$ dosing has been used on a large scale in greenhouse horticulture for many years to increase production. $CO_2$, being a natural component of the air, enters the plant through tiny openings in the leaves, known as stomata. In that respect, $CO_2$ is different from all other nutrients, which are all absorbed through the roots.

And again, as mentioned earlier, the effect of $CO_2$ on plant growth and production is somewhat different for different crops, but it is possible to generalize the relative effect of $CO_2$ on production as a $CO_2$ rule of thumb. The rule of thumb has been determined based on the experiments on vegetable crops using a base concentration of 350 ppm, but in principle they are valid for almost all crops. A rule of thumb is obviously not quite precise, but it is simple and generally applicable. This rule of thumb is expressed as follows:

$$X = \left(1{,}000/CO_2\right)^2 \times 1.5 \tag{5.10}$$

where $X$ is the percent production increase per 100 ppm $CO_2$ increase ($X$ is expressed as a % per 100 ppm). For instance, at 350 ppm increase (relative to the 350 ppm base concentration of $CO_2$) the effect of 100 ppm increase (up to 700 ppm) would be determined as follows:

$$\left(1{,}000/350\right)^2 \times 1.5 = 12\% \text{ per } 100\,\text{ppm}$$

At 550 ppm increase (it means 900 ppm in the greenhouse environment), the effect would be 5% and at 1,000 ppm, the increase would only be 1.5% per 100 ppm.

However, increasing the $CO_2$ concentration has its limits, and maintaining too high concentrations are not necessary. In summer time, when the light level is high on average, 1,000 ppm and a maximum of 1,200 ppm; and in winter time, when there is on average little light and no ventilation, 700 ppm and a maximum of 1,000 ppm are the recommended $CO_2$ concentrations. These levels are optimal levels from the crop perspective; however, economic side must also be considered when making decisions. In the summer months, a high $CO_2$ concentration in a greenhouse is desirable; however, it is not feasible due to high ventilation requirements. With a lot of light available and no $CO_2$ dosing, the concentration can quickly drop below the outside levels, and $CO_2$ then becomes the limiting factor and production is greatly reduced by the $CO_2$ deficiency. In such cases, it is always profitable to maintain the $CO_2$ concentration at least at the outside levels.

##### *5.4.3.3.2 Installation of $CO_2$ Dosing*

Greenhouse growers take a portion of the flue gases from their natural gas-fired boilers into the greenhouse to increase the $CO_2$ concentration in the greenhouse aerial environment. Such a $CO_2$ dosing installation consists of a PVC piping system that transport the flue gases from the boiler through the $CO_2$ pipes. $CO_2$ distribution lines, on the other hand, are plastic branches (layflats) with a diameter of 40 or 60 mm. Today, 90 mm diameter layflats are also available. These branches are provided with tiny holes through which the flue gases flow into the greenhouse. Because the flue gases do not flow the $CO_2$ into the greenhouse by itself, a blower is placed at the beginning of the piping system. The primary duty of the blower is to suck the desired amount of flue gases from the boiler vent stack and bring it into the $CO_2$ transport line with adequate pressure.

#### *Determining the Capacity of the $CO_2$ Blower*

For optimal $CO_2$ dosing, the blower must have sufficient capacity to bring the desired amount of flue gases through the piping system into the greenhouse. When choosing a blower, it is important to determine the correct quantity of dosing gas and what resistance this gas will undergo through the piping system. An adequate blower can be selected based on i) the amount of dosing gases and ii) the dosing value. The amount of dosing gases is determined by the desired dosing value and the temperature of the flue gases. And the dosing value, on the other hand, is the amount of natural gas that the grower wants to burn per hectare for the $CO_2$ dosing. The dosing value is strongly dependent on the type of crop and whether a heat buffer is present or not. For greenhouse vegetables, the dosing value without a heat buffer is 40–80 $m^3$ gas/(h.ha); and with a heat buffer, the dosing value is 80–120 $m^3$ gas/(h.ha). For floriculture, a dosing value of 40–60 $m^3$ gas/(h.ha) is used when no heat buffer is present; and if a heat buffer is present, then the dosing value is 40–80 $m^3$ gas/(h.ha). For potting plants, there is no specific guideline or recommendation. It should be kept in mind that, in order to provide $CO_2$ dosing, we can also operate burners at low speeds or burners can be operated with variable speed fans.

When there is a heat buffer, it is particularly important that the capacity of the installed $CO_2$ dosing system is accurate and aligned with the content of the heat buffer system. If it is not designed properly, the heat buffer will heat up either too quickly or too slowly. As a general rule, the shortest filling time used for the heat buffer is 6 hours; and based on the filling time and the buffer tank capacity, the desired dosing value can easily be calculated when a heat buffer is used.

#### *Temperature of Flue Gas*

A $CO_2$ dosing system is composed of PVC pipes; therefore, the temperature of the flue gases may not exceed 60°C. When there exists a flue gas condenser installed on the boiler, the heat recovered can be used for low grade heating (such as work spaces, walkways, and in-floor heating) and the flue gas temperature stays under 60°C; however, without a flue gas condenser, the flue gas temperature goes up to 180°C. The flue gases at these temperatures are definitely not suitable for $CO_2$ dosing. Cooling the flue gases down to 60°C is however still possible by using the outside air and then it will become suitable for $CO_2$ dosing. When flue gas temperature rises beyond 60°C, the amount of outside air to be mixed in obviously increases, and hence increasing the amount of air to be moved by the $CO_2$ blower. The quantity of the air to be moved can be calculated using the dosing value and the temperature of the flue gases.

#### *Resistance*

If the amount of dosing gases is known, then the resistance in the pipe system can be calculated. The head of the blower for the calculated quantity of dosing gases must be at least equal to the resistance in the piping system. The head of a blower depends on the temperature of the air to be moved. The higher the temperature, the lower the head is for a given volume of air. Since most blower manufacturers specify the delivery head at an air temperature of 20°C, the delivery head of the blower must be adjusted to 60°C.

### Blower Connection to the Boiler Vent Stack

The connection of the $CO_2$ blower to the boiler chimney is quite important when designing a $CO_2$ dosing system installation. The flue gases not only face resistance after the blower, but also before the blower itself. For an optimally functioning $CO_2$ dosing system installation, it is very important to keep the suction side resistance as small as possible, and this can be done in several ways: i) by placing the fan as close to the chimney as possible. The length of the suction side line (maximum 10 m) and the number of fittings is then kept at a minimum; ii) by preventing flow velocities in the suction pipe above 10 m/s; and iii) so-called guide plates can be placed in the boiler chimney. These partitions guide the flue gases from the chimney into the suction pipe of the $CO_2$ dosing system installation. It should be kept in mind that, typically, 100 mm $H_2O$ head loss is observed on the suction side for sucking the flue gas, and 70 mm $H_2O$ head loss is observed just through the layflat connection to the distribution piping; and this leaves the remaining blower head for the transport line and along the layflat. In order to have reasonable pressure drops, a maximum flue gas velocity of 10 m/s is used in the transport main and the distribution lines. The maximum velocity used in the layflats, on the other hand, is limited to 7.0 m/s.

Carbon monoxide (CO) concentration in the flue gas is continuously monitored as well; and if the concentration is 30 ppm or higher, it is not considered for $CO_2$ dosing. Typically, however, when the CO concentration at the $CO_2$ blower reaches 25 ppm, then the $CO_2$ dosing system automatically shuts off.

### $CO_2$ Piping System

The $CO_2$ piping system consists of three components: i) the transport pipes; ii) the distribution pipes; and iii) the branches. The transport pipe carries the $CO_2$ dosing gases from the blower to the distribution pipe. In order to keep the resistance in a transport line within acceptable values, the number of bends in the transport line should be kept at a minimum level. The diameter of a transport line PVC piping can be 125, 160, 200, 250, 315, or 400 mm (see Table A.28), and are available with different wall thicknesses. The wall thickness is indicated by a class designation, such as Class 65, 51, and 41. The lower the class is, the thicker the wall thickness. At least Class 51 must be used for $CO_2$ dosing system. PVC pipes of Class 65 are considered too thin for $CO_2$ dosing systems.

The distribution pipes, on the other hand, distribute the dosing gases evenly around the greenhouse, and in order to keep the resistance in the manifold within acceptable levels, a maximum dosing gas flow velocity of 10 m/s is used. Because the amount of dosing gases in the manifold decreases further along the line, smaller diameter pipes can be used as the amount of gas gets smaller. PVC pipes with a diameter of 90, 110, 125, 160, 200, 250, and 315 mm are used for the $CO_2$ manifold design, which will be shown in a design example shortly in this section. In terms of the PVC class, the same applies to distribution pipes as to transport pipes.

The branch forms the connection between the distribution pipe and the $CO_2$ dosing layflats. A branch consists of at least a layflat holder with a throttling plate with which the layflat holder is screwed into the distribution pipes. Possible options for branching are shown in Figure 5.34; however, exceptions to these branch connections are possible. Depending on the amount of dosing gases flowing through a branch, a 32 mm or a 40 mm branch connection can be used. The distance between the branches on the distribution pipe is usually 1.60 or 3.20 m; however, this distance can be modified depending on the type of cultivation system being used in a greenhouse.

The layflats connected to the distribution pipe are provided with holes through which the dosing gases are released into the greenhouse environment. The number of holes in the

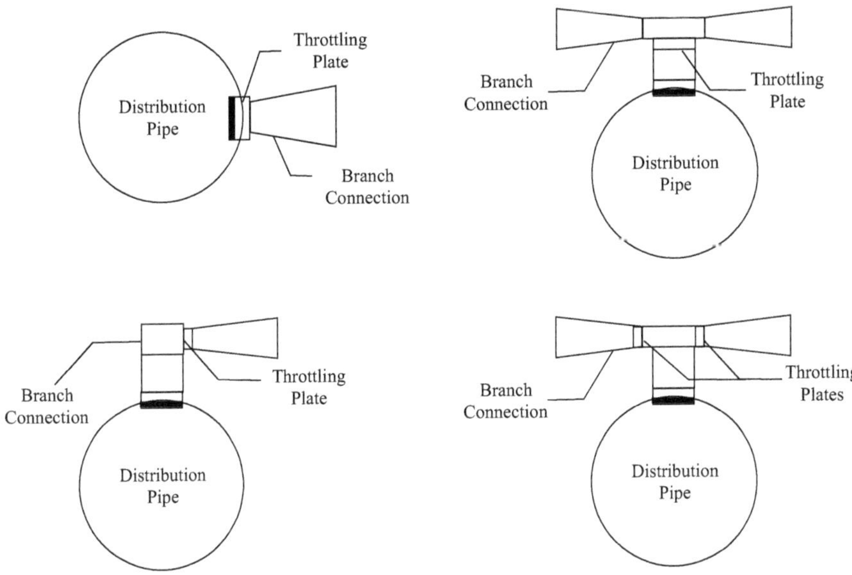


**FIGURE 5.34**
Possible options for branch connections.

$CO_2$ layflats determines the resistance that the dosing gases face in the layflats. Due to this resistance, a pressure builds up in the layflats, which is called the layflat pressure. The more holes there are on a layflat, the more dosing gases a layflat can release into the greenhouse at the same layflat pressure. It means, the greater the dosing value and/or the flue gas temperature, the greater the quantity of dosing gases, and hence the more holes the layflats must have.

The number of holes on the $CO_2$ layflat is determined by the prick (puncture) distance, which is the distance between the holes along the length of layflat. By default, 4 holes with a diameter of 0.8 mm are pricked per prick distance. Instead of $4 \times 0.8$ mm holes, we can also use $1 \times 2$ mm hole per prick distance. For an optimal $CO_2$ distribution over a layflat, the design of a $CO_2$ dosing installation is based on a layflat pressure of 70 mm $H_2O$ column. At this pressure, the $CO_2$ layflats are not squeezed by stems and/or fruits easily, and the condensate can easily get out of the layflats. Using Equations (5.14) and (5.15) provided at the end of this chapter, the prick distance on the $CO_2$ layflats can be determined for a layflat pressure of 70 mm $H_2O$ column.

### *Pressure Drop across the Layflat*

Pressure loss also takes place in a $CO_2$ layflat. As the amount of dosing gases through a layflat increases, the resistance increases as well. As a result, the pressure drop at the beginning of a layflat is larger than at the end of a layflat. When the pressure drop across a layflat exceeds 20% of the pressure at the beginning of the connection, the difference in $CO_2$ release between the beginning and the end of the layflat is not acceptable. The pressure drop can then be reduced by reducing the layflat distance or by using a larger diameter layflat. In practice, excess pressure differences across the layflat can be prevented by adhering to the following guideline values. With a branch distance of 3.20 m (c.t.c.) and a layflat diameter of 41 and 60 mm, the maximum layflat length is 40 and 75 m, respectively. With a branch distance of 1.60 m (c.t.c.) and a layflat diameter of 41 and 60 mm, the maximum layflat length is 60 and 120 m, respectively.

*Throttling Plates (Discs)*

The distribution of the dosing gases over the greenhouse is optimal if the pressure in all $CO_2$ lines is the same. However, the same line pressure is only possible if the dosing gases in the CO piping system experience the same resistance, which is in practice not possible. The dosing gases that flow into the $CO_2$ layflats at the end of the distribution pipe obviously travel a much longer distance than the dosing gases that enter the $CO_2$ layflat at the beginning of the distribution pipe. By using throttling plates, we can build in resistance and throttle the pressure to the desired layflat pressure; and as a consequence, an equal layflat pressure can be achieved everywhere. This throttling is done with so-called throttles or throttling plates. Throttling plates are round plates with a hole of a certain diameter. The ratio between the diameter of the hole in the throttle plate and the diameter of the branch determine the degree to which the pressure is throttled. The smaller this ratio is, the more pressure is throttled away. The available diameters of the holes in the throttle plates are for a 32 mm branch 11 to 18, 20, 22, and 24 mm. For a 40 mm branch, these are 20, 22, 24, 26, 28, 30, and 32 mm. Designing the correct throttle plates is not easy. Incorrectly designed throttle plates can have negative consequences on the dosing, and they should never be glued.

*Condensate Drainage*

One of the biggest problems in $CO_2$ dosing installations is flooding. The dosing gases in the $CO_2$ pipes have a higher temperature than the greenhouse air, which cools the dosing gases. Since the relative humidity of the dosing gases is around 100%, water condenses from the flue gases due to cooling. If this condensate water is not properly discharged, it will remain in the $CO_2$ pipes and result in poor $CO_2$ distribution. Flooding can be prevented by installing condensate discharge points in the pipes, laying the pipes with a slope and preventing the pipes from subsiding. Condensation water can be discharged from the $CO_2$ pipes in two ways: through holes drilled in the bottom of the pipes or through a "gooseneck" connection (see Figure 5.35). The first solution is quite simple and cheap.

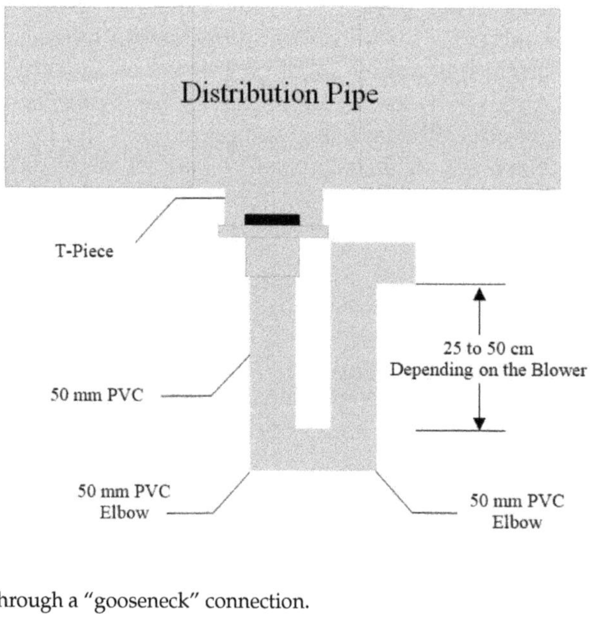


**FIGURE 5.35**
Condensate discharge through a "gooseneck" connection.

However, the disadvantage of this method is that significant leakages can occur across the holes. That is why, this condensation discharge method is not recommended. The second solution is discharging the condensate water via a gooseneck connection, which is a more expensive but a more secure way of condensate water removing. Since water remains in a gooseneck, no dosing gases can escape through the gooseneck. It is of course important that the gooseneck has a sufficient length. Otherwise, the blower will blow the water out of the gooseneck, causing dosing gases to escape. The gooseneck must be properly connected to the $CO_2$ line. This is usually done with a pointed piece, as a result of which a layer of water will always remain in the pipe. This can be prevented by connecting the gooseneck to the $CO_2$ pipe with the help of an insert inlet or a T-piece. With pipe diameters of 200 mm and larger, an insert inlet is used, for smaller diameters, however, a T-piece is used.

For optimal condensate discharge, it is necessary to install sufficient discharge points in the $CO_2$ piping system. It is therefore recommended to place a condensate discharge point at least at the end of a transport pipe and in a distribution pipe every 25 m. The condensate water from the discharge point can be discharged into the drainage via a gravel bed. It is also possible to drain the collected condensate water with a pipe to the recirculation system or the ditch. A minimum slope of 1 mm per meter of pipe length (0.1%) can be used safely. It is important that the slope is always in the flow direction of the dosing gases. Subsidence may occur in underground $CO_2$ pipes if they are not properly buried. As a result of the subsidence, water remains in the pipe, which will adversely affect the $CO_2$ distribution. Therefore, it is important to install the underground $CO_2$ pipes on a solid and level surface.

Designing a proper $CO_2$ piping system, which requires a thorough knowledge of fluid dynamics, is not an easy task. The $CO_2$ dosing system installation must therefore be carried out by an experienced engineer who has proper experience in fluid dynamics.

*Calculating the Dosing Value and Dosing Gases for a Heat Buffer*

The dosing value is the quantity of natural gas to be burned per hour in the floor surface area (ha) of the greenhouse.

$$D_s = \left[ V_{wb} \times C_w \times r_w \times (T_{max} - T_{min})_b \right] / \left( A_{gh} \times \Delta t_{fill} \times H_b \times \eta_{boiler} \right) \tag{5.11}$$

where $D_s$ is the dosing value in m$^3$ gas/h.ha; $V_{wb}$ is the water content of the heat buffer in m$^3$, $C_w$ is the specific heat of water, 4.2 kJ/kg.K; $\rho_w$ is the density of water in kg/m$^3$; $T_{max}$ is the maximum water temperature in buffer in °C; $T_{min}$ is the minimum water temperature in °C; $A_{gh}$ is the greenhouse floor surface area in ha; $\Delta t_{fill}$ is the minimum filling time of the heat buffer full load hours boiler (hour), $H_b$ is the upper calorific value of natural gas, 35,170 kJ/m$^3$ (varies regionally; therefore, local value must be used); and $\eta_{boiler}$ is the efficiency of the boiler in fraction.

The quantity of dosing gases $V_{dgas}$ can be calculated using the dosing value $D_s$ and the volume of combustion gases per m$^3$ of natural gas to be burned $V_{cg}$ as follows:

$$V_{dgas} \left( \text{m}^3/\text{h.ha} \right) = D_s \times V_{cg} \tag{5.12}$$

$$V_{cg} \left( \text{m}^3/\text{m}^3 \text{natural gas} \right) = \left[ (8.3871 \times n - 0.7171) \times T_{cg} \right] / 273 \tag{5.13}$$

where $n$ is the excess air in the combustion process. For $n = 1.17$ and the combustion gas temperature of $T_{cg} = 27°C$, $V_{cg} = 10$ m$^3$/m$^3$ natural gas.

*Prick Distance of $CO_2$ Layflat*

The prick distance depends on the amount of dosing gases and the layflat (branch) distance.

At a branch distance of 3.20 m (c.t.c.):

$$\text{Space}_{3.2} = 450/V_{dgas} \text{ with a maximum of } 1.2\,\text{m and a minimum of } 0.2\,\text{m} \tag{5.14}$$

At a branch distance of 1.6 m (c.t.c.):

$$\text{Space}_{1.6} = 900/V_{dgas} \text{ with a maximum of } 1.2\,\text{m and a minimum of } 0.2\,\text{m} \tag{5.15}$$

**Example 5.9: Calculating Prick Distance**

For the given case below, determine the prick distance:
The greenhouse floor surface area, $A_{gh}$ = 15,000 m = 1.5 ha
The buffer capacity, $V_{wb}$ = 150 m$^3$
The maximum water temperature in the buffer, $T_{max}$ = 95°C
The minimum water temperature in the buffer, $T_{min}$ =45°C
The minimum filling time of the heat buffer, $\Delta t_{fill}$ = 6 hours
The boiler efficiency, $\eta_{boiler}$ = 85%
The excess air in the combustion process, $n$ = 1.17
The branch distance = 1.6 m

**Solution**

Using Equation (5.11), we can first calculate the dosing value $D_s$ as follows:

$$D_s = \left[V_{wb} \times C_w \times \rho_w \times \left(T_{\max} - T_{\min}\right)_b\right] / \left(A_{gh} \times \Delta t_{fill} \times H_b \times \eta_{boiler}\right)$$

$$D_s = \left[150 \times 4.2 \times 1{,}000 \times (95 - 45)\right] / (1.5 \times 6 \times 35{,}170 \times 0.85)$$
$$= 117\ \text{m}^3\text{gas/h.ha} \approx 120\ \text{m}^3\text{gas/h.ha}$$

And then, using Equations (5.12) and (5.13), we can calculate the amount of dosing gases as follows:

$$V_{dgas}\left(\text{m}^3/\text{h.ha}\right) = D_s \times V_{cg}$$

$$V_{dgas} = 120 \times (8.3871 \times 1.17 - 0.7171) \times 300/273 = 1{,}200\ \text{m}^3/\text{h.ha}$$

And for 1.5 ha given greenhouse space, the required amount of dosing gases would be:

$$V_{dgas} = 1.5\ \text{ha} \times 1{,}200\ \text{m}^3/\text{h.ha} = 1{,}800\ \text{m}^3/\text{h}$$

And then finally, using Equation (5.15), the prick distance can be calculated as follows:

$$\text{Space}_{1.6} = 900/V_{dgas}$$

$$\text{Space}_{1.6} = 900/1{,}800 = 0.5\ \text{m} = 50\ \text{cm}$$

**Example 5.10: Designing a Carbon Dioxide Dosing System**

In this example, we will design a $CO_2$ dosing system for a vegetable greenhouse with the following dimensions in Canada.

| GREENHOUSE DIMENSIONS | | |
|---|---|---|
| **COMPONENT** | **(ft)** | **(m)** |
| House Width | 24 | 7.2 |
| House Length | 252 | 75.6 |
| Width of Center Walkway | 12 | 3.6 |
| Number of Bays | 17 | 17 |

For the given greenhouse:
The greenhouse floor surface area = 9,253 m = 0.925 ha
The $CO_2$ dosing standard in Canada (for vegetables) = 100 m$^3$ natural gas/h.ha
The dosing capacity = 0.925 ha × 100 m$^3$ natural gas/h.ha = 92.5 m$^3$/h
The flue gas temperature = 60°C (333 K)
The branch distance = 1.6 m
The distribution lines will be placed on the ground along the gable end walls.

**Solution**

Let's first determine the conversion factor for converting the amount of flue gas to the amount at 60°C:

$$\left(1\,\text{m}^3\text{gas} + 10\,\text{m}^3\text{air}\right) \text{x}\, 1.15\, \text{x} \left(273 + 60°\text{C}\right) \text{K} / 273\,\text{K} \approx 15$$

Then, the amount of flue gases at 60°C:

$$\left(92.5\,\text{m}^3/\text{h}\right) \text{x} \left[15\right] = 1,388 \approx 1,400\,\text{m}^3/\text{h}\left(@60°\text{C}\right)$$

And then, we will choose a $CO_2$ blower having a capacity of:

$$1,400\,\text{m}^3/\text{h}\left(\text{with let's say } 3,500\,\text{Pa} = 350\,\text{mm H}_2\text{O head}\right)$$

See the greenhouse layout below for the $CO_2$ dosing line lengths. Then, the maximum line length is:

$$= 5 + 45 + 36 + 36 + 50 = 172\,\text{m}$$

Remember that the blower sucking side pressure drop is 100 mm $H_2O$, the pressure drop at the layflat connection is 70 mm $H_2O$; then the blower head left for the rest of the transport and distribution lines is:

$$350\,\text{mm H}_2\text{O} - 100\,\text{mm H}_2\text{O} - 70\,\text{mm H}_2\text{O} = 180\,\text{mm H}_2\text{O}$$

So, we have to design the $CO_2$ dosing line based on this remaining 180 mm $H_2O$ head.

Then the allowable pressure drop per meter line length can be calculated as follows:

$$180\ \text{mm}\ H_2O/172\ \text{m} = 1.047\ \text{mm}\ H_2O/\text{m line}$$

However, typically 50% of the all resistance (pressure drop) is accounted for the special resistance; therefore. The actual allowable resistance for the pipe line is:

$$\left(1.047\ \text{mm}\ H_2O/\text{m}\right)/2 = 0.52\ \text{mm}\ H_2O/\text{m line}$$

Remember that the branch distance is given as 1.6 m; then the total number of layflats = 2 × 122 m/1.6 m = 152 (each side of the center walkway has 76 layflats). Then,

Flue Gas Flow Rate per Layflat, $\Phi$ = Amount of Flue Gas/Total Number of Layflats

Flue Gas Flow Rate per Layflat, $\Phi$ = (1,400 $m^3$/h)/152 = 9.2 $m^3$/h

Using the pressure drop per meter PVC line (0.52 mm $H_2O$/m line) for each flow rate, we can now go to Table A.28 and design the PVC piping sizes as follows:

| GABLE WALL SECTION (m) | | $CO_2$ Flow Rate (l/h) | PVC OD (mm) | GABLE WALL SECTION (m) | | $CO_2$ Flow Rate (l/h) | PVC OD (mm) |
|---|---|---|---|---|---|---|---|
| 0.00 | 3.20 | 18.4 | **75** | 60.80 | 64.00 | 368.0 | **160** |
| 3.20 | 6.40 | 36.8 | **75** | 64.00 | 67.20 | 386.4 | **160** |
| 6.40 | 9.60 | 55.2 | **75** | 67.20 | 70.40 | 404.8 | **160** |
| 9.60 | 12.80 | 73.6 | **90** | 70.40 | 73.60 | 423.2 | **160** |
| 12.80 | 16.00 | 92.0 | **90** | 73.60 | 76.80 | 441.6 | **160** |
| 16.00 | 19.20 | 110.4 | **90** | 76.80 | 80.00 | 460.0 | **160** |
| 19.20 | 22.40 | 128.8 | **110** | 80.00 | 83.20 | 478.4 | **160** |
| 22.40 | 25.60 | 147.2 | **110** | 83.20 | 86.40 | 496.8 | **160** |
| 25.60 | 28.80 | 165.6 | **110** | 86.40 | 89.60 | 515.2 | **160** |
| 28.80 | 32.00 | 184.0 | **110** | 89.60 | 92.80 | 533.6 | **160** |
| 32.00 | 35.20 | 202.4 | **110** | 92.80 | 96.00 | 552.0 | **160** |
| 35.20 | 38.40 | 220.8 | **125** | 96.00 | 99.20 | 570.4 | **160** |
| 38.40 | 41.60 | 239.2 | **125** | 99.20 | 102.40 | 588.8 | **200** |
| 41.60 | 44.80 | 257.6 | **125** | 102.40 | 105.60 | 607.2 | **200** |
| 44.80 | 48.00 | 276.0 | **125** | 105.60 | 108.80 | 625.6 | **200** |
| 48.00 | 51.20 | 294.4 | **125** | 108.80 | 112.00 | 644.0 | **200** |
| 51.20 | 54.40 | 312.8 | **160** | 112.00 | 115.20 | 662.4 | **200** |
| 54.40 | 57.60 | 331.2 | **160** | 115.20 | 118.40 | 680.8 | **200** |
| 57.60 | 60.80 | 349.6 | **160** | 118.40 | 121.60 | 699.2 | **200** |
| **Transport Main: 1,400 l/h → 250 mm OD PVC** | | | | | | | |

And the velocities in each PVC pipe have been checked and they are all less than 10 m/s. Note that when condensation in the pipes is considered, in reality, we may not want to use any pipe sizes less than 110 mm OD; but we will keep the piping design as is. The overall design layout then looks as follows:

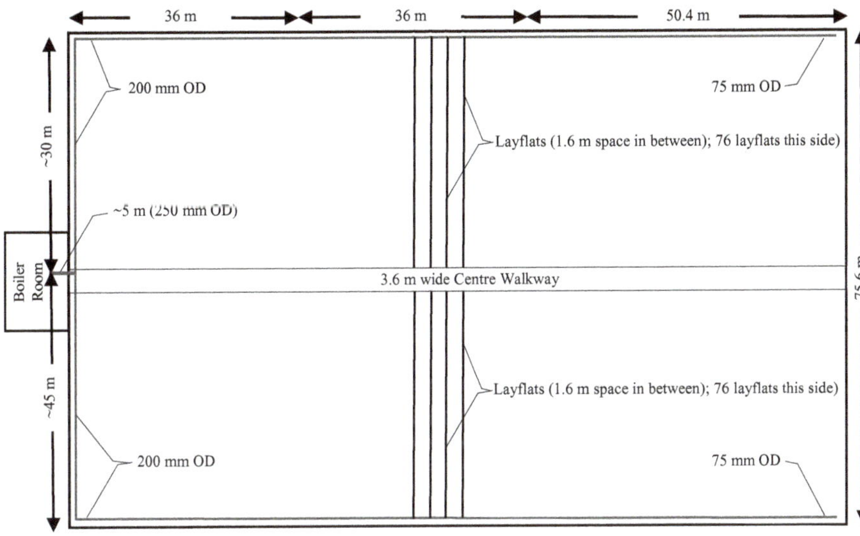


Before we conclude our design, let's determine the prick distance as well. We can determine the prick distance using Equation (5.15) as follows:

$$\text{Space}_{1.6} = 900 / V_{dgas} = 900 / 1,400 = 0.64 \text{ m} = 64 \text{ cm} = 640 \text{ mm}$$

However, it should be noted that Equation (5.15) was developed for a dosing standard of 60 $m^3$/h. And we have used the Canadian standard of 100 $m^3$/h in this example; therefore, in order to eliminate the effect of 60 $m^3$/h, we can use the following relationship to determine the prick space for a dosing norm other than 60 $m^3$/h:

$$\text{Space}_{1.6}(\text{mm}) = 96,000/[\text{Branch Distance}(\text{m}) \times \text{Dosing Norm}(\text{m}^3/\text{h})]$$

$$\text{Space}_{1.6}(\text{mm}) = 96,000/(1.6 \text{ m} \times 100 \text{ m}^3/\text{h}) = 600 \text{ mm}$$

This means that we need to use 4 × 0.8 mm holes or 1 × 2 mm hole every 600 mm along the layflat. And if we use the 60 mm OD layflats (standard for vegetables), then the maximum layflat length cannot exceed 120 m and our layflat length is less than 120 m.

#### *5.4.3.4 Heat Storage Tanks*

When heat is produced, but not consumed at the time of its production, the heat is stored in a tank. The stored heat is later used again when there is a heat demand in the next cycle. When the hot water is pumped to the top of the storage tank, a hot and a cold section are created at the top and the bottom of the tank, respectively. These sections are normally separated by a thin boundary layer in between, and this layer gets thicker when the hot water is stored longer. The heat storage tank is unloaded based on this line. And obviously, a heat loss occurs during the heat storage, depending on the method of filling, the storage time, and the insulation of the storage system. After a period of heat surplus, typically a period of heat demand follows; and heat is stored exactly in between these two periods that follow each other within 24 hours. Heat storage tanks can be either horizontal (maximum temperature is 100°C) or vertical (maximum temperature is less than 93°C) and are used in greenhouses in the following situations; and when they are used, they become the heart of whole system.

1. **$CO_2$ Dosing**
   As discussed earlier, $CO_2$ is an essential ingredient for crop growth and development. In summer months, crops can take up a lot of $CO_2$, and therefore, for a proper crop production, at least the outside $CO_2$ level has to be maintained in a greenhouse environment. In spring months, however, higher $CO_2$ levels can be maintained, and extra crop production can be achieved as a result. In a mass crop production facility, $CO_2$ can be generated in a cost effective way by simply burning natural gas. The heat generated during $CO_2$ dosing is, however, quite often not used entirely in the greenhouse in the spring and summer months. For instance, there is a heat surplus during the time greenhouses are being ventilated to get rid of the excess heat accumulated inside. The heat in the storage tank is easily unloaded at night in the summer months; however, even though it is hoped for, it is a challenge to empty the tank during the days in summer. A heat storage (buffer) system for $CO_2$ dosing only is an old and proven system (see Figure 5.36 for heat storage cycle (loading/filling cycle) and Figure 5.37 for heat recovery cycle (unloading cycle)), for which a design example will be executed shortly at the end of this section. In these systems, the tank capacity is a function of $CO_2$ dosing norm. If the dosing norm is 100 $m^3$ natural gas/h.ha, then the tank capacity is 100 $m^3$/ha; and if the norm is 125 $m^3$ natural gas/h.ha, then the tank capacity is 125 $m^3$/ha. The dosing time has to be 10 hours.

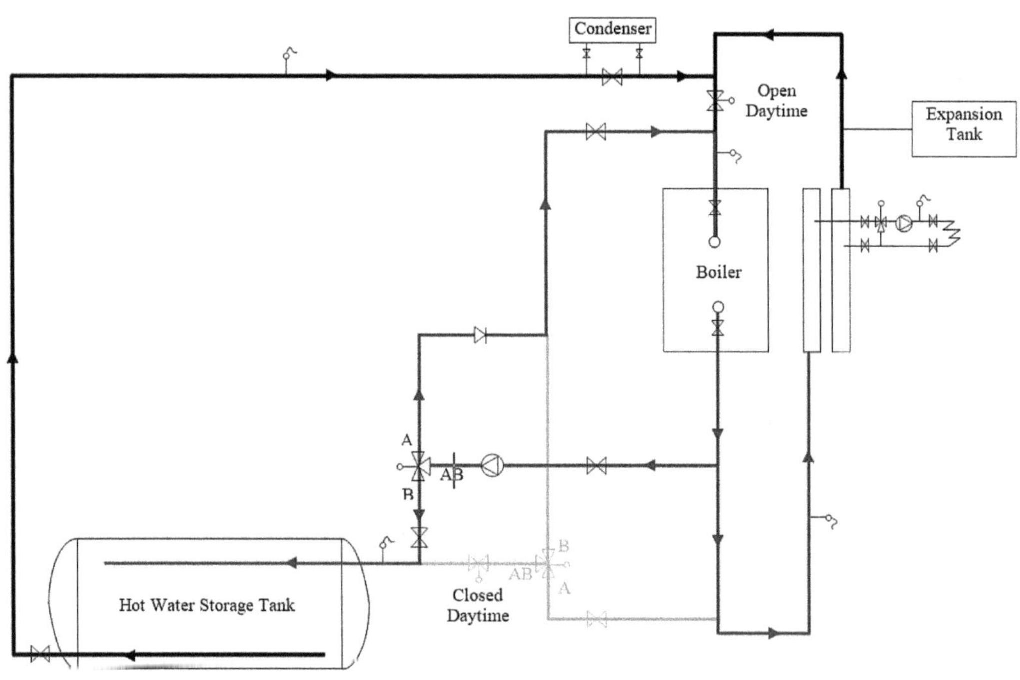


**FIGURE 5.36**
Hot water storage system with a horizontal tank (loading cycle).

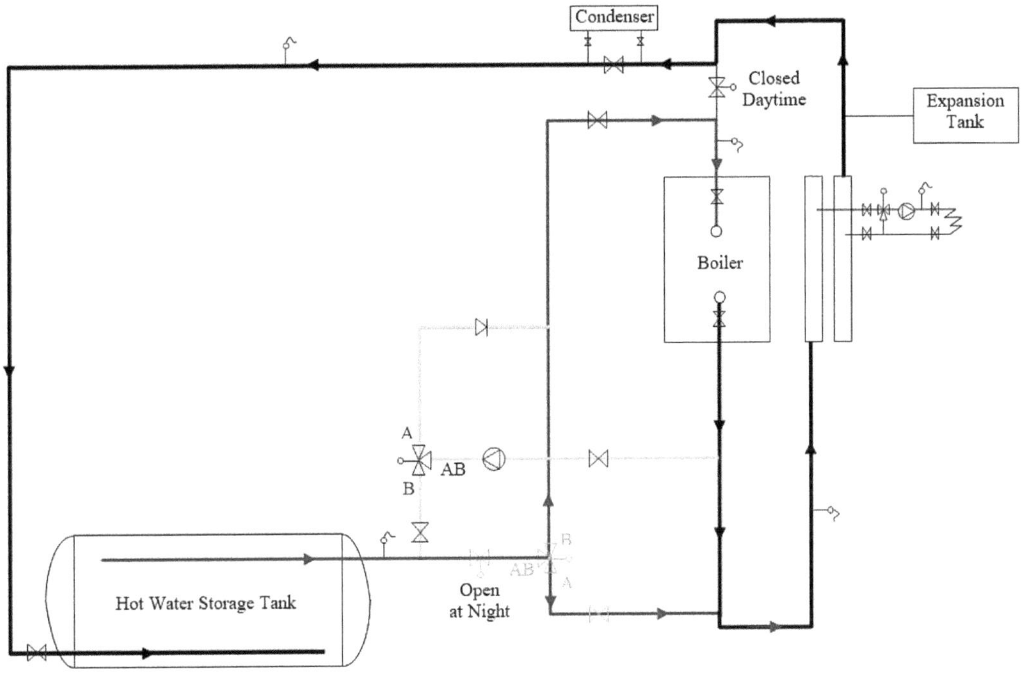


**FIGURE 5.37**
Hot water storage system with a horizontal tank (unloading cycle).

During the loading/filling cycle, the primary goal of the system is to perform $CO_2$ dosing in the greenhouse and store the hot water generated in the storage tank. This system must, at the very same time of $CO_2$ dosing, be able to respond to the heating needs of the greenhouse whether for primarily heating or humidity control. As shown in Figure 5.36, in the loading mode, hot water coming from the boiler is directed into the top of the storage tank while colder water is drawn from the bottom of the storage tank by the load/fill pump through a control valve. When the $CO_2$ dosing is finished, that is, the loading cycle is over, then the hot water stored in the storage tank be unloaded as shown in Figure 5.37.

During full unloading or recovery, the hot water storage tank essentially replaces the boiler and becomes the heart of the heating system. The hot water is drawn from the top of the tank; and if the unloading temperature is not hot enough, then the unloading valve is modulated to draw additional hot water from the boiler (see Figure 5.37). During unloading, the load/fill pump must be off, and the load/fill valve is fully closed in order to prevent hot water from being drawn into the return water destined for the bottom of the storage tank. Needless to remind, high return water temperatures reduce the hot water storage system capacity.

2. **Open Buffer Systems and Combined Heat and Power Generation**
Peak shaving is the prime reason for the development of open buffer systems, and we must have a high number of operation hours for such systems. In these systems, the boiler is operated during the off-peak hours when there is not much or no heat demand at all in the greenhouse and the generated heat is stored in heat storage tanks for later use; and in this way, the overall natural gas consumption at the greenhouse operation is reduced. Hourly natural gas usages with an open buffer system has a

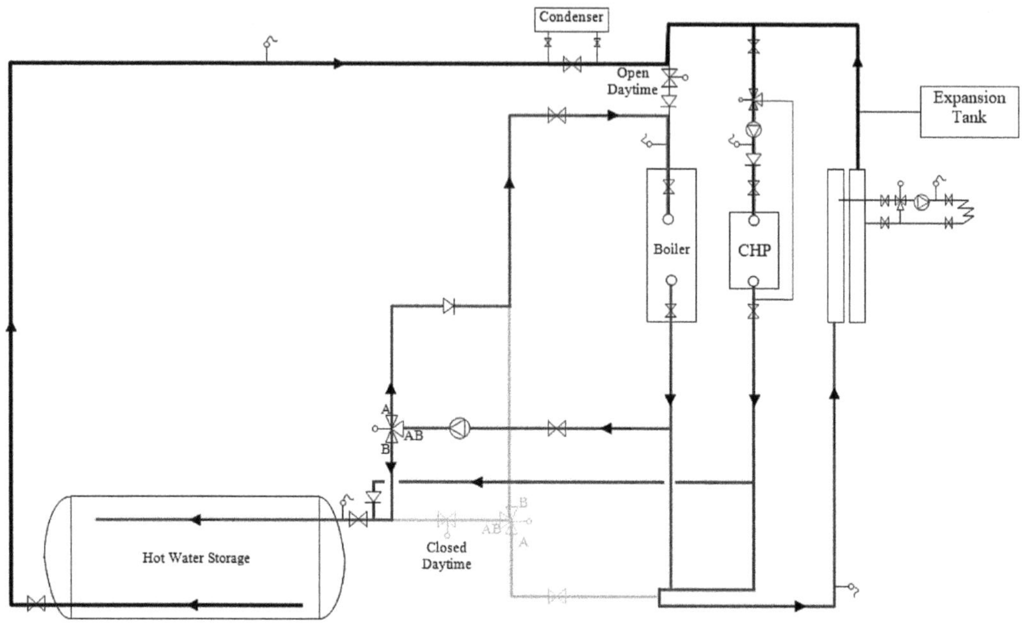


**FIGURE 5.38**
Hot water storage system with CHP in parallel connection (loading cycle).

totally different regime from those of a system without an open buffer system; therefore, it saves considerable amount of natural gas and help conserve our resources and make our greenhouse operations more sustainable. A design example will be executed for an open buffer system as well at the end of this section.

In an effort to conserve energy and make greenhouse production facilities more sustainable, greenhouse operations are using more and more CHP generation systems every passing year. These "localized" systems are more environmentally friendly, more flexible, more efficient, and more cost-effective than traditional centralized systems, especially when power transmission savings are taken into account. Conventional generation of electricity in large central power stations is normally only 30–40% energy efficient. More recent CHP cycle generation can improve this to 55%, excluding the transmission loses of electricity, which can be as great as 20% of the delivered efficiency. Through the utilization of the heat, the efficiency of cogeneration unit can reach up to 90% or more (see Figure 5.38 for loading and Figure 5.39 for unloading). Cogeneration of heat and power in greenhouse operations not only utilizes the heat and power generated on site, but also, essentially eliminates the $CO_2$ emissions to the environment. In essence, these systems serve as multigeneration systems in greenhouse operations, generating heat, power, $CO_2$, and water, all precious resources. CHP systems can be connected to the system either in series or parallel connection.

#### *5.4.3.4.1 Tank Size, Heat Surplus, and Demand*

A hot water storage tank must be used as efficiently as possible. In order to avoid unnecessary investments, the size of the tank will have to be sized and optimized in advance. Special attention must also be given to both filling and emptying the tank.

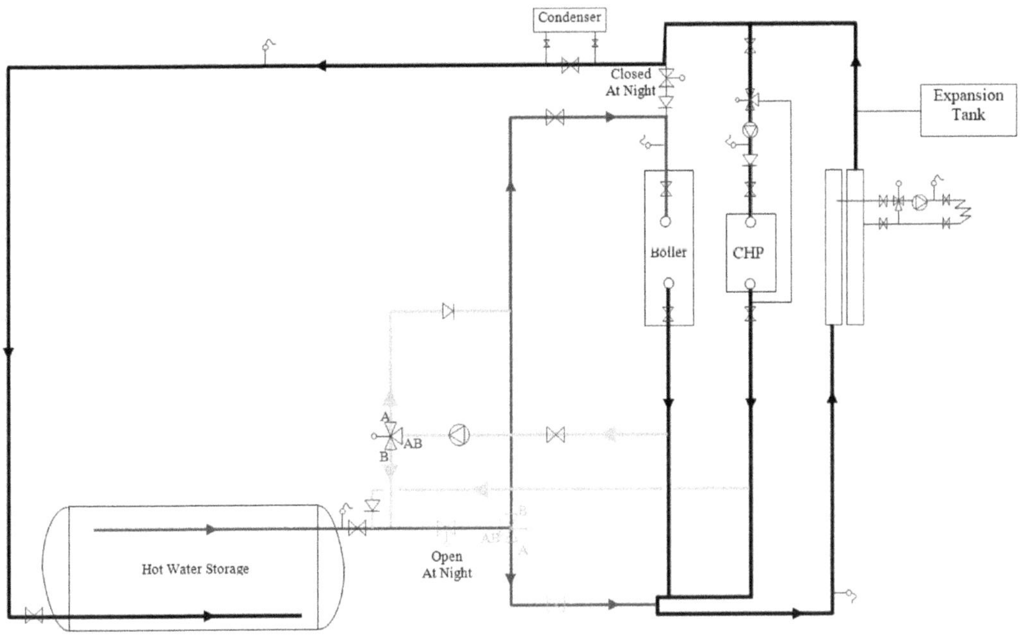


**FIGURE 5.39**
Hot water storage system with CHP in parallel connection (unloading cycle).

How long the burner only runs for $CO_2$ dosing can often be taken from the computer recordings in a greenhouse operation. The number of hours, in combination with the burner setting for $CO_2$ dosing (often the minimum setting), gives us the excess heat produced. The determination of heat surplus will be numerically shown in an example later in this section. Similarly, the exhausted heat (too large) from a CHP installation can also be calculated.

The stored heat has to be depleted in the following period with heat demand, usually at night. It is highly desirable to start the next day with an empty heat tank, especially when $CO_2$ dosing is practiced. In short, the heat demand during the night sets a limit for heat storage; that is, no further heat can be stored than what can be used in the short term. In spring months, the nights are still cold enough to be able to use much of the stored heat; hence, it is not a good practice to size the heat storage tank based solely on the heat levels in the summer nights.

### *5.4.3.4.2 Setting Up Heat Storage Tanks*

The heat storage efficiency decreases in proportion to the increase in the surface of the boundary layer. That is why, the vertical arrangement of storage tanks is particularly important for large storage tanks (see Figure 5.40). Furthermore, with a vertical arrangement, the load per $m^2$ of ground surface increases, which obviously increases the cost of building a tank foundation. This can be prevented by placing several tanks in series; however, the consequence of having more tanks is to have a larger exposed surface and hence more heat losses. A horizontal tank that can be quickly unloaded is therefore simply the best option where interfacial mixing can hardly occur (see Figure 5.41). With the use of more than just two supports, the load of the tank on the ground can no longer be an issue; hence, no special foundation design would be required. In order to reduce the load on the

**FIGURE 5.40**
A vertical hot water storage tank used in a greenhouse facility.

**FIGURE 5.41**
A horizontal hot water storage tank used in a greenhouse facility.

ground, it is recommended to use more supports, for instance every 2.5 m or so. It should also be kept in mind that considerable heat loss occurs through the supports.

In order to make optimum use of the heat storage tank, it is necessary to select the pipe diameters large enough. The stored heat must cover the entire heat demand of the greenhouse as much as possible. This means that in order to empty the heat tank, a pipe diameter that is at least the same size as the supply from the boiler is required. Both pipes to and from the heat storage tank can be the same size. As long as the pipe size is suitable for emptying, filling is never a problem. All the methods used in practice to connect the boiler and heat storage tank to the manifold are based on two methods: i) in series connection with the boiler; or ii) in parallel connection with the boiler.

Occasionally, hot water storage tanks are damaged due to high pressure, which is caused by the improper use of valves or pipes that are too small. Using a safety valve is required to prevent major damages to the tank; and in order to prevent corrosion, an air vent must also be installed on the tank.

#### *5.4.3.4.3 Practical Considerations*

We should always choose a tank, which has more capacity than the calculated capacity for summer. The larger tank will not be more expensive than the one for summer storage; and it will provide more room for $CO_2$ dosing, especially in the spring. And remember that the tank does not have to be completely full in the summer. Other useful considerations are provided below.

1. **Distance to Boiler Room**
   In practice, emptying the tank can often become a problem. Sometimes, even though there is a lot of heat demand in the greenhouse, the heat storage tank cannot be used effectively because the hot water from the storage tank is not released quickly enough. This is because of the chosen pipe diameters that are quite often too small with respect to their lengths. In the event of an immediate heat demand, sufficient heat can be supplied from the tank because the tank is full of heat, and the boiler is prevented from starting. The rule of thumb stating that the pipes to the heat storage tank should be the same size as the pipes to the boiler is only applicable when the heat storage tank is close to the manifold. The heat storage tank is, however, typically further away from the manifold; therefore, larger diameters (than those used for the boiler) are often recommended. Also, due to the potential energy loses, the heat storage tank is placed as close as possible to the manifold.
2. **Insulating Supports**
   The supports of the hot water storage tank can become quite hot. However, insulating these supports is very easy and reduces heat losses during the storage. Also, rather than insulating these supports, rubber pieces can be placed between the tank and supports.

#### *5.4.3.4.4 Heat Storage Tank Design for $CO_2$ Dosing*

Now let's design a hot water heat storage system shown in Figure 5.36 and 5.37.

**Example 5.11: Designing a Horizontal Hot Water Storage Tank System for $CO_2$ dosing**

Design a hot water heat storage system for a 15,000 m² greenhouse where the $CO_2$ dosing norm is 100 m³ gas/h.ha, and the calorific (heating) value of the locally available natural gas is 9,568 kcal/m³. The boiler capacity is given as 4,000,000 kcal/h, and the combined boiler and condenser efficiency as 90%.

**Solution**

Heat Storage Tank Capacity (in practice) = 1.5 ha × 100 m³ gas/h.ha = 150,000 liter = 150 m³

Burner Capacity = 1.5 ha × 100 m³ gas/h.ha = 150 100 m³ gas/h

Heat Supply during $CO_2$ Dosing = 150 × 100 m³ gas/h × 9,568 kcal/m³ × 0.9 = 1,291,680 kcal/h

$CO_2$ Dosing Time = 10 hours

Total Heat Supply during $CO_2$ Dosing = 10 h × 1,291,680 kcal/h = 12,916,800 kcal

For a horizontal heat storage tank, we can have a maximum water temperature of 100°C in the tank, while the minimum temperature is 40°C. Then, the water temperature difference over the storage tank would be:

$$\Delta T_{water\ over\ the\ tank} = 100°\text{C} - 40°\text{C} = 60°\text{C}$$

Then, in order to store all the heat generated, using Equation (2.24), the theoretical capacity of the hot water storage tank can be calculated as follows:

$$12{,}916{,}800\ \text{kcal}/60°\text{C} = 215{,}280\ \text{litre}$$

As it is seen here, the theoretical tank capacity (215,280 l) is greater than the capacity (150,000 l) determined based on the practical experience. We will use the capacity (150,000 l) based on practical experience because in reality the daily $CO_2$ dosing duration is not always 10 hours and sometimes there is an additional heat demand in the greenhouse.

And,

$$\text{Maximum Heat Supply}\left(\text{kcal/h}\right)\text{during } CO_2 \text{ Dosing} = \left(1{,}291{,}680\ \text{kcal/h}\right)/60°\text{C}$$
$$= 21{,}528\ \text{l/h}$$

This means that 21,528 l/h hot water at 100°C comes from the boiler and goes into the hot water storage tank while 21,528 l/h hot water at 40°C goes from the hot water storage tank to the boiler.

And, in order to avoid condensation of the flue gasses in the boiler fire tubes and material stresses, using the hot water from the boiler at 100°C, we have to increase the water temperature coming from the tank at 40–70°C. Then this means that the flow over the boiler is 21,528 l/h plus the hot water to increase the return temperature (let's call this as $x$ l/h). Then,

$$\left(21{,}528\ \text{l/h} + x\right)\left(70°\text{C}\right) = 21{,}528\ \text{l/h}\left(40°\text{C}\right) + x\left(100°\text{C}\right)$$

$$30\,x = 645{,}840\ \text{l/h}$$

$$x = 21{,}528\ \text{l/h}$$

Then the load (fill)/shunt pump capacity would be:

$$21{,}528\ \text{l/h} + 21{,}528\ \text{l/h} = 43{,}056\ \text{l/h}\left(\sim 190\ \text{gpm}\right)$$

Since the tank is close to the boiler room, the experience shows that the line resistance is 5,000 mm $H_2O$. If the storage tank is further away from the boiler room, then we need to add the line resistance (both supply and return) to this resistance. So, we will go ahead and use the typical resistance of 5,000 mm $H_2O$ (=16.4 ft ≈ 17 ft). And then the load pump will be 190 gpm @ 17 ft. We can go to the pump manufacturers catalog now, and choose a high efficiency pump, which will deliver 190 gpm at 17 ft head.

Now, using Table A.21, we can go ahead and determine the line size for 43,053 l/h. In boiler room only, we can use the standard pipe sizes (33 mm = 1″; 42 mm = 1 ¼″; 40 mm = 1 1/2″; 60 mm = 2″; 89 mm = 3″; 114 mm = 4″; 140 mm = 5″; 168 mm = 6″; 219 mm = 8″; 273 mm = 10″; 324 = 12″; etc.). Note that the sizes in mm is the outside diameter, and the sizes in inches is the inside diameter. Other sizes in mm are only used in distribution headers and for heating in greenhouses.

Then, for the line, we will choose 114 mm OD (4″ ID) pipe, which will give us a resistance of 14 mm $H_2O$/m and a velocity of approx. 1.30 m/s.

We can also determine the size of the 3-way mixing valve. A 100 mm 3-way mixing valve has a $K_{vs} = 160$, then

$$\Delta P = [(43\ m^3/h)/160]^2 \times 10{,}000\ mm\ H_2O = 722\ mm\ H_2O, \text{ which is perfectly fine.}$$

The load (fill) line size for 21,528 l/h can also be determined by using Table A.21; and the line size would be 89 mm OD (3″ ID), which would give us a line resistance of approx. 14 mm $H_2O$ and a velocity of 1.10 m/s.

Let's now work on the shunt line:

$$\Delta P_{boiler\ circuit} = 1{,}000\ mm\ H_2O$$

$$\Delta P_{mixing\ valve} = 722\ mm\ H_2O$$

$\Delta P_{fill/shunt\ line} \approx 100$ mm $H_2O$ (including the special resistance; just a short distance)
Total of the Above = 1,822 mm $H_2O$.

$$\Delta P_{tank\ circuit} = 5{,}000\ mm\ H_2O$$

Then,

$$\begin{aligned}\Delta P_{shunt\ line} &= \Delta P_{tank\ circuit} - 1{,}822\ mm\ H_2O \\ &= 5{,}000\ mm\ H_2O - 1{,}822\ mm\ H_2O = 3{,}178\ mm\ H_2O\end{aligned}$$

$$\begin{aligned}\text{Special Resistance Over the Shunt Line} &= 90° \text{ Elbow + Tee + Check Valve} \\ &= 0.5 + 1.5 + C/V\end{aligned}$$

(See Table 2.30 for the resistance coefficients of 90° elbow and the tee.)

Since the flow rate is 21,528 l/h (21.5 $m^3$/h); if we choose a 50 mm check valve, then $K_{vs} = 55$. And then the pressure drop over the check valve to be used would be:

$\Delta P = [(3.063 \times 5{,}572 l/h)/(40\ mm)^2]^2 \approx 114\ mm\ H_2O$, which is perfectly fine.

If we then, using Table A.21, choose a 60 mm OD pipe for the shunt line, it would give us a line resistance of 110 mm $H_2O$ and a velocity of approx. 2.43 m/s. Since we still have room in the shunt line resistance, this resistance is fine; and even though the velocity (2.43 m/s) is much larger than the maximum velocity (1.2 m/s) of 60 mm OD pipe, because the line is a pretty short line, we can still go ahead and use 60 mm OD pipe for the shunt line. However, we can always use larger pipe sizes in order to get the velocity within the limits.

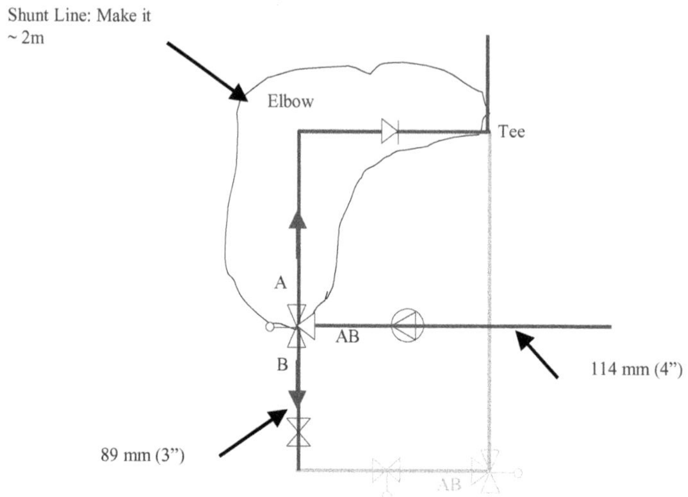


*Main Lines to the Heat Storage Tank*

The capacity of the main lines to the tank is 70% of the maximum flow rate over the boiler. And the maximum flow rate over the boiler is the boiler capacity (kcal/h) divided by the $\Delta T$ (15°C over the loop + additional 3°C). In this case, the boiler capacity was given as 4,000,000 kcal/h; then

The Maximum Flow Rate = (4,000,000 kcal/h)/18°C = 222,222 l/h or the sum of flow rates in all groups.

Then, 70% of the maximum flow rate (222,222 l/h) would be 155,555 l/h ≈ 156,000 l/h. We can take this flow rate and go to Table A.21, and determine the main line size as 219 mm OD (8″ ID), which would give us a line resistance of 5.6 mm $H_2O$ (the resistance limit for this line is 10 mm $H_2O$/m) and a velocity of 1.23 m/h, which is perfectly fine. See the sketch below for the hot water storage system details.

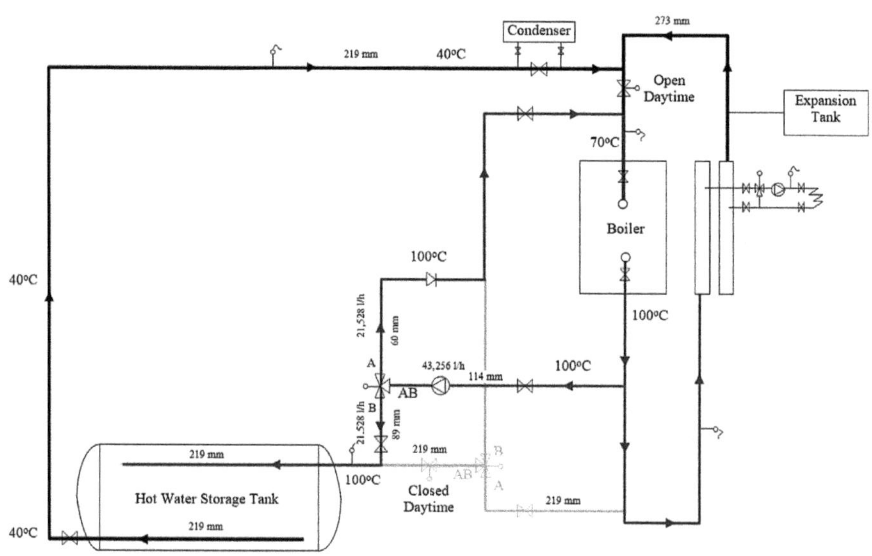

*Holes in the Distribution Headers in the Horizontal Heat Storage Tank*
In horizontal tanks, usually 40 mm holes are used in the headers. This would give us a cross-sectional area ($\pi D^2/4$) of 1,256 mm². The cross-sectional area of 219 mm OD (211.1 mm ID) pipe is 34,982 mm². Then,

$$\text{The Number of Holes} = 34{,}982/1{,}256 = 28$$

$$\text{The Flow Rate per Hole} = 156{,}000 \text{ liter}/28 \text{ holes} = 5{,}572 \text{ l/h}$$

Then, using $d = \sqrt{\dfrac{3.063\,Q}{\sqrt{\Delta P}}}$, we can determine $\Delta P$ across the holes as follows:

$$\Delta P = [(3.063 \times 5{,}572 \text{l/h})/(40 \text{ mm})^2]^2 \approx 114 \text{ mm } H_2O \text{ (which is perfectly fine)}$$

*Choosing the Tank*
Remember that the tank capacity is 150 m³, we now go to tank data provided by the tank manufacturers, and choose the largest diameter applicable because the bigger the tank is, the cheaper the insulation and the tank itself become. Therefore, we will choose a tank with a 4 m diameter and 11.5 m length, which provides a capacity of 152 m³.

Having the tank length information in hand, we can now determine the distance between the holes as follows:

$$\text{Distance between the Holes (c.t.c.)} = 11{,}500 \text{ mm}/28 \text{ holes} = 411 \text{ mm} \approx 400 \text{ mm}$$

It should be noted that the holes are placed upward on the supply pipe (a 10 mm downward hole is also provided on the supply pipe to drain the line), and downward on the reverse distribution (not on the reverse main). A 10 mm downward hole is provided on the reverse distribution as well to drain the line. The sketch of the storage tank is provided below.

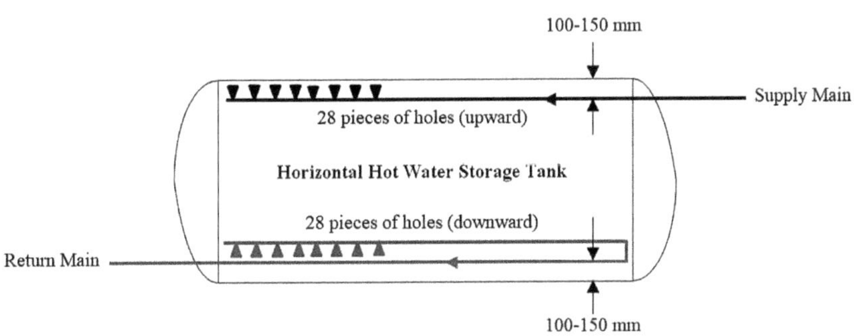


The storage tank is also equipped with a 2″ drain pipe, an auto air release and a vacuum relief valve, which opens at 25 mm $H_2O$ pressure (very close to the atmospheric pressure).

#### *5.4.3.4.5 Open Buffer System Design*

Now let's design a vertical tank open buffer system as shown in Figure 5.42.

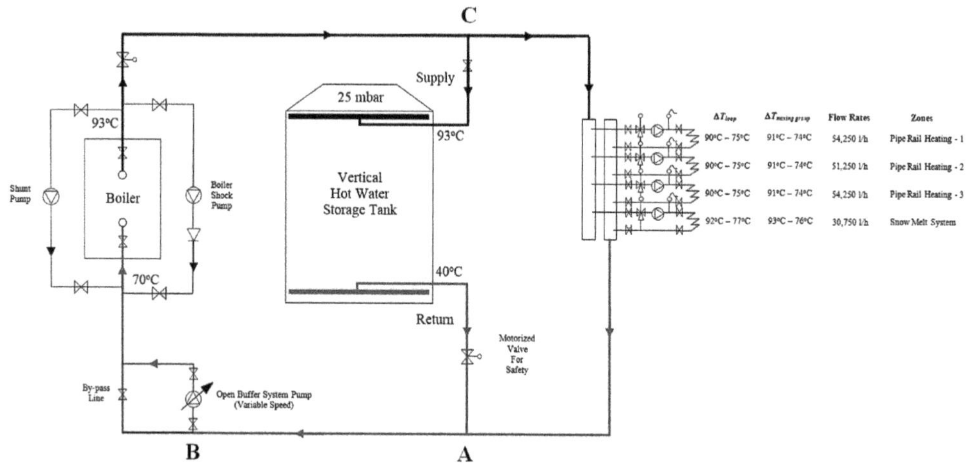


**Example 5.12: Designing a Vertical Tank Open Buffer System**

Design a vertical tank open buffer system for a 15,000 m² greenhouse where the $CO_2$ dosing norm is 100 m³ gas/h.ha, and the calorific (heating) value of the locally available natural gas is 9,568 kcal/m³. The boiler capacity is given as 4,000,000 kcal/h, and the combined boiler and condenser efficiency as 90%.

**Solution**

Burner Capacity = 1.5 ha × 100 m³ gas/h.ha = 150 100 m³ gas/h

Heat Supply during $CO_2$ Dosing = 150 × 100 m³ gas/h × 9,568 kcal/m³ × 0.9 = 1,291,680 kcal/h

Shunt Pump Capacity = 1.5% of the Boiler Capacity (kcal/h) = 0.015 × 4,000,000 = 60,000 l/h

For this facility, we could have used a horizontal tank as well; however, in order to provide diversity, we are going to use a vertical tank and design the open buffer system accordingly. In a vertical heat storage tank, we can have a maximum (supply) water temperature of 93°C in the tank, while the minimum (return) temperature is 40°C. Remember that the boiler return temperature is 70°C (as discussed in the previous example). Then, the water temperature difference over the storage tank would be:

$$\Delta T_{water\ over\ the\ tank} = 93°C - 40°C = 53°C$$

Then, in order to store all the heat generated, using Equation (2.24), the theoretical capacity of the hot water storage tank can be calculated as follows:

$$1{,}291{,}680\ \text{kcal}/53°C \approx 24{,}370\ \text{l/h (during } CO_2 \text{ dosing)}$$

$$\text{Shunt Pump Capacity} + \text{Boiler Shock Pump Capacity} = 24{,}370\ \text{l/h} \times \left(\frac{70-40}{93-70}\right)°C = 31{,}707\ \text{l/h}$$

We have earlier determined the shunt pump capacity as 60,000 l/h, which covers both the shunt pump and boiler shock pump capacities; therefore, no separate shock pump capacity is required. In cold regions, we have large capacity boilers and therefore high flow rate shunt pumps, which handle shock pump flow rate requirements as well.

If, however, we had a boiler return temperature of 80°C instead, then we would have a combined flow rate requirement as follows:

$$\text{Shunt Pump Capacity} + \text{Boiler Shock Pump Capacity} = 24{,}370\,\text{l/h} \times \left(\frac{80-40}{93-80}\right)^{\circ}\text{C}$$
$$= 74{,}985\,\text{l/h}$$

And then, a separate shock pump capacity would have been required as follows:

$$\text{Boiler Shock Pump Capacity} = 74{,}985\,\text{l/h} - 60{,}000\,\text{l/h} = 14{,}985\,\text{l/h}$$

We can now determine the tank/boiler flow rate from the provided information as follows:

$$\text{Total Flow Rates for Pipe Rail Heating} = 159{,}750\,\text{l/h and } \Delta T_{mixing\ group} = 91-74^{\circ}\text{C}$$

$$\text{Flow Rate for Snow Melting} = 159{,}750\,\text{l/h and } \Delta T_{mixing\ group} = 93-76^{\circ}\text{C}$$

Then, the flow rates from the tank/boiler to all the pipe rail groups would be calculated as follows:

$$159{,}750\,\text{l/h} \times \left(\frac{91-74}{93-74}\right)^{\circ}\text{C} = 142{,}934\,\text{l/h}$$

and the flow rate from tank/boiler to the snow melt group:

$$30{,}750\,\text{l/h} \times \left(\frac{93-76}{93-74}\right)^{\circ}\text{C} = 30{,}750\,\text{l/h}$$

Then, the total flow rate from the tank/boiler would be 142,934 l/h + 30,750 l/h = 173,684 l/h, which gives us the open buffer pump capacity.

Remember that the flow resistance over Points ABC must be equal to the resistance over CA, and based on practical experience, $\Delta P$ is 5,000 mm $H_2O$; therefore, the open buffer pump has a flow rate capacity of 173,684 l/h @5,000 $H_2O$ (765 gpm @16.5 ft). Using the pump manufacturers catalog, we will then go ahead and pick a proper pump with high efficiency, which will meet both the flow and head requirements.

Open Buffer Heat Storage Tank Capacity (in practice) = 200 m$^3$ gas/h.ha (in the Netherlands, for instance)

Open Buffer Heat Storage Tank Capacity (in practice) = 250 m$^3$ gas/h.ha (in Canada, for instance)

We can now go ahead and pick an adequate size of vertical tank for our 1.5 ha greenhouse operation, let's say in Canada (250 m$^3$ gas/h.ha × 1.5 ha = 375 m$^3$ vertical tank volume), and calculate the pipe resistances and piping for the mains and the pipes within the buffer tank.

The total flow rate to and from the tank/boiler 173,684 l/h. We can take this flow rate and go to Table A.21, and determine the main line size as 219 mm OD (8″ ID), which would give us a line resistance of 7.00 mm $H_2O$ (the resistance limit for this line is 10 mm $H_2O$/m) and a velocity of 1.38 m/h, which is perfectly fine. See the sketch below for the open buffer system details.

*Distribution Header Sizes and Holes in the Distribution Headers*

The cross in the buffer tank has four arms; then, the flow rate per arm would be 173,684 l/h/4 = 43,421 l/h. And based on experience, the flow resistance in the cross piping within the tank is approx. 5 mm $H_2O$/m. Again, we can take this flow rate and the flow resistance, and go to Table A.21, and determine the arm size as 140 mm OD (5″ ID), which would give us a line resistance of 5 mm $H_2O$ (as stated above) and a velocity of 0.87 m/h, which is perfectly acceptable.

And in buffer tanks, usually 40 mm holes are used in the distribution headers; and the flow resistance through the 40 mm holes must be less than 100 mm. This would give us a cross-sectional area ($\pi\ D^2/4$) of 1,256 mm². Then, we can determine the flow rate for each hole using the following relationship:

$$d = \sqrt{\frac{3.063\,Q}{\sqrt{\Delta P}}}$$

$$Q = \left[\left(40^2\right) \times \sqrt{100}\ \right] / 3{,}063 = 5{,}224\ \mathrm{l/h}$$

So, the total number of 40 mm holes would be (173,684 l/h)/5,224 l/h per hole) = 34. The cross in the buffer tank has four arms; then, the number of holes per arm would be 34/4 = 9 ≈ 10 holes. Since the holes are placed on both sides of the arm, we need to place 5 holes on each side of the arms, which would make a total of 40 holes.

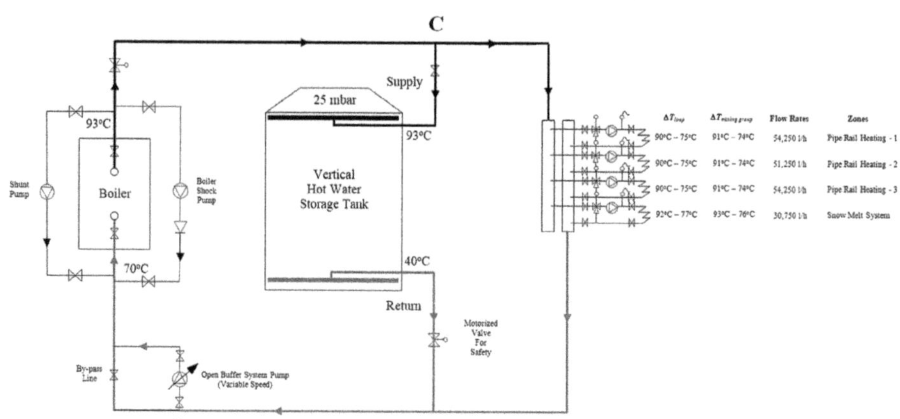


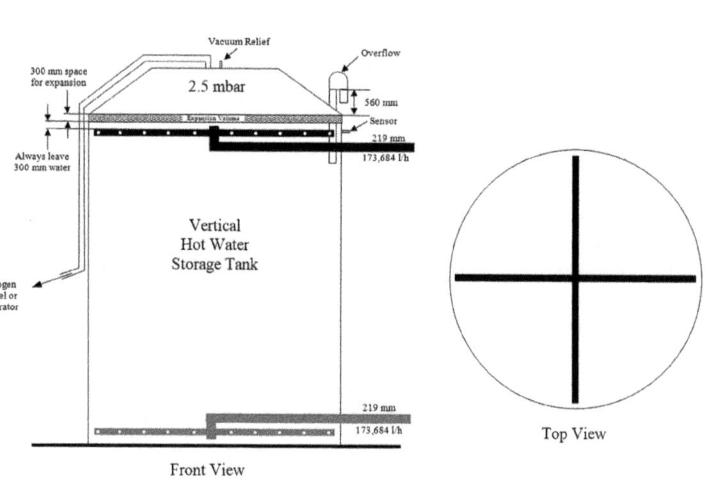

*Space for Expansion in the Buffer Tank*

It should be noted that a separate expansion vessel is not required in an open buffer system that is because the expansion space is provided at the top of the buffer tank. Using the predetermined water contents in different sections of the whole system, we will now perform a quick calculation for the required expansion space below.

| **Expansion** | $\Delta T$ | **Water Content** | **Expansion** | **Expanded Volume** |
|---|---|---|---|---|
| Boiler | 10–93°C | 17,000 liter | 3.6% | 615 liter |
| Manifold | 10–93°C | 2,500 liter | 3.6% | 90 liter |
| Vertical Tank | 10–93°C | 375,000 liter | 3.6% | 13,500 liter |
| Snow Melt | 10–93°C | 12,000 liter | 3.6% | 432 liter |
| Pipe Rail for Min.Temp. | 40°C | 60,000 liter | 1% | 600 liter |
| | | | | **Net Expansion Volume = 15,237 liter** |

From the available tanks in the manufacturer's catalog, we will choose an 8 m (=80 dm) diameter tank. And 1 dm (=10 cm) tank height will have a volume of $(\pi\, 80^2/4) \times 1 = 5{,}026$ liter. Then, the height for the expansion volume would be (15,237 liter)/(5,026 liter) = 3.03 dm ( = 30.3 cm). So, we can leave approx. 300 mm height in the tank for expansion (see the sketch of the buffer tank).

It should be noted that, since there is no separate expansion tank in the open buffer system, there would be no space for expansion when tank shuts off. As mentioned earlier, the upper section of the buffer tank is maintained at 25 mm $H_2O$, and this is maintained by either a separate nitrogen vessel or a nitrogen generator. The buffer tank is also equipped with a drain pipe, an overflow pipe, and a vacuum relief valve, which opens at 25 mm $H_2O$ pressure (very close to the atmospheric pressure).

#### *5.4.3.5 Irrigation Water Pre-Heating System Design*

In greenhouse crop production, irrigation water is typically pre-heated to reduce the shock of the cold water, and Figure 5.42 below shows the details of an irrigation water pre-heating system.

**Example 5.13: Designing an Irrigation Water Pre-Heating System**

See Examples 2.57 and 2.58 for irrigation water pre-heating.

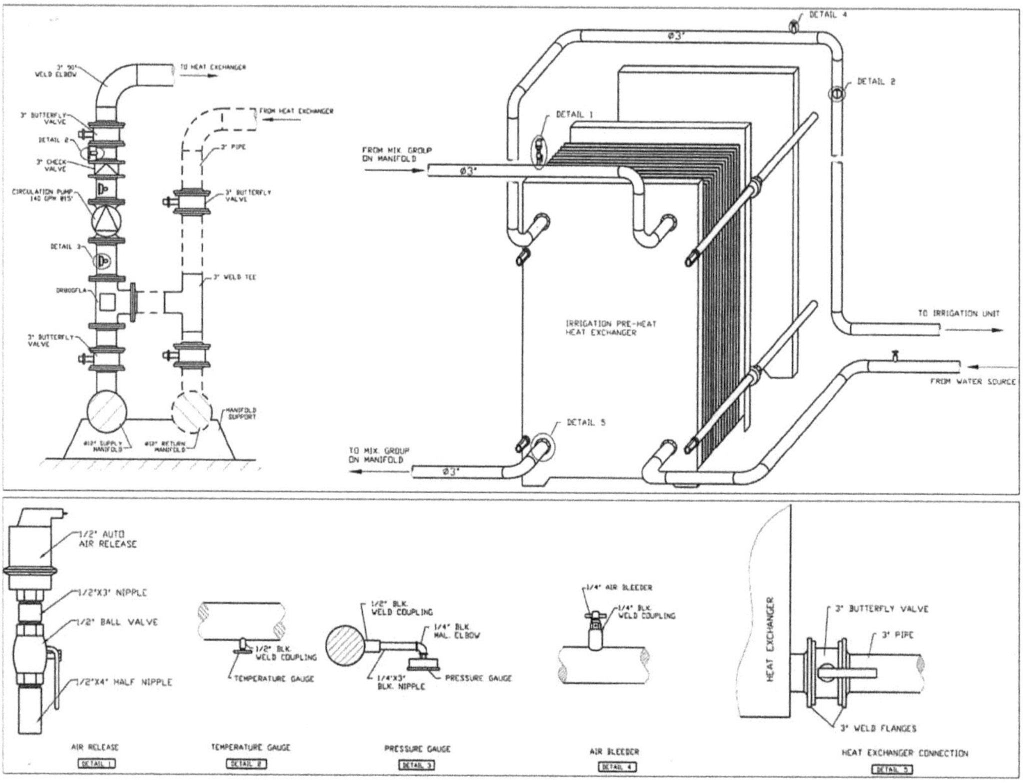


**FIGURE 5.42**
Details of an irrigation water pre-heating system.

## 5.4.4 Radiant Heating Systems

As it has been shown again and again, conventional greenhouse heating systems heat the greenhouse air first, and then the heated air heats the plants. A low infrared heater, on the other hand, operates with heated pipes – up to 595°C (1,100°F) – and first heat the objects in the greenhouse (including crops) directly through infrared radiation heat exchange, and then the objects exchange this heat with the environment via conduction, convection, and again radiation heat transfer. Since the greenhouse air is not heated directly, the temperature of the interior air is typically lower; and this low inside air temperatures result in less heat losses from the greenhouse environment. Furthermore, infrared heaters eliminate the pumping energy requirements and the heat losses through distribution lines typically seen in conventional greenhouse heating systems. As a consequence, when installed properly, infrared heating can provide uniform plant and vertical air temperatures and be a cost effective option for greenhouse heating.

## 5.4.5 Horizontal Air Circulation System Design

HAF fans, also known as circulation fans, provide a more uniform greenhouse environment and reduce the difference between internal air temperature and plant temperature

**FIGURE 5.43**
A horizontal air flow (HAF) fan installed in a greenhouse.

(see Figure 5.43). They not only improve uniform temperature distribution, but also humidity and carbon dioxide distributions costing marginal energy expenditure. The flowing air removes moisture and prevents condensation on the leaves, therefore reduce foliar fungal diseases.

Typically, a total air flow of 0.01 $m^3/s.m^2$ (2 $cfm/ft^2$) of floor area is provided keeping the greenhouse air mass moving at a velocity of 0.5 m/s (100 ft/min) to 0.7 m/s (140 ft/min) and not exceeding 1.0 m/s (200 ft/min) in the vicinity of crop canopy. HAF systems come with 30 cm (12 in), 40 cm (16 in), 45 cm (18 in), and 50 cm (20 in) diameter blades, and generally move the air approximately 30 times the blade diameter. A 30 cm (12 in) diameter HAF fan would be able to move the air 9 m (30 ft). In a well-designed HAF fan system, the fans are installed in a way that air flow is horizontally directed down the greenhouse bay and aligned with similar fans pushing the greenhouse air in the opposite direction down the next bay. In order to make sure that the air within the canopy is moved rather than the air above the canopy, fans are mounted at between 0.6 m (2 ft) to 0.9 m (3 ft) above the plants. The first and last fans are placed about 4.5 m (15 ft) to 6.0 m (20 ft) from the gable ends to pull the air that is coming from the adjacent bay. Crop canopy and hanging baskets resist HAF; therefore, the fans should be spaced properly. After the first fan, they are typically placed 9 m (30 ft) to 15 m (50 ft) apart to keep the greenhouse air mass moving smoothly.

**Example 5.14: Sizing Horizontal Air Flow Fans for a Greenhouse Complex**

Now let's consider the 10-bay gutter connected single glass gable roof greenhouse complex in Columbus, Ohio, USA, for which we designed a heating system using natural gas-fired unit heaters in Example 5.1.

**Solution**

Remember that the unit heaters chosen for the greenhouse complex were GHE200 and GHE230 types, which come with one 16 in. and two 16 in. diameter fans, respectively. And as discussed above, a 16 in. (40 cm) diameter fan can move the air approximately 30 times the blade diameter, that is, 40 ft (12 m). The remaining distance in each bay one way can be handled by two 18 in. (45 cm) diameter fans, each moving the air additional 45 ft (13.5 m). Then, for each bay we will use four GHE18 fans, two fans in each direction as shown below.

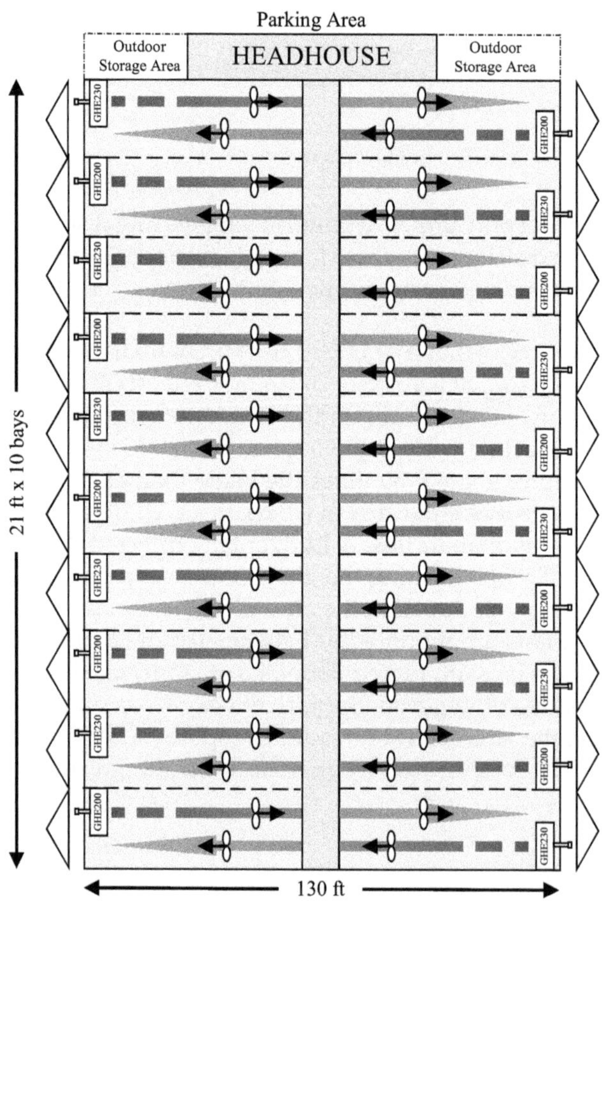

**Example 5.15: Sizing Horizontal Air Flow Fans for a Greenhouse Complex**

Now let's consider the 10-bay gutter connected, IR inhibited double poly curved roof greenhouse complex in Columbus, Ohio, USA, for which we designed a heating system using natural gas-fired unit heaters in Example 5.2.

**Solution**

Remember that the unit heaters chosen for the greenhouse complex were GHE100 and GHE115 types, which come with a 16 in. diameter fan each. And as discussed earlier, a 16 in. (40 cm) diameter fan can move the air approximately 30 times the blade diameter, that is, 40 ft (12 m). The remaining distance in each bay one way can again be handled by two 18 in. (45 cm) diameter fans, each moving the air additional 45 ft (13.5 m). Then, for each bay we will use four GHE18 fans, two fans in each direction as shown below.

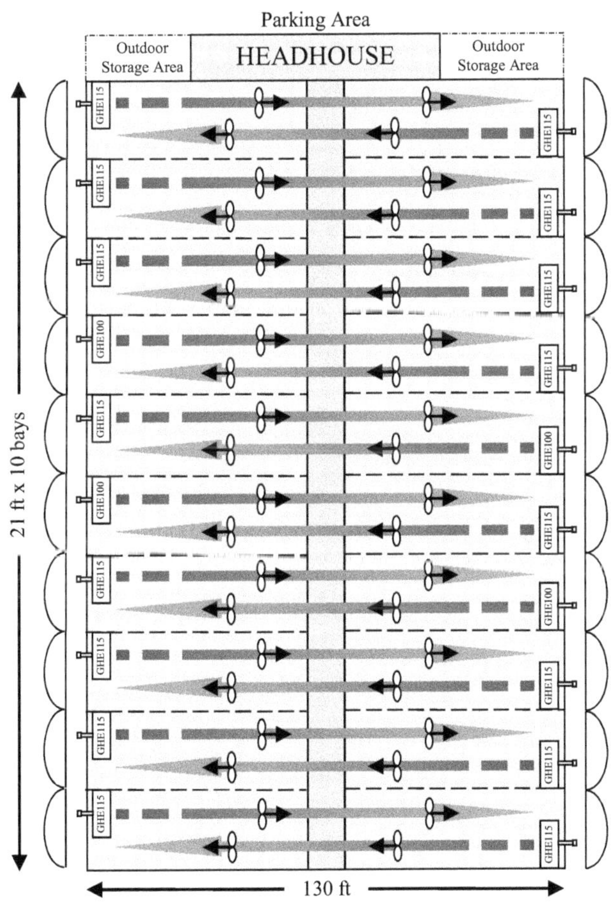

## Nomenclature

| | |
|---|---|
| $A$ | area |
| ACH | air change per hour |
| BHP | boiler horse power |
| $C$ | specific heat |
| COP | coefficient of performance |
| $d$ | diameter |
| $D$ | diameter; dosing value |
| $E$ | energy |
| $I$ | content of expansion vessel |
| $M$ | mixing group size |
| $n$ | excess air |
| $Q$ | heat transfer rate; flow rate |
| $R$ | return |
| $S$ | supply |
| $t$ | time |
| $T$ | temperature |
| $U$ | heat transmission rate |
| $V$ | volume |
| $x$ | variable |
| $X$ | percent increase in production |

### Greek letters

| | |
|---|---|
| $\Delta$ | difference |
| $\Phi$ | flue gas flow rate |
| $\eta$ | efficiency |
| $\rho$ | density |

### Subscripts

| | |
|---|---|
| $b$ | boiler; buffer |
| $c$ | conduction |
| $cg$ | combustion gas |
| $dgas$ | dosing gas |
| $g$ | natural gas |
| $gh$ | greenhouse |
| $t$ | total |
| $cond$ | conduction |
| $evap$ | conduction |
| $i$ | inside |
| $inf$ | infiltration |
| $max$ | maximum |
| $min$ | minimum |
| $o$ | outside |
| $p$ | pressure |
| $t$ | total |
| $w$ | water |

## Bibliography

Andrews JW (1978) Development of a cost effective solar assisted heat pump. *Proc. 1978 Ann. Meet. Am. Section Int. Solar Energy Soc. Inc.*, Vol. 2.1, pp. 281-287.

ASAE (1989) *Standards, Engineering Practices and Data developed and adopted by the American Society of Agricultural Engineers*. 36th edition. St. Joseph, MI, U.S.A.

Boersma L and Rykbost KA (1975) Soil warming with power plant waste heat greenhouses. *Hort Science*, 10:28–30.

Bosio RC and Suryanarayana NV (1975) Solar assisted heat pump system: a parametric study for space heating of a characteristic house in Madison, Wisconsin, U.S.A., ASME Paper 75.WA/Sol-8.

Casault B, Gauthier L, and Boily R (1989) Comparison of three gas-fired greenhouse heating systems in cold climates. ASAE Paper No: 89-4033, ASAE, St. Joseph, MI 49085, U.S.A.

Duncan RT and van Zuiden GJ (1979) Economic viability of the solar-assisted industrial heat pump system. Solar Indust. Process Heat Conf. Proc. Solar Energy research Institute/TP-333-429, San Fransisco, California, U.S.A.

Freeman TL, Mitchell JW, and Audit TE (1979) Performance of combined solar-heat pump systems. *Solar Energy*, 22:125–135.

Gosselin A and Trudel MJ (1982) L'utilisation du chauffage au dol dans les cultures de legume de serre. *Revue Horticole*, 232:21–25.

Knies P, van de Braak NJ, and Breuer JJG (1984) Infrared heating in greenhouses. *Acta Hort.*, 148:73–80.

Seager JCR (1982) Pot plants respond to root zone warming. *Farm and Food Res.*, 12:178–179.

Zeroni M and Gale J (1982) Effect of root temperature on the development, growth, and yield of 'Sonia' roses. *Scientia Hortic.*, 18:177–184.

Yildiz I and Stombaugh DP (2006) Dynamic modeling of microclimate and environmental control strategies in a greenhouse coupled with a heat pump system. *Acta Hort.* 718:331–340.

Yildiz I, Yue J, and Lowrey J (2014a) Spatiotemporal distributions of energy and water consumption in open, semi-closed and closed loop heat pump greenhouse systems, and associated management strategies in Canada. *Acta Hort.* 1037 (1): 119–128.

Yildiz I, Yue J, and Lowrey J (2014b) Spatiotemporal distributions of energy and water consumption in open, semi-closed and closed loop heat pump greenhouse systems, and associated management strategies in the Canadian North. *Acta Hort.* 1037 (1): 171–177.

# 6

# Greenhouse Ventilation and Cooling

## 6.1 Introduction

Plants are biological air-conditioners, which cool their own immediate surroundings. However, during hot and sometimes warm weather, heat builds up inside the greenhouse due to high solar radiation loads; and if quality crops are to be grown, it is required to provide greenhouses with appropriate cooling. Cooling may be provided by different means starting from the most cost-effective options such as whitewashing the greenhouse exterior, shading screens, and natural ventilation, and moving toward relatively costly options such as mechanical ventilation, evaporative cooling, and refrigerated air-conditioning.

This chapter will briefly go over estimating cooling loads in greenhouses first, and then move to the passive and active greenhouse cooling techniques. The techniques covered include whitewashing the greenhouse exterior, shading curtains, natural and mechanical ventilation, evaporative cooling, and heat pump cooling.

## 6.2 Estimating Cooling Loads in Greenhouses

Using the energy balance equation, Equation (5.2), and the other pertinent equations provided in Chapter 5, we can determine the cooling load in a transient greenhouse environment. Greenhouse cooling load can be reduced using permanent or retractable shading devices before taking other cooling measures as discussed below.

## 6.3 Passive Cooling of Greenhouses by Shading

Greenhouse shading is used mostly during warm and bright sunny days to reduce cooling loads and therefore indoor temperatures; hence, as will be discussed later in Chapter 11, it is also an effective energy conserving technique commonly used in commercial greenhouses. These include an external spray (also known as whitewash) applied on the exterior surface of the glazing (see Figure 6.1) and variable retractable shading (internal or less commonly external; see Figure 6.2) using different materials. Whitewashing is an inexpensive method to provide shading and reduce cooling loads; however, the proportion of shading it provides is relatively fixed and permanent even though it is washed away

**FIGURE 6.1**
Whitewash applied on the outside surface of the glazing.

**FIGURE 6.2**
Internal retractable shading system: open (*left*) and closed (*right*).

slowly in time. It can be applied on the whole or only a portion of the glazing in alternating strips. Whitewashing application's degree of shading depends on the coating thickness or the frequency of alternating strips. Also, the whitewash must be washed off when the solar radiation becomes limiting again in the fall.

Retractable shading, as shown in Figure 6.2, on the other hand, is a relatively expensive shading technology; however, when controlled properly, it dynamically and precisely reduces cooling load. When whitewashing or external shading curtains are used, solar radiation can be reflected before it enters and increases the cooling load in the greenhouse. These systems should not be controlled based on only inside temperatures, but also inside light levels for not to hinder crop growth and development. Both external and internal shading cloths come in different colors and materials having different costs and shading percentages. Another variable shading system, which reduces the cooling loads in greenhouses, is the one with a combination of polystyrene pellets that provides precise shading in greenhouses; however, it is not commercially available yet due to its relatively expensive channelized double acrylic glazing and control system.

## 6.4 Passive Cooling of Greenhouses by Natural Ventilation

Ventilation is a necessary component of conventional greenhouse systems. It is a fundamental instrument to control not only the indoor temperature – especially in mild and hot summer months, but also the humidity and other gas concentrations such as $CO_2$ throughout the year. Ventilation removes the excess heat and moisture from the greenhouse environment reducing both the cooling and moisture loads. It is also a vital environmental control component in terms of introducing the outside air, which has a higher $CO_2$ content compared to that of the inside air, if $CO_2$ enrichment is not being utilized. As will be discussed later in Chapter 11, in energy conserving greenhouses, it is desirable to reduce ventilation rates to reduce the heat loss due to ventilation in colder months. Other greenhouses may have entirely closed systems that require no ventilation at all, which will also be covered later in this chapter as well as in Chapter 11.

Greenhouse ventilation systems can essentially be classified into two systems: i) natural (passive) ventilation systems, and ii) mechanical (forced-air) ventilation systems. Obviously, a pressure difference must be created by wind or temperature gradient for natural ventilation to occur. And this type of cooling heavily depends on the crop's cooling themselves via transpirational cooling. In order to take advantage of wind-created pressure differences, it is recommended that vent openings be present on the sides and on the ridge of the greenhouse (see Figures 6.3, 6.4, 6.5, and 6.6). Most greenhouse operations depend only on natural ventilation even for summer cooling. In relatively dry climates

**FIGURE 6.3**
Sidewall (left) and ridge (right) ventilation at a Venlo greenhouse complex.

**FIGURE 6.4**
Roof ventilation with retractable shading system at a greenhouse complex.

**FIGURE 6.5**
Sidewall (left) and ridge (right) ventilation at a polyethylene greenhouse complex.

where wet-bulb depression permits efficient evaporative cooling, then evaporative cooling systems become handy to support both commercial and institutional greenhouse operations as a cost effective option.

Greenhouses with only sidewall vents depend on wind pressure to force air exchange and are usually ineffective, especially in large commercial greenhouse operations. Therefore, greenhouses typically have vent openings on both sides of the ridge and on both sidewalls running continuously the full length of the greenhouse. The total ridge vent area should be equal to the total sidewall vent area, and each should be designed to have at least 15–20% of the greenhouse floor area. And depending on the wind direction, the leeward (downwind side) vents are opened automatically to create a negative pressure at the top of the ridge so that effective natural ventilation can be achieved. Vent openings are adjusted depending on the rain and wind conditions as well.

**FIGURE 6.6**
Sidewall and roof ventilation with insect screen installed at a research facility.

Efficient greenhouse ventilation systems are desperately needed for providing comfortable and uniform greenhouse environments in different greenhouse structures. Much research efforts have been spent on providing uniform greenhouse ventilation; however, the outcomes of ventilation studies performed in research-scale greenhouses are many times questionable with respect to their effectiveness in much larger commercial greenhouses. Many thanks to the powerful CFD (computational fluid dynamics) modeling tools that we are now able to determine the weakest distribution areas of the traditional ventilation systems and increase their performances (see Figure 6.7).

## 6.5 Active Cooling of Greenhouses by Mechanical Ventilation

The other ventilation system is the mechanical, or fan, or forced-air ventilation system. The fans are usually located on the leeward side or on the end of the greenhouses (Figure 6.8), and incoming air is introduced into the greenhouse so that it mixes and is warmed by interior air before contacting the plants. If the fans need to be placed on the windward sides, then air exchange capacity should be increased at least 10%. Ventilation fans are sized to deliver required air exchange for a minimum static pressure of 0.015 kPa (0.060 in. $H_2O$); however, a static pressure of 0.025 kPa (0.10 in. $H_2O$) is typically used in designs. Also, the fans are placed no more than 7.6 m (25 ft) along the sidewalls or gable ends. It is recommended that mechanical ventilation rates should be 0.005–0.01 $m^3/s.m^2$ (1–2 $cfm/ft^2$) for minimum winter air exchange, and 0.06 $m^3/s.m^2$ (12 $cfm/ft^2$) for summer cooling. The fans should be activated in stages by the control system based on the air exchange needs of the greenhouse; and in the heating season, the first stage of ventilation should be activated at

**Example 6.1: Sizing Natural Ventilation for a Greenhouse**

Now let's consider a 3 bay single glass gable roof greenhouse with 24 ft (7.2 m) span, 150 ft (45 m) long, and 14 ft (4.2 m) wall height. Determine the required sidewall (both sides) and ridge (both sides) opening sizes for natural ventilation running continuously the full length of the greenhouse.

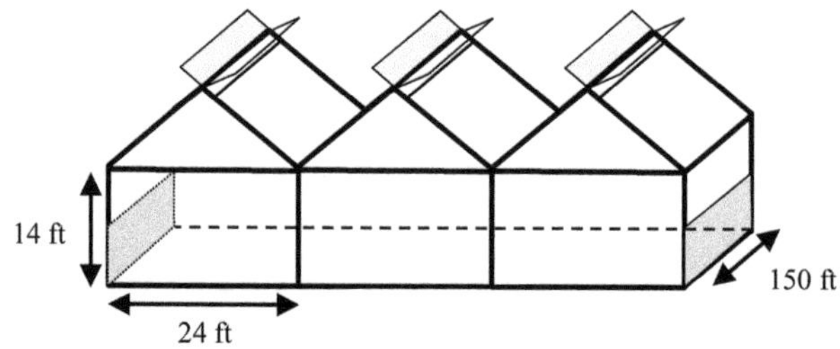


**Solution**

Remember that the total ridge vent area will be equal to the total sidewall vent area, and each will be designed to have at least 15–20% of the greenhouse floor area. Here, let's assume that we are going to apply 20%.

$$\text{Greenhouse Floor Area} = 3\text{ bays} \times 24\text{ ft} \times 150\text{ ft} = 10{,}800\text{ ft}^2$$

$$\text{Total Ridge Vent Area} = 0.20 \times 10{,}800\text{ ft}^2 = 2{,}160\text{ ft}^2 \left(\text{this is total sidewall vent area as well}\right)$$

$$\text{Ridge Vent Area per Bay} = 2{,}160\text{ ft}^2/3\text{ bays} = 720\text{ ft}^2$$

$$\text{Ridge Vent Area}\left(\text{each side}\right)\text{ per Bay} = 720\text{ ft}^2/2 = 360\text{ ft}^2$$

$$\text{Sidewall Vent Area}\left(\text{each side}\right) = 2{,}160\text{ ft}^2/2 = 1{,}080\text{ ft}^2$$

Both the ridge and sidewall openings will be running continuously the full length of the greenhouse; then,

$$\text{Ridge Vent Area}\left(\text{each side}\right)\text{ Dimension} = \mathbf{24\text{ ft} \times 150\text{ ft}} = 360\text{ ft}^2\left(\sim 34\text{ m}^2\right)$$

Sidewall Vent Area Dimension $= \mathbf{7.2\text{ ft} \times 150\text{ ft}} = 1{,}080\text{ ft}^2\left(\sim 100\text{ m}^2\right)$ which will be installed at the ground level on each side.

least 4–6°C (8–10°F) above the heating set-point so that the exhaust fans will not operate at the same time. In greenhouses where the air inlet-to-fan distance is less than 30 m (100 ft), ventilation flow rates should be increased to improve the heat transfer based on the following equation:

$$Q_{adj} = \frac{C.\ Q_v}{\sqrt{D}} \tag{6.1}$$

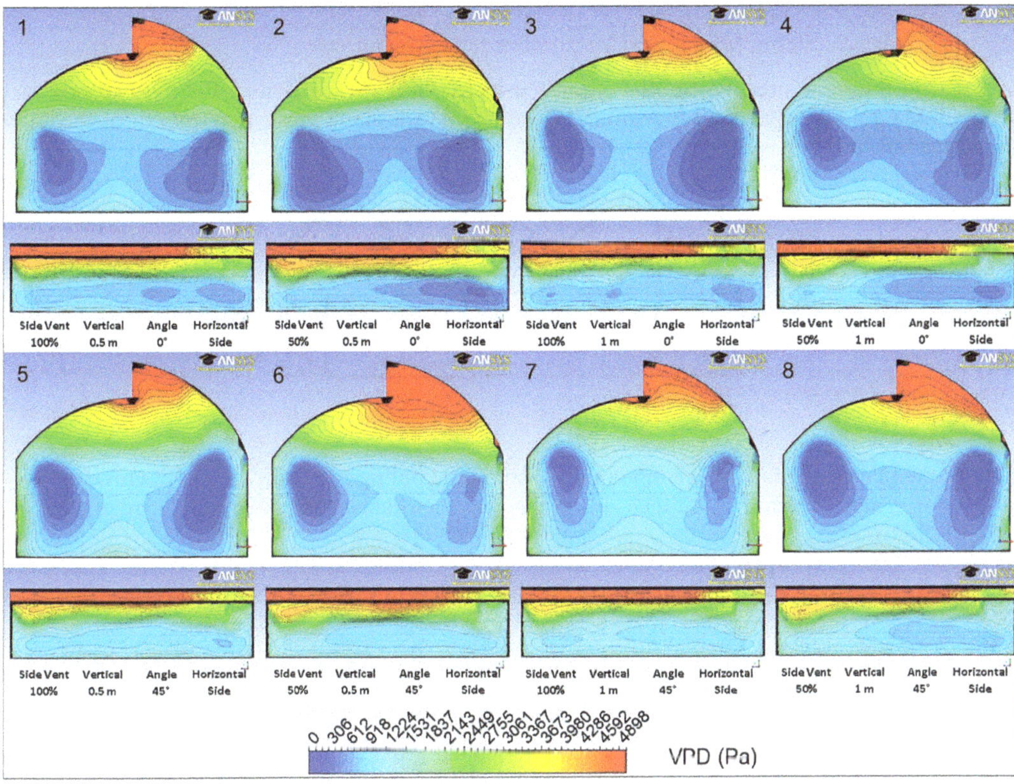


**FIGURE 6.7**
Uniformity of greenhouse climate under natural ventilation coupled with high pressure fogging system (cases with side nozzles). Courtesy of Dr. Murat Kacira, University of Arizona, Tucson, AZ, USA.

**FIGURE 6.8**
Mechanical ventilation systems installed on a gable end (left) and a sidewall (right).

**Example 6.2: Sizing Mechanical Ventilation for a Greenhouse**

Using the same greenhouse that was used in Example 6.1, determine the mechanical ventilation and the fan sizes required. Assume that the static pressure required for the fans is 0.10″.

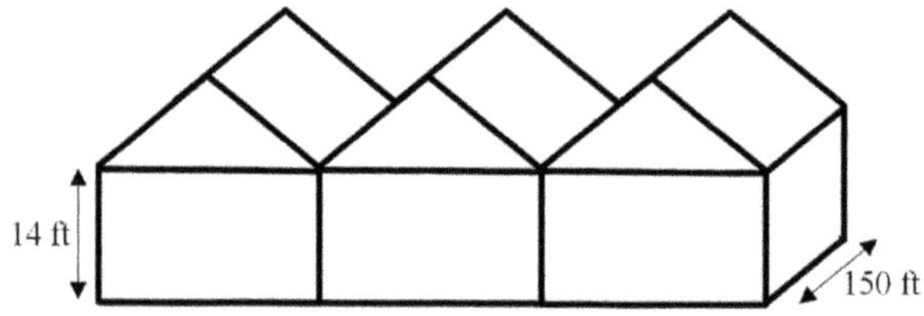


**Solution**

Remember that the air exchange rate for summer cooling is 0.06 $m^3/s.m^2$ (12 $cfm/ft^2$) while the minimum winter air exchange requirement is 0.005–0.01 $m^3/s.m^2$ (1–2 $cfm/ft^2$).

$$\text{Greenhouse Floor Area} = 3\text{ bays} \times 24\text{ ft} \times 150\text{ ft} = 10{,}800\text{ ft}^2$$

$$\text{Air Exchange Required for Summer Cooling} = 12\text{ cfm/ft}^2 \times 10{,}800\text{ ft}^2 = 129{,}600\text{ cfm}$$

Since the inlet-to-fan distance is 72 ft (21.6 m), which is less than 100 ft (30 m), we need to adjust the ventilation rate using Equation (6.1) as follows:

$$Q_{adj} = \frac{(10.0)(129{,}600\text{ cfm})}{\sqrt{72\text{ ft}}} = 152{,}735\text{ cfm}\left(\text{adjusted maximum summer ventilation rate}\right)$$

$$\text{Minimum Winter Air Exchange Requirement} = 2\text{ cfm/ft}^2 \times 10{,}800\text{ ft}^2 = 21{,}600\text{ cfm}$$

Again, we need to adjust the minimum ventilation rate as follows:

$$Q_{adj} = \frac{(10.0)(21{,}600\text{ cfm})}{\sqrt{72\text{ ft}}} = 25{,}456\text{ cfm}\left(\text{adjusted minimum winter ventilation rate}\right)$$

Minimum winter ventilation can uniformly be supplied by four fans installed each delivering 6,438 cfm at the assumed static pressure of 0.10″ (Model: GHE24; Fan Size: 24 in; Motor: 1/2 HP; Fan RPM: 891). Please note that while fan specifications are real, the manufacturer model used in this example is a generic model.

These four fans can deliver a total of 25,752 cfm when they operate all together; therefore, in order to determine the spring/fall and summer ventilation fans, this total needs to be subtracted from the grand total (152,735 cfm).

152,735 cfm – 25,752 cfm = 126,983 cfm more to be provided for spring/fall and summer ventilation

We may have other locally available options; however, for this example, we will pick six fans each delivering 21,283 cfm at the assumed static pressure of 0.10″ (Model: GHE54, Fan Size: 54 in; Motor: 1 HP; Fan RPM: 374). Again while fan specifications are real, the manufacturer model is a generic model.

| | | | Manufacturer Model No. |
|---|---|---|---|
| 1. | Winter Ventilation | 4 fans at 6,438 cfm each | GHE24 (single speed) |
| 2. | Spring/Fall Ventilation | 3 fans at 21,283 cfm each | GHE54 (two speed) |
| 3. | Summer Ventilation | 3 fans at 21,283 cfm each | GHE54 (two speed) |
| | **Total Fan Capacity:** | **153,450 cfm (installed)** | |

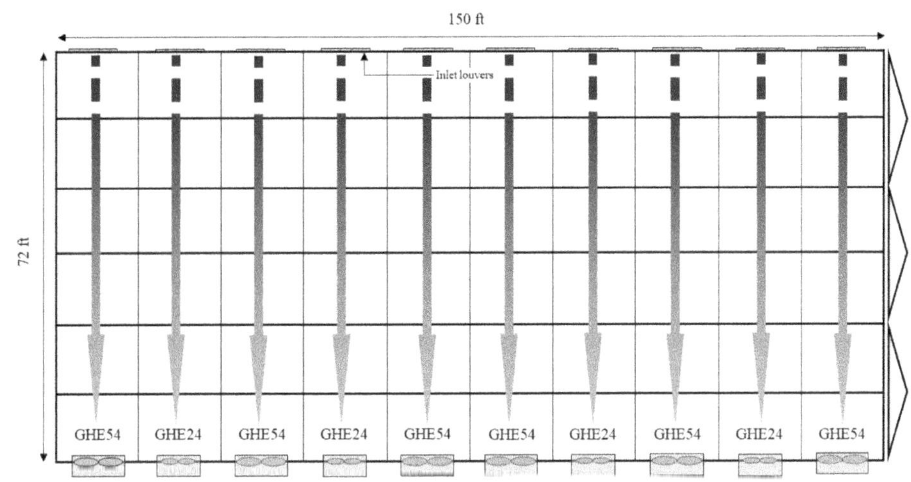


In the above design, the fans and inlet louvers are placed within the distance of 7.6 m (25 ft) along the sidewalls. Also, note that the inlet-to-fan distance is less than 100 ft; therefore, we have made an adjustment in the ventilation rate. In the following example, we are going to use five bays, which will give us an inlet-to-fan distance of 120 ft that is greater than 100 ft, where no adjustment will be required. For an efficient fan operation, keep in mind that a distance of 4–5 fan diameters should also be maintained between the fan discharge and any nearby obstruction such as another greenhouse or building.

Our design for this greenhouse has a total installed capacity of 153,450 cfm, which will provide little more than what is required (152,735 cfm); therefore, the overall ventilation system will be in good shape providing uniform air exchange with the provided configuration in the greenhouse as shown above. The fans are nicely sized and located to provide uniform air exchange and prevent oscillations when controlled properly in stages. Remember that we have used four fans (6,438 cfm each) for minimum (winter) ventilation distributed along the sidewall, which will operate continuously. These four fans can as well operate in steps or stages providing further (proportional) control over minimum ventilation when needed. Furthermore, we have split the total summer (the maximum) air exchange requirement into two stages (low and high) using two-speed fans, providing half the capacity in stages for spring/fall ventilation, and the other half to top up in stages for summer ventilation. It means, all the GHE24 fans (minimum winter ventilation fans) and all the GHE54 fans (low stage only) would be operating for spring/fall ventilation in stages as needed; and all the GHE24 and GHE54 fans (low and high stages) would be operating in stages and eventually providing the total air exchange of 153,450 cfm when needed in summer. This configuration will provide excellent spatial and temporal ventilation for this specific greenhouse. It should however be kept in mind that larger fan diameter and lower HP fans consume less operational energy for delivering the same air exchange; therefore, alternative fans with larger diameter and lower HP motors should be chosen, when possible. Reducing fan rotational speed is the most efficient way to reduce fan energy use in variable flow, and this can be efficiently achieved with frequency-controlled electric motors. For instance, if we reduce the fan rotational speed by 10%, based on the fan/pump affinity laws (this is further discussed in Chapter 11), the fan power will reduce 27%. It means, not only the initial costs but also the operational costs of fans should be considered for making a justified fan decision.

Here, we will pick motorized air inlet louvers based on 1.5 times the areas of the fans. They are controlled based on the number of active fans or in response to the pressure drop generated across the inlet louver. Total installed ventilation capacity in this example is 153,450 cfm. Using the inlet air velocity maintained at about 3.6 m/s (700 ft/min) at the inlet louver, we can calculate the cross-sectional area of inlet louver as follows:

$$\text{Total Cross-Sectional Area of Inlet Louver} = 1.5\,\text{x}\left(153{,}450\,\text{ft}^3/\text{min}\right) \div 700\,\text{ft/min} \approx 329\,\text{ft}^2$$

If a continuous inlet louver is to be installed on the opposite sidewall, since the total length of the greenhouse sidewall is 150 ft, then the width of the inlet louver becomes 329 $\text{ft}^2 \div 150\ \text{ft} = 2.2\ \text{ft} \approx 2\frac{1}{4}$ ft. Hence, the inlet louver dimension becomes 2 ¼ ft × 150 ft. We can also install multiple louvers distributed along the opposite sidewall for a uniform airflow distribution in the greenhouse. As shown on the figure above, for instance, ten inlet louvers can be used with a cross-sectional area of approximately 33 $\text{ft}^2$ each.

**Example 6.3: Sizing Mechanical Ventilation for a Greenhouse**

Maintaining the floor area of the greenhouse that was used in Example 6.2, however, this time with 5 bays, determine the required mechanical ventilation and the fan sizes. Again, assume that the static pressure required for the fans is 0.10″.

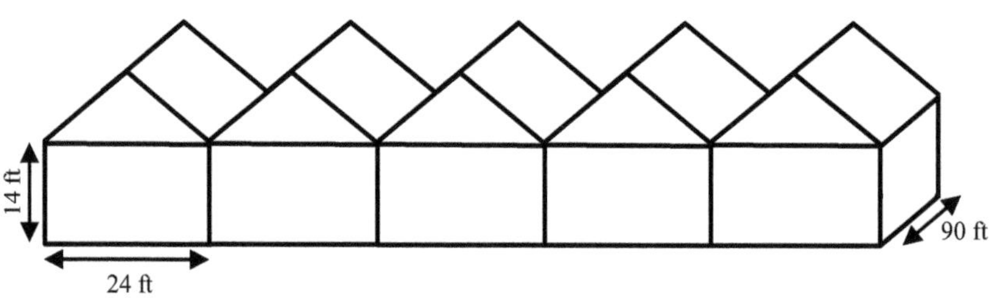


**Solution**

Again, remember that the air exchange rate for summer cooling is 0.06 $\text{m}^3/\text{s.m}^2$ (12 $\text{cfm/ft}^2$) while the minimum winter air exchange requirement is 0.005–0.01 $\text{m}^3/\text{s.m}^2$ (1–2 $\text{cfm/ft}^2$).

$$\text{Greenhouse Floor Area} = 5\,\text{bays} \times 24\,\text{ft} \times 90\,\text{ft} = 10{,}800\,\text{ft}^2$$

$$\text{Air Exchange Required for Summer Cooling} = 12\,\text{cfm/ft}^2 \times 10{,}800\,\text{ft}^2 = 129{,}600\,\text{cfm}$$

$$\text{Minimum Winter Air Exchange Requirement} = 2\,\text{cfm/ft}^2 \times 10{,}800\,\text{ft}^2 = 21{,}600\,\text{cfm}$$

In this case, minimum winter ventilation can uniformly be supplied by four fans each delivering 5,421 cfm at the assumed static pressure of 0.10″ (Model: GHE24; Fan Size:

24"; Motor: 1/3 HP; Fan RPM: 769). Again, the manufacturer model used in this example is a generic model.

These four fans can deliver a total of 21,684 cfm maximum when they operate all together; therefore, in order to determine the spring/fall and summer ventilation fans, this total need to be subtracted from the grand total, 129,600 cfm.

129,600 cfm − 21,684 cfm = 107,916 cfm more to be provided for spring/fall and summer ventilation

We may have other locally available options; however, for this example, we will pick six fans each delivering 18,238 cfm at the assumed static pressure of 0.10" (Model: GHE48; Fan Size: 48"; Motor: 1 HP; Fan RPM: 380). Again, the manufacturer model is a generic model.

| | | | Manufacturer Model No. |
|---|---|---|---|
| 1) | Winter Ventilation | 4 fans at 5,421 cfm each | GHE24 (single speed) |
| 2) | Spring/Fall Ventilation | 3 fans at 18,238 cfm each | GHE48 (two speed) |
| 3) | Summer Ventilation | 3 fans at 18,283 cfm each | GHE48 (two speed) |
| | **Total Fan Capacity:** | **131,112 cfm (installed)** | |

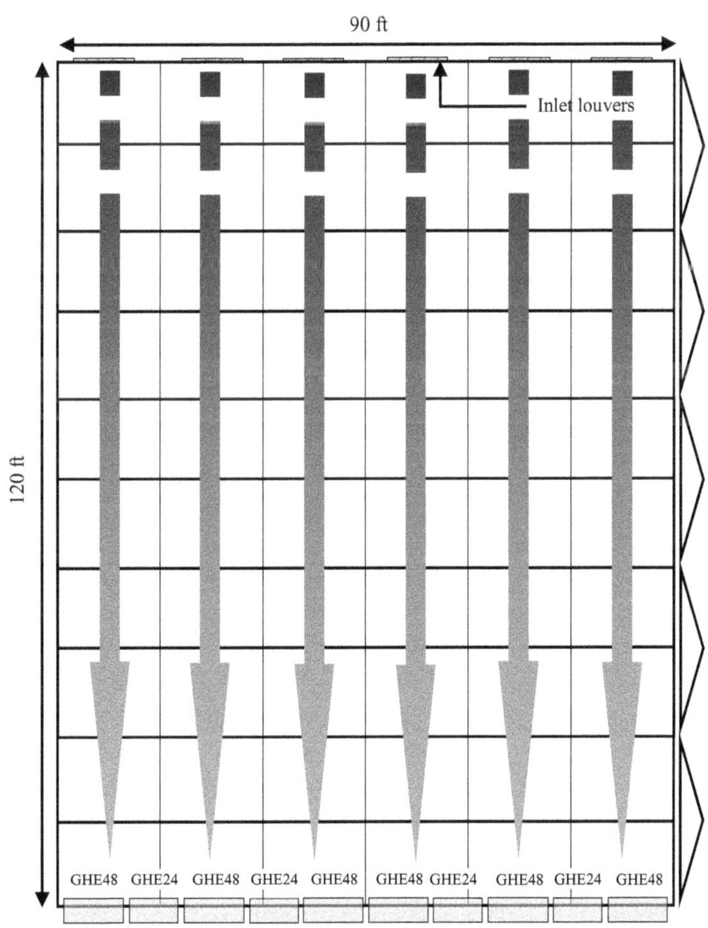

In the above design, the fans and inlet louvers are placed within the distance of 7.6 m (25 ft) along the sidewalls. Also, note that the inlet-to-fan distance is less than 100 ft; therefore, no adjustment was required in the ventilation rate. For an efficient fan operation, keep in mind that a distance of 4–5 fan diameters should also be maintained between the fan discharge and any nearby obstruction such as another greenhouse or building.

Our design for this greenhouse has the total installed capacity of 131,112 cfm, which will provide little more than what is required (129,600 cfm); therefore, the overall ventilation system will be in good shape providing uniform air exchange with the provided configuration in the greenhouse as shown above. The fans are nicely sized and located to provide uniform air exchange and prevent oscillations when controlled properly. Remember that we have used four fans (5,421 cfm each) distributed along the sidewall for minimum (winter) ventilation that will operate continuously. These four fans can operate in steps or stages providing further (proportional) control over minimum ventilation when needed. Furthermore, we have split the total summer (the maximum) air exchange requirement into two stages (low and high) using two-speed fans, providing half the capacity in stages for spring/fall ventilation, and the other half to top up in stages for summer ventilation. It means, all the GHE24 fans (minimum winter ventilation fans) and all the GHE48 fans (low stage only) would be operating for spring/fall ventilation in stages as needed; and all the GHE24 and GHE48 fans (low and high stages) would be operating in stages and eventually providing the total air exchange of 131,112 cfm when needed in summer. This configuration will provide excellent spatial and temporal ventilation for this specific greenhouse. It should however be kept in mind that larger fan diameter and lower HP fans consume less operational energy for delivering the same air exchange; therefore, alternative fans with larger diameter and lower HP motors should be chosen, when possible. Again, reducing the fan rotational speed is the most efficient way to reduce fan energy use in variable flow, and this can be efficiently achieved with frequency-controlled electric motors.

Here, we will again pick motorized air inlet louvers based on 1.5 times the areas of the fans. They are controlled based on the number of active fans or in response to the pressure drop generated across the inlet louver. Total installed ventilation capacity in this example is 131,112 cfm. Using the inlet air velocity maintained at about 3.6 m/s (700 ft/min) at the inlet louver, we can calculate the cross-sectional area of inlet louver as follows:

$$\text{Total Cross-Sectional Area of Inlet Louver} = 1.5 \times \left(131{,}112\,\text{ft}^3/\text{min}\right) \div 700\,\text{ft/min} \approx 281\,\text{ft}^2$$

If a continuous inlet louver is to be installed on the opposite sidewall, since the total length of the greenhouse sidewall is 150 ft, then the width of the inlet louver becomes 281 $\text{ft}^2 \div 90$ ft = 3.12 ft ≈ 3 ¼ ft. Hence, the inlet louver dimension is 3 ¼ ft × 90 ft. We can also install multiple louvers distributed along the opposite sidewall for a uniform airflow distribution in the greenhouse. As shown on the figure above, for instance, six inlet louvers can be used with a cross-sectional area of approximately 48 $\text{ft}^2$ each.

where $Q_{adj}$ is the adjusted ventilation rate in $m^3/s.m^2$ ($cfm/ft^2$), $C$ is coefficient 5.5 in SI units (10.0 in I-P units), and $D$ is the inlet to fan distance in m (ft).

For an efficient fan operation, a distance of 4–5 fan diameters should also be maintained between the fan discharge and any nearby obstruction such as another greenhouse or building, etc. Air inlet louvers or shutters, on the other hand, have an inlet area of minimum 1.25 times the fan area, and should open outward. In cold regions where outside temperatures go below 5°C (40°F), incoming air is mixed and warmed by the warmer interior air before reaching the plants. In such cases, continuous louvers can be installed high along the sidewall, and the inlet air velocity is maintained at 3.6 m/s (700 ft/min) at the required ventilation flow rate. They are motorized and controlled in proportion to the number of active fans or in response to the pressure drop generated across the inlet louver.

## 6.6 Active Cooling of Greenhouses by Evaporative Cooling

Evaporative cooling, using current technology and available equipment, is an energy efficient and cost-effective method of cooling. It cools air by evaporation of water. When water evaporates into the air being cooled, it creates direct evaporative cooling, which is the oldest and the most common form of cooling. As discussed in Chapter 2 earlier, evaporative cooling is essentially an adiabatic saturation process, from an initial state at low humidity ratio and high dry-bulb temperature to a final state at higher humidity and lower dry-bulb temperature. The quantity of heat associated with a change in phase from liquid water to water vapor, termed as *latent heat of vaporization*, $Q_L$, that is, the energy required to evaporate 1 kg liquid water without a change in temperature is about 2,430 kJ (at ~30°C). In imperial units, 1 lb of water at 86°F requires ~1,044 Btu of energy to evaporate. This energy in the form of sensible heat is extracted from the air and this is how the evaporative cooling system cools the air as water evaporates and transforms the sensible heat into the latent heat. When the evaporation occurs separately and the air is cooled without humidity gain, the process is called indirect evaporative cooling. There are a number of techniques to cool the air using direct evaporative cooling. These include i) fan and pad systems, ii) humidifiers such as fog systems, iii) spray-filled and wetted-surface air washers, and iv) sprayed coil units.

Some evaporative cooling methods can provide cost-effective greenhouse cooling. However, direct evaporative cooling adds moisture to the greenhouse air that may reduce plant transpiration and also increase the risk of diseases particularly fungal diseases if not managed adequately. The amount of useful evaporative cooling obtained depends largely on the wet bulb temperature of the air entering the equipment. High ambient relative humidity levels severely limit the effectiveness of evaporative cooling and fog systems because of the reduced vapor pressure differential (also known as wet-bulb depression). The *wet-bulb depression*, which is the difference between the dry-bulb and corresponding web-bulb temperatures, provides the maximum limit the air can be cooled down via evaporative cooling. The drier the air, the bigger the wet-bulb depression is, and therefore, the bigger the potential for evaporative cooling.

### 6.6.1 Fan and Pad Evaporative Cooling Systems

The fan and pad type evaporative cooling system (see Figure 6.9) has been the standard greenhouse cooling system for many years. These systems consist of exhaust fans on the sidewall or gable end of the greenhouse and the pad assembly on the opposite side of the greenhouse. When the exhaust fans operate, they dump the hot and moist greenhouse air outside pulling the cooler and drier outside air through the wetted pads where the water evaporates and the outside air is cooled through the process. Each kilogram of water removes 2,430 kJ (and each pound of water removes 1,044 Btu) at ~30°C (86°F) of heat from

**Example 6.4: Determining Wet-bulb Depressions on the Psychrometric Chart**

Determine the wet-bulb depressions for Tucson, Arizona, USA; Columbus, Ohio, USA; and Montreal, Québec, Canada, and show the evaporative cooling potentials on the psychrometric chart.

**Solution**

Using Tables A.11–A.14, we first determine the summer climatic conditions at the given sites (1% level) as follows:

| LOCATION | | °C | °F | Wet-bulb Depression (°C) | Wet-bulb Depression (°F) |
|---|---|---|---|---|---|
| Tucson, AZ | Design Dry-bulb Temperature (T) | 39 | 102 | 21 | 37 |
| | Coincident Wet-bulb Temperature ($T_w$) | 18 | 65 | | |
| Columbus, OH | Dry-bulb Temperature ($T$) | 31 | 88 | 8 | 15 |
| | Coincident Wet-bulb Temperature ($T_w$) | 23 | 73 | | |
| Montreal, QC | Dry-bulb Temperature ($T$) | 28 | 83 | 7 | 13 |
| | Coincident Wet-bulb Temperature ($T_w$) | 21 | 70 | | |

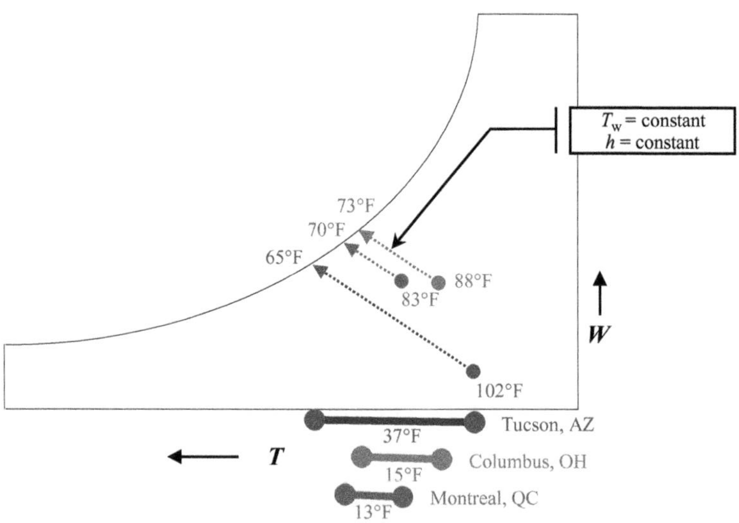


As shown above, Tucson, AZ, has very hot and dry summers, while Columbus, OH, has a milder and more humid summer climate. Montreal, QC, on the other hand, has relatively cooler and more humid summer climate compared to the other two locations. As it has been discussed earlier, the drier the air, the bigger the wet-bulb depression is, and therefore, the bigger the potential for evaporative cooling. Tucson, AZ, has the biggest wet-bulb depression (37°F) of all; and if we assume 80% efficiency for the evaporative cooler, then we can achieve 0.8 × 37°F = 29.6°F ≈ 30°F temperature drop with this cooler, while the same evaporative cooler can approximately provide 12°F and 10°F temperature drops in Columbus, OH, and Montreal, QC, respectively. This exercise clearly shows that evaporative cooling is quite an effective cooling method in hot and dry climates such as the one in Arizona.

**FIGURE 6.9**
A fan and pad evaporative cooling system. Only the pad system is shown here.

the air by changing the phase from liquid to vapor, hence cools the outside air, which is being pulled to the greenhouse.

A fan and pad cooling system, as shown in Figure 6.10, has a pad, a water pump with valves and a water filter, a water distribution pipe, a water collection gutter at the bottom of the fiber pad (connected to a covered sump via a return pipe with a 50 mesh inclined screen), and a covered sump tank, and fan(s) on the opposite wall of the greenhouse. The water is pumped and distributed horizontally through 3 mm (0.125 in.) holes spaced 7.5 cm (3 in.) from each other along the distribution pipe at the top of the vertically mounted pads. For pad lengths less than 15 m (50 ft), a 32 mm (1 ¼ in.) PVC pipe handles the task, while for lengths up to 18 m (60 ft), a 38 mm (1 1/2 in.) PVC pipe is typically used. Non-evaporated water is collected in the collection gutter at the bottom of the vertical pads and

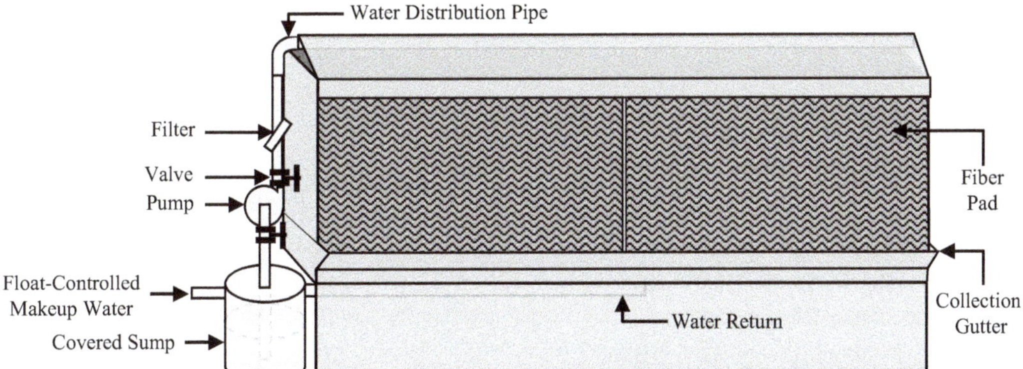


**FIGURE 6.10**
Schematic view of a fan and pad evaporative cooling system. Only the pad system is shown here.

returned to a sump tank to be circulated again. The fans mounted on the opposite sidewall or gable end determine the total evaporative cooling pad area and pulls in fresh air through the pads, and cools the greenhouse interior.

There are a number of commercially available pad materials such as corrugated cellulose (most widely used), aspen pads, and other pads made from aluminum or plastic fibers. Corrugated pads are relatively inexpensive and, when maintained properly, have a lifetime of approximately 10 years. Recommended air velocities through various evaporative cooling pads are presented in Table 6.1.

As mentioned earlier, the fans mounted on the opposite wall determine the total pad area, which can be calculated by dividing the volume of air exchange ($m^3/s$ or $ft^3/min$) by the face velocity of air (m/s or ft/min, respectively) (Table 6.1) that can be moved through the pad. The pads are typically mounted on the interior sides of sidewalls or gable ends extending the entire length of the greenhouse wall. On the exterior side of the pad, air inlets are installed so that outside air can be sealed off during cold weather. When the pad height is more than that of the inlet, the pad should have a minimum of half the height difference space from the inlet. And maximum air velocities through the inlet should be limited to 1.8 m/s (350 ft/min).

Obviously, the cooled outside air moves from the pad side wall toward the opposite wall where the fans are mounted. And as a consequence of this distance, a temperature gradient forms across the greenhouse. Depending on the temperature tolerance of crops – a temperature gradient of 4–5.5°C (7–10°F) is common, greater pad-to-fan distances can be used. However, a pad-to-fan distance of 30–45 m (100–150 ft) is typically utilized even though it is sometimes extended up to 60 m (200 ft).

Cooling pads are typically installed continuously or uniformly along the sidewall or gable end (when possible in the windward direction of prevailing summer winds) opposite the exhaust fans. For a uniform wetting, therefore, an efficient cooling, vertical pad heights are kept at or below 2.4 m (8.0 ft), but not less than 0.60 m (2.0 ft). Algae growth and salt buildup should be monitored and controlled. We need to ensure that the pads remain wet from top to bottom during the cooling process, and when the pads are sized based on Table 6.1, the recommended water flow rates given in Table 6.2 are good enough. When algae growth becomes a problem, then it can be controlled via injecting chemicals through the circulation water. Where high mineral content water is used for evaporative cooling, a continuous water bleed-off rate of 0.02 L/min per $m^3/s$ (0.05 gal/min per 1,000 cfm) of airflow is for minimizing mineral precipitation and buildup in the pads. The sump volumes given in Table 6.2 are sized to operate at the half depth of the tank, hence they will provide room for water returning from the pad when the system is turned off.

**TABLE 6.1**

Recommended Air Velocities Through Various Cooling Pad Materials (ASABE Standards, 2018).

| | Face Velocity of Air [a] | |
|---|---|---|
| **Material** | **(m/s)** | **(ft/min)** |
| Corrugated Cellulose, 100 mm (4 in) thick | 1.27 | 250 |
| Corrugated Cellulose, 150 mm (6 in) thick | 1.78 | 350 |
| Aspen Fiber, 50–100 mm (2–4 in) thick, vertical | 0.76 | 150 |
| Aspen Fiber, 50–100 mm (2–4 in) thick, horizontal | 1.00 | 200 |

[a] Where space is limiting, these velocities may be increased by no more than 25%

**TABLE 6.2**

Recommended Water Flow Rates and Sump Capacities for Vertically Mounted Pads (ASABE Standards, 2018).

| | Minimum Water Flow Rate per Unit Length of Pad | | Minimum Sump Capacity per Unit Pad Area | |
|---|---|---|---|---|
| **Material** | **(L/min.m)** | **(US gal/min.ft)** | **(L/m²)** | **(US gal/ft²)** |
| Corrugated Cellulose, 100 mm (4 in) | 6.2 | 0.50 | 33 | 0.80 |
| Corrugated Cellulose, 150 mm (6 in) | 9.9 | 0.80 | 40 | 1.00 |
| Aspen Fiber, 50–100 mm (2–4 in) | 3.7 | 0.30 | 20 | 0.50 |
| Aspen Fiber, 50–100 mm (2–4 in), desert conditions | 5.0 | 0.40 | 20 | 0.50 |

As mentioned earlier, exhaust fans are sized to provide the required airflow rate at an overall static pressure of 0.030 kPa (0.12 in. $H_2O$) and mounted on the opposing greenhouse wall. And they should not exhaust directly into the pads of the neighboring greenhouse unless the distance between fan and inlet is more than 15 m (50 ft). Also, the fans should not exhaust directly into each other unless they are more than 4–5 fan diameters apart.

### 6.6.2 Fog Cooling Systems

Fog cooling uses exactly the same principle that is used in fan and pad evaporative cooling. Fog cooling systems however employ high-pressure pumps typically operating at 6.9–10.3 MPa (1,000–1,500 psi), copper, stainless steel, or reinforced flexible poly-tubes for water piping, and high-pressure ceramic, plastic, or stainless steel nozzles to form 0.5–50 μm (microns) water droplets for cooling the greenhouse (see Figure 6.11). The water consumption per nozzle is approximately 0.00105–0.00126 L/s (1.0–1.2 gallons per hour). The pipes and nozzles are located over aisles in order to prevent water dripping directly over plants. The water droplets almost vaporize immediately and cool the greenhouse air; however, fog-cooling systems are not as effective as the fan and pad systems because these systems typically operate on the greenhouse interior air. Therefore, they must be monitored carefully with respect to the inside relative humidity levels. Even though they can be uniformly spaced and used in greenhouses with natural ventilation systems, for an effective fog cooling system, the conventional exhaust fans need to replace the humid air inside

**Example 6.5 Fan and Pad Type Evaporative Cooling**

Let's now consider a fan and pad type evaporative cooling system used for air-conditioning an institutional greenhouse. Assume that the outside air enters the evaporative cooling pad at 35°C dry-bulb temperature and 25% relative humidity with a flow rate of 4 $m^3/s$. Assuming that the fan and pad evaporative cooler has a cooling efficiency of 80%, determine the supply $T$, $\phi$, total cooling capacity provided, and the rate of water used for this evaporative cooling. Note: 1 ton of refrigeration equals a heat removal rate of 12,000 Btu/h.

**Solution**

See Example 2.83.

**FIGURE 6.11**
Fog cooling system installed together with a retractable shading system.

**Example 6.6: Fan and Pad Type Evaporative Cooling**

Now, using the same greenhouse that was used in Example 6.3, design a fan and pad type evaporative cooling system using 100 mm (4 in.) thick cellulose pads. Remember that the static pressure required for the fans is 0.030 kPa (0.12 in. $H_2O$). Assuming that the greenhouse is located in Tucson, Arizona, USA, and the fan and pad evaporative cooler has a cooling efficiency of 80%, determine the supply $T$, $\phi$, total cooling capacity provided, and the rate of water used for this evaporative cooling. Note: 1 ton of refrigeration equals a heat removal rate of 12,000 Btu/h.

**Solution**

Remember that in Example 6.3 the total fan capacity, 131,112 cfm, was installed for a static pressure of 0.10 $H_2O$. Since the static pressure required for the evaporative fans is 0.12 in. $H_2O$, we now need to pick new fans as follows:

$$\text{Minimum Winter Air Exchange Requirement} = 21{,}600\ \text{cfm}$$

Minimum Winter Air Exchange Requirement = 21,600 cfm

In this case, minimum winter ventilation can uniformly be supplied by four fans each delivering 5,648 cfm at a static pressure of 0.125″ (Model: GHE30; Fan Size: 30″; Motor: 1/4 HP; Fan RPM: 491). Again, the manufacturer model used in this example is a generic model.

These four fans can deliver a total of 22,592 cfm maximum when they operate all together; therefore, in order to determine the spring/fall and summer ventilation fans, this total need to be subtracted from the grand total, 129,600 cfm.

$$129{,}600\ \text{cfm} - 22{,}592\ \text{cfm} = 107{,}008\ \text{cfm more to be provided for spring /fall and summer ventilation}$$

We may have other locally available options; however, for this example, we will pick five fans each delivering 17,456 cfm at a static pressure of 0.125″ (Model: GHE48; Fan Size: 48″; Motor: 1 HP; Fan RPM: 380); and another fan delivering 20,480 cfm again at the required static pressure (Model: GHE54; Fan Size: 54″; Motor: 1 ½ HP; Fan RPM: 414). Both manufacturer models are generic.

| | | | **Manufacturer Model No.** |
|---|---|---|---|
| 1. | Winter | 4 fans at 5,648 cfm each | GHE24 (single speed) |
| 2. | Spring/Fall/Summer | 5 fans at 17,456 cfm each | GHE48 (two speed) |
| 3. | Spring/Fall/Summer | 1 fan at 20,480 cfm | GHE54 (two speed) |
| | **Total Fan Capacity:** | **130,352 cfm (installed)** | |

Now, using the total installed fan capacity of 130,352 cfm, we can calculate the face area of the 6 in. pad as follows:

Face Area of the Pad = 130,352 cfm ÷ 250 ft/min (see Table 6.1 for 4 in. thick cellulose) = 521 $ft^2$

Now, calculate the pad height if pad extends full length of 90 ft sidewall:

$$\text{Pad Height} = \text{Area of the Pad} \div \text{Length of the Greenhouse}$$

$$\text{Pad Height} = 521\ \text{ft}^2 \div 90\ \text{ft} \approx 6\ \text{ft}$$

Then, the overall 4 in. pad dimension is 6 ft × 90 ft.

Now, we can calculate the pump flow rate as follows:

Water Flow Rate = Pad Length × Flow Rate per Unit Length of Pad (see Table 6.2)

Water Flow Rate = 90 ft × 0.5 gal/min . ft (see Table 6.2 for 4 in. thick cellulose pad)

$$\text{Water Flow Rate} = 45\ \text{gpm}$$

And, the minimum sump size as follows:

Minimum Sump Capacity = Pad Face Area x Required Sump Capacity per Unit Pad Area (see Table 6.2)

Minimum Sump Capacity = (6 × 90) $ft^2$ × 0.80 gal/$ft^2$ (see Table 6.2 for 4 in. thick cellulose pad)

$$\text{Minimum Sump Capacity} = 432\ \text{gal}$$

Note that in greenhouses where the air inlet-to-fan distance is less than 30 m (100 ft), ventilation flow rates should be increased to improve the heat transfer based on Equation (6.1). In our example, the distance is 36 m (120 ft); therefore, no adjustment was needed.

Now considering the above fan and pad type evaporative cooling system has a cooling efficiency of 80%, and assuming that the greenhouse is located in Tucson, AZ, and the outside air enters the evaporative cooling pad at the summer design dry-bulb temperature of 102°F (39°C) (Tables A.11 and A.12) and the coincident wet-bulb temperature of 65°F (18°C) (Tables A.11 and A.12) with a flow rate of 130,352 cfm (~61.3 $m^3/s$), we can determine the supply $T$, $\phi$, and total cooling capacity provided as follows:

State 1 is the outside air condition. If the cooling system were 100% efficient, then the temperature at State 2 (after the cooling pad) would be the $T_w$ of State 1 (102°F), that is, State 1 is located at the intersection of $T = 102°F$ and $T_w = 65°F$, which would at the same time be the temperature at State 2. With 80% cooling efficiency, however, $T_2$ would be $T_1$

minus the 80% of ($T_1 - T_w$), that is, $T_2$ = 102°F – 0.8 (102°F – 65°F) = 72.4°F, which is the supply temperature to the greenhouse.

Using the intersection of $T_2$ = 72.4°F and the constant wet-bulb temperature line of $T_w$ = 65°F on the psychrometric chart or using psychrometric equations (Chapter 2), we can now determine the humidity ratio at State 2 (after the pad) as $W_2 = 0.01150\ \text{lb}_{\text{water}}/\text{lb}_{\text{da}}$, the specific volume $v_2 = 13.66\ \text{ft}^3/\text{lb}_{\text{da}}$, and the relative humidity at State 2 as $\phi \approx 69\%$.

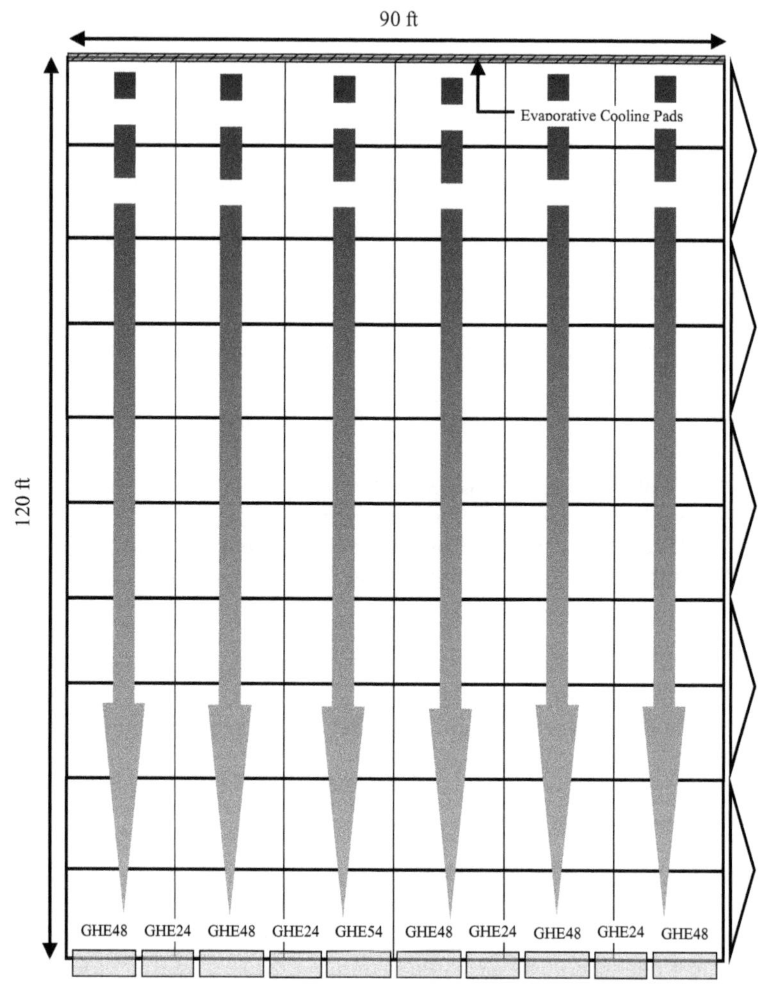


Now, we can first calculate the mass flow rate of dry air using the specific volume $v_2$ at State 2 as the fan is positioned on the opposite wall (negative pressure or suction type ventilation). The mass flow rate of dry air is then:

$$\dot{m}_a = \left(130{,}352\frac{\text{ft}^3}{\text{min}}\right) / \left(13.66\frac{\text{ft}^3}{lb_{da}}\right) = 9{,}543\ \text{lb}_{\text{da}}/\text{min}$$

And now using Equation (2.138) and the specific heat of air 0.24 $\text{Btu}/\text{lb}_{\text{da}}\cdot°\text{F}$ (Table 2.12),

$$\dot{Q} = \dot{m}_a c_p (T_2 - T_1) = \left(9{,}543 \frac{\text{lb}_{\text{da}}}{\text{min}}\right)\left(60 \frac{\text{min}}{\text{h}}\right)(0.24\ \text{Btu/lb}_{\text{da}}{}^\circ\text{F})(72.4 - 102)^\circ\text{F}$$

$$\dot{Q} = -4{,}067{,}608\ \text{Btu/h}\ (-1{,}192\ \text{kW})$$

And since 1 ton of refrigeration is equal to 12,000 Btu/h, the amount of cooling provided by this fan and pad evaporative cooling system is approximately equal to 340 ton.

And using Equation (2.134), the theoretical rate of water use for the evaporative cooling process can be calculated as:

$$\dot{m}_w = \dot{m}_a (W_2 - W_1) = (9{,}543\ \text{lb}_{\text{da}}/\text{min})(0.01150 - 0.00543)\ \text{lb}_{\text{water}}/\text{lb}_{\text{da}} = 57.93\ \text{lb}_{\text{water}}/\text{min}$$

$$\dot{m}_w = (57.93\ \text{lb}_{\text{water}}/\text{min})\left[0.12\ \text{gpm}/(\text{lb}/\text{min})\right] = 6.95\ \text{gpm} \approx 7\ \text{gpm}$$

**Example 6.7: Fan and Pad Type Evaporative Cooling**

Now, using the same greenhouse that was used in Examples 6.3 and 6.6, design a fan and pad type evaporative cooling system, this time, using 150 mm (6 in.) thick cellulose pads.

**Solution**

Remember that the total fan capacity is 130,352 cfm. Now we can calculate the face area of the 6 in. pad as follows:

Face Area of the Pad = 130,352 cfm ÷ 350 ft/min (see Table 6.1 for 6 in. thick cellulose) = 372 ft²

Now, calculate the pad height if pad extends full length of 90 ft sidewall:

$$\text{Pad Height} = \text{Area of the Pad} \div \text{Length of the Greenhouse}$$

$$\text{Pad Height} = 372\ \text{ft}^2 \div 90\ \text{ft} \approx 4\ ½\ \text{ft}$$

Then, the overall 6 in. cellulose pad dimension is 4 ½ ft × 90 ft.

Now, we can calculate the pump flow rate as follows:

Water Flow Rate = Pad Length × Flow Rate per Unit Length of Pad (see Table 6.2)

Water Flow Rate = 90 ft × 0.8 gal/min . ft (see Table 6.2 for 6 in. thick cellulose pad)

$$\text{Water Flow Rate} = 72\ \text{gpm}$$

And, the minimum sump size as follows:

Minimum Sump Capacity = Pad Face Area x Required Sump Capacity per Unit Pad Area (see Table 6.2)

Minimum Sump Capacity = (4 ½ × 90) ft² × 1.00 gal/ft² (see Table 6.2 for 6 in. thick cellulose pad)

$$\text{Minimum Sump Capacity} = 405\ \text{gal}$$

See the difference between Examples 6.6 and 6.7. The total amount of cooling is 329 ton here as well.

periodically with the drier outside air. In fan-cooled greenhouses, most of the nozzles are located near the inlet with the remaining distributed evenly over rest of the greenhouse space. Fog nozzles can also be installed inside the air inlets when mechanical ventilation is used.

Fogging can be controlled with a computer or simply by a two-stage thermostat. Temperature settings of 3–5.5°C (5–10°F) apart would provide good enough control with this two-stage control. Since the nozzles have quite tiny openings, good quality water and filtration is extremely important for the performance of fog cooling systems.

## 6.7 Active Cooling of Greenhouses by Heat Pumps

Several heat pump cycles are considered feasible: the reversed Rankine cycle, reversed Brayton cycle, reversed Stirling cycle, and Ericson cycle. Among these, the most commonly used cycle is the reversed Rankine cycle, which is also known as the vapor-compression cycle (see Figure 6.12). The driving energy of these cycles can be electricity generated in a power plant, an internal combustion engine, a Rankine or Brayton engine cycle, or an absorption system.

The vapor-compression cycle is the most common residential refrigeration/heat pump cycle used today. Current electric heat pump systems use a vapor-compression cycle. The refrigerant removes heat from the environment in the evaporator; it is, then, compressed in an electrically driven compressor or an internal combustion engine, and then sent to a condenser. The condenser rejects heat to the heat sink, and then the refrigerant flows back to the evaporator through an expansion valve. The energy consumer of this cycle is the electric motor driving the compressor. The COP (coefficient of performance) of heat pumps with respect to the compressor work varies in cooling and heating modes depending on the refrigerant used. However, if the energy source were considered, real efficiency is then depended on the electric plant or the engine efficiency. Colosimo (1987) reported that the typical efficiency of an electric power plant is about 32%, and approximately 3% (even more) of the electricity is lost during the transmission process, leaving only 29% of the

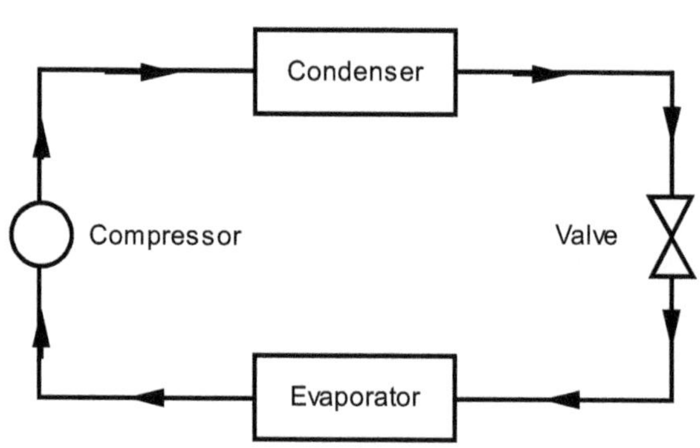


**FIGURE 6.12**
Vapor-compression cycle.

energy supplied by fuel to be utilized as electrical energy. Consequently, the actual COP value of the electric heat pump system is less than a third of the value obtained with respect to the compressor work.

In order to improve COP, alternative cycles have been developed utilizing the energy source directly (oil or natural gas). These are the Brayton or Rankine cycle engine-driven heat pumps. The power cycles in these systems generate energy to drive the compressor for the vapor-compression cycle. Since both of these systems do not have an acceptable COP in both heating and cooling, Christensen and Santoso (1990) developed an alternative configuration of a Rankine cycle engine-driven heat pump system, which uses a hydraulically connected rolling diaphragm piston-cylinder device as the motor, compressor and pump. Initial evaluations of the system gave heating and cooling COPs of 2.01 and 1.06, respectively.

Another commonly used refrigeration cycle is the absorption refrigeration cycle (see Figure 6.13). Usually, natural gas is used as a source of energy, which supplies some heat to the generator. This cycle uses an absorption system (instead of a compressor), which consists of an absorber, a pump, and a generator to drive the whole system. Ammonia-water and water-lithium bromide solutions are the most commonly used refrigerants. Absorption refrigeration systems have several advantages compared to the vapor compression cycle. These are the quietness and the high efficiency at reduced load operations. However, they have disadvantages (such as lower COPs, more components, and ammonia toxicity) when compared to the vapor-compression systems. Santoso (1989) compared an absorption heat pump to a Rankine cycle engine-driven heat pump for the same operating conditions, and determined that the absorption system using ammonia-water had COPs of 0.73 and 0.63 in the heating and the cooling modes, respectively. For the same operating conditions, a Rankine power cycle, engine-driven heat pump system had better COP predictions using R22 and R113 refrigerants. The COPs were 2.01 and 1.06 in the heating and the cooling modes, respectively.

Yildiz (1993) developed a simulation model for a gas-fired Rankine cycle engine-driven heat pump system, which was used for winter heating and warm weather cooling

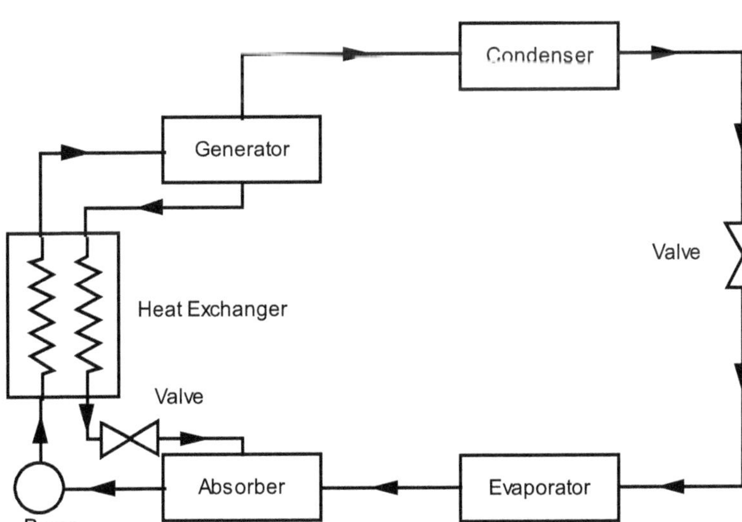


**FIGURE 6.13**
Absorption cycle.

requirements of a greenhouse. Three energy sources are typically used for heat pump heating applications. These are the ambient air, ground water, and the earth. The earth is considered to be the best energy source; however, since most heat pump applications today use ambient air as an energy source or sink, ambient air was used as the energy source or sink for this heat pump system. The heat pump system was a 10,548 W (3 ton, 36,000 Btu/hr) unit consisting of a Rankine cycle and a vapor compression cycle. In the system, the power generated by the Rankine cycle replaces the electric power used to drive the compressor in Figure 6.12. Natural gas was used as the energy source, which supplied heat to the evaporator of the Rankine cycle. A fixed burner rate of 12,306 W (42,000 Btu/hr) and a burner efficiency of 0.9 were used. The vapor compression cycle removed the energy from the heat source through an evaporator attached to a hydronic cycle. The Rankine cycle used R123 (dichlorotrifluoroethane) as the working fluid; and the vapor compression cycle used R22 (chlorodifluoromethane) as its refrigerant.

Using this heat pump system, Yildiz & Stombaugh (2006a,b) studied heating and cooling of a greenhouse at Delaware (latitude 40° 17′ N, longitude 83° 05′ W), Ohio, USA. January, April, and July weather files were used to represent winter, spring, and summer in the simulations, respectively. Simulations were performed starting at the beginning of the fifth day and ended at the end of 29th day of the month providing a 25-day simulation.

Table 6.3 presents some important greenhouse characteristics. The heat pumps evaluated were 3-ton gas-fired units, and in order to provide multiposition proportional control, it was assumed that three heat pump units were used in each greenhouse. Along with variable shading and ventilation, an evaporative cooler was employed for cooling the conventional (CON) greenhouse. Outdoor air was used in the evaporative cooling system, and it was assumed that the air at the evaporative cooler outlet was fully saturated. In the open heat pump (OHP) and closed heat pump (CHP) systems, however, in addition to the variable shading system, heat pump units provided the cooling requirements. Indoor air was circulated and introduced back to the inside at a lower temperature using an overhead plastic tube for the air distribution. Two shading clothes with transmissivities of 0.75 and 0.50 were used to reduce the cooling loads. The use of these shading clothes provided shading levels of 25%, 50%, and 62.5% by using them individually or together. In the open systems (CON and OHP), ventilation was provided by two fans, one with a fixed flow rate

**TABLE 6.3**

Greenhouse and Crop Characteristics used in the Simulation Model.

| | |
|---|---|
| Greenhouse length | 7.50 m (CON & OHP) and 25.0 m (CHP) |
| Greenhouse width | 7.50 m |
| Greenhouse height at eaves | 2.50 m |
| Greenhouse height at ridges | 4.50 m |
| Greenhouse glazing | Double polyethylene |
| Floor surface | Reflective mulch with 60% reflectivity |
| Crop | Cucumber |
| Crop row orientation | North–South |
| Distance between plant rows | 0.75 m |
| Average leaf length and width | 0.30 m and 0.75 |
| Plant height | 2.0 m |

to provide a minimum level (0.01 $m^3/s.m^2$) of air exchange at all times, and the other one with a variable flow rate (with a maximum rate of 0.08 $m^3/s.m^2$).

Based on the indoor air temperature, the control system operated in either the heating or cooling mode. If the system was in heating mode and if heating was required, the ventilation rate was set to the minimum rate. The cooling mode operated in two steps. The first step was to reduce the cooling load using a variable shading system and to cool the inside air by increasing ventilation rates. If the first step in cooling could not handle the cooling load, then the second step was activated, in which the heat pump units (OHP) or evaporative cooling system (CON) provided the cooling. Inside relative humidity in the conventional system was controlled indirectly by the temperature control. In the OHP system, however, additional relative humidity control was provided. When the inside relative humidity levels exceeded 80%, additional ventilation was introduced to decrease inside relative humidity. In the cooling mode of the CHP system, there was only one step unlike the conventional and OHP systems, which had two-step cooling systems. Here, no cooling was provided by ventilation; instead, it was done by the three heat pump units providing a multiposition proportional control, when possible, after reducing the cooling load using the variable shading system. The operation of the shading system was the same as in the other two systems. Here, either the cooling units or the dehumidifier (the first heating unit) controlled inside relative humidity.

Figure 6.14 shows the diurnal changes in solar radiation, relative humidity, and temperature regimes in conventional, open, and CHP greenhouse systems on a spring day. Figure 6.15a,b, on the other hand, shows the operations of control systems in the conventional greenhouse system on a summer day and the confined greenhouse system on a spring day to illustrate the wide diversity in operational strategies used in this system.

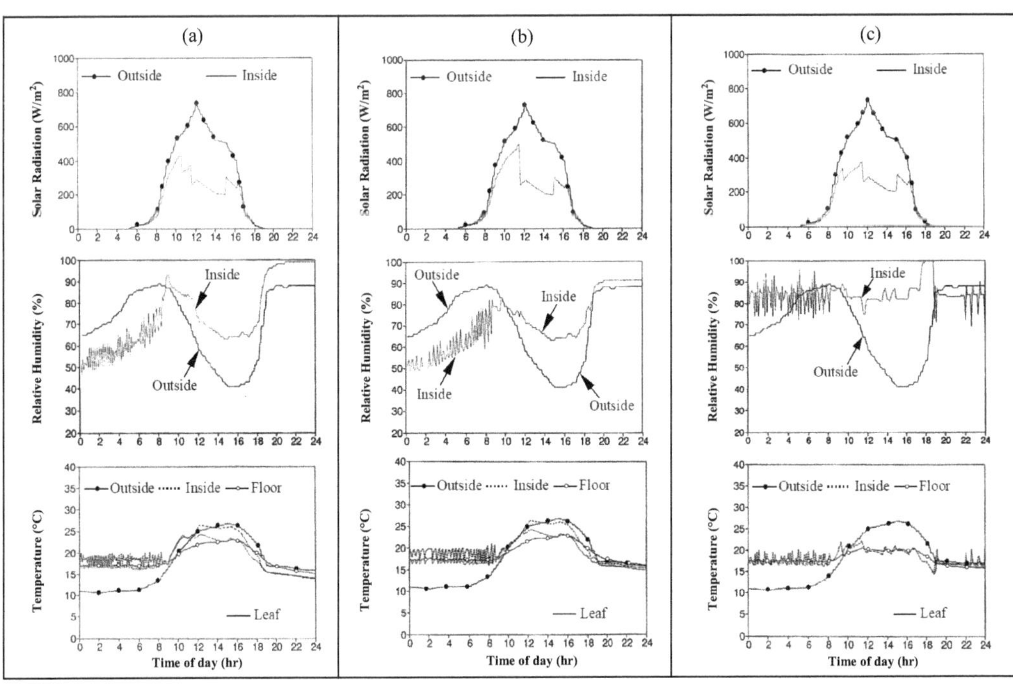


**FIGURE 6.14**
Diurnal changes in predicted greenhouse climatological quantities in the (a) conventional, (b) open heat pump, and (c) closed heat pump greenhouse systems on a spring day.

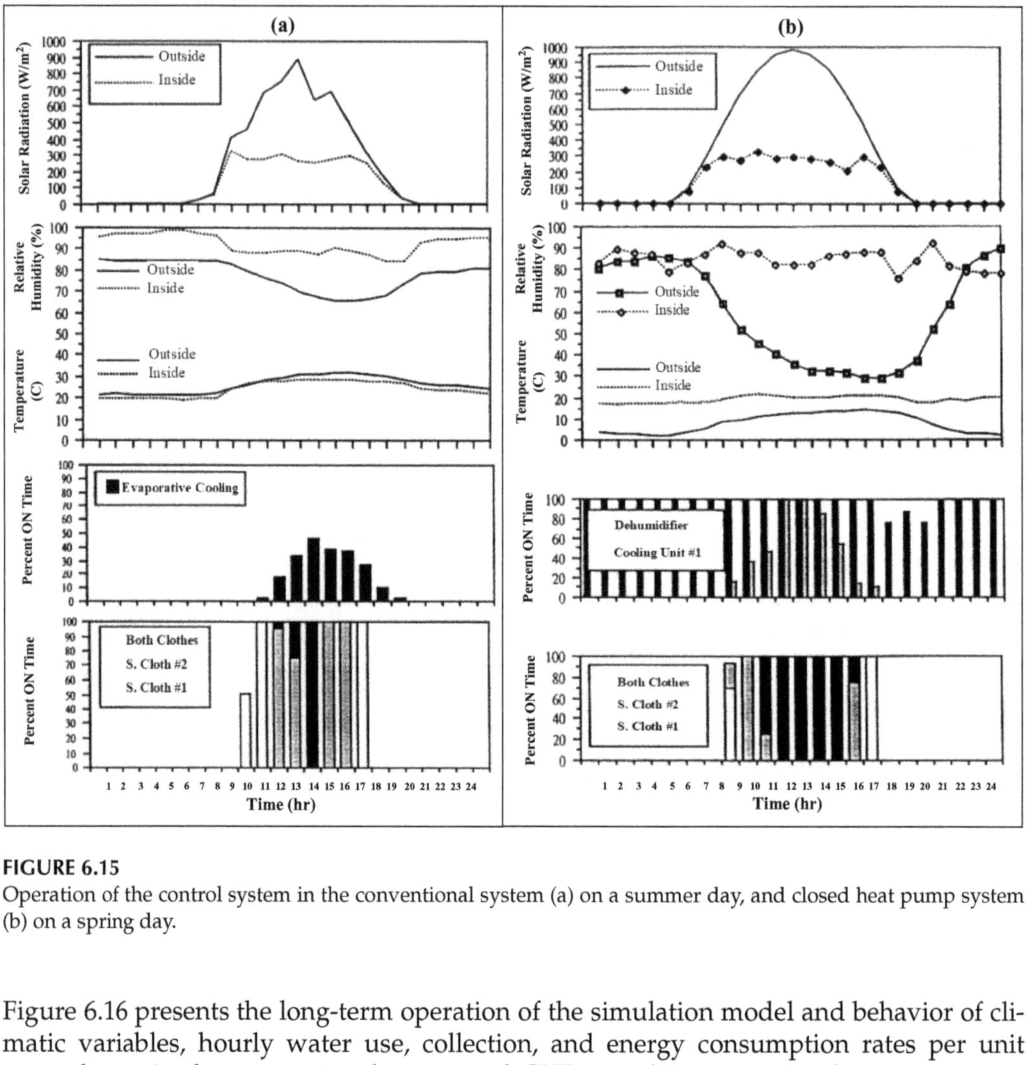


**FIGURE 6.15**
Operation of the control system in the conventional system (a) on a summer day, and closed heat pump system (b) on a spring day.

Figure 6.16 presents the long-term operation of the simulation model and behavior of climatic variables, hourly water use, collection, and energy consumption rates per unit ground area in the conventional, open, and CHP greenhouse systems for 10 successive days in summer. In the conventional system, water was used for both transpiration and evaporative cooling (Figure 6.16a), whereas in the heat pump greenhouse systems water was used for transpiration only, some of it was recovered in the open system and almost all in the confined system (Figure 6.16b).

Table 6.4 summarizes the daily averages of energy consumption for cooling and water consumption in conventional, open, and CHP greenhouse systems in winter, spring, and summer. When both heating and cooling seasons are considered, it was determined that 99.5% of the energy consumption in the conventional system was used for heating, while the remaining portion (0.5%) was used for ventilation at this particular location. In the OHP system, 94.2% of the total energy consumption was used for heating, while 3.1% and 2.7%, respectively, were used for ventilation and heat pump cooling. In the CHP system, 70% of the total energy consumption was used for heating, while 30% was used for cooling. The energy used for cooling was much less than the energy used for heating, because the shading system functioned very well and reduced the cooling loads. Also, the plants

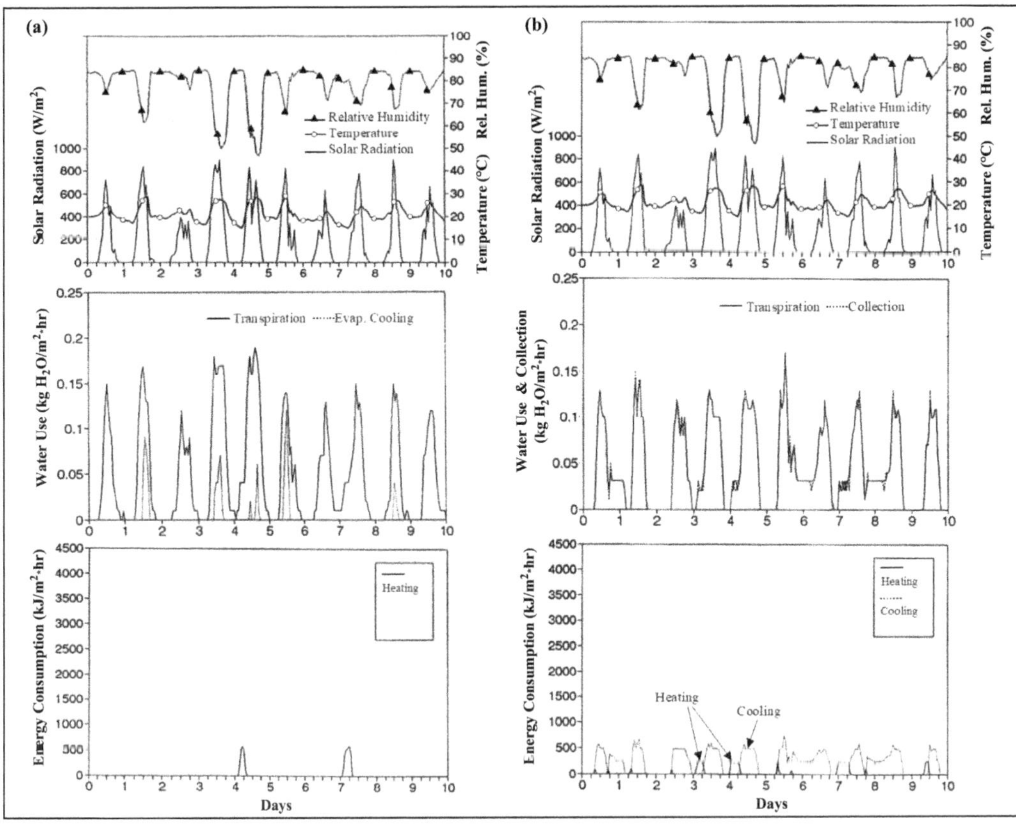


**FIGURE 6.16**
Predicted water use, collection, and energy consumption per unit ground area in the (a) conventional and (b) closed heat pump greenhouse systems for 10 successive days in summer.

themselves are biological air-conditioners and it should be remembered that they can provide much of their own cooling.

Table 6.4 shows that, in addition to the transpiration, water was also used for evaporative cooling in the conventional system. In the CHP system, however, almost all the transpired water was recovered on the condensing coils, making the overall water consumption in this system essentially zero. Thus, the CHP system was the most water conserving system, while the conventional system used the most water of all the systems in this study. When all three systems are compared with respect to the total energy consumption (both heating and cooling), the CHP system was the most energy conserving system, while the conventional system used the most energy of all the systems. If the three systems are to be compared with respect to the energy consumption for cooling only, based on the average daily energy consumption (winter, spring, and summer values averaged together), the CHP system had the highest energy consumption for cooling (2.48 MJ/day.m$^2$), while the OHP and conventional greenhouse systems had the consumptions of 0.66 MJ/day.m$^2$ and 012 MJ/day.m$^2$, respectively.

Yildiz and Stombaugh (2006a,b) concluded that greenhouse operation with the heat pump system has a great potential for enormous energy and water savings, as well as for dehumidification. Closed greenhouse operation with a heat pump system would also make it easier to control the humidity and keep high $CO_2$ levels in greenhouses, while

**TABLE 6.4**

Energy Consumption with Respect to Minimum Ventilation Rate in Winter, Spring, and Summer.

| | WINTER | | | SPRING | | | SUMMER | | |
|---|---|---|---|---|---|---|---|---|---|
| | CHP | OHP | CON | CHP | OHP | CON | CHP | OHP | CON |
| **ENERGY CONSUMPTION FOR COOLING (MJ/d.m²)** | | | | | | | | | |
| Natural Gas | 0.02 | -- | -- | 1.94 | -- | -- | 5.19 | 0.91 | -- |
| Fan (Electric) | * | -- | -- | 0.05 | -- | -- | 0.14 | 0.02 | -- |
| Pump (Electric) | * | -- | -- | 0.03 | -- | -- | 0.08 | 0.01 | -- |
| **Subtotal** | 0.02 | -- | -- | 2.02 | -- | -- | 5.40 | 0.94 | -- |
| Ventilation | -- | 0.11 | 0.11 | -- | 0.23 | 0.11 | -- | 0.71 | 0.13 |
| TOTAL | **0.02** | **0.11** | **0.11** | **2.02** | **0.23** | **0.11** | **5.40** | **1.65** | **0.13** |
| **WATER CONSUMPTION (kg $H_2O$/d.m²)** | | | | | | | | | |
| Transpiration | 0.85 | 2.52 | 2.54 | 1.05 | 2.22 | 2.17 | 1.17 | 1.66 | 1.28 |
| Water Recovered | 0.83 | -- | -- | 1.04 | -- | -- | 1.17 | 0.35 | -- |
| Evaporative Cooling | -- | -- | -- | -- | -- | -- | -- | -- | 0.29 |
| TOTAL | **0.02** | **2.52** | **2.54** | **0.01** | **2.22** | **2.17** | **0.00** | **1.31** | **1.57** |

* Negligible amount. CON, OHP, and CHP: Conventional, open, and closed heat pump greenhouse systems, respectively.

reducing the risk of insects and diseases within the confined greenhouse environment. The heat pump system performed very well with a variable shading system significantly conserving both energy and water when both heating and cooling are considered. A combination of open and CHP systems would be the most energy and water conserving system, operating the system as a closed system in fall, winter, and spring and as an open system in summer. However, when only the cooling season is considered, the commercial feasibility of using a heat pump system would depend on the market conditions of the produce and the other savings such as savings in water and $CO_2$ use.

It should be kept in mind that greenhouses can also be cooled by earth or water coupled heat pumps; and cold-water reservoirs can be underground water, oceans, lakes, and rivers. Furthermore, we can directly cycle cold water from these reservoirs and provide ground cooling using PE tubes in greenhouses.

## 6.8 Active Cooling of Greenhouses by the Watergy Concept

For hot and arid regions, Buchholz et al. (2004) developed the Watergy concept – a novel, closed greenhouse system. The Watergy concept follows the development of a technology platform for decentralized basic supply of energy, water, and food, which can be used within a number of different applications. The system has no external aeration and has almost no need for further water supply. The technological principle is actually proven at two different

**FIGURE 6.17**
First prototype of Watergy greenhouse built in Almeria, Spain.

prototypes that were built within a European Union (Framework Program Promoting Energy, Environment, and Sustainable Development) funded research project. In 2004, the first prototype was built for the purpose of greenhouse horticulture in Almeria, Spain (see Figure 6.17), which holds great promise for heating and cooling greenhouses. The second prototype was built in Berlin, Germany, as a living or office building with an attached façade-greenhouse. It has a great potential to reduce the demand for or to totally displace natural gas, to reduce the load on power grid by electricity savings, and the demand for irrigation water in greenhouse operations. It also allows plant protection without chemical insecticides and the accumulation of carbon dioxide without any aeration losses. The process generates heat, clear water, and cooled/dehumidified air at minimal need for primary energy.

The main focus of Watergy is the decrease of the temperature peaks in the greenhouse by using water based thermal storage for alternate heating and cooling. A day–night balance of the temperatures is aimed at greenhouse daytime cooling for times of strong solar irradiation and for heating during winter nights. A day–night storage is a cheap system and can economically provide heating and cooling with temperature differences from about 10 K. Daytime cooling during the summer is provided from the storage that is unloaded through the night. In the cooling duct, the circulated air has to reach the dew-point temperature, as the greenhouse air has to be continuously dehumidified to enable a sufficient transpiration from the vegetation for providing enough evaporative cooling capacity in the closed system. The accumulated condensate in this process can be used for further irrigation.

Besides the production of food, closed greenhouses can also establish a basic module for decentralized systems of combined water supply and wastewater treatment. In order to exploit the full economical potential of the system, it is necessary to go beyond the concept of just recovering the evaporated water for reuse within the same greenhouse. Graywater can be used for irrigation in the greenhouse, and following the above-described heating/cooling process, can also supply drinking water. The idea is to substitute drinking water with cheaper and easier available sources of rainwater, graywater, and saltwater. Rainwater is already commonly used in horticulture when water resources are limited. Through a further humidification device in the roof area of the greenhouse, even additional sea or brackish water can be evaporated to provide desalination within the same process. In this way, greenhouse horticulture can turn from a water-consuming activity into a water-producing technology.

The ability of air to absorb water vapor is however dependent on its temperature. Due to the temperature gradients within the air cycle of the Watergy system, humidification is achieved within two steps: The first step is done through the transpiration of the vegetation. While using graywater, the waste components in the water are also used as plant nutrition. In the second step at higher temperatures, the water gets into direct contact with the air by using sprinklers or sprayers. Here, rainwater, pretreated graywater, or saltwater can be used. In the latter case, the brine can be concentrated within the component to the limit of solubility and after that, sent back to the sea or to an exterior system for complete desalination. This does not only have a cleaning effect, but also a passive cooling effect in the greenhouse. Evaporation acts as an interface between the water and energy management systems.

## Nomenclature

| | |
|---|---|
| $C$ | coefficient |
| **cfm** | cubic feet per minute |
| **CHP** | closed heat pump greenhouse system |
| **CON** | conventional greenhouse system |
| **COP** | coefficient of performance |
| $D$ | distance |
| $h$ | enthalpy |
| **OHP** | open heat pump greenhouse system |
| $Q$ | ventilation rate |
| $T$ | temperature |

### Greek letters

| | |
|---|---|
| $v$ | specific volume |
| $\phi$ | relative humidity |

### Subscripts

| | |
|---|---|
| $adj$ | adjusted |
| $v$ | ventilation |
| $w$ | wet bulb |
| **1,2** | state numbers |

## Bibliography

Aldrich RA and Bartok, JW (1994) Greenhouse Engineering. *NRAES-33. Northeast Regional Agricultural Engineering Service*. Cornell University, Ithaca, NY, USA.

ASABE Standards (2018) Standards, Engineering Practices, and Data. *V21(1)*. St. Joseph, MI, USA.

ASHRAE (1987) Handbook, HVAC Syst. & Applic. *American Society of Heating, Refrigeration and Air Conditioning Engineers, Inc.* Atlanta, GA 30329. pp. 10.25–10.29.

Baeza EJ (2007) Optimización del diseño de los sistemas de ventilación en invernaderos tipo parral. *Tesis doctoral. Escuela Politécnica Superior. Departamento de Ingeniería Rural.* Universidad de Almería, Almería, Spain.

Boulard T (2006) Greenhouse natural ventilation modelling: a survey of the different approaches. *Acta Hort.* 719:29–40.

Buchholz M, Jochum P, and Zaragoza G (2004) *Basic Water, Heat and Food Supply from a Closed Greenhouse – The Watergy Project, International Symposium GREENSYS 2004 – Sustainable Greenhouse Systems, Leuven, Belgium, 12–16 September 2004*, published in International Society of Horticultural Science (ISHS) und Wageningen University and Research Center, Wageningen 2004 ISBN: 907601924X, S. 133.

Buchholz M, Zaragoza G, Buchholz R and Pérez-Parra J (2006) Temperature and humidity control in the watergy greenhouse. *Acta Hort.* 719:401–408.

Castilla N and Montero JI (2008) Environmental control and crop production in Mediterranean greenhouses. *Acta Hort.* 797:25–36.

Christensen RN and Santoso M (1990) An evaluation of a Rankine cycle driven heat pump. *Heat Recovery Systems and CHP*, 10 (2): 161–175.

Colosimo DD (1987) Introduction to engine driven heat pump: concepts, approach and economics. *ASHRAE Transaction*, 93 (2): 987–996.

de Zwart HF and Kempkes FLK (2008) Characterizing of cooling equipment for closed greenhouses. *Acta Hort.* 801:409–416.

Friedman D (1977) *Light commercial Bryton/Rankine space conditioning system. 12th Intersociety Energy Conversion Engineering Conference*, Washington, D.C., U.S.A., p.172.

Goldammer T (2019) *Greenhouse Management: A Guide to Operations and Technology*. Apex Publishers, Centreville, Virginia, USA

Handarto Hayashi M and Kozai T (2006) Air and leaf temperatures and relative humidity in a naturally ventilated single-span greenhouse with a fogging system for cooling. *Acta Hort.* 710:165–170.

Hellickson ML and Chen CF (1987) Simulated and tested performance of a rockbed heat sink for cooling summer ventilation air. *Transactions of the ASAE*, 30 (1): 207–214.

Jordan RC and Preister GB (1956) *Refrigeration and Air Conditioning*. Prentice-Hall, Inc., New York, NY, USA.

Kacira M, Sase S and Okushima L (2004) Effects of side vents and span numbers on wind-induced natural ventilation of a gothic multi-span greenhouse. *JARQ* 38:227–233.

Kim K, Yoon JY, Hahn JH, Kwon HJ, Son JE, Nam SW, Giacomelli GA and Lee IB (2008) 3-D CFD analysis of relative humidity distribution in greenhouse with fog cooling system and refrigerative dehumidifiers. *Biosyst. Eng.* 100:245–258.

Lee I, Lee S, Kim G, Sung J and Yoon Y (2005) PIV verification of greenhouse ventilation air flows to evaluate CFD accuracy. *Trans. ASAE* 48:2277–2288.

Mistriotis A, Bot GPA, Picuno P and Scarascia-Mugnozza G (1997) Analysis of the efficiency of greenhouse ventilation using computational fluid dynamics. *Agr. Forest Meteorol.* 85:217–228.

Montero JI (2006) Evaporative cooling in greenhouses: effect on microclimate, water use efficiency and plant response. *Acta Hort.* 719:373–384.

Montero JI, Short TH, Curry RB and Bouerle WL (1981) Influence of evaporative cooling systems on greenhouse environment. *ASAE Paper No: 81-4027*, ASAE, St. Joseph, MI 49085, USA.

Montero JI, van Henten EJ, Son JE, Castilla N (2011) Greenhouse Engineering: New Technologies and Approaches. *Acta Hort.* 893:51–64.

Nam YI, Yu IH, Kim TY, Roh MY and Cho MW (2007) Effect of newly developed fan and mist cooling system on greenhouse cooling and growth of cucumber. *Acta Hort.* 761:49–53.

Nielsen OF (2002) Natural ventilation of a greenhouse with top screens. *Biosys. Eng.* 81:443–452.

Okusima L, Sase S and Nara M (1989) A support system for natural ventilation design of greenhouses based on computational aerodynamics. *Acta Hort.* 284:129–136.

Ryou YS, Kang YK, Kang KC, Kim YJ, and Paek Y (2008) Cooling performance of horizontal geothermal heat pump system for protected horticulture. *J. Bio-Environ. Control* 17:90–95.

Santoso M (1989) An alternative configuration of Rankine cycle engine-driven heat pump system. *M. S. Thesis*. The Ohio State University, Columbus, OH, USA.

Sase S (1989) The effects of plant arrangement on airflow characteristics in a naturally ventilated glasshouse. *Acta Hort.* 245:429–435.

Sase S (2006) Air movement and climate uniformity in ventilated greenhouses. *Acta Hort.* 719:313–324.

Swensen PF and Rose RK (1977) *Development of a high seasonal performance factor gas heat pump. Proc. 12th Intersociety Energy Conversion Engineering Conference*, Washington, D.C., U.S.A., p.390.

Voogt J and van Weel P (2008) Climate control based on stomatal behavior in a semi closed greenhouse system 'AIRCOKAS'. *Acta Hort.* 797:151–156.

Yildiz I (1993) A performance study of a Rankine cycle driven heat pump system. *M.S. Thesis*. The Ohio State University, Columbus, Ohio, U.S.A.

Yildiz I, Christensen RN, and Drake MS (1996) Energy consumption of a Rankine cycle heat pump system in a residential building. Süleyman Demirel University, Isparta, *Turkey, Mechanical Engineering Journal* 1(9):57–66.

Yildiz I, Lin C, and Christensen RN (1993a) A performance study of a Rankine cycle driven heat pump system. AES-Vol. 29, *Heat Pump and Refrigeration Systems Design, Analysis, and Applications*. ASME 1993: 9–18.

Yildiz I, Lin C, and Christensen RN (1993b) A performance study of a Rankine cycle driven heat pump system. *Heat Pump and Refrigeration Systems: Design, Analysis and Applications* 29: 9–18. ASME-Publications-AES (Advanced Energy Systems).

Yildiz I and Stombaugh DP (2006a) Dynamic modeling of microclimate and environmental control strategies in a greenhouse coupled with a heat pump system. *Acta Hort.* 718:331–340.

Yildiz I and Stombaugh DP (2006b) Heat pump cooling and greenhouse microclimates in open and confined greenhouse systems. *Acta Hort.* 719: 255–262.

Yu, I.H. 2008. *Optimum design of air-circulation fans using CFD model in air-heated greenhouse for chrysanthemum*. Ph.D. Dissertation. Seoul National University, Seoul, Korea.

# 7

# Greenhouse Lighting

## 7.1 Introduction

Environmental factors such as light and radiation, temperature, relative humidity, carbon dioxide ($CO_2$), water, nutrition, insect, and disease stresses are the key environmental components affecting optimum plant growth and yield. And greenhouse growers have been successfully using a number of mechanisms for modulating each and every one of these components for better plant growth and yield including artificial lighting. As discussed in depth earlier in Chapter 4, light energy is required directly to sustain the growth of all green plants in the first place while light intensity (quantity), light quality (wavelength), and light duration (photoperiod) all influence plant development, and plants respond primarily to the intensity and duration of light within photosynthetically active radiation (PAR) range, the visible portion of the spectrum. Light has very strong interactions on photosynthesis as well particularly with ambient temperature and $CO_2$ concentrations.

The greenhouse location, orientation, and glazing determine the amount of free natural light that is available for plant growth and development in a greenhouse. The best greenhouse location therefore is the location where the free solar energy enters the greenhouse all day long with no shadows cast on the greenhouse. When the intensity and/or duration of natural light is not sufficient under normal operating conditions (temperature, $CO_2$ concentration, water, nutrients), then artificial lighting might be considered for increasing the intensity of light and extending the duration of light. Some of the lighting sources include incandescent (INC) bulbs, fluorescent and compact fluorescent lamps (CFLs), high-intensity discharge (HID) lamps such as high pressure sodium (HPS) and metal halide (MH) lamps, and light-emitting diodes (LEDs).

This chapter covers essential light definitions and units used in lighting systems first, and then moves to the lighting strategies and different light sources used in greenhouses and other controlled environments. And finally, the chapter will cover the future perspectives of greenhouse lighting before the chapter is concluded.

## 7.2 Some Definitions, Light Units, and Conversions

As discussed earlier in Chapter 2, light is considered as a stream of particles of radiant energy, also known as *photons*, and light waves require no medium to propagate. Photons have no mass, occupy no space, and travel at the speed of light. The amount of energy carried by an individual photon is, on the other hand, extremely small and inversely related to wavelength. The shorter the light's wavelength is, the higher the energy content of its photons. In the visible spectrum, for instance, blue light photons have more energy content than those of red light photons.

While sunlight is the most common and free source of radiation for plant growth and yield, the intensity and/or duration of this natural light may not be sufficient. In such cases, artificial lighting is provided. In order to understand what's happening under either natural or artificial lightings, we need to know and understand the following concepts pertinent to controlled environment lighting for plants.

First of these concepts is PAR, which stimulates photosynthesis and was defined earlier as the visible portion (400–700 nm) of the spectrum. And basically, three measuring systems are used for measuring radiation intensity:

1. Radiometric Units (Irradiance): W/m$^2$ or W/ft$^2$ with specific wavelength intervals (e.g. 400–700 nm)
2. Quantum Units (Photon Flux Density): μmol/s.m$^2$ (400–700 nm unless otherwise specified)
3. Photometric Units (Illuminance): lumen/m$^2$ (lux) or lumen/ft$^2$ (footcandle); 1 footcandle = 10.8 lux

Only one of the above three, the quantum units, is appropriate for photosynthesis and therefore for plant production; and it will be discussed further below and becomes more meaningful. Also, since the spectral power distribution varies, it is important to use proper conversions (see Table 7.1) between individual light sources for the wavelength interval considered so that meaningful comparisons can be made.

*Photosynthetic photon flux* (PPF) is the amount (μmol) of PAR (amount of photons available between 400 and 700 nm) emitted from a lamp per second. The unit of PPF is μmol/s. PPF is typically measured in laboratory settings with an integrating sphere that captures all the photons emitted from a particular lamp, for instance.

**TABLE 7.1**

Light Conversion Factors (Thimijan and Heins, 1983).

| Light Source | μmol/s.m$^2$ Per W/m$^2$ (400–700 nm) | Footcandle Per μmol/s.m$^2$ (400–700 nm) | Lux Per μmol/s.m$^2$ (400–700 nm) |
|---|---|---|---|
| Sun and sky, daylight | 4.57 | 5.0 | 54 |
| Blue sky only | 4.24 | 4.8 | 52 |
| High-pressure sodium lamps | 4.98 | 7.6 | 82 |
| Metal halide lamps | 4.59 | 6.6 | 71 |
| Warm-white fluorescent lamps | 4.67 | 7.0 | 76 |
| Cool-white fluorescent lamps | 4.59 | 6.9 | 74 |
| Incandescent lamps | 5.00 | 4.6 | 50 |
| Low-pressure sodium lamps | 4.92 | 9.8 | 106 |

*Photosynthetic photon flux density* (PPFD) is the PPF (μmol/s) per unit surface area ($m^2$); therefore, it has an overall unit of μmol/s.$m^2$.

Engineers and plant scientists have different preferences with respect to the PPF and PPFD; however, since the PPFD refers to the intensity of PAR at a surface and is typically measured at the top of a plant canopy by a quantum sensor, it is more meaningful than the PPF, which refers to the total amount of light emitted by a lamp spherically, and not necessarily the intensity arriving on a surface such as plant canopy. And there are three groups of PPFD provided as follows:

1. Low-light Intensity : 100 μmol/s.$m^2$
2. Medium-light Intensity : 300 μmol/s.$m^2$
3. High-light Intensity : 600 μmol/s.$m^2$

Microgreen crops are considered to be low-light intensity (100–200 μmol/s.$m^2$) crops while tomatoes, for instance, are considered as high-light intensity (600 μmol/s.$m^2$) crops. During peak hours, daylight PPFD can easily reach about 1,500 μmol/s.$m^2$ and 2,000 μmol/s.$m^2$ levels in mild and hot desert climates, respectively.

*Photosynthetic photon efficacy* (PPE) is the photosynthetic photon (400–700 nm) output (μmol/s) per unit electrical power input (W = J/s). In essence, PPE expresses how effectively a lighting fixture converts electrical power into photosynthetic light useful for photosynthesis, and the overall unit of PPE is therefore μmol/s.W or μmol/J.

*Luminous efficiency* is the number of lumens emitted per unit electrical power input (W). This efficiency is relevant when comparing the efficiency of lamps for human applications of visible light and not for plant applications such as greenhouses and other controlled environments for plant production.

*Daily light integral* (DLI) is the integrated PPFD (400 and 700 nm) received on one square meter area per day. The unit is therefore mol/$m^2$·d and the overall equation can be expressed as follows:

$$\mathrm{DLI}\left(\frac{\mathrm{mol}}{\mathrm{m^2.d}}\right) = \left(\frac{\text{Photosynthetic Photon Flux Density}\left(\frac{\mu\mathrm{mol}}{\mathrm{s.m^2}}\right)\left(\text{Photoperiod}\left(\frac{\mathrm{h}}{\mathrm{d}}\right)\right)\left(3{,}600\,\frac{\mathrm{s}}{\mathrm{h}}\right)}{\left(1{,}000{,}000\,\frac{\mu\mathrm{mol}}{\mathrm{mol}}\right)}\right) \tag{7.1}$$

Table 7.2 was generated using Equation (7.1), and the readers may find it useful for quick visualization of the DLI as a result of the combination effect of PPFD and photoperiod. Using this table, one can quickly decide how long of a photoperiod must be provided for a given DLI requirement and PPFD value, or what sort of a photoperiod and PPFD combination can be used for providing a required optimum DLI value for a particular plant considering the plant's minimum and maximum PPFD limits.

There exists a correlation between the average DLI and plant growth, which increases almost linearly with increasing DLI. As a rule of thumb, a 1% increase in production is estimated for each 1% increase in DLI. Number of fruits and fruit sizes increase with the increasing DLI as well for fruiting vegetables such as tomatoes and cucumbers. And the DLI in a greenhouse is obviously influenced by the greenhouse glazing, overhead obstructions, shading curtains, and high-intensity supplemental lighting (and distance

**TABLE 7.2**

Daily light integral values obtained using Equation (7.1) for corresponding PPFDs and photoperiods.

| DLI | Photosynthetic Photon Flux Density (μmol/s.m²) | | | | | | | | | | | | | | | | | | | |
|---|---|---|---|---|---|---|---|---|---|---|---|---|---|---|---|---|---|---|---|---|
| **Photoperiod (h/d)** | **50** | **100** | **150** | **200** | **250** | **300** | **350** | **400** | **450** | **500** | **550** | **600** | **650** | **700** | **750** | **800** | **850** | **900** | **950** | **1,000** |
| **1** | 0.2 | 0.4 | 0.5 | 0.7 | 0.9 | 1.1 | 1.3 | 1.4 | 1.6 | 1.8 | 2.0 | 2.2 | 2.3 | 2.5 | 2.7 | 2.9 | 3.1 | 3.2 | 3.4 | 3.6 |
| **2** | 0.4 | 0.7 | 1.1 | 1.4 | 1.8 | 2.2 | 2.5 | 2.9 | 3.2 | 3.6 | 4.0 | 4.3 | 4.7 | 5.0 | 5.4 | 5.8 | 6.1 | 6.5 | 6.8 | 7.2 |
| **3** | 0.5 | 1.1 | 1.6 | 2.2 | 2.7 | 3.2 | 3.8 | 4.3 | 4.9 | 5.4 | 5.9 | 6.5 | 7.0 | 7.6 | 8.1 | 8.6 | 9.2 | 9.7 | 10.3 | 10.8 |
| **4** | 0.7 | 1.4 | 2.2 | 2.9 | 3.6 | 4.3 | 5.0 | 5.8 | 6.5 | 7.2 | 7.9 | 8.6 | 9.4 | 10.1 | 10.8 | 11.5 | 12.2 | 13.0 | 13.7 | 14.4 |
| **5** | 0.9 | 1.8 | 2.7 | 3.6 | 4.5 | 5.4 | 6.3 | 7.2 | 8.1 | 9.0 | 9.9 | 10.8 | 11.7 | 12.6 | 13.5 | 14.4 | 15.3 | 16.2 | 17.1 | 18.0 |
| **6** | 1.1 | 2.2 | 3.2 | 4.3 | 5.4 | 6.5 | 7.6 | 8.6 | 9.7 | 10.8 | 11.9 | 13.0 | 14.0 | 15.1 | 16.2 | 17.3 | 18.4 | 19.4 | 20.5 | 21.6 |
| **7** | 1.3 | 2.5 | 3.8 | 5.0 | 6.3 | 7.6 | 8.8 | 10.1 | 11.3 | 12.6 | 13.9 | 15.1 | 16.4 | 17.6 | 18.9 | 20.2 | 21.4 | 22.7 | 23.9 | 25.2 |
| **8** | 1.4 | 2.9 | 4.3 | 5.8 | 7.2 | 8.6 | 10.1 | 11.5 | 13.0 | 14.4 | 15.8 | 17.3 | 18.7 | 20.2 | 21.6 | 23.0 | 24.5 | 25.9 | 27.4 | 28.8 |
| **9** | 1.6 | 3.2 | 4.9 | 6.5 | 8.1 | 9.7 | 11.3 | 13.0 | 14.6 | 16.2 | 17.8 | 19.4 | 21.1 | 22.7 | 24.3 | 25.9 | 27.5 | 29.2 | 30.8 | 32.4 |
| **10** | 1.8 | 3.6 | 5.4 | 7.2 | 9.0 | 10.8 | 12.6 | 14.4 | 16.2 | 18.0 | 19.8 | 21.6 | 23.4 | 25.2 | 27.0 | 28.8 | 30.6 | 32.4 | 34.2 | 36.0 |
| **11** | 2.0 | 4.0 | 5.9 | 7.9 | 9.9 | 11.9 | 13.9 | 15.8 | 17.8 | 19.8 | 21.8 | 23.8 | 25.7 | 27.7 | 29.7 | 31.7 | 33.7 | 35.6 | 37.6 | 39.6 |
| **12** | 2.2 | 4.3 | 6.5 | 8.6 | 10.8 | 13.0 | 15.1 | 17.3 | 19.4 | 21.6 | 23.8 | 25.9 | 28.1 | 30.2 | 32.4 | 34.6 | 36.7 | 38.9 | 41.0 | 43.2 |
| **13** | 2.3 | 4.7 | 7.0 | 9.4 | 11.7 | 14.0 | 16.4 | 18.7 | 21.1 | 23.4 | 25.7 | 28.1 | 30.4 | 32.8 | 35.1 | 37.4 | 39.8 | 42.1 | 44.5 | 46.8 |
| **14** | 2.5 | 5.0 | 7.6 | 10.1 | 12.6 | 15.1 | 17.6 | 20.2 | 22.7 | 25.2 | 27.7 | 30.2 | 32.8 | 35.3 | 37.8 | 40.3 | 42.8 | 45.4 | 47.9 | 50.4 |
| **15** | 2.7 | 5.4 | 8.1 | 10.8 | 13.5 | 16.2 | 18.9 | 21.6 | 24.3 | 27.0 | 29.7 | 32.4 | 35.1 | 37.8 | 40.5 | 43.2 | 45.9 | 48.6 | 51.3 | 54.0 |
| **16** | 2.9 | 5.8 | 8.6 | 11.5 | 14.4 | 17.3 | 20.2 | 23.0 | 25.9 | 28.8 | 31.7 | 34.6 | 37.4 | 40.3 | 43.2 | 46.1 | 49.0 | 51.8 | 54.7 | 57.6 |
| **17** | 3.1 | 6.1 | 9.2 | 12.2 | 15.3 | 18.4 | 21.4 | 24.5 | 27.5 | 30.6 | 33.7 | 36.7 | 39.8 | 42.8 | 45.9 | 49.0 | 52.0 | 55.1 | 58.1 | 61.2 |
| **18** | 3.2 | 6.5 | 9.7 | 13.0 | 16.2 | 19.4 | 22.7 | 25.9 | 29.2 | 32.4 | 35.6 | 38.9 | 42.1 | 45.4 | 48.6 | 51.8 | 55.1 | 58.3 | 61.6 | 64.8 |
| **19** | 3.4 | 6.8 | 10.3 | 13.7 | 17.1 | 20.5 | 23.9 | 27.4 | 30.8 | 34.2 | 37.6 | 41.0 | 44.5 | 47.9 | 51.3 | 54.7 | 58.1 | 61.6 | 65.0 | 68.4 |
| **20** | 3.6 | 7.2 | 10.8 | 14.4 | 18.0 | 21.6 | 25.2 | 28.8 | 32.4 | 36.0 | 39.6 | 43.2 | 46.8 | 50.4 | 54.0 | 57.6 | 61.2 | 64.8 | 68.4 | 72.0 |
| **21** | 3.8 | 7.6 | 11.3 | 15.1 | 18.9 | 22.7 | 26.5 | 30.2 | 34.0 | 37.8 | 41.6 | 45.4 | 49.1 | 52.9 | 56.7 | 60.5 | 64.3 | 68.0 | 71.8 | 75.6 |
| **22** | 4.0 | 7.9 | 11.9 | 15.8 | 19.8 | 23.8 | 27.7 | 31.7 | 35.6 | 39.6 | 43.6 | 47.5 | 51.5 | 55.4 | 59.4 | 63.4 | 67.3 | 71.3 | 75.2 | 79.2 |
| **23** | 4.1 | 8.3 | 12.4 | 16.6 | 20.7 | 24.8 | 29.0 | 33.1 | 37.3 | 41.4 | 45.5 | 49.7 | 53.8 | 58.0 | 62.1 | 66.2 | 70.4 | 74.5 | 78.7 | 82.8 |
| **24** | 4.3 | 8.6 | 13.0 | 17.3 | 21.6 | 25.9 | 30.2 | 34.6 | 38.9 | 43.2 | 47.5 | 51.8 | 56.2 | 60.5 | 64.8 | 69.1 | 73.4 | 77.8 | 82.1 | 86.4 |

from the canopy) when employed. Natural DLI in a commercial greenhouse rarely reaches 30 mol/m².d and based on their DLI requirements, crops can be classified as follows:

1. Low-light crops : 3–6 mol/m².d
2. Medium-light crops : 6–12 mol/m².d
3. High-light crops : 12–18 mol/m².d
4. Very high-light crops : 18+ mol/m².d

Furthermore, Table 7.3 provides a list of simplified DLI targets for different plants; however, the provided figures might change as a function of ambient air temperatures, $CO_2$ concentrations, and other environmental and economic conditions.

**TABLE 7.3**
DLI targets for some plants grown in controlled environments.

| Plant Type | Target DLI (mol/m²·d) | Source |
|---|---|---|
| Shade (annuals and perennials) and foliage plants | 6–10 | |
| Potted bulbs | 6–15 | |
| Stock plants (for cuttings) | 10–20 | |
| Annual bedding plants | 10+ | |
| Leafy greens and herbs | 12+ | Runkle (2019) |
| Potted flowering plants | 12+ | |
| Herbaceous perennials and shrubs | 12+ | |
| Cut flowers | 15+ | |
| Fruiting vegetables | 15+ | |
| Butterhead lettuce | 14–16 | |
| Cucumber | 22–30 | |
| Capsicum | 22–30 | Morgan (2013) |
| Eggplant | 22–30 | |
| Tomatoes | 22–30 | |

As it will be worked on numerically later in this section, for instance, having the long-term natural DLI figures available, one can easily determine how much extra light is required for maximizing crop yields in a greenhouse at a particular location; and then using this information, it can easily be decided whether installing and operating artificial lighting would be feasible or not. Knowing DLI also makes it possible to determine the intensity of artificial lighting and when and how long the artificial light sources need to be operated for providing optimal light conditions for maximum production in greenhouses where artificial lighting systems already exist. DLI knowledge in and outside of a greenhouse also lets a greenhouse grower decide whether a shading system needs to be used or removed, or whether the time for cleaning or replacing a greenhouse glazing has come or not.

Greenhouse lighting systems designers and growers may find the DLI maps very useful, especially the monthly maps as more of those become publicly available around the world. Figures 7.4–7.11 provide available outdoor annual DLI maps around the world, and the outdoor monthly average DLI maps for the USA can be found in Korczynski et al. (2002).

Now, it is a good time to perform some numerical examples for the concepts and conversions discussed above.

Table 7.4 also shows that the PPE is an important parameter for comparing lighting fixtures; however, it also shows that PPF and measured powers are also important. In addition to these considerations, other factors should also be considered before investing in a particular technology, such as light spectrum, spatial uniformity of light intensity, lighting fixture shading, durability, and service.

Table 7.5, on the other hand, compares the photosynthetic photon efficacy (PPE) (for plants) and the luminous efficiency (for people) of four light sources. The table shows that high PPE fixtures (blue and red LEDs) for plants do not necessarily have high luminous

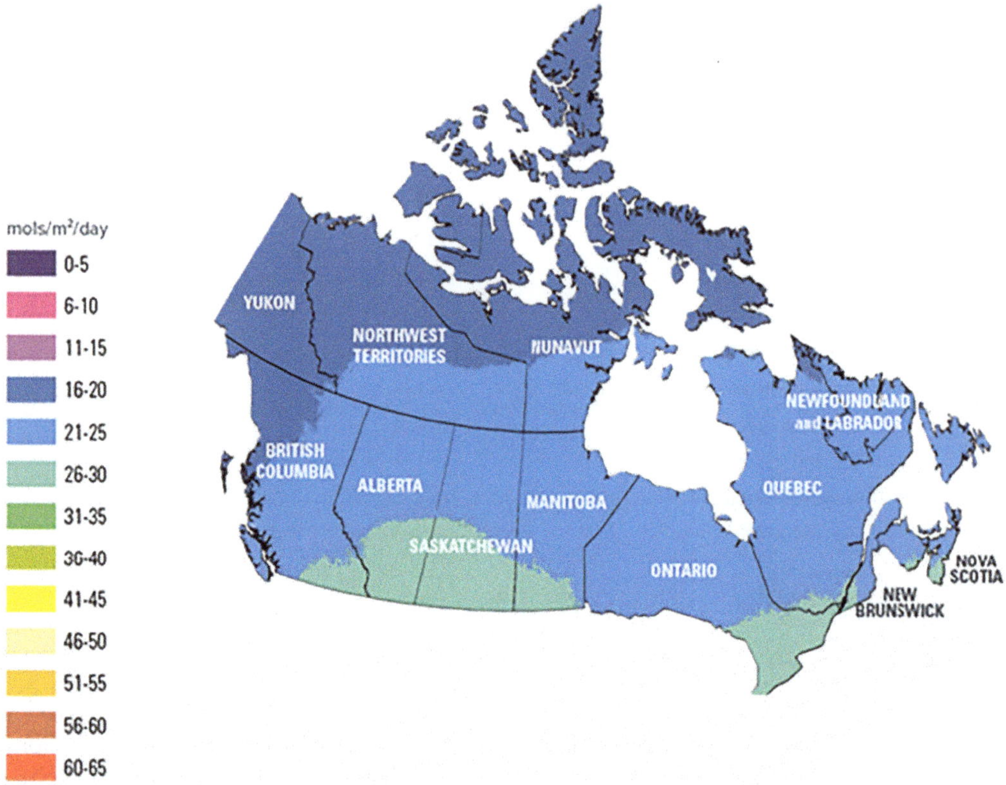


**FIGURE 7.1**
Outdoor annual average DLI map for Canada (Faust & Logan, 2018). (Courtesy of Dr. James E. Faust, Clemson University and Dr. Joanne Logan, The University of Tennessee).

efficiency (which is appropriate for people) or high luminous efficiency fixtures do not necessarily have high photosynthetic photon efficacy (which is appropriate for plants). So, it is obvious that the measurement system based on photometric units (lux and footcandle) is quite misleading and inappropriate for plant applications while the measurement system based on quantum units is the most appropriate system for photosynthesis and therefore for plant production.

A double-ended 1,000-W HPS lamp with electronic ballast, for instance, has a PPE of approx. 1.7 μmol/J while the PPEs for LEDs vary quite a bit. The PPEs of many new LEDs today commonly exceed 2.0 μmol/J and even 3.0 μmol/J (see Table 7.4). Obviously, the higher the PPE value of a light bulb is, the better it is at converting electric energy into photosynthetic photons.

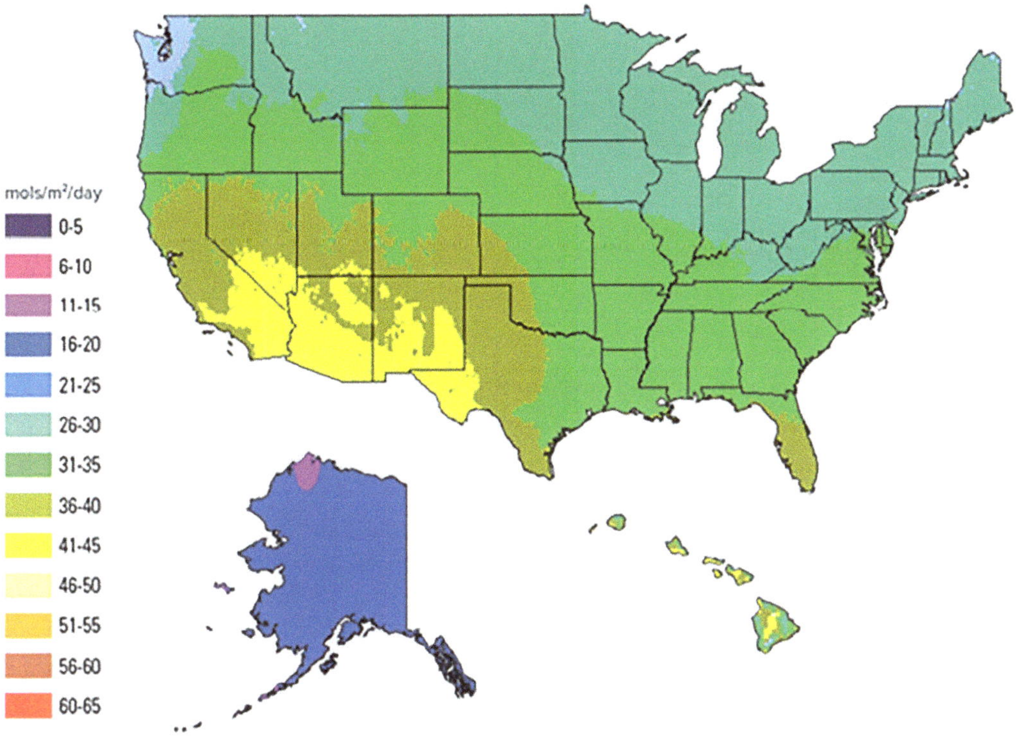


**FIGURE 7.2**
Outdoor annual average DLI map for the USA (Faust & Logan, 2018). (Courtesy of Dr. James E. Faust, Clemson University and Dr. Joanne Logan, The University of Tennessee).

**FIGURE 7.3**
Outdoor annual average DLI map for Mexico (Faust & Logan, 2018). (Courtesy of Dr. James E. Faust, Clemson University and Dr. Joanne Logan, The University of Tennessee).

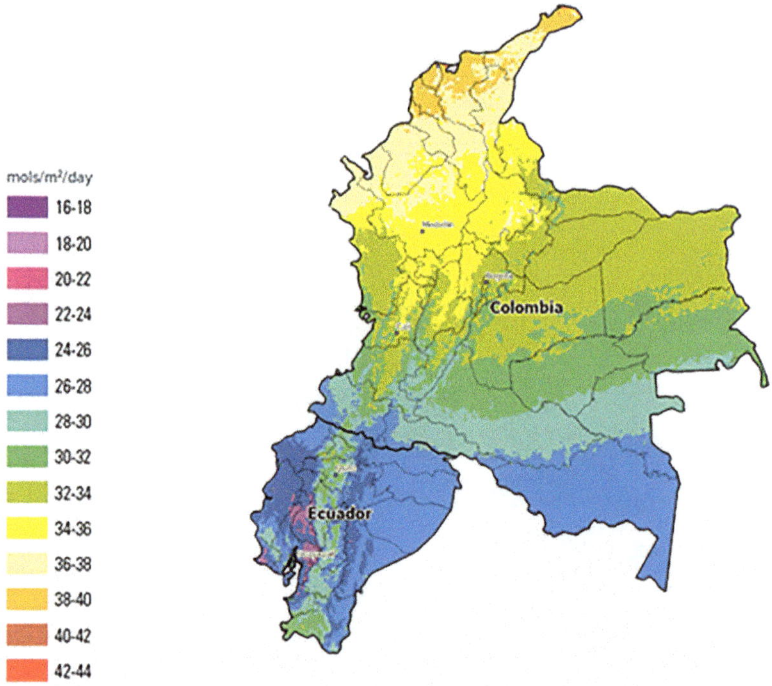


**FIGURE 7.4**
Outdoor annual average DLI map for Colombia and Ecuador (Faust & Logan, 2018). (Courtesy of Dr. James E. Faust, Clemson University and Dr. Joanne Logan, The University of Tennessee).

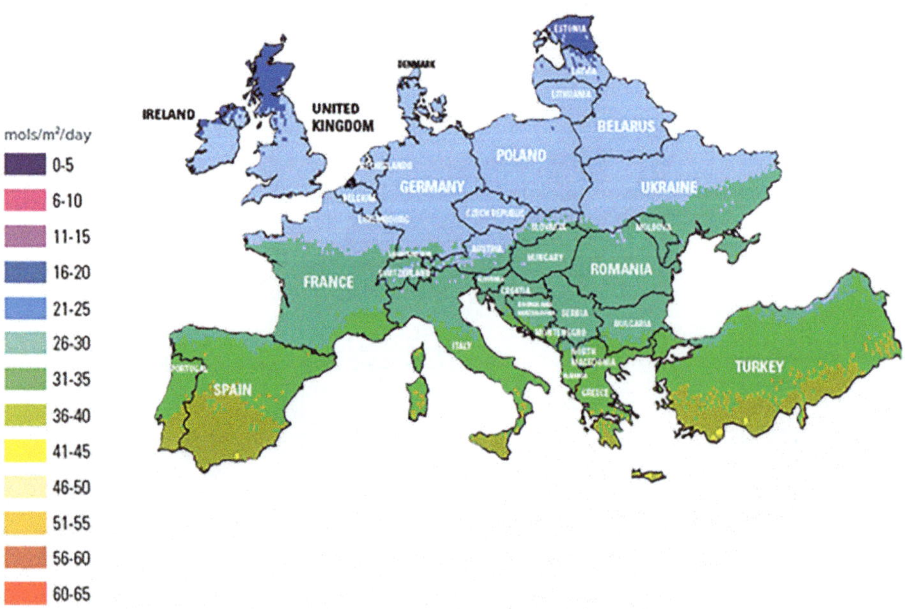


**FIGURE 7.5**
Outdoor annual average DLI map for Europe (Faust & Logan, 2018). (Courtesy of Dr. James E. Faust, Clemson University and Dr. Joanne Logan, The University of Tennessee).

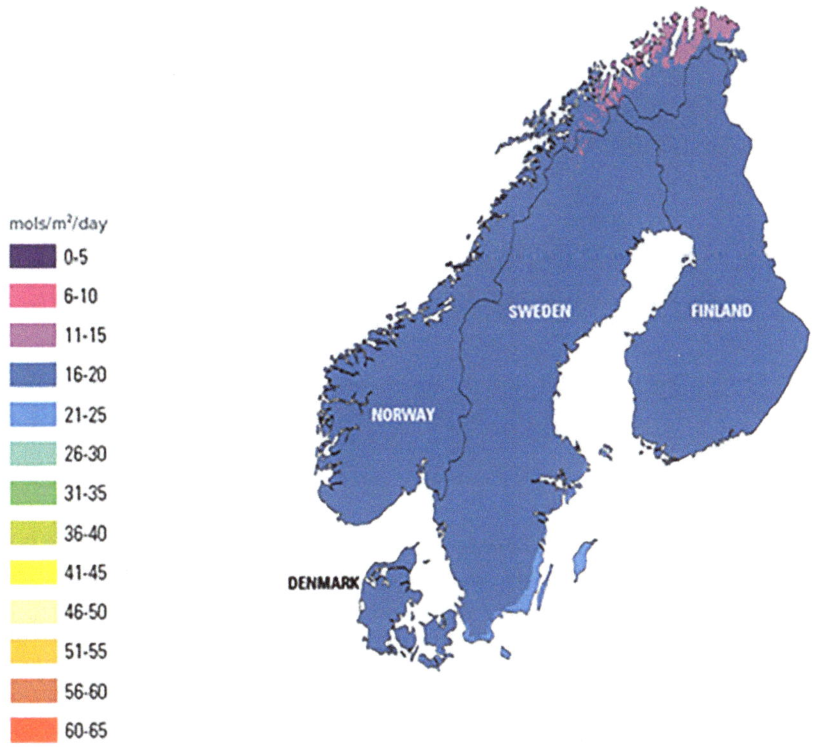


**FIGURE 7.6**
Outdoor annual average DLI map for Scandinavia (Faust & Logan, 2018). (Courtesy of Dr. James E. Faust, Clemson University and Dr. Joanne Logan, The University of Tennessee).

## 7.3 Artificial Lighting Strategies in Controlled Environments

Artificial lighting is increasingly being used and now a common practice in the commercial production of plants grown in controlled environments such as greenhouses. Different practices of greenhouse lighting have totally different goals, such as increasing the DLIs and/or extending the photoperiod. Photoperiodic lighting needs a low-intensity lighting (typically 10 footcandle or 110 lux or somewhere between 1 and 2 μmol/s.m$^2$), which is only used in greenhouses after sunset or before sunrise, whereas supplemental (photosynthetic) lighting requires a high-intensity lighting (usually 400–500 footcandle or 4,320–5,400 lux or 60–120 μmol/s.m$^2$) and is generally used in greenhouses from October through March in mid and high latitudes on cloudy days and during the night (Table 7.6). Sole-source lighting, on the other hand, is used for indoor growing of plants, and typically requires 150–200 μmol/s.m$^2$ (Table 7.6). For a majority of crops grown indoors, plants are usually provided light for 18–24 hours per day. Lighting is not a cheap investment, especially when the purchase, installation, and operational cost of such systems are considered. Therefore, it is important to understand the differences between different lighting strategies in the first place and then install an adequate lighting system that can deliver the target DLI and/or the photoperiod to the plants in a greenhouse or other controlled

**FIGURE 7.7**
Outdoor annual average DLI map for Russian Federation (Faust & Logan, 2018). (Courtesy of Dr. James E. Faust, Clemson University and Dr. Joanne Logan, The University of Tennessee).

**FIGURE 7.8**
Outdoor annual average DLI map for China (Faust & Logan, 2018). (Courtesy of Dr. James E. Faust, Clemson University and Dr. Joanne Logan, The University of Tennessee).

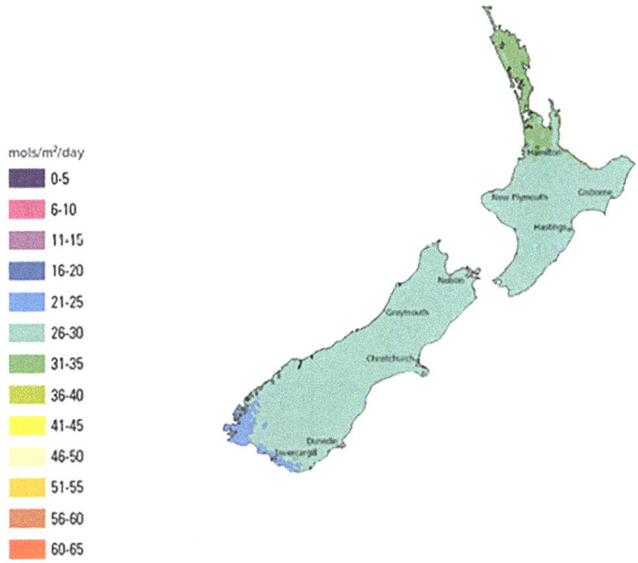


**FIGURE 7.9**
Outdoor annual average DLI map for New Zealand (Faust & Logan, 2018). (Courtesy of Dr. James E. Faust, Clemson University and Dr. Joanne Logan, The University of Tennessee).

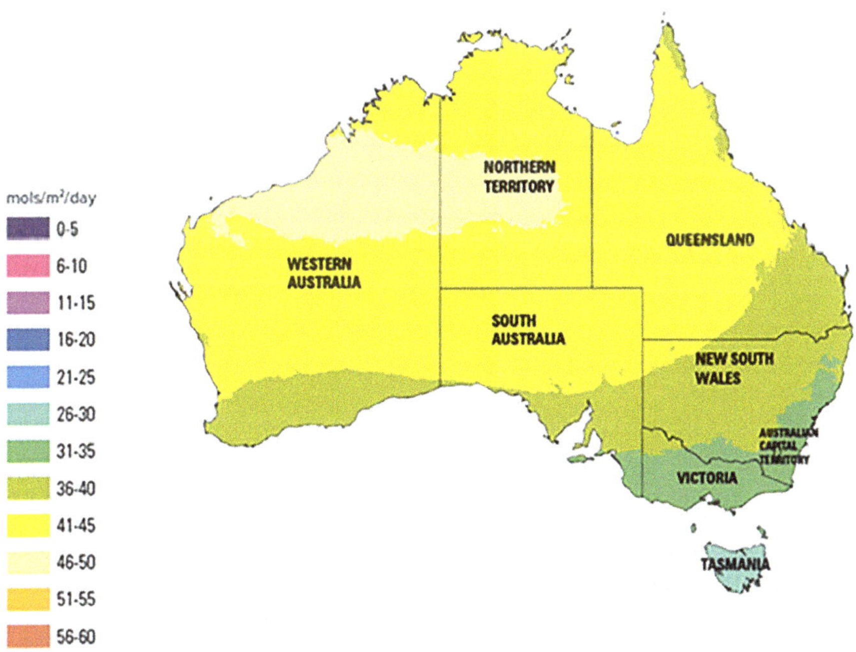


**FIGURE 7.10**
Outdoor annual average DLI map for Australia (Faust & Logan, 2018). (Courtesy of Dr. James E. Faust, Clemson University and Dr. Joanne Logan, The University of Tennessee).

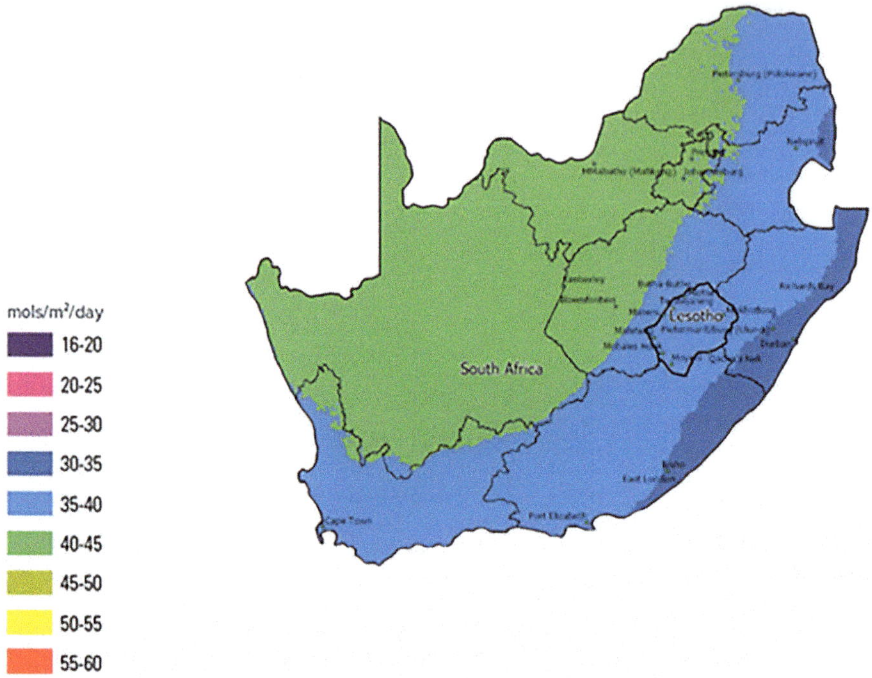


**FIGURE 7.11**
Outdoor annual average DLI map for South Africa (Faust & Logan, 2018). (Courtesy of Dr. James E. Faust, Clemson University and Dr. Joanne Logan, The University of Tennessee).

**Example 7.1: Radiometric to quantum and quantum to photometric units conversions and light comparisons**

Determine and compare the photosynthetic photon flux densities (PPFD), footcandle per μmol/s.m², and lux per μmol/s.m² for 400 W/m² daylight (400–700 nm) and 400 W/m² HPS lamp light (400–700 nm).

**Solution**

Using the conversion factors provided in Table 7.1, we can easily determine all the conversions as follows:

Photosynthetic Photon Flux Density Conversions:

$$\text{Daylight:}\left(400\,\frac{\text{W}}{\text{m}^2}\right)\left(\frac{4.57\ \mu\text{mol/s.m}^2}{\text{W/m}^2}\right)=1{,}828\ \mu\text{mol/s.m}^2$$

$$\text{High-pressure sodium lamp:}\left(400\,\frac{\text{W}}{\text{m}^2}\right)\left(\frac{4.98\ \mu\text{mol/s.m}^2}{\text{W/m}^2}\right)=1{,}992\ \mu\text{mol/s.m}^2$$

As we see here, the light coming from the HPS lamp provides higher PPFD compared to the daylight at the same radiation intensity per unit surface area.

Footcandle Conversions:

$$\text{Daylight:}\left(1{,}828\,\frac{\mu\text{mol}}{\text{s.m}^2}\right)\left(\frac{5.0\text{ footcandle}}{\mu\text{mol/s.m}^2}\right)=9{,}140\text{ footcandle}=9{,}140\text{ lumen/ft}^2$$

$$\text{High-pressure sodium lamp:}\left(1{,}992\,\frac{\mu\text{mol}}{\text{s.m}^2}\right)\left(\frac{7.6\text{ footcandle}}{\mu\text{mol/s.m}^2}\right)$$
$$=15{,}139\text{ footcandle}=15{,}139\text{ lumen/ft}^2$$

Lux Conversions:

$$\text{Daylight:}\left(1{,}828\,\frac{\mu\text{mol}}{\text{s.m}^2}\right)\left(\frac{54\text{ lux}}{\mu\text{mol/s.m}^2}\right)=98{,}712\text{ lux}=98{,}712\text{ lumen/m}^2$$

$$\text{High-pressure sodium lamp:}\left(1{,}992\,\frac{\mu\text{mol}}{\text{s.m}^2}\right)\left(\frac{82\text{ lux}}{\mu\text{mol/s.m}^2}\right)$$
$$=163{,}344\text{ lux}=163{,}344\text{ lumen/m}^2$$

And further comparisons between different light sources can be made easily using the conversion factors provided in Table 7.1.

**Example 7.2: Photometric light (footcandle of light) from high-pressure sodium lamp into quantum conversion**

Determine the amount of PPFD provided by 1,500 footcandle of light from HPS lamp (400–700 nm).

**Solution**

Again, using the proper conversion factor provided in Table 7.1, we can easily perform this conversion as follows:

$$\left(\frac{1{,}500\text{ footcandle}}{7.6\,\dfrac{\text{footcandle}}{\mu\text{mol/s.m}^2}}\right)=197\,\mu\text{mol/s.m}^2$$

**Example 7.3: Photometric light (lux of light) from metal halide lamp into quantum conversion**

Determine the amount of PPFD provided by 14,000 lux of light from metal halide lamp (400–700 nm).

**Solution**

Once again, using the proper conversion factor provided in Table 7.1, we can perform this task as follows:

$$\left(\frac{14{,}000\text{ lux}}{71\,\dfrac{\text{lux}}{\mu\text{mol/s.m}^2}}\right)=197\,\mu\text{mol/s.m}^2$$

**Example 7.4: Quantum to radiometric conversion**

Convert 1,992 μmol/s.m² of HPS light (400–700 nm) to radiometric units (400–700 nm).

**Solution**

The quantum unit (μmol/s.m²) is based on the number of photons, and photons at different wavelengths have different energy contents. Therefore, in order to be able to calculate total energy, we have to know both the number of photons available at each wavelength and the energy content of photon at each wavelength. In reality, it is a complex calculation; however, Table 7.1 makes this process much simpler for known light sources. For this example, the light source is HPS lamp; and for this lamp, Table 7.1 provides a conversion factor of 4.98. Therefore, we can perform this task easily as follows:

$$\left(\frac{1{,}992\,\frac{\mu\text{mol}}{\text{s.m}^2}}{\frac{4.98\,\mu\text{mol/s.m}^2}{\text{W/m}^2}}\right) = 400\ \text{W/m}^2$$

One should also keep in mind that it is impossible to convert 24-hour average solar radiation values into, for instance, 8-, 16-, or 20-hour average solar power as the time distribution of solar power throughout the day is typically not known. Hence, we must convert lamp light data into 24-hour average values first, and then compare them with sunlight values. Let's now assume that the above 400 W/m² HPS light is instantaneous and the irradiance is constant, and the lamp is used in a greenhouse lighting for 8 hours, and would like to convert this radiometric light output to 24-hour basis:

$$(\text{Irradiance})\left(\frac{\text{Hours of Irradiation}}{24\ \text{hours}}\right) = \left(400\ \text{W/m}^2\right)\left(\frac{8\ \text{h}}{24\ \text{h}}\right) = 133\ \text{W/m}^2\,(24\text{-hour basis})$$

Furthermore, Table 7.4 shows the power consumption, quantity of photosynthetic photons (400–700 nm) (PPF) emitted per lighting fixture, and photosynthetic photon efficacy (PPE) for a number of selected lighting fixtures, that is, HPS lamps and LEDs.

environment plant production facilities. Hence, this section covers different lighting strategies and characteristics used in such plant production facilities.

Table 7.6 compares the primary characteristics of each lighting strategy. Controlled environment growers sometimes use artificial lighting to create longer days (photoperiods) or to increase the DLI (supplemental lighting) or both. For instance, if we provide, let's say 16 hours of artificial lighting using HPS lamps or any other artificial light sources on an overcast winter day, we will not only increase the DLI (supplemental lighting) but at the same time also provide a long-day photoperiod (photoperiodic lighting).

### 7.3.1 Photoperiodic Lighting

As discussed earlier in Chapter 2, daily path of the Sun is not constant throughout the year; rather it shifts slightly and steadily every day. And this has a huge impact on insolation (intensity) and daylight hours (natural photoperiods). The latitude and the season are a result of the solar geometry. The Northern Hemisphere, for instance, has more daylight

**TABLE 7.4**

Power Consumption, Quantity of Photosynthetic Photons (PPF) Emitted Per Lighting Fixture, and Photosynthetic Photon Efficacy (PPE) for Selected Lighting Fixtures.

| Light Source | Manufacturer and Model | Electrical Power Input (W) | Photosynthetic Photon Flux (μmol/s) | Photosynthetic Photon Efficacy (μmol/J) |
|---|---|---|---|---|
| HPS[1] | Sunlight Supply Sun Star, 400 W (magnetic ballast) | 443 | 416 | 0.94 |
| HPS[2] | PARsource GLXI (magnetic ballast) | 1,004 | 1,165 | 1.16 |
| HPS[2] | PARsource GLXII (electronic ballast) | 1,026 | 1,334 | 1.30 |
| HPS[3] | P.L. Light SON-T PIA, 600 W (electronic ballast) | 690 | 926 | 1.34 |
| HPS[3] | Gavita Pro 1000e, 1,000 W (double-ended bulb, electronic ballast) | 1,069 | 1,837 | 1.72 |
| LED[3] | Heliospectra LX601C, 630 W | 595 | 673 | 1.13 |
| LED[2] | LumiGrow Pro 325 | 304 | 392 | 1.29 |
| LED[3] | Illumitex PowerHarvest W, 300 W | 268 | 475 | 1.77 |
| LED[3] | Lumigrow Pro 325e, 300 W | 300 | 540 | 1.80 |
| LED[3] | Hubbell Cultivaire, 425 W | 358 | 736 | 2.06 |
| LED[3] | P.L. Light HortiLED TOP 150° Full Spectrum, 320 W | 330 | 696 | 2.11 |
| LED[5] | Illumitex DHW7S12NLX6 [H,U]DWH10 | 620 | 1,359 | 2.20 |
| LED[5] | Lumigrow TL-2-T-03-PT-XX-SP | 620 | 1,442 | 2.33 |
| LED[5] | Sunritek GX330A FSG 26 D1 XXXXX | 318 | 769 | 2.42 |
| LED[4] | P.L. Light HortiLED TOP 80° Red-Blue, 320 W | 313 | 798 | 2.55 |
| LED[5] | Philips (Signify) 9290 009 80406, 200 W | 199 | 510 | 2.56 |
| LED[3] | Philips (Signify) GreenPower Deep Red/White Low Blue, 200 W | 195 | 504 | 2.59 |
| LED[5] | California LightWorks GHPRO 340 | 341 | 890 | 2.61 |
| LED[5] | Philips (Signify) GP LED Toplighting 2.1 DR/W MB 277-400V | 285 | 774 | 2.72 |
| LED[5] | GE Current GEHTL-HPKB4-(2,3)NX1 | 606 | 1,681 | 2.77 |
| LED[5] | Philips (Signify) GPL toplight_2.1 DR/W MB 200–400V RO | 178 | 506 | 2.90 |
| LED[5] | Philips (Signify) GPL toplight_2.1 DR/B LB 200–400V RO | 170 | 550 | 3.20 |

*Source:* In: Runkle (2018a):
1 Nelson and Bugbee (2014);
2 http://cpl.usu.edu/files/publications/factsheet/pub__4338884.pdf.
3 Radetsky (2018);
4 Johnson, Kusuma, and Bugbee (2017). In: Runkle (2014):
5 https://www.designlights.org/horticultural-lighting

**TABLE 7.5**

Comparison of the Photosynthetic Photon Efficacy (For Plants) and the Luminous Efficiency (For People) of Four Light Sources (Runkle & Bugbee, 2013).

| Light Source | Photosynthetic Photon Efficacy (μmol/J) | Luminous Efficiency (lumen/W) |
|---|---|---|
| Cool-white fluorescent | 1.2 | 90 |
| High-pressure sodium | 1.6 | 130 |
| Blue LEDs (peaking at 455 nm) | 1.8 | 47 |
| Red LEDs (peaking at 655 nm) | 1.7 | 17 |

**Example 7.5: Calculating the DLI deficiency for supplemental lighting**

Let's now consider that we have leafy greens in a greenhouse, which have a DLI requirement of 12 mol/m$^2$·d (Table 7.3) and the natural light provides an average DLI of 7 mol/m$^2$·d inside the greenhouse. Determine the supplemental lighting intensity required to reach the DLI target of 12 mol/m$^2$·d if the grower wants to provide 20 hours of lighting per day.

**Solution**

We first need to calculate the DLI deficiency (the required supplemental DLI):

$$\text{Required Supplemental DLI} = \left[(\text{Plant DLI Requirement}) - (\text{Natural DLI Available})\right]$$

$$\text{Required Supplemental DLI} = \left[(12\,\text{mol/m}^2\cdot\text{d}) - (7\,\text{mol/m}^2\cdot\text{d})\right]$$

$$\text{Required Supplemental DLI} = 5\,\text{mol/m}^2\cdot\text{d}$$

The grower wants to provide 20 hours of lighting per day; so, the light intensity required to reach the DLI target can be calculated after rearranging Equation (7.1) as follows:

$$\text{Supplemental PPFD} = \left(\frac{(\text{Required Supplemental DLI, mol/m}^2\cdot\text{d})\left(1{,}000{,}000\,\frac{\mu\text{mol}}{\text{mol}}\right)}{\text{Photoperiod}\left(\frac{\text{h}}{\text{d}}\right)\left(3{,}600\,\frac{\text{s}}{\text{h}}\right)}\right)$$

$$\text{Supplemental PPFD} = \left(\frac{(5\,\text{mol/m}^2\cdot\text{d})\left(1{,}000{,}000\,\frac{\mu\text{mol}}{\text{mol}}\right)}{\left(20\,\frac{\text{h}}{\text{d}}\right)\left(3{,}600\,\frac{\text{s}}{\text{h}}\right)}\right) = 69\,\mu\text{mol/s.m}^2$$

And exactly the same magnitude (69 μmol/s.m$^2$) of supplemental PPFD can be found using Table 7.2 and the method of interpolation for a DLI of 5.0 mol/m$^2$·d and a photoperiod of 20 hours per day.

Moreover, using this required light density (supplemental PPFD), we can now determine the number of particular lighting fixtures required for the high-density supplemental lighting taking into consideration the greenhouse dimension and the distance between the light source and the plants.

**Example 7.6: Calculating how long to operate supplemental lighting lamps for DLI deficiency**

Now, let's determine how long we need to operate the supplemental lighting to achieve a target supplemental DLI of 5 mol/m²·d assuming that the supplemental lighting is provided by HPS lamps that provide 70 µmol/s.m² (532 footcandle or 5,740 lux, see Table 7.1 for conversion factors) at plant height.

**Solution**

Rearranging Equation (7.1) for photoperiod this time, we can easily calculate the photoperiod, that is, how long we need to operate the supplemental lighting:

$$\text{Photoperiod} = \left(\frac{\left(\text{Required Supplemental DLI, mol/m}^2\cdot\text{d}\right)\left(1{,}000{,}000\,\dfrac{\mu\text{mol}}{\text{mol}}\right)}{\text{Provided Light Intensity}\left(\dfrac{\mu\text{mol}}{\text{s.m}^2}\right)\left(3{,}600\,\dfrac{\text{s}}{\text{h}}\right)}\right)$$

$$= \left(\frac{\left(5\,\text{mol/m}^2\cdot\text{d}\right)\left(1{,}000{,}000\,\dfrac{\mu\text{mol}}{\text{mol}}\right)}{\left(70\,\dfrac{\mu\text{mol}}{\text{s.m}^2}\right)\left(3{,}600\,\dfrac{\text{s}}{\text{h}}\right)}\right) = 20\,\text{h/d}$$

As we did in the previous example, we can achieve 20 h/d photoperiod using Table 7.2 and the method of interpolation for a DLI of 5 mol/m²·d and a PPFD of 70 µmol/s.m².

**Example 7.7: Calculating how much of total DLI is coming from supplemental lighting lamps**

Now, let's determine how much of DLI is coming from the supplemental lighting lamps in a greenhouse if metal halide (MH) lamps are operated from 6 a.m. to 11:00 a.m. and from 6 p.m. to 11:00 p.m. (10 hours of operation in total) providing 70 µmol/m²·s.

**Solution**

Using Equation (7.1), now we can easily calculate how much of total DLI is coming from supplemental lighting lamps:

$$\text{DLI} = \left(\frac{\text{Provided PPFD}\left(\dfrac{\mu\text{mol}}{\text{s.m}^2}\right)\left(\text{Photoperiod}\left(\dfrac{\text{h}}{\text{d}}\right)\right)\left(3{,}600\,\dfrac{\text{s}}{\text{h}}\right)}{\left(1{,}000{,}000\,\dfrac{\mu\text{mol}}{\text{mol}}\right)}\right)$$

$$= \left(\frac{70\left(\dfrac{\mu\text{mol}}{\text{s.m}^2}\right)\left(10\,\dfrac{\text{h}}{\text{d}}\right)\left(3{,}600\,\dfrac{\text{s}}{\text{h}}\right)}{\left(1{,}000{,}000\,\dfrac{\mu\text{mol}}{\text{mol}}\right)}\right) = 2.5\,\text{mol/m}^2\text{.d}$$

And again, we can determine exactly the same DLI of 2.5 mol/m².d using Table 7.2 and the method of interpolation for a PPFD of 70 µmol/m²·s and a photoperiod of 10 hours per day.

**Example 7.8: Calculating the DLI from footcandle or lux measurements**

Now, let's consider that the footcandle and lux measurements provided in the following table were made over a 24-hour period inside a greenhouse. Determine the DLI for each measurement units.

| Time | Footcandle | Lux | Time | Footcandle | Lux |
|---|---|---|---|---|---|
| 01:00 | 0 | 0 | 13:00 | 4,100 | 44,280 |
| 02:00 | 0 | 0 | 14:00 | 2,350 | 25,380 |
| 03:00 | 0 | 0 | 15:00 | 1,200 | 12,960 |
| 04:00 | 0 | 0 | 16:00 | 900z | 9,720 |
| 05:00 | 70 | 756 | 17:00 | 750 | 8,100 |
| 06:00 | 200 | 2,160 | 18:00 | 400 | 4,320 |
| 07:00 | 450 | 4,860 | 19:00 | 210 | 2,268 |
| 08:00 | 980 | 10,584 | 20:00 | 80 | 918 |
| 09:00 | 1,500 | 16,200 | 21:00 | 0 | 0 |
| 10:00 | 2,350 | 25,380 | 22:00 | 0 | 0 |
| 11:00 | 2,580 | 27,864 | 23:00 | 0 | 0 |
| 12:00 | 3,500 | 37,800 | 24:00 | 0 | 0 |

**Solution**

First, we need to calculate the daily total and 24-hour average footcandle and lux measurements, and then we can perform the following calculations as discussed below.

| | Footcandle | Lux |
|---|---|---|
| Daily Total | 21,625 | 233,550 |
| 24-hour Average (per hour) | 901.0 | 9,731.3 |
| PPFD (µmol/s.m$^2$) | 180.2 | 180.2 |
| DLI (mol/m$^2$.d) | 15.6 | 15.6 |

In the above calculations, 24-hour footcandle and lux averages were converted to PPDF using the conversion factors (1/5.0 = 0.2 µmol/s.m$^2$ per footcandle and 1/54 = 0.0185185 µmol/s.m$^2$ per lux, respectively) for daylight provided in Table 7.1. And then, DLIs were calculated using Equation (7.1).

hours in the summer, reaching a maximum value on summer solstice, June 21 (see Figure. 2.33). On summer solstice, the daily path is highest in the sky, and the length of daylight is the longest of the year. In the Southern Hemisphere, on the other hand, the situation is the reverse, and the hemisphere, on or about June 21, receives the least amount of sunlight and onset of winter. In the Northern Hemisphere's winter, the Sun is directly overhead at latitude 23.5°S, which is called the winter solstice (December 21); and at this time, the Southern Hemisphere starts enjoying the beginning of summer with long hours of the Sun high in the sky. At the extreme, the noontime Sun is directly overhead at either the Tropic of Cancer (summer solstice), or Tropic of Capricorn (winter solstice). At winter solstice, the areas north of the Arctic Circle have 24-hour darkness, while the areas south of

**TABLE 7.6**

Lighting Strategies and Characteristics used in Controlled Environment Production of Horticultural Crops (Runkle, 2018b).

| | Photoperiodic Lighting | Supplemental Lighting | Sole-Source Lighting |
|---|---|---|---|
| **General Description** | Low-intensity lighting to regulate flowering of crops grown in greenhouses | Moderate-intensity lighting to increase photosynthesis of crops grown in greenhouses | High-intensity lighting for indoor production of high-value specialty crops |
| **Primary Objective** | Promote or inhibit flowering | Increase growth, yield, and quality attributes | Precise regulation of plant growth and quality attributes |
| **Major Crops** | Ornamental crops | Plugs, liners, small fruits, vegetables, and other high-value crops | Leafy greens, herbs, young plants, small fruits, and other high-value crops |
| **Operational Use** | When the natural photoperiod is less than 13 hours, and only during the night | When the daily light integral is low, and only at night and on cloudy days | Every day |
| **Typical PPFD (μmol/s.m²)** | 1–2 μmol/s.m² | 60–120 μmol/s.m² | 150–200 μmol/s.m² |
| **Typical Spectrum** | Red ± Far-red | Red ± Blue/White or White | Red ± Blue/White or White |
| **Number of Hours Used per Day** | Usually up to 4 | Usually up to 20 | Usually 18–24 |
| **Control of Plant Height and Leaf Size** | Slight to moderate | Slight | Strong |

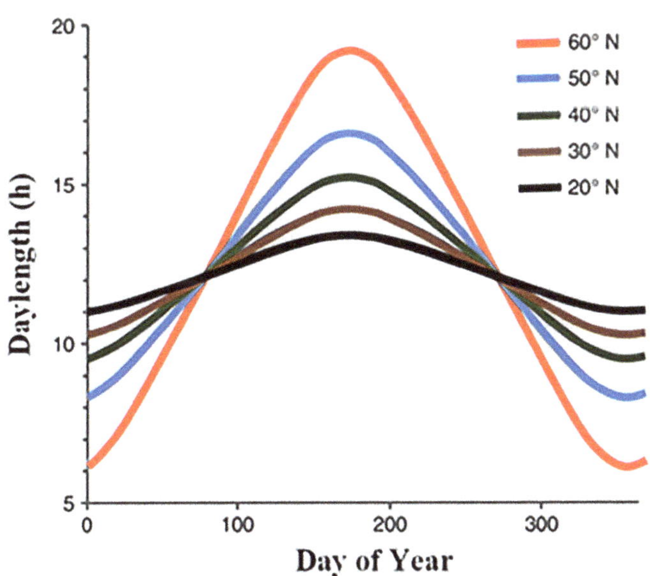


**FIGURE 7.12**
Calculated natural photoperiods by latitude and Julian day of year.

the Antarctic Circle have 24-hour daylight. When the Sun is directly overhead at the equator (spring and autumnal equinoxes – on March 21 and September 21, respectively), everywhere on the Earth, there are 12 hours of daylight Figure 7.12); that is, all latitudes have equal amounts of day and night.

Figure 2.34 shows that as summer moves into fall, and then into winter, the sunrise and sunset times of the Sun's motion across the sky gradually move southward. As a result, the day lengths (natural photoperiods) get shorter and the solar path gets lower in the sky. So, the natural photoperiod changes by latitude and time of the year as shown in Figure 7.12. For instance, as we move from the equator toward higher latitudes, the annual range of natural photoperiods get much larger, and this phenomenon requires intervention in greenhouses so that the crops grown can have adequate photoperiods for their optimal growth and yields.

Photoperiodic lighting is called as the delivery of low-intensity lighting to artificially generate long days when the natural day lengths are short as discussed above. Once the natural photoperiod is long enough, there is no reason to provide artificial photoperiodic lighting. Typically, a light intensity of 1–2 μmol/s.m$^2$ is provided (Table 7.6), which is not enough or negligible for plants to perform photosynthesis; but, it is bright enough for plants to sense and take proper action. Hence, this low-intensity (photoperiodic) lighting does not increase plant growth and development; rather, it is only used to regulate flowering, that is, to accelerate and induce flowering of long-day plants (many annuals and perennials) so that the produce can make the target market dates, and to inhibit flowering of short-day plants (chrysanthemums, kalanchoes, poinsettias, etc.) again for marketing purposes.

We can extend the day length by simply turning on a HID light source such as HPS, MH, high-intensity LID lamps before the Sun sets, and keeping it on until the proper day length is achieved for a particular crop. Growers typically keep their stationary HPS lamps on for day-extension lighting, which indeed serves as both photoperiodic and supplemental lighting (increasing the DLI) depending on the light intensity and duration provided. Night interruption is another lighting strategy that has no effect on the DLI, and employs the photoperiodic lighting in the middle of night by interrupting the dark period for a 4-hour period, typically from 10 p.m. to 2 a.m. because plants are extremely sensitive to light during this time. It is well known that continuous light during night interruption period is not absolutely required; hence an alternative lighting strategy called as cyclic or intermittent photoperiodic lighting for generating longer days can be employed and energy consumption can be reduced considerably. During the cyclic lighting period, lights are turned on for 10 minutes and off for the following 20 minutes (or lights are kept on 5 minutes minimum every 30 minutes). Since both extending the day and night interruption strategies are similarly effective for plant response, using the day-extension strategy with HPS lamps may cost additional energy if only operated for providing low-intensity photoperiodic lighting. The simplest way to deliver photoperiodic lighting is by installing low-output, screw-in lighting fixtures (CFL, INC, or LED bulbs) and turning them on for 5 minutes minimum every half an hour. Using high-intensity fixtures (HPS, MH, or high-intensity LED bulbs) with rotating reflectors or on a moving boom above crop canopy and moving back and forth for about four hours during night does the same job. Fluorescent lamps, red LEDs, and white LEDs have been reported quite effective for inhibiting flowering of short-day plants. Lamps that emit red and far-red lights are, on the other hand, reported as the most effective lighting for promoting flowering of long-day plants.

### 7.3.2 Photosynthetic (Supplemental) Lighting

High-intensity photosynthetic (supplemental) lighting provides much higher light intensity compared to that of low-intensity photoperiodic lighting, typically around 60–120 μmol/s.m$^2$ (Table 7.6) from HID lamps such as HPS and MH lamps or high-intensity LEDs, and increases photosynthesis and therefore plant growth and production. As a result of the increased DLI, plants having supplemental lighting typically flower one to two weeks earlier. As discussed earlier in the previous section, photosynthetic lighting can also be used to extend daylength (photoperiod); however, compared to low-intensity photoperiodic lighting, the installation and operational costs of high-intensity supplemental lighting are much higher. Therefore, supplemental lighting is generally used for high-value crops and when the harvestable yield of crops increases with light quantity (e.g. fruiting vegetables and cut flowers), and there is no need to have separate photosynthetic and photoperiodic lighting systems in the same greenhouse growing space. The choice of lighting technology is, on the other hand, primarily based on economics; therefore, the decision of providing lighting and the choice of lighting source will vary from one greenhouse operation to another.

### 7.3.3 Sole-Source Lighting

Sole-source lighting is the delivery of light in controlled environment plant production facilities that do not receive natural light from the Sun, that is, artificial lighting is provided for indoor production of plants and provides the whole light (photosynthetic and photoperiodic) required by plants as the sole source of light. Inside plant factories and growth room facilities, environmental conditions are optimized for basically any desired crop characteristics. And the most important environmental factor controlling plant growth and operational production costs in such kind of production facilities is lighting, that is, the sole-source lighting.

Relatively speaking, much higher intensity lighting (150–200 μmol/s.m$^2$) is provided for most plants grown in these production facilities, and the lighting fixtures operate for 18–24 hours per day mostly year round (Table 7.6). As lighting fixtures generate a lot of heat, such facilities require properly designed cooling systems for providing and maintaining an adequate thermal environment for plants. Therefore, the lighting efficacy of light sources is critically important. New LEDs have relatively high lighting efficacies (see Table 7.4); hence, they generate less heat. As a consequence, we see plant factories today fully equipped with sole-source LED lighting systems, especially red LEDs with blue or white LEDs, or only a type of white LED.

## 7.4 Artificial Lighting Sources in Controlled Environments

In the previous sections, we have discussed the fundamental concepts of natural and artificial lighting, their relationships with plant growth and development as well as manipulating different lighting strategies for better target plant attributes and maximum plant production. However, the ultimate question is how the light (natural, artificial, or both) required by plants (preferably slightly over) will be deployed at the plant height. The answer to this question requires an adequate engineering design so that the desired light intensity (PPFD) at the crop level (as a function of the crop type, the natural DLI, and the target DLI), uniformity, and the duration of lighting can be designed effectively and sustainably. In their designs, engineers should also take the light depreciation, spectral changes, and material degradation into account so that the light output from the designed lighting systems can be maintained over their life cycles. Since each and every controlled environment lighting system is case-specific, it is extremely important for growers to work with a reputable engineer who has proper experience in controlled environment plant production lighting systems and can perform custom designs rather than using "a cookie cutter" approach.

This section covers different artificial lighting sources used successfully in controlled environment plant production facilities, and their characteristics.

### 7.4.1 High Intensity Discharge Lamps

HID lamps, such as HPS and MH lamps have been used in controlled environments effectively for a quite a long time. They are employed for providing daylight extension and night-interruption lighting. HPS and MH lamps are commonly suspended over the canopy to provide crop lighting. However, as mentioned earlier in the previous section, they are sometimes mounted on booms that move back and forth above the canopy, and still provide the required lighting. Sometimes, these lamps are used with oscillating reflectors, a relatively more recent approach. Table 7.4 shows the PPE of different HPS bulbs. The higher the PPE is, the more effective the fixture is at converting electrical energy into photosynthetic photons. For instance, a 400-W single-ended HPS lamp with magnetic ballast has a PPE value of 0.94 μmol/J while a double-ended 1,000-W HPS lamp with electronic ballast has a PPE of 1.72 μmol/J. Table 7.4 also shows that the PPE value for high-intensity LED products varies considerably, and many new LEDs now even approach and exceed 3.0 μmol/J.

### 7.4.2 Light-Emitting Diodes

LEDs are solid-state, semiconducting diodes, which can emit light from a wavelength of approximately 250 nm to over 1,000 nm. They are a wonderful light source with extremely promising controlled environment plant production applications. There are several features of LEDs that make them attractive alternatives to HPS lamps. First, LEDs can produce quite a narrow-spectrum light at target wavebands useful for plant growth and development, including blue (450 nm) and red (660 nm). Additionally, LEDs currently have a luminous efficiency of 38% (red) to 50% (blue) in converting energy to light with an estimated life of approximately 50,000 luminous hours or more. The theoretical maximum efficiency obviously occurs when all input energy is converted to energy in photosynthetic photons. Table 7.7 shows that blue LEDs can be 93% efficient, phosphor-converted "whites" 76% efficient, and red LEDs 81% efficient. LED efficiency describes the ratio of the optical power output (W) to the electrical power input (W), using the same units. Therefore, the term efficiency is expressed as a percentage. As discussed earlier, the term efficacy on the other hand applies to the same ratio using different units, that is, μmol photon output per second, per watt of electrical power input, and therefore having an overall unit of μmol/J. Also, as discussed earlier in Chapter 2, the relationship between photon energy content and wavelength is expressed in Equation (2.210), the Planck's relationship. This equation

**TABLE 7.7**

Efficiency and Efficacy of Common LEDs at 100 mA per mm² (Near-Optimal Efficacy) and a 25°C Junction Temperature (Kusuma et al., 2020).

| Light Source | Peak Wavelength (nm) or Correlated Color Temperature (K) | Efficiency (W/W) | Photosynthetic Photon Efficacy (µmol/J) |
|---|---|---|---|
| Blue | 450 nm | 0.93 | 3.5 |
| Green | 530 nm | 0.42 | 1.9 |
| Red | 660 nm | 0.81 | 4.5 |
| Far-red | 730 nm | 0.77 | 4.7 |
| Cool-white | 6,500 K | 0.76 | 2.9 |
| Warm-white | 2,700 K | 0.69 | 2.6 |

is used to calculate the maximum possible photosynthetic photon efficacy for a given spectrum, and it is employed to convert between efficiency and efficacy. Based on Planck's relationship, energy content of photon is inversely proportional to wavelength.

In addition to efficacy, initial cost of lighting fixture per photon/s of output capacity is an important criterion. HPS fixtures (1 kW) range from US$200 to US$350 per kW (5–9 µmol/s per $) while LED fixtures ranging from US$1,000 to US$3,000 per kW (0.5–1.5 µmol/s per $). On a photon flux basis, the initial cost of LEDs is therefore reported to be somewhere between 3 and 18 times higher compared to those of HPS fixtures. The initial cost of both HPS and LED technologies has decreased, but the cost of LED fixtures is expected to decrease further and faster than HPS fixtures. While the initial cost is higher, LED fixtures have lower operational costs compared with those of HPS fixtures. Depending on the hours of usage and the cost of electricity, a payback period of 3–5 years for sole-source lighting and 5–8 years for supplemental lighting of LEDs can be expected (Nelson & Bugbee, 2014). The PPE of LEDs is increasing and initial costs are decreasing, and this makes the payback periods more acceptable every passing year. Their energy efficiency, long lifespan, and ability to target specific wavelengths make them an excellent option particularly for precision managing of photoperiod and some crop attributes. LEDs also provide the opportunity to manipulate the ratio of red (R) and far-red (FR) light for desired plant responses. As it was discussed earlier in Chapter 4, low red/far-red ratio promotes stem elongation (shade avoidance response) in many plant species. The best combinations of different light mixtures and ratios are currently being studied and producing quite exciting findings.

### 7.4.3 Incandescent Lamps and Compact Fluorescent Lamps

INC lamps have been used effectively in greenhouse crop production for decades, and even though they are being dropped out of greenhouse production due to their low energy efficiency, they are still commonly employed for providing daylength extension and/or night-interruption lighting. Many growers have replaced most or all INC lamps with attractive alternative, CFLs, which consume about one-fourth the amount of electricity as INC lamps. Since the frequent on-off operation required in cyclic lighting does not affect bulb life or fixture longevity, INC lamps are easily and adequately used for such lighting strategy. However, while CFLs have better energy efficiency and are quite effective for use in greenhouses, their spectrum is not as effective at controlling flowering of some long-day

**TABLE 7.8**

The Percentage Distribution of Light within the Photosynthetic Range (Blue, Green, Red), Far-Red, and the Red to Far-Red Ratio (Runkle, 2007).

| Light Source | Light Distribution Percentage | | | | Red to Far-red Ratio |
|---|---|---|---|---|---|
| | Blue | Green | Red | Far-red | |
| Cool-white fluorescent lamp | 21 | 52 | 24 | 2 | 10.7 |
| High-pressure sodium lamp | 5 | 51 | 38 | 6 | 6 |
| Incandescent lamp | 2 | 13 | 34 | 52 | 0.7 |
| Metal halide lamp | 18 | 49 | 25 | 8 | 3 |
| Sun and sky, daylight | 23 | 26 | 26 | 25 | 1 |

plants. CFLs are particularly low in far-red light, which is required for rapid flowering of some long-day crops (Table 7.8).

It is, however, possible to alternate CFL with INC bulbs to provide the required light quality for effective photoperiodic lighting so that some energy savings can still be achieved. A considerably more expensive alternative LEDs presented above, on the other hand, are much more energy efficient than both INC bulbs and CFLs, and last much longer.

## 7.5 Future Perspectives

The moderately efficient HPS lamp is currently the most widely used light source for high-intensity supplemental (photosynthetic) lighting in greenhouses. Efficiency, intensity, and longevity of LEDs are being enhanced rapidly while INC lamps are being dropped out of greenhouse production due to their low energy efficiency. CFLs are cost-effective substitutes; however, as discussed earlier, they delay flowering in some long-day crops due to their very little far-red light emissions. Low-intensity LEDs are, on the other hand, slowly replacing INC lamps for low-intensity photoperiodic lighting.

As many greenhouses use supplemental lighting for only 3 or 4 months of the year, the payback periods are typically longer. However, as the production and sales of LEDs increase, their initial investment costs will drop further and faster. As they advance more and their costs decrease further, it will be a common practice to use them first in combination with HPS lamps, and then, they eventually become the only light source for both low-intensity and high-intensity lighting. As well, LED lighting fixture arrays will become much more compact for minimizing crop shading and making in-canopy lighting possible (for penetrating light deeper in the canopy) as they emit much less heat compared with HPS lamps. The flexibility of designing the spectral output of LEDs for specific plant requirements will enable extremely exciting customized precision lighting applications for specific crops and crop attributes paving the way for crop engineering further.

Other lighting sources such as plasma lamps are also being developed for controlled environment plant production lighting; and companies are mostly focusing on high-intensity lighting to increase plant photosynthesis. We see a very bright light at the end of the tunnel, but more diverse and exciting developments are yet to come.

## Nomenclature

**CFL** compact fluorescent lamp
**DLI** daily light integral
**HID** high intensity discharge
**HPS** high-pressure sodium
**INC** incandescent
**LED** light-emitting diode
**MH** metal halide
**PAR** photosynthetically active radiation
**PPE** photosynthetic photon efficacy
**PPF** photosynthetic photon flux
**PPFD** photosynthetic photon flux density

## Bibliography

ASHRAE (1997) Handbook of Fundamentals. American Society of Heating, Refrigeration and Air Conditioning Engineers, Inc. Atlanta, GA 30329. pp. 10.25–10.29.

Currey C and Lopez R (2013) Comparing LED lighting to high-pressure sodium lamps. *Greenhouse Grower*, October 2013, Meister Media Worldwide, Willoughby, OH, USA. pp. 34–40.

Currey C and Lopez R (2014) Managing photoperiodic lighting. *GrowerTalks*, March 2014, Ball Publishing, Chicago, ILUSA. pp. 72–73.

DesignLights Consortium (2020) https://www.designlights.org/horticultural-lighting Retrieved 07-06-2020

Dutta Gupta S (2017) *Light Emitting Diodes for Agriculture*. Springer Singapore.

Faust JE and Logan J (2018) Daily light integral: A research review and high resolution maps of the United States. *HortScience* 53(9):1250–1257.

Goldammer T (2019) *Greenhouse Management: A Guide to Operations and Technology*. Apex Publishers, Centreville, VI, USA.

Grevstad FS and Coop LB (2015) The consequences of photoperiodism for organisms in new climates. *Ecological Applications* 25(6):1506–1517.

Korczynski PM, Logan J, and Faust JE (2002) Mapping monthly distribution of daily light integrals across the contiguous United States. *HortTechnology* 12:12–16.

Kozai T, Fujiwara K, Runkle ES (2016) LED Lighting for Urban Agriculture. Springer Singapore.

Kusuma P, Pattison PM, and Bugbee B (2020) From physics to fixtures to food: current and potential LED efficacy. *Hortic Res* **7**, 56 (2020). doi:10.1038/s41438-020-0283-7

Lopez R and Currey C (2014) Understanding light. *GrowerTalks*, February 2014, Ball Publishing, Chicago, ILUSA. pp. 58–61.

Lopez R and Runkle ES (2017) *Light Management in Controlled Environments*. Meister Media Worldwide, Willoughby, OH, USA.

Morgan L (2013) Daily Light Integral (DLI) and greenhouse tomato production. *The Tomato Magazine*, Winter 2013. p. 15.

Nelson J and Bugbee B (2014) Economic analysis of greenhouse lighting: light emitting diodes vs. high intensity discharge fixtures. *PLoS ONE* 9, e99010 (2014).

OMAFRA (2001) Growing Greenhouse Vegetables. Ontario Ministry of Agriculture, Food and Rural Affairs. Publication 371. ISSN: 1492-6601. Toronto, Canada. p.77.

Runkle ES (2007) Light quality defined. *Greenhouse Product News Magazine*, June 2007. Sparta, MI, USA. p. 82.

Runkle ES (2010) The future of greenhouse lighting. *Greenhouse Product News Magazine*, September 2010. Sparta, MI, USA. p. 66.

Runkle ES (2018a) An update on LED lighting efficacy. *Greenhouse Product News Magazine*, July 2018. Sparta, MI, USA. p. 58.

Runkle ES (2018b) Horticultural Lighting Applications. *Greenhouse Product News Magazine*, June 2018. Sparta, MI, USA. p. 58.

Runkle ES (2019) DLI 'Requirements'. *Greenhouse Product News Magazine*, May 2019. Sparta, MI, USA. p. 50

Runkle ES and Bugbee B (2013) Problems with foot-candles, lux and lumens. *Greenhouse Product News Magazine*, September 2013. Sparta, MI, USA. p. 78.

Thimijan RW and Heins RD (1983) Photometric, radiometric and quantum light units of measure: A review of procedures for interconversion. *Hort Science* 18(6):818–822.

# 8

# *Greenhouse Environmental Control Systems*

## 8.1 Introduction

Greenhouses are extremely complex and non-linear living systems whose behaviors change in time depending on numerous disturbances, generally weather-related. It is therefore difficult to control greenhouse environments with classical controllers; hence, complex and smarter approaches are required in order to bring the indoor and outdoor stakeholders such as whether, the cost of energy, water, $CO_2$, and others all together and make proper adjustments in greenhouse environment instantly for healthy crop development and therefore a profitable operation. Smart digital greenhouse environmental control systems are available today, and they control every single parameter in an integrated fashion in modern commercial greenhouses providing smart decision-making processes for optimum growth conditions. Further precision controls and therefore savings can be achieved via incorporating smarter strategies with artificial intelligence (AI) algorithms into the environmental control systems. For instance, through reducing ventilation a bit or maintaining little higher indoor relative humidity levels without harming the crop growth and development (and therefore the eventual economic gains), achieving special savings available through power companies, supplying power from co-generation plants and alternate power supply plans, etc., and performing $CO_2$ enrichment and storing the heat in hot water storage tanks for later use in heating, we can reduce energy losses and save dollars escaping through greenhouse vents or our pockets. The consequence of every management step and strategy needs to be evaluated and dynamically optimized in an integrated fashion with continuous environmental monitoring before it is actuated. Today, the remote monitoring and control systems are on duty day and night continuously; and with the help of AI, computer vision, and enhanced sensor integration, our greenhouses will soon become more flexible and autonomous operations, and therefore more universal and sustainable.

Before something is controlled, its position needs to be sensed first somehow, so that proper adjustments can be made in its position. And we call these sensing devices as sensors, which are essential components of any greenhouse monitoring and control systems. These sensors measure conditions of different environmental variables such as temperature, relative humidity, light intensity, $CO_2$ concentration, and wind speed. Today, both hardwired and wireless monitoring and control systems are available, and commercial greenhouse operations have either one of these types or sometimes the combination of both. It should be kept in mind that control systems are limited by the accuracy and responsiveness (time constant) of their sensors; that is, they not only need to sense the environment accurately, but also fast enough so that the control system can take proper actions in a timely manner. Sensor accuracy and reliability under greenhouse conditions are particularly important due to the unfavorable environmental conditions in greenhouses.

This chapter briefly covers control theory and classification first, then moves to classical and modern control theories and system classifications, and concludes with future evolution of control systems.

## 8.2 Control Theory

A *control system* has four main functions termed as measuring, comparing, computing, and correcting functions. These four functions are realized by a detector, a transducer, a transmitter, a controller, and a final control element. The detector, transducer, and transmitter, all contained in a single unit (e.g. a resistance thermometer) do the measuring task. Comparing and computing tasks are performed by the controller, which may be implemented electronically by a proportional (P) control, integral (I) control, a proportional-integral (PI) controller, a proportional-derivative (PD) controller, a proportional-integral-derivative (PID) controller, or programmable logic controller (PLC). Older controller units have been mechanical devices, for instance, as in a carburetor. Correcting task is, on the other hand, performed by a final control element, the actuator. The final control element changes an input or output in the control system that influences the adjusted or controlled variable such as temperature, $CO_2$ concentration, and light intensity.

*Control theory* is an interdisciplinary branch of engineering and mathematics that deals with influencing the behavior of dynamic systems, and how their behavior is modified by feedback. The usual objective of control theory is to control a system or process so its output tracks a desired control signal (the reference), which may have a fixed or changing value. In order to perform this task, a controller is designed, which monitors the output and compares it with the reference value. The difference between actual and desired outputs (a.k.a. the error signal) is fed back as the input to the system in order to bring the actual output closer to the set reference.

A system or process to be controlled is represented by a *block diagram* in Figure 8.1. The input–output relationship represents the cause-and-effect relationship of the process, which in turn represents a processing of the input signal to provide an output signal variable, often with power amplification.

There are two main types of control systems. When there is no feedback as shown in Figure 8.2, the control system is called as an *open-loop controller* or a *non-feedback controller*

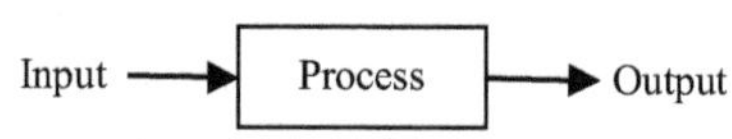


**FIGURE 8.1**
A process to be controlled.

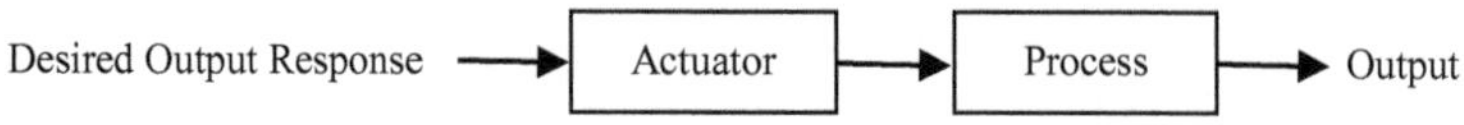


**FIGURE 8.2**
Open-loop control system (without feedback).

because the system output is not measured, fed back, and used to change the position of the control. As a consequence of this configuration, the controller cannot compensate for changes acting on the system or process. If the process transfer function, which theoretically models the output of the process for each possible input, is known and there is very small disturbance, an open-loop controller, as shown in Figure 8.2, would satisfy the most design requirements. A manual control system is an open-loop control system. Examples of an open-loop control system can be listed as an electric toaster, automatic coffee maker, light switch, automatic washing machine, timer based clothes drier, etc., we use every day.

In a *closed-loop control system*, as shown in Figure 8.3, on the other hand, data from a sensor monitoring the system output enter a controller, which continuously subtracts the quantity representing the actual output from the reference quantity representing the desired output response. The measure of the output is called the *feedback signal*. An open-loop control system can as well be easily converted into a closed-loop system by providing a feedback.

With an accurate sensor, the measured output is equal to the actual output of the system. The difference between the actual (measured) and the desired output (i.e. the error) determines the position of the control. The result is to continuously and dynamically match the actual or measured output to the reference response in order to maintain the desired system output. As the sensed output drops below the reference, obviously the error increases; hence the throttle opens. These control systems using the *feedback loop*, as shown schematically in Figure 8.3, affect the system output, which in turn is measured and fed back to the controller dynamically. Quite often, the error is amplified and used to control the process so that the difference is continually reduced. The output of the control device causes the actuator to modulate the process in order to reduce the error. The feedback concept has been the foundation for control system analysis and design. If the feedback signal is positive, then the system is called a *positive feedback system*, and if the feedback signal is negative, then the system is called a *negative feedback system*. With dynamic variations and significant disturbance in the process, feedback control is the only choice, whose simplicity is quite appealing. Significant steady-state error and overshoot are the most common problems with this proportional controller, a control response in proportion to how far away the measured output is from the desired output response.

A common example of a closed-loop control system is a person steering a car by visually inspecting the car's location on the road, and based on this visual "feedback," making proper adjustments and controlling the car's position. The system shown in Figure 8.3 is called a *negative feedback* control system; because the measured output is subtracted from the desired input and the difference (error) is used as the input signal to the power amplifier. Many other examples can be given for the closed-loop feedback system such as automatic electric iron, water level controller, and an air conditioner.

Increasing complexity of systems and interest in achieving optimal performance has forced control systems to grow tremendously in the past several decades. Furthermore, as

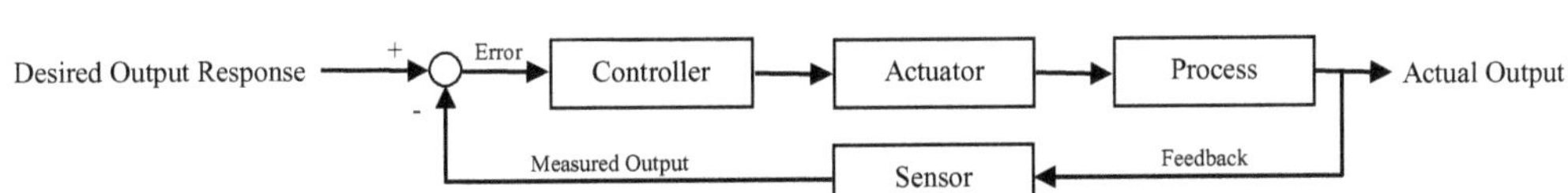


**FIGURE 8.3**
Closed-loop feedback control system.

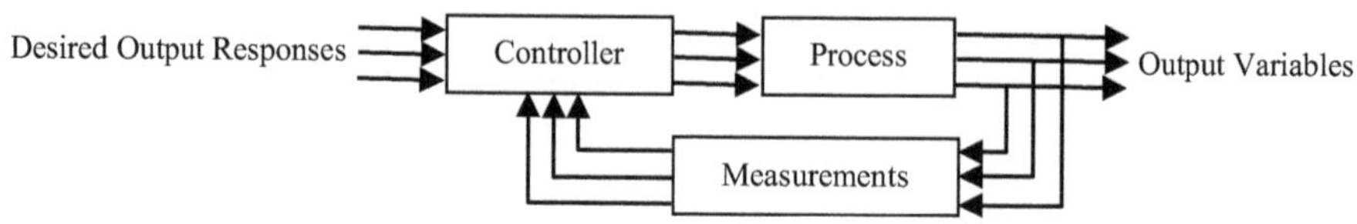


**FIGURE 8.4**
Multivariable control system.

the systems become more complex, the interrelationship of many controlled variables such as those in greenhouses need to be considered in the control scheme. A block diagram schematically demonstrating a multivariable control system is shown in Figure 8.4.

## 8.3 Classification

There are numerous types of control systems, which are broadly categorized and discussed below.

### 8.3.1 Linear Versus Nonlinear Control Theory

The field of control theory can be divided into two branches termed as *linear control theory* and *nonlinear control theory*. In order to comprehend the linear control system, we first need to cover the principle of superposition. The principle of superposition has two important properties termed as homogeneity and additivity. A system is said to be homogeneous, if the input is multiplied by a constant, then the output as well is multiplied by the same constant. Now, let's suppose there is a system, which is fed by the input $a_1$ and provides the output $b_1$ for the input $a_1$ for the first time; and the next time, when the system is fed by the input $a_2$, the system provides the output $b_2$ for the input $a_2$. When the sum of the previous inputs (i.e. $a_1 + a_2$) is provided as the input to the system and we receive the output as the sum of previous outputs as well (i.e. $b_1 + b_2$), then it is said that the system follows the property of additivity. Now, the linear control systems can be defined as the systems, which obey the superposition principle, that is, the principle of homogeneity and additivity (only exists in theory). This roughly means that the output is proportional to the input, and these systems are governed by linear differential equations.

A nonlinear control system is, on the other hand, defined as a control system that does not obey the principle of homogeneity. In practice, all control systems are nonlinear systems where we cannot apply the principle of superposition. Sometimes, it is even desirable to introduce a nonlinearity intentionally in order to improve the performance of a system or make its operation safer and more economical. One of the simplest examples of a system with an intentionally introduced nonlinearity is a relay controlled or ON/OFF system. For instance, a furnace in a home heating system is turned ON when the temperature falls below a specified temperature and OFF when the temperature exceeds this time another specified temperature. Analysis of different nonlinear control systems is usually very difficult due to their nonlinear behavior. These systems are often governed by nonlinear differential equations, and such systems are quite often analyzed using numerical methods on computers by simulating their operation using a simulation language, for instance.

### 8.3.2 Frequency Domain Versus Time Domain

Mathematical techniques for analyzing and designing control systems, on the other hand, fall into two different domains called as the *frequency domain* and the *time domain*.

In frequency domain, the state variables, that is, the input, output, and feedback are all represented as functions of frequency. The input signal and the transfer function are converted from time functions to functions of frequency by a transform function such as the Fourier, Laplace, or Z transform. In this technique, the differential equations representing the system are replaced by algebraic equations in the frequency domain, so that the mathematical solutions become much simpler; however, it should be noted that frequency domain techniques cannot be used with nonlinear systems, rather they only be used with linear systems.

As frequency domain techniques can only be used with linear systems, time domain is extensively used for analyzing the real-world nonlinear systems. In time domain type design, the state variables are represented as functions of time using differential equations, which are obviously more difficult to solve compared to algebraic equations; however, they have turned into routine analyses with the help of computer simulation techniques today.

### 8.3.3 SISO Versus MIMO Systems

Based on the number of inputs and outputs, on the other hand, control systems are divided into two different categories: *single-input-single-output* (SISO) and *multiple-input-multiple-outputs* (MIMO).

SISO is the simplest and most common type, in which one output is controlled by one control signal. Simple example is an audio system where the input audio signal is the control input, and the sound waves from the speaker is the system's output. Other examples include temperature control and position control system.

MIMO are, however, found in more complicated systems. For example, modern greenhouse facilities today are composed of many segments, each of which is controlled by a separate actuator. Complex systems such as nuclear reactors, greenhouses, and many other biological systems are handled by MIMO control systems such as PLCs.

### 8.3.4 Analog Versus Digital Systems

Control systems can be classified into two different categories depending on the input signal: *analog systems* and *digital systems*.

Analog control systems, also known as continuous control systems, have a continuous signal such as 4–20 mA or 0–10 VDC as the input to the system to control different equipment. These signals are continuous input signals such as sinusoidal, square, or triangle, and that's why they are called continuous control systems. This type of control is preferred where a faster equipment response is needed, and is the standard method of control for certain equipment such as proportioning devices and variable speed motors.

Digital control systems, also known as discrete control systems, have a discrete or a pulse input signal to the system. And these signals have a discrete interval of time. Various sources of continuous (analog) input signals such as sinusoidal or square type can be converted into a discrete signal. Simply, digital control is a simple on/off control and used to control on/off devices such as louvers, motors, and solenoid valves. Proportional devices such as curtains, motorized valves, vents, and other on/off devices can be controlled with a digital signal as well. Digital control is achieved by activating an open or close relay for a period of time to move the device to the desired position. Then, proportional devices are controlled based on the movement time.

Digital systems handle nonlinear control systems more effectively, and are more reliable and accurate compared to the analog control systems.

### 8.3.5 Lumped Versus Distributed Parameter Systems

*Lumped parameter* control systems have the various active (current sources, voltage sources, etc.) and passive (resistors, capacitors, etc.) components concentrated at singular points in space; hence, the spatial derivatives are neglected and we only have the time derivatives to consider. Therefore, the analyses of such systems include ordinary differential equations and are relatively easy to handle. An electrical circuit with passive elements like resistor and capacitor, is a common example for such systems, and the physical quantities current and voltage are functions of *time only*.

*Distributed parameter* control systems, on the other hand, have various active and passive components distributed in space, so that physical quantities *depend on both time and space*. And therefore, the analyses of such systems include partial differential equations and are slightly difficult compared to lamped parameter control systems. The classical example is the electrical line where inductance, capacitance, and resistance are not constant but functions of length. This leads to partial derivatives in time and space (length).

## 8.4 Classical Control Theory

In order to overcome the above-discussed limitations of the open-loop controller, feedback is introduced in control theory. As mentioned earlier, a closed-loop controller provides feedback to adjust outputs of a dynamic system. Process inputs such as voltage applied to an electric motor have an effect on the process outputs (e.g. speed or torque of a pump or a fan motor) measured with sensors and processed by the controller; the result (the control signal) is fed back as input to the process and closes the loop; that's why, it is called as a closed-loop controller.

It is possible for some systems to use closed-loop and open-loop controls simultaneously. In such systems, the open-loop control is termed *feed-forward* and serves to further improve the already improved reference tracking performance of a closed-loop controller. This type of dual loop control provides a higher degree of precision than controls based on closed-loop control only because the response can be predicted ahead of time.

### 8.4.1 Closed-loop Transfer Function

In a closed-loop or feedback control system, as discussed earlier, the output of the system is fed back to the desired output response (set-point) via a sensor measurement. Then, the controller takes the difference (error) between the set-point and the output to change the inputs $u$ to the system under control. This is shown in Figure 8.3. This closed-loop control system is called a SISO control system; and as discussed earlier, MIMO systems with more than one input and output are common as well, especially in modern greenhouse operations.

If we assume that the controller, the process, and the sensor are linear and not time-dependent, then the abovementioned systems can be analyzed using the Laplace transform. Its simplest form can be given as $u = Ke$, where $K$ is a constant. It means that a proportional controller provides a control signal, which is proportional to the error ($e$), that is, a proportional controller returns its input multiplied by the proportional gain $K$. Significant steady-state error and overshooting are the common problems with this type of controller.

### 8.4.2 PID Controller

In order to correct the abovementioned steady-state error and overshooting problems with the constant gain (proportional) controller, two additional terms, the integral and the derivative terms, are added as shown below in Figure 8.5. All together, these three terms operate on the error signal $e(t)$ to produce a control signal $u(t)$.

The end result is the well-known *proportional-integral-derivative* (PID) controller, which is the most-used feedback control system by engineers today. The proportional term usually plays the key role ensuring the stability, with the integral and derivative terms added to reduce or eliminate the steady-state error by rejecting step disturbance and the overshoot by providing damping, respectively. If an integral action is executed, the integral of the error is added to the control signal as well. In the case the error signal is large, then the control signal increases quickly; but when the error signal is small, then the control signal increases slowly. When a derivative control action is executed, on the other hand, a signal proportional to the signal error change rate is provided only when a change in the error signal is present. The main problems with the PID controller are that the integrator introduces a phase lag that might lead to stability problems and the differentiator makes the controller sort of sensitive to noise. The desired closed-loop dynamics is mainly obtained by adjusting the three parameters, $K_p$, $K_i$, $K_d$, and often iteratively and with no specific knowledge of a process model. PID controllers are the most established class of control systems; however, they cannot be used in some complicated cases, especially if MIMO

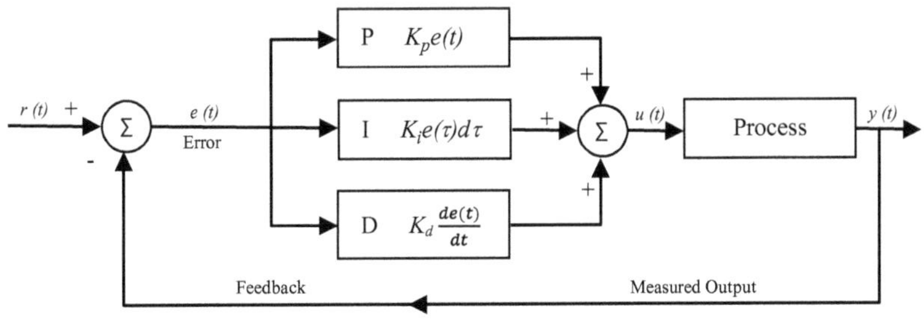


**FIGURE 8.5**
Block diagram of a PID controller in a feedback system. $r(t)$ is the desired output response or set-point and $y(t)$ is the actual (measured) output.

systems are considered. Since greenhouse systems are quite complex mathematical models with many nonlinearity and MIMO systems, PID controllers are not the most appropriate systems for solving the pertinent problems in greenhouses. As a consequence, this has led to the development of new control techniques.

### 8.4.3 Nonlinear PID Controller

Nonlinear PID is an improved PID controller for better performance. Maintaining the simplicity and intuition of PID, nonlinear PID makes the PID stronger with nonlinear gains. In a nonlinear PID controller, when the error is small, the need for the integral control is reduced by making the proportional gain bigger. The differential gain is also reduced for small errors to reduce sensitivities to sensor noise.

## 8.5 Modern Control Theory

In this chapter, we have covered some fundamental analysis and design methods in classical control theory, primarily the feedback control, the development of which was primarily driven by engineering practice and needs of modern industry and society. With an accurate sensor, the measured output is equal to the actual output; and the difference between the desired output and the actual output is equal to the error, which is then adjusted by the control device as covered earlier. Over the last few decades or so, huge efforts in control research have led to the creation of modern mathematical control theory or advanced control paving the way for intelligent control, which uses a number of AI computing methods such as artificial neural networks (ANN), Bayesian probability, fuzzy logic systems (FLS), machine learning systems (MLS), expert systems (ES), evolutionary computation, and genetic algorithms (GA), to control a dynamic system. This development originally started with *optimal control theory* in the late 1950s and early 1960s to study the optimality of control design. Briefly, a cost function is minimized in optimal control theory; and for deriving optimal controllers, analytical or computational techniques are employed. Examples of optimal control include minimized cost, fuel or energy use, and water use situations, which are quite applicable in greenhouse operations. As they can guarantee closed-loop stability, the most widely used optimal control design approaches are Model Predictive Control (MPC) and linear-quadratic-Gaussian control (LQG). Along with PID controllers, MPC systems are the most widely used control techniques in process control. In optimal control of greenhouse environments, it is required to consider aspects like greenhouse energy consumption, water consumption, and crop response, as input parameters of the control system, for which ES and MPC are the most commonly used techniques. Other AI-based approaches are also considered as optimal control when they consider input parameters such as energy consumption, water consumption, and crop responses. Figure 8.6 schematically shows the general classification of greenhouse control systems.

Other major achievements in modern control theory include, for instance, MIMO control systems as discussed earlier, and this achievement extended the well-known SISO concepts to MIMO systems; adaptive control is another example of modern control approaches, which is primarily designed to extend the operating range of a controller by modulating automatically the controller parameters based on the estimated dynamic changes in the processes, analysis and design of nonlinear control systems, and so on. Modern control

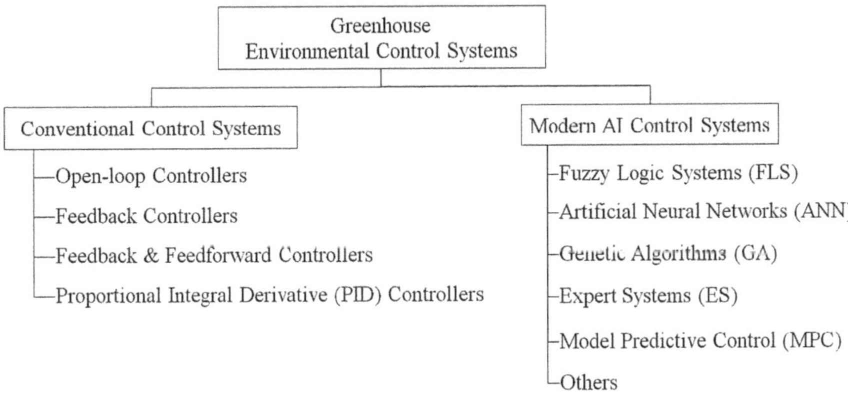


**FIGURE 8.6**
Classification of greenhouse control systems.

theory employs the time-domain state-space representation, a mathematical model of a physical system as a set of input, output, and state variables related by first-order differential equations, whereas the frequency domain analysis forms the basis of the classical control theory. In contrast to the frequency domain approach, the utilization of the state-space representation is not limited to systems with linear components and zero initial conditions.

## 8.6 System Classifications

Control systems can further be classified based on the characteristics of systems being controlled such as linear systems control, nonlinear systems control, decentralized systems control, and centralized systems control.

Linear and nonlinear systems have already been discussed, and they can even be elaborated further and we can talk about, for instance, the linearization of control theory, and then the application of linear techniques, etc.; however, such topics are outside the scope of this manuscript.

Furthermore, when the multiple controllers control the system, the control system is called as *decentralized control*. Decentralization is helpful in many ways; for instance, it helps control systems to operate over a larger geographical area. The agents in decentralized control systems can interact with each other using communication channels and coordinate their actions. Computer zone controllers in greenhouses, for instance, are examples of decentralized controllers; and they bring much greater control flexibility and programmability for different zones, and reduce the wiring requirements compared to centralized controls if wireless systems are not in operation or an option.

*Centralized* or *integrated computer controls* provide coordinated control of distributed systems such as all greenhouse environmental functions in a single, integrated computer control system. Such systems bring together the capability of individual decentralized controllers (e.g. zone controllers) along with the individual control devices such as timers.

**FIGURE 8.7**
Robotics research for harvesting greenhouse tomatoes (Courtesy of Dr. Peter Ling, The Ohio State University, Wooster, OH, USA).

## 8.7 Future Perspectives

In the past, the application of modern control technologies was quite limited due to the limited availability of computational power. Controllers based on AI such as ANNs, FLS, GAs, and others alike could not be implemented in the old technological platforms. Modern techniques based on computational intelligence or algorithms on new technological platforms are now becoming more popular than ever in almost every industry.

Control systems and the automatic control of a production process are used to achieve improved performance and high-quality products along with increased productivity in systems such as greenhouses. Today, our overall goal by developing and utilizing modern control systems is to provide higher level of flexibility and autonomy. And modern control theory deals with systems that typically have adaptive, robust, self-organizing, learning, and optimum qualities. Therefore, in its modern usage, automation can be defined as a technology that uses algorithms to operate a process with supplied feedback of information to ensure that the programmed commands executed properly. Automated processes operate without any human interference; and indeed, most automated systems are able to perform their functions with much greater accuracy and precision, and in less time.

Two system concepts (control systems and robotics) are approaching the goal of providing higher level of flexibility and autonomy through different evolutionary pathways. When it is programmed properly, today's industrial robot is quite autonomous and normally requires no further human intervention. However, due to the sensory limitations, robotic systems (see Figure 8.7) today have sort of limited flexibility in adapting to the changes in work environment, which indeed provides the motivation for computer/machine vision (see Figure 8.8) research. The control system is highly adaptable, but it still

**FIGURE 8.8**
Computer/machine vision research for detecting greenhouse crop anomalies at the University of Arizona.

relies on human supervision. Advanced robotic systems are trying to achieve work adaptability through further enhanced sensory feedback. Sensor integration, particularly crop physiological sensing via computer/machine vision, and AI will soon make our systems, including the commercial greenhouses, more flexible and autonomous, and therefore more universal and sustainable. Improvements in supervisory control and human–machine interface technologies will make the lives of operators much easier, while big data management via powerful computers will improve the efficiency of operators and intelligent control systems. Involving in improved communication techniques and advanced programming languages, both robotics and control systems research are aimed toward reducing the cost of implementation and eventually reaching the universal application.

## Nomenclature

| | |
|---|---|
| **AI** | artificial intelligence |
| **ANN** | artificial neural networks |
| **ES** | expert systems |
| *e* | error |
| **FLS** | fuzzy logic systems |
| **GA** | genetic algorithm |
| **I** | integral |
| **LQG** | linear-quadratic-Gaussian control |
| **MIMO** | multiple-input multiple-output |
| **MLS** | machine learning systems |
| **MPC** | Model Predictive Control |
| **P** | proportional |
| **PD** | proportional-derivative |

**PI** proportional-integral
**PID** proportional-integral-derivative
**PLC** programmable logic controller
$r$ desired output response
**SISO** single-input single-output
$t$ time
$u$ transfer function
$y$ actual/measured output

### Greek letters

$\tau$ time

### Subscripts

$d$ derivative
$i$ integral
$p$ proportional

## Bibliography

Ali IA, Abdalla AM (1993) A microcomputer-based system for all-year-round temperature control in greenhouses in dry arid lands. *Comput Electron Agr* 8(3):195–210.

Ang KH, Chong GCY, and Li Y (2005) PID control system analysis, design, and technology. *IEEE Trans Control Systems Tech*, 13(4): 559–576.

Antsaklis PJ, and Michel AN (1997) *Linear Systems*. McGraw-Hill, New York.

Arvanitis KG, Paraskevopoulos PN, Vernardos AA (2000) Multirate adaptive temperature control of greenhouses. *Comput Electron Agr* 26(3):303–320.

Åström KJ and Murray RM (2008) *Feedback Systems: An Introduction for Scientists and Engineers.* Princeton University Press, New Jersey. ISBN 0-691-13576-2.

Baptista FJ, Bailey BJ, Meneses JF, Navas LM (2010) Greenhouses climate modelling. Tests, adaptation and validation of a dynamic climate model. *Span J Agric Res* 8(2):285–298.

Belforte G, Deboli R, Gay P, Piccarolo P and Ricauda Aimonino D (2006) Robot design and testing for greenhouse applications. *Biosyst. Eng.* 95:309–321.

Bellman R (1964) Control Theory, *Scientific American* 211(3):186–200.

Bennis N, Duplaix J, Enéa G, Haloua M, Youlal H (2008) Greenhouse climate modeling and robust control. *Comput Electron Agr* 61(2):96–107.

Blasco X, Martínez M, Herrero JM, Ramos C, Sanchis J (2007) Model-based predictive control of greenhouse climate for reducing energy and water consumption. *Comput Electron Agr* 55(1):49–70.

Briat C (2015) *Linear Parameter-Varying and Time-Delay Systems. Analysis, Observation, Filtering & Control.* Springer Verlag Heidelberg, Berlin. ISBN 978-3-662-44049-0.

Caponetto R, Fortuna L, Nunnari G, Occhipinti L, Xibilia MG (2002) Soft computing for greenhouse climate control. *IEEE T Fuzzy Syst* 8(6):753–760.

Castañeda-Miranda R, Ventura-Ramos E, Peniche-Vera RR, Herrera-Ruiz G (2006) Fuzzy greenhouse climate control system based on a field programmable gate array. *Biosyst Eng* 94(2):165–177.

Chalabi ZS, Bailey BJ, Wilkinson DJ (1996) A real-time optimal control algorithm for greenhouse heating. *Comput Electron Agr* 15(1):1–13.

Coelho JP, De Moura Oliveira PB, Boaventura Cunha J (2005) Greenhouse air temperature predictive control using the particle swarm optimisation algorithm. *Comput Electron Agr* 49(3): 330–344.

Contreras-Medina LM, Torres-Pacheco I, Guevara-Gonzalez RG, Romero-Troncoso RJ, Terol-Villalobos IR, Osornio-Rios RA (2009) Mathematical modeling tendencies in plant pathology. *Afr J Biotechnol* 8(25):7391–7400.

Davis PF, Hooper AW (2002) Improvement of greenhouse heating control. In: *Control theory and applications, IEE Proceedings D* 138, IET:249–255.

De Baerdemaeker J, Munack A, Ramon H, Speckmann H (2002) Mechatronic systems, communication, and control in precision agriculture. *Control Systems Magazine, IEEE,* 21(5):48–70.

De Koning ANM (1990) Long-term temperature integration of tomato. Growth and development under alternating temperature regimes. *Sci Hortic* 45(1–2): 117–127.

Duarte-Galvan C, Torres-Pacheco I, Guevara-Gonzalez RG, Romero-Troncoso RJ, Contreras-Medina LM, Rios-Alcaraz MA and Millan-Almaraz JR (2012) Review. Advantages and disadvantages of control theories applied in greenhouse climate control systems. *Spanish Journal of Agricultural Research* 10(4):926–938

El Ghoumari MY, Tantau HJ, Serrano J (2005) Non-linear constrained MPC: Real-time implementation of greenhouse air temperature control. *Comput Electron Agr* 49(3):345–356.

Fan X, Zuo-Hua T (2006) Application of a genetic simulated annealing algorithm in the greenhouse system. *Computer Simulation* 12: 045.

Fang L, Zhen-Xiao L (2008) Research on the control mode of agriculture greenhouse control system in China. *J Agr Mechaniz Res* 10:223–226.

Fitz-Rodríguez E, Giacomelli GA (2009) Yield prediction and growth mode characterization of greenhouse tomatoes with neural networks and fuzzy logic. *T ASABE* 52(6):2115–2128.

Fourati F, Chtourou M (2007) A greenhouse control with feed-forward and recurrent neural networks. *Simul Model Pract Th* 15(8):1016–1028.

Franklin GF, Powell JD, and Emami-Naeimi A (1994) *Feedback Control of Dynamic Systems.* 3rd edition, Addison-Wesley, Reading, MA.

Freeman JA, Skapura DM (1991) *Neural Networks: Algorithms, Applications, and Programming Techniques*, Addison Wesley, New York.

Gauthier L (1992) A smalltalk-based platform for greenhouse environment control. *Part I. Modeling and managing the physical system. T ASABE* 35(6):2003–2009.

Gauthier L, De Halleux D, Boisvert A, Trigui M, Zehrouni A (1995) A control strategy for the operation of a reversible heat-pump in greenhouses. *Appl Eng Agr* 11(6): 873–879.

Gauthier L, Guay R (1990) An object-oriented design for a greenhouse climate control system. *T ASABE* 33(3):999–1004.

Goggos V, King RE (2000) Qualitative-evolutionary design of greenhouse environment control agents. *Comput Electron Agr* 26(3):271–282.

Goodwin GC, Graebe SF, and Salgado ME (2001) *Control System Design*. Prentice Hall, Upper Saddle River.

Gutman PO, Lindberg PO, Ioslovich I, Seginer I (1993) A non-linear optimal greenhouse control problem solved by linear programming. *J Agr Eng Res* 55(4):335–351.

Hashimoto Y, Murase H, Morimoto T, Torii T (2002) Intelligent systems for agriculture in Japan. *IEEE Contr Syst Mag* 21(5): 71–85.

Hellerstein JL, Tilbury DM, and Parekh S (2004) *Feedback Control of Computing Systems.* John Wiley and Sons, Piscataqay. ISBN 0-471-26637-X.

Hooper AW (1988) Computer control of the environment in greenhouses. *Comput Electron Agr* 3(1):11–27.

Hooper AW, Davis PF (1988) An algorithm for temperature compensation in a heated greenhouse. *Comput Electron Agr* 2(4):251–262.

Hunt KJ, Sbarbaro D, Zbikowski R, Gawthrop PJ (1992) Neural networks for control systems – a survey, *Automatica* 28 (1992) 1083–1112.

Ibrahim IA, Sørensen CG, 2010. A more energy efficient controller for the greenhouses climate control system. *Appl Eng Agr* 25(3):491–498.

Ioslovich I, Gutman P, Linker R (2009) Hamilton-Jacobi-Bellman formalism for optimal climate control of greenhouse crop. *Automatica* 45(5):1227–1231.

Ioslovich I, Gutman P, Seginer I (1996) A non-linear optimal greenhouse control problem with heating and ventilation. *Optim Contr Appl Met* 17(3):157–169.

Jacobson BK, Jones PH, Jones JW, Paramore JA (1989) Real-time greenhouse monitoring and control with an expert system. *Comput Electron Agr* 3(4):273–285.

Jewett TJ, Short TH (1992) Computer control of a five-stage greenhouse shading system. *T ASABE* 35(2):651–658.

Kilian C (2005) *Modern Control Technology*. Thompson Delmar Learning, Clifton Park. ISBN 1-4018-5806-6.

Körner O, Challa H (2003) Process-based humidity control regime for greenhouse crops. *Comput Electron Agr* 39(3):173–192.

Körner O, Van Straten G (2008) Decision support for dynamic greenhouse climate control strategies. *Comput Electron Agr* 60(1):18–30.

Kuo BC (1995) *Automatic Control Systems*. 7$^{th}$ edition, Prentice Hall, Englewoods, Cliff.

Kurata K, Eguchi N (1990) Machine learning of fuzzy rules for crop management in protected cultivation. *T ASAE* 33(4):1360–1368.

Lafont F, Balmat JF (2004) Fuzzy logic to the identification and the command of the multidimensional systems, *Int J Comput Cog* 2:21–47.

Levine WS (1996) *The Control Handbook*. CRC Press, New York. ISBN 978-0-8493-8570-4.

Linker R, Gutman PO, Seginer I (2000) Robust model-based failure detection and identification in greenhouses. *Comput Electron Agr* 26(3):255–270.

Linker R, Seginer I, Gutman P (1998) Optimal $CO_2$ control in a greenhouse modeled with neural networks. *Comput Electron Agr* 19(3):289–310.

Narendra KS, Parthasarathy K (1990) Identification and control of dynamical systems using neural networks, *IEEE Transactions on Neural Networks* 1:4–27.

Nielsen OF (1995) Climate computer algorithms for peak shaving of greenhouse heating demand. *Comput Electron Agr* 13(4):315–335.

Ogata K (1997) *Modern Control Engineering*. 3rd edition, Prentice Hall, New Jersey.

Pasgianos GD, Arvanitis KG, Polycarpou P, Sigrimis N (2003) A nonlinear feedback technique for greenhouse environmental control. *Comput Electron Agr* 40(1–3):153–177.

Pekkeriet EJ and van Henten EJ (2011) Current developments of high-tech robotic and mechatronic systems in horticulture and challenges for the future. *Acta Hort.* 893:85–94.

Pham DT, Liu X (1995) *Neural Networks for Identification, Prediction and Control*, Springer, London.

Pinon S, Camacho EF, Kuchen B, Peña M (2005) Constrained predictive control of a greenhouse. *Comput Electron Agr* 49(3):317–329.

Rohrs CE, Melsa JL, and Schultz DG (1993) *Linear Control Systems*. McGraw-Hill, New York.

Seginer I (1997) Some artificial neural network applications to greenhouse environmental control. *Comput Electron Agr* 18(2–3):167–186.

Seginer I, McClendon RW (1992) Methods for optimal control of the greenhouse environment. *T ASAE* 35(4):1299–1307.

Setiawan A, Albright LD, Phelan RM (2000) Application of pseudo-derivative-feedback algorithm in greenhouse air temperature control. *Comput Electron Agr* 26(3):283–302.

Sigrimis N, Antsaklis P, Groumpos PP (2002) Advances in control of agriculture and the environment. *IEEE Contr Syst Mag* 21(5):8–12.

Sigrimis N, King RE (2000) Advances in greenhouse environment control. *Comput Electron Agr* 26(3): 217–219.

Sigrimis N, Rerras N (1996) A linear model for greenhouse control. *T ASABE* 39(1):253–261.

Soto-Zarazua GM, Rico-Garcia E, Ocampo R, Guevara-Gonzalez RG, Herrera-Ruiz G (2010) Fuzzy-logic-based feeder system for intensive tilapia production (*Oreochromis niloticus*). *Aquacult Int* 18(3):379–391.

Stengel RF (1994) *Optimal Control and Estimation*. Dover Publications, Newburyport. ISBN 0-486-68200-5.

Takakura T, Manning TO, Giacomelli GA, Roberts WJ (1994) Feedforward control for a floor heat greenhouse. *T ASABE* 37(3):939–945.
Trabelsi A, Lafont F, Kamoun M, Enea G (2007) Fuzzy identification of a greenhouse. *Appl Soft Comput* 7(3):1092–1101.
Van Straten G, Challa H, Buwalda F (2000) Towards user accepted optimal control of greenhouse climate. *Comput Electron Agr* 26(3):221–238.
Van Straten G, Van Willigenburg LG, Van Henten EJ, Van Ooteghem RJC (2011) *Optimal control of greenhouse cultivation*. CRC Press, NY, USA.
Yildiz I (1993) Simulation of greenhouse microclimates and environmental control strategies using a Rankine cycle heat pump. *Dissertation*. The Ohio State University, Columbus, Ohio, USA.
Yildiz I and Stombaugh DP (2006a) Dynamic modeling of microclimate and environmental control strategies in a greenhouse coupled with a heat pump system. *Acta Hort.* 718:331–340.
Yildiz, I. and Stombaugh, D.P. (2006b) Simulated performances of a heat pump system for energy and water conservation in open and confined greenhouse systems. *Acta Hort.* 718: 341–350.
Yildiz, I. and Stombaugh, D.P. (2006c) Heat pump cooling and greenhouse microclimates in open and confined greenhouse systems. *Acta Hort.* 719: 255–262.
Yildiz I and Stombaugh DP (2009) Simulated effects of various environmental management practices on energy consumption in open and confined greenhouse systems. *Acta Hort.* 801:1317–1324.
Yildiz I, Yue J, and Lowrey J (2014a) Spatiotemporal distributions of energy and water consumption in open, semi-closed and closed loop heat pump greenhouse systems, and associated management strategies in Canada. *Acta Hort.* 1037 (1):119–128.
Yildiz I, Yue J, and Lowrey J (2014b) Spatiotemporal distributions of energy and water consumption in open, semi-closed and closed loop heat pump greenhouse systems, and associated management strategies in the Canadian North. *Acta Hort.* 1037 (1):171–177.

# 9

# *Energy Estimating in Greenhouses*

## 9.1 Introduction

Energy requirements and fuel consumption of heating, ventilation, and air-conditioning (HVAC) systems directly affect the operational cost of a greenhouse and indirectly affect the greenhouse internal environment. This chapter discusses methods for estimating energy use for two distinct purposes: i) modeling for greenhouse and HVAC system design and pertinent design optimization (also known as *forward modeling*), and ii) modeling energy use of existing greenhouses for calculating retrofit savings (also known as *data-driven modeling*).

A mathematical model of a greenhouse describes the behavior of a greenhouse system, and is basically made up of three main components: i) input (forcing) variables – controllable or uncontrollable (e.g. outside solar radiation, temperature, relative humidity), ii) system structure and parameters/properties, which provide the necessary physical description of the greenhouse system (e.g. thermal mass), iii) output (response) variables, which describe the reaction of the system to the input variables (e.g. energy use, water use, indoor temperature, leaf temperature, relative humidity levels, photosynthesis, respiration, overall production; see Figures 9.1 and 9.2).

The objective or purpose of the modeling study dictates what type of approach needs to be used, that is, the forward (classical) approach or the data-driven (inverse) approach.

And the objective of the forward modeling approach is to predict the output variables of a specified greenhouse model with known structure and parameters when subject to specified input variables. With the availability of powerful computers and their easy accessibility, the greenhouse models today have become increasingly complex with greater accuracy levels. This approach requires detailed knowledge of the various natural phenomena affecting system behavior along with the magnitude of various interactions (e.g. effective thermal mass, heat and mass transfer coefficients).

Forward modeling approach starts with a physical description of the greenhouse system. For example, geographical location, greenhouse orientation and geometry, greenhouse glazing, type of equipment and operational strategies, type of HVAC system, are all specified (see Figure 9.2). And then, the energy, water, and $CO_2$ use profiles of such a greenhouse can for instance be simulated by the forward simulation model. The main advantage of this method is that it is completely based on sound engineering principles; and therefore, the system need not be physically built to predict its behavior. Consequently, the approach is the most commonly used approach by mathematical modelers and has gained widespread acceptance by the scientific and professional community. Figure 9.3, on the other hand, illustrates the ordering of the analysis typically performed by a greenhouse energy simulation model.

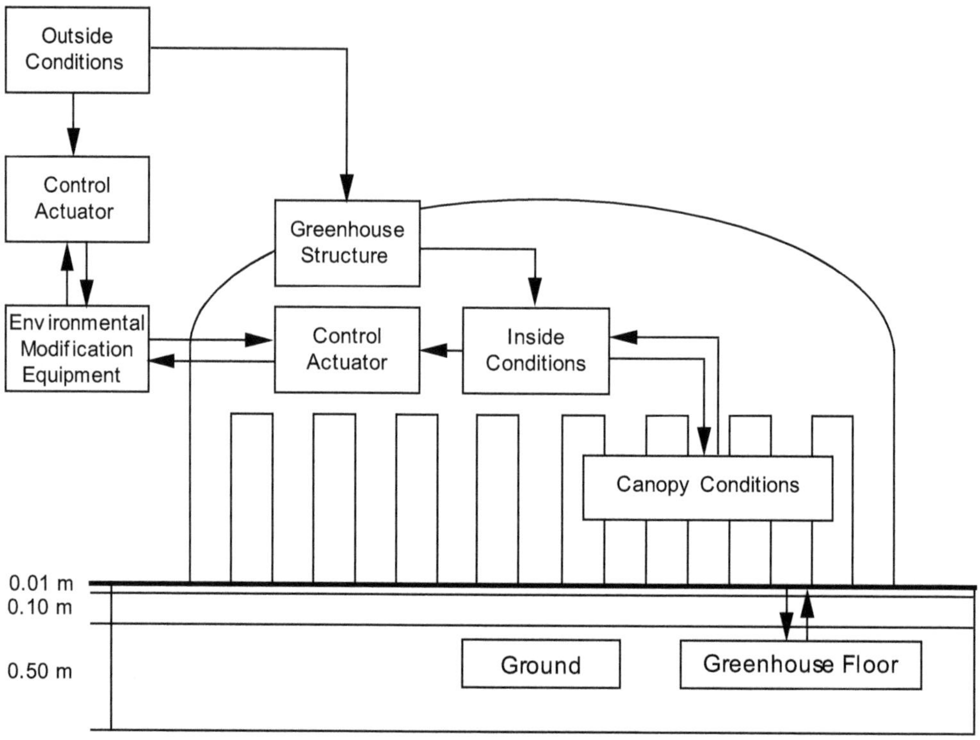


**FIGURE 9.1**
Overall structure of a greenhouse mathematical model (Yildiz, 1993).

In the case of data-driven approach, on the other hand, input and output variables are measured and therefore known; and the objective of developing such a model is then to determine a mathematical description of the system and estimate parameters of the system when the system has already been built and actual performance data are available for model development and/or identification. Data-driven models often provide more accurate predictors of future system performance than forward models. Choosing or developing a greenhouse energy analysis program, however, depends on its application, the frequency of its use, the experience or knowledge of users, and the available hardware to run the simulation program. Since detailed models are almost all the time computationally intensive, simplified methods such as the degree-day, bin, and correlation methods have been developed for energy estimating. The most accurate methods for determining greenhouse energy consumption are the most complex ones due to their intense computational requirements and the expertise required. Simulation programs that assemble component models into system models and then exercise those models with weather and crop data are preferred by experts for determining energy, water, and $CO_2$ use in greenhouses.

## 9.2 Characteristics of Energy Estimating Procedures

Procedures for energy estimating requirements range quite a bit in their complexity; however, they mostly have the following common elements: i) load calculation for the

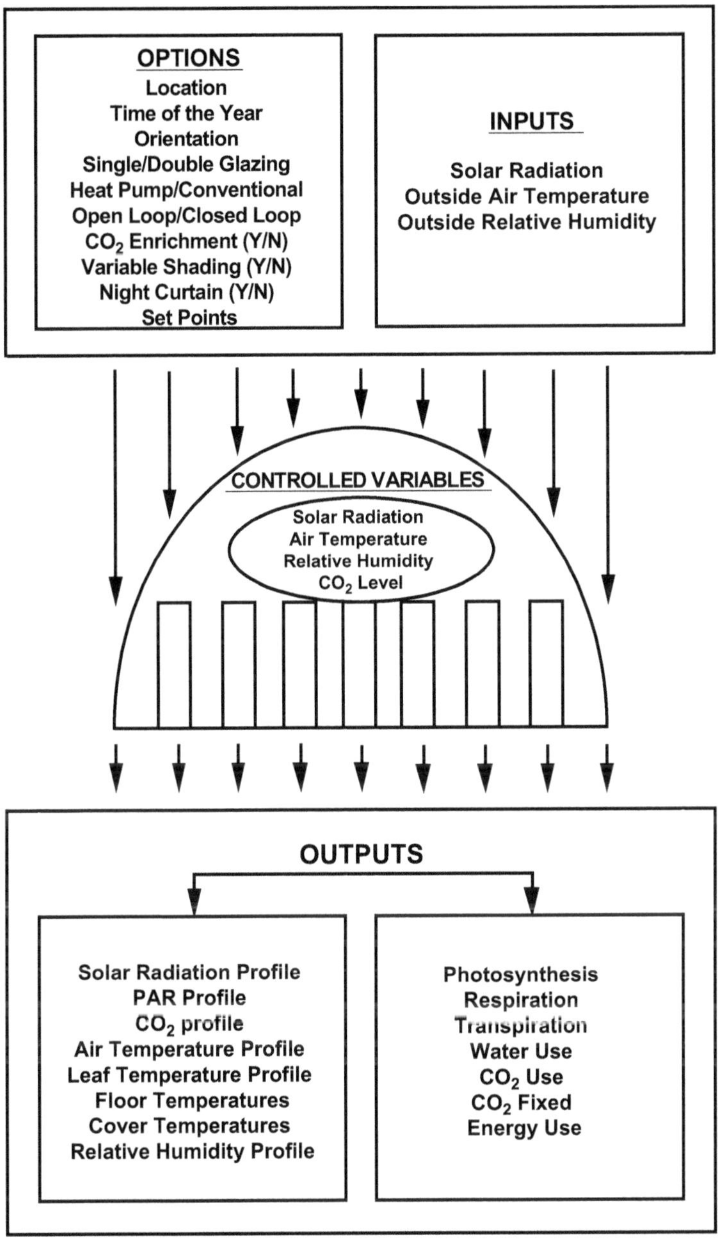


**FIGURE 9.2**
Physical description of a forward greenhouse simulation model (Yildiz, 1993).

controlled greenhouse space, ii) load calculation for the secondary equipment, and iii) calculation of primary equipment energy necessities. Secondary equipment means the equipment that distributes the heating, cooling, or ventilating medium to controlled greenhouse spaces, whereas primary equipment are the central plant (e.g. boiler room) equipment that converts fuel or electric energy to thermal (heating or cooling) effect.

Based on the procedure provided above, in order to estimate energy requirements in a greenhouse, we first need to calculate the greenhouse space load, which is the amount of

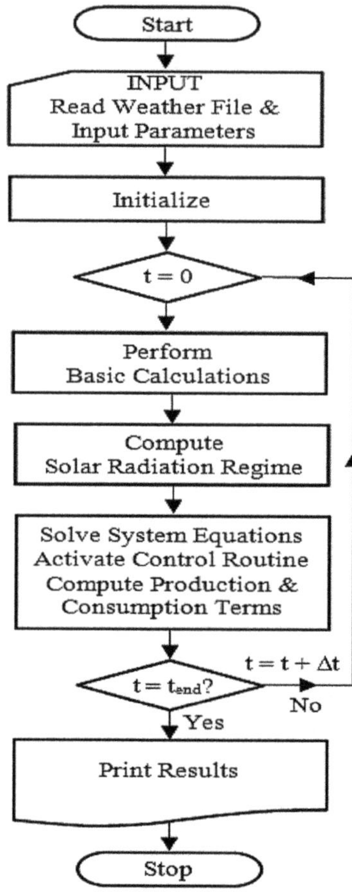


**FIGURE 9.3**
Typical flowchart of a forward greenhouse simulation model (Yildiz, 1993).

energy that must be added to or removed from a controlled greenhouse space for maintaining thermal comfort for plants and personnel. The simplest procedures typically assume that the energy required for providing and maintaining thermal comfort is a function of the outdoor dry-bulb temperature only. And relatively more detailed and accurate approaches, on the other hand, consider the solar effects, crop effects, internal gains (e.g. artificial lighting), and heat storage in the floor, the inside air, the crop canopy and other interior surfaces, along with the effects of shading, night curtains, insect screens, and wind, on the heat loss from the greenhouse envelope and infiltration, etc. Calculations for energy estimating are similar to the heating and cooling load calculations used for sizing primary and secondary equipment as performed in Chapters 5 and 6; but, they are not the same as we will see shortly. Energy estimating calculations are based on typical weather conditions rather than worst-case weather scenarios. Sophisticated procedures usually use hourly climatic conditions and operational characteristics for a number of typical meteorological days of the year or on 8,760 h of operation per year.

After determining the space load, we then take the calculated space load and translate it to the secondary (piping, ducts, pumps, fans, etc.) equipment load. This can only be a simple estimate of duct or piping losses or gains or a complex hour-by-hour, minute-by-minute, or even second-by-second simulation of a greenhouse air system. This step includes all forms of energy required by all the secondary equipment (e.g. electrical energy to operate fans and/or pumps, including the energy in the form of heated or chilled water, such as irrigation water).

And then finally, we calculate the fuel and energy required by the primary equipment to meet all of the abovementioned loads and the peak demand considering the efficiencies and partial load characteristics of the equipment. It is quite often necessary to keep track of the form of energy consumptions, such as electrical, natural gas, biomass, and oil, so that they can be used in economic analyses or business models for new greenhouse complexes and/or the cost-effectiveness of energy conservation measures.

Before going into further details of the most elaborate *detailed simulation* or *complex modeling* approaches, we will spend some time in the following sections on the abovementioned simpler methods for estimating energy consumption in greenhouses. One of these simpler methods is what's called *single-measure* procedure, which uses only one measure such as annual degree-days (*degree-day* method). The other procedure is called *simplified multiple-measure* methods in which more information is used and therefore accuracy is improved. The best known of such kind of methods is the *bin* method where occurrence hours are used for each temperature interval called as temperature bin.

## 9.3 Balance Point Temperature

The *balance point temperature* ($T_{bal}$) of a building is defined as the outdoor temperature $T_o$ at which, for the specified value of the interior temperature $T_i$, the total heat losses from a building equal the heat gains in the building (in our case, the building is the greenhouse).

For a steady-state method, heat gains must be the average for the period in question, not for the peak values. In particular, solar radiation must be based on averages, not peak values. Applying the energy balance equation (Equation (2.2)), developed in Chapter 2, to the internal air, we can write the following equation for internal heat gains:

$$\dot{Q}_{internal} = \dot{Q}_{equipment} + \dot{Q}_{lighting} + \dot{Q}_{solar} \tag{9.1}$$

Only when the outdoor temperature $T_o$ drops below the balance point temperature $T_{bal}$, supplemental heat is required to maintain the indoor design temperature $T_i$. Above the balance point temperature $T_{bal}$, internal heat gains (solar, equipment and lights, when available) are able to maintain the indoor temperature at or above the indoor design temperature $T_i$. At the balance point temperature $T_{bal}$, the total heat loss by conduction and ventilation (or infiltration) equals the internal heat gains $\dot{Q}_{internal}$:

$$\dot{Q}_{internal} = \Sigma UA \times (T_i - T_{bal}) + \text{ACH}\, V \rho c_p (T_i - T_{bal}) \quad \text{(I-P)} \tag{9.2a}$$

$$\dot{Q}_{internal} = \Sigma UA \times (T_i - T_{bal}) + \text{ACH}\left(\frac{1}{3{,}600}\right) V \rho c_p (T_i - T_{bal}) \quad \text{(SI)} \tag{9.2b}$$

where the first and second terms on the right hand side of Equations (9.2 a,b) are the total heat loss terms by conduction and the heat loss terms due to infiltration, respectively.

Then, the balance point temperature $T_{bal}$ in I-P units becomes

$$T_{bal} = T_i - \left( \frac{\dot{Q}_{internal}}{(\Sigma UA) + \text{ACH} V \rho c_p} \right) \quad \text{(I-P)} \tag{9.3a}$$

where $T_{bal}$ is the balance point temperature (°F), $\dot{Q}_{internal}$ is the total internal sensible heat gain (Btu/h) from lights, equipment, etc., $\Sigma UA$ is the total conduction heat loss term (Btu/h.°F), ACH is the number of air changes per hour, $V$ is the volume of the greenhouse (ft³), $\rho$ is the density (lb/ft³), and $c_p$ is the specific heat of greenhouse inside air (Btu/lb.°F).

And the balance point temperature $T_{bal}$ in SI units becomes

$$T_{bal} = T_i - \left( \frac{\dot{Q}_{internal}}{(\Sigma UA) + \text{ACH}\left(\frac{1}{3{,}600}\right) V \rho c_p} \right) \quad \text{(SI)} \tag{9.3b}$$

where $T_{bal}$ is the balance point temperature (°C), $\dot{Q}_{internal}$ is the total internal sensible heat gain (W) from lights, equipment, etc., $\Sigma UA$ is the total conduction heat loss term (W/°C), and ACH is the number of air changes per hour, $V$ is the volume of the greenhouse (m³), 1/3,600 is for unit conversion (1 h = 3,600 s), $\rho$ is the density (kg/m³) and $c_p$ is the specific heat of greenhouse inside air (J/kg.°C).

## 9.4 Annual Degree-Day Method

Degree-day (single-measure) methods are the simplest methods for energy estimating and are adequate if building use and HVAC equipment efficiency are constant. Greenhouse heating and cooling are responsible for a large fraction of energy consumption in greenhouse production around the world; and heating degree-days (HDDs) and cooling degree-days (CDDs) are proxies for the energy demand needed to heat or cool, respectively, a greenhouse, a home or a business. Both variables are derived from outside air temperature measurements. HDDs or degree-hours for a balance point temperature of 18.3°C (65°F) have been widely tabulated (this temperature represents average conditions in typical residential buildings in the past). The 18.3°C (65°F) base is assumed whenever $T_{bal}$ is not indicated explicitly. The heating and cooling requirements for a given building such as a greenhouse at a specific location are, to some degree, proportional to the number of HDDs and CDDs at that location. HDDs and CDDs are calculated using daily mean outside temperature ($T_{mean} = [T_{max} - T_{min}]/2$) and a base temperature – the outside temperature, below or above which a building requires heating or cooling for thermal comfort. The baseline (balance point) temperatures today are quite different from what was developed for residential buildings in the 1940s – still in use – as buildings are much tighter, more insulated and have more internal heat sources today. Therefore, in order to accurately determine energy demands and gauge the impacts of climate change, new baseline temperatures are required for both heating and cooling. Annual and monthly HDD and CDD values have been developed for selected base temperatures in certain geographical jurisdictions; however, further coverage is still required. Outside daily mean temperatures are used in determining HDD and CDD calculations even though cooling a space depends more on the maximum rather than the mean temperature, while the minimum temperature is more relevant for heating. Further, long-term temperature data closer to present time should be used for more accurate calculations of HDD and CDD and hence for future energy assessments of buildings including greenhouses. Overall, it can be said that climate change is reducing the demand for space heating and increasing the demand for space cooling globally. In any case, even if a specified base temperature HDDs or CDDs may not be available for the location in question, we can still use the tabulated annual HDD and CDD values determined for classical baseline temperatures for energy estimating, which will be shown numerically later in this section.

We have defined the balance point temperature $T_{bal}$ of a building, and appropriate equations were also developed earlier in this section. It was stated that heating is only needed when outside temperature $T_o$ drops below the balance point temperature $T_{bal}$. Therefore, the rate of energy consumption for a heating system can be expressed as follows:

$$E = \left( \frac{(24)(\mathrm{HDD})(\dot{Q}_t)}{\mathrm{HV}\,\eta\,\Delta T} \right) \quad \text{(I-P)} \tag{9.4a}$$

where $E$ is the fuel or energy consumption (in trade units), the constant 24 is for unit conversion (i.e. 24 h/d), HDD is the mean annual heating degree days for the location (Table 9.1), $\dot{Q}_t$ is the total heat loss from the greenhouse (Btu/h), HV is the heating value of fuel (unit consistent with $\dot{Q}_t$ and $E$), $\eta$ is the efficiency of heating system (in decimals), $\Delta T$ the design temperature difference (°F) for $\dot{Q}_t$.

**TABLE 9.1**

Degree-Day Data for Various Locations in the United States (Engineering ToolBox, 2009).

| | Variable-Base Heating Degree-Day, °C.days (°F.days) | | |
|---|---|---|---|
| **LOCATION** | **7.2°C (45°F)** | **12.8°C (55°F)** | **18.3°C (65°F)** |
| New Haven, CN | 983 (1,769) | 1798 (3,237) | 3276 (5,897) |
| Denver, CO | 860 (1,548) | 1911 (3,440) | 3069 (5,524) |
| Chicago, IL | 1094 (1,969) | 2079 (3,743) | 3268 (5,882) |
| Lansing, MI | 1409 (2,537) | 2469 (4,444) | 3838 (6,909) |
| New York, NY | 784 (1,412) | 1716 (3,089) | 2706 (4,871) |
| Columbus, OH | 889 (1,600) | 1808 (3,255) | 3144 (5,660) |
| Portland, OR | 207 (373) | 1062 (1,911) | 2283 (4,109) |
| Nashville, TN | 233 (419) | 932 (1,678) | 1988 (3,578) |
| Seattle, WA | 227 (408) | 1214 (2,185) | 2458 (4,424) |
| Madison, WI | 1704 (3,067) | 2694 (4,850) | 4368 (7,863) |

**Example 9.1: Heating Energy Estimating in a Greenhouse via Degree-Day Method**

Now, let's consider a single glass greenhouse in Columbus, Ohio, USA, whose dimensions are given below.

| GREENHOUSE DIMENSIONS and CHARACTERISTICS | | |
|---|---|---|
| **GREENHOUSE COMPONENT** | **Dimension, ft (m)** | |
| House wall height | 8 (2.4) | |
| House width | 24 (7.2) | |
| House length | 192 (57.6) | |
| Rafter length | 13.4 (4.0) | |
| Lower wall height | 1.5 (0.45) | |
| Upper wall height | 6.5 (1.95) | |
| Gable wall height | 6 (1.8) | |
| Number of bays | 4 (1.2) | |
| | **Construction Material** | **Reference** |
| Lower wall | 6″ (0.15m) poured concrete | See Table 3.3 |
| Upper wall | Single glass | See Table 3.3 |
| End area | Single glass | See Table 3.3 |
| Roof | Single glass | See Table 3.3 |
| Perimeter | Uninsulated perimeter | See Table 3.3 |

Estimate the annual heating energy requirement and the costs of heating using fuel oil and natural gas furnaces for this greenhouse. Information on heating values and fuel prices can be found in Chapter 2.

**Solution**

We first need to calculate the UA values as follows:

| UA VALUES and AIR CHANGE PER HOUR | | | | |
|---|---|---|---|---|
| **GREENHOUSE COMPONENT** | ***A* (ft²)** | ***U* (Btu/h.ft².°F)** | **Reference** | ***UA* (Btu/h.°F)** |
| Lower Wall Area | 864 | 0.75 | Table 3.3 | 648 |
| Upper Wall Area | 3,744 | 1.1 | | 4,118 |
| Gable or Curved Area | 576 | 1.1 | | 634 |
| Roof Area | 20,582 | 1.1 | | 22,640 |
| Perimeter Length | 576 | 0.8 | | 461 |
| $\Sigma UA$ | | | | 28,501 |
| ACH | | 1.0 | | |

Then, using Equation (5.4), we can easily determine the greenhouse envelop and infiltration heat losses as well as the fuel oil and natural gas consumption calculations as shown in the following table:

| | | | | Annual Cost of Heating (US$/season) |
|---|---|---|---|---|
| $T_i$ | 65°F | | | |
| $T_o$ | 6°F (99% level; Table A.11) (ASHRAE, 1997) | | | |
| ΔT | 59°F | | | |
| HDD | 5,660 °F.days at 65°F base temp. (Table 9.1 or consult your local Weather Bureau) | | | |
| Furnace Efficiency | 80% | | | |
| HV of fuel oil (Btu/gal) | 139,000 | Price of fuel oil (US$/gal) | 2.45 | |
| HV of natural gas (Btu/ft³) | 1,036 | Price of natural gas (US$/ft³) | 0.005 | |
| $Q_{cond}$ (Btu/h) | $UA \times \Delta T$ | | 1,681,559 | |
| $Q_{inf}$ (Btu/h) | $0.02 \times \Delta T \times V \times ACH$ | | 243,302 | |
| $Q_t$ (Btu/h) | $Q_{cond} + Q_{inf}$ | | 1,920,806 | |
| $E$ (gal oil/year) | $24 \times HDD \times Q_t/(HV \times \eta \times \Delta T)$ | | 39,770 | 97,436 |
| $E$ (ft³ natural gas/year) | $24 \times HDD \times Q_t/(HV \times \eta \times \Delta T)$ | | 5,335,922 | 26,680 |

As shown in the above table, the greenhouse heating energy consumption was estimated using the trade units for fuel oil and natural gas, and then they were converted to US$ per season. Based on our calculations, annual natural gas consumption for heating this greenhouse would cost $26,680 while it would cost $97,436 per year if fuel oil is used to heat the same greenhouse facility. For more accurate monetary estimates though, up-to-date local fuel prices should be used.

**Example 9.2: Heating Energy Estimating in a Greenhouse with Variable Set-Point Temperatures**

Now, using the same greenhouse that was used in Example 9.1, a single glass greenhouse in Columbus, Ohio, USA, estimate the annual heating energy requirement and the costs of heating using fuel oil and natural gas furnaces for different day- and night-time indoor set-point temperatures. Assume that a daytime set-point temperature of 65°F is used, while the night-time set-point temperature is 55°F.

**Solution**

See Example 11.3.

**Example 9.3: Estimating the Reduction in Heating Energy Requirement When Double-Poly Glazing Used**

Let's now consider the same greenhouse facility that was used in Example 9.1; however, this time a double-poly glazing was used instead of single glass, and the lower wall (6″ poured concrete + 2″ foamed urethane board) and the perimeter are insulated as shown in the following table. And determine the energy savings.

**Solution**

See Example 11.1.

**Example 9.4: Estimating the Reduction in Heating Energy Requirement When Thermal Blanket Used**

Let's now consider the same greenhouse facility that was used in Example 9.3; however, this time, in addition to using a double-poly glazing instead of single glass, and an insulated lower wall (6″ poured concrete + 2″ foamed urethane board) and perimeter, a thermal blanket is used as well. Determine the additional energy savings.

**Solution**

See Example 11.2.

$$E = \left( \frac{(24)(3{,}600)(\text{HDD})(\dot{Q}_t)}{\text{HV}\,\eta\,\Delta T} \right) \quad (\text{SI}) \tag{9.4b}$$

where $E$ is the fuel or energy consumption (in trade units), the constant 24 is for unit conversion (i.e. 24 h/d), the constant 3,600 is for another unit conversion (i.e. 3,600 s/h) HDD is the mean annual HDDs for the location (Table 9.1), $\dot{Q}_t$ is the total heat loss from the greenhouse (W), HV is the heating value of fuel (unit consistent with $\dot{Q}_t$ and $E$), $\eta$is the efficiency of heating system (in decimals), $\Delta T$ the design temperature difference (°C) for $\dot{Q}_t$.

## 9.5 Bin Method

The degree-day method provides a simple estimate of annual heating or cooling loads, which are pretty accurate if the indoor temperature and internal gains are relatively constant, and if the HVAC systems operate for a complete heating or cooling season. However, since the HVAC system efficiency or the balance point temperature $T_{bal}$ may not be sufficiently constant, there is a legitimate concern with the use of degree-day method even with the variable-base method. For instance, if a heat pump is used, the heat pump efficiency changes with outdoor temperature, and the efficiency of other HVAC equipment (boilers, chillers, etc.) may also change as a function of varying load. Furthermore, in greenhouses, free heating energy such as solar and artificial lighting (when used) heat gains have pronounced patterns, which influence the overall greenhouse heat gains, indoor temperature, and therefore ventilation rate. In such cases, steady-state calculation might yield pretty good estimates for annual energy consumption if different temperature intervals (bins) and time periods are evaluated separately. This approach is called as the *bin* method, a simplified multiple-measure method, because consumption is calculated for several outdoor temperatures and multiplied by the number of occurrence hours ($N_{bin}$) in the temperature interval (bin) centered around that temperature:

$$E = \left( \frac{(N_{\text{bin}})(\dot{Q}_t - \dot{Q}_{internal})}{\text{HV}\,\eta} \right) \tag{9.5}$$

where $E$ is the annual (seasonal) fuel or energy consumption (in trade units), $N_{bin}$ is the number of occurrence hours (see Table 9.2), $\dot{Q}_t$ is the total heat loss from the

**TABLE 9.2**

Hourly weather occurrences per year for selected locations.

| LOCATION | Outdoor (Bin) Temperature, °F (°C) | | | | | | | | | | | | | | | | | | |
|---|---|---|---|---|---|---|---|---|---|---|---|---|---|---|---|---|---|---|---|
| | 72 (22.2) | 67 (19.4) | 62 (16.7) | 57 (13.9) | 52 (11.1) | 47 (8.3) | 42 (5.6) | 37 (2.8) | 32 (0) | 27 (−2.8) | 22 (−5.6) | 17 (−8.3) | 12 (−11.1) | 7 (−13.9) | 2 (−16.7) | −3 (−19.4) | −8 (−22.2) | −13 (−25) | −18 (−27.8) |
| Albany, NY | 588 | 733 | 740 | 708 | 652 | 625 | 647 | 769 | 793 | 574 | 404 | 278 | 184 | 110 | 63 | 32 | 10 | 5 | 4 |
| Bakersfield, CA | 831 | 898 | 966 | 977 | 908 | 746 | 541 | 247 | 77 | 7 | | | | | | | | | |
| Columbus, OH | 744 | 820 | 720 | 648 | 622 | 603 | 653 | 772 | 730 | 502 | 280 | 169 | 94 | 40 | 20 | 10 | 4 | 1 | |
| Chicago, IL | 762 | 769 | 653 | 592 | 569 | 543 | 591 | 800 | 822 | 551 | 335 | 196 | 117 | 85 | 59 | 25 | 12 | 3 | |
| Dallas, TX | 831 | 795 | 693 | 656 | 629 | 576 | 504 | 371 | 231 | 91 | 34 | 17 | 4 | 1 | | | | | |
| Denver, CO | 549 | 684 | 783 | 731 | 678 | 704 | 652 | 717 | 721 | 553 | 359 | 216 | 119 | 78 | 36 | 22 | 6 | 1 | 1 |
| Grand Rapids, MI | 634 | 739 | 712 | 647 | 571 | 565 | 554 | 742 | 938 | 690 | 469 | 293 | 172 | 78 | 31 | 10 | 1 | 1 | |
| Hartford, CT | 617 | 755 | 751 | 752 | 649 | 575 | 683 | 807 | 825 | 552 | 370 | 233 | 153 | 77 | 33 | 11 | 3 | 2 | |
| Jacksonville, FL | 1,334 | 975 | 879 | 692 | 530 | 355 | 288 | 154 | 83 | 24 | 2 | | | | | | | | |
| Las Vegas, NV | 651 | 644 | 699 | 786 | 769 | 716 | 591 | 396 | 194 | 44 | 7 | 1 | | | | | | | |
| Los Angeles, CA | 881 | 1,654 | 2,193 | 1,904 | 1,054 | 428 | 107 | 10 | | | | | | | | | | | |
| Minneapolis, MN | 621 | 690 | 695 | 602 | 588 | 482 | 500 | 560 | 632 | 609 | 514 | 383 | 311 | 246 | 186 | 119 | 62 | 31 | 16 |
| New York, NY | 926 | 877 | 754 | 745 | 722 | 796 | 838 | 603 | 330 | 188 | 82 | 26 | 10 | 1 | | | | | |
| Phoenix, AZ | 762 | 776 | 767 | 769 | 659 | 540 | 391 | 182 | 57 | 8 | | | | | | | | | |
| Portland, OR | 407 | 627 | 780 | 808 | 760 | 748 | 722 | 839 | 820 | 599 | 408 | 293 | 190 | 109 | 60 | 29 | 15 | 5 | 1 |
| Raleigh, NC | 1,087 | 937 | 848 | 762 | 707 | 672 | 638 | 527 | 410 | 236 | 103 | 38 | 11 | 1 | | | | | |
| San Francisco, CA | 285 | 665 | 1,264 | 2,341 | 2,341 | 1,153 | 4[illegible]9 | 99 | 10 | | | | | | | | | | |
| Seattle, WA | 258 | 448 | 750 | 1,272 | 1,462 | 1,445 | 1,408 | 914 | 427 | 104 | 39 | 20 | 3 | | | | | | |
| Washington, DC | 960 | 766 | 740 | 673 | 690 | 684 | 790 | 744 | 542 | 254 | 138 | 54 | 17 | 2 | | | | | |

greenhouse, $\dot{Q}_{internal}$ is the free internal heat gains, HV is the heating value of fuel (unit consistent with $\dot{Q}_t$ and $E$), η is the efficiency of heating system (in decimals), $\Delta T$ is the design temperature difference, and the others are as developed in Chapter 5 as follows:

$$\dot{Q}_t = \dot{Q}_{cond} + \dot{Q}_{inf}$$

$$\dot{Q}_{cond} = \Sigma UA \times (T_i - T_o) \quad \text{(I-P or SI)}$$

$$\dot{Q}_{inf} = \text{ACH}\, V \rho c_p (T_i - T_o) \quad \text{(I-P)}$$

$$\dot{Q}_{inf} = \text{ACH}\left(\frac{1}{3,600}\right) V \rho c_p (T_i - T_o) \quad \text{(SI)}$$

and $\dot{Q}_{internal}$ is as expressed in Equation (9.1).

No heating is needed when $T_o$ is above $T_{bal}$. Equation (9.5) is evaluated for each bin, and the total consumption is the sum of the $E_{bin}$ over all bins. In the United States, for instance, the necessary weather data are available in ASHRAE (1986) and USAF (1978). The bins usually have 2.8°C (5°F) ranges and are typically collected in three daily 8 h shifts. Hourly weather occurrences can easily be developed when they are not readily available.

And as discussed earlier, the main advantage of the bin method is that we can analyze the performance of partly loaded heating equipment, weather related variations in capacity and performance of heat pumps, effects of changing crop density and canopy in the greenhouse, UA value, minimum winter ventilation, or any other energy saving techniques. Even though easily accessible computers can calculate the energy consumption of a greenhouse via complex modeling, the concepts of degree-days and bin approach are nevertheless simple and quite valuable tools.

**Example 9.5: Estimating Heating Energy in a Greenhouse via Bin Method**

Now, let's consider the same greenhouse, which was used in Example 9.1, a single glass greenhouse in Columbus, Ohio, USA, and estimate the annual heating energy requirement and the costs of heating using fuel oil and natural gas furnaces for this greenhouse. Assume that internal heat gains are zero.

**Solution**

We need to follow the steps provided below:

- First, we will calculate the balance point temperature $T_{bal}$ using either Equation (9.3a) or (9.3b) depending on the units used.
- Remember that the internal gains are neglected in this example; therefore, $T_{bal} = T_i = 65°\text{F}$. So, no supplemental heating is required at or above 65°F outside temperatures.
- See Table 9.2 for the number of occurrence hours in each bin.
- Since $T_{bal} = 65°\text{F}$, the highest temperature bin to be considered is the 67°F bin (64.5°F and 69.5°F).
- 67°F bin occurs for 820 h; but, supplemental heating is needed only when $T_o < 65°\text{F}$.
- The outside temperature is between 64.5°F and 67.9°F for about:

$$820\ \text{h} \times (65°\text{F} - 64.5°\text{F})/5°\text{F} = 82\ \text{h}$$

(Continued)

- During this 82-h period, the average outside (bin) temperature is:

$$\left(65°\mathrm{F} + 64.5°\mathrm{F}\right)/2 = 64.75°\mathrm{F}$$

So, 64.75°F is the highest temperature interval (bin) and the corresponding occurrences of 82 h need to be used first in our calculations. And then, the remaining (lower) temperature bins and the corresponding occurrence hours will be incorporated fully for estimating the seasonal heating energy consumption.

Finally, using Equation (9.5) and its components, we will then perform the full energy estimating calculations as presented in the following table:

**BIN METHOD**

| **Bin Temp** | **ΔT** | $Q_{inf}$ | $Q_{cond}$ | $Q_{internal}$ | $Q_{inf} + Q_{cond} - Q_{internal}$ | **Occurrence** | *E* |
|---|---|---|---|---|---|---|---|
| **(°F)** | **(°F)** | **(Btu/h)** | **(Btu/h)** | **(Btu/h)** | **(Btu/h)** | **(h)** | **(Btu/year)** |
| **64.75** | 0.25 | 1,014 | 7,125 | 0 | 8,139 | 82 | 667,399 |
| **62** | 3 | 12,165 | 85,503 | 0 | 97,668 | 720 | 70,321,046 |
| **57** | 8 | 32,440 | 228,008 | 0 | 260,448 | 648 | 168,770,511 |
| **52** | 13 | 52,716 | 370,513 | 0 | 423,229 | 622 | 263,248,139 |
| **47** | 18 | 72,991 | 513,018 | 0 | 586,009 | 603 | 353,363,258 |
| **42** | 23 | 93,266 | 655,523 | 0 | 748,789 | 658 | 492,703,109 |
| **37** | 28 | 113,541 | 798,028 | 0 | 911,569 | 730 | 665,445,458 |
| **32** | 33 | 133,816 | 940,533 | 0 | 1,074,349 | 772 | 829,397,675 |
| **27** | 38 | 154,092 | 1,083,038 | 0 | 1,237,130 | 502 | 621,039,019 |
| **22** | 43 | 174,367 | 1,225,543 | 0 | 1,399,910 | 280 | 391,974,722 |
| **17** | 48 | 194,642 | 1,368,048 | 0 | 1,562,690 | 169 | 264,094,596 |
| **12** | 53 | 214,917 | 1,510,553 | 0 | 1,725,470 | 94 | 162,194,191 |
| **7** | 58 | 235,192 | 1,653,058 | 0 | 1,888,250 | 40 | 75,530,013 |
| **2** | 63 | 255,468 | 1,795,563 | 0 | 2,051,031 | 20 | 41,020,610 |
| **−3** | 68 | 275,743 | 1,938,068 | 0 | 2,213,811 | 10 | 22,138,107 |
| **−8** | 73 | 296,018 | 2,080,573 | 0 | 2,376,591 | 1 | 2,376,591 |
| **−13** | 78 | 316,293 | 2,223,078 | 0 | 2,539,371 | 1 | 2,539,371 |
| **−18** | 83 | 336,568 | 2,365,583 | 0 | 2,702,151 | 0 | 0 |
| | | | | | | Σ | **4,426,823,817** |

And then, the heating fuel oil and natural gas consumptions were determined as follows:

| | | | **Annual Cost of Heating (US$/season)** |
|---|---|---|---|
| *E* (gal oil/year) | Equation (9.5) | 39,810 | 97,533 |
| *E* ($ft^3$ natural gas/year) | Equation (9.5) | 5,341,245 | 26,706 |

*(Continued)*

Our calculations show that both the annual HDD (Example 9.1) and the bin method approaches provided quite similar estimates. We can use the same energy conservation measures that were used in Examples 9.3 and 9.4, and determine similar savings this time using the bin method. One can easily introduce varying heat pump performance (below the balance point temperature) as a function of outside temperature or some sort of internal heat gains and determine new balance point temperature and therefore estimate the heating energy consumption due to the varying heat pump efficiency therefore the heat pump capacity (which must be adjusted for the changing efficiency) or the use of free energy (internal heat gains) available, respectively. We will now introduce a fixed internal heat gain in the following example and see what happens.

**Example 9.6: Estimating Heating Energy in a Greenhouse with a Fixed Internal Heat Gain via Bin Method**

Let's now consider the greenhouse, which was used in Example 9.5, a single glass greenhouse in Columbus, Ohio, USA, and estimate the annual heating energy requirement and the costs of heating using fuel oil and natural gas furnaces for this greenhouse. Assume an average internal heat gain of 200,000 Btu/h.

**Solution**

First, using Equation (9.3a) or (9.3b) we have to calculate the balance point temperature $T_{bal}$ for this greenhouse complex:

$$T_{bal} = T_i - \left( \frac{\dot{Q}_{internal}}{(\Sigma UA) + \text{ACH}\, V\, \rho\, c_p} \right)$$

$$= 65°\text{F} - \left( \frac{200,000\ \text{Btu/h}}{\left(28,501 \frac{\text{Btu}}{\text{h.°F}}\right) + 1.0\left(202,752\ \text{ft}^3\right)0.02} \right) = 58.86°\text{F} = 58.9°\text{F}$$

- Remember that no supplemental heating is required at or above 58.9°F outside temperatures. Even though the inside design temperature $T_i = 65°\text{F}$, the difference ($\Delta T = T_i - T_{bal}$) is handled by free internal heat gains rather than the furnace.
- See Table 9.2 for the number of occurrence hours in each bin. Since $T_{bal} = 58.9°\text{F}$, the highest temperature bin to be considered is the 57°F bin (59.5°F and 54.5°F).
- 57°F bin occurs for 648 h; but, supplemental heating is needed only when $T_o < 58.9°\text{F}$.
- The outside temperature is between 54.5°F and 58.9°F for about:

$$648\ \text{h} \times \left(58.9°\text{F} - 54.5°\text{F}\right)/5°\text{F} = 564.6\ \text{h}$$

- During this 564.6-h period, the average outside (bin) temperature is:

$$\left(58.9°\text{F} + 54.5°\text{F}\right)/2 = 56.7°\text{F}$$

So, 56.7°F is the highest temperature interval (bin) and the corresponding occurrences

*(Continued)*

of 565 h need to be used first in our calculations. And then, the remaining (lower) temperature bins and the corresponding occurrence hours will be incorporated fully for estimating the seasonal heating energy consumption.

Finally, using Equation (9.5) and its components, we will perform the full energy estimating calculations as presented in the following table:

| BIN METHOD | | | | | | | |
|---|---|---|---|---|---|---|---|
| **Bin Temp** | **ΔT** | $Q_{inf}$ | $Q_{cond}$ | $Q_{internal}$ | $Q_{inf} + Q_{cond} - Q_{internal}$ | **Occurrence** | $E$ |
| **(°F)** | **(°F)** | **(Btu/h)** | **(Btu/h)** | **(Btu/h)** | **(Btu/h)** | **(h)** | **(Btu/year)** |
| **56.7** | 8.3 | 33,657 | 236,558 | 200,000 | 70,215 | 565 | 39,645,877 |
| **52** | 13 | 52,716 | 370,513 | 200,000 | 223,229 | 622 | 138,848,139 |
| **47** | 18 | 72,991 | 513,018 | 200,000 | 386,009 | 603 | 232,763,258 |
| **42** | 23 | 93,266 | 655,523 | 200,000 | 548,789 | 658 | 361,103,109 |
| **37** | 28 | 113,541 | 798,028 | 200,000 | 711,569 | 730 | 519,445,458 |
| **32** | 33 | 133,816 | 940,533 | 200,000 | 874,349 | 772 | 674,997,675 |
| **27** | 38 | 154,092 | 1,083,038 | 200,000 | 1,037,130 | 502 | 520,639,019 |
| **22** | 43 | 174,367 | 1,225,543 | 200,000 | 1,199,910 | 280 | 335,974,722 |
| **17** | 48 | 194,642 | 1,368,048 | 200,000 | 1,362,690 | 169 | 230,294,596 |
| **12** | 53 | 214,917 | 1,510,553 | 200,000 | 1,525,470 | 94 | 143,394,191 |
| **7** | 58 | 235,192 | 1,653,058 | 200,000 | 1,688,250 | 40 | 67,530,013 |
| **2** | 63 | 255,468 | 1,795,563 | 200,000 | 1,851,031 | 20 | 37,020,610 |
| **−3** | 68 | 275,743 | 1,938,068 | 200,000 | 2,013,811 | 10 | 20,138,107 |
| **−8** | 73 | 296,018 | 2,080,573 | 200,000 | 2,176,591 | 1 | 2,176,591 |
| **−13** | 78 | 316,293 | 2,223,078 | 200,000 | 2,339,371 | 1 | 2,339,371 |
| **−18** | 83 | 336,568 | 2,365,583 | 200,000 | 2,502,151 | 0 | 0 |
| | | | | | | Σ | **3,326,310,737** |

And then, the heating fuel oil and natural gas consumptions were determined as follows:

| | | | **Annual Cost of Heating (US$/season)** |
|---|---|---|---|
| $E$ (gal oil/year) | Equation (9.5) | 29,913 | 73,287 |
| $E$ (ft$^3$ natural gas/year) | Equation (9.5) | 4.013,406 | 20,067 |

The above calculations show that using the available free energy (internal heat gains) reduces the heat provided by the furnace approximately 25% in this example.

## 9.6 Energy Estimating via Complex Modeling

Mathematical models are needed to accurately predict and aid in the optimization of the greenhouse thermal and radiant environments. Accurate prediction models for greenhouse and plant growth performance are also used as a design tool and in economic feasibility analyses. And, simulation of greenhouse environments requires an accurate prediction of greenhouse energy and moisture exchanges as a function of dynamic environmental factors such as solar energy, wind speed, outside temperatures and moisture levels, auxiliary heating, plant moisture and energy exchanges, and excess heat removal or storage. Both steady-state and dynamic (transient) analyses techniques have been used in simulating greenhouse thermal environments. Since a steady-state analysis does not account for energy storage in the system and does not permit accurate predictions of the instantaneous thermal and biological responses of greenhouse systems, a dynamic analysis is required for more accurate prediction and control of greenhouse thermal environments, and therefore energy use estimations.

Steady-state models use one or more variables for energy estimating in greenhouses, and do not account for the effects such as thermal mass or capacitance that cause short-term temperature transients; hence, they are often used for baseline model development. Dynamic models, however, capture the effects of greenhouse peak loads, and warm-up or cool-down periods, and are especially useful for greenhouse heating and cooling load control. A dynamic simulation of a greenhouse thermal environment should also incorporate the dynamics of the plant canopy in a reasonably accurate manner. The treatment of the plant canopy varies tremendously from model to model. While some simulation models incorporates only a simple plant canopy for energy and moisture exchanges within the greenhouse environment, others do not account for the plant canopy at all. One can easily recognize whether a model is steady-state or dynamic by looking for the existence of time-lagged variables, either in the output or input variables. Steady-state models simply do not have any time-lagged variables.

Several types of steady-state greenhouse models have been developed for estimating baseline energy consumption: single- or multivariate, polynomial, etc. Single-variate (i.e. one input variable only) energy estimating models are the easiest of steady-state modeling approaches. They estimate energy use in a greenhouse based on one forcing function, that is, the input variable. Many studies have shown that the outdoor air dry-bulb temperature is the most important forcing function. The simplest steady-state data-driven model is the 18.3°C (65°F) degree-day model. As mentioned above, the main disadvantage of steady-state single-variate data-driven models is their insensitivity to dynamic effects (e.g. thermal mass) and to variables other than temperature (e.g. humidity, solar gain, and artificial lighting).

Besides, there also exist two types of steady-state, multivariate models in general: i) standard multiple-linear model where the data set is treated without having the time-series type data; and ii) Fourier series models having the time-series type data and capturing the diurnal and seasonal cycles, based on which greenhouses are operated. Multivariate models are naturally an extension of single-variate models, and the goal of such type of energy use modeling is to characterize greenhouse energy use with a few readily available and reliable input variables. The environmental variables for modeling heating and cooling energy use include outdoor air dry-bulb temperature, solar radiation, and outdoor specific humidity. In commercial greenhouses, internally generated loads, such as the heat given off by lights and other electrical equipment, also affect heating and cooling energy use. Breaking up energy use data into hourly bins corresponding to each hour and then identifying 24 individual hourly models obviously lead to considerably greater accuracy.

Dynamic data-driven greenhouse models, on the other hand, require solving a set of differential equations, and are typically used with hourly or sub-hourly data in greenhouses where the thermal mass is significant enough to delay heat gains or losses. Dynamic models usually require a high level of knowledge of the greenhouse or the overall system being modeled. Dynamic models based on pure statistical approaches have also been developed for the past couple of decades; and such kind of approaches include machine learning and artificial neural networks.

Neural networks have the ability to "learn," and they learn by example rather than by programmed rules. The weights of a network are initiated with small random numbers; and then, the weights are adjusted (trained) iteratively with a training data set that consists of many input/output pairs, so that applying a set of inputs eventually produces the desired set of outputs. Neural networks have been useful in modeling energy use in commercial greenhouses for: i) predicting what an adequately operating greenhouse should be doing compared to the actual operation; and if there is a difference, it can be used in an expert system to produce early diagnoses of greenhouse operation problems; and ii) predicting what sort of energy a greenhouse would have consumed under present conditions, before a retrofit; and when compared to the measured consumption of the retrofitted greenhouse, the difference represents quite a reasonable estimate of the energy savings due to the retrofit, for instance.

### 9.6.1 Model Validation and Testing

A developed simulation model's accuracy must be validated and tested before it is used for reliable estimations, controls, or any other strategies. There exist three methods to evaluate the accuracy of a whole-greenhouse energy simulation model: i) *empirical validation*, which compares simulated results to monitored data from a real greenhouse; ii) *analytical verification*, which compares simulated outputs to results from a known analytical solution or a generally accepted numerical method calculation for isolated heat transfer under very simple, highly constrained boundary conditions; and iii) *comparative testing*, which compares a simulation model to itself or to other simulation models. All of these three validation and testing procedures are fully acceptable and have been used extensively in model development.

### 9.6.2 Dynamic Greenhouse Simulation Model I for Energy Estimating

Yildiz (1993) developed and validated a dynamic simulation model for accurate prediction of greenhouse energy, moisture, and $CO_2$ exchanges as a function of dynamic environmental factors such as solar energy, outside temperatures and moisture levels, plant moisture and energy exchanges, and heat removal or storage. The model has options to evaluate the effects of location, time of the year, orientation, single and double polyethylene glazings, conventional and heat pump heating and cooling systems, open and closed greenhouse systems, $CO_2$ enrichment, variable shading, and the use of night curtains. Conventional gas furnace and evaporative cooling, respectively, provided heating and cooling in the conventional system. In the heat pump systems, gas-fired heat pump units provided both heating and cooling. The heat pump systems were operated both as an open and a completely confined systems. Outputs of the simulation model included both temporal and vertical distribution of air, leaf, floor and cover temperatures, $CO_2$, relative humidity, solar radiation, and photosynthetically active radiation in addition to the dynamics of photosynthesis, respiration, transpiration, energy, and $CO_2$ use and fixation. Comparison of experimental and predicted results showed that the compared microclimatological parameters were in fairly good agreement (Yildiz & Stombaugh, 2006a,b). The model is useful for plant scientists and engineers to evaluate individual or combined effects of various forcing functions on greenhouse environment and plant responses; and to develop control strategies for different parameters.

#### 9.6.2.1 Physical Model, Weather File, and Greenhouse Characteristics

The developed simulation model was optimized and validated. A description of cucumber plant canopy as a series of parallel rows with rectangular cross sections and variable architectural features was extended to an overall greenhouse simulation model having single or double plastic glazings and three greenhouse floor layers. In the energy and mass exchanges of the greenhouse system, the outside weather conditions and the deep ground temperature served as boundary conditions. Finite difference methods were used to solve the set of differential equations. Five vertical nodes were used for the canopy stand to reduce the computational time without disrupting the solution accuracy. Integration over space was performed when dealing with the radiative heat transfer. The integral equation was approximated using the Composite Trapezoidal Rule for the solar radiation. However, the more complicated closed Newton–Cotes formula with Simpson's Rule was employed when the spatial integration was performed for other selected basic variables such as cumulative leaf area index in describing heat and mass exchanges. This resulted in better accuracy. The ordinary differential equations with an initial value were solved using Euler's Method, and no stability problems were observed. A time interval of 10 s was employed. A small time interval was required due to the rapid response of plants to dynamic environmental parameters. This small time interval caused a considerable increase in computational time, but it represented plant responses in a reasonably accurate manner.

January, April, and July weather files for Delaware (latitude 40° 17′ N, longitude 83° 05′ W), Ohio, USA, were used to represent winter, spring, and summer climates in the simulations. The heat pumps evaluated for open (OHP) and closed (CHP) greenhouse systems were 3-ton (based on system heat removal capacity) gas fired units, and to provide multiposition proportional control it was assumed that three units were used in each greenhouse. The heat pump consisted of a Rankine power cycle and a vapor compression cycle, which uses a novel hydraulically connected rolling diaphragm piston cylinder device as motor, compressor, and pump (Yildiz, 1993; Yildiz et al., 1993). R123 (dichlorotrifluoroethane) and R22 (chlorodifluoromethane) refrigerants were used for the power and refrigeration cycles, respectively. One of the heat pump units in the CHP greenhouse system was used as a dehumidifier while operating as a heater. The only difference from the original heating unit was circulating the inside greenhouse air through the outdoor coil instead of the outside air. This prevented moisture build up in the confined greenhouse system. During the winter season, 0.943 $m^3/s$ of inside air passed through the condensers, and 2.829 $m^3/s$ of inside air was passed through the outside evaporator, which provided the dehumidification. For the CON greenhouse system simulations, it was assumed that the conventional gas-fired furnace provided 24,612 W of heat input each. A furnace efficiency of 0.7 was assumed for the conventional heating system. An evaporative cooler was employed for cooling the conventional greenhouse. Outdoor air was used in the evaporative cooling system, and it was assumed that the air at the evaporative cooler outlet was fully saturated. No heat storage facility was used in this study. An overhead plastic tube was used for the hot and the cold air distribution in both heating and cooling modes. Two shading clothes with transmissivities of 0.75 and 0.50 were used to reduce the cooling loads. The use of these shading clothes provided shading levels of 25%, 50%, and 62.5% by using them individually or together. An aluminized (both sides) night curtain was used at

**TABLE 9.3**

Greenhouse and crop characteristics used in the simulation model.

| | |
|---|---|
| Greenhouse length | 7.50 m (CON and OHP) and 25.0 m (CHP) |
| Greenhouse width | 7.50 m |
| Greenhouse height at eaves | 2.50 m |
| Greenhouse height at ridges | 4.50 m |
| Crop row orientation | North–South |
| Distance between plant rows | 0.75 m |
| Floor surface material | Reflective mulch |
| Glazing | Double polyethylene |
| Distance between plant rows | 0.75 m |
| Average leaf length | 0.30 m |
| Average leaf width | 0.25 m |

night to reduce the heat loss due to long-wave radiation exchanges between the inside greenhouse components and the sky. In the open systems (CON and OHP), ventilation was provided by two fans, one with a fixed flow rate to provide a minimum level of air exchange at all times, and the other one with a variable flow rate. The greenhouse characteristics are presented in Table 9.3.

#### *9.6.2.2 Operational and Control Strategies*

The day or nighttime greenhouse temperature set points were based on the solar position. Based on the indoor air temperature, the control system operated in either the heating or cooling mode. If the system was in heating mode and when heating was required, the ventilation rate was first set to the minimum rate. The control system turned on other heating units based on the difference between the indoor and set point temperatures, providing a multiposition proportional control. If no heating was required in this mode, no heating unit operated; but the system remained in the heating mode until it was switched to the cooling mode.

The cooling mode operated in two steps. The first step was to reduce the cooling load using a variable shading system and to cool the inside air by increasing ventilation rates. Two shading clothes with transmissivities of 0.75 and 0.50 provided shading levels of 25%, 50%, and 62.5% by using them individually or together. The minimum and maximum ventilation rates were 0.01 $m^3/s.m^2$ and 0.08 $m^3/s.m^2$, respectively. If the first step in cooling could not handle the cooling load, then the second step was activated, where the heat pump units or evaporative cooling provided the cooling. In the CON system, introducing an outside airflow rate of 0.08 $m^3/s.m^2$ when the second step was activated in the cooling mode provided evaporative cooling. Relative humidity levels in the conventional system were controlled indirectly by the temperature control. In the OHP system, however, additional relative humidity control was provided. When the inside relative humidity levels exceeded 80%, additional ventilation was introduced to decrease inside relative humidity. In the CHP system, the same criterion was used to prompt the heating mode. However, the cooling mode was activated at lower inside temperatures than those used in the other two systems. The operation of the heating system was the same as in the other two systems. However, the minimum ventilation rate was used in the open system while no ventilation was used in this closed system. In the cooling mode of the closed system, there was only one step unlike the conventional and open systems, which had two-step cooling systems. Here, no cooling was provided by ventilation; instead, the cooling was provided by the three heat pump units providing a multiposition proportional control, after reducing the cooling load using the variable shading system. The operation of the shading system was the same as in the other two systems. Either the cooling units or the dehumidifier (the first heating unit) controlled inside relative humidity. When the inside relative humidity exceeded 80% level, this heating unit operated as a dehumidifier to prevent excess moisture within the closed loop heat pump system.

#### *9.6.2.3 Energy and Mass Balances*

Without going into any further details, only the major balance equations will be provided here. For instance, energy balance of the internal air for a single layer greenhouse was expressed as:

$$\rho_a V_a c_{pa} dT_a / dt = Q_{sr} - Q_{ac} - Q_{vio} - Q_{flux} \tag{9.6}$$

where $\rho_a$ was the density of the bulk air (kg/m$^3$), $V_a$ was the volume of the bulk air above the canopy stand (m$^3$), $c_{pa}$ was the specific heat of the air (J/kg.C), $T_a$ was the bulk air temperature (°C), and t was the time (s). $Q_{sr}$ was the heat input rate by the heating source into the internal air (W), $Q_{ac}$ was the heat transfer rate from the internal air to the structural cover (W/m$^2$), $Q_{vio}$ was the rate of heat transfer from the inside air to the outside air due to ventilation (W), and $Q_{flux}$ was the rate of heat transfer from the bulk air to the top layer of the canopy stand (W).

Moisture balance of the internal air was:

$$(1/\nu_a)V_a dw/dt = -M_{conc} - M_{vio} - M_{coil} - M_{flux} \tag{9.7}$$

where $\nu_a$ was the specific volume of the air (m$^3$/kg dry air), and $w$ was the humidity ratio (kg $H_2O$/kg dry air), $M_{conc}$ was the rate of moisture condensed on or evaporated from the inside surface of the cover, $M_{vio}$ was the amount of moisture transferred from the inside air to the outside air via ventilation (kg $H_2O$/s), $M_{coil}$ was the rate of moisture condensed on the cooling coil (kg $H_2O$/s), and $M_{flux}$ was the rate of moisture transferred from the bulk air to the top layer of the canopy stand (kg $H_2O$/kg dry air).

Thornley and Johnson (1990) indicated that, in describing $CO_2$ concentrations, the units of parts per million (ppm) has two deficiencies. First, it is not clear whether the definition is kilograms per million kilograms or molecules per million molecules. The second problem is that photosynthesis depends on the absolute number of $CO_2$ molecules per unit volume, and not just the proportion of $CO_2$ molecules in the air. Therefore, they suggest that the use of ppm should be avoided. Instead, they recommend the following definition of concentration to be used in any model. From the gas laws, the concentration of any gas at an arbitrary temperature and pressure is given by:

$$\text{Concentration} = (273.15/T)(P/101325.0)(0.044618) \tag{9.8}$$

$$\text{Density} = (\text{Concentration})(\text{Relative Molecular Mass}) \tag{9.9}$$

where the concentration is in kmole $CO_2/m^3$, $T$ is the temperature (K) and $P$ is the pressure (Pa), and 0.044618 is the concentration of pure $CO_2$ (or any other gas) in kmole/$m^3$ at normal temperature (273.15 K) and pressure (101,325.0 Pa). The relative molecular mass of $CO_2$ is 44.0098 kg $CO_2$/kmole $CO_2$. Using these relations, the following equation was derived for the $CO_2$ balance of the inside air:

$$dCO_2/dt(\text{kg}\,CO_2/\text{s}) = C_{inj} - (C_{vio} + C_{flux}) \tag{9.10}$$

where $C_{inj}$ was the $CO_2$ injection rate (kg $CO_2$/s), $C_{vio}$ was the $CO_2$ exchange rate due to ventilation (kg $CO_2$/s), and $C_{flux}$ was the $CO_2$ exchange rate between the bulk air above the canopy stand and the top layer of the canopy (kg $CO_2$/s). $CO_2$ concentrations for inside and outside air were in ppm. The instantaneous gross rate of canopy photosynthesis was defined according to Thornley and Johnson (1990), whereas dark respiration was expressed using the $Q_{10}$ factor.

In dealing with the energy and mass exchanges of the structural cover, it was assumed that the exchanges occurred homogeneously on the cover, the heat storage capacity of the cover material was small compared to the existing fluxes, and no condensation or evaporation occurred on or from the cover. The steady-state energy balance equation for a single cover greenhouse was

$$0 - Q_{ac} + Q_{oc} + \alpha_{s,c} A_c IG_s + LW_c \tag{9.11}$$

where $Q_{ac}$ was the convective heat transfer rate between the internal air and the cover (W), $Q_{oc}$ was the corresponding term between the outside air and the cover (W), $\alpha_{s,c}$ was the short-wave absorptivity of the cover, $A_c$ was the cover area ($m^2$), $IG_s$ was the rate of solar radiation on the cover (W/$m^2$), and $LW_c$ was the net long-wave radiation on the cover (W).

It was assumed that the floor was covered with a polyethylene film; however, an option was provided so that bare soil could also be used. A one-dimensional heat conduction equation was used in dealing with the energy balance of the greenhouse floor, by dividing the floor into three layers (0.01, 0.10, and 0.50 m) with the assumption of homogeneous thermal and hydraulic properties within each layer. It was also assumed that the deep ground temperature was constant at 15°C, and no condensation or evaporation occurred on or from the floor surface.

The solar radiation was treated by splitting it into direct, diffuse, and scattered components and assuming that all the radiation reflected by and/or transmitted through foliage elements contributed only to the diffuse component. The expression widely used in microclimatological studies for the penetration function of direct solar radiation for uniformly distributed plant canopies was expanded to a row plant stand whose foliage area distribution varied both vertically and horizontally. It was assumed that the scattering distributions (both upward and downward) were uniform horizontally.

A resistance concept was used in dealing with the thermal radiation as outlined by Incropera and DeWitt (1985). A parallel plane analysis was employed whenever it was applicable. For the other cases, a multiple surface radiation exchange analysis using the same approach (resistance concept) was employed.

#### 9.6.2.4 Validation and Testing

The comparisons of simulation results with experimental findings were made using the following inputs for the conventional greenhouse system (CON) (the shading and evaporative cooling systems were not operational for these comparison simulations): a plant height of 2.00 m, a distance of 0.86 m between rows, a north–south row direction, and a greenhouse floor of uncovered soil. Figure 9.4 shows the diurnal changes in predicted and measured air and leaf temperatures for 2 successive days. The predicted and the measured temperatures were very close to each other. Generally, the predicted air temperatures were slightly lower than the measured temperatures. The predicted inside relative humidity levels were also compared to the measured levels, and plotted together with the predicted and the measured transpiration rates (Figure 9.4).

The model consistently overestimated (~7%) the daytime inside relative humidity levels while underestimating (~10%) at night. Figure 9.5 shows the comparison between predicted and measured transpiration rates per unit ground area for the 2 successive days, plotted together with measured solar radiation. The highest transpiration rate (365 g/h.m$^2$) was observed on the second day when high solar radiation existed. The lowest transpiration rate (20 g/h.m$^2$) was observed at night. Transpiration rates closely followed the changes in the solar radiation, and the values were in fairly good agreement during the day. The transpiration rates at night, however, were overestimated. This was due to the stomatal resistance expression, which was derived from daytime data only, not counting for the effects of climatic variables other than solar radiation. Since the nighttime transpiration rates account for a very small portion of the total transpiration, the absolute magnitude of

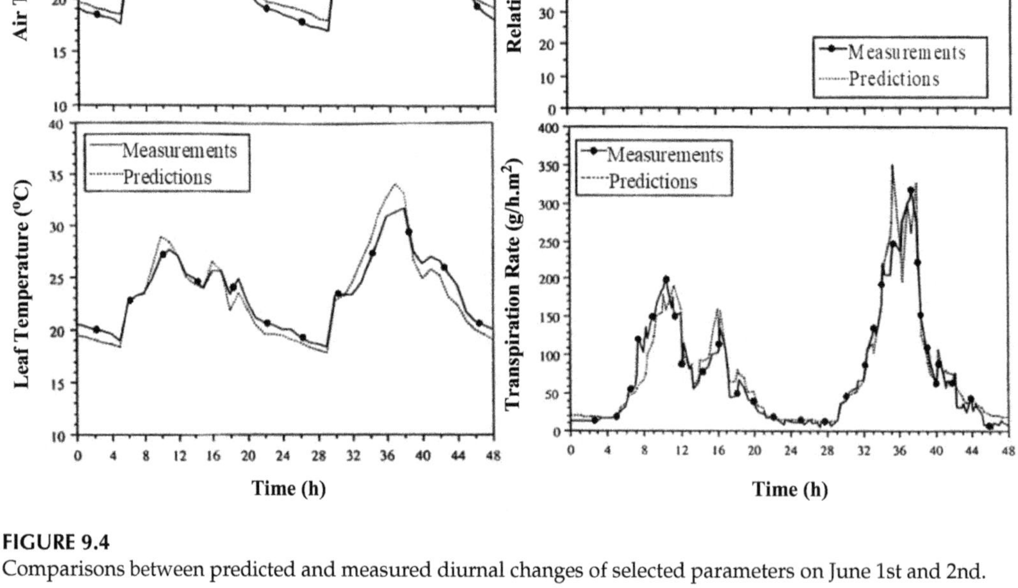


**FIGURE 9.4**
Comparisons between predicted and measured diurnal changes of selected parameters on June 1st and 2nd.

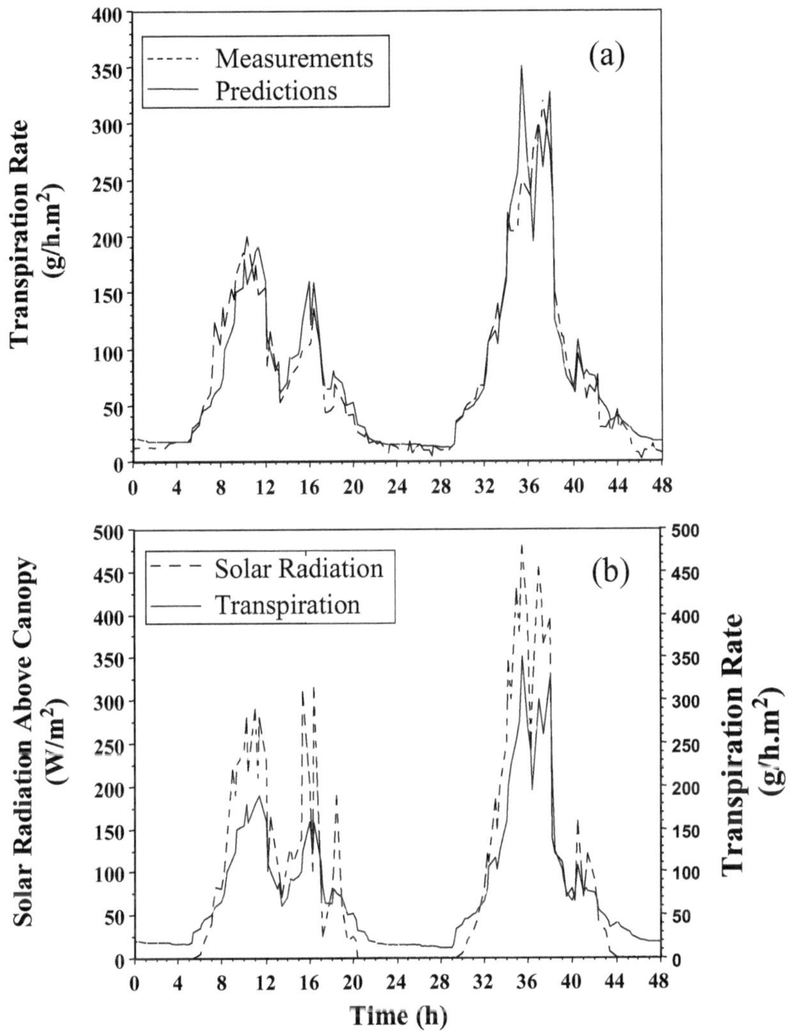


**FIGURE 9.5**
Comparisons between predicted and measured diurnal changes of transpiration rates (a) on June 1st and 2nd, and plotted together with measured solar radiation (b).

the error due to overestimation at night was not significant. Predicted and measured relative humidity, and air and leaf temperatures were also in fairly good agreement. The predicted air temperatures were slightly lower than the measured temperatures.

Figure 9.6a shows the diurnal solar radiation, temperature, and humidity regimes in response to the outside climatic conditions within the OHP system in winter (January). Only small temperature fluctuations were observed in this greenhouse during the day, because the two heating units were operating continuously. Temperature fluctuations were observed again in the late afternoon. This time, the third heating unit was turned on due to the reduced solar radiation and the decrease in the outside air temperature. Leaf temperatures were affected by the variations in the air temperature. The fluctuations in the inside relative humidity caused fluctuations in the transpiration rates due to the resulting changes

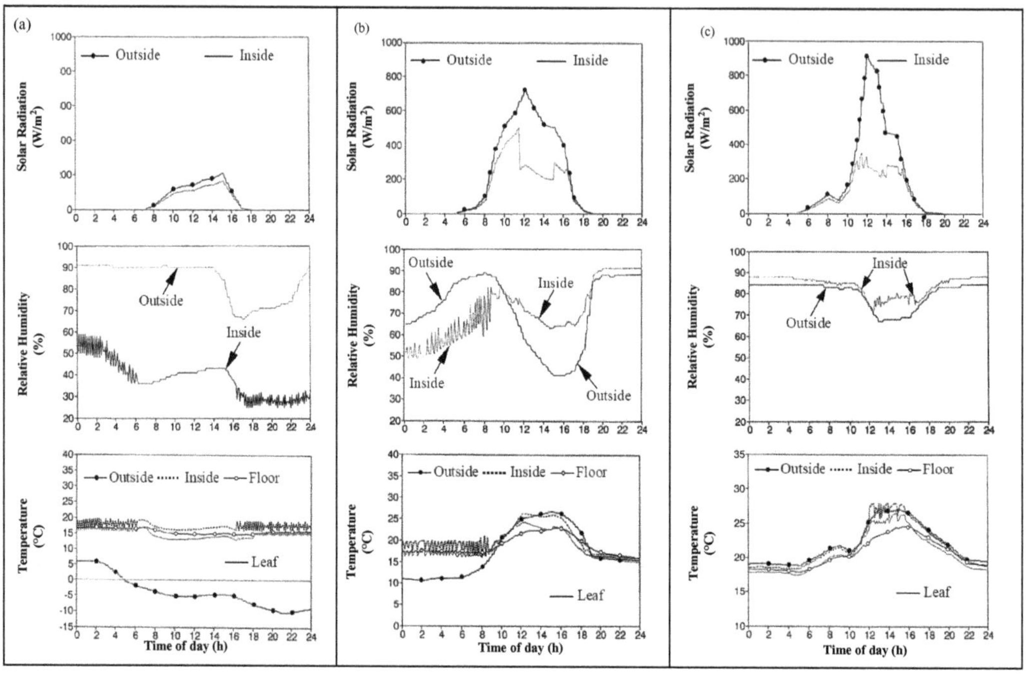


**FIGURE 9.6**
Diurnal changes in predicted greenhouse environmental quantities in the OHP greenhouse system in winter (a), spring (b), and summer (c).

in vapor pressure deficit. Temperature fluctuation of 3°C resulted in fluctuations of 1°C, 5%, and 3 mg $H_2O/s.m^2$ in the leaf temperature, inside relative humidity, and transpiration rates, respectively. Figure 9.6b shows the diurnal changes in solar radiation, temperature, and moisture regimes in the same greenhouse system in spring (April). Additional control (increased ventilation rates) was employed to control the relative humidity at about 80%. Figure 9.6c shows the diurnal changes in solar radiation, temperature, and moisture regimes in the same greenhouse system in summer (July). Inside air temperatures were very close to the outside air temperatures because of the increased ventilation rates due to the increased inside relative humidity levels. No heating or cooling was required at night. The increase in the air temperature resulted in the use of the shading system between noon and 16:00. However, it was still required to do some mechanical cooling between the hours of 13:00 and 15:00. The difference between the inside air and leaf temperatures was about 1.6°C at night, and it was 5.0°C at noon. This was because of the low leaf temperatures due to increased transpiration rates (44 mg $H_2O/s.m^2$) as a result of the increased solar radiation and inside air temperature. The inside relative humidity followed the variations in the outside relative humidity levels due to the increased ventilation rates.

Generally speaking, regardless of the individual greenhouse system and the season, when high relative humidity levels prevailed when there was no or low solar radiation, then the leaf temperatures were lower than, but very close to the air temperatures. However, high solar radiation along with high relative humidity levels resulted in leaf temperatures that were higher than the air temperature. At this point, the plants could not transpire much, and hence they did not remove the excess solar energy on their surfaces. This resulted in increased leaf temperatures.

This shows that the simulation model with all its components performed very well, and the greenhouse with the heat pump system can be operated as a closed system, because the dehumidifier handled the excess moisture adequately. The study also showed that proportional control was an essential part of the control system. A multiposition proportional control, as employed in this study, would provide a reasonable control for the greenhouse environment.

#### *9.6.2.5 Performance of the Simulation Model*

Figure 9.7 presents the predicted hourly transpiration and energy consumption rates per unit ground area in the CONs, open greenhouse heat pump systems (OHPs), and closed greenhouse heat pump systems (CHPs) for 10 successive days in winter. In Figure 9.7, only 10-day simulation results were presented to show the long-term operation of the simulation model and behavior of the greenhouse environmental quantities. These 10-day simulations were obtained from 25-day simulation runs ignoring the first 6 and last 9 days of the simulations.

Figure 9.7a shows predicted hourly transpiration and energy consumption rates per unit ground area for the outside climatic conditions shown in a CON in winter. The highest transpiration rate occurred on the fifth day (at noon) with a rate of approx. 0.24 kg/h.m$^2$. This was due to the high outside solar radiation and the low outside relative humidity levels on that particular day. The low outside temperature and relative humidity resulted in low relative humidity levels and bigger vapor pressure deficits inside. As a result of this combination, higher transpiration rates were observed on the fifth day compared to the other days. At night however, the transpiration rates were at approximately 0.10 kg/h.m$^2$ except for the ninth day on which both night and daytime transpiration rates were

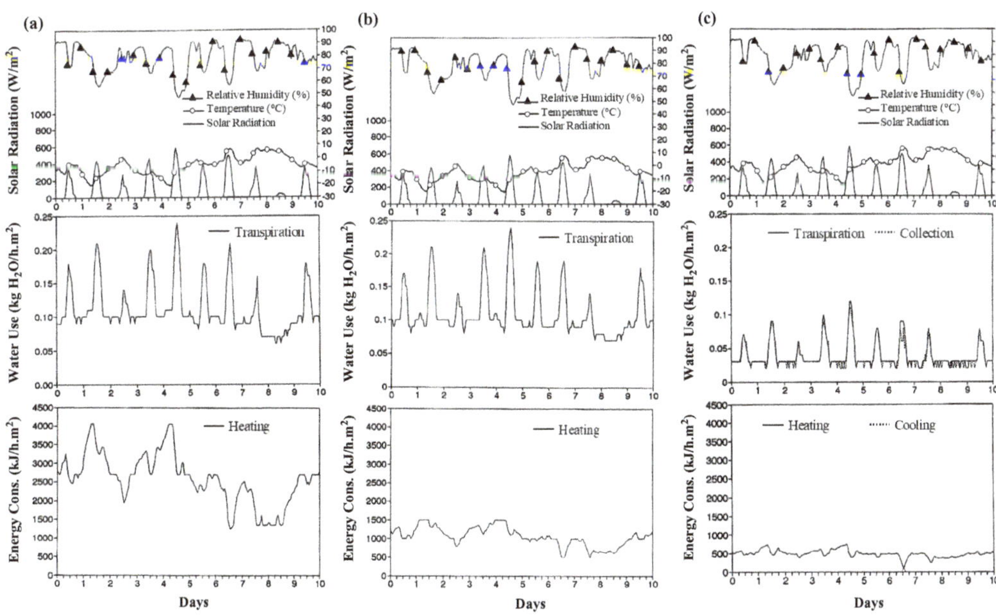


**FIGURE 9.7**
Predicted transpiration rate and energy consumption per unit ground area in the conventional system (a), open greenhouse heat pump system (b), and closed greenhouse heat pump system (c) for 10 successive days in winter.

considerably lower than the rates on the other days. This was partly due to the very low outside solar radiation (about 50 W/m$^2$) for the duration of only 2–3 h. Another cause for these low rates was the high outside relative humidity and air temperatures during the day. Figure 9.7a also shows that the highest energy consumption rates occurred at very early hours on the second and fifth days due to the low outside air temperatures. The low outside temperatures resulted in lower inside temperatures increasing the inside relative humidity levels. This high relative humidity levels called for heating, eventually increasing the energy consumption. The energy consumption rates were around 4,000 kJ/h.m$^2$ on both days. The minimum energy consumption occurred on the seventh day, on which the air temperatures were comparatively higher.

Figure 9.7b shows that the transpiration rates in the OHP greenhouse system were almost the same as in the conventional system. This was due to the fact that the indoor conditions were very similar, and nearly identical control strategies were used for both systems in cold weather. While the trends in energy consumption rates were the same as in the conventional system, the magnitudes were considerably lower. The highest energy consumption for heating in the conventional system was about 4,000 kJ/h.m$^2$, whereas it was about 1,500 kJ/h.m$^2$ in the OHP greenhouse system. This resulted from the efficiency differences between the two heating systems. The average COP's (coefficient of performance) for the heating units were 0.70 and 1.78 for the conventional furnace and open loop heat pump systems, respectively. That is, the conventional gas furnace consumed about 2.5 times more energy than the proposed heat pump system to heat the same size greenhouse.

Figure 9.7c shows that the only differences between the CHP greenhouse system and the other two systems were the magnitudes of the transpiration rates and the fact that almost all the water transpired was condensed on either the dehumidifier coil or the cooling unit coil. The highest transpiration rate was about 0.12 kg/h.m$^2$ on the fifth day while the minimum rate was about 0.02 kg/h.m$^2$, primarily at night. Transpiration rate was lower in the CHP greenhouse system because the inside relative humidity levels were approximately 20% higher. High inside humidity levels reduced the vapor pressure deficit, and therefore the transpiration rates. Since almost all the water transpired by plants was collected on the coils, water use in the CHP system was essentially zero. Figure 9.7c also shows that the CHP system had the lowest heating energy consumption of all three systems. The highest and lowest energy consumption rates occurred on the same days as in the other two systems due to the same outside climatic conditions. However, the highest heating energy consumption rate was only about 750 kJ/h.m$^2$ observed early in the morning on the second and fifth days, whereas the lowest energy consumption rate was observed on the seventh day with a rate of about 100 kJ/h.m$^2$. The reason for the CHP system having low water use (transpiration) and energy consumption was due to the fact that there was no ventilation in this greenhouse system. This resulted in a moisture build up and high humidity levels (between 70% and 90%) within the greenhouse. No ventilation also meant that no heat loss or gain occurred via ventilation. Therefore, although both the OHP and CHP greenhouse systems used the same heating systems, the energy consumption rates were considerably different in these two systems.

Table 9.4 summarizes the daily averages of energy consumption and transpiration rates in CON, OHP, and CHP greenhouse systems in winter, spring, and summer. The conventional system used considerably more energy than the other two systems in winter and spring. In winter, average daily heating energy consumption per unit ground area were 56.9, 23.5, and 11.3 MJ/d.m$^2$ in the CON, OHP, and CHP greenhouse systems, respectively. The average consumptions in the same greenhouse systems in spring,

**TABLE 9.4**

Daily energy consumption and water use in conventional (CON), open loop (OHP), and closed loop heat pump (CHP) systems in winter, spring, and summer.

| | WINTER | | | SPRING | | | SUMMER | | |
|---|---|---|---|---|---|---|---|---|---|
| | CHP | OHP | CON | CHP | OHP | CON | CHP | OHP | CON |
| ENERGY CONSUMPTION (MJ/d.m²) | | | | | | | | | |
| **Heating** | | | | | | | | | |
| Natural Gas | 10.75 | 22.22 | 56.72 | 5.04 | 8.36 | 20.54 | 0.66 | 0.09 | 0.44 |
| Fan (Electric) | 0.44 | 0.91 | 0.20 | 0.20 | 0.36 | 0.07 | 0.03 | * | * |
| Pump (Electric) | 0.16 | 0.32 | -- | 0.07 | 0.12 | -- | 0.01 | * | -- |
| **Subtotal** | **11.34** | **23.45** | **56.92** | **5.31** | **8.82** | **20.61** | **0.70** | **0.10** | **0.44** |
| **Cooling** | | | | | | | | | |
| Natural Gas | 0.02 | -- | -- | 1.94 | -- | -- | 5.19 | 0.91 | -- |
| Fan (Electric) | * | -- | -- | 0.05 | -- | -- | 0.14 | 0.02 | -- |
| Pump (Electric) | * | -- | -- | 0.03 | -- | -- | 0.08 | 0.01 | -- |
| **Subtotal** | **0.02** | -- | -- | **2.02** | -- | -- | **5.40** | **0.94** | -- |
| **Ventilation** | -- | **0.11** | **0.11** | -- | **0.23** | **0.11** | -- | **0.71** | **0.13** |
| TOTAL | **11.36** | **23.56** | **57.03** | **7.33** | **9.04** | **20.73** | **6.10** | **1.75** | **0.57** |
| WATER CONSUMPTION (kg $H_2O$/d.m²) | | | | | | | | | |
| **Transpiration** | 0.85 | 2.52 | 2.54 | 1.05 | 2.22 | 2.17 | 1.17 | 1.66 | 1.28 |
| **Water Collected** | 0.83 | -- | -- | 1.04 | -- | -- | 1.17 | 0.35 | -- |
| **Evaporative Cooling** | -- | -- | -- | -- | -- | -- | -- | -- | 0.29 |
| TOTAL | **0.02** | **2.52** | **2.54** | **0.01** | **2.22** | **2.17** | **0.00** | **1.31** | **1.57** |

*Negligible amount.

however, were 20.6, 8.8, and 5.3 MJ/d.m², respectively. In summer, however, the average daily energy consumption were considerably less than those in winter and summer, and the CHP system had the highest energy consumption. The major portion of the total energy consumption was for heating in all the greenhouse systems (Table 9.4 and Figure 9.8). The energy used for cooling and ventilation was of much less importance, because the shading reduced the required cooling loads. Also, the plants themselves are biological humidifiers and can provide much of their own cooling.

Figure 9.8 shows the average daily energy consumption in winter, spring, and summer when all three were averaged together. The conventional system (26.1 MJ/d.m²) consumed about 2.3 and 3.2 times more energy than the OHP (11.5 MJ/d.m²) and the CHP (8.3 MJ/d.m²) greenhouse systems, respectively. In the CON, 99.5% of the energy consumption was used for heating, while the remaining was used for ventilation. Evaporative cooling in the conventional system provided the cooling in summer. In the OHP greenhouse system, 94.2% of the total energy consumption was used for heating, while 3.1% was used for ventilation, and 2.7% was used for cooling. In the CHP greenhouse system, however, 70% of the total energy consumption was used for heating, while 30% was used for cooling.

Table 9.4 also summarizes the water consumption and collection rates in different greenhouse systems. Higher transpiration rates were predicted in winter than in spring and

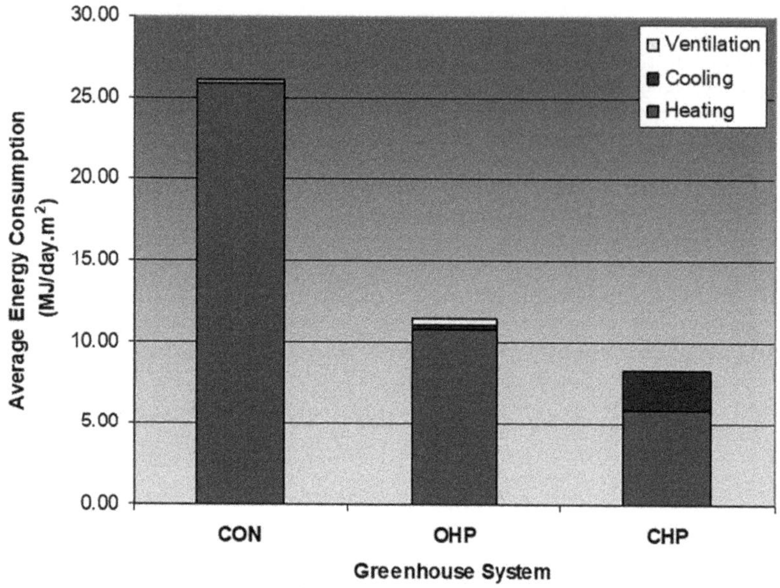


**FIGURE 9.8**
Average of winter, spring, and summer energy consumption in conventional greenhouse (CON), open greenhouse heat pump (OHP), and closed greenhouse heat pump (CHP) systems.

summer in the conventional and OHP greenhouse systems, even though lower solar radiation levels were observed in winter. However, low outside and inside relative humidity levels occurred in winter and this was the driving force that caused the increased transpiration rates. The transpiration rates in the conventional and OHP greenhouse systems were about the same in winter and spring. In summer, however, the transpiration rates in the OHP greenhouse system were higher than those in the conventional system. This was due to the high relative humidity levels and small vapor pressure deficits in the conventional system as a result of evaporative cooling. The difference in transpiration rates in the CHP greenhouse system with respect to different months, however, was mostly the result of the increased solar radiation rather than the vapor pressure deficit, because the inside relative humidity levels were kept at about the same levels. Since the inside relative humidity was maintained at higher levels (80–95%) in the CHP greenhouse system compared to the other two systems, the transpiration rates in this system were lower than those in the other two systems. Average daily transpiration rates of winter, spring, and summer were 2.09, 2.02, and 1.02 kg/d.m$^2$ in the conventional, OHP, and CHP greenhouse systems, respectively.

In addition to the transpiration, water was also used for evaporative cooling in the conventional system (Table 9.4). On the other hand, in the closed greenhouse system, almost all the transpired water was recollected on the coils, making the overall water consumption in the closed greenhouse system essentially zero. Thus, the closed greenhouse system was the most energy and water conserving system, while the conventional system used the most energy and water of all the systems in this study.

This study concluded that greenhouse operation with the heat pump system has a great potential for enormous energy and water savings, as well as cooling and dehumidification. Closed greenhouse operation with a heat pump system would make it easier to control the humidity and keep high $CO_2$ levels, while reducing the risk of insects and diseases within the confined greenhouse space. The heat pump system performed very well with a variable shading system significantly conserving both energy and water. The cooling system COP varied from 0.93 to 1.03, whereas the heating system COP varied from 1.64 to 1.81. A combination of open and closed greenhouse systems, that is, a semi-closed greenhouse would be the most energy and water conserving system, operating the system as a closed system in fall, winter, and spring months and as an open system in hot summer months. We will address this system again in Chapter 11, and discuss the performance of a semi-closed greenhouse system in depth.

### 9.6.3 Dynamic Greenhouse Simulation Model II for Energy Estimating

Mehlitz (2009) developed a dynamic greenhouse simulation model (see Figure 9.9) to provide an accurate prediction of heating and cooling energy requirements of a horizontal tubular photobioreactor (PBR) for *Chlorella* spp. production at any desired location as a function of dynamic environmental factors such as solar energy, outside temperature, and wind speed.

#### 9.6.3.1 Physical Model, Weather File, and Greenhouse Characteristics

This dynamic prediction model was developed from basic energy conservation and heat transfer principles for a single glass glazing greenhouse, and was based on a PBR system also known as "BioFence." The BioFence system was developed by Varicon Aqua Solutions Ltd., UK. Horizontal plastic tubes are stacked in a rack to allow a high surface-to-volume ratio. The expandability of the system allows a scale-up to basically all desired sizes. According to Varicon Aqua Solutions Ltd., the BioFence can produce an equivalent of 2,000 L of bag-grown algae per day in an array of tubes of 10 m × 1.8 m. Each meter of transparent tube has an internal volume of 0.66 liters and an internal surface area of 0.1 $m^2$. It is expandable in blocks of 16 × 5 m tubes. A light-to-dark ratio of at least 50% is recommended. The tank size required is therefore equal or greater than the internal tube volume. A description of the PBR as a series of parallel rows was extended to an overall greenhouse model for indoor operation, whereas the same parallel rows were modeled without a greenhouse for the outside operation, as schematically shown in Figure 9.9. In dealing with the energy exchanges, the outside weather conditions and the deep ground temperature served as boundary conditions. The bulk air was analyzed by adopting the assumption of "perfect mixing." All dimensions of the PBR were kept as variables in the model. This feature allowed optimization and flexibility for different PBR configurations.

The developed simulation model was optimized and validated. The model incorporated a dynamic simulation of *Chlorella* spp. growth. The PBR and greenhouse characteristics used in the simulation model for this study are presented in Table 9.5. The model was used to predict heating and cooling loads and the benefits associated with heating and cooling. Hourly weather data (temperature, solar radiation, wind speed, relative humidity) were downloaded from the California Irrigation Management Information System (CIMIS) weather station for San Luis Obispo (35° 17′N; 120° 39′W), CA, USA, for the month of April 2007. However, the simulation model is flexible enough to simulate any other location, any other month, or even any year. Simulations were performed starting at the beginning of the 5th day and ended at the end of 30th day of the month providing 26-day simulations. Operating the model with hourly weather data, heating and cooling loads can be calculated early in the planning stage of a project. Simulation of different PBR row

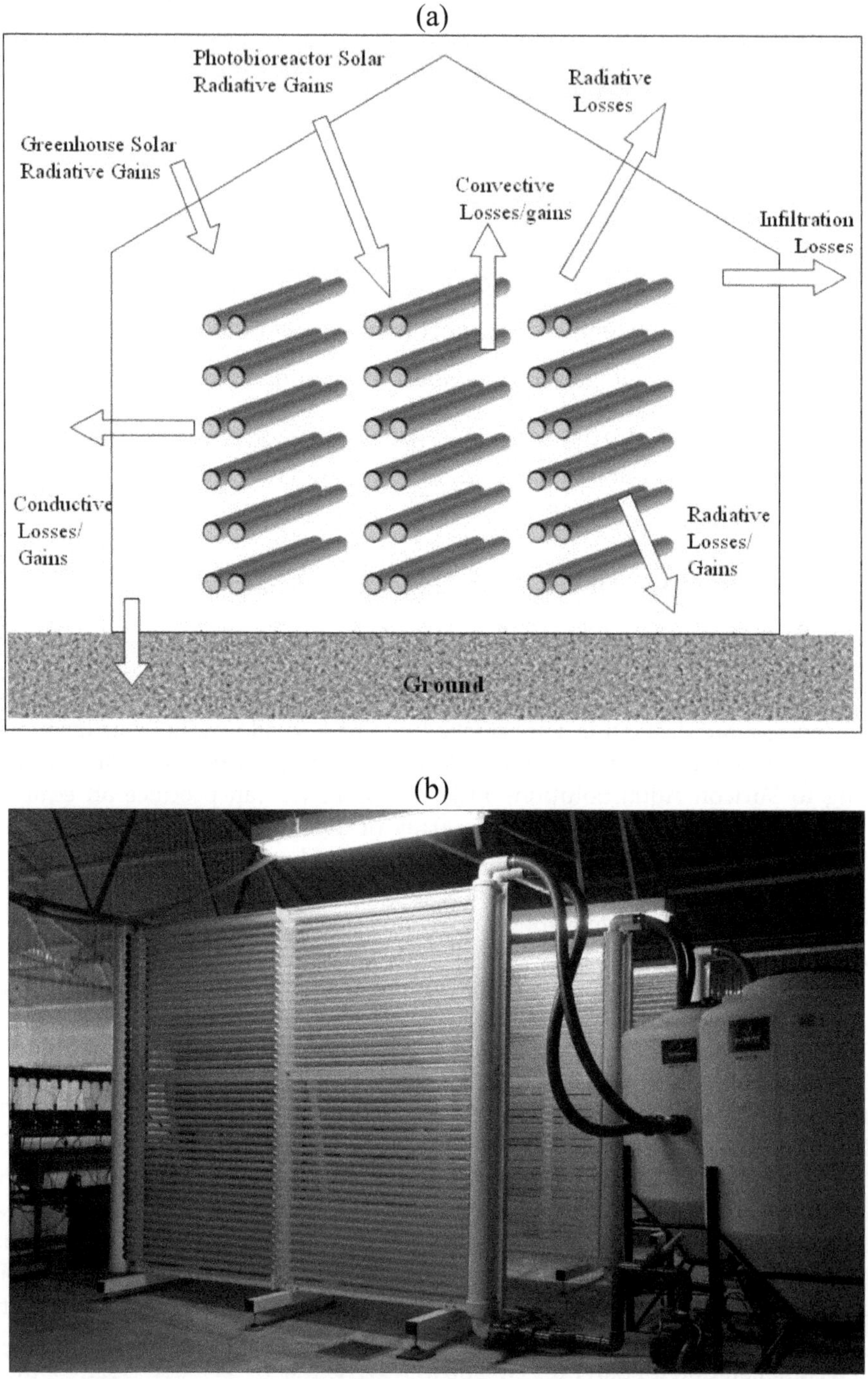


**FIGURE 9.9**
Schematic view of the PBR greenhouse model (a) and the view of actual PBR in the greenhouse (b).

**TABLE 9.5**
Photobioreactor and greenhouse characteristics.

| Parameter | Description | Value |
|---|---|---|
| $IR_{AIR}$ | Index of refraction n for the air | [1]1 |
| $IR_{TUBE}$ | Index of refraction n for the glass tube | [1]1.526 |
| $\Phi$ | Latitude for location (San Luis Obispo) | 35.23° |
| $\gamma$ | Surface azimuth angle | [3]0° |
| $\beta$ | Slope of virtual surface for beam incidence | 45° (assumed) |
| $R_{GROUND}$ | Reflectivity of the ground | 20% (assumed) |
| $R_{ALGAE}$ | Reflectivity of the algae algae | 5% (assumed) |
| $\tau_{GH}$ | Transmissivity of the greenhouse cover | [2]60% |
| $Cp_{WAT}$ | Specific heat of water | [4]1.16278 Wh/kg.K |
| $Cp_{AIR}$ | Specific heat of air | [4]0.2811 Wh/kg.K |
| $\rho_{WAT}$ | Density of water | [4]996 kg/m$^3$ |
| $\rho_{AIR}$ | Density of air | [4]1.2 kg/m$^3$ |
| $\varepsilon$ | Emissivity of the glass/Plexiglass tubes | [3,4]94% |
| $AEH$ | Infiltration rate | [2]1.5 (air exchanges per hour) |
| $U_G$ | Heat transfer coefficient – single layer glass | [2]0.00628 kW/m$^2$.K |
| $U_{FL}$ | Heat transfer coefficient – poured concrete 152 mm | [2]0.000872 kW/m$^2$.K |
| $T_{OPT}$ | Temperature at max growth rate | 29°C |
| $\mu_{max}$ | Maximum growth rate | 1.44/day |
| $T_{SUP}$ | Temp. at 10% of max growth rate – upper | 35°C |
| $T_{INF}$ | Temp. at 10% of max growth rate – lower | 11°C |
| $nc$ | Number of PBR parallel columns | 19 |
| $na$ | Number of PBR columns axial | 10 |
| $d$ | PBR tube diameter | 0.03 m |
| $lt$ | PBR tube length | 10 m |
| $dt$ | Distance between PBR tube centers | 0.18 m |
| $dtt$ | Distance between PBR twin tube centers | 0.06 m |
| $b$ | Distance between ground and first PBR tube | 0.3 m |
| $x$ | Distance between PBR columns (parallel) | 1.5 m |
| $xa$ | Distance between PBR columns (axial) | 1.5 m |
| $t$ | Number of tubes per PBR column | 32 |
| $GHW$ | Greenhouse width | 7.2 m |
| $GHL$ | Greenhouse length | 57.6 m |
| $GHWH$ | Greenhouse wall height | 2.4 m |
| $GHG$ | Greenhouse gable height | 1.8 m |
| $GHN$ | Number of greenhouse bays | 12 |

[1] Duffie and Beckman, 2006;
[2] Aldrich and Bartok, 1992;
[3] Iqbal, 1983;
[4] Incropera et al., 2007.

spaces, tube diameters, lengths, and distances between the tubes provides essential information for energy studies as well as efficient space use. The best configuration for a specific location can be evaluated easily. Preliminary results proved the importance of a simulation model and the impact of a sophisticated heat management on economics. Energy savings due to an optimized energy management system will eventually increase proficiency of the systems, and support a sustainable industry to grow further. The model was exemplary tested for a hypothetical 100,000 liter PBR located in San Luis Obispo, CA, USA.

#### 9.6.3.2 *Energy Balances*

Many factors influence the heat losses and gains of tubular PBRs. The simulation model is able to determine the heating requirements for an outdoor operation as well as for an indoor (greenhouse) operation. Different heat transfer modes occur under the two different conditions. In the case of an outdoor operation, heat losses due to forced convection (wind) and radiation (ground and sky), and gains due to solar radiation are present. The PBR and ground have thermal masses; therefore, heat storage has to be considered.

Considering the heat storage term, the energy balances for the greenhouse air (Equation (9.12)) and PBR (Equation (9.13) in the greenhouse were expressed as:

$$\rho_A V_{GH} Cp_A dT_{GH}/dt = Q_{GHHG} - Q_{GHI} - Q_{GHC} \tag{9.12}$$

$$\rho_W V_{PBR} Cp_W dT_{PBR}/dt = Q_{GSRA} - Q_{GHLR} + Q_{GHCL} \tag{9.13}$$

where $\rho_A$ was the density of air (kg/m$^3$); $V_{GH}$ was the greenhouse volume (m$^3$); $Cp_W$ was the specific heat capacity of air (Wh/kg.K); d$T_{GH}$ was the greenhouse air temperature change (K) over a period of time dt (1 h); $Q_{GHHG}$ was the greenhouse heat gains due to solar radiation (kW); $Q_{GHI}$ was the heat transfer due to infiltration (kW); $Q_{GHC}$ was the convective heat transfer; $\rho_w$ was the density of the algae water (kg/m$^3$); $V_{PBR}$ was the total PBR volume (m$^3$); $Cp_W$ was the specific heat capacity of water (Wh/kg.K); d$T_{PBR}$ was the PBR temperature change (K) over a period of time dt (1 h); $Q_{GSRA}$ was the total solar radiation absorbed by the PBR (kW); $Q_{OS}$ was the long-wave radiation exchange with the sky and the ground (kW); and $Q_{GHCL}$ was the convective heat transfer.

It was assumed that the greenhouse air was well mixed and had negligible temperature gradients. Moreover, a one-dimensional heat conduction equation was used in dealing with the floor for inside and outside conditions, by dividing the floor into three layers with the assumption of homogeneous thermal and hydraulic properties within each layer (Kindelan, 1980; Yildiz, 1993). The ground layer thicknesses were 0.05, 0.10, and 0.50 m for the top, middle, and bottom layers, respectively. It was assumed that the deep ground temperature was constant at 15°C (Takakura et al., 1971). The readers of this book are referred to Mehlitz (2009) or Yildiz et al. (2013) for further details of this simulation model.

#### 9.6.3.3 *Sensitivity Analyses, Validation, and Model Performance*

After developing the simulation model, sensitivity analyses were performed on the indoor PBR system, which had a glass greenhouse glazing. The shading system was not operational. Each parameter (outside temperature, solar radiation, ground reflectivity, PBR tube distance, etc.) was varied individually while all other parameters were held at their standard values. In addition to the sensitivity analyses, responses of some key parameters to step inputs in outside air temperature and outside solar radiation were also determined. For the temperature sensitivity analysis, the solar radiation was kept constant at 600 W/m² for 9 h each day. The outside temperature was kept the same for 5 consecutive days at each temperature starting at −20°C and ending at 40°C. The greenhouse air temperature and the PBR temperature responded accordingly.

Figure 9.10 shows the behavior at a step change. It took roughly 2 days for the responding temperatures to adjust to the new climatic condition as seen in Figure 9.10. Focusing on the critical region, where the outside temperature step change took place, it was obvious to see that the maximum temperatures for the greenhouse and the PBR had a gap of about 2–3 h due to the difference in their thermal masses. The distribution of the PBR temperature as well was shown in detail. Due to shading effects within the PBR columns, the temperature rose rapidly until the critical sun angle (where tubes shade each other) and slower afterward due to less direct sun light reaching the PBR tubes. The temperature differences at the maximums were between 8°C and 10°C.

For the solar radiation sensitivity analysis, the outside air temperature was kept constant at 18°C for day and night. The solar radiation was varied in 300 W/m² increments between 0 and 1,200 W/m² for 5 days each. It was assumed that the Sun was shining at a constant radiation rate for 9 h every day. Figure 9.11 shows the behavior at a step change. The greenhouse air temperature and the PBR temperature responded accordingly, and increased after each step. Due to the difference in thermal masses, the maximum temperature of the PBR was reached about 1–2 h after the greenhouse air temperature reached its maximum. Also, the temperature variability of the PBR was due to higher thermal mass less than the one of the greenhouse temperature. The shape of the temperature distribution looked similar to the previous sensitivity analysis, and was again a result of the shading function, and therefore, of the solar angle. However, the magnitudes were different. The temperature differences at the maximums were between 5°C and 18°C, depending on the level of incoming solar radiation.

For the ground reflectivity sensitivity analysis, the original weather file for San Luis Obispo, CA, USA, for April 2007 was used. The simulations were performed for ground reflectivity values of 20%, 40%, 60%, and 80%. Associated energy requirements and

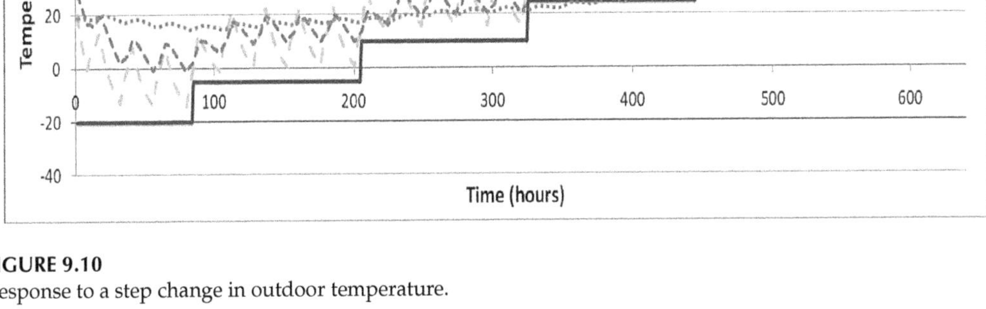


**FIGURE 9.10**
Response to a step change in outdoor temperature.

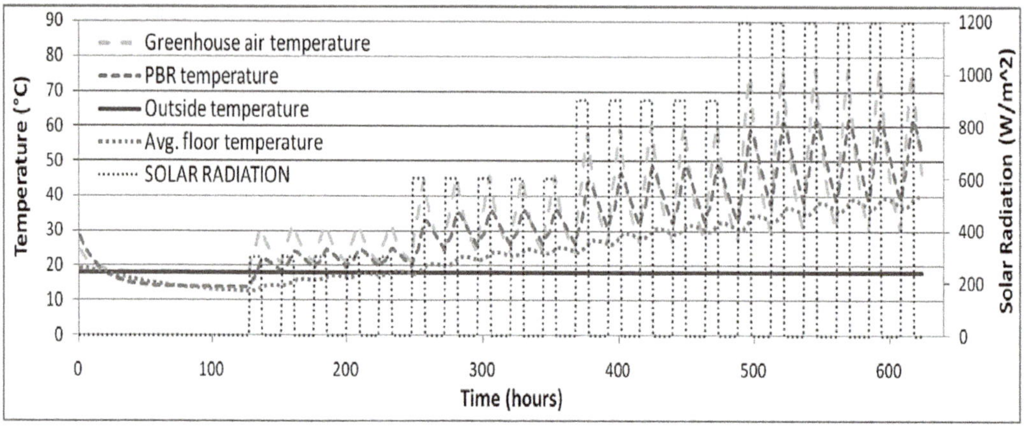


**FIGURE 9.11**
Response to a step change in solar radiation.

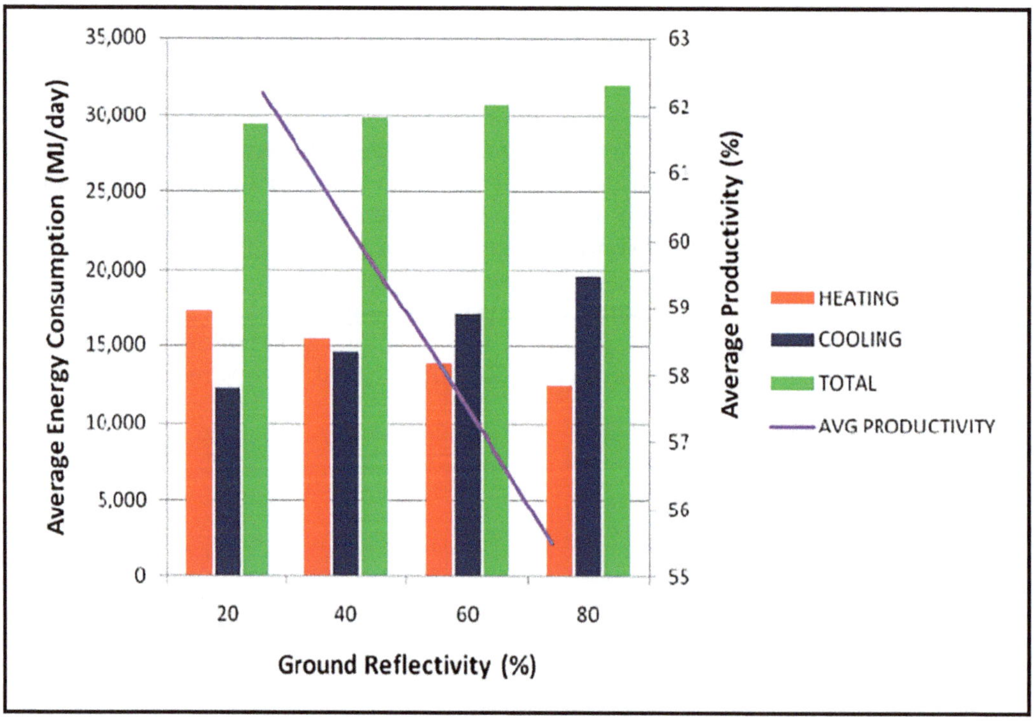


**FIGURE 9.12**
Energy requirements and average ***Chlorella*** spp. productivity as a function of ground reflectivity.

average productivities were monitored (Figure 9.12). The results showed that the higher the reflectivity was, the lower the average productivity was (a function of the PBR temperature). Also, the higher the reflectivity was, the lesser the heating and the more was the cooling required. The total energy needs for heating and cooling increased with increasing ground reflectivity.

For the PBR column distance sensitivity analysis, the original weather file for San Luis Obispo, CA, USA, for April 2007 was used. The simulations were performed for PBR column distances of 0.25, 0.5, 0.75, 1.0, 1.5, and 2.0 m. In each simulation, the greenhouse size was kept the same, which means that there was the same number of PBRs in the greenhouse, but they were closer or wider apart. The distance between the PBR columns had a major impact on the energy requirements and the average productivity (Figure 9.13). The most heating (18,331 MJ per day) was required at a distance of 0.75 m, while the least (16,474 MJ per day) was observed at 2 m distance. The most cooling (12,974 MJ per day) was required for a distance of 2 m, while the least (10,531 MJ per day) was observed at 0.75 m distance. The total energy needs increased with larger distances and peaked at the 2 m distance with a 29,448 MJ per day energy requirement. However, the algal productivity was a function of the PBR temperature and its distribution during the day. Too high and too low temperatures had direct impact on the productivity. Shading had a major impact on the temperature regime of the PBR. The distance between the PBRs influenced the shading factor as previously described. Apparently, an optimal distance between columns existed, where heat gains due to solar radiation and the shading at midday (avoids overheating) were in a most advantageous ratio. For the evaluated PBR, the best distance seemed to be about 0.75 m. However, the total productivity (biomass output and price per area) is also a measure of the price of greenhouse space. A smaller distance will eventually save space. Whether space savings outweigh the higher productivity rate or not can now be investigated.

The simulations were performed for greenhouse shading material transmissivities of 20%, 40%, 60%, and 80% as well as for a greenhouse without shading (100% transmissivity) (Figure 9.14). The transmissivity of the shading material had a major impact on the energy requirements and the average productivity. The most heating (59,678 MJ per day) was required for a transmissivity of 20%, the least (15,830 MJ per day) for 100% (without

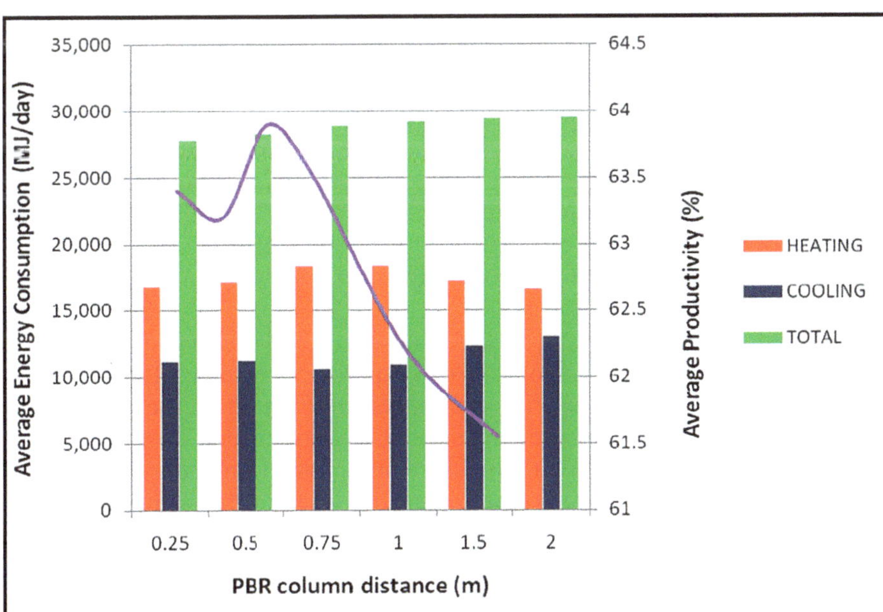


**FIGURE 9.13**
Energy requirements and average ***Chlorella*** spp. productivity as a function of PBR column spacing.

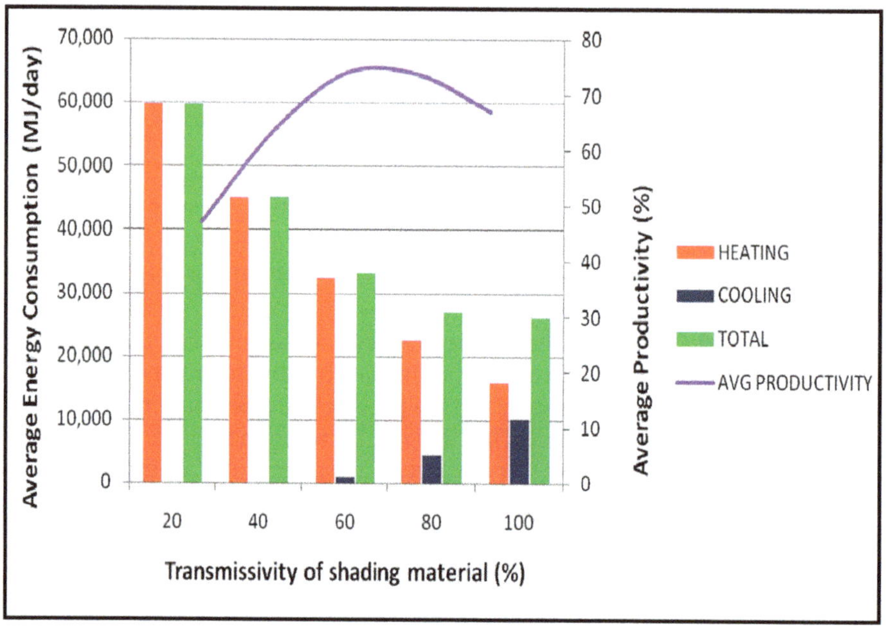


**FIGURE 9.14**
Energy requirements and average ***Chlorella*** spp. productivity as a function of shading.

shading material). The most cooling (10,243 MJ per day) was required for no shading material (100% transmissivity), the least (0 MJ per day) for a transmissivity of 20%. The total energy needs increased with lesser transmissivity and was the least for 100% transmissivity with 26,073 MJ per day. However, the *Chlorella* spp. productivity as a function of the PBR temperature and its distribution during the day could not be correlated to the energy consumption. The highest productivity (73.9%) was determined at 60% transmissivity, whereas the lowest productivity (47.1%) was determined for a transmissivity of 20%. Important for further considerations was the productivity rate of 66.9% for a transmissivity of 100% (no shading). Apparently, an optimal shading transmissivity existed, where heat gains due to solar radiation and the shading at midday (avoid overheating) were in a most advantageous ratio. The productivity increased by 7% by using a shading material with 60% transmissivity. The next question is whether more energy input (using a shading material with less transmissivity) would outweigh the higher *Chlorella* spp. productivity or not.

The developed energy-estimating model delivered good results for the temperature regime in greenhouses and PBRs, comparable to experiences made for the same location at the same time of the year on the Cal Poly campus greenhouse. From the temperature data, the heat management requirements for any specific location could be estimated. Using the model could in advance (i.e. in a planning stage of a project) precisely estimate its future heating or cooling loads, or estimate the productivity without heating or cooling. This simulation model is a fundamental tool and guidance in the economical decision-making processes related to algae growth in PBRs.

A simulation of a hypothetical tubular PBR in San Luis Obispo, CA, USA, with a volume of about 100,000 liter capacity showed the potential of the model exemplary. Actual energy loads (heating and cooling) needed to maintain the PBR at optimal temperature were determined and compared as well. With the knowledge of the actual energy requirements, PBR operators will be able to estimate their operational costs in advance and see what changes in their system would decrease the costs. In addition to the presented analyses, many other parameters could be examined. Since all greenhouse and PBR dimensions are variable, energy requirements for different greenhouse types or PBRs could be compared easily. The model could later on be extended to estimate all energy needs, for instance, the pumping energy requirements for the specific PBR, the energy and productivity of artificial lighting, or the energy required for harvesting (centrifuge, drying, ultrasonic, pressing, etc.). All operational costs could be estimated and optimized accordingly. Eventually, a complete production facility could be simulated with all inputs and outputs.

## Nomenclature

| | |
|---|---|
| *A* | area |
| ACH | air change per hour |
| *c* | specific heat |
| CDD | cooling degree-days |
| CHP | closed greenhouse heat pump system |
| CON | conventional greenhouse system |
| *E* | energy |
| HDD | heating degree days |
| HV | heating value |
| HVAC | heating, ventilation, and air-conditioning |
| *IG* | flux of solar radiation on the cover |
| *LW* | net long-wave radiation on the cover |
| *M* | moisture transfer rate |
| *N* | number |
| OHP | open greenhouse heat pump system |
| PBR | photobioreactor |
| ppm | parts per million |
| $\dot{Q}$ | heat transfer rate |
| t | time |
| *T* | temperature |
| *U* | conduction heat transmission coefficient |
| *V* | volume |
| *w* | humidity ratio |

### Greek letters

| | |
|---|---|
| $\alpha$ | absorptivity |
| $\Delta$ | difference |
| $\eta$ | efficiency |
| $\nu$ | specific volume |
| $\rho$ | density |
| $\Sigma$ | total |

**Subscripts**

| | |
|---|---|
| *a* | air |
| *ac* | from internal air to structural cover |
| *bal* | balance point |
| *c* | cover |
| *conc* | condensation/evaporation to/from the structural cover |
| *cond* | conduction |
| *coil* | cooling coil |
| *flux* | flux from the bulk air to the top layer of the canopy stand |
| *GH* | greenhouse |
| *GHC* | heat loss due to convection |
| *GHHG* | greenhouse heat gains due to solar radiation |
| *GHCL* | heat loss due to convection |
| *GHI* | greenhouse heat loss due to infiltration |
| *GSRA* | total solar radiation absorbed by the PBR in the greenhouse |
| *i* | internal |
| *inf* | infiltration/ventilation |
| *inj* | injection |
| *oc* | from internal air to structural cover |
| *OS* | radiation exchange with the sky and ground |
| *p* | pressure |
| *pbr* | photobioreactor |
| *s* | short-wave |
| *sr* | heating source |
| *t* | total |
| *TSRA* | total solar radiation |
| *vio* | ventilation |
| *w* | water |

## Bibliography

Aldrich R and Bartok JW (1992) Greenhouse Engineering. Northeast Regional Agricultural Engineering Service. Ithaca, New York, U.S.A.

Arinze EA, Schoenau GJ, Besant RW (1984) A dynamic thermal performance simulation model of an energy conserving greenhouse with thermal storage. *Transactions of the ASAE* 27:508–519.

ASHRAE (1986) Bin and degree-hour weather data for simplified energy calculations. American Society of Heating, Refrigeration and Air Conditioning Engineers, Inc. Atlanta, GA 30329.

ASHRAE (1989a) Handbook of Fundamentals. American Society of Heating, Refrigeration and Air Conditioning Engineers, Inc. Atlanta, GA 30329. pp. 24.1-24-22.

ASHRAE (1989b) Handbook of Fundamentals. American Society of Heating, Refrigeration and Air Conditioning Engineers, Inc. Atlanta, GA 30329. pp. 28.1-28-42.

ASHRAE (1997) Handbook of Fundamentals. American Society of Heating, Refrigeration and Air Conditioning Engineers, Inc. Atlanta, GA 30329.

ASHRAE (2005) Handbook of Fundamentals. American Society of Heating, Refrigeration and Air Conditioning Engineers, Inc. Atlanta, GA 30329. pp. 32.1–32.39.

Duffie JA and Beckman WA (2006) Solar Engineering of Thermal Processes. John Wiley and Sons, Inc. Hoboken, New Jersey, U.S.A.
Engineering ToolBox, (2009). US Heating Degree-Days. [online] Available at: https://www.engineeringtoolbox.com/us-degree-days-d_1569.html (Accessed 12.08.220).
Hinrichs R and Kleinbach M (2006) *Energy: Its Use and the Environment*. Thompson Higher Education. Belmont, U.S.A.
Incropera FP and DeWitt DP (1985) *Introduction to Heat Transfer*. John Wiley & Sons, Inc. New York, U.S.A.
Incropera F, Dewitt D, Bergman T, and Lavine A (2007) *Introduction to Heat Transfer*. John Wiley & Sons, Inc. New York, U.S.A.
Iqbal M (1983) *An Introduction to Solar Radiation*. Academic Pub. Toronto, Canada.
Kindelan M (1980) A dynamic modeling of greenhouse environment. *Transactions of the ASAE* 23 (5): 1232-1239.
Mehlitz T (2009) Temperature and Heat Management Requirements of Microalgae Cultivation in Photobioreactors. Master Thesis. GRIN Verlag Publishing, Germany. Archive No.: V132058. ISBN (E-book): 978-3-640-38214-9. ISBN (Book): 978-3-640-38225-5.
Mehlitz T and Yildiz I (2011) Dynamic simulation of microalgae growth and microclimate in large-scale closed photobioreactors. *Acta Hort.* 893: 413-420.
Navas LM, de la Plaza S, Garcia JL, Duran JM, and Retamal N (1998) Formulation and sensitivity analysis of a dynamic model of the greenhouse climate validation for a mild Mediterranean climate. *Acta Hort.* 456: 305-312.
NOAA (1973) *Degree-days to selected bases*. U.S. National Climatic Data Center, Asheville, NC, USA.
Rijsdijk AA and Houter G (1993) Validation of a model for energy consumption, $CO_2$ consumption and crop production (ECP-Model). *Acta Hort.* 328: 125-131.
Takakura TK, Jordan A, and Boyd LL (1971) Dynamic simulation of plant growth and environment in the greenhouse. *Transactions of the ASAE* 14 (5): 964-971.
Thornley JHM and Johnson IR (1990) *Plant and Crop Modelling. A Mathematical Approach to Plant and Crop Physiology*. Clarendon Press, Oxford, U.K. pp. 48–49 and 128.
USAF (1978) Engineering Weather Data. Air Force Manual AFM 88-29, U.S. Government Printing Office 725-243/511, July.
Yang X, Short TH, Fox RD, and Bauerle WL (1990) Dynamic modeling of the microclimate of a greenhouse cucumber row-crop. Part I. Theoretical Model. *Transactions of the ASAE* 33 (5): 1701-1709.
Yildiz I (1993) Simulation of greenhouse microclimates and environmental control strategies using a Rankine cycle heat pump. Dissertation. The Ohio State University, Columbus, Ohio, USA.
Yildiz I, Lin C, and Christensen RN (1993) A performance study of a Rankine cycle driven heat pump system. *Heat Pump and Refrigeration Systems: Design, Analysis and Applications* 29: 9.1. ASME-Publications-AES (Advanced Energy Systems).
Yildiz I and Stombaugh DP (2006a) Dynamic modeling of microclimate and environmental control strategies in a greenhouse coupled with a heat pump system. *Acta Hort.* 718: 331–340.
Yildiz I and Stombaugh DP (2006b) Simulated performances of a heat pump system for energy and water conservation in open and confined greenhouse systems. *Acta Hort.* 718: 341–350.
Yildiz I, Nguyen-Quang T, Mehlitz T and Brooker B (2013) Algae, Biofuels, and Modeling. In: Dincer, I., Colpan, C.O., and Kadioglu, F. (Eds.), *Causes, Impacts, and Solutions to Global Warming*. DOI 10.1007/978 1-45614-7588-0_3, pp.528-628. Springer: New York, USA.

# 10

# *Common Fuels, Renewable and Alternative Energy Sources*

## 10.1 Introduction

Whether it is renewable or nonrenewable, many different forms of energy sources are used in greenhouse operations around the world. Of course, it is preferable to use locally available, clean, and renewable energy sources; however, as mentioned earlier, fossil fuels have always had the major share in the global primary energy consumption. Overall fossil fuel consumption increased approximately 51% in the period of 1995–2015, and it is predicted that the consumption will increase approximately 18% more in the period of 2015–2035. It was observed that, in 1995, oil, natural gas, and coal had total primary energy consumption shares of 38%, 22%, and 26% (87% in total), respectively; that, in 2015, they had total primary energy consumption shares of 32%, 24%, and 29% (85% in total), respectively; whereas non-fossil energy consumption in 2015 included hydroelectric 6.8%, nuclear 4.4%, wind 1.4%, solar 0.45%, and other renewables amounting to 0.89% (Yildiz, 2018a). World energy consumption is growing, and it is predicted that, by 2035, the shares will be 29%, 25%, and 24% (78% in total), respectively, having natural gas second place after oil, and that a 7% decrease in fossil fuels share in consumption is predicted in the global primary energy consumption. Even though the share of fossil fuel consumption has decreased and is predicted to decrease further, they will continue to hold the major share in the primary energy mix in the foreseeable future as more unconventional fossil fuels are explored.

## 10.2 Oil

Oil is formed by the anaerobic decomposition of organic matter including phytoplankton and zooplankton, which were settled in large quantities in deposition basins such as a seabed or lakebed over geological time. This organic matter, after being mixed with mud, remained buried under increasing layers of sediment over millions of years. As a result, the organic matter and mud mixture was exposed to high pressures and temperatures over millions of years causing the organic matter to chemically change, first into kerogen (a waxy material, which is found in oil shales), and then, with more exposure to further heat, into liquid and gaseous hydrocarbons in a process known as catagenesis. Throughout all these processes, the energy density of hydrocarbons might increase; however, the origin of the energy is still the ancient photosynthesis.

Currently, the world's primary fuel source for transportation is oil, which also exists in shale and tar sands; however, most oil is pumped out of underground reservoirs. Oil is extracted, and then it is processed in refineries to create fuel oil, gasoline, liquefied petroleum gas, and other products like fertilizers, pesticides, pharmaceuticals, and plastics. However, oil causes serious environmental problems and the heavy reliance on oil particularly for transportation makes it difficult to reduce its consumption. Furthermore, the environmental degradation caused by extraction and oil spills, combustion of oil emits fine particulates, which can lead to serious health problems for humans. As we keep depleting conventional oil sources from underground reservoirs, unconventional sources such as tar sands and oil shale attract more attention. Heavier crude oils extracted from such sources result in more emissions and environmental disturbance compared to conventional oil because their extraction is involved in the use of energy intensive methods.

### 10.2.1 Sources of Oil

Oil comes from both conventional and unconventional sources. The terms conventional and unconventional sources are however not clearly defined in the literature and the definitions vary considerably. We can give a 2013 study, as an example, which predicted an early peak, excluded the oil such as deepwater oil, tight oil, and oil close to the poles, all of which were defined as non-conventional sources (White & Cordell, 2008).

Oil extracted on land and offshore using standard techniques is defined as conventional. And with respect to grade, though the exact definitions of the grades vary based on where the oil comes from, these sources are categorized as light, medium, heavy, or extra heavy. Light oil naturally flows to the surface and can simply be extracted by pumping it to the surface. Heavy oil obviously refers to the oil that does not flow easily because of its higher density. While conventional techniques are partly used for producing this heavy oil, the use of unconventional methods however proves to provide better recovery rates (Palaniappan & Gleick, 2008).

Currently, the following oil sources are considered unconventional; however, it does not mean that the following list will be considered as unconventional forever:

- Oil sands: These are unconsolidated sandstone deposits containing large amounts of viscous crude bitumen or extra-heavy crude oil. These formations are quite often called "tar sands," such as Canadian tar sands; however, the material exists there is actually not tar, it is an extra-heavy oil known as bitumen (Energyseer, 2017).
- Oil shale: This term is used as a common term for sedimentary rock such as shale or marl, which contains kerogen that is yet to be transformed into crude oil by the high pressures and temperatures.
- Tight oil: This is the oil extracted via hydraulic fracturing from low-permeability rock deposits, shale deposits, and many times from other rock types as well. It should not be confused with shale oil, which is, as mentioned above, the oil manufactured from the kerogen contained in an oil shale.
- Coal liquefaction or gas-to-liquids product: These are synthesized liquid hydrocarbons from the conversion of coal or natural gas by a number of processes such as Fischer--Tropsch, Bergius, or Karrick process.
- Minor sources: These include thermal depolymerization, which could indefinitely be used to produce oil out of waste feedstocks such as agricultural waste, garbage, and sewage.

### 10.2.2 Oil Reserves

Flows or daily production of carbon-based fossil energy sources are realized from these reserves. When we talk about oil reserves, we specifically mean the amount of crude oil, which can technically be recovered in a financially feasible manner at the present price of oil (Society of Petroleum Engineers, 2017). The total estimated amount of oil in an oil reservoir, including both producible and non-producible oil, is called oil-in-place. Due to the limitations of oil extraction technologies at the time and reservoir characteristics, not the whole but only a portion of this oil can be recovered and brought to the surface of the Earth. And this producible portion is considered as reserves. Oil resources however include all oil that can technically be recovered at any price, not only at the present price. Therefore, as it is clearly seen and understood, unlike oil resources, oil reserves vary as a function of the price. Reserves for instance typically indicate the reserves for a point of interest, such as a single oil well, a single oil reservoir, a single oil field, a single country, or the whole world. The ratio of reserves to the total amount of oil in a specific reservoir is called the recovery factor. Thence, the method of oil recovery used, the operation itself, and the relevant technological developments, all together determine the recovery factor for a given field (BP, 2017). Most early estimates of the oil field reserves are generally considered conservative and tend to grow with time (which is called reserves growth) as a function of new technological developments (Morehouse, 1997). Figure 10.1 shows the global oil reserves in billion barrels (bbl) in 2013.

Oil-producing nations do not typically reveal their engineered reservoir data due to their national security concerns, and instead, provide manipulated data for political reasons (Goodman, 2011). However, British Petroleum (BP) at the end of 2016 reported (Table 10.1) that the highest proved oil reserves (including non-conventional oil deposits) are located in Venezuela (17.6% of global reserves), Saudi Arabia (15.6% of global reserves), Canada (10% of global reserves), Iran (9.3%), and Iraq (9.0%), with the highest reserves-to-production (R/P) ratios belonging to Venezuela and Canada, 341.1 and 105.1 years, respectively, and the global ratio is 50.6 years. Organization of the Petroleum Exporting Countries (OPEC) data at the end of 2016 showed similar reserve numbers (OPEC, 2017).

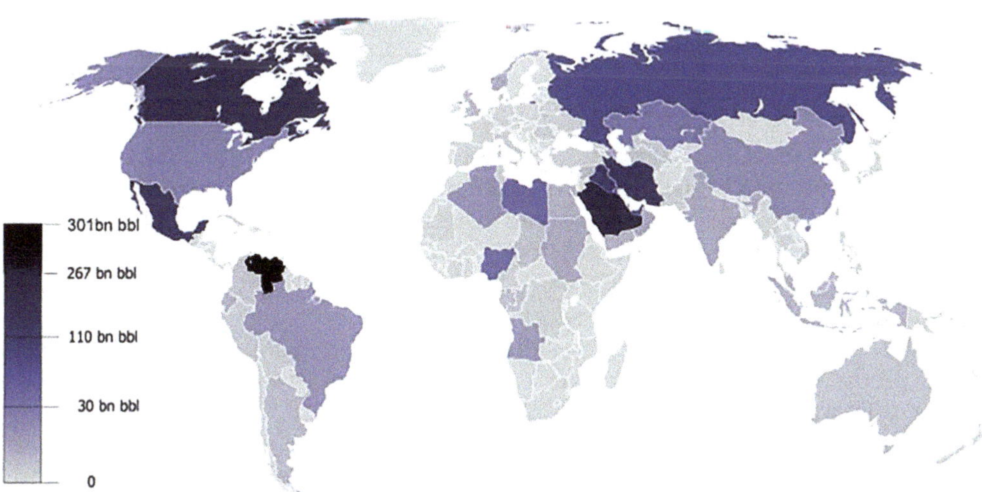


**FIGURE 10.1**
Global map of oil reserves in 2013. Source: Wikipedia (2017a).

**TABLE 10.1**

Proved Oil Reserves in Top Five Countries Between the Years 1996 and 2016 (Reserve Levels are in Thousand Million Barrels).

| COUNTRY | 1996 | 2006 | 2016 | Share of total (%) | R/P ratio |
|---|---|---|---|---|---|
| Venezuela | 72.7 | 87.3 | 300.9 | 17.6 | 341.1 |
| Saudi Arabia | 261.4 | 264.3 | 266.5 | 15.6 | 59.0 |
| Canada | 48.9 | 179.4 | 171.5 | 10.0 | 105.1 |
| Iran | 92.6 | 138.4 | 158.4 | 9.3 | 94.1 |
| Iraq | 112.0 | 115.0 | 153.0 | 9.0 | 93.6 |
| **Total World** | **1,148.8** | **1,388.3** | **1,706.7** | **100.0%** | **50.6** |

Data from BP (2017).

The oil reserves were suspected to be running out in the coming decades. However, with the help of advanced technologies, more oil reserves have been discovered and became viable. As a result, the global oil reserves increased by 60% within the past 20 years with a 25% growth in oil production (WEC, 2013), now having South and Central America with the highest (120 years) oil reserves-to-production ratios rather than the Middle East (BP, 2017). It is presently expected to expand further as unconventional oil sources are taken into account. We know that oil sands are mixture compounds that contain clay, water, sand, and bitumen. (Gosselin et al., 2010) indicated that the Alberta oil sands, for instance, have great competitiveness in Canada that accounted for approximately 50% of the total Canadian oil production in 2009. Also, there is no oil in the oil shale; however, it contains a unique material called kerogen, which is a precursor of crude oil (Allen et al., 2012). The world resources of oil shale are estimated to be around 730 trillion liters, which is reported to be four times greater than crude oil reserves currently (WEC, 2016). Both oil sands and oil shale must be refined before they can work as conventional crude oil (Khatib & Verbeek, 2003). Therefore, they not only represent an increased oil reserve, but also an increased environmental burden. The more technologies employed for extraction and processing, the more severe is the impact on air, water, land, and energy resources.

### 10.2.3 Oil Production, Demand, and Trade

Figure 10.2 shows the global distribution of oil producing nations in 2013 (Wikipedia, 2017c). BP (2017) provided a detailed report for the current and past 20 year oil production data as well as the projections for the next 20 years. Based on this report, annual production in million tonnes of oil equivalent (Mtoe), oil, coal, and natural gas provided 85.4% of primary energy production in 2015 (32.4% + 29.2% + 23.9%, respectively). Currently, oil is the world's largest primary source of energy.

Table 10.2, on the other hand, presents the regional oil productions in 1995, 2015, projected oil productions in 2035, as well as the relevant changes in regional productions. Based on this report, the Middle East, North America, and the Commonwealth of Independent States (CIS) will continue to preserve their leading shares of global oil production with further production increases through 2035, while the South and Central America sustaining its share with increased production, and other regions observing a decline in oil production. The major decline took place in Europe between 1995 and 2015, and will continue to drop further through 2035. Overall, there has been an annual increase

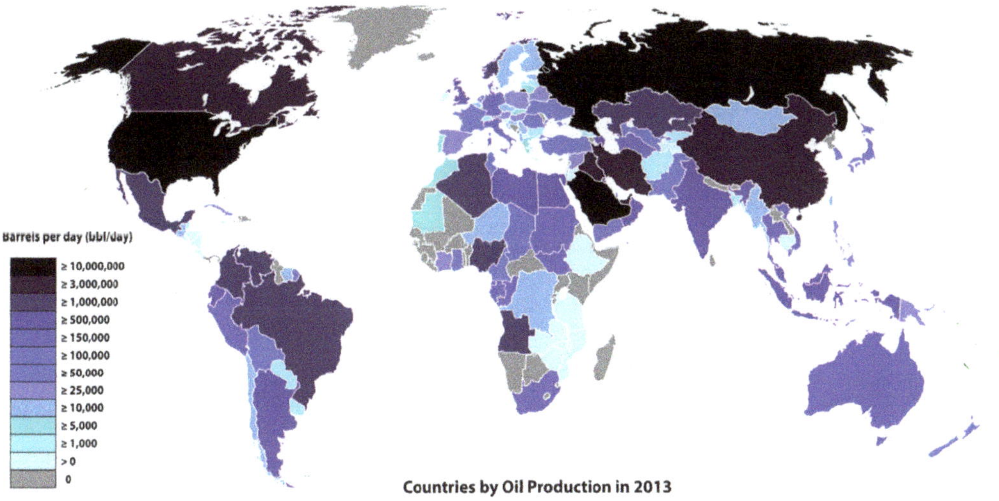


**FIGURE 10.2**
Global map of oil producing countries in 2013. Source: Wikipedia (2017b). The map is Ali Zifan's work, which was prepared using data from CIA World Factbook.

**TABLE 10.2**

Changes in Regional Oil Production for the Years 1995–2015 and 2015–2035 (Production Levels are in Million Tonnes of Oil Equivalent).

| REGION | 1995 | 1995 Total Share (%) | 2015 | 2015 Total Share (%) | 2035 | 2035 Total Share (%) | 1995–2015 Change (%) | 1995–2015 Annual Change (%) | 2015–2035 Change (%) | 2015–2035 Annual Change (%) |
|---|---|---|---|---|---|---|---|---|---|---|
| North America | 645.8 | 19.7 | 910.3 | 20.9 | 1,004.1 | 21.1 | 41.0 | 2.0 | 10.3 | 0.5 |
| S & C America | 300.1 | 9.1 | 396.0 | 9.1 | 454.0 | 9.5 | 31.9 | 1.5 | 14.7 | 0.7 |
| Europe | 311.1 | 9.5 | 164.7 | 3.8 | 66.6 | 1.4 | −47.1 | −2.2 | −59.6 | −2.8 |
| CIS | 358.3 | 10.9 | 682.0 | 15.6 | 730.2 | 15.3 | 90.4 | 4.3 | 7.1 | 0.3 |
| Middle East | 979.2 | 29.8 | 1,412.4 | 32.4 | 1,818.7 | 38.1 | 44.2 | 2.1 | 28.8 | 1.4 |
| Africa | 339.6 | 10.3 | 398.0 | 9.1 | 382.7 | 8.0 | 17.2 | 0.8 | −3.8 | −0.2 |
| Asia Pacific | 352.2 | 10.7 | 398.3 | 9.1 | 311.9 | 6.5 | 13.1 | 0.6 | −21.7 | −1.0 |
| **Total** | **3,286.3** | **100.0** | **4,361.6** | **100.0** | **4,768.1** | **100.0** | **32.7** | **1.6** | **9.3** | **0.4** |

Data from BP (2017).

(1.6%) with 32.7% total change in global oil production in the period of 1995–2015, and the annual increase is expected to happen around 0.4% with 9.3% total change for the period of 2015–2035.

Table 10.3 shows the top oil producing nations in the past 10 years with their relevant growth rates and shares of global oil production. Saudi Arabia has consistently been the top producer of oil followed by the Russian Federation and US. Iraq, on the other hand, has experienced a major increase (>8.0%) in oil production in the past 10 years and moved to the fourth place in production.

**TABLE 10.3**

Oil Production in Top Five Countries from 2006 to 2016 (Production Levels are in Million Tonnes).

| | | | | Annual Growth Rate | | |
|---|---|---|---|---|---|---|
| COUNTRY | 2006 | 2011 | 2016 | 2016 | 2005–2015 | Share of 2016 |
| Saudi Arabia | 508.9 | 525.9 | 585.7 | 2.9 | 0.4 | 13.4 |
| Russian Fed. | 485.6 | 518.8 | 554.3 | 2.2 | 1.3 | 12.6 |
| US | 304.5 | 344.9 | 543.0 | −4.2 | 6.2 | 12.4 |
| Iraq | 98.0 | 136.7 | 218.9 | 10.8 | 8.2 | 5.0 |
| Canada | 150.6 | 169.8 | 218.2 | 0.9 | 4.2 | 5.0 |
| **Total World** | **3,964.8** | **4,007.9** | **4,382.4** | **0.3** | **1.0** | **100.0** |

Data from BP (2017).

**TABLE 10.4**

Changes in Regional Oil Consumption for the Years 1995–2015 and 2015–2035 (Consumption Levels are in Million Tonnes of Oil Equivalent).

| REGION | 1995 | 1995 Total Share (%) | 2015 | 2015 Total Share (%) | 2035 | 2035 Total Share (%) | 1995–2015 Change (%) | 1995–2015 Annual Change (%) | 2015–2035 Change (%) | 2015–2035 Annual Change (%) |
|---|---|---|---|---|---|---|---|---|---|---|
| North America | 951.4 | 28.9 | 1,036.3 | 23.9 | 894.9 | 17.8 | 8.9 | 0.4 | −13.6 | −0.6 |
| S & C America | 208.9 | 6.3 | 322.7 | 7.5 | 372.2 | 7.4 | 54.5 | 2.6 | 15.3 | 0.7 |
| Europe | 738.5 | 22.4 | 673.6 | 15.6 | 552.9 | 11.0 | −8.8 | −0.4 | −17.9 | −0.9 |
| CIS | 210.4 | 6.4 | 188.5 | 4.4 | 226.5 | 4.5 | −10.4 | −0.5 | 20.1 | 1.0 |
| Middle East | 215.8 | 6.5 | 425.7 | 9.8 | 550.0 | 11.0 | 97.3 | 4.6 | 29.2 | 1.4 |
| Africa | 106.2 | 3.2 | 183.0 | 4.2 | 284.8 | 5.7 | 72.3 | 3.4 | 55.7 | 2.7 |
| Asia Pacific | 864.2 | 26.2 | 1,501.4 | 34.7 | 2,140.4 | 42.6 | 73.7 | 3.5 | 42.6 | 2.0 |
| **Total** | **3,295.5** | **100.0** | **4,331.3** | **100.0** | **5,021.8** | **100.0** | **31.4** | **1.5** | **15.9** | **0.8** |

Data from BP (2017).

Table 10.4, on the other hand, reports the regional oil consumptions in 1995 and 2015, projected oil consumption in 2035, and the relevant changes in regional consumption. The report indicates that Asia Pacific and North America will continue to preserve their leading shares of global oil consumption with the United States having some drop in consumption and Asia Pacific having a further consumption increase (42.6%) through 2035. The Middle East as well has doubled its oil consumption in the past 20 years, and is expected to increase it further through 2035, reaching a total share of 11% of the global consumption with a 29.2% change between 2015 and 2035. Europe, on the other hand, has had the third largest share in global oil consumption in the past 20 years with a drop of 8.8%; however, it is expected to drop (17.9%) further between 2015 and 2035 reaching the same level of the Middle East. The CIS experienced a 10.4% drop in oil consumption between the years of 1995 and 2015; however, it is forecasted that it will have a 20.1% increase in the next

**TABLE 10.5**

Oil Consumption in Top Five Countries Between the Years 2006 and 2016 (Production Levels are in Million Tonnes).

| COUNTRY | 2006 | 2011 | 2016 | Annual Growth Rate | | |
|---|---|---|---|---|---|---|
| | | | | 2016 | 2005–2015 | Share of 2016 |
| US | 930.7 | 834.9 | 863.1 | 0.5 | −0.9 | 19.5 |
| China | 353.1 | 465.1 | 578.7 | 2.7 | 5.5 | 13.1 |
| India | 128.3 | 163.0 | 212.7 | 8.3 | 4.9 | 4.8 |
| Japan | 238.0 | 203.7 | 183.4 | −2.8 | −2.6 | 4.2 |
| Saudi Arabia | 98.4 | 139.1 | 167.9 | 0.5 | 5.9 | 3.8 |
| **Total World** | **3,984.2** | **4,125.7** | **4,418.2** | **1.5** | **1.0** | **100.0** |

Data from BP (2017).

20 years. Overall, there has been an annual increase (1.5%) with a 31.4% total change in global oil consumption in the period of 1995–2015, and the annual increase is expected to take place around 0.8% again with a 15.9% total change in the next 20 years.

Table 10.5 presents the top oil consuming nations in the past 10 years with their relevant growth rates and shares of global oil consumption. The United States has consistently been the top consumer of oil followed by China and India (replaced Japan in 2016). India is highly dependent on imported crude oil. Japan, on the other hand, has experienced a major decline (2.6%) in annual growth of oil consumption in the past 10 years, with an annual drop of 2.8% in 2016 alone. In many other Asian countries, the oil consumption is still rising at high speeds to give access to urbanization and industrialization. The reason that oil has remained the leading energy supplier is that the consumption growth in developing economies exceeds the declined demands in the Organisation for Economic Co-operation and Development (OECD) countries (REN21, 2016; Tverberg, 2012). The OECD countries have already benefited from energy efficiency transition and upgrading strategies that can satisfy the needs (IEA, 2012b).

## 10.3 Coal

Coal is a fossil fuel, which originates from a combustible sedimentary rock occurring in rock strata in layers or veins called coal beds or coal seams. Coal forms when dead plant matter is converted into peat, which is then converted into lignite, then sub-bituminous coal, and then bituminous coal, and finally anthracite. The whole process involves biological and geological processes that take place over geological time. The harder forms (e.g. anthracite) are regarded as metamorphic rocks due to exposure to high temperatures and pressures. Coal is primarily composed of carbon, along with variable quantities of other elements, mainly hydrogen, sulfur, oxygen, and nitrogen (Blander, 2017).

Coal is now primarily burned for the production of electricity and heat (responsible for 39% of the electric power supply in the United States only in 2014 (EESI, 2017), and also used for industrial purposes, such as refining metals. Coal is the largest source of energy for the generation of electricity and $CO_2$ emissions globally. The extraction of coal, its use, and by-products all have environmental and health effects including climate change.

### 10.3.1 Sources of Coal

At a number of different times in the geologic past, the Earth had quite dense forests (USDE, 2017) in wetland areas. These forests were buried under the soil in time due to a number of natural processes including flooding. And they were compressed further as more soil was deposited over. As the dead plant material moved deeper in the Earth due to further deposition of soil, the temperature and pressure increased as a result. It is considered that the mud or acidic conditions preserved the buried dead plant matter from biological degradation and oxidation, and eventually the dead material was converted to coal through slow carbonization process under high pressures and high temperatures. And as mentioned earlier, under proper conditions, as ancient plant material are exposed to high temperatures and pressures over geological time, metamorphic grade of the dead material increases successively into peat, sub-bituminous, bituminous, steam coal, anthracite, and graphite.

### 10.3.2 Coal Reserves

Coal, of the three fossil fuels, has the most widely distributed reserves. It is mined in many different countries (over 100) and on all continents except Antarctica. The 948 billion short tons of recoverable coal reserves estimated by the US Energy Information Administration (EIA) are equal to about 4,196 billion barrels of oil equivalent (EIA, 2012). British Petroleum, in its 2017 report, estimated that there was 153 years reserves-to-production ratio based on proved coal reserves worldwide. Other researchers predicted that the deposits can last more than 250 years, which is considerably longer than those of oil and natural gas reserves (Ritchie & Dowlatabadi, 2015). Note that this figure only includes reserves classified as proved; explorations by mining companies, specifically in underexplored areas, are providing new reserves of coal every passing year. The largest reserves are found in the United States, China, Russian Federation, Australia, and India (Figure 10.3 and Table 10.6).

North America has the highest (>350 years) coal reserves-to-production ratio (BP, 2017); however, the reserves are expected to expand further as coal resources in underexplored regions are taking part in the overall picture.

### 10.3.3 Coal Production, Demand, and Trade

Following oil (32%), coal with a share of 29% is currently the world's second largest source of primary energy. Currently, about 40% of global electricity production is contributed by coal, but this number is expected to decline due to increased shares from other renewable energy sources and natural gas.

Table 10.7 shows the regional coal productions in 1995 and 2015, projected coal production in 2035, and the relevant changes in regional production. The main changes have been and will be in Asia Pacific, the South and Central America, and Africa, while there is a major decline in coal production in Europe and North America. Coal production in the Middle East has always been negligible. Based on this report, the Asia Pacific, North America, and the CIS will continue to preserve their leading shares of global coal production having Asia Pacific and the CIS with further production increases through 2035, while North America decreasing its share with further reduction in coal production. The major

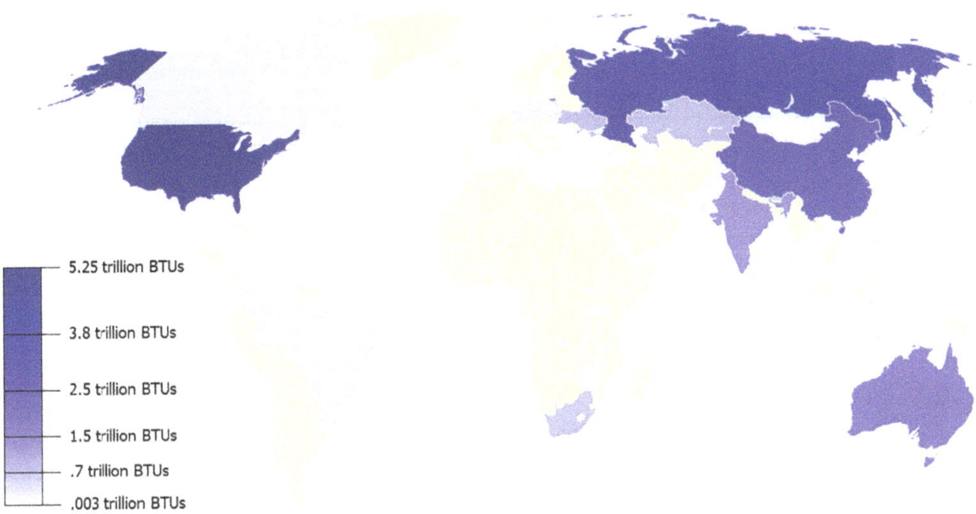


**FIGURE 10.3**
Global map of coal reserves in 2007. Source: BP (2010).

**TABLE 10.6**
Proved Coal Reserves in Top Five Countries in 2016 (Reserve Levels are in Million Tonnes).

| COUNTRY | Anthracite & bituminous | Sub-bituminous & lignite | Total | Share of total (%) | R/P ratio |
|---|---|---|---|---|---|
| US | 221,400 | 30,182 | 251,582 | 22.1 | 381 |
| China | 230,004 | 14,006 | 244,010 | 21.4 | 72 |
| Russian Federation | 69,634 | 90,730 | 160,364 | 14.1 | 417 |
| Australia | 68,310 | 76,508 | 144,818 | 12.7 | 294 |
| India | 89,782 | 4,987 | 94,769 | 8.3 | 137 |
| **Total World** | **816,214** | **323,117** | **1,139,331** | **100.0** | **153** |

Data from BP (2017).

decline took place in Europe and North America and will continue to drop further through 2035. Overall, there has been an annual increase (3.3%) with 69.3% total increase in global coal production in the period of 1995–2015, and the annual increase is expected to occur around 0.4% with 8.4% total change in the next 20 years.

Table 10.8 presents the top coal producing nations in the past 10 years with their relevant growth rates and shares of global coal production. China and the United States have consistently been the top producers of coal. China alone, in 2016, produced 1,685.7 Mtoe coal, 44.1% of global coal production. Other major producers in the same year were the United States – Wyoming, West Virginia, Kentucky, Pennsylvania, and Texas – leading in production, Australia, and India.

Table 10.9, on the other hand, reports the regional coal consumptions in 1995 and 2015, projected coal consumption in 2035, and the relevant changes in regional consumptions. As in the past 20 years, Asia Pacific, North America, and the CIS will continue to preserve their leading shares of global coal consumption with North America having some drop in

**TABLE 10.7**

Changes in Regional Coal Production for the Years 1995–2015 and 2015–2035 (Production Levels are in Million Tonnes of Oil Equivalent).

| REGION | 1995 | 1995 Total Share (%) | 2015 | 2015 Total Share (%) | 2035 | 2035 Total Share (%) | 1995–2015 Change (%) | 1995–2015 Annual Change (%) | 2015–2035 Change (%) | 2015–2035 Annual Change (%) |
|---|---|---|---|---|---|---|---|---|---|---|
| North America | 602.3 | 26.6 | 494.3 | 12.9 | 307.0 | 7.4 | −17.9 | −0.9 | −37.9 | −1.8 |
| S & C America | 23.5 | 1.0 | 61.3 | 1.6 | 81.3 | 2.0 | 161.2 | 7.7 | 32.5 | 1.5 |
| Europe | 304.6 | 13.5 | 170.4 | 4.4 | 66.2 | 1.6 | −44.1 | −2.1 | −61.1 | −2.9 |
| CIS | 199.7 | 8.8 | 249.4 | 6.5 | 291.4 | 7.0 | 24.9 | 1.2 | 16.8 | 0.8 |
| Middle East | 0.7 | 0.0 | 0.7 | 0.0 | 0.6 | 0.0 | 0.2 | 0.0 | −18.2 | −0.9 |
| Africa | 121.7 | 5.4 | 151.4 | 4.0 | 186.9 | 4.5 | 24.4 | 1.2 | 23.4 | 1.1 |
| Asia Pacific | 1,009.8 | 44.6 | 2,702.6 | 70.6 | 3,217.5 | 77.5 | 167.6 | 8.0 | 19.1 | 0.9 |
| **Total** | **2,262.2** | **100.0** | **3,830.1** | **100.0** | **4,150.9** | **100.0** | **69.3** | **3.3** | **8.4** | **0.4** |

Data from BP (2017).

**TABLE 10.8**

Coal Production in Top Five Countries Between the Years 2006 and 2016 (Production Levels are in Million Tonnes of Oil Equivalent).

| | | | | Annual Growth Rate | | |
|---|---|---|---|---|---|---|
| COUNTRY | 2006 | 2011 | 2016 | 2016 | 2005–2015 | Share of 2016 |
| China | 1,329.4 | 1,851.7 | 1,685.7 | −7.9 | 3.9 | 46.1 |
| US | 595.1 | 556.1 | 364.8 | −19.0 | −2.5 | 10.0 |
| Australia | 220.4 | 245.1 | 299.3 | −2.4 | 3.6 | 8.2 |
| India | 198.2 | 255.0 | 280.9 | 2.4 | 4.0 | 7.9 |
| Indonesia | 114.2 | 208.2 | 255.7 | −6.2 | 11.7 | 7.0 |
| **Total World** | **3,194.7** | **3,897.3** | **3,656.4** | **−6.2** | **2.5** | **100.0** |

Data from BP (2017).

consumption, and Asia Pacific having a further consumption increase. The CIS has maintained its share and is expected to continue to do so in the next 20 years. Europe, on the other hand, had the third largest share (13.5%) in global coal consumption in 1995; however, the share dropped to 4.4% in 2015 moving it to the fourth place; and it is predicted to drop further in the period of 2015–2035 reaching 1.6% only. Overall, there has been an annual increase (3.3%) with 69.3% total change in global coal consumption between the years 1995 and 2015, and the annual increase is expected to be around 0.4% with a total change of 8.4% in the next 20 years.

Table 10.10 presents the top coal consuming nations in the past 10 years with their relevant growth rates and shares of global coal consumption. Based on this report, China has consistently been the top consumer of coal followed by India (recently changed the place with the United States), and the United States, with their shares of 2016 as 50.6%, 11.0%,

**TABLE 10.9**

Changes in Regional Coal Consumption for the Years 1995–2015 and 2015–2035 (Consumption Levels are in Million Tonnes of Oil Equivalent).

| REGION | 1995 | 1995 Total Share (%) | 2015 | 2015 Total Share (%) | 2035 | 2035 Total Share (%) | 1995–2015 Change (%) | 1995–2015 Annual Change (%) | 2015–2035 Change (%) | 2015–2035 Annual Change (%) |
|---|---|---|---|---|---|---|---|---|---|---|
| North America | 602.3 | 26.6 | 494.3 | 12.9 | 307.0 | 7.4 | −17.9 | −0.9 | −37.9 | −1.8 |
| S & C America | 23.5 | 1.0 | 61.3 | 1.6 | 81.3 | 2.0 | 161.2 | 7.7 | 32.5 | 1.5 |
| Europe | 304.6 | 13.5 | 170.4 | 4.4 | 66.2 | 1.6 | −44.1 | −2.1 | −61.1 | −2.9 |
| CIS | 199.7 | 8.8 | 249.4 | 6.5 | 291.4 | 7.0 | 24.9 | 1.2 | 16.8 | 0.8 |
| Middle East | 0.7 | 0.0 | 0.7 | 0.0 | 0.6 | 0.0 | 0.2 | 0.0 | −18.2 | −0.9 |
| Africa | 121.7 | 5.4 | 151.4 | 4.0 | 186.9 | 4.5 | 24.4 | 1.2 | 23.4 | 1.1 |
| Asia Pacific | 1,009.8 | 44.6 | 2,702.6 | 70.6 | 3,217.5 | 77.5 | 167.6 | 8.0 | 19.1 | 0.9 |
| **Total** | **2,262.2** | **100.0** | **3,830.1** | **100.0** | **4,150.9** | **100.0** | **69.3** | **3.3** | **8.4** | **0.4** |

Data from BP (2017).

**TABLE 10.10**

Coal Consumption in Top Five Countries Between the Years 2006 and 2016 (Production Levels are in Million Tonnes).

| COUNTRY | 2006 | 2011 | 2016 | Annual Growth Rate 2016 | Annual Growth Rate 2005–2015 | Share of 2016 |
|---|---|---|---|---|---|---|
| China | 1,454.7 | 1,903.9 | 1,887.6 | −1.6 | 3.7 | 50.6 |
| India | 219.4 | 304.8 | 411.9 | 3.6 | 6.5 | 11.0 |
| US | 565.7 | 495.4 | 358.4 | −8.8 | −3.8 | 9.6 |
| Japan | 112.3 | 109.6 | 119.9 | −0.2 | 0.5 | 3.2 |
| Russian Federation | 97.0 | 94.0 | 87.3 | −5.5 | −0.3 | 2.3 |
| **Total World** | **3,293.9** | **3,807.2** | **3,732.0** | **−1.7** | **1.9** | **100.0** |

Data from BP (2017).

and 9.6%, respectively. It should be noted that China is burning almost half of the global coal production (more than what it produces) and the trend is expected to continue due to the dramatic growth in energy demand. In many other Asian countries, the coal consumption is still increasing at high speeds to give access to urbanization and industrialization.

## 10.4 Natural Gas

Natural gas is a hydrocarbon gas mixture mainly consisting of methane ($CH_4$), varying amounts of other higher alkanes, and sometimes a small portion of carbon dioxide, nitrogen, hydrogen sulfide, and helium. Natural gas forms when layers of decomposing biogenic material are exposed to very high temperatures and pressures under the surface of the Earth over millions of years. It is often referred to as "gas" when compared to other energy forms such as oil or coal; and is a non-renewable fossil fuel resource that is used as a source of energy for heating, cooking, and electricity generation, as fuel for vehicles, and as a chemical feedstock for manufacturing fertilizer, plastics, paints, and other chemicals. Natural gas comprised 24% of world's primary energy use in 2015 (BP, 2017) and is most commonly used to produce heat or electricity for buildings or industrial processes. It is typically transported by pipeline; however, it is increasingly being transported by ships in a liquefied form (liquefied natural gas [LNG]).

Natural gas usually contains a significant amount of ethane, propane, butane, and pentane, which are removed for commercial use prior to the methane being sold as a fuel or chemical feedstock. Non-hydrocarbons such as carbon dioxide, nitrogen, helium, and hydrogen sulfide are also removed before the natural gas is transported. In petroleum production, natural gas is often burnt as flare gas, which is estimated by the World Bank that over 150 $km^3$ of natural gas is flared or vented annually (World Bank, 2017).

Natural gas, which is primarily composed of methane, is also generated by the decomposition of municipal waste in landfills and manure from livestock production. Methane is a greenhouse gas that is approx. 22 times more potent than carbon dioxide. Capturing and burning the gas to produce usable heat and power prevents the methane from being released from the landfill or feedlot into the atmosphere directly. Before this type of natural gas can be used as a fuel, most of the time, but not always, is processed to remove impurities such as water vapor to meet the specifications of marketable natural gas. The by-products of such processing include ethane, propane, butanes, pentanes, and higher molecular weight hydrocarbons, hydrogen sulfide, carbon dioxide, water vapor, and sometimes helium and nitrogen (EESI, 2017).

### 10.4.1 Sources of Natural Gas

Natural gas posed a disposal problem in the active oil fields, and was burned off at oil fields. However, unwanted gas associated with oil extraction today is often returned to the reservoir with injection wells and wait for a possible market in future. When it becomes economically feasible to transport natural gas from a well site to an end user, pipelines are built in high natural gas demand regions, and mainly used in power generation, while other uses for natural gas include exporting as LNG or conversion of natural gas into other liquid products such as gasoline, diesel, or jet fuel. Gas liquefaction is one of the advanced technologies that make natural gas transportation more convenient. Pollutants such as heavy metals, $CO_2$, and $H_2S$ need to be removed, and then the natural gas is compressed into LNG through refrigeration processing (EIA, 2016; Ebenezer & Gudmunsson, 2005). Nowadays, more and more countries are using LNG and the market is expected to expand in the near future (Jalilvand, 2013). Many studies predict that natural gas will account for a larger portion of electricity generation and heat in the future (Kanellos, 2017).

### 10.4.2 Natural Gas Reserves

Figure 10.4 shows the global map of natural gas reserves in 2014. The gas reserves are broadly, however, unevenly distributed around the world. And there is no agreement on which country has the largest proved natural gas reserves. The US CIA (47,600 $km^3$) (CIA, 2013), the US Energy Information Administration (47,800 $km^3$) (EIA, 2013), and OPEC (48,700 $km^3$) (OPEC, 2013), consider that Russia has the largest proved reserves. However,

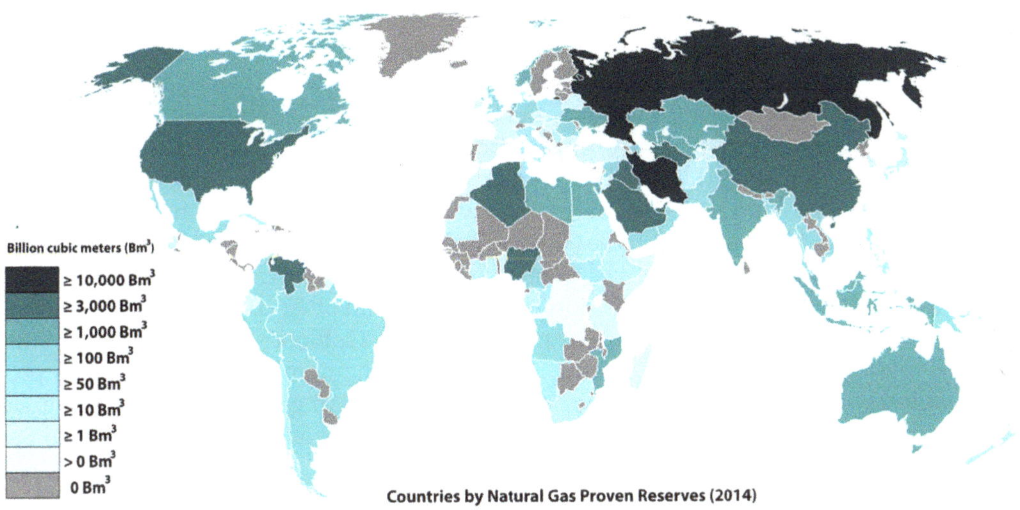


**FIGURE 10.4**
Global map of natural gas reserves in 2014. Source: Wikipedia (2017d). The map is Ali Zifan's work, which was prepared using data from CIA World Factbook.

**TABLE 10.11**
Proved Natural Gas Reserves in Top Five Countries Between the Years 1996 And 2016 (Reserve Levels are in Trillion Cubic Meters).

| COUNTRY | 1996 | 2016 | Share of total (%) | R/P ratio |
|---|---|---|---|---|
| Iran | 23.0 | 33.5 | 18.0 | 165.5 |
| Russian Federation | 30.9 | 32.3 | 17.3 | 55.7 |
| Qatar | 8.5 | 24.3 | 13.0 | 134.1 |
| Turkmenistan | n/a | 17.5 | 9.4 | 261.7 |
| US | 4.7 | 8.7 | 4.7 | 11.6 |
| **Total World** | **123.5** | **186.6** | **100.0** | **52.5** |

Data from BP (2017).

BP credits Russia with only 32,900 $km^3$, which would put it in the second place, just behind Iran (33,100–33,800 $km^3$, depending on the source; Table 10.11). Major proven resources (in $km^3$) at the end of 2016 are world 186,600, Iran 33,500 (18.0% of total world), Russian Federation 32,300 (17.3%), Qatar 24,300 (13.0%), Turkmenistan 17,500 (9.4%), and the United States 8,700 (4.7%) (Table 10.11). The world's largest gas field is the offshore South Pars/North Dome Gas-Condensate field, shared between Iran and Qatar. It is estimated to have 51,000 $km^3$ of natural gas and 50 billion barrels (6 $km^3$) of natural gas condensates. In 2015, natural gas had already accounted for 21.5% of global power generation with the largest growth in total primary energy. This number is expected to maintain its share through 2035 (BP, 2017). Moreover, natural gas is regarded as the cleanest energy resource among the three types of fossil fuels. According to BP (2017), with the development of exploitation technology, the natural gas reserves have intensively increased by 51.1% over the past 20 years.

The Middle East having the highest natural gas reserves-to-production ratios (>120 years) (BP, 2017), it is presently expected to expand further as unconventional natural gas sources are taking part in the overall picture. There are three main types of unconventional natural gas, including coalbed methane, tight natural gas, and shale gas (Hoyos, 2007). Currently, the global resources of unconventional natural gas reservoirs have not been fully calculated, so it is hard to determine their total proved volumes. However, it is estimated that there are about 900,000 $km^3$ of unconventional natural gas such as shale gas, of which 180,000 $km^3$ may be recoverable (Knight, 2010). Shale gas has become a major source of natural gas in the United States and Canada since 2000 (Mouawad, 2009). After the success in the United States, shale gas exploration has begun in other countries such as China, Poland, and South Africa (Financial Times, 2012; Yang, 2012; Bloomberg, 2012). Due to this increase in shale production, the United States is now the number one natural gas producer in the world (Beschloss, 2014). Unconventional natural gas resources have a great prospect in natural gas markets because of limited resources of conventional natural gas (EIA, 2010; Jackson et al., 2011).

### 10.4.3 Natural Gas Production, Demand, and Trade

According to the data provided by BP released in 2017 (BP, 2017), following oil (32%) and coal (29%), natural gas was the world's third largest (24%) source of primary energy in 2015. However, the share of natural gas, with a projected change of 38% between the years 2015 and 2035, is expected to rise to 25%, making natural gas the world's second largest source of primary energy in 2035.

Figure 10.5 shows the global distribution of natural gas producing nations in 2014. According to BP Energy Outlook 2017, as mentioned earlier, annual production of oil, coal, and natural gas in Mtoe provided 85.4% of primary energy production in 2015 (32.4% + 29.2% + 23.9%, respectively).

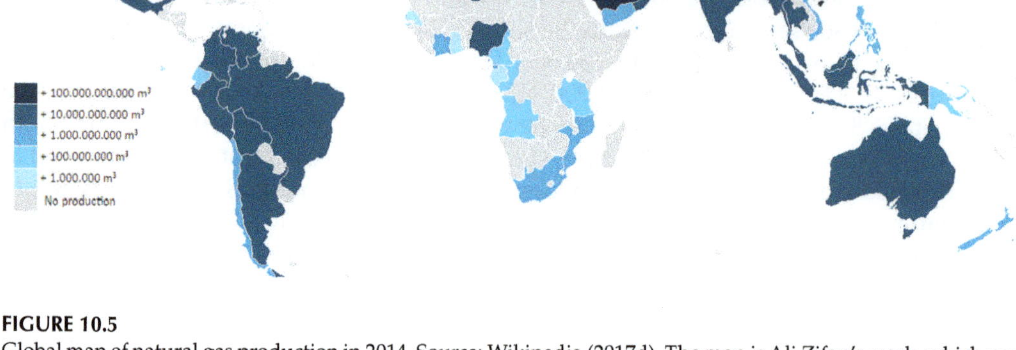


**FIGURE 10.5**
Global map of natural gas production in 2014. Source: Wikipedia (2017d). The map is Ali Zifan's work, which was prepared using data from CIA World Factbook.

Table 10.12, on the other hand, shows the regional natural gas productions in 1995 and 2015, projected natural gas production in 2035, and the relevant changes in regional production. The main changes have been and will be in the Middle East, Asia Pacific, and North America, while there is a major decline in natural gas production in Europe. North America, the CIS, and the Middle East will continue to preserve their leading shares of global natural gas production, having Asia Pacific with further production increases through 2035, while Europe decreasing its share with major reduction in natural gas production. Overall, there has been an annual increase (3.3%) with a total increase of 67.9% in global natural gas production in the period of 1995–2015, and the annual increase is expected to occur around 1.7% with a total change of 35.1% in the period of 2015–2035.

Table 10.13 presents the top natural gas producing nations in the past 10 years with their relevant growth rates and shares of global natural gas production. The United States, Russian Federation, and Iran (after replacing Canada) have consistently been the top

**TABLE 10.12**

Changes in Regional Natural Gas Production for the Years 1995–2015 and 2015–2035 (Production Levels are in Million Tonnes of Oil Equivalent).

| REGION | 1995 | 1995 Total Share (%) | 2015 | 2015 Total Share (%) | 2035 | 2035 Total Share (%) | 1995–2015 Change (%) | 1995–2015 Annual Change (%) | 2015–2035 Change (%) | 2015–2035 Annual Change (%) |
|---|---|---|---|---|---|---|---|---|---|---|
| North America | 651.7 | 34.2 | 900.4 | 28.1 | 1,330.7 | 30.8 | 38.2 | 1.8 | 47.8 | 2.3 |
| S & C America | 68.1 | 3.6 | 160.6 | 5.0 | 168.8 | 3.9 | 135.9 | 6.5 | 5.1 | 0.2 |
| Europe | 218.6 | 11.5 | 214.4 | 6.7 | 111.6 | 2.6 | −1.9 | −0.1 | −47.9 | −2.3 |
| CIS | 569.1 | 29.9 | 676.5 | 21.1 | 881.4 | 20.4 | 18.9 | 0.9 | 30.3 | 1.4 |
| Middle East | 134.1 | 7.0 | 556.1 | 17.4 | 792.8 | 18.3 | 314.5 | 15.0 | 42.6 | 2.0 |
| Africa | 76.8 | 4.0 | 190.6 | 6.0 | 280.3 | 6.5 | 148.3 | 7.1 | 47.0 | 2.2 |
| Asia Pacific | 187.4 | 9.8 | 501.0 | 15.7 | 756.0 | 17.5 | 167.4 | 8.0 | 50.9 | 2.4 |
| **Total** | **1,905.7** | **100.0** | **3,199.5** | **100.0** | **4,321.5** | **100.0** | **67.9** | **3.2** | **35.1** | **1.7** |

Data from BP (2017).

**TABLE 10.13**

Natural Gas Production in Top Five Countries Between the Years 2006 and 2016 (Production Levels are in Billion Cubic Meters).

| | | | | Annual Growth Rate | | |
|---|---|---|---|---|---|---|
| COUNTRY | 2006 | 2011 | 2016 | 2016 | 2005–2015 | Share of 2016 |
| US | 520.0 | 648.5 | 749.2 | −2.5 | 4.1 | 21.1 |
| Russian Federation | 595.2 | 607.0 | 579.4 | 0.5 | −0.1 | 16.3 |
| Iran | 111.5 | 159.9 | 202.4 | 6.6 | 6.4 | 5.7 |
| Qatar | 50.7 | 145.3 | 181.2 | 1.3 | 14.6 | 5.1 |
| Canada | 171.7 | 144.4 | 152.0 | 1.7 | −1.3 | 4.3 |
| **Total World** | **2,876.7** | **3,290.2** | **3,551.6** | **0.3** | **2.4** | **100.0** |

Data from BP (2017).

producers of natural gas. The United States alone, in 2016, produced 749.2 billion cubic meters of natural gas, 21.1% of 3,551.6 billion cubic meters of world natural gas production. Other major producers in the same year were Russian Federation with 579.4 billion cubic meters of production (16.3% of the world's total), Iran with 202.4 billion cubic meters (5.7% of the world's total), Qatar with 181.2 billion cubic meters (5.1% of the world's total), and Canada with a production of 152.0 billion cubic meters (4.3% of the world's total).

Table 10.14, on the other hand, reports the regional natural gas consumptions in 1995 and 2015, projected natural gas consumption in 2035, and the relevant changes in regional consumptions. North America and the CIS (although dropped from 24.5% in 2005 to 15.7% in 2015) continued to preserve their leading shares of global natural gas consumption in the past 20 years, with Europe having some drop in consumption (18.2% in 1995 to 13.1% in 2015) and Asia Pacific having a further consumption increase (9.9% in 2005 to 20.1% in 2015). Europe had the third largest share in global natural consumption in 1995; however, the share dropped in 2015 moving it to the fifth place, and is expected to drop further between 2015 and 2035 reaching 11.4%. It is also expected that North America (with a total share of 26.0%) and Pacific Asia (with 25.9%) will maintain their leading consumptions, while the Middle East (with 15.8%) taking the third place and replacing the CIS (with 11.5%) by 2035. Overall, an annual increase in global natural gas consumption (3.0%) with a 63.0% total change in consumption was observed between the years 1995 and 2015, and the annual increase is expected to take place around 1.8% with a 37.7% total change in the next 20 years.

Table 10.15 presents the top natural gas consuming nations in the past 10 years with their relevant growth rates and shares of global oil consumption. The United States has consistently been the top consumer of natural gas followed by Russian Federation, and China, with their shares of 2016 as 22.0%, 11.0%, and 5.9%, respectively. It should be noted that the United States is burning almost quarter of the global natural gas production (more than what it produces) and the trend is expected to continue due to its high energy demand.

**TABLE 10.14**

Changes in Regional Natural Gas Consumption for the Years 1995–2015 and 2015–2035 (Consumption Levels are in Million Tonnes of Oil Equivalent).

| REGION | 1995 | 1995 Total Share (%) | 2015 | 2015 Total Share (%) | 2035 | 2035 Total Share (%) | 1995–2015 Change (%) | 1995–2015 Annual Change (%) | 2015–2035 Change (%) | 2015–2035 Annual Change (%) |
|---|---|---|---|---|---|---|---|---|---|---|
| North America | 673.8 | 35.0 | 880.7 | 28.1 | 1,123.6 | 26.0 | 30.7 | 1.5 | 27.6 | 1.3 |
| S & C America | 67.7 | 3.5 | 157.3 | 5.0 | 186.5 | 4.3 | 132.4 | 6.3 | 18.6 | 0.9 |
| Europe | 350.6 | 18.2 | 412.2 | 13.1 | 492.0 | 11.4 | 17.6 | 0.8 | 19.4 | 0.9 |
| CIS | 472.2 | 24.5 | 490.9 | 15.7 | 494.6 | 11.5 | 3.9 | 0.2 | 0.8 | 0.0 |
| Middle East | 126.9 | 6.6 | 441.2 | 14.1 | 682.7 | 15.8 | 247.8 | 11.8 | 54.8 | 2.6 |
| Africa | 42.7 | 2.2 | 121.9 | 3.9 | 220.0 | 5.1 | 185.4 | 8.8 | 80.4 | 3.8 |
| Asia Pacific | 189.9 | 9.9 | 631.0 | 20.1 | 1,119.1 | 25.9 | 232.3 | 11.1 | 77.3 | 3.7 |
| **Total** | **1,923.8** | **100.0** | **3,135.2** | **100.0** | **4,318.5** | **100.0** | **63.0** | **3.0** | **37.7** | **1.8** |

Data from BP (2017).

**TABLE 10.15**

Natural Gas Consumption in Top Five Countries Between the Years 2006 and 2016 (Production Levels are in Million Tonnes).

| | | | | Annual Growth Rate | | |
|---|---|---|---|---|---|---|
| COUNTRY | 2006 | 2011 | 2016 | 2016 | 2005–2015 | Share of 2016 |
| US | 614.4 | 683.1 | 778.6 | 0.4 | 2.2 | 22.0 |
| Russian Federation | 415.0 | 424.6 | 390.9 | −3.2 | 0.2 | 11.0 |
| China | 59.3 | 137.1 | 210.3 | 7.7 | 15.0 | 5.9 |
| Iran | 112.0 | 162.2 | 200.8 | 5.0 | 6.4 | 5.7 |
| Saudi Arabia | 73.5 | 92.3 | 109.4 | 4.4 | 3.9 | 3.1 |
| **Total World** | **2,850.6** | **3,245.9** | **3,542.9** | **1.5** | **2.3** | **100.0** |

Data from BP (2017).

## 10.5 Nuclear Energy

While there are a number of different nuclear power generation designs the method in which power is generated is largely the same. Heavy and nearly unstable elements such as Uranium 235 are impacted with neutrons. These neutrons cause the uranium to fall out of stability and split, releasing energy in the form of radiation. This radiation heats water in the core whose steam is then used to turn a turbine. The condensed water is then cycled back through the system to the core. This controlled nuclear chain reaction is known as nuclear fission (Abu-Khader, 2009).

The share of nuclear power in global electricity generation is expected to grow from 2.3 trillion kWh in 2012 to 3.1 trillion kWh in 2020 and continue to expand to 4.5 trillion kWh by 2040 (EIA, 2016). Currently, nuclear power represents the fastest growing energy sources among all the renewables, with the consumption rate rising by 2.3% per year over the period from 2012 to 2040 (WEC, 2016). According to WEC, uranium is the main fuel source of nuclear reactors, and the demand grew by 40% from 2004 to 2013 (WEC, 2016). By the end of 2015, the International Atomic Energy Agency (IAEA) announced that 65 new nuclear reactors were under construction and 40 of them were under establishment in China, India, Russia, and South Korea (Schneider & Froggatt, 2012). International Energy Outlook 2016 reported that almost all the projected nuclear power plants are concentrated in non-OECD countries. China is attributed as having the primary increased nuclear capacity, with approximately 139 GW increment from 2012 to 2040. In contrast, even though South Korea shows a steady increase of nuclear capacity, other OECD countries such as Japan, Europe, and Canada tend to have reduced the budget for nuclear power plants (EIA, 2016). The Fukushima Daiichi nuclear disaster occurred in March 2011 and has had prolonged impacts on the global energy security. This nuclear accident probably hinders the expansion of nuclear power to some extent. What is more, in many OECD countries, the competitiveness of nuclear power has declined, and has been replaced by natural gas to fill the share of the market (WEC, 2013).

The cooling system in nuclear power plants consumes a vast amount of water, which is estimated about 3 billion liters volume of water per day for a large nuclear plant. Generally, the water used for cooling is discharged back to the adjacent aquatic system, and water temperatures can be elevated by 8°C depending on the operations and settings (Laws, 1993). For nuclear power plants that are located in tropical regions, the water temperatures might increase to 30°C or even higher in summer time (Teixeira et al., 2009; Teixeira et al., 2012). As a result, biodiversity and population dynamics in adjacent waters eventually change due to thermal pollution (Teixeira et al., 2009). The waste heat discharged by cooling waters provides great opportunities for greenhouse heating and moisture control within crop canopy.

On the other hand, Generation IV nuclear reactors are proposed nuclear systems, which take into account at least one of the following considerations: increased efficiency, generation and capture of process heat to be used in other thermal applications such as the production of hydrogen, increased safety, and waste reduction and handling (Zinkle & Was, 2013). Much higher operating temperatures in these reactors place more of a burden on reaction materials than traditional systems. This means that the construction material is of utmost importance when designing these new reactors (Hansen, 2015). The names of most of these systems are derived from the material used to cool the reactors. As the temperatures are too high for traditional coolants, materials such as molten salt, high-pressure helium gas, liquid sodium, and lead are all being used to cool the reaction. There are still a number of other issues associated with the construction of these reactors. Despite all the technological challenges associated with Generation IV nuclear reactors, they still represent a promising way toward cleaner energy production. By taking into account the aspects of efficiency and safety, the hope is that these systems can propel nuclear past's negative public image and into a more energy secure future.

## 10.6 Bioenergy

The World Energy Council reported that bioenergy is the largest renewable source, which accounts for 78% of renewable energy and contributes 10% of the global energy supply (Shimizu et al., 2006). A broad definition of bioenergy production is the use of conversion technologies to change various kinds of biological feedstocks into heat, power, and biofuel. Generally, bioenergy contains traditional biomass (fuel wood, charcoal, and agricultural waste), modern biomass, and biofuels (Baxter, 2005). With a growing demand for energy, in 2017, 55.6 EJ of biomass was utilized for energy purposes – 86% of the use was in the form of primary solid biofuels including wood chips, wood pellets, and fuelwood for cooking and heating, and 7% of the biomass was used as liquid biofuels. Biogas, municipal waste, and industrial waste had almost equal share at 2–3% (WBA, 2019). The principal merit of bioenergy is that domestic resources, either from natural biomass or the organic waste, are produced in association with resultant products from industry or agriculture to be transformed and utilized as a bioenergy source. For instance, tobacco, pulp, and paper production use woody biomass as a raw material and create huge amounts of organic residues, including wood chips and bark, which can be collected to provide secondary sources for bioenergy (Olsson, 2012).

### 10.6.1 Sources and Supplies of Biomass

As mentioned above, bioenergy consists of traditional biomass (fuelwood, charcoal, and agricultural waste), modern biomass, and biofuels. And domestic supply of biomass is the highest in Asia followed by Africa and Americas. Asia and Africa together have a high share of biomass due to the extensive use of fuelwood and charcoal for mainly heating and cooking purposes. For liquid biofuels, Americas (particularly the United States and Brazil) has the highest share due to high production of bioethanol mainly from corn and sugarcane, respectively. It is interesting to note that the region owns more than 70% of the global biofuels supply; and Europe, on the other hand, leads the way in biogas supply and has more than 50% of the global biogas supply (WBA, 2019). Due to extensive use of waste to energy technologies including incineration and gasification, Europe also leads the world in the supply of municipal waste for bioenergy having two thirds of the global supply.

#### *10.6.1.1 Forestry Products and Supplies*

The forestry sector is the largest global contributor to the bioenergy mix. WBA (2019) stated that forestry products including charcoal, fuelwood, pellets, and wood chips all together has a share of more than 85% of all the biomass used for energy purposes. Mostly, the forestry products is in the form of residues from pulp, paper, and sawmill industries while a significant portion is in the form of fuelwood (a.k.a. traditional biomass) for cooking and heating purposes in Africa and Asia.

There is an abundant forest biomass resource in the world, totaling close to 4,000 million hectares with woody biomass amounting over 400 billion tonnes (Table 10.16). Among continents, majority of the forestland occurs in Americas (40%) followed by Europe (25%) and equal share in Africa (15%) and Asia (15%). This is due to the enormous area of primary forest in the Amazon in South America which accounts for almost half of all primary forests globally. Canada alone has 10% of all the world's forests (about 400 million hectares). Forests are a major source of wealth for Canadians, providing a wide range of economic, social, and environmental benefits for the country. In 2013, forestry production (lumbers, pulp/paper, etc.) contributed $19.8 billion (or 1.25%) to Canada's real gross domestic product (GDP), the world's largest forest product trade balance.

For many decades, the forestry sector has consumed a substantial amount of forestry residues (bark, sawdust, and lignin) for generating process steam and electricity for the pulp, paper, and lumber manufactures. Using biomass for heat and electricity generation is currently developing rapidly with the potential to drastically reduce dependency on fossil fuels. While the sources of biomass energy are highly variable, the technology can best

**TABLE 10.16**

Global Forest Area and Woody Biomass Resources. Modified after Parikka (2004) (Courtesy of Elsevier. Used with permission).

| Region | Forest Area (ha) ($10^6$) | Woody Biomass (tonne/ha) | Woody Biomass (tonne) ($10^9$) |
|---|---|---|---|
| Africa | 649 | 109 | 70 |
| Asia | 547 | 82 | 44 |
| Europe | 1,039 | 59 | 61 |
| North and Central America | 549 | 95 | 52 |
| Oceania | 197 | 64 | 12 |
| South America | 885 | 203 | 179 |
| World | 3,869 | 109 | 421 |

be described as the combustion of biological material for the purpose of capturing and utilizing the generated heat (Petrus & Noordermeer, 2006). The heat generated from the combustion is used to boil water, whose steam is then used to turn a turbine and generate electricity. Many systems are designed as cogeneration systems – combined heat and power (CHP) generation – in that the steam which passes through the turbine is then utilized to heat nearby buildings including greenhouses or is recirculated back to the boiler to keep the temperature of the water high (Barreiro et al., 2013).

On a smaller scale, biomass is also consumed in direct combustion for heat generation such as in woodstoves and fireplaces. Conventionally, hardwoods were burned in these instances; however, with the growing shortage of large hardwoods globally, there is a need to find new solutions. One such avenue comes in the form of grass pellets and briquettes. Grass pellets and briquettes are produced by heat pressing selected grasses in order to increase their energy density. The pellets and briquettes can then be burned as a suitable alternative to burning hardwood. Products like the ones described can be seen in Figure 10.6.

##### *10.6.1.1.1 Woodfuel*

One of the main products from forests that are used for bioenergy production is woodfuel. Most of the woodfuel is used for cooking and heating purposes in Asia and Africa. Globally, 1.9 billion $m^3$ of woodfuel was reported to be used for energy purposes. It should be noted that woodfuel does not consist of the wood residues from industrial processing of round wood, which forms a major share of bioenergy in Europe. Asia and Africa together has approximately 75% of all woodfuel production globally (WBA, 2019).

##### *10.6.1.1.2 Pellets*

Pellets are a solid biomass fuel, mainly produced from wood residues, grasses, and agricultural residues such as straw. Specific advantages of pellets as compared to unprocessed biomass include standardized properties, high energy content, and high density, and therefore, reduced costs for transport, storage, and handling. Pellets are used for heating purposes in pellet stoves and pellet boilers as well as for the generation of heat, steam, and electricity for manufacturing and service industry. The wood pellet sector has shown

**FIGURE 10.6**
Biomass pellets produced from switchgrass, ***Panicum virgatum*** (left), and biomass briquettes produced from dried lawn clippings (right).

tremendous global growth in recent years to become one of the largest biomass commodities. WBA (2019) reported that, in 2018 alone, 35.4 million tonnes of wood pellets were produced – 55% of the production occurred in Europe while Americas accounted for 32%.

##### 10.6.1.1.3 Charcoal

Charcoal is mainly produced from woody biomass as well as from agricultural products and biological fraction of municipal and industrial solid waste. Charcoal is mainly used for cooking and heating predominately in Asia, Africa, and South America. Even though the global production of charcoal is about 1.5 times the production of wood pellets globally, it is quite neglected and there is significantly much less attention drawn to the challenges and opportunities of charcoal. The global production of charcoal in 2017 was 51.6 million tonnes (WBA, 2019).

#### 10.6.1.2 Crops and Supplies

The agriculture sector accounted for less than 10% of the overall supply of biomass for bioenergy globally. The main source for energy from agricultural land is in the form of crops for biofuels and residues for biogas as well as use in the form of heating and cooking. In terms of area harvested, cereal food crops such as maize, rice, and wheat together account for more than 580 million ha of land use and together account for more than 80% of the area harvested for major crops. Crops which show a tremendous growth in area harvested include maize (45%), soybean (66%), and cassava (55%). It is important to note that the use of crops for biofuels is still a very small share of the overall use of crops for food production (WBA, 2019).

One of the most promising sectors for growth in bioenergy production is in the form of residues from agriculture sector. Currently, the sector contributes less than 3% to the total bioenergy production. However, due to the increasing demand for replacing fossil fuels for generating sustainable heat and electricity, agricultural residues such as straw and husk can easily take a major share of the bioenergy generation.

#### 10.6.1.3 Municipal and Industrial Waste

Another source of bioenergy supply is municipal and industrial waste, which is primarily used for energy production in urban areas. The proper technologies to produce energy from forestry residues and agricultural wastes are also applicable for the conversion of any other organic matter in solid wastes. The technologies used for this purpose include gasification and combustion of biogenic waste to produce steam for turbines and hence producing electricity and heat for domestic and commercial use. Municipal waste primarily comes from households, industries, etc., while industrial waste typically comes from non-renewable origins and is composed of solid and liquid products (e.g. rubber tires) that are combusted in specialized plants for generating heat and/or power. WBA (2019) reports that, in 2017, domestic supply of waste to bioenergy was 2.51 EJ, of which 58% was in the form of municipal waste while the remaining was industrial waste. In 2017, Europe had more than half of the domestic supply of waste while Asia had 30%.

Based on the information provided by the World Bank (Table 10.17), the three classes of low, lower middle, and lower upper income levels produce the highest carbon-based municipal solid wastes (MSW), which are considered suitable for the waste-to-bioenergy production processes.

**TABLE 10.17**

World Bank Estimates for global MSW Composition by Income, for the Year 2025.

| | Waste Composition | | | | | |
|---|---|---|---|---|---|---|
| Income Level | Organic (%) | Paper (%) | Plastic (%) | Glass (%) | Metal (%) | Other (%) |
| Low | 62 | 6 | 9 | 3 | 3 | 17 |
| Lower Middle | 55 | 10 | 13 | 4 | 3 | 15 |
| Upper Middle | 50 | 15 | 12 | 4 | 4 | 15 |
| High | 28 | 30 | 11 | 7 | 6 | 18 |

*Sources:* Hoornweg and Bhada-Tata (2012) and DeCicco (2013).

**TABLE 10.18**

Moisture Contents and Heating Values of Municipal Solid Wastes.

| | Moisture Content (%wt) | | Heating Value[a] (MJ/kg) |
|---|---|---|---|
| Type of Waste | Range | Typical | Range |
| Mixed Food Wastes | 50–80 | 70 | 11–19 |
| Mixed Paper | 4–10 | 6 | 13–21 |
| Mixed Agricultural wastes | 40–80 | 50 | 16–17 |
| Yard Wastes | 30–80 | 60 | 10.16 |
| Mixed Wood Waste | - | 20 | 11–16 |
| Mixed Plastics | - | 2 | 22–42 |
| Textiles | - | 10 | 17 |
| Leather | - | 10 | 17.5 |
| Plastic Composite | - | <1 | 37 |
| Polystyrene (PS) | - | <1 | 40 |
| Polyvinyl chloride (PVC) | - | <1 | 25 |
| Polyethylene terephthalate (PT) | - | <1 | 25 |
| Polypropylene (PP) | - | <1 | 44 |
| Rubber | - | 2 | 28–33 |
| Glass | - | 2 | 0.14 |
| Metal | - | 3 | 0.7 |

*Sources:* Horttanainen et al. (2013) and Zhou et al. (2014).
[a] Heating (calorific) value of dried materials.

As mentioned earlier, these waste fuels can be handled via a range of available technologies. Each process is considered based on the waste's moisture content, the calorific value (Table 10.18), the composition, and the consistency to ensure that the relevant byproducts and the emissions from the waste processing are identified as non-hazardous and stable over time (Słania & Sacharz–Szewczyk, 2016; Egorov & Strizhak, 2013).

**Example 10.1: Determining Energy Content of Municipal Solid Waste in a High-Income Country**

Determine the energy and specific energy contents of a typical municipal solid waste with an average composition in a high-income country.

**Solution**

Using Table 10.17 and the typical moisture contents and heating values of MSWs provided in Table 10.18, we can first prepare the following table:

| Component | Wet Weight (kg) | Moisture Content (%) | Dry Weight (kg) | Heating Value (MJ/kg) | Energy Content (MJ) |
|---|---|---|---|---|---|
| Organic | 46 | 48.0 | 23.9 | 14.2 | 339.7 |
| Paper | 30 | 6.0 | 28.2 | 17.0 | 479.4 |
| Plastic | 11 | 2.0 | 10.8 | 32.0 | 345.0 |
| Rubber | 0 | 2.0 | 0.0 | 30.5 | 0.0 |
| Glass | 7 | 2.0 | 6.9 | 0.14 | 1.0 |
| Metal | 6 | 3.0 | 5.8 | 0.70 | 4.1 |
| **Total** | **100** | | | | **1,169.1** |

And then we can calculate the specific energy content as follows:

$$Specific\ Energy\ Content = \frac{\left[Energy\ Content\,(\mathrm{MJ})\right]}{\left[Total\ Solid\ Waste\,(\mathrm{kg})\right]} = 1{,}169.1\,\mathrm{MJ}/100\,\mathrm{kg}$$

$$Specific\ Energy\ Content = 11.7\,\mathrm{MJ}/\mathrm{kg}$$

Please note that in this example, organic waste included the food waste, cardboard, yard waste, wood, textiles, and leather. Therefore, the relevant moisture content and calorific value represent the approximate average of the individual compositions, respectively.

**Example 10.2: Determining Energy Content of Municipal Solid Waste in a Low-Income Country**

Determine the energy and specific energy contents of a typical municipal solid waste with an average composition in a low-income country.

**Solution**

Using Table 10.17 and the typical moisture contents and heating values of MSWs provided in Table 10.18, as in the previous example, we can first prepare the following table:

| Component | Wet Weight (kg) | Moisture Content (%) | Dry Weight (kg) | Heating Value (MJ/kg) | Energy Content (MJ) |
|---|---|---|---|---|---|
| Organic | 77.6 | 48.0 | 40.4 | 14.2 | 573.0 |
| Paper | 6 | 6.0 | 5.6 | 17.0 | 95.9 |
| Plastic | 9 | 2.0 | 8.8 | 32.0 | 282.2 |

*(Continued)*

| Component | Wet Weight (kg) | Moisture Content (%) | Dry Weight (kg) | Heating Value (MJ/kg) | Energy Content (MJ) |
|---|---|---|---|---|---|
| Rubber | 1.4 | 2.0 | 1.4 | 30.5 | 41.8 |
| Glass | 3 | 2.0 | 2.9 | 0.14 | 0.4 |
| Metal | 3 | 3.0 | 2.9 | 0.70 | 2.0 |
| **Total** | **100** | | | | **995.4** |

And then we can calculate the specific energy content as follows:

$$Specific\ Energy\ Content = \frac{\left[Energy\ Content\,(\text{MJ})\right]}{\left[Total\ Solid\ Waste\,(\text{kg})\right]} = 995.5\ \text{MJ}/100\ \text{kg}$$

$$Specific\ Energy\ Content = 10.0\ \text{MJ}/\text{kg}$$

Please note that in this example as well, organic waste included the food waste, cardboard, yard waste, wood, textiles, and leather. Therefore, the relevant moisture content and calorific value represent the approximate average of the individual compositions, respectively.

The above examples show that the specific energy contents of typical MSWs in high- and low-income countries are different (higher in high-income countries). In the above examples, we calculated the energy contents based on the average waste compositions in both high- and low-income countries. In reality, we would determine the energy contents of the MSWs based on the site-specific waste compositions and relevant moisture contents. Therefore, the specific energy contents, and the total energy contents (considering the total amount of solid waste generated) of the MSW would be different at any specific site.

#### Example 10.3: Comparing Mixed Wood Waste with Natural Gas

What should be the price of mixed wood waste per ton to deliver the same amount of heat to a greenhouse, if natural gas costs $1.00 per therm? Assume that no $CO_2$ dosing is performed in the greenhouse.

#### Solution

Switching to wood waste for providing space heating obviously depends on the cost of competing fuels. The wood has to cost less than the natural gas to provide comparable heat output for the same price. This example assumes that the efficiency of the wood furnace is 50%, and the efficiency of the gas furnace is 75%. Our calculations below assume that the wood furnace's attractiveness, the need to cut and carry wood to the furnace, and the removal of the ash are ignored.

From Table 10.18, we know that mixed wood waste has a typical moisture content of 20%, and a heating value of 11–16 MJ/kg dry wood (for this example, we will assume that it is 13 $\text{MJ}/\text{kg}_{dw}$). A therm is a measurement of energy content and is approximately equal to 100,000 British thermal units (BTU) and 105.506 MJ.

Then,

$$Cost\ of\ Natural\,Gas = \$1.00 \div 105.506\ \text{MJ} = \$9.48 \times 10^{-3}/\text{MJ}$$

Considering the 75% natural gas furnace efficiency,

$$Cost\ per\ \text{MJ}\ Delivered\ Heat = (100 \div 75) \times Cost\ of\ Natural\ Gas(\$/\text{MJ})$$

$$Cost\ per\ \text{MJ}\ Delivered\ Heat = 1.33 \times \$9.48 \times 10^{-3}/\text{MJ}$$

$$Cost\ per\ \text{MJ}\ Delivered\ Heat = \$0.0126/\text{MJ}$$

And considering 50% wood furnace efficiency and 20% moisture content, the amount of heat delivered by the moist mixed wood waste becomes:

$$0.5 \times (13\ \text{MJ}/\text{kg}_{\text{dw}}) \times (100 - 20) \div 100 = 5.2\ \text{MJ}/\text{kg}$$

This tells us that 1 MJ of heat is provided by 1 MJ ÷ (5.2 MJ/kg) = 0.192 kg moist mixed wood waste. Therefore, the maximum cost of mixed wood waste can be calculated as: \$0.0126 ÷ 0.192 kg = \$0.066 per kilogram, or approx. \$66 per ton. In order to be competitive, the price of mixed wood waste per ton should be less than \$66 delivered.

**Example 10.4: Comparing Rubber Waste with Natural Gas**

As in the previous example, if natural gas costs \$1.00 per therm, what should be the price of rubber waste per ton to deliver the same amount of heat? Assume that no $CO_2$ dosing is performed in the greenhouse.

**Solution**

Similarly, switching to rubber for providing space heating depends on the cost of competing fuels. The rubber has to cost less than that of the natural gas to provide comparable heat output for the same price. This example assumes that the efficiency of the rubber furnace is 50%, and the efficiency of the gas furnace is 75%. Our calculations below assume that the rubber furnace's attractiveness, the need to shred and carry rubber to the furnace, and the removal of the residue are ignored.

From Table 10.18, we know that rubber waste has a typical moisture contentment of 2%, and a heating value of 28–33 MJ/kg dry rubber (for this example, we will assume that it is 30 MJ/$kg_{dw}$). As mentioned earlier, therm is a measurement of energy content and is approximately equal to 100,000 BTU and 105.506 MJ. In the previous example, the natural gas cost was determined as \$9.48 × $10.^{3}$/MJ. Considering a natural gas furnace efficiency of 75%, the cost per MJ delivered heat was also calculated as \$0.0126/MJ.

And considering 50% rubber furnace efficiency and 2% moisture content, the amount of heat delivered by the moist rubber waste would be:

$$0.5 \times (30\ \text{MJ}/\text{kg}_{\text{dw}}) \times (100 - 2) \div 100 = 14.7\ \text{MJ}/\text{kg}$$

This means that 1 MJ of heat is provided by 1 MJ ÷ (14.7 MJ/kg) = 0.068 kg moist rubber waste. Therefore, the maximum cost of rubber waste can be calculated as \$0.0126 ÷ 0.068 kg = \$0.1853 per kilogram, or approx. \$185 per ton. In order to be competitive, the price of rubber waste per ton should be less than \$185 delivered. When this price is compared to that of the wood in the previous example, it becomes clear that we can afford higher prices of rubber waste due to its higher energy and lower moisture contents.

**Example 10.5: Electricity Generation from Municipal Solid Waste**

Determine the net electric power generation potential for City A if total MSW collected from the city is 500 tons per day, and the heat generation potential of the waste is 15 MJ/kg. Assume that the available steam energy is 70% of heat energy, and also consider that the conversion efficiency is 31% in a power plant, and the net electric power generated by solid waste is determined after accounting station service allowance (5%) and heat losses (another 5%).

**Solution**

Let's first determine the steam energy available for running the turbine:

$$\textit{Available Steam Energy} = 70\%\ \textit{of heat energy}$$

$$\textit{Available Steam Energy} = 0.70 \times 15\ \text{MJ/kg} = 10.5\ \text{MJ/kg}$$

This available steam energy is used to turn the turbines, which are coupled with generators that eventually produce electricity. We also know that heat rate is the heat input required to produce one unit of electricity (kWh) and 1 kW is equal to 3.6 MJ/h.

Therefore, considering the conversion efficiency of 31% in the power plant,

$$\text{Heat Input Required} = 3.6\ \text{MJ/h} \div 31\% = 11.6\ \text{MJ/kWh}$$

In order to produce 1 kWh electrical energy, 11.6 MJ of steam energy is required. Then,

$$\textit{Generated Electric Power} = \frac{\textit{Available Steam Energy}}{\textit{Heat Input Required}}$$

$$\textit{Generated Electric Power} = \frac{10.5\ \text{MJ/kg}}{11.6\ \text{MJ/kWh}} = 0.905\ \text{kWh/kg}$$

Total Electricity Generation = 0.905 kWh/kg × 500,000 kg/day

Total Electricity Generation = 452,500 kWh/day

We assumed a station service allowance and heat loss of 5% each of the total electric power generation, then

Station Service Allowance = 0.05 × 452,500 kWh/day = 22,600 kW/h

Heat Loss (unaccounted) = 0.05 × 452,500 kWh/day = 22,600 kW/h

Net Power Generation = (Electric Power Generation)
− (Service Allowance + Heat Loss)

Net Power Generation = 452,500 − (22,600 + 22,600)

Net Power Generation = 406,800 kWh/day = 406.8 MWh/day

The above generated electricity is for 1 day (i.e. 24 h); therefore, using this net electric power, we can calculate the power for per hour basis as follows:

$$\textit{Net Electric Power Generation} = (406.8\ \text{MWh})/(24\ \text{h}) = 16.95\ \text{MW} \approx 17\ \text{MW}$$

Remember that a capacity of 17 MW power could supply power for approximately 17,000 average homes. Furthermore, in addition to the generation of electricity, we can use the heat lost for heating and moisture control purposes in a greenhouse complex that could be built in the proximity of this waste power plant.

### 10.6.2 Combustion of Solid Biomass

Direct combustion of biomass is the most common method for generating heat from biomass for cooking, heating, or generating electricity (Nussbaumer, 2003). Among different biomass combustion technologies, fluidized bed combustion is considered as one the most promising technologies as it can provide sufficiently high temperature, strong turbulence of the air-gas mixture, and a long residence time (the three T rule). Fluidized beds are widely used in various industrial processes, including fluid catalytic cracking (FCC), drying, coating on solid items, and solid fuel (such as biomass) combustion/gasification for generation of steam, electricity, and hydrogen.

Many fluidized bed combustors have been constructed in the United States, Europe, and Southeast Asia using a wide range of biomass as fuel. Fluidized bed combustion has many advantages such as fuel flexibility (different shapes, sizes, and moisture contents), high efficiency, and low emission levels. Fluidized bed boilers are used in pulp and paper mills and sawmill to produce high-pressure steam and electric power, primarily in Europe and North America. Fluidized bed combustors can be classified into two main types: circulating fluidized bed combustor (CFBC) and bubbling fluidized bed combustor (BFBC). CFBC is preferred in large units while BFBC uses simpler technologies in smaller units for burning biomass (Hupa, 2012). Bubbling fluidized bed and chain stoker grate boilers are the two main types of biomass boilers commonly used in different industries. Older industries mostly use stoker grate boilers while the newer operations typically use bubbling fluidized boilers as they are larger and more efficient (Montes, 2014). Despite all the advantages of fluidized bed combustion, there are some shortcomings such as sintering, deposition of ash particles, and bed material agglomeration, a major problem (Montes, 2014; Shimizu et al., 2006).

Co-combustion (or co-firing) technology, on the other hand, has proven to be a cost-effective technology for achieving the goal of increased use of biomass for heat and power generation, thereby significantly reducing emissions of greenhouse gases and other toxic gases such as $SO_x$ and sometimes $NO_x$ (Baxter, 2005; McIlveen-Wright et al., 2007).

#### *10.6.2.1 Biomass to Electricity*

Electricity generation from biomass is a sustainable and renewable option for reducing fossil fuel consumption and demand. In 2017, globally 596 TWh of electricity was produced from biomass, having an increase of 3.6 folds compared to the figures in year 2000 (WBA, 2019). About 65% of the produced electricity was from primary solid biofuels including wood chips and wood pellets, used in large-scale electricity only and CHP plants. Electricity is generated using the biogas as well produced from municipal and industrial waste. Liquid biofuels form a tiny part of bioelectricity generation, and are better suited for transportation sector instead.

Bioelectricity is produced either in electricity-only plants or CHP plants that simultaneously produce heat and electricity. In electricity-only plants, in 2017, 5.15 EJ of biomass was used for electricity production, which translates to approximately 429 TWh (WBA, 2019). Asia is the global leader in producing bioelectricity in electricity-only plants. Asia used approximately 64% of globally used biomass for electricity generation in electricity-only plants while Europe and Americas together used 18% of the biomass in electricity-only plants. The conversion efficiency of such plants is quite low and they have huge losses as heat; therefore, it would be better to invest in CHP plants for generating more useful energy per unit biomass used. And greenhouses can easily be integrated into such plants for increasing the overall efficiency. In CHP plants – having one of the most efficient technologies for converting biomass to energy, the biomass is combusted to produce heat, which converts water to steam to run a turbine for generating electricity. In 2017, 2.65 EJ of biomass was used in CHP plants to produce heat and steam, and therefore, electricity. This is approximately the half of all biomass used in electricity-only plants, and most of the biomass used in CHP plants is in the form of solid biofuels. Europe being the global leader in biomass CHP technology used 63% of the globally used biomass in its CHP facilities while Americas used only 35% (WBA, 2019).

#### *10.6.2.2 Biomass To Heat*

Heating covers both space heating and hot water needs of residential, industrial, and commercial establishments around the world. Compared to power generation, the options for heating sector is relatively limited. Geothermal, solar thermal, and biobased heating are the most important renewable energy options for heating including greenhouse heating. In 2017, 1.08 EJ of biobased heat was generated globally, and solid biofuels such as wood pellets and wood chips covered almost half of the biobased heat, which was followed by both municipal and industrial waste, biogas, and liquid biofuels. Even though Asia uses the most biomass for residential heating, Europe is the global leader in using biobased heat. About 87% of the biobased heat generated globally is generated in Europe mainly because of the extensive use of all biomass feedstock including municipal waste, solid biofuels, and industrial waste.

In heat-only plants, only biobased heat is generated without any electricity generation. In heat-only facilities, globally, 0.53 EJ of biomass was used to produce heat in 2017. Approximately, 84% of this biomass was used for generating heat in Europe. In CHP facilities, the heat (steam) after running the turbine is typically used for district heating. In 2017, approximately 1.06 EJ of biobased heat was generated in CHP plants globally, 63% of which was produced in Europe – about 0.67 EJ, 34% was produced in Americas – about 0.36 EJ, and the remaining 3% biobased heat was produced in rest of the world (WBA, 2019).

### 10.6.3 Liquid and Gaseous Biofuels

Bioenergy development is currently focused on biofuel (biogas, biodiesel, bioethanol, etc.) production particularly in the developed world. Classified as several generation feedstocks, biofuels have been perceived as an alternative source of fossil fuel and are gradually being the most sustainable substitution to break the oil monopoly in the transportation sector.

#### *10.6.3.1 First Generation Biofuels*

A first generation biofuel refers to any biofuel produced from a conventional food crop. While yields for many first generation biofuels are quite high, the issue, which often arises, is the food versus fuel debate. This debate is anchored on the notion that conventional food crops are more valuable as food crops than as biofuel crops. As first generation biofuels reduce the amount of food produced, their value needs to be evaluated as either a food or fuel crop. In the United States, the leading ethanol producer, corn is used as the dominant feedstock for bioethanol production. Brazil is the world's largest sugar producer as well as the second largest fuel ethanol producer where sugarcane is a favorable feedstock. Half of the sugarcane produced goes into ethanol, resulting in an increased sugarcane price (Olsson, 2012). The Energy Independence and Security Act (EISA) announced that biofuel production should be increased to at least 136 billion liters by 2022 to improve energy productivities (Energy Independence and Security Act, 2007). World Water Development Report (2014) introduced the bioenergy status in Asia, which is recently developing into an important market for biofuels, as well as becoming a biofuel producer, due to its potentially enormous biomass resources (Zhou & Thomson, 2009). The report pointed out that India, the Philippines, and Thailand are promoted as burgeoning industrial biofuel generators, and China is now the third-largest producer in the world (BIOFUELSTP, 2009; PR Newswire, 2009). Komarudin et al. (2010) reported that Indonesia contributed 45% of the global palm oil market and nearly 8% of the land was designated for biofuel production, which has the benefit of gaining extra income for local farmers and workers. As more croplands are requisitioned for biofuel production, food crisis might occur due to decreased arable land available for food production. Hence, it is preferable that biofuel crops be grown on agriculturally marginal land (Daniel et al., 2010; Lal & Pimentel, 2007).

#### *10.6.3.2 Second Generation Biofuels*

A second generation biofuel refers to any biofuel produced from the lignocellulosic material of a plant and aim to directly combat the food versus fuel issue. In using the "non-food" parts of the plant along with crops not considered as food crops, second generation biofuels do not encounter the same food versus fuel issues as first generation biofuels. Various plants have been confirmed as potential feedstocks for renewable fuels, and range from dedicated herbaceous bioenergy crops (e.g. switchgrass and miscanthus – a.k.a. silvergrass) to woody species (e.g. poplar and willow) (Wilhelm et al., 2010). Wright et al. (2010) and McLaughlin & Kszos (2005) reported that switchgrass is one of the most suitable feedstocks that has many positive environmental attributes, including wide geographical growth range, low water requirement, high nutrient efficiency and relatively high productivity (Wright & Turhollow, 2010).

#### *10.6.3.3 Third Generation Biofuels*

In addition to plants, microalgae (third generation biofuel) have also become a prevalent area of interest for the production of biofuels in recent years. A number of algal species contain high amounts of lipids, and can be cultivated in mass with the proper system. This makes them an attractive option as production costs decline and fuel demand increases. Microalgae neither compete with food crops nor compete for the same land resources as second generation biofuels. Instead, they are grown in ponds or photobioreactors – sometimes housed in greenhouses. These aspects in combination help to distinguish microalgae as the most promising feedstock for biofuel conversion. The major issue, which arises with many second and third generation biofuels, is that the yields are not high enough to justify production.

#### *10.6.3.4 Biogas*

Biogas is an important component of the whole bioenergy industry. It is produced by anaerobic fermentation of different forms of organic matter and is composed of mainly (60–70%) methane ($CH_4$) and (30–40%) carbon dioxide ($CO_2$). One of the most common implementation of anaerobic digesters is in agricultural operations. Organic solids are often mixed with agricultural manures in order to improve the digestion process. Typical feedstocks are manure and sewage, residues of crop production, the organic fraction of the waste from households and industry, as well as energy crops such as maize and grass silage. The resultant methane is either captured or burned off as it is produced. Some larger digesters incorporate a generator, which uses the produced methane to generate electricity, reducing operational and labor costs. Capturing methane in this way not only provides a valuable biofuel but also helps in mitigating its effects as a greenhouse gas. Methane has a heat-trapping capacity about 21 times higher than that of carbon dioxide, making mismanagement highly detrimental to the environment. When produced in this way, methane is thought of as a valuable renewable resource. Biogas is used for a variety of purposes such as electricity generation, heat generation, and transportation fuel. Europe is the global leader in the production and use of biogas. In 2017, the global domestic supply of biogas was 1.33 EJ, and Europe provided half of this, followed by Asia, which provided 1/3rd of the supply (WBA, 2019). Overall, biogas accounts for only 2% of the bioenergy supply; however, it has much more potential to contribute.

**Example 10.6: Calculating Methane and Electricity Production from Manure-Bedding Mixture**

Consider that 10 short tons of manure-bedding mixture with the characteristics provided below goes into an anaerobic digester every day, and it is combined with adequate dilution water to reduce ammonia levels below 1,200 mg/l. The volatile solids (VS) molecular formula of this manure-bedding mixture is provided as $C_7H_{11}O_4N$.

Manure:
85% $H_2O$ (wet basis)
15% Total Solids (*TS*) (wet basis)
90% Volatile Solids (*VS*) (dry basis)
Reactor:
Influent total solids = 5% (dry basis)
Digester operating temperature = 35°C (mesophilic)

Determine the stoichiometric methane ($CH_4$) production from the degradation of volatile solids (VS) assuming that 60% of the total solids (TS) degrade during the 20-day hydraulic retention time. Using the methane produced, also calculate the amount of electricity that can be generated in kWh per day.

**Solution**

Let's first calculate the total solids and volatile solids produced, and then the volatile solids degraded per day so that we can determine the amount of methane produced per day. Remember that 1 short ton = 2,000 lb.

Amount of total solids (TS) produced per day:

$$\left(\frac{10\text{ ton wet manure}}{\text{day}}\right)\left(\frac{2{,}000\text{ lb}}{\text{ton}}\right)\left(\frac{\text{kg}}{2.205\text{ lb}}\right)\left(\frac{0.15\text{ TS}}{\text{wet manure}}\right)=\frac{1{,}360.5\text{ kg TS}}{\text{day}}$$

Amount of volatile solids (VS) produced per day:

$$\left(\frac{1{,}360.5\text{ kg TS}}{\text{day}}\right)\left(\frac{0.9\text{ VS}}{\text{TS}}\right)=\frac{1{,}224.5\text{ kg VS}}{\text{day}}$$

Amount of volatile solids degraded per day:

$$\left(\frac{1{,}224.5\,\text{kg VS}}{\text{day}}\right)\left(\frac{0.6\,\text{TS degraded}}{\frac{0.9\,\text{VS}}{\text{TS}}}\right)=\frac{816.3\,\text{kg VS degraded}}{\text{day}}$$

The chemical reaction for the degradation can be written on a molar basis (Buswell equation) as:

$$C_7H_{11}O_4N + 2.25\,H_2O \rightarrow 3.125\,CO_2 + 3.875\,CH_4$$

Note that $NH_3$ and other minor reactants are not included in the Buswell equation.

In order to use the stoichiometric equation for converting volatile solids (VS) degradation to methane, we need to change the chemical equation from molar basis (provided above) to mass basis (provided below). We can calculate this directly keeping in mind that the atomic weights of H = 1, C = 12, N = 14, and O = 16. Note that all of these elements are volatile (i.e. they burn off as gases at 550°C), and the mineral (ash) components of the manure (P, Na, K, Ca, Mg, etc.) were neglected in the formula given.

$$173\,C_7H_{11}O_4N + 40.5\,H_2O \rightarrow 137.5\,CO_2 + 62\,CH_4$$

$$\left(\frac{816.3\,\text{kg VS degraded}}{\text{day}}\right)\left(\frac{62\,\text{kg}\,CH_4}{173\,\text{kg manure VS}}\right)=\frac{292.6\,\text{kg}\,CH_4}{\text{day}}$$

The density of methane is 0.715 $kg/m^3$ at standard temperature and pressure (STP). And at 35°C, the density of methane is about 0.654 $kg/m^3$. We could use any of these densities in our estimations; however, we will use the latter value in our calculation:

$$\left(\frac{292.6\,\text{kg}\,CH_4}{\text{day}}\right)\left(\frac{m^3}{0.654\,\text{kg}\,CH_4}\right)=\frac{447.4\,m^3CH_4}{\text{day}}$$

Here, the critical assumption was the percent volatile solids degradation (60%). Testing the accuracy of this assumption (and thus improving the estimates) could be done by either measuring methane production or volatile solids degradation and the elemental composition of manure. Measuring methane production would probably be much easier for most farm digesters.

The average energy content of methane can reasonably be assumed as 35.5 $MJ/m^3$ (Table 2.18) at STP.

$$\left(\frac{447.4\,m^3CH_4}{\text{day}}\right)\left(\frac{35\,\text{MJ}}{m^3CH_4}\right)\left(\frac{0.278\,\text{kWh}}{\text{MJ}}\right)=\frac{4{,}353.2\,\text{kWh}}{\text{day}}$$

Note that the conversion of this biogas energy to electricity will not be 100% efficient (typical conversion efficiency can be considered as 20%). The remainder is lost from the cogenerator's engine as waste heat, and some of that heat can be captured and used to heat the anaerobic digester itself, a greenhouse, water, etc. We also need to remember that combustion of methane releases $CO_2$, which can easily be captured and used for $CO_2$ dosing in a nearby greenhouse complex for increased crop production.

Now, considering the typical conversion efficiency of 20%, we can calculate the power generated as:

$$0.2\times\left(\frac{447.4\,m^3CH_4}{\text{day}}\right)\left(\frac{35\,\text{MJ}}{m^3CH_4}\right)\left(\frac{\text{day}}{86{,}400\,\text{s}}\right)=0.0362\,\text{MW}=36.2\,\text{kW}$$

## 10.7 Hydro Energy

Hydropower utilizes the energy generated from moving water to create electric power; the water is discharged downstream via turbines. The application of this excellent power source can be dated to ancient times, and as the main force in renewables, hydropower has developed rapidly and been able to compete with other primary energy resources in recent years. The operational costs of hydropower are considerably low, and the technologies are proven; however, hydropower requires high capital expenditures. It is estimated that the total installed capacity has grown by 35% from 2005 to 2015 with a 4% annual growth rate (EIA, 2016). Most of the hydropower projects are concentrated in developing countries, and they are expected to continue to expand in the near future. The World Energy Council reported that, in 2012, approximately 27–30 GW of new hydropower and 2–3 GW of pumped storage capacity were built around the world (WEC, 2013). As a leading renewable energy source for global electricity production, 1,064 GW of installed hydropower capacity was reached in 2016, which accounted for 16.4% of the total electricity supply (EIA, 2016). Today, there are more than 100 countries with hydropower plants, and they contribute approximately 15% of the global electricity production in the world. Typically, compared with other areas, Asia and South America regions have larger potential installed capacity of hydropower. These countries are emerging markets due to easier access to electricity and developing requirements to improve the local economy (WEC, 2013; WEC, 2016). The number of high dams is growing rapidly, particularly in China in the past few decades. The top six largest markets for potential hydropower capacity in 2015 were China, USA, Brazil, Canada, India, and Russia (Table 10.19) (WEC, 2013).

There were 5,300 dams in 1950, and only 2 of them built in China. After 30 years, China now owns half of the total 36,500 dams in the world. Apparently, China has significantly exceeded the others, contributing 26% of global installed capacity, and well ahead of the USA, Brazil, and Canada, which accounted for 8.4%, 7.6%, and 6.5%, respectively (WEC, 2016; REN21, 2015). It has been reported that in many countries, such as Canada, Iceland, Nepal, and Mozambique, hydropower also contributed more than half of total electricity production in domestic energy markets (WEC, 2013). The global energy resource survey indicated that

**TABLE 10.19**

Top Hydropower Capacity as of 2015.

| Country | Total Capacity End of 2015 (GW) | Added Capacity in 2015 (GW) | Production (TWh) |
|---|---|---|---|
| China | 319 | 19 | 1,126 |
| USA | 102 | 0.1 | 250 |
| Brazil | 92 | 2.5 | 382 |
| Canada | 79 | 0.7 | 376 |
| India | 52 | 1.9 | 120 |
| Russia | 51 | 160 | 120 |

Data from REN21 (2015).

the application of hydropower is expected to rise from 3,000 TWh to nearly 5,500 TWh during the period from 2010 to 2050, which is more than an 80% improvement (Olsson, 2012).

### 10.7.1 Water Energy to Electricity Conversion

Water energy to electricity conversion, as with any energy conversion process, is inherently reliant on the first law of thermodynamics, which states that energy can neither be created nor destroyed, and it can only change forms. In this way, water turbines convert the energy inherent to water into electricity using a hydraulic turbine. However, the maximum hydropower power output entirely depends on how much head and flow is available at the site. Therefore, we can use the following equation for determining how much electric power we could generate from the energy of water:

$$P = \dot{m}\, g\, H_{net}\, \eta \tag{10.1}$$

where $P$ is the power in Watts, $\dot{m}$ is the mass flow rate of water in kg/s (numerically the same as the flow rate in liters per second), $g$ is the gravitational acceleration (9.81m/s$^2$), $H_{net}$ is the net head (which is the gross head physically measured at the site, less any head losses; and it can be assumed to be 10%, so $H_{net} = 0.9\, H_{gross}$), and $\eta$ is the product of all component efficiencies (typically the turbine, drive system, and generator efficiencies). For a typical small hydropower plant, we can use a turbine efficiency of 85%, a drive efficiency of 95%, and a generator efficiency of 93%, so the overall system efficiency would be: $0.85 \times 0.95 \times 0.93 = 0.751$, that is, 75.1%.

We should keep in mind that approximately 1 kW hydropower system could meet the needs of the annual electrical energy needs of an average home, whereas a utility-scale 100 MW system could supply approximately 100,000 average homes.

**Example 10.7: Conversion of Water Energy to Electricity**

Determine the maximum power output if the gross head is 2.0 m, and the maximum flow rate is 5 m$^3$/s.

**Solution**

Let's first convert the gross head into the net head:

$$H_{net} = 0.9\, H_{gross} = 0.9\,(2.0\text{ m}) = 1.8\text{ m}$$

The flow rate 5 m$^3$/s = 5,000 l/s; and remember that 1 liter of water = 1 kg water; therefore, $\dot{m}$ is numerically the same as the flow rate in l/s, hence 5,000 kg/s.

We can now calculate the hydropower power using Equation (10.1) as follows:

$$P = \dot{m}\, g\, H_{net}\, \eta = (5{,}000)(9.81)(2.0)(0.751) = 73{,}673\text{ W} = 73.7\text{ kW}$$

Now, let's do the same calculation for a high-head hydropower plant whose gross head is 50 m and maximum flow rate through the turbine is 225 l/s.

As we did before, let's first convert the gross head into the net head:

$$H_{net} = 0.9\,(50\text{ m}) = 45\text{ m}$$

*(Continued)*

And the flow rate is 250 l/s; therefore,

$$P = \dot{m}\, g\, H_{net}\, \eta = (225)(9.81)(45)(0.751) = 74{,}594\ \text{W} = 74.6\ \text{kW}$$

In this example, it is quite interesting to observe that the hydropower plants with a net head of 1.8 m and 45 m can generate about the same amount of power. This is because the low-head power plant has much higher flow compared to the high-head power plant. This example clearly shows how the power output is proportional to the head multiplied by the flow. The low-head plant would need a physically large turbine, for instance, Kaplan turbine – the size of a large garage to discharge such a large volume of water with a relatively low head (pressure) across. The high-head plant, on the other hand, would only need a small turbine, for instance, Pelton type turbine – the size of a fridge to discharge about 5% of the flow rate of the low-head system and under a much higher pressure.

## 10.8 Wind Energy

While the idea of utilizing wind energy is far from a recent one, the utilization of wind for electricity generation certainly is. A wind turbine can be thought of as any device, which converts the energy of wind into electricity. It is as important to understand the distinction that these are not windmills, which convert wind energy into mechanical work. While the majority of wind turbines on earth are quite small, ranging from 1 to 10 kW, the majority of electricity generation comes from much larger turbines in the range of 1.5–5 MW. These larger turbines are generally oriented in equally large wind farms, which can incorporate hundreds of these massive turbines. Europe and North America have been the quickest in adopting this new technology with both having numerous wind farms located throughout their continents (Manwell et al., 2010).

The majority of modern wind turbines are known as horizontal wind axis turbines. These wind turbines are generally comprised of three angled blades, rotating parallel to the ground, which take advantage of the aerodynamic force of lift to rotate and generate torque on a shaft. This shaft is used to drive a generator which produces electricity. Unlike other renewable energy sources such as hydroelectric, wind turbines are only able to produce energy from what wind is immediately available. There is no capacity to store it for later use during peak demand periods. As such, it would be inconceivable to have a grid, which generated 100% of its electricity from wind as the output is highly variable. Based on this, the best implementation of wind turbines is as a secondary producer, serving to lighten the load on whatever other method of electricity generation is being employed (Manwell et al., 2010).

In looking at Figure 10.7, it can easily be observed where the areas of greatest average wind speeds are globally. As can be expected, those areas located closer to shore as well as those areas of greater elevation generally observe higher average wind speeds. Coastal areas such as Atlantic Canada, Southern Argentina, Greenland, Great Britain, Western Sahara, Somalia, Japan, and New Zealand see drastically higher average wind speeds than those nearby inland areas. In the same way, major mountain regions such as the Rockies, Andes, and Himalayas also observe the same patterns of increased wind speeds.

While Figure 10.7 shows global average wind speeds, the same sort of wind maps can be produced for more localized areas when progressing toward turbine siting. By using a

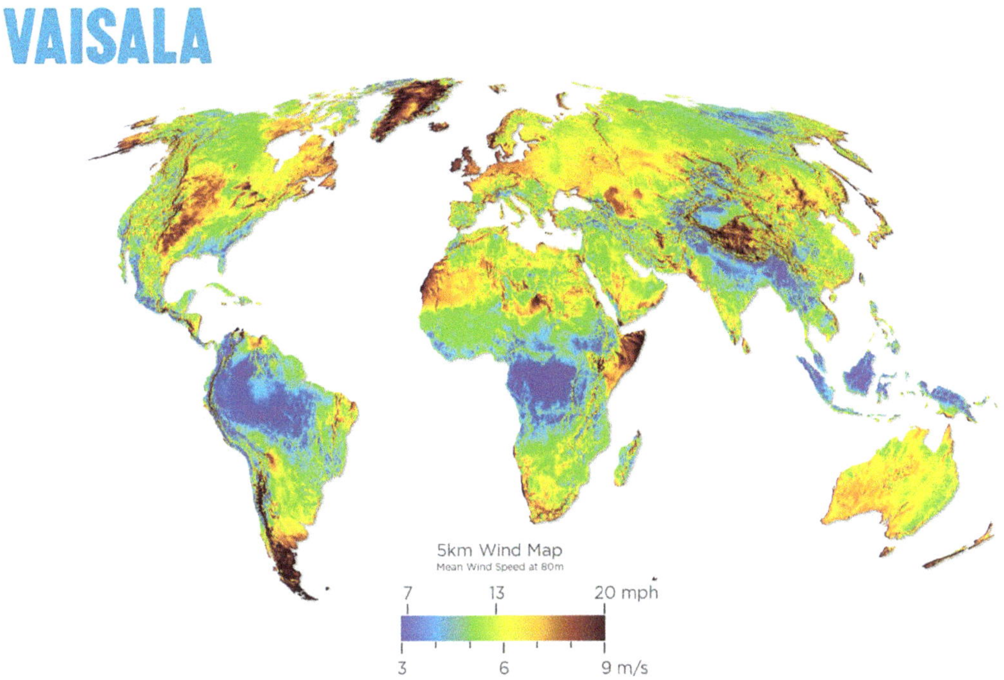


**FIGURE 10.7**
Global wind speed map showing average on shore wind speeds (Arent et al., 2012).

more strict scale, small differences may become apparent, which can lead to more efficient micro-siting. In the same way that wind maps can be produced for on shore applications, they can also be produced for offshore. Figure 10.8 shows a similar map, which has been produced based on offshore average wind speeds. This map is used in exactly the same way by engineers when selecting the best sites for offshore wind projects. As it can quickly be seen, the general trend is that the further away from shore we go, the greater the wind speeds become. This is due largely to the lack of surface obstructions but it does create an interesting challenge versus reward scenario for engineers. By siting turbines in deeper waters, the turbines will experience greater average wind speeds; and therefore, they will be able to convert more wind energy into electricity. With that being said, the project costs will be drastically increased as constructing deeper turbines in often much harsher environments can be a major challenge. Balancing the greatest efficiencies, with cost mitigation remains one of the most important tasks in siting wind turbines.

Wind energy has quickly become one of the leading renewable energy sources employed globally. The wide-scale availability of the resource coupled with increases in efficiency has led to more and more developments and installations throughout the world. Due largely in part to the high infrastructural costs associated with wind energy, this is a technology led largely by the world's most developed countries. Table 10.20 shows the top ten countries having installed wind energy capacity globally.

China has clearly separated itself when it comes to wind energy and in recent years have become the global leader in the technology. Not only do they currently control almost 35% of the global installed wind energy capacity, but also, in 2016, they had the new installation of a further 23,328 MW of wind turbines. This represented almost 43% of all wind turbines installed in 2016. A breakdown of the top ten countries by installed megawatt can

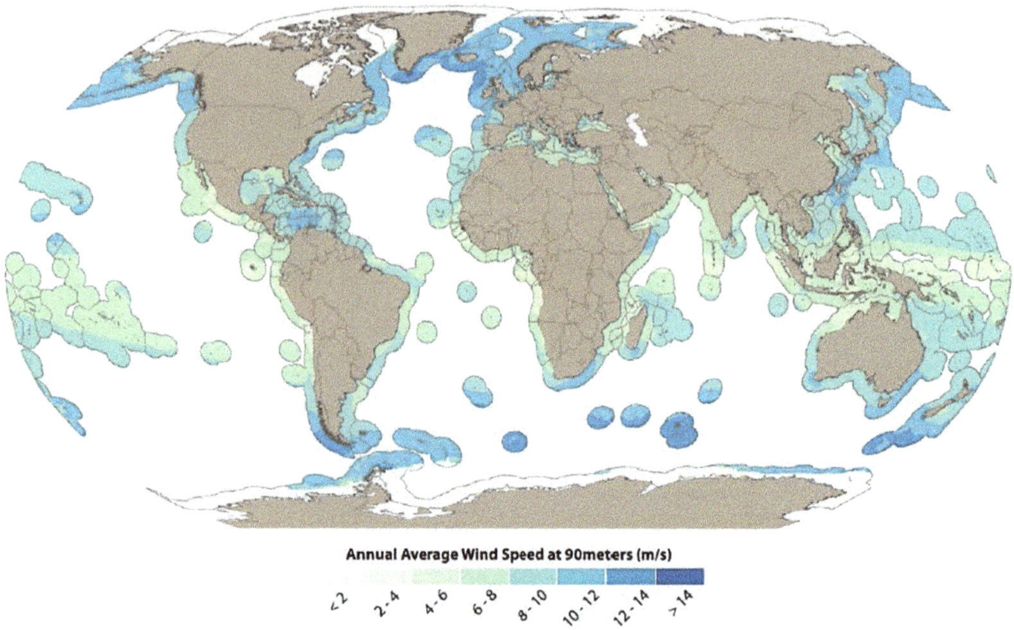


**FIGURE 10.8**
Global wind speed map showing average off shore wind speeds adjusted for a 90 m hub height (Arent et al., 2012).

**TABLE 10.20**
Total Installed Wind Energy Capacity.

| Country | Installed Capacity (MW) | Global Share (%) |
|---|---|---|
| China | 168,690 | 34.66 |
| USA | 82,184 | 16.88 |
| Germany | 50,018 | 10.28 |
| India | 28,700 | 5.90 |
| Spain | 23,074 | 4.74 |
| United Kingdom | 14,543 | 2.99 |
| France | 12,066 | 2.48 |
| Canada | 11,900 | 2.44 |
| Brazil | 10,740 | 2.21 |
| Italy | 9,257 | 1.90 |
| Rest of World | 75,557 | 15.52 |
| **Total World** | 486,749 | – |

*Source:* Thompson et al. (2010).

**TABLE 10.21**

Total Newly Installed Wind Energy for 2016.

| Country | New Installed (MW) | Global Share (%) |
|---|---|---|
| China | 23,328 | 42.73 |
| USA | 8,203 | 15.02 |
| Germany | 5,443 | 9.97 |
| India | 3,612 | 6.62 |
| Brazil | 2,014 | 3.69 |
| France | 1,561 | 2.86 |
| Turkey | 1,387 | 2.54 |
| Netherlands | 887 | 1.62 |
| United Kingdom | 736 | 1.35 |
| Canada | 702 | 1.29 |
| Rest of World | 6,727 | 12.32 |
| **Total World** | 54,600 | – |

*Source:* Thompson et al. (2012).

be seen in Table 10.21. While China is leading the way in onshore wind, the United Kingdom and Germany have taken major strides when it comes to offshore wind energy. Together, they account for over 64% of global offshore wind. This is due largely in part to their proximity to the North Sea and English Channel regions of Europe. These waters represent some of the best locations of offshore wind projects globally. This has led to the surge of offshore wind projects seen in countries that also share proximity to these waters such as the Netherlands, Denmark, and Belgium. Table 10.22 shows the complete list of offshore wind energy installations.

On a global scale, there has been a drastic increase in wind energy installations. Over the past 15 years, there has been an over 20-fold increase in the global installed wind energy capacity placing it amongst the fastest growing and most relevant renewable energy sources. It can quickly be ascertained that the prevalence and impact of wind energy is on the rise globally. The technology has seen particular implementation in developed countries; and as infrastructural and construction costs fall, the rest of the world is likely to follow suit. The applicability of the technology in most regions of the world coupled with the high potential for vast amounts of clean energy generation make the technology impossible to ignore. It is likely that this technology will see even further implementation globally moving into the future.

### 10.8.1 Wind Energy to Electricity Conversion

Wind energy to electricity conversion as well, is inherently reliant on the first law of thermodynamics. In this way, wind turbines convert the energy inherent to wind into electricity using a turbine. Wind energy, also known as kinetic energy, $E_k$, can be represented by the following equation:

$$E_k = \frac{1}{2}mv^2 \tag{10.2}$$

**TABLE 10.22**

Total Installed Offshore Wind Energy.

| Country | Installed Capacity (MW) | Global Share (%) |
|---|---|---|
| United Kingdom | 5,156 | 35.85 |
| Germany | 4,108 | 28.56 |
| China | 1,627 | 11.31 |
| Denmark | 1,271 | 8.84 |
| Netherlands | 1,118 | 7.77 |
| Belgium | 712 | 4.95 |
| Sweden | 202 | 1.40 |
| Japan | 60 | 0.42 |
| South Korea | 35 | 0.24 |
| Finland | 32 | 0.22 |
| USA | 30 | 0.21 |
| Ireland | 25 | 0.17 |
| Rest of the World | 8 | 0.06 |
| **Total World** | 14,384 | – |

*Source:* Thompson et al. (2012).

where, $m$ is the mass and $v$ is the velocity. To appropriate this better for wind, a few aspects of the equation must be altered. Firstly, mass of air is difficult to quantify directly; and for that reason, mass in the sense of wind can be thought of as density of air multiplied by the wind speed, the rotor area, and the time (the air is flowing through that area). In order to conceptualize this, consider wind blowing through an open window. Obviously, if the wind is blowing faster, then more wind (i.e. more mass) makes its way through the window. Much in the same way, for a consistent wind speed, a window with a larger opening will see more wind pass through it. Finally, if the window is open for a longer time with all other variables held constant, then more wind makes its way through the window. Taking all of this into consideration, the kinetic energy equation for wind on a wind turbine can then be rewritten as:

$$E_k = \frac{1}{2}\rho_a A_T v^3 t \qquad (10.3)$$

where $\rho_a$ represents the density of air, $A_T$ represents the turbine rotor area, $v$ is the velocity of air, and $t$ is the time (Mathew, 2006). While this equation paints a nice picture of energy production for a wind turbine, it is often more useful to calculate power generation as opposed to energy production when considering electrical energy conversion. And power, $P$, is simply the energy produced over time, so to get the equation we simply move the time component and receive the following (Mathew, 2006):

$$P = \frac{1}{2}\rho_a A_T v^3 \qquad (10.4)$$

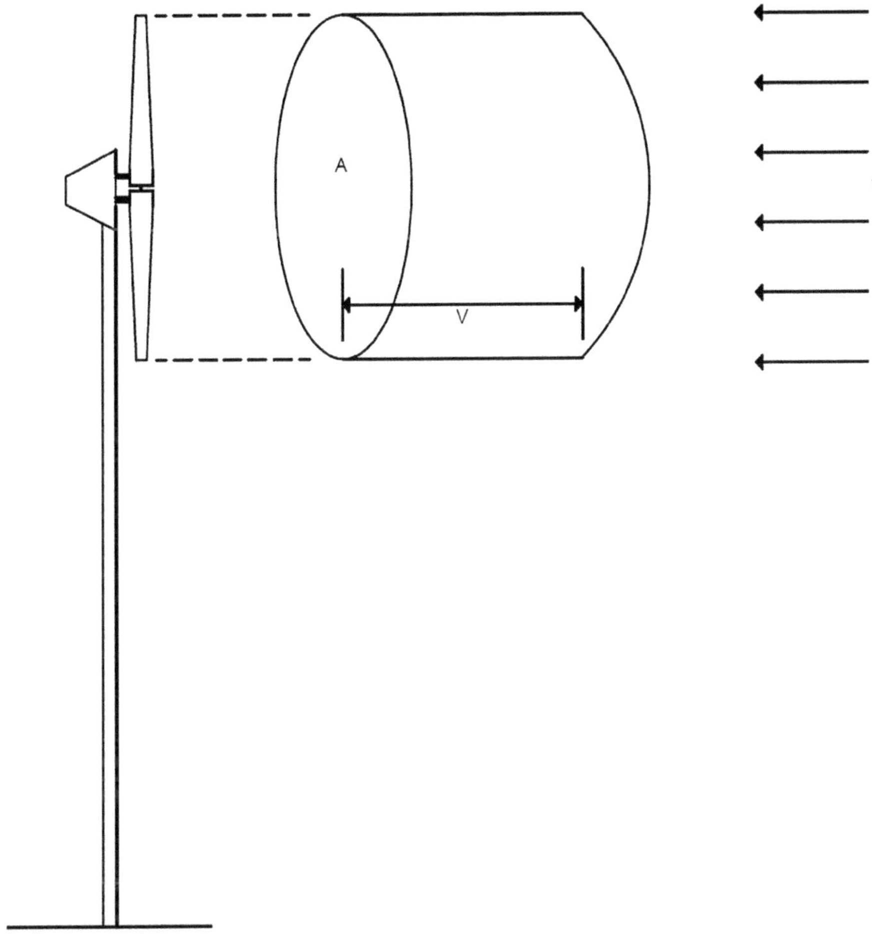


**FIGURE 10.9**
Air parcel moving toward a turbine (MacEachern & Yildiz, 2018).

In applying this equation to a practical scenario, what can be seen is that, engineers truly have influence only over one of the factors contributing directly to power generation. Air density is not something that can be controlled, and while turbines can be positioned in areas of greater average wind velocity, this still cannot be considered as a truly controllable factor. As such, the major way in which turbine power generation can be increased is by increasing the rotor area, which can be seen in Figure 10.9. From the power equation, it can simply be observed that a larger swept area will result in greater power generation, and while this may be the case theoretically, in practice, it is not that simple. This is because a greater rotor area comes with the need for increased wind speeds to efficiently turn the larger rotor. This puts emphasis on the need to design turbines on a location-to-location basis, considering aspects such as average wind velocity, peak wind velocity, elevation, and diurnal wind fluctuations to name a few. We must also remember the *Betz limit*, which states that no turbine can capture more than 16/27 (59.3%) of the kinetic energy in wind, and adjust Equation (10.4) accordingly as done in the following numerical examples.

**Example 10.8: Conversion of Wind Energy to Electricity**

Calculate the maximum achievable power output of a wind turbine of blade radius 2 m when the wind speed is 10 m/s.

**Solution**

Assuming the air density of 1.225 kg/m$^3$ at sea level and 15°C, using Equation (10.4) and remembering the Betz limit, which states that no turbine can capture more than 16/27 (59.3%) of the kinetic energy in wind, the maximum achievable power output can be calculated as follows:

$$P = \frac{1}{2}\rho_a A_T v^3$$

$$P = \frac{1}{2}\rho_a \left[\left(\pi D^2\right)/4\right] v^3$$

where $\rho_a$ represents the density of air, $A_T$ represents the turbine rotor area, $v$ is the velocity of air, and $t$ is the time (Mathew, 2006). Taking the Betz limit into account:

$$P = (0.59)\frac{1}{2}\left(1.225\,\text{kg/m}^3\right)\left(\pi\left[4\,\text{m}\right]^2/4\right)\left(10\,\text{m/s}\right)^3$$

$$P = 4{,}539\ \text{W} = 4.5\ \text{kW}$$

And this capacity would be enough to provide electricity to about four average homes.

**Example 10.9: Conversion of Wind Energy to Electricity**

Calculate the maximum achievable power output of a wind turbine of blade radius 4 m (double the size in the previous example) and when the wind speed is exactly the same as in the previous example, which is, 10 m/s.

**Solution**

Again, assuming the air density of 1.225 kg/m$^3$ at sea level and 15°C, using Equation (10.4) and taking the Betz limit into account, the maximum achievable power output for the doubled blade radius can be calculated as follows:

$$P = (0.59)\frac{1}{2}\left(1.225\,\text{kg/m}^3\right)\left(\pi\left[8\,\text{m}\right]^2/4\right)\left(10\,\text{m/s}\right)^3$$

$$P = 18{,}156\ \text{W} = 18.2\ \text{kW}$$

Note: Compared to Example 10.8, an increase in blade diameter from 4 to 8 m will yield $(2)^2$ or 4 times more power from the wind turbine.

**Example 10.10: Conversion of Wind Energy to Electricity**

Calculate the maximum achievable power output of a wind turbine of blade radius 6 m (triple the size in Example 10.8) and when the wind speed is exactly the same as in the previous example, which is, 10 m/s.

**Solution**

Assuming the air density of 1.225 kg/m³ at sea level and 15°C, using Equation (10.4) and taking the Betz limit into account, the maximum achievable power output for the doubled blade radius can be calculated as follows:

$$P = \frac{1}{2}\rho_a\left(\pi\left[D^2\right]/4\right)v^3$$

Taking the Betz limit into account:

$$P = 40{,}850\ \text{W} = 40.9\ \text{kW}$$

Note: Compared to Example 10.8, an increase in blade diameter from 4 to 12 m will yield $(3)^2$ or 9 times more power from the wind turbine.

### Example 10.11: Conversion of Wind Energy to Electricity

Calculate the maximum achievable power output of a wind turbine of blade radius 2 m and when the wind speed is double the speed as in Example 10.8, which is, 20 m/s.

**Solution**

Again, assuming the air density of 1.225 kg/m³ at sea level and 15°C, using Equation (10.4) and taking the Betz limit into account, the maximum achievable power output for the doubled wind speed can be calculated as follows:

$$P = (0.59)\frac{1}{2}\left(1.225\ \text{kg/m}^3\right)\left(\pi\left[4\ \text{m}\right]^2/4\right)\left(20\ \text{m/s}\right)^3$$

$$P = 36{,}311\ \text{W} = 36.3\ \text{kW}$$

Note: Compared to Example 10.8, an increase in wind speed from 10 to 20 m/s will yield $(2)^3$ or 8 times more power from the wind turbine.

### Example 10.12: Conversion of Wind Energy to Electricity

Calculate the maximum achievable power output of a wind turbine of blade radius 2 m and when the wind speed is triple the speed as in Example 10.7, which is, 30 m/s.

**Solution**

Assuming the air density of 1.225 kg/m³ at sea level and 15°C, using Equation (10.8) and again taking the Betz limit into account, the maximum achievable power output for the doubled wind speed can be calculated as follows:

$$P = (0.59)\frac{1}{2}\left(1.225\ \text{kg/m}^3\right)\left(\pi\left[4\ \text{m}\right]^2/4\right)\left(30\ \text{m/s}\right)^3$$

$$P = 122{,}550\ \text{W} = 122.6\ \text{kW}$$

Note: Compared to Example 10.8, an increase in wind speed from 10 to 30 m/s will yield $(3)^3$ or 27 times more power from the wind turbine.

The calculated maximum theoretical outputs are for the wind turbines converting 59% of the winds energy into usable output. In reality, due to the aerodynamic imperfections, and mechanical and electrical losses, the calculated numbers have to be multiplied by about 0.5–0.7. The examples show the importance of blade diameter and turbine siting; that is, the turbine blade should be large, and the support tower for the turbine must be as high as possible.

## 10.9 Solar Energy

Solar energy usage can be broadly categorized into two categories, active and passive. Active solar systems have the active components such as pumps and fans, and also take advantage of the photovoltaic (PV) effect to directly convert sunlight into electricity. The photovoltaic effect can be best described as the appearance of an electric voltage between two electrodes as a result of introducing a light into the system. The electrodes can be attached to either a solid or a liquid medium depending on the system (Goetzberger et al., 2002). The result of this phenomenon is usable electricity, whose generation is achieved through the use of PV cells, comprised of a semiconductor. The design of PV cells was first created by Bell Laboratories in 1954 and this original PV cell had a conversion efficiency of 6%, though the design was quickly improved upon to reach efficiencies of up to 10%. These efficiencies were achieved through the use of a silicon semiconductor, which, when exposed to sunlight, give up electrons from their atom. By attaching conductors to the terminals in the cell, these electrons can be captured as an electrical current. Depending on the semiconductor material, the reaction can be anywhere from 7% to 40% efficient (Guney, 2016). Even today, with modern PV cells, the most often employed semiconductor is silicon due to its ability to be worked easily with minimum defects (Mayer, 2016). While silicon might not be the best semiconductor, it is cheaper, more available, and more reliable than its competitors. Modern solar cells arranged in solar panels and subsequently organized into a solar array. For smaller operations, one array may suffice or multiple arrays ranging into the thousands can be constructed for larger active solar farms.

Passive solar is a method of designing objects and buildings so as to take best advantage of the Sun's heat and light. Wide varieties of passive solar techniques have been used for thousands of years for a variety of purposes. Even today, passive solar has a large effect on many construction and design choices such as greenhouses, which are perfect passive solar systems. For example, in the northern hemisphere, houses with more south facing windows can utilize the Sun's energy to provide home heating and eliminate some of the cost of supplemental heating methods. Another example of passive solar would be painting an outdoor water storage tank black to preheat the water and reduce the amount of heat and energy required for producing hot water (Guney, 2016).

While details of the invention of solar thermal energy conversion are sparse, the consensus is that the process was invented in 1615 by the French engineer Solomon de Caux describes a device he created which was able to pump water by concentrating solar energy to cause a desired expansion of air within the system. This air would then, as a result of the expansion, move water throughout the pumping system. While this was the start of solar thermal energy conversion, it was not until around 1854 that solar thermal energy conversion took the form we know today. Modern solar thermal energy conversion to electricity is generally performed under specific conditions. Solar thermal energy conversion can be performed by either concentrating or non- concentrating collectors. With that being said, concentrating collectors are able to achieve far higher thermodynamic efficiencies. Concentrating collectors use mirrors to focus sunlight onto an absorber tube or a central tower collector. Regardless of the collection unit, the principle is largely the same. The concentrated sunlight heats a fluid (often synthetic oil) whose heat is then used to boil water. The steam generated from the boiling water is used to turn turbines and subsequently generate electricity (Tian & Zhao, 2013).

Characteristics of solar radiation, global solar energy resources, numerical examples of solar energy and solar power applications are provided in Chapter 2 earlier. For more information, the readers are referred to Chapter 2.

## 10.10 Geothermal Energy

Geothermal energy can most simply be defined as the heat energy contained within the Earth. Geothermal energy usage truly took off following the Second World War. The technology was seen as being economically competitive as well as being readily available in areas where other technologies were not feasible. Today, geothermal energy is used in 23 countries for electricity generation and 58 for use in other non-electrical applications including greenhouse heating.

Geothermal energy can capture heat from the natural earth, and it is regarded as an environmentally friendly energy source (Rybach, 2003). The wells utilized in geothermal production are usually 2–3 km deep, and the temperatures in the well bottom can easily reach to 90–100°C (WEC, 2013). According to the temperature of the heat, the geothermal resource can be classified into several types, and they differ in their functions. For instance, the electric power is typically contributed by the resources above 150°C, while those resources below 150°C can primarily be used in heating or cooling systems (Lund, 2006; White & Williams, 1975). A geothermal system is comprised of three primary elements: a heat source, a reservoir, and a fluid to transfer the heat. The heat source can either be the Earth's heat itself – obtained by reaching an appropriate drilling depth or a magmatic intrusion, which can be found in some cases at shallower depths. The reservoir is an area of permeable rocks, which transfer heat to the fluid as the fluid passes through the voids in the rock. The fluid works in a cycle, being heated underground and bringing that heat to the surface, the cooled fluid then makes its way back underground to be heated once more and the cycle continues (Dickson & Fanelli, 2013). This setup is often seen in homes or greenhouses, which are heated by geothermal energy. In addition to heating, when high temperature resources are available, slight alterations can be made to the design so that the generated steam drives a turbine attached to a generator. This process can be used to generate renewable electricity at a low cost.

As geothermal activity varies globally, select locations are better than others are for electricity generation. Some of the world's largest facilities are located in California, Iceland, Indonesia, Mexico, the Philippines, and Turkey. However, geothermal energy only shares a small market of the world's primary energy consumption compared with other renewables, such as wind and solar energy discussed earlier. The reason is likely because geothermal energy requires higher installation cost and heavily relies on government subsidies (UN, 2014). WEC (2016) reported that global investment in geothermal energy was US$2 billion, with an additional 315 MW new energy capacity in 2015. Typically, about half of the new capacity was contributed by Turkey, followed by the United States and Mexico.

## 10.11 Ocean Energy

### 10.11.1 Wave Energy

Wave energy has the largest share (80%) of marine energy, and it would be sufficient to supply renewable energy to the whole world if efficiently converted into electricity. Wave energy is the most promising form of marine energy, and widely available around the world (Figure 10.10), and currently, the wave energy harnessing technologies are under active research and development. The uneven heating of surface and therefore air by the

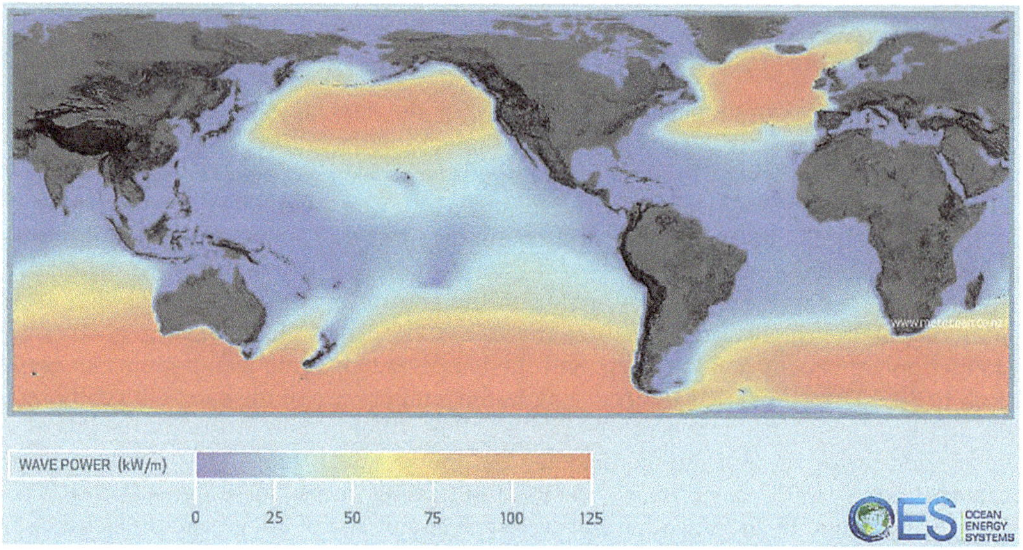


**FIGURE 10.10**
Global annual average wave energy distribution. Courtesy of International Renewable Energy Agency (IRENA, 2014).

Sun causes winds that whip over the surface of the oceans, creating a wide range of waves with a lot of energy. Therefore, in essence, wave energy is an indirect form of solar energy (USDI, 2006; Muetze & Vining, 2006). Waves have very high power density; typically 30 kW/m, which is about 10 and 5 times greater than those of solar and wind, respectively. Furthermore, the loss of wave energy while they propagate across the oceans is quite small (Clément et al., 2002).

Wave energy is widely available across the oceans; the annual wave energy extraction potential is somewhere between 8,000 and 80,000 TWh (IEA, 2016). The near-shore areas with the most abundant wave energy resources include the west coast of the tip of South America, the south coast of Australia, the west coast of Europe, and the west coast of North America (WorldWave, 2017). In the case of Europe, for instance, 290 GW of wave power can potentially be harnessed off the west coast of Europe, and another 30 GW of potential exists along the Mediterranean coast. The wave energy potential is about 25 kW/m near the Canary islands, 75 kW/m off the coast of Ireland and Scotland, 30 kW/m off the Norwegian coast, 10–21 kW/m in the North Sea, and 4–11 kW/m in the Mediterranean (Clément et al., 2002).

The wave energy flux (rate of transfer) $P$ (W/m) is defined as follows (Falnes, 2002):

$$P = \frac{\rho g^2 T_e H_{mo}^2}{64\pi} \tag{10.5}$$

where $\rho$ is the density of sea water (kg/m$^3$), $g$ is the gravitational acceleration (9.81 m/s$^2$), $T_e$ is the wave energy period (s), and $H_{mo}$ is the significant wave height (m).

### 10.11.2 Tidal Energy

Tidal energy is a clean and renewable form of energy that is expected to be widely exploited in the coming years (see Figure 10.11). The practice of trapping tidal waters behind a

**Example 10.13: Calculating the Energy Flux of a Wave**

Determine the energy flux of a wave with an energy period of 5 seconds and significant wave height of 1.0 m.

Assume that the density of sea water is 1,020 kg/m³.

**Solution**

$$P = \frac{\rho g^2 T_e H_{mo}^2}{64\pi} = \frac{1{,}020 \times 9.81^2 \times 5 \times 1.0^2}{64\pi} = 2{,}442\ \text{W/m} = 2.4\ \text{kW/m}$$

**Example 10.14: Calculating $CO_2$ Emission Reduction by a Wave Power Plant**

Now consider a greenhouse operation located in a coastal area. The operation's average electric energy consumption is reported to be 300 MWh per month. The utility company that provides the electricity to the greenhouse operation is using fossil fuel sources in the power generation mix. The coal-fired power plants have a 35% share and the natural gas fired power plants have a 30% share in their power generation mix. The remaining power comes from clean energy sources without any emissions. Assume that the coal and natural gas power plants have 30% and 40% efficiencies, respectively; and that the average heating values of coal (C) and natural gas ($CH_4$) are 30 MJ/kg and 50 MJ/kg, respectively.

1. Ignoring transmission losses, determine the total amount of $CO_2$ released per year by the utility company due to the regular electricity consumption of this greenhouse complex.
2. Due to the abundant ocean wave resources in the proximity, the greenhouse complex decides to install a clean wave energy conversion plant itself rated at 500 kW. Determine the amount of $CO_2$ emissions reduced per year due to the operation of this new plant at 50% capacity factor.

**Solution**

1. Let's first determine the carbon footprint of the greenhouse complex:
   Total electric energy demand of the greenhouse complex in a year is 12 × 300 MWh = 3,600 MWh
   Coal's share of the total demand is 35% of 3,600 MWh = 1,260 MWh
   With 30% power plant efficiency, the total amount of coal energy used is 1,260/0.3 = 4,200 MWh

Since 1 MWh = 3,600 MJ; then, the total amount of coal used can be determined as follows:

$$(4{,}200\ \text{MWh})(3{,}600\ \text{MJ/MWh})/(30\ \text{MJ/kg}_{\text{coal}}) = 504{,}000\ \text{kg}_{\text{coal}} = 504\ \text{ton}_{\text{coal}}$$

$CO_2$ has a molecular weight of 44 g/mol, of which 12 g is C. Hence, based on the stoichiometric equation ($C + O_2 \rightarrow CO_2$), burning 12 g of coal (C) will generate 44 g of $CO_2$. Then, the total amount of $CO_2$ generated by burning coal is:

$$(2{,}700\ \text{MWh})(3{,}600\ \text{MJ/MWh}) / (50\ \text{MJ/kg}_{\text{gas}}) = 194{,}400\ \text{kg}_{\text{gas}} = 194.4\ \text{ton}_{\text{gas}}$$

Natural gas's share of the total demand is 30% of 3,600 MWh = 1,080 MWh

With 40% power plant efficiency, the total amount of natural gas energy used is 1,080/0.4 = 2,700 MWh

The total amount of natural gas used can then be calculated as follows:

$$(2{,}700\ \text{MWh})(3{,}600\ \text{MJ/MWh})/(50\ \text{MJ/kg}_{gas}) = 194{,}400\ \text{kg}_{gas} = 194.4\ \text{ton}_{gas}$$

In case of natural gas ($CH_4$), the molecular weight is 16 g/mol, of which 12 g is C. Hence, based on the stoichiometric equation ($CH_4 + 2O_2 \rightarrow CO_2 + 2H_2O$), burning 16 g of natural gas will generate 44 g of $CO_2$. Then, the total amount of $CO_2$ generated by burning natural gas is:

$$194.4\ \text{ton of natural gas} \times (44\ \text{g}\ CO_2/16\ \text{g natural gas}) = 534.6\ \text{ton of}\ CO_2$$

Then,
Total $CO_2$ generated by both fossil energy sources = 1,848 ton + 534.6 ton = 2,382.6 ton

2. For this part, we need to calculate the reduction in $CO_2$ emissions:
   Total energy produced per year using the wave energy conversion plant can be calculated as follows:

$$500\ \text{kW} \times (24\ \text{h/day}) \times (365\ \text{days/year}) \times (0.5) = 2{,}190{,}000\ \text{kWh} = 2{,}190\ \text{MWh}$$

If the calculations above for the greenhouse operation are repeated for 2,190 MWh rather than 3,600 MWh, then the total amount of reduction in $CO_2$ emissions due to coal burning would be 1,124.2 ton, and the total amount of reduction in $CO_2$ emissions due to natural gas burning would be 325.2 ton. Then, the total amount of reduction in $CO_2$ emissions if the greenhouse complex starts using a wave plant would be:

$$1{,}124.2(\text{coal}) + 325.2(\text{natural gas}) = 1{,}449.4\ \text{ton of}\ CO_2$$

barrier in order to capitalize on the potential energy is far from a modern concept. What is a modern practice, however, is the exploitation of tidal waters for the generation of electricity. Due to the immense potential and reliability of tidal waters, they make an excellent candidate for the implementation of tidal energy systems. Tidal range power generation functions by taking advantage of the head difference generated between two bodies of water. The difference is created by separating the two water bodies by means of a wall. As the tide moves in and out, the wall is closed to detain water, generating the head difference. Once an optimal difference has been achieved, port holes in the wall are opened, allowing water to make its way back toward the lower potential. With two tidal cycles per day, the process can be performed four times. As the water passes through the portholes, it is used to turn turbines, which in turn generate electricity. The process is then repeated with the tidal cycles (Waters & Aggidis, 2016). In general, the velocities of ebb and flood should be higher than 2.5 m/s to get sufficient tidal energy.

Tidal stream turbines function in practice quite similarly to wind turbines. A dynamic fluid is used to turn a turbine and as a result generate electricity. The difference between them is that for tidal stream turbines the dynamic fluid is water, not wind. These turbines are either fastened to the seafloor or tethered by a cable in areas of elevated tidal activity. As the tide moves in and out, the turbine is turned. Tidal stream turbines take up less space

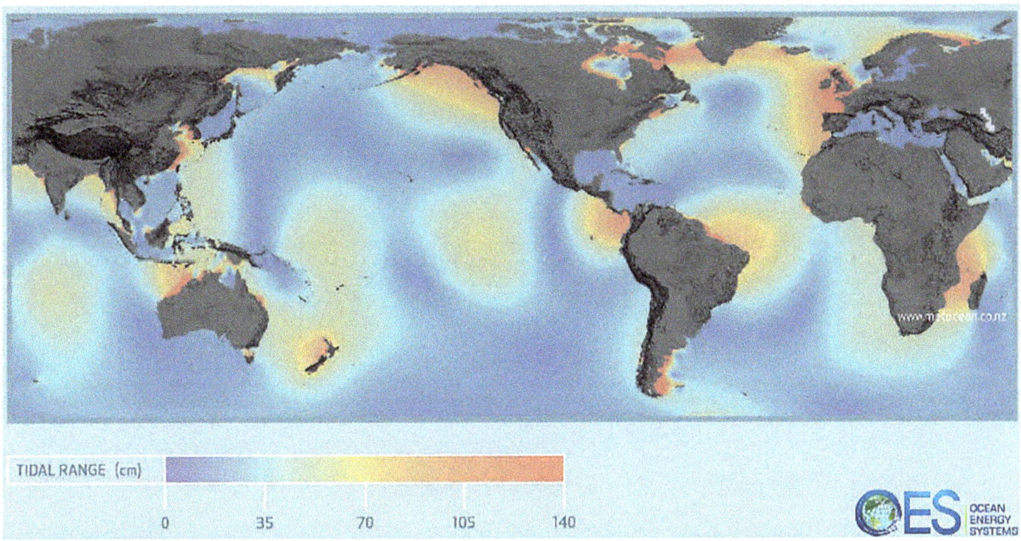


**FIGURE 10.11**
Global tidal range energy distribution. Courtesy of International Renewable Energy Agency (IRENA, 2014).

and have a far lower impact than tidal range; however, they do not generate as much electricity. There are currently three employed and conceptualized designs for tidal stream generators, horizontal axis turbines (similar to wind turbines), vertical axis turbines, and hydrodynamic lift force energy devices. Currently, only the horizontal axis turbines are employed on any large-scale grid connected designs (Lewis et al., 2015).

The technology for tidal range energy is relatively mature and a number of tidal range power plants in different parts of the world are operating very well. However, large investment required by tidal range power plants prevents them from being more widely used (Rourke et al., 2010). If the economic challenge could be solved effectively, more tidal range power plants could be built in many bay areas to harness the enormous energy in ocean tides.

### 10.11.3 Other Ocean Energy Resources and Technologies

Ocean thermal energy (see Figure 10.12), ocean current energy, and salinity gradient are other ocean energy resources.

Quite a significant portion of solar energy reaching the ocean surface is absorbed and stored as thermal energy in the upper layers of the ocean and as shown in Figure 10.12, the resource is primarily distributed in the tropics. The temperature gradient between the ocean surface and the deep ocean water can be harnessed using different ocean thermal energy conversion (OTEC) technologies that can provide especially island populations with continuous clean and renewable energy sources.

Ocean currents are, on the other hand, driven by spatial distributions of atmospheric and thermohaline ocean circulations. Compared to tidal currents, they are much slower; however, as shown in Figure 10.13, they are more continuous renewable energy sources around the world.

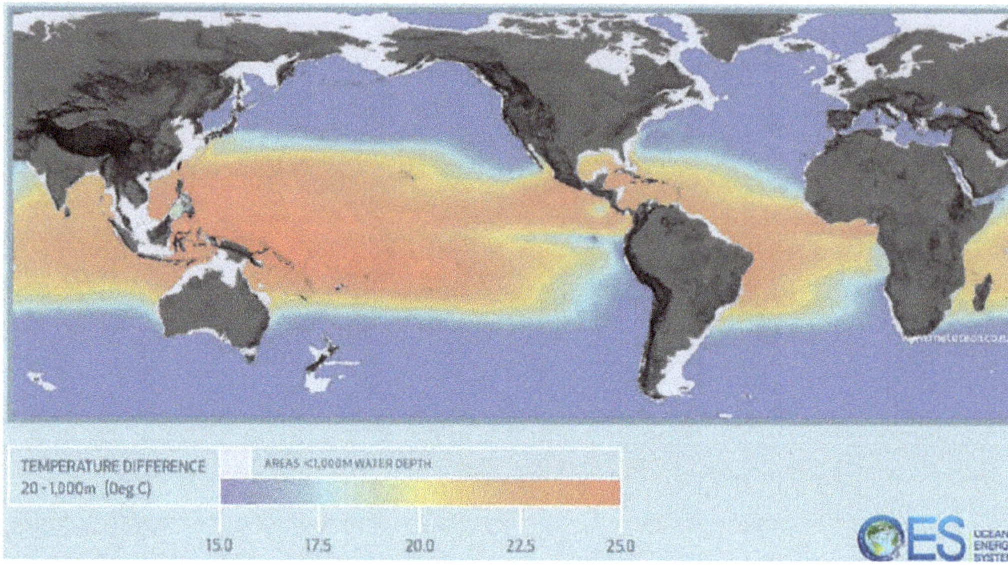


**FIGURE 10.12**
Global ocean thermal energy resource distribution. Courtesy of International Renewable Energy Agency (IRENA, 2014).

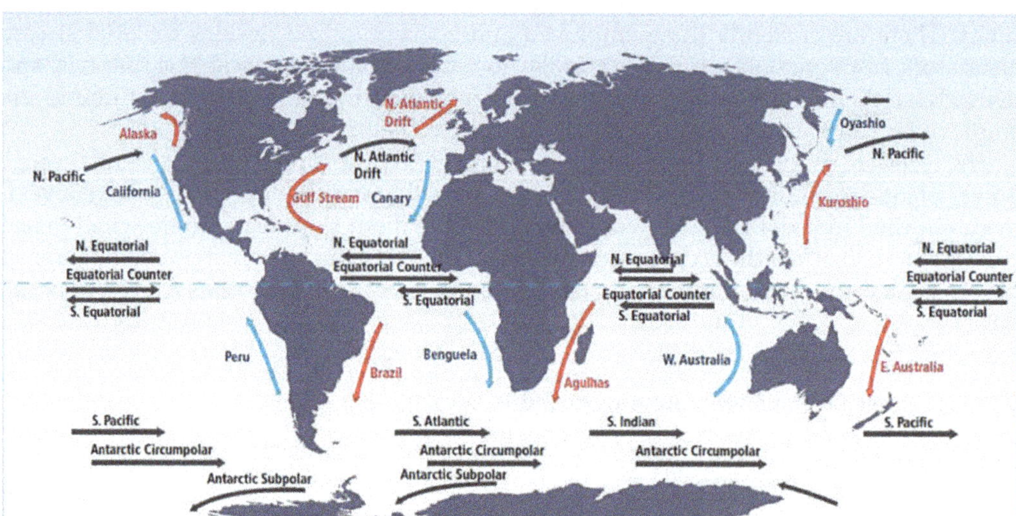


**FIGURE 10.13**
Ocean surface currents showing warm (red) and cold (blue) current systems. Courtesy of International Renewable Energy Agency (IRENA, 2014)

Salinity gradient is another globally distributed potential ocean energy resource located where freshwater from rivers meets saline seawater as shown in Figure 10.14, which can be cultivated using osmotic power technologies. The main obstacle standing against this energy resource is the cost of membranes, which is currently preventing the commercialization of salinity gradient power plants. When cost-effective conversion technologies

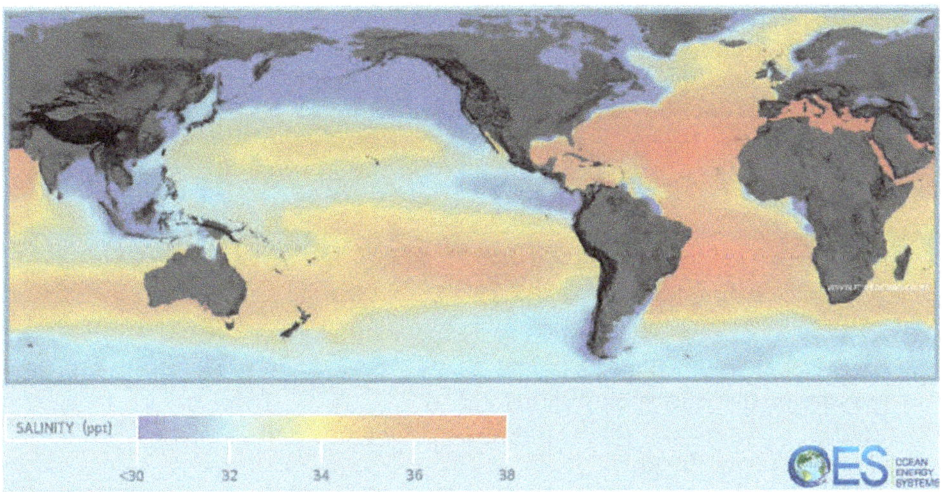


**FIGURE 10.14**
Global salinity gradient potential energy resource distribution. Courtesy of International Renewable Energy Agency (IRENA, 2014).

become available, this local energy resource can as well start providing our communities with renewable energy.

## 10.12 Heat Pumps

A heat pump operates by transferring heat from a source to a sink through a refrigeration cycle. The refrigeration cycle functions by changing the pressure at certain points within the loop in order to vaporize and condense a refrigerant. This change in state either releases or absorbs heat depending on the direction of the state change. This process has a dual benefit of being able to provide heating in the winter or cooling in the summer simply by reversing the flow of the system (Hepbasli & Kalinci, 2009). As discussed earlier, heat pumps have a great potential for conserving energy and water resources as well as insect-free operations in closed or semi-closed greenhouse operations.

The first heat pumps were actually designed as icemakers and for cooling practices in the food industry. The first practical heat pump was designed and built for just this purpose in 1834 by Jacob Perkins. From this point until 1875, a number of small improvements were made to the system including the implementation of various refrigerants including, ammonia, methyl ether, and carbon dioxide. In 1875, the Polytechnic Society of Munich began its comparative analysis of refrigeration technologies, which led to most of the significant advancements with the technology. By the year 1918, ammonia had separated itself as the best refrigerant and a number of compressor manufacturers emerged globally. It was at this time when refrigeration for space heating (heat pumps) began to garner attention. This was largely due to the advancements made with electric motors, which made running the compressors more economically viable for heating larger spaces (Zogg, 2008). Heat pumps continued to develop throughout the twentieth century though for a variety of reasons never saw truly wide-scale implementation. In recent years, however, the technology has seen a dramatic increase in installations. Domestic scale heat pumps have seen an upturn in sales for a variety of reasons including improved design, more reliable equipment, energy security concerns, environmental concerns, and the increased price of fossil fuels.

## Nomenclature

| | |
|---|---|
| $A$ | area |
| $E$ | energy |
| $g$ | gravitational acceleration (9.81m/s$^2$) |
| $H$ | head |
| $H_{mo}$ | significant wave height |
| $m$ | mass |
| $\dot{m}$ | mass flow rate |
| $P$ | power; wave energy flux (rate of transfer) |
| $T_e$ | wave energy period |
| $TS$ | total solids |
| $v$ | velocity |
| $VS$ | volatile solids |

### Greek letters

| | |
|---|---|
| $\rho$ | density |
| $\eta$ | efficiency |

### Subscripts

| | |
|---|---|
| a | air |
| k | kinetic |
| t | time |
| $T$ | turbine |

## Bibliography

Abu-Khader MM (2009) Recent advances in nuclear power: A review. *Progress in Nuclear Energy*, 51(2):225–235.

Allen L, Cohen MJ, Abelson D, Miller B (2012) Fossil fuels and water quality, In *The World's Water, Island Press/Center for Resource Economics*, 4: 74–96.

Arent D, Sullivan P, Heimiller D, Lopez A, Eurek K, Badger J, et al. (2012) Improved offshore wind resource assessment in global climate stabilization scenarios. *Contract*; 303:275–3000.

Barreiro DL, Prins W, Ronsse F, Brilman W (2013) Hydrothermal liquefaction (HTL) of microalgae for biofuel production: State of the art review and future prospects, *Biomass and Bioenergy*, 53 (2013) 113–127.

Baxter L (2005) Biomass-coal co-combustion: Opportunity for affordable renewable energy. *Fuel*, 84(10):1295–1302.

Beschloss M (2014) U.S. Now World's Leading Natural Gas Producer, Dessert Sun, 2 September 2014.

BIOFUELSTP (2009) European Biofuels Technology Platform, Global Biofuels: An Overview, 2009.

Blander M (2017) *Calculations of the Influence of Additives on Coal Combustion Deposits, Argonne National Laboratory*, p. 315, http://web.anl.gov/ Retrieved 2017-10.21.

Bloomberg (2012) *South Africa Allows Exploration of Shale Gas Resources*, www.bloomberg.com Retrieved 2012-10.18.

BP (2010) *BP Statistical Review of World Energy June 2010*, https://commons.wikimedia.org/wiki/File:2007_Coal_Reserves_in_CTUs.png Retrieved 2010.09-28.

BP (2013) *Oil reserve definitions*, https://www.bp.com, Retrieved 2013-12-04.

BP (2017) http://bp.com/statisticalreview Retrieved 21-10.2017.

CIA (2013) Natural gas – proved reserves, https://www.cia.gov Retrieved 2013-12-01.

Clément A, McCullen P, Falcão A, et al. (2002) Wave energy in Europe: current status and perspectives, *Renewable and Sustainable Energy Reviews*, 6(5): 405–431.

Daniel G, He L, Jensen KL, English BC (2010) Expanded ethanol production: Implications for agriculture, water demand, and water quality, *Biomass and Bioenergy*, 34(11): 1586–1596.

DeCicco JM (2013) Biofuel's carbon balance: Doubts, certainties and implications. *Climatic Change*, 121(4):801–814.

Dickson MH and Fanelli M (2013) *Geothermal energy: Utilization and technology*. London: Routledge.

Ebenezer SA, Gudmunsson JS (2005) *Removal of carbon dioxide from natural gas for LNG production*, Semester project work.

EESI (2017) *Data*. http://www.eesi.org/ Retrieved 2017-10.21.

Egorov RI and Strizhak PA (2013) The light-induced gasification of waste-derived fuel. *Fuel*. ;197:28–30.

EIA (2010) *International Energy Outlook 2010*, Washington DC, 2010.

EIA (2012) *International Energy Statistics. Coal: Recoverable Reserves*, U.S. Department of Energy, Washington, D.C., https://www.eia.gov Retrieved 2012-05-31.

EIA (2013) *International Statistics*, U.S. Department of Energy, Washington, D.C., 2013.

EIA (2016) *United States Energy Information Administration, International Energy Outlook 2016*. Washington, DC.

Energy Independence and Security Act (2007) *Energy Independence and Security Act of 2007*. Public Law, pp. 10.140.

Energyseer (2017) *Strategic Energy & Economic Research Inc.*, http://www.energyseer.com/ Retrieved 2017-10.21.

Falnes J (2002) *Ocean Waves and Oscillating Systems*, Cambridge University Press.

FAOSTAT (2019) Data. Retrieved from FAOSTAT, http://www.fao.org/faostat/en/?#home Retrieved 2019-09-10.

Financial Times (2012) Poland Seeks to Boost Shale Gas Industry, 18 October 2012.

Goetzberger A, Luther J, Willeke G (2002) Solar cells: past, present, future. *Solar Energy Materials and Solar Cells*, 74(1):1–1.

Goodman LM (2011) *The Asylum: The Renegades Who Hijacked the World's Oil Market*. William Morrow and Company, New York.

Guney MS (2016) Solar power and application methods. *Renewable and Sustainable Energy Reviews*, 57:776–785.

Hansen MO (2015) *Aerodynamics of wind turbines*. London, Routledge.

Hepbasli A and Kalinci Y (2009) A review of heat pump water heating systems. *Renewable and Sustainable Energy Reviews*, 13(6):1211–1229.

Hoornweg D and Bhada-Tata P (2012) What a waste: a global review of solid waste management. *World Bank Urban development series knowledge papers*,15:1–98.

Horttanainen M, Teirasvuo N, Kapustina V, Hupponen M, Luoranen M (2013) The composition, heating value and renewable share of the energy content of mixed municipal solid waste in Finland. *Waste Manage*, 33(12):2680-2686.

Hoyos C (2007) Study sees harmful hunt for extra oil, Financial Times, 18 February 2007.

Huckerby JA, Jeffrey H, and Moran B (2011) *An International Vision for Ocean Energy. Ocean Energy Systems Implementing Agreement.* http://www.oceanrenewable.com/wp-content/uploads/2011/05/oes_vision_brochure_2011.pdf (Retrieved 10.02-2014)

Hupa M (2012) Ash-related issues in fluidized-bed combustion of biomasses: Recent research highlights. *Energy and Fuels*, 26: 4–14.

IEA (2012a) *International Energy Agency, Key World Energy Statistics*, OECD, Paris.

IEA (2012b) *International Energy Agency, World Energy Outlook 2012*, OECD, Paris.

IEA (2013) *International Energy Agency, World Energy Outlook 2013*, OECD, Paris.

IEA (2016) International Energy Agency – Ocean Energy Systems, Implementing agreement on ocean energy systems, Annual Report 2007. http://pt21.ru/docs/pdf/28_e.pdf Retrieved 2016-10.10.

IEA (2019) *Statistics*. Retrieved from https://www.iea.org/statistics Retrieved 2019-12-15.

IRENA (2014) Ocean Energy: Technology Readiness, Patents, Deployment Status and Outlook.http://www.irena.org/DocumentDownloads/Publications/IRENA_Ocean_Energy_report_2014.pdf Retrieved 2016-10.10.

IRENA (2019) *Renewable Energy and Jobs – Annual Review 2019*. Abu Dhabi: IRENA.

Jackson RB, Pearson BR, Osborn GS, Warner NR, Vengosh A (2011) *Research and policy recommendations for hydraulic fracturing and shale-gas extraction, Center on Global Change*, Duke University, Durham, NC

Jalilvand DR (2013) *Iran's Gas Exports: Can Past Failure Become Future Success?* Oxford Institute for Energy Studies.

Kanellos M (2017) In Natural Gas, U.S. Will Move From Abundance to Imports, www.greentechmedia.com Retrieved 2017-10.21.

Khademi F and Yildiz I (2018) Energy and Solid Wastes. In: Dincer, I. (ed.), *Comprehensive Energy Systems*. Vol. 1, Oxford: Elsevier. pp. 980–1020.

Khatib Z and Verbeek P (2003) Water to value-produced water management for sustainable field development of mature and green fields, *Journal of Petroleum Technology*, 55: 26–28.

Knight H (2010) Wonderfuel: Welcome to the age of unconventional gas, New Scientist, (June 2010) 44–7.

Komarudin H, Obidzinski K, Andrianto A (2010) Biofuel Development in Indonesia: Progress and Challenges, Bogor, Indonesia, Center for International Forestry, Forests and Governance Programme (CIFOR).

Lal R and Pimentel D (2007) Biofuels from crop residues, *Soil and Tillage Research*, 93(2): 237–238.

Laws EA (1993) *Aquatic Pollution-An Introductory Text*, Wiley, New York.

Lewis A, Estefen E, Huckerby J, Musial W, Pontes T and Torres-Martinez J (2011) Ocean Energy. In *IPCC Special Report on Renewable Energy Sources and Climate Change Mitigation* [Edenhofer O, Pichs-Madruga R, Sokona Y, Seyboth K, Matschoss P, Kadner S, Zwickel T, Eickemeier P, Hansen G, Schlömer S and Von Stechow C (eds)], Cambridge University Press, Cambridge, United Kingdom and New York, NY, USA. http://srren.ipcc-wg3.de/report/IPCC_SRREN_Ch06. pdf. Retrieved 2014-10-02.

Lewis M, Neill SP, Robins PE, Hashemi MR (2015) Resource assessment for future generations of tidal-stream energy arrays. *Energy*, 83:403–415.

Liu Y and Yildiz I (2018) Energy and Water Pollution. In: Dincer, I. (ed.), *Comprehensive Energy Systems*. Vol. 1, Oxford: Elsevier. pp. 950–979.

Lund JW (2006) Chena Hot Springs, Geo-Heat Center Quarterly Bulletin, Vol. 27, No. 3 (September), Klamath Falls, OR, USA, pp. 2–4.

MacEachern C and Yildiz I (2018) Wind Energy. In: Dincer, I. (ed.), *Comprehensive Energy Systems*. Vol. 1, Oxford: Elsevier. pp. 665–701.

Manwell JF, McGowan JG, Rogers AL (2010) *Wind Energy Explained: Theory, Design and Application*. Chichester, John Wiley & Sons.

Mathew S (2006) *Wind Energy: Fundamentals, Resource Analysis and Economics*. Berlin, Springer.

Mayer M (2016) *Why are solar cells made of silicon? Berkley Energy and Resource Collaborative.* http://berc.berkeley.edu/why-are-solar-cells-made-of-silicon_1/ (accessed 09.06.16).
McIlveen-Wright DR, Huang Y, Rezvani S, Wang Y (2007) A technical and environmental analysis of co-combustion of coal and biomass in fluidised bed technologies. *Fuel*, 86(14): 2032–2042.
McLaughlin SB, Kszos LA (2005) Development of switchgrass (Panicum virgatum) as a bioenergy feedstock in the United States, *Biomass and Bioenergy*, 28(6): 515–535.
Montes A (2014) Factors Affecting Bed Agglomeration in Bubbling Fluidized Bed Biomass Boilers. Western University MSc. Thesis.
Morehouse DF (1997) The Intricate Puzzle of Oil and Gas Reserves Growth, U.S. Energy Information Administration, Natural Gas Monthly, (July 1997) vii–xx
Mouawad J (2009) Estimate Places Natural Gas Reserves 35% Higher, New York Times, 17 June 2009.
Muetze A and Vining JG (2006) Ocean wave energy conversion – a survey, 41st IEEE IAS Annual Meeting Conference Record 3, pp. 1410–1417.
NaturalGas.org (2017a) *Composition of natural gas*, http://www.naturalgas.org/overview/background Retrieved 2017-10.21.
NaturalGas.org (2017b), *Natural gas overview*, https://web.archive.org/web/20110101063224/http://naturalgas.org/ Retrieved 2017-10.21.
Nussbaumer T (2003) Combustion and Co-combustion of Biomass: Fundamentals, Technologies, and Primary Measures for Emission Reduction. *Energy and Fuels*,17:1510–1521.
Olsson, G (2012) *Water and Energy: Threats and Opportunities*, IWA Publishing, London, UK.
Omer A (2008) Energy, environment and sustainable development, *Renewable and Sustainable Energy Reviews*, 12(9): 2265–2300.
OPEC (2013) *Table 3.2 World Proven Natural Gas Reserves by Country*, http://www.opec.org/ Retrieved 2013-12-01.
OPEC (2017) OPEC *Share of World Oil Reserves 2016*, http://www.opec.org/opec_web/en/data_graphs/330.htm Retrieved 2017-10.21.
Palaniappan M and Gleick PH (2008) Peak Water, The World's Water 2008–2009, Ch 1., *Pacific Institute* (2008), pp.1–16.
Parikka, M (2004) Global biomass fuel resources. *Biomass and Bioenergy*, 27: 613–620.
Petrus L and Noordermeer MA (2006) Biomass to biofuels, a chemical perspective. *Green Chemistry*, 8(10): 861–867.
PR Newswire (2009) *Global biofuels growth to double by 2015, challenges remain for second generation technologies*, Press Release, Houston, PR Newswire Association.
REN21 (2015) *Renewable Energy Policy Network for the 21st Century, Renewables 2015*, Global Status Report, Paris.
REN21 (2016) *Renewable Energy Policy Network for the 21st Century*, Renewables 2016 Global Status Report, Paris.
Ritchie J and Dowlatabadi H (2015) Fossil Fuel Divestment: Reviewing Arguments, Implications and Policy Opportunities, *Pacific Institute for Climate Solutions*, 6 (2015) 51–74.
Rourke FO, Boyle F, Reynolds A (2010) Tidal energy update 2009, *Applied Energy*, 87(2): 398–409.
Rybach L (2003) Geothermal Energy: Sustainability and the Environment, *Geothermics*, 32(4): 463–470.
Scarlat N, Dallemand JF, Ferrario FM, Nita V (2015) The role of biomass and bioenergy in a future Bioeconomy: policies and facts. *Environmental Development*, 15: 3–34.
Schneider M and Froggatt A (2012) 2011-2012 World Nuclear Industry Status Report, *Bulletin of the Atomic Scientists*, 68(5): 8–22.
Shimizu T; Han J; Choi S; Kim L; Kim H (2006) Fluidized-Bed Combustion Characteristics of Cedar Pellets by Using an Alternative Bed Material. *Energy & Fuels*, 20: 2737–2742.
Słania J and Sacharz–Szewczyk A (2016) Protective Coatings in the Power Boilers Used to Incinerate Waste–Fuel Characteristics of Waste as the Source of Energy. *Archives of Metallurgy and Materials*, 61(1):103–108.
Society of Petroleum Engineers (2017) *Petroleum Reserves and Resources Definitions*, http://www.spe.org/industry/reserves.php Retrieved 2017-10.21.

Teixeira TP, Neves LM, Araújo FG (2009) Effects of a nuclear power plant thermal discharge on habitat complexity and fish community structure in Ilha Grande Bay, Brazil, *Marine Environmental Research*, 68(4): 188–195.

Teixeira T, Neves P, Araújo L (2012) Thermal impact of a nuclear power plant in a coastal area in Southeastern Brazil: Effects of heating and physical structure on benthic cover and fish communities, *Hydrobiologia*, 684(1): 161–175.

Thompson PM, Lusseau D, Barton T, Simmons D, Rusin J, Bailey H (2010) Assessing the responses of coastal cetaceans to the construction of offshore wind turbines. *Marine Pollution Bulletin* 60(8):1200–1208.

Tian Y, Zhao CY (2013) A review of solar collectors and thermal energy storage in solar thermal applications. *Applied Energy*, 104: 538–553.

Tverberg GE (2012) Oil supply limits and the continuing financial crisis, *Energy*, 37(1): 27–34.

U.S. Census Bureau (2012) International Programs, World Population.

UN (2014) *U.N. World Water Assessment Program, The United Nations World Water Development Report 2014: Water and Energy*, Paris.

UN (2015) *U.N. World Water Assessment Program, The United Nation World Water Development Report: Water for a Sustainable World*, UNESCO, Paris.

USDE (2017) Office of Fossil Energy, *How Coal Is Formed, Washington, D.C.*, https://energy.gov Retrieved 2017-10.21.

USDI (2006) *Technology white paper on wave energy potential on the U.S. outer continental shelf.* http://www.boem.gov/Ocean-Current-White-Paper-2006/ Retrieved 2016-10.11.

Waters S and Aggidis G (2016) Tidal range technologies and state of the art in review. *Renewable and Sustainable Energy Reviews*, 59: 514–529.

WBA (2019) *World Bioenergy Statistics 2019*. Retrieved from https://worldbioenergy.org Retrieved 2017-10.10.

WEC (2013) *World Energy Council, 2013 Survey of World Energy Resources*, London, UK.

WEC (2016) *World Energy Council, 2016 Survey of World Energy Resources*, London, UK.

White S and Cordell D (2008) *Peak Phosphorus: the sequel to Peak Oil. Global Phosphorus Research Initiative (GPRI)* http://phosphorusfutures.net/peak-phosphorus/ Retrieved 2017-10.21.

White DE and Williams DL (1975) Assessment of Geothermal Resources of the United States – 1975.

Wikipedia (2017a) https://en.wikipedia.org/wiki/Hubbert_peak_theory Retrieved 2017-10.21.

Wikipedia (2017b) https://en.wikipedia.org/wiki/Oil_reserves Retrieved 2017-10.21.

Wikipedia (2017c), https://en.wikipedia.org/wiki/Peak_oil#cite_note-102 Retrieved 2017-10.21.

Wikipedia (2017d) https://commons.wikimedia.org/wiki/File:Countries_by_Natural_Gas_Proven_Reserves_(2014) Retrieved 2017-10.21

Wilhelm WW, Hess JR, Karlen DL, Johnson JM, Muth DJ (2010) Review: Balancing limiting factors and economic drivers for sustainable Midwestern US agricultural residue feedstock supplies, *Industrial Biotechnology*, 6(5): 271–287.

World Bank (2017) *World Bank, GGFR Partners Unlock Value of Wasted Gas, World Bank Group*, http://web.worldbank.org/ Retrieved 2017-10.21.

World Bank (2019) *Data*. Retrieved from https://data.worldbank.org Retrieved 2019-10.10.

WorldWave (2017) *Global Offshore Database*, http://www.oceanor.no/Services/Worldwaves/WW_database Retrieved 2017-04-02.

WRA (2010) *Fossil Foolishness: Utah's Pursuit of Tar Sands and Oil Shale*. Western Resource Advocates.

Wright L and Turhollow A (2010) Switchgrass selection as a "model" bioenergy crop: a history of the process, *Biomass and Bioenergy*, 34(6): 851–868.

Yang CT (2012) China Drills Into Shale Gas, Targeting Huge Reserves Amid Challenges, National Geographic, 9 August 2012.

Yildiz I (2018a) Fossil Fuels. In: Dincer, I. (ed.), *Comprehensive Energy Systems*, Vol. 1, Elsevier, Oxford, UK, pp. 521– 567.

Yildiz I (2018b) Solar Energy. In: Dincer, I. (ed.), *Comprehensive Energy Systems*. Vol.1, Oxford: Elsevier. pp. 638– 664.

Yildiz I and MacEachern C (2018) Historical Aspects of Energy. In: Dincer, I. (ed.), *Comprehensive Energy Systems*. Vol. 1, Oxford: Elsevier. pp. 24–48.

Zhou H, Meng A, Long Y, Li Q, Zhang Y (2014) An overview of characteristics of municipal solid waste fuel in China: Physical, chemical composition and heating value. *RSER Renewable and Sustainable Energy Reviews*, 36:107–122.

Zhou A and Thomson E (2009) The development of biofuels in Asia, *Applied Energy*, 86(1): 11–20.

Zinkle SJ and Was GS (2013) Materials challenges in nuclear energy. *Acta Materialia*, 61(3):735–758.

Zogg M (2008) History of heat pumps. Swiss Contributions and International Milestones. Oberburg: Process and Energy Engineering CH-3414, Switzerland. p. 114.

# 11

# *Energy Conservation and Environmental Considerations*

## 11.1 Introduction

Integrated energy management entails energy conservation, energy mixture, and environmental considerations. Since it is already available in hand, we start with energy conservation. Energy conservation is the practice of reducing energy use as a way of extending the lifetime of our fossil-fuel supplies, of being less wasteful, and of reducing our environmental impact. Personal choice and increased efficiency are the two parallel pathways for effective energy conservation. While using more efficient lighting fixtures, cogeneration systems or high efficiency boilers in our greenhouses are some of the examples of energy conservation via personal choice, developing more efficient lighting fixtures, cogeneration systems, or high efficiency boilers form the technical fix, that is, the increased efficiency.

As mentioned earlier, in order to make their operations sustainable, commercial greenhouse operations have to reduce the two largest expenses: labor and energy. Every passing day, the use of mechanization and automation systems is becoming more common in plant production facilities for combating the labor issues. Particularly in colder climates, heating energy is a major expense category in commercial greenhouses and energy prices have been rising. Consequently, considerable effort is expended to conserve energy and look for alternative energy sources, especially environmentally friendly energy sources and technologies. However, greenhouse operations in certain countries have not been fast enough to adopt energy conservation and environmentally friendly renewable energy technologies.

In order to provide economically optimal microenvironments for plant growth, greenhouse operations can use or control the number of glazing layers, infrared inhibited double poly glazing, sidewall, end wall, and foundation insulation, energy curtains to reduce long-wave radiation losses at night, different shading devices to control incoming solar radiation, energy efficient lighting systems, reduced ventilation rates, evaporative coolers, and high efficiency condensing boilers or flue gas condensers to extract waste heat that would otherwise be released to the environment and wasted. In addition, heat pumps hold great promise for reducing winter heating and warm weather cooling energy requirements, especially through semi-closed or closed greenhouse operations.

Greenhouses will keep using conventional energy sources for the foreseeable future while incorporating more alternative energy sources in their energy mix and improved energy collection and storage technologies. Paying special attention to energy conservation measures certainly help growers to maximize energy savings and to make their operations more sustainable. However, conservation and sustainability require a multifaceted approach, and there is an additional dimension to energy use and conservation, the environmental dimension. In the years to come, energy investments will increasingly be determined by three factors: using energy more efficiently, improving competitiveness, and protecting the environment. Hence, this chapter not only covers the conservation techniques and technologies available for greenhouse growers supported with numerical figures but also dives deeper into the environmental dimension of commercial greenhouse operations in an effort to convert the global greenhouse industry to a fully sustainable net energy generating industry.

## 11.2 Energy Conservation Techniques

New energy efficient greenhouse designs should incorporate features that take maximum advantage of solar energy and are compatible with energy conservation. A number of techniques have been developed for energy conservation in greenhouses and are currently being used by many commercial greenhouse operations especially in larger greenhouse operations. Some of these techniques or technologies can be listed as the use of increased insulation on foundation, sidewalls, and end walls, the use of inflated double or triple plastic alone or over glass, the use of energy curtains at night and different shading devices to control incoming solar radiation, energy efficient lighting systems, and high efficiency condensing boilers or flue gas condensers to extract waste heat that would otherwise be released to the environment and wasted. Other approaches for energy conservation include the use of evaporative cooling, reduced ventilation, and preheating the incoming air.

### 11.2.1 Transparent and Opaque Insulation

Short and Bauerle (1980) studied energy conservation techniques for commercial greenhouses. They reported that double-layer-air-separated plastic provided a double glazing with low initial costs and a 30–35% energy savings over most single glazings (see Example 11.1 below for similar energy saving estimates). Similar findings on the use of double glazing instead of single glazing were published by Roberts and Mears (1969). In inland Australian climatic conditions, Garzoli and Blackwell (1987) reported an average of 32% energy savings by introducing double layer plastic glazing over single layer glazing. Short and Bauerle (1977) as well reported that glass houses can be modified for energy conservation by lap-sealing or using air-inflated-double-plastic-over-glass or rigid-double-wall-acrylic to save up to 20%, 60%, and 60%, respectively. In another report, Short et al. (1978) indicated that, in Ohio, a polystyrene pellet nighttime insulation system could reduce greenhouse nighttime energy requirements by 80–90%. Miyagawa (1977) reported similar findings in Japan. In addition to reducing nighttime heating requirements, the same system can provide variable shading, which can reduce the cooling requirements during hot sunny days.

Polyethylene film is the most common glazing material used in greenhouses today. Two or sometimes three layers of polyethylene films are used in most heated greenhouses to reduce heat loss. In earlier sections, it was discussed that the introduction of each glazing layer impacts light transmission negatively; however, energy losses in an inflated double-poly greenhouse, as shown below, is much less than those of a single layer glass

**Example 11.1: Estimating the Reduction in Heating Energy Requirement When Double-Poly Glazing Used**

Let's now consider the same greenhouse facility that was used in Example 9.1; however, this time, a double-poly glazing is used instead of single glass, and the lower wall (6″ poured concrete + 2″ foamed urethane board) and the perimeter are insulated as shown in the following table. Determine the energy savings.

**Solution**

We first need to calculate the UA values as follows:

| UA VALUES and AIR CHANGE PER HOUR | | | | |
|---|---|---|---|---|
| **GREENHOUSE COMPONENT** | **$A$ (ft²)** | **$U$ (Btu/h.ft².°F)** | **Reference** | **$UA$ (Btu/h.°F)** |
| Lower Wall Area | 864 | 0.07 | Table 3.3 | 60 |
| Upper Wall Area | 3,744 | 0.7 | | 2,621 |
| Gable or Curved Area | 576 | 0.7 | | 403 |
| Roof Area | 20,582 | 0.7 | | 14,407 |
| Perimeter Length | 576 | 0.4 | | 230 |
| $\Sigma UA$ | | | | 17,722 |
| ACH | | 1.0 | | |

And then, using Equations (5.2 and 5.4), we can determine the greenhouse envelop and infiltration heat losses as well as the fuel oil and natural gas consumption calculations as shown in the following table:

| | | | **Annual Cost of Heating (US$/season)** |
|---|---|---|---|
| **$Q_{cond}$ (Btu/h)** | $UA \times \Delta T$ | **1,045,615** | |
| **$Q_{inf}$ (Btu/h)** | **$0.02 \times \Delta T \times V \times$ ACH** | **239,247** | |
| **$Q_t$ (Btu/h)** | $Q_{cond} + Q_{inf}$ | **1,284,862** | |
| $E$ (gal oil/year) | $24 \times HDD \times Q_t/(HV \times \eta \times \Delta T)$ | 26,603 | 65,177 |
| $E$ (ft³ natural gas/year) | $24 \times HDD \times Q_t/(HV \times \eta \times \Delta T)$ | 3,569,294 | 17,846 |

As shown in the above table, just by changing the glass glazing to double-poly and insulating the lower wall and the perimeter, we have saved approximately 33% in heating energy.

greenhouse. Greenhouse operators need to make sure that outside air is used for inflating the space between the polyethylene film layers to reduce the risk of condensation.

Furthermore, installing an infrared (IR) polyethylene film, which has quite a short payback period (just a few months), can block long-wave (infrared) radiation as well and therefore reduce heat losses at night lowering the heating requirements in the winter and help reduce cooling requirements in the summer. IR polyethylene can be used in different ways: i) IR poly as an exterior layer – where outside IR radiation is partly blocked from entering the greenhouse; ii) IR poly as an interior layer – where inside IR radiation is partly blocked from leaving the greenhouse and trapped inside; iii) IR poly as both interior and exterior layer – where both layers of a double-poly glazing are IR poly films preventing both outside infrared radiation entering the greenhouse and inside infrared radiation leaving the greenhouse.

As discussed in Chapter 3 earlier, condensation is quite a common problem in greenhouses, and water droplets on the glazing material reduces light transmission and commonly lead to disease problems particularly when they drip onto plants. AC films prevent early morning and late afternoon fog formation in the greenhouse; hence, fewer water droplets fall on plants and less solar radiation is blocked compared to a condensing plastic film. The anti-condensation agent reduces the surface tension and allows condensate to flow rather than forming droplets. However, it should be kept in mind that the additives are slowly washed out by water in time; and therefore, the AC effect lasts for up to 2 years or so.

Not only insulating the roof of a greenhouse with transparent materials such as double or triple polyethylene glazings, but also insulating the sidewalls and end walls with the same type glazing materials is extremely important. Depending on the sidewall and end wall materials, opaque insulation may need to be used as discussed in Chapter 3. Same type of insulation is required for simple and further energy savings on foundations as well.

Special attention must also be given to cracks and other unintended openings, especially where the glazing is attached to the foundation, sidewalls, and gable end walls, as well as around doors, fans, fan louvers, and vents. It is the easiest energy conservation measure with great savings and an extremely quick payback period.

### 11.2.2 Retractable Energy Curtains, Shading Systems, and Reflective Floors

Another energy conservation technique, the use of retractable energy curtains, is a highly effective method in reducing nighttime heat requirements for greenhouses (see Figure 11.1). Energy curtains are usually used during the winter in higher latitude colder climates for keeping the warm air closer to the plant canopy and reduce heat losses due to the extra

**FIGURE 11.1**
Retractable energy curtains reduce nighttime heat requirements.

insulation layer introduced as well as reduced radiational cooling. The plants under the energy curtains do not have direct interaction with the cold sky temperature any longer; therefore, a couple degrees higher plant temperatures are observed compared to the temperatures of plants directly exposed to the sky in the absence of these curtains. As a result, plants develop faster under an energy curtain, particularly in colder climates.

Night curtains are used in the summer months as well in order to provide shading for plants and reduce cooling loads on particularly hot summer days. The savings using energy curtains have been shown to be as much as 50% (Simpkins et al., 1976; Rebuck et al., 1977). However, Albright et al. (1978) reported that a 90% savings could be achieved using a highly insulated, reflective night curtain. Greenhouse operators report energy savings up to 30% with the use of energy curtains depending on the climate and greenhouse location as well as the curtain material and management, and these figures are in line with our energy savings estimates shown in Example 11.1. It should also be kept in mind that the effect of the curtain is naturally greater on colder nights and northern climates; however, special attention is required in opening the curtains in the morning for preventing the cold air from rapidly sinking down to the crop level.

Furthermore, as discussed earlier in other chapters, light levels inside greenhouses drop considerably in late fall, winter, and early spring months at or beyond certain latitudes. Therefore, providing adequate light for plants might be challenging while trying to take advantage of the energy conserving potential of the energy curtains. It is not that uncommon to see a second energy curtain (with high light transmissivity) installed in greenhouses in an effort to find a solution to this dilemma. This second energy curtain reduces nighttime energy losses further; and when used during daytime, it transmits more solar radiation compared to the conventional aluminized curtain. It means it serves as a shading cloth optimizing the light during the day when needed. Or sometimes, energy curtains are kept closed until the outside solar radiation exceeds a crop specific level not hindering the plant growth and development.

### Example 11.2: Estimating the Reduction in Heating Energy Requirement When Thermal Blanket Used

Let's now consider the same greenhouse facility that was used in Example 11.1; however, this time, in addition to using a double-poly glazing instead of single glass, and an insulated lower wall (6″ poured concrete + 2″ foamed urethane board) and perimeter, a thermal blanket is used as well. Determine the additional energy savings.

**Solution**

Let's first calculate the UA values as follows:

| **UA VALUES and AIR CHANGE PER HOUR** | | | | |
|---|---|---|---|---|
| **GREENHOUSE COMPONENT** | ***A*** **($ft^2$)** | ***U*** **($Btu/h.ft^2.°F$)** | **Reference** | ***UA*** **($Btu/h.°F$)** |
| Lower Wall Area | 864 | 0.07 | | 60 |
| Upper Wall Area | 3,744 | 0.44 | | 1,647 |
| Gable or Curved Area | 576 | 0.44 | Table 3.3 | 253 |
| Roof Area | 20,582 | 0.44 | | 9,056 |
| Perimeter Length | 576 | 0.4 | | 230 |
| ΣUA | | | | 11,248 |
| ACH | | 1.0 | | |

And then, using Equations (5.2) and (5.4), we determine the greenhouse envelop and infiltration heat losses as well as the fuel oil and natural gas consumption calculations as shown in the following table:

| **$Q_{cond}$ (Btu/h)** | **$UA \times \Delta T$** | **663,618** | **Annual Cost of Heating (US$/season)** |
|---|---|---|---|
| **$Q_{inf}$ (Btu/h)** | **$0.02 \times \Delta T \times V \times ACH$** | **239,247** | |
| **$Q_t$ (Btu/h)** | **$Q_{cond} + Q_{inf}$** | **902,865** | |
| $E$ (gal oil/year) | $24 \times HDD \times Q_t/(HV \times \eta \times \Delta T)$ | 18,694 | 45,799 |
| $E$ (ft$^3$ natural gas/year) | $24 \times HDD \times Q_t/(HV \times \eta \times \Delta T)$ | 2,508,123 | 12,541 |

As shown in the above table, after changing the glass glazing to double-poly and insulating the lower wall and the perimeter, by introducing an internal thermal blanket, we have saved 30% more in heating energy consumption.

Greenhouse shading is another energy conserving technique, used mostly during warm and bright sunny days, to reduce indoor temperatures, and therefore cooling loads. Several shading systems are available and currently used by many commercial greenhouse growers. These include an external spray (whitewash) applied on the outside surface of the glazing (see Figure 11.2) and variable retractable shading (internal or less commonly external) (see Figure 11.3) using different materials. There are advantages and disadvantages of both systems. For instance, whitewash is quite an inexpensive technique to provide shading for plants; however, the proportion of shading it provides is relatively fixed and permanent, and it is washed away slowly with rainfall. Also, it must be washed off when the solar radiation becomes limiting again in the fall. Retractable shading, on the other hand, is a relatively expensive technology, but provides controlled shading when it is needed. Another variable shading system developed by Kuack (1986) is the one with a combination of polystyrene pellets, which provides precise shading in greenhouses; however, it is not commercially adopted yet due to its relatively expensive channelized double acrylic glazing and control system.

**FIGURE 11.2**
Whitewash applied on the outside surface of the glazing.

**FIGURE 11.3**
An internal retractable shading system: open (*left*) and closed (*right*).

Reflective floors (see Figure 11.4) have also been reported to alter crop environmental factors by increasing the light intensity by the reflection of incoming light (Roberts and Anderson, 1994), reducing the soil temperature (Schalk and Robbins, 1987), and increasing the air temperature above the reflective surface (Zalom, 1981). In their study, Mehlitz et al. (2008) focused on reducing energy and water consumption in semi-arid California greenhouses and studied the influences of mulch (having different reflectivities) on energy and water conservation in San Luis Obispo (35° 17′ N and 120° 39′ W), California, USA. They developed a dynamic computer simulation model to compare different mulch reflectivities, cucumber plant heights, and leaf dimensions to draw a conclusion about energy and water conservation. The effect of floor surface reflectivity was studied together with four plant heights and three leaf dimension sets (Table 11.1). The study concluded that the greenhouse energy consumption was dependent on the reflectivity of mulch (soil) used in the greenhouse. The higher the reflectivity was, the more energy was consumed. Mulch with a reflectivity of 20% instead of 80% saved as high as 4.2% energy (Table 11.2). The water consumption for transpiration was also dependent on the reflectivity. It was also shown that significant savings in water consumption (as much as 8.6%) could be realized if 80% reflective mulch was used rather than 20% (Table 11.2). However, the effect of mulch on water use faded away as the shade from canopy stand increased. In order to make a conclusion about total energy and water consumption, the greenhouse has to be treated as a complete system. Potential energy savings due to lower reflectivity have to be seen in conjunction with a larger amount of water use. They concluded that total energy and water conservation analysis has to be performed for each case separately, considering the specific infrastructure, local water and energy prices, and the crop type.

### 11.2.3 Reduced Ventilation Rates

While lower ventilation rates reduce the heating requirements, they may cause high relative humidity levels in greenhouses and low $CO_2$ levels if $CO_2$ enrichment is not provided. Consequently, low ventilation rates may adversely affect plant responses unless carefully controlled. If conventional ventilation rates are used to maintain reasonable relative

**FIGURE 11.4**
Reflective floor surface in a greenhouse.

humidity levels within greenhouses, this time, they result in higher heating energy requirements, and relatively high operating costs.

Yildiz and Stombaugh (2009) studied the effects of minimum ventilation rate on energy and water conservation in a greenhouse heated and cooled with a gas-fired heat pump in Delaware (latitude 40° 17′ N, longitude 83° 05′ W), Ohio, USA. Table 11.3 shows the greenhouse and crop characteristics used in the dynamic greenhouse simulation model.

In their assessment, a minimum ventilation rate of 0.005 $m^3/s.m^2$ was used and compared to the conventional rate of 0.01 $m^3/s.m^2$. Table 11.4 and Figure 11.5 show that energy consumption was significantly affected by the variation in minimum ventilation rate. A 50% reduction in the ventilation rate resulted in 26%, 21%, and 1.5% decreases in total energy consumption in winter, spring, and summer, respectively. In winter, 26% and 50% drops were observed in heating and ventilation energy consumption, respectively. This showed that a 50% reduction in minimum ventilation rate resulted in a 50% reduction in energy consumption for ventilation in winter. This was due to the nearly continuous

**TABLE 11.1**
Greenhouse and Crop Characteristics Used in the Simulation Model.

| | |
|---|---|
| Greenhouse length | 15.0 m |
| Greenhouse width | 7.50 m |
| Greenhouse height at eaves | 2.50 m |
| Greenhouse height at ridges | 4.50 m |
| Greenhouse glazing | Double polyethylene |
| Floor surface | Mulch with reflectivity of 20%, 40%, 60%, or 80% |
| Crop | Cucumber |
| Crop row orientation | North–South |
| Distance between plant rows | 0.75 m |
| Average leaf length × width | Set 1: 0.30 × 0.25 m; Set 2: 0.23 × 0.18 m; Set 3: 015 × 0.10 m |
| Average plant height | 0.5 m, 0.75 m, 1.0 m, 2.0 m |

operation of minimum ventilation rates in winter. A 50% reduction in the minimum ventilation rate for spring, however, resulted in only a 15% decrease in energy consumption. This indicated that ventilation rates higher than the minimum were used for some time due to increased relative humidity levels within the greenhouse. The effect of minimum ventilation rate in summer was insignificant for the same reasons that were determined for spring.

### 11.2.4 Crop Temperature Management

As discussed earlier, temperature is one of the main environmental variables that influences the growth and development of greenhouse crops, which require crop-specific minimum, optimum, and maximum temperatures for their growth. The minimum temperature is the temperature at which a plant stops growing. The optimum temperature is the temperature at which plant develops most rapidly. Maximum temperature, on the other hand, is the temperature that exerts stress on crops when the day, night, or both temperatures are above some crop-specific temperature.

Temperature is an extremely valuable tool for regulating crop timing by changing temperatures during the day and night (DIF) or by controlling the average daily temperature (ADT). Plants integrate temperatures over a 24-hour period, and plant growth and development are strongly influenced by this integrated temperature, that is, the average temperature. DIF is the temperature difference between daytime and nighttime, which influences plant height, chlorophyll content, etc.; therefore, it is used in greenhouses to control plant height and reduce the use of chemical growth retardants. Plant height can be decreased by decreasing the daytime set-point temperature or increasing the nighttime set-point temperature or both by achieving a zero or negative DIF. In order to increase plant height, however, we can increase the daytime temperature or decrease the nighttime temperature or both by maintaining a zero or positive DIF. Generally speaking, a nighttime temperature beyond 24°C (75°F) is not desirable because many crops are negatively influenced by high temperatures. Another method for manipulating temperature to control plant elongation is to drop the temperature (also known as DIP) by approx. 2.8–8.3°C (5–15°F) before sunrise for a period of 2–3 hours. Therefore, when greenhouse temperature is managed based on the crop requirements and the crop finish date, further savings would be possible in greenhouse energy consumption.

**TABLE 11.2**

Comparisons of Energy and Water Consumption Means With Respect to Different Mulch Reflectivities Using Tukey's Simultaneous Test at a 95% Confidence Level.

| | Energy use (all values in $MJ/m^2/day$) | | | | | Water use (all values in $kg/m^2/day$) | | | |
|---|---|---|---|---|---|---|---|---|---|
| | **Reflectivity (Mean)** | **Compared to Reflectivity (Mean)** | **Difference of Means** | **Statistically Significant at a 95% level?** | | **Reflectivity (Mean)** | **Compared to Reflectivity (Mean)** | **Difference of Means** | **Statistically Significant at a 95% level?** |
| WINTER | 0.2 (32.05) | 0.4 (32.65) | 0.6 | YES | WINTER | 0.2 (2.49) | 0.4 (2.45) | 0.034 | YES |
| | | 0.6 (32.91) | 0.86 | YES | | | 0.6 (2.44) | 0.046 | YES |
| | | 0.8 (33.07) | 1.02 | YES | | | 0.8 (2.43) | 0.057 | YES |
| | 0.4 (32.65) | 0.6 (32.91) | 0.26 | YES | | 0.4 (2.45) | 0.6 (2.44) | 0.011 | YES |
| | | 0.8 (33.07) | 0.42 | YES | | | 0.8 (2.43) | 0.023 | YES |
| | 0.6 (32.91) | 0.8 (33.07) | 0.17 | YES | | 0.6 (2.44) | 0.8 (2.43) | 0.012 | YES |
| SPRING | 0.2 (25.39) | 0.4 (25.95) | 0.56 | YES | SPRING | 0.2 (2.509) | 0.4 (2.45) | 0.055 | YES |
| | | 0.6 (26.20) | 0.8 | YES | | | 0.6 (2.44) | 0.074 | YES |
| | | 0.8 (26.51) | 1.12 | YES | | | 0.8 (2.41) | 0.098 | YES |
| | 0.4 (25.95) | 0.6 (26.20) | 0.25 | YES | | 0.4 (2.454) | 0.6 (2.44) | 0.017 | YES |
| | | 0.8 (26.51) | 0.56 | YES | | | 0.8 (2.41) | 0.043 | YES |
| | 0.6 (26.20) | 0.8 (26.51) | 0.32 | YES | | 0.6 (2.435) | 0.8 (2.41) | 0.024 | YES |
| SUMMER | 0.2 (13.17) | 0.4 (13.52) | 0.35 | YES | SUMMER | 0.2 (2.433) | 0.4 (2.36) | 0.07 | NO |
| | | 0.6 (13.59) | 0.41 | YES | | | 0.6 (2.36) | 0.072 | NO |
| | | 0.8 (13.79) | 0.62 | YES | | | 0.8 (2.22) | 0.214 | YES |
| | 0.4 (13.52) | 0.6 (13.59) | 0.07 | YES | | 0.4 (2.363) | 0.6 (2.36) | 0.002 | NO |
| | | 0.8 (13.79) | 0.27 | YES | | | 0.8 (2.22) | 0.144 | YES |
| | 0.6 (13.59) | 0.8 (13.79) | 0.21 | YES | | 0.6 (2.361) | 0.8 (2.22) | 0.142 | YES |

**TABLE 11.3**

Greenhouse and Crop Characteristics Used in the Simulation Model.

| | |
|---|---|
| Greenhouse length | 7.50 m |
| Greenhouse width | 7.50 m |
| Greenhouse height at eaves | 2.50 m |
| Greenhouse height at ridges | 4.50 m |
| Greenhouse glazing | Double polyethylene |
| Floor surface | Reflective mulch with 60% reflectivity |
| Crop | Cucumber |
| Crop row orientation | North–South |
| Distance between plant rows | 0.75 m |
| Average leaf length and width | 0.30 m and 0.75 m |
| Plant height | 2.0 m |

**TABLE 11.4**

Energy Consumption With Respect to Minimum Ventilation Rate in Winter, Spring, and Summer.

| VENTILATION RATE ($m^3/s.m^2$) | WINTER 0.01 | WINTER 0.005 | SPRING 0.01 | SPRING 0.005 | SUMMER 0.01 | SUMMER 0.005 |
|---|---|---|---|---|---|---|
| **ENERGY ($MJ/d.m^2$)** | | | | | | |
| Heating | 23.45 | 17.41 | 8.82 | 6.95 | 0.10 | 0.10 |
| Cooling | --- | --- | --- | --- | 0.94 | 0.93 |
| Ventilation | 0.11 | 0.06 | 0.23 | 0.19 | 0.71 | 0.71 |
| TOTAL | **23.56** | **17.47** | **9.04** | **7.15** | **1.75** | **1.73** |

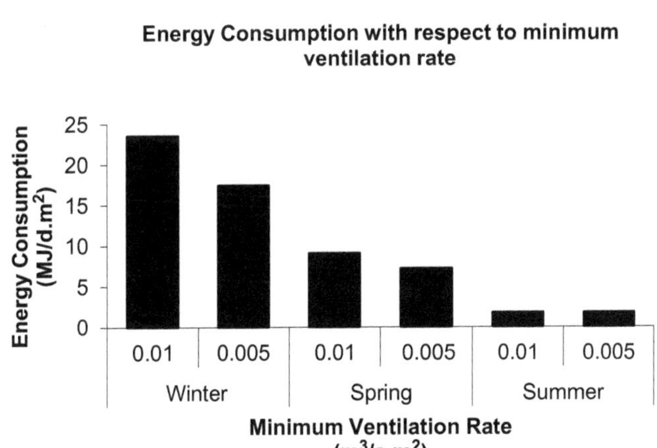


**FIGURE 11.5**
Energy consumption with respect to minimum ventilation rate in winter, spring, and summer.

For example, the estimations performed in the following example show that using variable set-point temperatures in a single glass greenhouse reduces the heating energy supplied by the furnace approximately 29%.

**Example 11.3: Heating Energy Savings in a Greenhouse with Variable Set-Point Temperatures**

Using the same greenhouse that was used in Example 9.1, a single glass greenhouse in Columbus, Ohio, USA, estimate the annual heating energy requirement and the costs of heating using fuel oil and natural gas furnaces for different daytime and nighttime indoor set-point temperatures. Assume that a daytime set-point temperature of 65°F is used, while the nighttime set-point temperature is 55°F.

**Solution**

The energy estimation using the degree-day method for the indoor set-point temperature of 65°F was performed in Example 9.1. Now, we need to do the same, but this time for a variable set-point temperature, that is, a daytime set point temperature of 65°F and a nighttime set point temperature of 55°F. Let's keep in mind that 75–80% of heating is done during night; so, we can easily calculate the energy savings (by adjusting the HDD) that could be achieved by employing a variable daytime and nighttime set-point temperatures rather than a continuous set-point temperature as follows:

$$\text{Adjusted HDD} = 0.75(5{,}660)(55°\text{F}/65°\text{F}) + 0.25(5{,}660)(65°\text{F}/65°\text{F}) = 5{,}007°\text{F.day}$$

And in Example 9.1, the energy consumption $E$ for the continuous set-point temperature of 65°F was calculated as 39,770 gal oil/year. Then, the new energy consumption $E$ for the variable set-point temperature would be:

$$E = (39{,}770\text{ gal oil/year})(5{,}007°\text{F.day})/(5{,}660°\text{F.day})$$

$$E = 35{,}182\text{ gal oil/year}$$

And then, the percent energy savings would be calculated as:

$$\left(\frac{(39{,}770\text{ gal oil/year}) - (35{,}182\text{ gal oil/year})}{(39{,}770\text{ gal oil/year})}\right)(100) = 11.5\%$$

The calculations performed in Example 9.1 and above show that using the variable set-point temperature reduces the heating energy provided by the furnace by approximately 11.5% in this particular greenhouse and at this particular location.

Energy consumption for heating indeed decreases as the set-point temperature is lowered. However, the impact of a lower daily average temperature on plant finish time must also be considered when using this strategy. The bottom line is, as mentioned above, we have to manage greenhouse temperature based on the crop requirements, that is, providing them with the required daily average temperatures.

It is also a known fact that, when plants are grown at colder temperatures, it takes longer for them to flower. Therefore, the energy consumption per plant basis may sometimes be higher when plants are grown at lower temperatures (due to the heating required for the extended growth period) compared to those grown at higher temperatures in order to make the finish time shorter.

### 11.2.5 Floor Heating, Infrared Heating, and Horizontal Air Flow

Main goal of greenhouse heating is to provide cost effective uniform heating where and when it is needed. Floor heating in greenhouses is a common practice, and hot-water floor heating systems (see Figure 11.6) today provide the most uniform and quite effective heat distribution for greenhouse crops grown directly on the floor or in the soil such as bedding plants, ornamentals, and bag-cultured vegetables.

As discussed earlier, conventional greenhouse heating systems (CONs) heat the greenhouse air first, and then the heated air heats the plants. A low infrared heater, on the other hand, operates with heated pipes – up to 595°C (1,100°F) – and first heat the objects in the greenhouse (including crops) directly through infrared radiation heat exchange, and then the objects exchange this heat with the environment via conduction, convection, and again radiation heat transfer. Since the greenhouse air is not heated directly, the temperature of the interior air is typically lower; and this low inside air temperatures result in less heat

**FIGURE 11.6**
A hot-water in-floor heating system installed in a concrete greenhouse floor. Crops are grown directly on the floor.

losses from the greenhouse environment. Furthermore, infrared heaters eliminate the pumping energy requirements and the heat losses through distribution lines typically seen in CONs. As a consequence, when installed properly, infrared heating can provide uniform plant and vertical air temperatures and be a cost effective option for greenhouse heating.

Horizontal air flow (HAF) fans, also known as circulation fans, provide a more uniform greenhouse environment and reduce the difference between internal air temperature and plant temperature (see Figure 11.7). They not only improve uniform temperature distribution, but also humidity and carbon dioxide distributions costing marginal energy expenditure. The flowing air removes moisture and prevents condensation on the leaves, therefore reduce foliar fungal diseases.

Typically, a total air flow of 0.01 $m^3/s.m^2$ (2 $cfm/ft^2$) of floor area is provided keeping the greenhouse air mass moving at a velocity of 0.5 m/s (100 ft/min) to 0.7 m/s (140 ft/min) and not exceeding 1.0 m/s (200 ft/min) in the vicinity of crop canopy. HAF systems come with 30 cm (12 in), 40 cm (16 in), 45 cm (18 in), and 50 cm (20 in) diameter blades, and generally move the air approximately 30 times the blade diameter. A 30 cm (12 in) diameter HAF fan would be able to move the air 9 meters (30 ft). In a well-designed HAF fan system, the fans are installed in a way that air flow is horizontally directed down the greenhouse bay and aligned with similar fans pushing the greenhouse air in the opposite direction down the next bay. In order to make sure that the air within the canopy is moved rather than the air above the canopy, fans are mounted at between 0.6 m (2 ft) to 0.9 m (3 ft) above the plants. The first and last fans are placed about 4.5 m (15 ft) to 6.0 m (20 ft) from the gable ends to pull the air that is coming from the adjacent bay. Crop canopy and hanging baskets resist HAF; therefore, the fans should be spaced properly. After the first fan, they are typically placed 9 m (30 ft) to 15 m (50 ft) apart to keep the greenhouse air mass moving smoothly.

**FIGURE 11.7**
A horizontal air flow (HAF) fan installed in a greenhouse.

### 11.2.6 Dynamic Row Spacing

The dynamic row spacing as well might play an important role in energy and water conservation, and have an influence on the plant growth. Papadopoulos and Pararajasingham (1997) reported that greater yields of greenhouse crops could be produced by using narrow spacings (high plant density) compared with wide spacings. Increased fruit yield in narrow spacings result from greater crop biomass generated by the effect of increased light interception and canopy photosynthesis. Increased high plant density also reduces the energy consumption on per plant basis.

Moya et al. (2008) developed and employed a dynamic computer simulation model to compare different row spacings, plant heights, and leaf dimensions of a cucumber crop to draw a conclusion about energy and water conservation. Table 11.5 shows the greenhouse and crop characteristics used in their simulation model. Their findings showed that using smaller spacings between cucumber crop rows (for instance, 0.5 m instead of 0.75 m) reduced energy consumption per unit floor area in average 14.4%. With a decrease in row spacing, the total amount of surface for radiation exchange decreases, and consequently plant canopy shading within the canopy increases. This leads to less radiational and evaporative cooling in smaller row spacings, and hence, results in lower heating requirements during the heating season. By changing the row spacing from 0.75 to 0.5 m, average water savings (adjusted to the whole greenhouse area) of 27.8% occurred. A complete system analysis, however, is necessary to be able to make a viable conclusion in total energy and water conservation.

The mean energy consumption for heating differed with the season and different row spacings. The colder the outside air, obviously the more heating energy was needed. The wider the row spacing was, the more energy was required in all seasons (i.e. 0.5 m the least, 0.75 m the most). For the row spacing of 0.5 m, the mean heating energy consumptions were 28.3, 21.5, and 11.8 MJ/m$^2$/day for winter, spring, and summer, respectively. For the row spacing of 0.75 m, however, the mean energy uses for heating were 32.7, 25.9, and 13.5 MJ/m$^2$/day for winter, spring, and summer, respectively. Statistical comparisons between all possible pairs were conducted using Tukey's Simultaneous Test at a 95% confidence level and Table 11.6 shows that all the presented means were statistically different from each other. The radiation exchange surfaces and airflow increased within wider row

**TABLE 11.5**

Greenhouse and Crop Characteristics Used in the Simulation Model.

| | |
|---|---|
| Greenhouse length | 15.0 m |
| Greenhouse width | 7.50 m |
| Greenhouse height at eaves | 2.50 m |
| Greenhouse height at ridges | 4.50 m |
| Greenhouse glazing | Double polyethylene |
| Floor surface | Reflective mulch with 60% reflectivity |
| Crop | Cucumber |
| Crop row orientation | North–South |
| Distance between plant rows | 0.50 and 0.75 m |
| Average leaf length × width | Set 1: 0.30 × 0.25 m; Set 2: 0.23 × 0.18 m; Set 3: 015 × 0.10 m |
| Average plant height | 0.5 m, 0.75 m, 1.0 m, 1.5 m, 2.0 m |

**TABLE 11.6**

Comparisons of Energy and Water Consumption Means With Respect to Different Row Spacings Using Tukey's Simultaneous Test at a 95% Confidence Level.

| | Energy use (all values in MJ/m²/day) | | | | Water use (all values in kg/m²/day) | | | |
|---|---|---|---|---|---|---|---|---|
| | Row Spacing (Mean) | Compared to row spacing (Mean) | Difference of Means | Statistically Significant at a 95% level? | Row Spacing (Mean) | Compared to row spacing (Mean) | Difference of Means | Statistically Significant at a 95% level? |
| WINTER | 0.5 (28.3) | 0.75 (32.7) | 4.4 | YES | 0.5 (1.72) | 0.75 (2.46) | 0.74 | YES |
| SPRING | 0.5 (21.5) | 0.75 (25 9) | 4.4 | YES | 0.5 (1.87) | 0.75 (2.45) | 0.58 | YES |
| SUMMER | 0.5 (11.8) | 0.75(13.5) | 1.7 | YES | 0.5 (1.64) | 0.75 (2.38) | 0.74 | YES |

spacings; as a result, radiational, convective, and evaporative cooling increased. Also, canopy shading decreased in wider row spacings; more plant surfaces were exposed to solar radiation inducing transpiration rates and evaporative cooling per unit floor area as a consequence. Therefore, the supplemental heating requirements in 0.75 m row spacing were higher. The row spacing could be kept just enough for the canopy stand, and increased dynamically as the crop grows and canopy stand becomes larger. Their findings showed that such a practice could reduce supplemental heating requirement of a greenhouse, in average of 14.4% (Table 11.6).

Not only the row spacing, but also the plant height and the leaf dimensions were accountable for energy consumption. The most energy was needed when the cucumber plant stands were still short (0.5–1.0 m). As the plants grew, energy needs for heating went down depending on the season and final plant height. The energy consumption also depended on the leaf dimensions, and increased linearly with decreasing leaf area. These results are valuable to estimate different energy consumptions for specific plants, but are less important given that a greenhouse operator or designer usually has no choice of which crop to grow with respect to energy consumption. However, this study showed that a greenhouse with tall plants had significantly less heating demand than a greenhouse with small plants. This was because the taller plants overall had much less exposed surface area for radiational heat exchange.

Simulation results of this dynamic model also showed that, due to increased plant density within the whole greenhouse space, the water consumption increased in average of 8.5% as row spacing decreased (i.e. 0.50 m, the most, and 0.75 m, the least water consumption). This is in agreement with the findings of Bryla and Strik (2007). In reality, for a dynamic row spacing system, plant density is kept the same throughout the growing season. Same number of plants exists no matter what the row spacing is. As crops grow, the row spacing expands covering a larger greenhouse space during the growing season. Considering the 0.75 m spacing as the final row spacing covering the whole greenhouse floor area, the 0.5 m spacing transpiration figures can be adjusted for the whole greenhouse space (i.e. to the 0.75 m spacing) reducing the water consumption used in the whole greenhouse space (Table 11.6). Then, for the row spacing of 0.5 m, the space-adjusted mean water consumption for transpiration (of 3 years) was 1.72, 1.87, and 1.64 kg/m²/day for winter, spring, and summer, respectively. For the row spacing of 0.75 m, however, the mean water consumption for transpiration (of 3 years) was 2.46, 2.45, and 2.38 kg/m²/day for winter, spring, and summer, respectively. Summer means show that less transpiration occurred in summer due to higher air moisture levels maintained inside the greenhouse (due to operational and control strategies employed) compared to those in winter and spring. Comparisons between all possible pairs were also made using Tukey's Simultaneous Test at a confidence level of 95%. All of the presented means (of 0.5 and 0.75 m row spacings) for winter, spring, and summer were statistically different from each other. From 0.75 to 0.50 m row spacing, the overall decrease in water consumption was in average of 27.8%. The effect of row spacing definitely changes the dynamics of airflow in the canopy and it affects the canopy resistances, especially, the aerodynamic and stomatal resistances.

The plant heights and leaf dimensions also played an important role with respect to the water consumption for transpiration. In general, the smaller the plants and the larger the leaves were, the less the water consumed. This observation is in agreement with the findings of a previous study by Claypool et al. (1997). In their study, they observed that large-leaved alfalfa plants had 10% to 16% lower transpiration than small (normal) leaves. The yield of large-leaved plants was also 17% to 22% higher than small-leaved. These findings indicate that crop transpiration can be potentially improved by genetically changing leaf size to conserve water.

In conclusion, the findings in this study showed that the energy consumption is influenced by the crop row spacing. The smaller the row spacing is, the less energy is consumed. If a smaller space requirement for smaller row spacing is considered, then it will be clear that less greenhouse space needs to be heated, increasing the potential to reduce energy consumption further. The growing space could be adjusted dynamically as the row spacing changes throughout the growing season, that is, as the crop grows, the spacing must increase. A sophisticated design or modification of greenhouses using dynamic row spacing can therefore save both energy and water during the production period of the greenhouse. It seems that using smaller row spacing when the canopy stand is small could save both energy and water. Instead of using fixed row spacing throughout the growing season, one could employ dynamic row spacing using smaller row spacing when the canopy stand is small, and extending dynamically as the canopy stand gets bigger. Also, the smaller the plants and the larger the leaves were, the less water was consumed. These findings indicate that crop transpiration can potentially be improved by genetically modifying crop leaf sizes and canopy stands. However, to be able to make a conclusion about total energy and water conservation, the greenhouse has to be treated as a complete system. Potential energy savings due to dynamic row spacing have to be seen in conjunction with the energy expenditure, automation, or labor required for providing such a system. Therefore, a total energy and water conservation analysis has to be conducted for each case in hand individually, considering the infrastructure, local water, and energy and labor costs, along with the crop type.

### 11.2.7 High Efficiency Furnaces, Cogeneration, and Heat Storage Systems

Recent developments in heating systems technologies help greenhouse growers to reduce their energy losses and consumptions further. These technologies now include high-efficiency direct-fired unit heaters, high efficiency condensing boilers, and boilers coupled with flue gas condensers (see Figures 11.8 and 11.9) that condense the exhaust gas and capture both waste heat and $CO_2$ (see Example 2.59). The overall efficiencies of such boilers can go up to 93–95%, and high-efficiency furnaces can save greenhouse growers up to 40–45% on energy costs, of course, based on the efficiency of the existing furnace, which is being replaced.

**FIGURE 11.8**
A greenhouse boiler room equipped with high efficiency boilers.

There is a clear trend toward local energy production, a trend which is being accelerated by the liberalization of gas and electricity markets. In parallel, technological development also points in the direction of more diversification and small-scale power production. Caring for the environment, on the other hand, requires more sustainable power production. Technological development, financial support, and government incentives will, in the future, all contribute to more sustainable power production. The integration of sustainable power supply and the highly efficient use of fossil fuels will be emphasized.

For years, we have been used to thinking of electricity production based on large-scale power stations that are usually remote from electricity consumers, and that require long-distance, high-voltage transmission networks. This is often known as a "centralized" system. Recent technological developments have introduced a whole range of much-smaller scale generation systems into the energy market, including engines, turbines, fuel cells, and photovoltaics. These are sometimes known as "decentralized," "distributed," or "embedded" systems because they can produce electricity near to the consumer, or in the consumer's house, office or production area. These "localized" systems are more environmentally friendly, more flexible, more efficient, and more cost-effective than traditional centralized systems – especially when transmission savings are taken into account. Conventional generation of electricity in large central power stations is normally only 30–40% energy efficient. More recent combined heat and power (CHP) cycle generation can improve this to 55%, excluding the transmission loses of electricity, which can be as great as 20% of the delivered efficiency. Hence, conventional electricity-only stations release large amounts of energy as waste heat, normally via large cooling towers or cooling with sea or river water. They also release large amounts of greenhouse gases (mainly $CO_2$) to the environment. Through the utilization of the heat, the efficiency of cogeneration unit can reach 90% or more. Cogeneration of heat and power in greenhouse operations not only

**FIGURE 11.9**
Flue gas condensers recover heat by allowing condensation of the exhaust gases.

utilizes the heat and power generated on site, but also, essentially eliminates the $CO_2$ emissions to the environment (see Figure 11.10). In essence, these systems serve as multigeneration systems in greenhouse operations, generating heat, power, $CO_2$, and water, all precious resources.

It should also be kept in mind that the energy efficiency of greenhouses can be increased further by the addition of heat storage systems to, for instance, a CHP unit or a boiler (during $CO_2$ dosing in a greenhouse) so that the excess heat can be stored and used later when needed. Typically, these storage systems are the insulated hot water storage tanks (see

**FIGURE 11.10**
A microcogeneration (CHP) system.

**FIGURE 11.11**
A vertical heat storage system.

Figure 11.11) that has already been covered in Chapter 5; however, underground aquifers can also be used as heat storage facilities.

### 11.2.8 Heat Pumps, Semi-Closed and Closed Greenhouse Systems

Heat pumps are reversible refrigerators that can heat or cool the greenhouse environment with a single system depending on the flow direction of the refrigerant. The use of heat pumps for heating and cooling greenhouses makes it possible to use local and renewable energy sources while reducing or totally eliminating $CO_2$ emissions. Especially, the earth-coupled or geothermal heat pumps are highly efficient and stable systems for heating and cooling greenhouses. Heat pumps also make innovative confined greenhouse operations possible, which would totally eliminate the energy, $CO_2$, and water losses due to ventilation conserving all of these precious resources, and would make an insect-free greenhouse operation possible.

Yildiz et al. (2014a,b) studied the spatiotemporal distributions and spatial correlations of energy and water consumption in open, semi-closed, and completely closed heat pump (CHP) greenhouse systems in continental Canada in order to help enhance the competitive position of greenhouse and agri-food industries by introducing economically, environmentally and socially sustainable technologies and management strategies. The investigated heat pump technology and operational modes (especially semi-closed and closed) bring increased yield, and energy and water efficient solutions to the doorstep of greenhouse producers. The overall goal of this study was to help reduce the load on power grid, demand for fossil fuels and irrigation water, and also supply $CO_2$ for the greenhouse production, as well as transform the unfavorable regions in the Canadian North into potential greenhouse regions and open up northern population's limited access to fresh vegetables and locally produced greens for better public health and local sustainability. The greenhouse systems with heat pump units in this study were operated as open heat pipe (OHP), semi-closed heat pipe (SHP), and completely closed heat pipe (CHP) systems. Simulations were performed starting at the beginning of 1st day of January and ending at the end of 31st day of December for full 365-day simulations except the Canadian North where the seasons varied between seven months (March through September) and nine months (February through October). In order to make reasonable spatial comparisons in the Canadian North, a 7-month growing season was used for all locations, and simulations were performed under natural light conditions starting on the 1st day of March and ending on the 30th day of September for 214-day simulations. The readers of this book are referred to Yildiz et al. (2014a,b) for further information on the details of weather file, greenhouse characteristics, operational strategies, and data analyses.

The findings showed that the energy and water consumptions were both spatially and system dependent. Latitude, longitude, and altitude all significantly predicted total energy and water consumptions. SHP and CHP greenhouse systems were the most energy and water conserving systems, respectively (Table 11.7). Highest energy consuming locations of SHP system were still a lot more efficient than the least energy consuming locations of conventional (CON) systems. Closed systems had the lowest energy consumptions in colder months, but the consumption increased in warmer months due to increased cooling loads; as a result, the overall annual energy consumptions in closed systems were much higher than those in SHPs. Locations close to the Atlantic and Pacific Coasts as well as the Great Lakes region had the lowest energy consumptions in all systems (Figure 11.12). SHPs and CHPs at lower latitudes had more water consumptions (transpiration) due to increased

**TABLE 11.7**

Comparisons of Means With Respect to Different Greenhouse Systems Using Tukey's Simultaneous Test at a 95% Confidence Interval.

| SYSTEM | Mean Annual Energy Consumption (GJ/m²) | Mean Annual Water Consumption (kg/m²) | Mean Annual Water Collection (kg/m²) |
|---|---|---|---|
| CON | 11.32 a | 660.4 p | 0.0 w |
| OHP | 5.77 b | 700.4 q | 6.4 w |
| SHP | 5.27 c | 540.7 r | 125.3 x |
| CHP | 7.60 d | 420.0 s | 420.0 y |

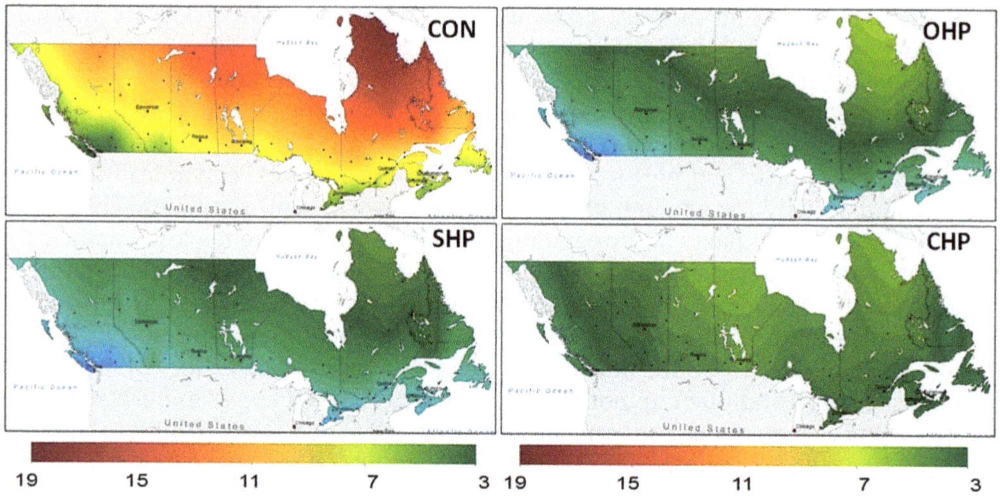


**FIGURE 11.12**
Spatial distributions of annual total energy consumptions (GJ/m$^2$) in CON, OHP, SHP, and CHP systems in Canada (Contour Line Intervals: 250 MJ/m$^2$). CON: Conventional greenhouse; OHP, SHP, and CHP: Open, semi-open, and closed heat pump greenhouse systems, respectively.

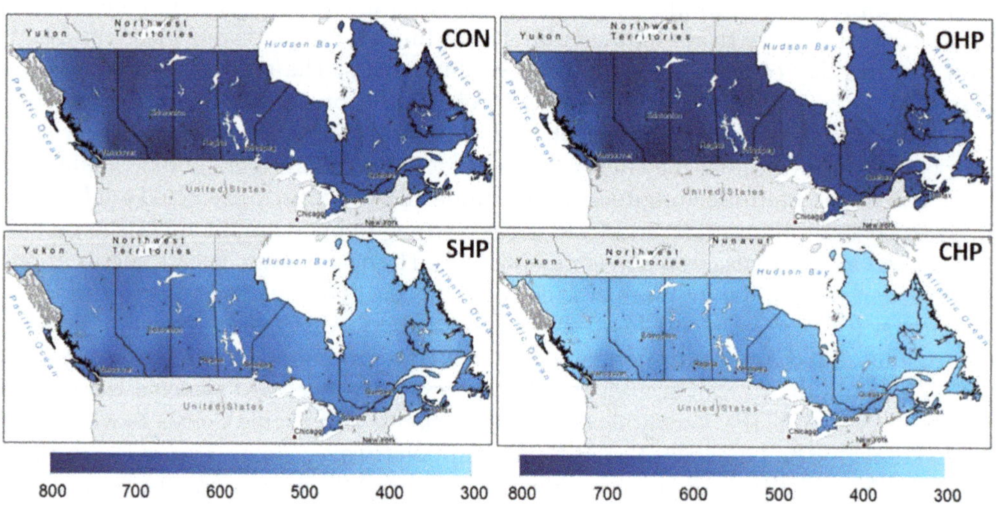


**FIGURE 11.13**
Spatial distributions of annual amount of transpiration (kg/m$^2$) in CON, OHP, SHP, and CHP systems in Canada (Contour Line Intervals: 15 kg/m$^2$). CON: Conventional greenhouse; OHP, SHP, and CHP: Open, semi-open, and closed heat pump greenhouse systems, respectively.

amount of available solar radiation (Figure 11.13). OHPs at higher latitudes had more water consumptions than those at lower latitudes as the vapor pressure deficits became larger due to decreased outside air temperatures in higher latitudes. Locations close to the Atlantic and Pacific Coasts as well as the Great Lakes region had lower water consumptions due to increased outside relative humidity levels hence reduced vapor pressure deficits. SHP greenhouse system transforms the non-energy-efficient, unfavorable regions under CON greenhouse systems into highly energy and water efficient potential greenhouse regions in Canada.

**TABLE 11.8**

Comparisons of Means With Respect to Different Greenhouse Systems in The Canadian North Using Tukey's Simultaneous Test at a 95% Confidence Interval.

| SYSTEM | Mean Annual Energy Consumption ($MJ/m^2.d$) | Mean Annual Water Consumption ($kg/m^2.d$) | Mean Annual Water Collection ($kg/m^2.d$) |
|---|---|---|---|
| CON | 33.3 a | 2.45 p | 0.0 w |
| OHP | 16.6 b | 2.49 p | 0.0 w |
| SHP | 15.1 b | 2.08 q | 0.34 x |
| CHP | 21.3 c | 1.33 r | 1.33 y |

CON: Conventional greenhouse; OHP, SHP, and CHP: Open, semi-open, and closed heat pump greenhouse systems, respectively. Means in the same column not followed by the same letter are significantly different.

The results in the Canadian North as well showed that the energy and water consumptions were both spatially and system dependent. Both latitude and longitude significantly predicted total energy consumptions. Latitude was significant in SHPs and OHPs water consumptions as well. SHP and OHP greenhouse systems proved to be the most energy conserving systems, while the CHP system was the most water conserving system (Table 11.8).

Even though the heat pump seasonal performance factors dropped at more northern locations due to colder outside temperatures, the highest energy consuming locations of SHP and OHP greenhouse systems were almost as efficient as the least energy consuming locations using the CON greenhouse system (southwest) (Figure 11.14). CHP systems had the lowest energy consumptions in colder months, but the consumption increased in warmer months especially in the southwest due to increased cooling loads; as a result, the overall annual energy consumptions in CHP systems were much higher than those in SHP systems. Locations in the southwest had the lowest energy consumptions in all systems. Water consumptions in all systems increased toward the southwest due to relatively increased available solar radiation (Figure 11.15). SHP system transforms the non-energy-efficient, unfavorable regions under CON systems in the Canadian North into potential greenhouse regions, and opens up the northern population's limited access to fresh vegetables and locally produced greens for better public health and local sustainability.

In hot and arid regions, on the other hand, greenhouses require large quantities of water for irrigation. During the summer months in these regions, inside temperatures rise above the acceptable limits, thence the greenhouses become quite unusable during this time of the year. Using proper technologies and environmental management systems can however significantly change the energy and moisture dynamics of greenhouse production systems. Hardin et al. (2008) focused on reducing natural gas, electricity, and water consumption in semi-arid California greenhouses introducing renewable energy heat pump technologies to both open and confined greenhouses in California. The simulations were performed for three consecutive years 2005, 2006, and 2007. The confined system had no external aeration and therefore need no further water supply. They demonstrated how water and energy consumptions inside the greenhouse changed under CON, OHP, and CHP greenhouse

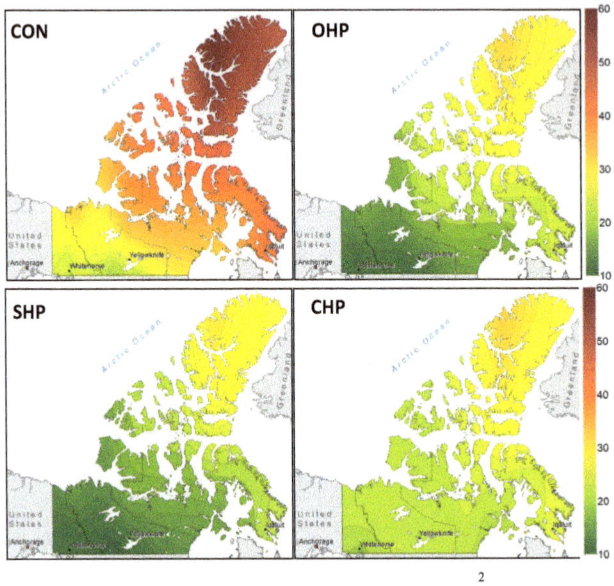


**FIGURE 11.14**
Spatial distributions of annual total energy consumptions (GJ/m²) in CON, OHP, SHP, and CHP systems in the Canadian North (Contour Intervals: 1 MJ/m².d). CON: Conventional greenhouse; OHP, SHP, and CHP: Open, semi-open, and closed heat pump greenhouse systems, respectively.

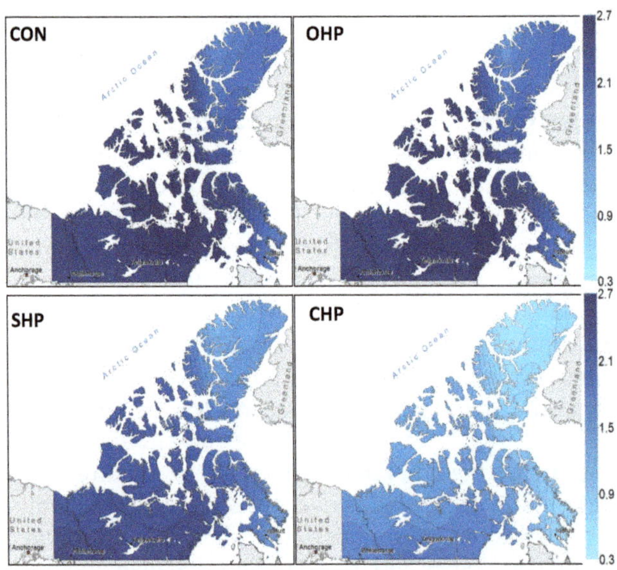


**FIGURE 11.15**
Spatial distributions of annual amount of transpiration (kg/m²) in CON, OHP, SHP, and CHP systems in the Canadian North (Contour Intervals: 100 g/m².d). CON: Conventional greenhouse; OHP, SHP, and CHP: Open, semi-open, and closed heat pump greenhouse systems, respectively.

**TABLE 11.9**

Comparisons of Energy and Water Consumption Means With Respect To Different Greenhouse Systems Using Tukey's Simultaneous Test at a 95% Confidence Level (CON: Conventional System; OHP: Open Heat Pump System; CHP: Closed Heat Pump System).

| Total Energy Consumption (all values in $MJ/m^2/day$) | | | |
|---|---|---|---|
| **System (Mean)** | **Compared to System (Mean)** | **Difference of Means** | **Statistically Significant at a 95% level?** |
| CON (19.29) | OHP (8.85) | 10.45 | YES |
| | CHP (11.20) | 8.09 | YES |
| OHP (8.25) | CHP (11.20) | −2.35 | YES |

| Water Consumption for Transpiration (all values in $kg/m^2/day$) | | | |
|---|---|---|---|
| **System (Mean)** | **Compared to System (Mean)** | **Difference of Means** | **Statistically Significant at a 95% level?** |
| CON (2.17) | OHP (2.25) | −0.08 | NO |
| | CHP (0.99) | 1.18 | YES |
| OHP (2.25) | CHP (0.99) | 1.26 | YES |

systems (Table 11.9). Outside environment had the most prevailing influence in the CON system and the least in the CHP system. The OHP system depended heavily on the outside weather as well, but it was much better than the CON greenhouse. The confined system, on the other hand, used only a heat pump to regulate humidity and temperature within the greenhouse and had slightly higher energy consumption than the OHP system during summer months. Almost all the transpired water in the CHP system was collected through dehumidification. Water is fully conserved in the confined system, essentially making the water consumption zero. However, additional energy was required to provide cooling in the confined system. Their findings suggest that a combination of CHP system and OHP system has the greatest benefit. That is, the greenhouse should be operated as a SHP in semi-arid California conditions. During the winter and spring months, the greenhouse system would be operated as a closed system conserving all the water and consuming much less energy for keeping the thermal environment at desirable levels. As mentioned earlier, closed greenhouse operation with a heat pump system would also make it easier to keep high $CO_2$ levels inside the greenhouse, while reducing the risk of insects and diseases within the confined greenhouse environment. During the summer months, however, the greenhouse would be operated as an OHP introducing ventilation into the greenhouse, hence allowing outside air to remove excess humidity within the greenhouse and reducing the cooling load over the heat pump.

Furthermore, for hot and arid regions, Buchholz et al. (2004) developed the Watergy concept – a novel, closed greenhouse system. Watergy follows the development of a technology platform for decentralized basic supply of energy, water, and food that can be used within a number of different applications. The system has no external aeration and has almost no need for further water supply. The technological principle is actually proven at two different prototypes, which were built within a European Union (Framework Program Promoting Energy, Environment, and Sustainable Development) funded research project. In 2004, the first prototype was built for the purpose of greenhouse horticulture in Almeria, Spain (see Figure 6.17). The second prototype was built in Berlin, Germany, as a living or office building with an attached façade-greenhouse. It has a great potential to reduce the demand for or to totally displace natural gas, to reduce the load on power grid by electricity savings, and the demand for irrigation water in greenhouse operations. It also allows plant protection without chemical insecticides and the accumulation of carbon dioxide without any aeration losses. In total, the process generates heat, clear water, and cooled/dehumidified air with minimum need of primary energy.

### 11.2.9 Variable Speed Pumps

Considering the whole pumping work that takes place all around the world, not only in greenhouses but also in industrial plants, domestic heating systems, etc., then it would probably be easier to see that the pumping energy consumption accounts for approximately 20% of the world's primary energy consumption. This shows that the global pumping industry has quite a large room to make significant contributions toward conserving our energy resources. In pumping applications such as greenhouses where the required pumping is not constant, installing a variable-speed pump will result in significant energy savings. Variable-speed pumps have a higher initial cost; however, they do not require mixing valves and their operational costs are lower than those of the equivalent fixed-speed pumps; therefore, they have more savings over time. Variable-speed pump systems cost more due to the added complexity of the variable frequency drive; however, this cost is outweighed by the long-term energy cost savings. When greenhouse operations replace their constant-speed pumps (see Figure 11.16) with variable-speed pumps, they typically recover the cost of the new pump through energy savings in less than a year.

In a variable speed pump system, when more heat is called for, the pumps run faster to provide this required heat. When less heat is required, on the other hand, the pump slows down. Since variable-speed pumps run below their maximum speeds, there is a reduction in wear and tear as well.

**FIGURE 11.16**
Constant-speed transport pumping and mixing group installed on manifolds.

**Example 11.4 Calculating Energy Savings with a Variable-Speed Pump**

Now, let's consider a 3 hp single (constant) speed pump, which is used for 5,000 hours annually. Calculate the energy savings if the greenhouse operator switches to a variable-speed pump. Remember that when we pay for electricity we pay for the kWh of energy used. Assume that the average cost of electricity is $0.12 per kWh.

**Solution**

*3 HP Constant-Speed Pump:*
From Chapter 2, one horsepower is equal to 746 W, and power is equal to the current times the voltage. Therefore, this constant-speed 3 hp pump uses 3 hp (746 W/1 hp) = 2,338 W, which is approximately 9.7 Amp at 230 V. In reality, this pump will draw more current than this given amount due to pump and plumbing system inefficiencies; hence, close to approximately 14 Amp at 230 V is a more realistic figure. However, for this simplified example, we will use this extremely conservative number of 9.7 Amp.

In order to determine the energy consumption (kWh) of this pump during 5,000 hours of operation, we simply multiply the power (Watts) by the operational hours. Then, the amount of energy this pump would use:

$$(2{,}338\text{W})(5{,}000\text{h}) = 11{,}690{,}000\text{ Wh or } 11{,}690\text{ kWh}$$

And, the annual cost of electricity would be (11,690 kWh) ($0.12/kWh) = $1,403 per year.

Now, let's do this again; but, this time using an equivalent variable-speed pump.

*3 HP Variable-Speed Pump:*
First of all, the pump Affinity Laws tells us three things: i) the flow changes proportionally to the speed (RPM); that is, double the RPM, double the flow; ii) the pressure changes by the square of the difference; that is, double the RPM, multiply the pressure by four; iii) the power changes by the cube of the difference; that is, double the RPM, multiply the power by eight. Hence, if we reduce the speed of the pump's motor, then based on the Affinity Laws, power is reduced as the cube of the speed. This is important because on a pump curve, the flow and head typically change simultaneously. As the flow increases, pressure capacity drops, and vice versa. In a variable-speed application, however, as the flow decreases and the pump pressure increases, the drive reduces the speed in order to match a constant discharge pressure by slowing the pump speed down. Then again, based on the Affinity Laws, as the pump slows down, the energy consumed by the pump is reduced by a factor of eight. This means that if we reduce the pump motor RPM (speed) by half, then we reduce the electricity consumption by eight times, that is, Power = $([1/2]\text{ RPM})^3 = 0.125$. Actually, this is the backbone of the efficient technology that variable-speed pumps are based on, and in a constant-speed system, this does not happen because the pump does not change speed.

A pump that uses 2,338 W per hour at 3,450 RPM (single-speed pump RPM) will use only 292 W at 1,725 RPM (half the speed). Then, in order to simplify our calculations, let's assume that the pump operates at the half the speed for performing the same job during the season. Then, the amount of energy this variable-speed pump would use:

$$(292\text{W})(5{,}000\text{ h}) = 1{,}460{,}000\text{ Wh or } 1{,}460\text{ kWh}$$

And, the annual cost of electricity would be (1,460 kWh) ($0.12/kWh) = $175 per year.

So, this variable-speed pump would save ($1,403 – $175) = $1,228 per year.

Again, even though the initial costs of variable-speed pump systems are higher, considering the energy savings we have just calculated above, this initial cost can easily be recovered in less than a year.

### 11.2.10 Energy Efficient Lighting Systems

As discussed earlier in Chapter 7, supplemental lighting is quite often used in commercial greenhouses in order to enhance crop quality and growth. In addition, photoperiodic lighting is employed to control flowering of certain crops. Most greenhouses use the electricity generated at central power plants, while increasing number of greenhouse facilities now generate their own electricity locally in a distributed fashion, that is, via cogeneration on site, conserving the waste heat and eliminating the transmission losses and $CO_2$ emissions as discussed earlier. Central power plants waste a lot of energy even before the power is transmitted to the consumer's site. And further energy is lost through the transmission lines. Efficiencies of some energy conversion systems and devices are presented in Table 11.10, while a simple case study is provided in Example 11.5 for determining overall "well to wheel" conversion efficiency. Table 11.11, on the other hand, provides a comparison of LED, incandescent (INC) light bulb, and compact fluorescent light (CFL) bulbs with respect to their energy efficiencies, environmental impacts, some important facts, and light outputs.

As Tables 11.10 and 11.11 show, quite a significant amount of energy can be saved by simply replacing INC lamps with more efficient lighting fixtures. CFLs, for instance, use approximately a quarter of the energy of INC lamps (e.g. a 15-W CFL replaces a 60-W INC lamp) because they generate less heat. Greenhouse growers can easily replace the INC lamps in work and other greenhouse areas where lighting is not being used for photoperiodic lighting. Michigan State University's Extension Service recommends installing CFL's for every other INC lamp in the areas where photoperiodic lighting is required due to the fact that flowering of some photoperiodic crops can be delayed if only CFLs are used. Moreover, LED flowering lamps can help greenhouse growers to save further energy and the environment. LEDs consume approximately one tenth of the energy of INC lamps and one half of the energy CFLs; therefore, they can easily replace INC and CFL bulbs for supplemental lighting and provide significant year-round energy savings in greenhouses where artificial lighting is used during winter months. In addition, using alternative lighting strategies such as cyclic night-interruption lighting instead of 4-hour continuous lighting every night can save further energy and provide one more step toward sustainability.

### 11.2.11 Efficient Environmental Control Systems

Smart digital greenhouse environmental control systems are available today, and they control every single parameter in mostly an integrated fashion. Further precision controls and therefore savings can be achieved via incorporating smarter strategies with artificial

**TABLE 11.10**

Efficiencies of Some Energy Conversion Systems and Devices.

| Energy Conversion System | Efficiency (%) |
|---|---|
| Gas furnace | 70–95 |
| Wind turbine | 35–50 |
| Fossil fuel power plant | 30–40 |
| LEDs | 50 |
| CFLs | 20 |
| Incandescent lamp | 5 |

**TABLE 11.11**

Comparison of LED, Incandescent (INC), and CFL Bulbs.

| | LEDs | INC | CFLs |
|---|---|---|---|
| **Energy Efficiency** | | | |
| Life span (hours) | 50,000 | 1,200 | 8,000 |
| Watts of electricity used (Equivalent to 60 W bulb) | 6–8 | 60 | 13 15 |
| **Environmental Impacts** | | | |
| Contains toxic mercury? | No | No | Yes |
| $CO_2$ emissions (30 bulbs per year) | 1 unit | 10 units | 2.3 units |
| Sensitivity to humidity | No | Some | Yes |
| **Important Facts** | | | |
| On/Off cycling | No | Some | Yes |
| Durability | Very durable | Not very durable | Not very durable |
| Heat emitted (Btu/h) | 3.4 | 85 | 30 |
| **Light Output** | **Watts** | | |
| 400 lumens | 4–5 | 40 | 9–13 |
| 800 lumens | 6–8 | 60 | 13–15 |
| 1,100 lumens | 9–13 | 75 | 18–25 |
| 1,600 lumens | 16–20 | 100 | 23–30 |
| 2,600 lumens | 25–28 | 150 | 30–55 |

*Source:* USAI Lighting (2020).

**Example 11.5: Calculating the Overall Efficiency for Chemical Energy to Light Energy Conversion**

Calculate the overall efficiency if the power plant efficiency is 35%, transmission efficiency is 90% and the light bulb is an INC bulb. What would be the overall efficiency if a CFL bulb replaces the INC bulb?

**Solution**

The overall efficiency is the product of individual efficiencies, and INC lamp efficiency is given as 5% in Table 11.10; therefore,

$$\text{Overall Efficiency} = \eta_1 \times \eta_2 \times \eta_3 = 0.35 \times 0.90 \times 0.05 = 0.016, \text{ which is } 1.6\%.$$

If we replace the INC bulb with a CFL, then

$$\text{Overall Efficiency} = \eta_1 \times \eta_2 \times \eta_3 = 0.35 \times 0.90 \times 0.20 = 0.063, \text{ which is } 6.3\%.$$

Furthermore, if we replace the CFL bulb with an LED, then the overall efficiency moves up to approximately 16%, saving not only our wallets but also the environment further.

**Example 11.6: Estimating the Savings in Operational Cost with Different Lighting Fixtures**

Determine the seasonal savings when a 100-W INC bulb is replaced with an equivalent CFL or LED bulb if the fixture is used for supplemental heating for an average of 10 hours a day from October through March, and the electricity costs $0.12 per kWh.

**Solution**

From October through March, we have 6 months; therefore, we have an average of 10 h/day × (180 days) = 1,800 hours of operation. Hence, the cost of electricity consumed per season

With a 100 W INC bulb:

$$100\text{W}\left(1\text{kW}/1{,}000\text{W}\right)\left(1{,}800\right)\left(\$0.12/\text{kWh}\right) = \$21.6$$

With a 25 W CFL (equivalent to 100 W INC bulb):

$$25\text{W}\left(1\text{kW}/1{,}000\text{W}\right)\left(1{,}800\text{h}\right)\left(\$0.12/\text{kWh}\right) = \$5.40$$

With a 10 W LED (equivalent to 100 W INC bulb):

$$10\text{W}\left(1\text{kW}/1{,}000\text{W}\right)\left(1{,}800\text{h}\right)\left(\$0.12/\text{kWh}\right) = \$2.16$$

This example shows that significant cost savings can be achieved just replacing a single 100-W INC bulb. It should also be kept in mind that the life span of an INC light bulb is 1,200 hours, which does not even last for the lighting season and has to be replaced before the season ends.

Examples 11.5 and 11.6 clearly show that, in order to save our funds and the environment, we need to replace our conventional lighting fixtures immediately, including those heavily used in our greenhouses, with more energy-efficient alternatives such as CFL and LED bulbs.

intelligence (AI) algorithms into the environmental control systems. For instance, reducing ventilation a bit or maintaining little higher indoor relative humidity levels without harming the crop growth and development (and therefore the eventual economic gains), heating water when the energy costs less, and storing the heat in hot water storage tanks for later heating, we can reduce the energy losses and save some dollars escaping through the greenhouse vents or our pockets. Greenhouses are extremely complex living structures; therefore, complex and smart approaches are required in order to bring all the stakeholders (e.g. whether, the cost of energy, water, $CO_2$, and others) together instantly for healthy crop development and therefore a profitable operation. The cost of every control step and strategy should be evaluated in an integrated fashion before it is actuated; and with today's AI technology, it is perfectly achievable.

### 11.2.12 Renewable Energy Based Greenhouse Systems

Before anything else, greenhouses are passive solar systems, that is, they are naturally renewable energy based systems. With the recent improvements in other renewable energy technologies and their dropping expenses, we are getting closer to net energy generating greenhouse complexes every passing year. While natural gas, a fossil fuel, is the predominant fuel used to heat commercial greenhouses in northern climates, renewable biomass, solar, wind, and heat pump technologies are increasingly incorporated to commercial greenhouse operations. Soon, we will be able to see greenhouse operations not only selling their green produce, but also selling the net green energy they are generating in a distributed fashion to the utility companies. In essence, each greenhouse facility is an energy generating station. Dropping prices of renewable energy technologies and various incentives will enable greenhouse growers to easily adopt a number of renewable energy systems in their operations. That's why; various renewable energy technologies and a number of numerical examples are intentionally introduced and provided in different sections of this book.

## 11.3 Environmental Considerations

The story of humankind's success and his eventual downfall is one that rests largely on the shoulders of our creative exploitation and reimagining of energy and its uses. Throughout history, humankind has been able to utilize energy in ways other species have been unable to grasp. Ingenious use of energy has led to increased brain development, enabled us to travel great distances, allowed for the manufacturing of a variety of products at tremendous speed, and helped to power the machines, which influence everything in life from healthcare to communication, and to science, research and development. The rate at which energy has been exploited has however left humankind in a compromising position. Finite resources are rapidly being depleted; carbon emissions and competing water demands and pollution continue to cause large-scale environmental issues, increasing the attention to sustainability in all contexts. Renewable energy sources offer one answer to the problem and with increased implementation, they may one day power the planet. This is the history of how humankind and society evolved alongside energy and how we have arrived at the current situation, full of challenges and opportunities. From the earliest uses of fire to the steam engine, to massive fossil fuel dependence, and to the beginning of the major shift toward renewable energy sources, humankind has continually innovated and adapted to his situation through desire or necessity. Advancements in energy production and usage improved social and economic aspects of life. Humankind has grown alongside energy and in a world where more and more is demanded, a continued growth and innovation will be required to sustain our societies and their ever growing needs.

### 11.3.1 Energy-Related Air Pollution

Air pollution refers to the pollution, which can harm human health and pollute the atmosphere. When pollutant concentrations in the air reach levels higher than the environmental quality standards, then we call the situation as air pollution. Air pollution can occur due to many different reasons and substances, which may be gaseous, solid, or liquid. One of the main sources of air pollution is burning fossil fuels such as oil, gasoline, and coal. During the combustion process, pollutants such as $CO_2$, particulate matter (PM), $SO_2$, and $NO_x$ are released. The evaporation or incomplete combustion of fuels release Volatile Organic Compounds (VOCs) as well. In addition, when unburned methane ($CH_4$) is released into the atmosphere, it contributes in the greenhouse effect due to its relatively high global warming potential. Burning fossil fuels also release heavy metals into the environment such as As, Cd, Cr, Ni, Mn, and Pb causing further air pollution. We, as energy consumers, need to be conscious about the sources of these pollutants and what they cause, so that we can make proper informed decisions with respect to our preferences in different energy sources.

### 11.3.2 Energy-Related Water Pollution

Water resources play influential roles in the energy production process because water is needed in almost all forms of energy production during either extraction or processing. For instance, thermal power and hydropower are the two primary sectors of total electricity production that consume huge amounts of water, accounting for 80% and 15%, respectively (UN, 2014). The IEA's World Energy Outlook 2012 published that 583 billion cubic meters of water was withdrawn for energy production in 2010 alone, which was up to 15% of global water withdrawals (IEA, 2012). Energy–Water nexus was first introduced in a Department of Energy (DOE) National Laboratory project in the USA in 2004 (Olsson, 2012). From that time, the link between energy and water resources has been more comprehensive and much more profound than initially thought.

The UN (2014) reported the global water withdrawal per year by sectors in different regions as shown in Figure 11.17. The total water consumption in different regions shows that more than 60% of water was withdrawn in Asia, more than 20% was withdrawn in the Americas, and around 10% and 5% withdrawals were made in Europe and Africa, respectively. It is obvious that more water is used for agriculture, especially in developing countries. In these countries, although a small fraction accounted for agricultural, over half of the total water withdrawals were consumed by the cooling system from power plants (Omer, 2008; Olsson, 2012; UN, 2014a). Taking the thermal power sector as an example, it was responsible for approximately 45% of freshwater withdrawals in Europe and 50% of total water withdrawal in the USA, which is much more than the water supply used in the agricultural sector (Kenny et al., 2009; Eurostat, 2010). From the global perspective, agriculture is the largest consumer, responsible for 70% water withdrawals in total for irrigation practice (UN, 2014a). According to the New Policies Scenario stated by the IEA, water withdrawals

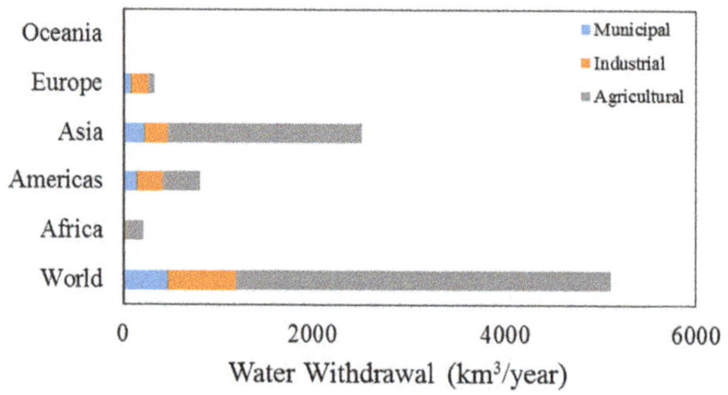


**FIGURE 11.17**
Total water withdrawals per year by sectors in different regions. Data from UN (2014).

**TABLE 11.12**
Water Requirements for Primary Energy Production and Water Quality Impacts (USDOE, 2006).

| Process | Energy Element | Water Quantity | Water Quality |
|---|---|---|---|
| **Energy Extraction and Production** | Oil and gas exploration | Water for drilling; completion and fracturing | Impact on shallow groundwater quality |
| | Oil and gas production | Large volume of produced impaired water | Produced water can impact surface and groundwater |
| | Coal and uranium mining | Large quantities water required | Tailings and drainage can impact surface water and groundwater |
| **Refining and Processing** | Traditional oil and gas refining | Water need to refine oil/gas | End-use can impact water quality |
| | Biofuels and ethanol | Water for growing and refining | Refinery wastewater treatment |
| | Synfuels and hydrogen | Water for synthesis or steam reforming | Wastewater treatment |
| **Energy Transportation and Storage** | Energy pipelines | Water for hydrostatic testing | Wastewater requires treatment |
| | Coal slurry pipelines | Water for slurry transport; water not returned | Final water is of poor quality; requires treatment |
| | Barge transport of energy | Fuel delivery impacted by river flows and stages | Spills or accidents can impact water quality |
| | Oil and gas storage caverns | Large quantities of water required for slurry mining of caverns | Slurry disposal impacts water quality, and ecology |
| **Electric Power Generation** | Thermoelectric (fossil, biomass, nuclear) | Surface water and groundwater for cooling and scrubbing | Thermal and air emissions impact surface waters and ecology |
| | Hydroelectric | Reservoirs lose large quantities to evaporation | Impacts water temperatures, water quality, and ecology |
| | Solar PV and wind | None during operation; minimal water use for panel and blade washing | |

are expected to increase by 20% together with an 85% consumption improvement by 2035 (IEA, 2015). However, higher power plant efficiencies could be achieved by then, with advanced cooling technologies and increased biofuels production.

Energy production and consumption have extraordinary positive impacts on social life, but negative environmental and social consequences of energy production cannot be ignored, including the water issues. Water is an already scarce resource from many aspects, and the expansion of food and energy production is adding further pressures on water resources and exacerbating the water scarcity further (Olsson, 2012). Water shortage will become a serious problem in many areas or the world such as India, Africa, and the Middle East, and New Zealand (WEC, 2013; IEA, 2013; UN, 2015). The United Nations Department of Economic and Social Affairs reported that the available water resources are decreasing, and approximately two-thirds of the population in the world will suffer water scarcity by 2025 (UN, 2014b). Consequently, water scarcity will, in turn, have dramatic impacts on pertinent production sectors and result in regional conflicts.

#### 11.3.2.1 Energy–Water–Pollution Nexus

The footprints of energy projects have significant effects on water quality. Obviously, better management can bring substantial benefits for energy savings and alleviate water pollution. Water requirements for primary energy and related pollution impacts are summarized in Table 11.12. According to the US Department of Energy (USDOE) (2006), all forms of energy production pollute water in varying degrees, including clean energy sources. For instance, the extraction of fossil fuels results in a vast volume of produced water discharge. Coal-mining activities further destroy the surrounding ecosystems, and produced water that contains toxic substances has critical risk to the aquatic system and human health (Olsson, 2012; UN, 2014a; UN, 2015). The construction of hydropower plants has a serious impact on the physical and chemical parameters of water and brings unexpected impacts on agriculture and nature. In addition, water used for cooling systems in power plants would be returned to the adjacent water bodies at higher temperatures, which may be detrimental to resident flora and fauna (Olsson, 2012; UN, 2014a). Energy can, on the other hand, be used for wastewater treatment and the waste thermal energy can be recovered. Solar energy, for instance, is widely used in wastewater treatment plants; and bioenergy production can as well be integrated with wastewater treatment systems (Argaw, 2003).

#### 11.3.2.2 Impacts of Oil Production on Water Quality

Almost each and every step in the oil production process has significant impacts on water resources. For example, steam injection and hydraulic fracturing are the two common technologies used for extracting oil from reservoirs, which contaminate water (Olsson, 2012). Oil fields usually contain a large volume of water and gas, so water or steam are frequently injected into reservoirs to maintain the pressure. After that, more steam is injected and the pressure is utilized to push oil up through the well (US EPA, 2011). Extraction of oil requires a lot of water withdrawal; and meanwhile, it contaminates the water. The polluted water that comes from the oil extraction process is called produced water. The composition of the produced water is quite complex and includes heavy metals, hydrocarbons, hydrogen sulfide, organic acids, and other chemicals added through the extraction and separation processes. The amount of produced water that is generated in oil production fields has been estimated about three times as much as the crude oil produced (Khatib & Verbeek, 2003). In the past few decades, most of the produced water is discharged directly into surrounding aquatic systems without any wastewater treatment, resulting in groundwater and even ocean contamination (Stephenson, 1992). Oil spills are the other direct environmental impacts that happen regularly during transportation of oil from place to place. Similarly, a large amount of oil may leak from underground storage tanks and flow into underground water resources. Oil production also causes indirect impacts on water quality. For instance, pollutants are released into the atmosphere during combustion of fuels. However, these particles can return to the water systems through precipitation. Furthermore, unconventional oil mining requires a massive amount of water and energy to refine and separate crude oil (Reynolds & Kiker, 2003). This clearly shows that the generation of produced water is huge, thereby causing severe water pollution.

#### 11.3.2.3 Impacts of Coal Production on Water Quality

Water consumption is tightly linked with coal mining, like oil production, as it is used in extraction, processing, and transportation. Water quality issues frequently accompany mining activities (Palmer et al., 2010). A large volume of water is used to wash the coal to limit the undesirable chemical components such as sulfur, mercury, and others. Then, these elements accumulate in wastewater, causing water pollution as they are discharged or leak into aquatic systems (Mielke et al., 2010). A high load of these detrimental elements, especially heavy metals, persistently accumulate in the environment; and even worse, these heavy metals can eventually incorporate into food chain. Coal spills during transportation can, on the other hand, lead to indirect water contamination. In 2008, a coal ash spill caused terrible impacts on the Emory and Chinch River system in Harriman, Tennessee, USA (Deonarine et al., 2013). Approximately 4.1 million $m^3$ of fly and bottom coal ash was spilled, and they were responsible for high mercury concentrations in river sediments. The water quality was not significantly impacted by the presence of mercury; however, the total mercury concentration in sediment was 3–4 times higher than non-polluted areas (Deonarine et al., 2013). Furthermore, people may have to use coal-contaminated water for daily requirements, such as drinking, cooking, and washing. Even worse, it could eventually force local residents to leave their homes and relocate.

#### *11.3.2.4 Impacts of Natural Gas Production on Water Quality*

A large amount of produced water is generated from the extraction process, usually containing heavy metals, hydrocarbon residues and other chemicals. Clark and Veil (2009) reported that the amount of produced water is about six times the gas production. Fontenot et al. (2013) illustrated that water parameters including arsenic, selenium, strontium, and TDS (total dissolved solids) exceeded the drinkable water quality standards in some wells near a natural gas extraction site. These wells were located within 3 km of the active natural gas extraction site. Lower concentrations have also been detected in the drinking water wells beyond 3 km radius from the gas extraction site. Moreover, high levels of methanol and ethanol were also observed in about one-third of all drinking water wells.

#### *11.3.2.5 Impacts of Bioenergy Production on Water Quality*

Agriculture is one of the main polluter of freshwater resources due to the use of fertilizers and pesticides that eventually end up in the aquatic system. Biofuels production expands the agricultural activities and increases the demand for the majority of traditional crops, resulting in water competition and pollution that typically cause what is called the eutrophication phenomenon, which is associated with algal blooms, dissolved oxygen depletion, and degraded surface water bodies. Water quality impacts can be tracked throughout the whole bioenergy feedstock production chain. However, the main problem is attributed to the expansion of agricultural activities in order to meet the bioenergy demand. Fertilizer and pesticide usage and other agricultural practices such as tillage and irrigation have damaging effects on water quality.

#### *11.3.2.6 Impacts of Hydropower on Water Resources*

The WEC (2013) indicated that, even though the global installed hydropower capacity has increased by 8% since the last survey, the global electricity generation dropped by 14%, mainly caused by water shortages. Hydropower generation is a large water consumer. However, most of the water used is returned to the river downstream as the reservoir is filled or when the water runs through the turbines. An increased temperature in dams may lead to an increased temperature in the river, which negatively influences the water quality downstream. As water temperature and flow discharges are altered, the life cycle of aquatic animals are threatened further. For instance, migration, spawning and hatching phases of salmon was found to be changed as a result of hydropower operations (International Commission for the Protection of the Rhine, 2010). Furthermore, water-borne diseases and malaria have infected more than 200 million people after the construction of dams and reservoirs, and it is a common health problem in the tropical regions of the world (Olsson, 2012). Detrimental water quality situations can however be mitigated by well-organized and well-managed hydropower plants before implementation.

#### 11.3.2.7 Impacts of Nuclear Power Plants' Cooling Water on Aquatic Systems

The cooling systems in nuclear power plants consume a vast amount of water. The water used for cooling is typically discharged back to the adjacent aquatic system. The water temperature can be elevated by 8°C, depending on the operations and settings (Laws, 1993). The structure of the marine ecosystem and population distribution are tightly linked with water temperature as it affects the physiological processes of fish and others. Therefore, increased temperature in the water bodies can have adverse impacts on aquatic life. Although some resident organisms can tolerate the warm water up to certain extent, most of the organisms are sensitive to environmental changes. As a result, biodiversity and population dynamics eventually change due to thermal pollution in water (Teixeira, 2012).

#### 11.3.2.8 Impacts of Nuclear Accidents on Water Resources and Aquatic Systems

The Chernobyl accident occurred on April 26, 1986, at the Chernobyl nuclear power plant in Ukraine. The bioaccumulation of radiocaesium can still be traced in the aquatic food chain. Some can even be found on lakes in Scandinavia and Germany. People who were living close to the lakes ingested the $^{137}Cs$ from fish. Besides, the major surface contamination around the accident region is dominated by the groundwater contaminations. A slow declining rate of $^{137}Cs$ in freshwater and fish has been reported, but it is expected to remain at a high level for decades. Nowadays, the surface contaminations from radioactivity have been dramatically reduced by good countermeasures and natural weather impacts (Alexakhin et al., 2006). Furthermore, on March 11, 2011, the devastating Tohoku earthquake and enormous tsunamis caused serious damages in Fukushima Daiichi Nuclear Power Plant in Japan, and tons of contaminated seawater was dumped back to the Pacific Ocean, which worsened the water pollution.

The international debate on the prospects of nuclear energy development has reignited after the Fukushima nuclear accident. Wittneben (2011) reported that each country has a different response and decision in Europe, mainly due to geographical and political reasons. For instance, the UK maintained its decision to increase nuclear share in the coming years, while the German government decided to shut down the commissioned nuclear reactors and make a comprehensive test to eliminate the potential safety hazard. In reality, the Fukushima accident reduced the public acceptability of nuclear power plants in many developing countries, while increasing the demands for fossil fuels, particularly natural gas (Wittneben, 2011; WEC, 2013).

#### 11.3.2.9 Impacts of Geothermal Energy Production on Water Resources

Recent studies reported that geothermal power plants have adverse impacts on water bodies and soils. Geothermal plants require the withdrawal of huge amounts of water for cooling and reinjection procedures (Macknick et al., 2011). Geothermal fluids contain various heavy metals and other toxic components such as Cd, Hg, As, Mn, F, and B, due to contact with hot fluids and rocks (Armienta et al., 2014). The geothermal fluid may contaminate the surface water as well if there are spills and if it is released into water ecosystems. They not only damage the aquatic ecosystems but also threaten water security for irrigation and drinking. Heat extraction and cooling by aquifers may, on the other hand, lead to temperature alteration in natural groundwater systems. Similar to the case in nuclear energy, the elevated temperature in water bodies might disturb physical and chemical compositions of water body, causing severe alterations in ecosystem functions. Actually, it is not necessary to use clean freshwater for cooling and reinjection purposes; hence, many geothermal plants recycle the geothermal fluids or wastewater for operating.

#### *11.3.2.10 Impacts of Other Energy Production on Water Resources*

Tidal energy, for instance, is a renewable and clean ocean energy that is expected to be broadly exploited in the coming years (Wang et al., 2015). A three-dimensional coastal ocean model showed that the water quality impacts were strangely related to the amount of power extracted from the tidal power plant system. The effects on water quality were minor under a lower level tidal energy extraction, while there were adverse impacts, such as low dissolved oxygen content in bottom waters when a larger amount of energy was extracted (Wang et al., 2015). Additionally, antifouling coating in turbines may have potential impacts on water quality (Polagye et al., 2011). It seems that a long-term extraction of tidal energy might alter both physical and biological parameters of water.

### 11.3.3 Indoor Air Pollution

The most common method used for $CO_2$ enrichment in greenhouses is burning hydrocarbons such as natural gas, propane or low sulfur kerosene. Air pollution in greenhouses due to $CO_2$ enrichment by burning these hydrocarbons sometimes causes visible leaf injuries, which include chlorosis, necrosis, and curling of the leaves, and invisible injuries (growth-reduction). The research findings explain the injurious effects of high $CO_2$ concentrations by: i) too high leaf temperatures at high light levels as affected by reduced transpiration at high $CO_2$ concentrations; ii) accumulation of starch, which breaks down the chlorophyll, promoted by high light levels and low temperatures; iii) reduced nutrient uptake because of reduced transpiration, particularly at high air humidity levels.

When $CO_2$ is supplied by burning kerosene, then ethylene ($C_2H_4$) pollution might be observed due to incomplete combustion. Ethylene is a plant hormone, which stimulates senescence of the plants, causing leaf chlorosis and abscission, and flower-drop; therefore, it has a potential to cause substantial crop losses. The concentration in the greenhouse environment should not exceed 0.05 ppm, above which plant injuries might be observed. These long-term plant exposures to even quite low ethylene concentrations usually inhibit overall plant growth and can harm small flower buds. While new kerosene $CO_2$ dosing equipment have pretty good burners, adequate burner maintenance and care is still vital.

Carbon monoxide (CO) is another pollutant (which can be deadly for people at certain concentrations and exposure times; see Table 2.15) produced by incomplete combustion and causes injuries similar to those caused by ethylene, but thousands of times higher concentrations of CO must be present in order to cause comparable damages. In practice, the CO concentration in greenhouses is somewhere between 0 ppm and 5 ppm, which creates no problems. Sometimes, kerosene contains sulfur that might generate enough $SO_2$ to cause damage as well; however, using low-sulfur kerosene eliminates this problem.

Other hydrocarbons such as propane might also cause pollution. Commercial propane leakage has caused severe plant injuries in some greenhouse operations in the past as propane contains propylene, which is an analogue of ethylene. With the modern equipment and current safety measures, however, the probability of leakage is quite low. Natural gas, on the other hand, consists mainly of methane; and with complete combustion, it does not pose any harm to plants.

Along with ethylene, nitrogen oxides ($NO_x$) are other important pollutants sourcing from the burning of hydrocarbons. $NO_x$ is produced during the combustion when $O_2$ reacts with the atmospheric $N_2$, and in greenhouses, concentrations up to 1 ppm levels have been measured with kerosene as well as with propane burners. At a $CO_2$ dosing of 1,000 ppm, the $NO_x$ concentration is usually about 0.5 ppm. Nitrogen oxide (NO) normally makes up approx. 75% of the oxides and concentrations above 250 ppb sustained over 10 days is considered harmful to plants; however, some propane and kerosene burners produce mainly $NO_2$ and concentrations above 132 ppb sustained over 10 days is considered harmful. The injuries of these oxides are similar and species specific, ranging from visible to invisible injuries.

Since burning hydrocarbons might cause injuries to plants, probably the safest source is pure liquid $CO_2$ from containers, recommended for general use for greenhouse crops even though it may also contain some undesirable impurities such as ethylene. Today, along with liquid $CO_2$, the flue gas $CO_2$ coming out of natural gas boilers is commonly and successfully used for $CO_2$ supplementation in commercial greenhouses.

### 11.3.4 Light Pollution

Today, more greenhouse operations are employing supplemental lighting to increase plant growth and development; however, this electric lighting can have undesirable and unintended consequences to neighboring areas at night, which is considered as light pollution. And these neighboring areas might be other greenhouse sections within the same operation, or nearby greenhouse complexes or residences. Some plants are quite sensitive to extremely low light intensities; therefore, the stray light can have sufficient intensity to reach other sections or nearby operations to modify the photoperiod of plants grown there. Light pollution may come not only from lighting fixtures used in greenhouses but also from other nearby buildings and street lighting, etc. Under such conditions, in order to be able to prevent light pollution, greenhouse operators can use retractable blackout screens and curtains along the greenhouse sidewalls and roof. These systems might be expensive to install; however, a blackout curtain not only prevents light pollution but also serves as an excellent night curtain that reduces heat losses from the greenhouse at night especially during the winter. Therefore, it is a justifiable investment.

### 11.3.5 Greenhouse Plastics and Organic Waste

Every year, millions of tons of plastic are used around the world for covering greenhouse structures, mulching greenhouse floors, and making all sorts of containers and plug flats. The uses of plastic are extremely convenient due to its low cost and ease of handling; however, the eventual environmental consequences are not so convenient and friendly due to the vast quantity of plastics generated. Managing plastic waste is a global challenge as landfills and incineration option many times do not provide economically and/or environmentally feasible solutions. Greenhouse glazing and mulching films cannot be reused as they deteriorate under ultraviolet radiation; however, plastic containers can be reused,

**TABLE 11.13**
Plastics Commonly Used in Greenhouses.

| Type of Plastic | Typical Uses | Reusability | Recyclability |
|---|---|---|---|
| High-Density Polyethylene (HPDE) | Nursery containers | Yes | Yes |
| Low-Density Polyethylene (LPDE) | Greenhouse glazing and plastic mulch | No | Yes |
| Polypropylene (PP) | Pots | Yes | Yes |
| High Impact Polystyrene (HIPS) | Carrier trays, plug and liner trays, and flats | Yes | Yes |

even though cleaning, sorting, and disease control might still be challenging. In order to be able to maintain the quality of the recycled end-product, greenhouse plastic waste ought to be sorted and handled separately. Contaminants and structural degradation of plastics, particularly soil and organic matter, reduce the recycled plastic quality dramatically. For effective recycling, a lot of waste plastic material must be collected from many greenhouse operations. Furthermore, waste plastic film cannot be transported very effectively due to its low bulk density. In order to reduce transport costs, bailers are used to compact the plastic film on the facility.

Today, several kinds of plastic are used in greenhouse industry, which includes high-density polyethylene (HDPE), low-density polyethylene (LDPE), polypropylene (PP), and high-impact polystyrene (HIPS) (Table 11.13).

*High-Density Polyethylene:*
Due to its chemical and thermal resistance, HDPE is commonly used in making nursery containers, particularly for trees, shrubs, pesticides, and herbicides. Since they are quite durable, they can conveniently be reused especially after sterilizing them. Plastic recycling companies recycle waste HDPE, which are typically used for making plastic timber for landscaping, picnic tables, etc., rather than being reformed into containers.

*Low-Density Polyethylene:*
LDPE is extensively used for making greenhouse glazing and mulching material for greenhouse plants. Due to its ultraviolet degradation during its lifecycle, LDPE greenhouse glazings cannot usually be reused. Waste LDPE requires washing (particularly mulching plastic) and baling due to its low bulk density before it is recycled. And most plastic recycling companies do not have the specialized cleaning systems particularly for removing the trapped soil; however, some recycling companies provide the baling equipment for handling the mulching plastic and greenhouse glazings.

*Polypropylene:*
PP is a durable, lightweight, and resistant plastic, which is commonly used for making plant containers for greenhouse production of herbs, potted flowering plants and bedding plants, etc. They can be reused especially after sterilization. And some plastic recycling companies recycle polypropylene by mixing it with HDPE and make plastic timber for landscaping, picnic tables, etc.

*High-Impact Polystyrene:*
HIPS is extremely lightweight plastic, difficult to compress, and used in making flats and plug trays for seedlings and small plants. HIPS plug and liner trays are mostly discarded

or recycled, but they are sometimes reused. Plastic recycling companies typically grind, melt and mold HIPS directly back into trays.

Every passing year, when they are provided with options, greenhouse operations use further technologies including environmental technologies in order to make their operations fully sustainable and turn to plastic recycling companies for managing their plastic waste as well rather than dumping the waste into landfills.

Moreover, tremendous amounts of produced greenhouse organic waste should also be considered for codigesting with livestock and other organic municipal waste to produce both electricity and thermal energy locally using CHP generators. The produced heat, power, and $CO_2$ can be used directly in greenhouse operations and make the operations and our rural communities much more resilient and sustainable.

## Nomenclature

| | |
|---|---|
| $A$ | area |
| **ACH** | air change per hour |
| **ADT** | average daily temperature |
| **CHP** | closed heat pump greenhouse system; combined heat and power |
| **CON** | conventional greenhouse system |
| **CFL** | compact fluorescent light |
| **DIF** | daytime and nighttime temperature difference |
| $E$ | energy |
| **HIPS** | high-impact polystyrene |
| **HDD** | heating degree-days |
| **HDPE** | high-density polyethylene |
| **HV** | heating value |
| **INC** | incandescent |
| **LDPE** | low-density polyethylene |
| **LED** | light-emitting diode |
| **OHP** | open heat pump greenhouse system |
| **PP** | polypropylene |
| $\dot{Q}$ | heat transfer rate |
| **SHP** | semi-closed heat pump greenhouse system |
| $T$ | temperature |
| $U$ | conduction heat transmission coefficient |

### Greek letters

| | |
|---|---|
| $\eta$ | efficiency |

### Subscripts

| | |
|---|---|
| *cond* | conduction |
| *inf* | infiltration |

## Bibliography

Albright LD, Novak S, and Wolfe DW (1986) *A preliminary evaluation of the effect of row cover material properties on the thermal environment under the cover*. Project Report. Dept. of Agric. Engng., Cornell University, Itacha, NY, U.S.A., p.114.

Albright LD, Reines RG, Anderson SE, Chandra P, Price DR, Langhans RW, and Cerilli RV (1978) *Experimental results of solar heating a Brace Institute style greenhouse. Proceedings, Third Annual Conference on Solar Energy for Heating Greenhouses and Greenhouse-Residence Combinations*. Fort Collins, CO, U.S.A.

Alexakhin R, Anspaugh L, Balonov M (2006) Environmental consequences of the Chernobyl accident and their remediation: Twenty years of experience. Report of the Chernobyl Forum Expert group "Environment", International Atomic Energy Agency.

Argaw N (2003) Renewable Energy in Water and Wastewater Treatment Applications, National Renewable Energy Laboratory.

Armienta MA,Rodriguez R, Ceniceros N (2014) Groundwater quality and geothermal energy. The case of Cerro Prieto Geothermal Field, México, *Renewable Energy*, 63: 236254.

ASABE Standards (2018) Standards, Engineering Practices, and Data. V21(1). St. Joseph, MI, USA.

Badger PC and Poole HA (1979) Conserving Energy in Greenhouses. OARDC Special Cir. 102, Extension Bul. 651, p. 32. Ohio Agricultural Research and Development Center, Wooster, Ohio, U.S.A.

Bryla DR and Strik BC (2007) Effects of cultivar and plant spacing on the seasonal water requirements of highbush blueberry. *J. Amer. Soc. Hort. Sci.* 132 (2): 270–277.

Buchholz M, Jochum P, and Zaragoza G (2004) Basic Water, Heat and Food Supply from a Closed Greenhouse – The Watergy Project, International Symposium GREENSYS 2004 – Sustainable Greenhouse Systems, Leuven, Belgium, 12–16 September 2004, published in International Society of Horticultural Science (ISHS) und Wageningen University and Research Center, Wageningen 2004 ISBN: 907601924X, S. 133.

Cameron A (2009) Horticultural plastics recycling – The future is brighter. *Greenhouse Product News Magazine*, June 2007. Sparta, MI, USA. pp. 20–24.

Clark CE, Veil JA (2009) Produced Water Volumes and Management Practices in the United States, Argonne National Laboratory ANL/EVS/R-09/1.

Claypool D, Delaney R, Ditterline R, and Lockerman R (1997) Genetic improvement of alfalfa to conserve water. Poster Presentation, Wyoming Water 1997 Conference, April 21–23, Casper, WY.

Deonarine A, Bartov G, Johnson T, Ruhl L, Vengosh A, Hsu-Kim H (2013) Environmental impacts of the Tennessee Valley Authority Kingston coal ash spill. 2. Effect of coal ash on methylmercury in historically contaminated river sediments, *Environmental Science and Technology*, 47(4): 2100–2108.

Ebru KA and Arif H (2007) A comparative study on exergetic assessment of two ground-source (geothermal) heat pump systems for residential applications. *Building and Environment* 42:2004–2013.

Eurostat (2010) Eurostat database. Brussels, European Commission.

Fontenot B, Hunt L, Hildenbrand Z, Carlton D, Walton, HO (2013) An evaluation of water quality in private drinking water wells near natural gas extraction sites in the Barnett Shale formation, *Environmental Science and Technology*, 47(17): 10032–10040.

Garzoli KV and Blackwell J (1987) An analysis of the nocturnal heat loss from a double skin plastic greenhouse. *Journal of Agricultural Engineering Research* 36: 75–85.

Hamada Y, Nakamura M, Saitoh H, Kubota H, and Ochifuji K (2007) Improved underground heat exchanger by using no-dig method for space heating and cooling. *Renewable Energy* 32:480–495.

Hardin C, Mehlitz T, Yildiz I, and Kelly SF (2008) Simulated performance of a renewable energy technology – Heat pump systems in semi-arid central California greenhouses. *Acta Hort.* 797: 347–352.

IEA (2012) *International Energy Agency, World Energy Outlook 2012*, OECD, Paris.

IEA (2013) *International Energy Agency, World Energy Outlook 2013*, OECD, Paris.

IEA (2015) *International Energy Agency, World Energy Outlook 2015*, OECD, Paris.

International Commission for the Protection of the Rhine (2010) Master Plan Migratory Fish Rhine. Report No. 179. Koblenz, Germany.

Karlsson F and Fahle´n P (2007) Capacity-controlled ground source heat pumps in hydronic heating systems. *International Journal of Refrigeration* 30:221–229.

Kenny JF, Barber NL, Hutson SS (2009) *Estimated Use of Water in the United States in 2005, USGS Circular 1344*, United States Geological Survey, Reston, VA.

Khademi F and Yildiz I (2018) 1.25 Energy and Solid Wastes. In: Dincer, I. (ed.), *Comprehensive Energy Systems*. vol. 1, pp. 980–1020. Oxford: Elsevier.

Khatib Z, Verbeek P (2003) Water to value-produced water management for sustainable field development of mature and green fields, *Journal of Petroleum Technology*, 55: 26–28.

Kuack DL (1986) Shading techniques utilize the unusual. *Greenhouse Grower* (May), pp. 29–30.

Laws EA (1993) *Aquatic Pollution-An Introductory Text*, Wiley, New York.

Liu Y and Yildiz I (2018) 1.24 Energy and Water Pollution. In: Dincer, I. (ed.), *Comprehensive Energy Systems*. vol. 1, pp. 950–979. Oxford: Elsevier.

Macknick J, Newmark R, Heath G, Hallett KC (2011) A Review of Operational Water Consumption and Withdrawal Factors for Electricity Generating Technologies, *Golden, CO, National Renewable Energy Laboratory*, 303:275–300.

Mehlitz T, Yildiz I, Hardin C, and Rahman S (2008) Simulated effects of reflective mulch on energy and water conservation in semi-arid central California greenhouses. *Acta Hort.* 797: 353–359.

Mielke E, Anadon LD, Narayanamurti V (2010) *Water consumption of energy resource extraction, processing, and conversion*, Belfer Center for Science and International Affairs.

Miyagawa I (1977) Studies on the pellet house. J. of Agricultural Laboratory Central Research Inst. of Electric Power Industry, Abiko, Chiba Perfecture, Japan, No. 8, February.

Moya A, Mehlitz T, Yildiz I, Kelly SF, and Hardin C (2008) Simulated effects of dynamic row spacing on energy and water conservation in semi-arid central California greenhouses. *Acta Hort.* 797: 307–313.

Olsson G (2012) *Water and Energy: Threats and Opportunities*, IWA Publishing, London.

Omer A (2008) Energy, environment and sustainable development, *Renewable and Sustainable Energy Reviews*, 12(9): 2265–2300.

Palmer MA, Bernhardt ES, Schlesinger WH (2010) Mountaintop mining consequences, *Science*, 327(5962):48–149.

Papadopoulos AP and Pararajasingham S (1997) The influence of plant spacing on light interception and use in greenhouse tomato (*Lycopersicon esculentum* Mill.): A review. *Scientia Horticulturae* (69): 1–29.

Polagye B, Van Cleve B, Copping A, Kirkendall K (2011) Environmental Effects of Tidal Energy Development: Proceedings of a Scientific Workshop, March 22–25, 2010. U.S. National Oceanographic and Atmospheric Administration.

Rebuck SM, Aldrich RA, and White JW (1977) Internal curtains for energy conservation in greenhouses. *Transactions of the ASAE* 20 (4): 732–734.

Reynolds RR, Kiker RD (2003) Produced Water and Associated Issues, Oklahoma Geological Survey.

Roberts BW and Anderson JA (1994) Canopy shade and soil mulch affect yield and solar injury of bell pepper. *HortScience* 29:258–260.

Roberts WJ and Mears DR (1969) Double covering a film greenhouse using air to separate film layers. *Transactions of the ASAE* 12 (1): 32–33, 38.

Schalk JM and Robbins ML (1987) Reflective mulches influences plant survival, production and insect control on fall tomatoes. *HortScience* 22:30–32.

Short TH, Badger PC, and Roller WL (1978) A solar pond-polystyrene bead system for heating and insulating greenhouses. Proc. of International Symposium on Potential Productivity in Protected Cultivation, Kyoto and Tokyo, Japan, August, 1978.

Short TH and Bauerle WL (1977) A double plastic heat conservation system for glass greenhouses. Proceedings of an International Symposium on Controlled Environment Agriculture, April 7–8, Tucson, AZ, U.S.A.

Short TH and Bauerle WL (1980) Energy conservation techniques for commercial glass greenhouses. ASAE Paper No: 80-4532. ASAE, St. Joseph, MI 49085, U.S.A.

Simpkins JC, Mears DR, and Roberts WJ (1976) Reducing heat losses in polyethylene covered greenhouses. *Transactions of the ASAE* 19 (4): 714–719.

Stephenson MT (1992) A survey of produced water studies. In Produced Water, Springer US, pp. 1–11.

Teixeira T, Neves P, Araújo L (2012) Thermal impact of a nuclear power plant in a coastal area in Southeastern Brazil: Effects of heating and physical structure on benthic cover and fish communities, *Hydrobiologia*, 684(1):161–175.

The University of Maine Cooperative Extension (2020) *Ag Plastic Recycling in Maine: Current Challenges and New Opportunities.* www.extension.umaine.edu/agriculture/greenhouse-plastic-recycling Retrieved 2020-07-10.

UN (2014a) World Water Assessment Program, The United Nations World Water Development Report 2014: Water and Energy, Paris.

UN (2014b) Department of Economic and Social Affairs, World Urbanization Prospects: The 2014 Revision, Highlights, (ST/ESA/SER.A/352), New York.

UN (2015) *World Water Assessment Program, The united nation world water Development report: Water for a sustainable world*, UNESCO, Paris.

US EPA (2011) Draft Plan to Study the Potential Impacts of Hydraulic Fracturing on Drinking Water Resources, Office of Research and Development, Washington, D.C.

USAI Lighting (2020) https://www.usailighting.com/ Retrieved 2020-07-07.

USDOE (2006) Energy demand on water resource: Report to congress on the interdependencies of energy and water. Washington, DC.

Wang Z (2018) 1.23 Energy and Air Pollution. In: Dincer, I. (ed.), *Comprehensive Energy Systems*. vol. 1, pp. 909–949. Oxford: Elsevier.

Wang T, Yang Z, Copping A (2015) A modeling study of the potential water quality impacts from in-stream tidal energy extraction, *Estuaries and Coasts*, 38(1):173–186.

WEC (2013) World Energy Council, Survey of World Energy Resources, London.

Willits DH and Gurjer YR (2004) Heat pumps for the heating and night-cooling of greenhouse crops: a simulation study. *Transactions of the ASAE* 47 (2):575–584.

Wittneben B (2011) The impact of the Fukushima nuclear accident on European energy policy. Environmental Science and Policy, Environmental Science and Policy.

Yildiz I and Stombaugh DP (2006a) Dynamic modeling of microclimate and environmental control strategies in a greenhouse coupled with a heat pump system. *Acta Hort.* 718:331–340.

Yildiz, I and Stombaugh, DP (2006b) Heat pump cooling and greenhouse microclimates in open and confined greenhouse systems. *Acta Hort.* 719: 255–262.

Yildiz I and Stombaugh DP (2009) Simulated effects of various environmental management practices on energy consumption in open and confined greenhouse systems. *Acta Hort.* 801:1317–1324.

Yildiz I, Yue J, and Lowrey J (2014a) Spatiotemporal distributions of energy and water consumption in open, semi-closed and closed loop heat pump greenhouse systems, and associated management strategies in Canada. *Acta Hort.* 1037 (1): 119–128.

Yildiz I, Yue J, and Lowrey J (2014b) Spatiotemporal distributions of energy and water consumption in open, semi-closed and closed loop heat pump greenhouse systems, and associated management strategies in the Canadian North. *Acta Hort.* 1037 (1): 171–177.

Zalom FG (1981) Effects of aluminum mulch on fecundity of apterous *Myzus persicae* on heat lettuce in a field planting. *Ent. Exp. and Appl.* 30:227–230.

# Index

Page numbers in *italics* refer to figures and those in **bold** refer to tables.

## D



## E

For Product Safety Concerns and Information please contact our EU
representative GPSR@taylorandfrancis.com
Taylor & Francis Verlag GmbH, Kaufingerstraße 24, 80331 München, Germany

www.ingramcontent.com/pod-product-compliance
Ingram Content Group UK Ltd.
Pitfield, Milton Keynes, MK11 3LW, UK
UKHW050756180926
475701UK00007B/1074

* 9 7 8 0 3 6 7 7 5 4 1 3 6 *